Acoustics of Fluid–Structure Interactions

Acoustics of Fluid–Structure Interactions deals with the theory of the production and absorption of noise and vibration by fluid flow. This includes the theory of aerodynamics sound, as well as more conventional areas of acoustics and structural vibration. After providing the background material in fluid mechanics, acoustics, and structural vibrations, it proceeds to more advanced topics, suited to a graduate level course on the theory of acoustics and aerodynamic sound. Theoretical concepts are illustrated and extended by numerous examples, many of which include complete worked solutions.

This book will be useful as a reference for analytical methods for modeling acoustic problems, as a repository of known results and methods in the theory of aerodynamic sound, and as a graduate level textbook.

Dr. Howe is Professor of Theoretical Mechanics at Boston University. He received his PhD in Contiuum Mechanics from the Imperial College in London. Dr. Howe has over 25 years experience of research in fluid mechanics, acoustics, random vibration, and structural mechanics.

CAMBRIDGE MONOGRAPHS ON MECHANICS

Acoustics of Fluid–Structure Interactions

M. S. HOWE

CAMBRIDGE
UNIVERSITY PRESS

CAMBRIDGE UNIVERSITY PRESS
Cambridge, New York, Melbourne, Madrid, Cape Town, Singapore, São Paulo

Cambridge University Press
The Edinburgh Building, Cambridge CB2 8RU, UK

Published in the United States of America by Cambridge University Press, New York

www.cambridge.org
Information on this title: www.cambridge.org/9780521633208

First published 1998
This digitally printed version 2008

A catalogue record for this publication is available from the British Library

Library of Congress Cataloguing in Publication data

Howe, M. S.
Acoustics of fluid-structure interactions / M. S. Howe.
p. cm. – (Cambridge monographs on mechanics)
Includes bibliographical references.
ISBN 0-521-63320-6
1. Fluid-structure interaction. 2. Structural dynamics.
3. Aeodynamic noise. 4. Acoustic models. I. Title. II. Series.
TA357.5.F56H69 1998
624.1′71–dc21 98-6119
CIP

ISBN 978-0-521-63320-8 hardback
ISBN 978-0-521-05428-7 paperback

Contents

Preface

This book deals with that branch of fluid mechanics concerned with the production and absorption of sound occuring when unsteady flow interacts with solid bodies. Problems of this kind are commonly known under the heading of *aerodynamic sound* but often include more conventional areas of *acoustics* and *structural vibration*. Acoustics is here regarded as a branch of fluid mechanics, and an attempt has therefore been made in Chapter 1 to provide the necessary background material in this subject. Elementary concepts of classical acoustics and structural vibrations are also reviewed in this chapter. Constraints of space and time have required the omission or the curtailed discussion of several important subareas of the acoustics of fluid–structure interactions, including in particular many problems involving supersonic flow. The book should be of value in one or more of the following ways: (i) as a reference for *analytical methods* for modeling acoustic problems; (ii) as a repository of known results and methods in the theory of aerodynamic sound and vibration, which have tended to become scattered throughout many journal and review articles over the past forty or so years; and (iii) as a graduate level textbook. Chapter 1 and selected topics from Chapters 2 and 3 have been used for several years in teaching an advanced graduate level course on the theory of acoustics and aerodynamic sound.

Theoretical concepts are illustrated and sometimes extended by numerous examples, many of which include complete worked solutions. Every effort has been made to ensure the accuracy of formulae, both in the main text and in the examples. The author would welcome notification of errors detected by the reader and more general suggestions for improvements.

Acknowledgement must be made to the Board of Directors of Bolt Beranek and Newman, Inc. (BBN) for the granting of a sabbatical leave during which the first half of this book was drafted. Much of the material is the outcome of many years of contractual research support by the National Aeronautics and Space Administration Langley Research Center (NASA LaRC), the Carderock Division of the Naval Surface Warfare Center (NSWC – formerly the David Taylor Model Basin), and the Air Force Office of Scientific Research (AFOSR). Particular thanks are due to Erich Bender and Stephen Breit (of BBN); Jay C. Hardin, Feri Farrasat, and Lucio Maestrello (NASA LaRC); Maurice M. Sevik, William K. Blake, and Theodore M. Farabee (NUWC); and Spencer Wu and Brian Sanders (AFOSR).

M. S. Howe
Boston University

1

Introduction

1.1 Fluids, Structures, and Acoustics

Sound and structural vibrations are an inevitable product of unsteady flow and of its interactions with structures immersed in the flow. These interactions, and the manner in which they affect the production and dissipation of sound, are the subject of this book. This first chapter is a review of several fundamental topics in fluid mechanics, acoustics, and the theory of elasticity.

1.1.1 Notation

The mathematical notation is developed in this chapter. For future reference, however, it is convenient to note here the following conventions.

Latin suffixes i, j, and so forth denote vector or tensor components in the direction of the $i-$, $j-$, ... axis; $\mathbf{i}_j$ is a unit vector in the j-direction.

Green's Functions

The frequency-domain Green's function $G(\mathbf{x}, \mathbf{y}; \omega)$ and the time-domain Green's function $G(\mathbf{x}, \mathbf{y}, t-\tau)$ are defined such that

$$G(\mathbf{x}, \mathbf{y}, t-\tau) = \frac{-1}{2\pi}\int_{-\infty}^{\infty} G(\mathbf{x}, \mathbf{y}; \omega)\mathrm{e}^{-i\omega(t-\tau)}\, d\omega.$$

However, the Fourier time transform $f(\mathbf{x}, \omega)$ of a function $f(\mathbf{x}, t)$ of position $\mathbf{x}$ and time t are related by

$$f(\mathbf{x}, t) = \int_{-\infty}^{\infty} f(\mathbf{x}, \omega)\mathrm{e}^{-i\omega t}\, d\omega.$$

For problems in the frequency domain, the notation $f(\mathbf{x}, \omega)$ implies a time dependence $e^{-i\omega t}$.

1.2 Equations of Motion of a Fluid

The analysis of fluid–structure interactions involves a consideration of equations governing the motion of fluids and solids and the boundary conditions that relate the solutions of these equations at the fluid–structure interface. A complete description of the state of a moving fluid is given at each time t and position $\mathbf{x} = (x_1, x_2, x_3)$ when the velocity $\mathbf{v}$ and any two thermodynamic variables are specified. These quantities generally vary with $\mathbf{x}$ and t, and five scalar equations are required to determine the motion in terms of the boundary conditions and a prescribed initial state. The equations are statements of the conservation of mass, momentum, and energy.

1.2.1 Equation of Continuity

Conservation of mass requires the rate of increase of the fluid mass within a fixed region of space V to be equal to the net influx due to convection across the boundaries of V. This is expressed in terms of $\mathbf{v}$ and the fluid density ρ by the equation of continuity

$$\partial\rho/\partial t + \operatorname{div}(\rho\mathbf{v}) = 0, \tag{1.2.1}$$

which may also be written

$$\frac{1}{\rho}\frac{D\rho}{Dt} + \operatorname{div}\mathbf{v} = 0, \tag{1.2.2}$$

where $D/Dt = \partial/\partial t + \mathbf{v}\cdot\nabla \equiv \partial/\partial t + v_j\partial/\partial x_j$ is the material derivative, which measures rate of change following the motion of a fluid particle, and a repeated suffix j implies summation over $j = 1, 2, 3$. By writing Equation (1.2.2) in the form $\operatorname{div}\mathbf{v} = \rho D(1/\rho)/Dt$, we see that $\operatorname{div}\mathbf{v}$ is just the rate of change of fluid volume per unit volume following the motion of the fluid. This vanishes when a fluid is incompressible, when the equation of continuity becomes

$$\operatorname{div}\mathbf{v} = 0. \tag{1.2.3}$$

1.2.2 Momentum Equation

The momentum equation for a viscous fluid is called the *Navier–Stokes equation* and expresses the rate of change of momentum of a fluid particle in terms of the pressure p, the *viscous stress tensor* σ_{ij}, and body forces (such as gravity) $\mathbf{F}$ per unit volume:

$$\begin{aligned}\rho Dv_i/Dt &= -\partial p/\partial x_i + \partial \sigma_{ij}/\partial x_j + F_i \\ &\equiv -\partial p_{ij}/\partial x_j + F_i,\end{aligned} \tag{1.2.4}$$

where

$$p_{ij} = p\delta_{ij} - \sigma_{ij} \tag{1.2.5}$$

is the *compressive stress tensor*. Viscous stresses are caused by frictional forces between neighboring fluid elements moving at different velocities. For an isotropic, *Newtonian* fluid (such as dry air and water)

$$\sigma_{ij} = 2\eta_{ij} + \left(\eta' - \frac{2}{3}\eta\right)\delta_{ij}e_{kk}, \tag{1.2.6}$$

where

$$e_{ij} = \frac{1}{2}(\partial v_i/\partial x_j + \partial v_j/\partial x_i) \tag{1.2.7}$$

is the rate of strain tensor and η and η' are, respectively, the *shear* and *bulk* coefficients of viscosity. These coefficients are functions of the pressure and temperature, T, and in general vary throughout the flow, although in many cases the variations are sufficiently small that they can be neglected. Equation (1.2.6) is a phenomenological relation that is applicable provided the time scales of unsteady fluid motions are much larger than the relaxation times of the internal degrees of freedom of the fluid molecules [1].

The bulk coefficient of viscosity η' vanishes for monatomic gases, and in this case (and for most liquids, such as water), the fluid is said to be Stokesian, with

$$\sigma_{ij} = 2\eta\left(e_{ij} - \frac{1}{3}e_{kk}\delta_{ij}\right). \tag{1.2.8}$$

When the variation of η is neglected, the momentum equation for a Stokesian fluid is

$$D\mathbf{v}/Dt = (-1/\rho)\nabla p + \nu\left[\nabla^2\mathbf{v} + \frac{1}{3}\nabla(\operatorname{div}\mathbf{v})\right] + F/\rho, \tag{1.2.9}$$

where

$$\nu = \eta/\rho \tag{1.2.10}$$

is the kinematic viscosity.

Table 1.2.1. *Air and water values*

Element	ρ kg/m^3	η kg/ms	ν m^2/s
Air	1.23	$1.764\ 10^{-5}$	$1.433\ 10^{-5}$
Water	1,000	$1.284\ 10^{-3}$	$1.284\ 10^{-6}$

Values of ρ, η, and ν for air and water at 10°C and one-atmosphere pressure are given in the Table 1.2.1.

When viscous stresses are ignored, equations (1.2.1) and (1.2.4) describe the motion of an ideal fluid and are called *Euler's equations.* These are of lower order in the spatial derivatives of velocity than the full Navier–Stokes equations, and this implies that it will not normally be possible to satisfy all of the conditions that a real fluid must satisfy at the physical boundaries of the flow (see Section 1.5).

1.2.3 Thermodynamic Relations

Apart from the pressure, density, and temperature, we shall also need to introduce the specific internal energy e, specific enthalpy $w \equiv e + p/\rho$, and specific entropy s, where a *specific* quantity denotes the value per unit mass of the fluid. For a single-component system in a state of equilibrium, any thermodynamic variable can be expressed in terms of any two given variables by making use of the equation of state and certain thermodynamic identities.

When a quantity of heat dQ is supplied to unit mass of fluid, as a result of which its volume changes by $d(1/\rho)$, the internal energy e increases by $de = dQ - pd(1/\rho)$, the second term on the right being the work done on the fluid element by the pressure because of the volume change. If the change occurs sufficiently slowly, the fluid element is always in thermodynamic equilibrium, and the heat input and the change in specific entropy are related by $dQ = T\,ds$, which leads to the fundamental identity

$$de = T\,ds - pd(1/\rho). \qquad (1.2.11)$$

This "quasi-static" relation does not necessarily imply that the motion is thermodynamically *reversible* because the fluid element, although internally in equilibrium, need not be in equilibrium with its surroundings. When it is written in terms of the enthalpy w, we have

$$dw = T\,ds + dp/\rho. \qquad (1.2.12)$$

The specific heats at constant volume c_v and constant pressure c_p are equal to the amounts of heat absorbed per unit mass of fluid per unit rise in temperature when the heating occurs at constant volume and constant pressure, respectively. Equations (1.2.11) and (1.2.12) supply

$$c_v = T(\partial s/\partial T)_\rho = (\partial e/\partial T)_\rho, \qquad c_p = T(\partial s/\partial T)_p = (\partial w/\partial T)_p. \tag{1.2.13}$$

The specific heats are related by

$$c_p - c_v = T\beta^2/\rho \mathrm{K}_T, \tag{1.2.14}$$

where β and K_T are, respectively, the coefficients of expansion and isothermal compressibility of the fluid, defined by

$$\beta = (-1/\rho)(\partial \rho/\partial T)_p, \quad \mathrm{K}_T = (1/\rho)(\partial \rho/\partial p)_T. \tag{1.2.15}$$

The equation of state of an ideal gas is usually expressed in the form

$$p = \rho \mathrm{R} T, \tag{1.2.16}$$

where the gas constant $\mathrm{R} = 8{,}314/(\text{mean molecular weight})$ in SI units ($\mathrm{R} \approx 287$ Joules/kg °K for dry air at normal temperatures and pressures). For a nonreacting mixture of gases in equilibrium at temperature T, Dalton's law of partial pressures states that the total pressure $p = \sum_j p_j$, where $p_j = \rho_j \mathrm{R}_j T$ is the pressure of the jth constituent gas of density ρ_j and gas constant R_j.

The specific heats of an ideal gas composed of rigid molecules each with N rotational degrees of freedom, satisfy $c_p - c_v = \mathrm{R}$, and $\gamma = c_p/c_v = (N+5)/(N+3)$. When c_p, c_v are *constant*, we also have

$$\begin{aligned} &e = c_v T = p/(\gamma - 1)\rho, \quad w = c_p T = \gamma p/(\gamma - 1)\rho, \\ &s = c_v \ln(p/\rho^\gamma). \end{aligned} \tag{1.2.17}$$

Equation (1.2.11) implies that the excess thermal energy required to raise the temperature of a fluid particle at constant pressure, as opposed to constant volume, is expended as work $\int p\,d(1/\rho)$ performed on its environment during expansion and that the relevant thermodynamic function for describing such processes is the enthalpy w.

1.2.4 Thermodynamic Variables

When velocity gradients are present the fluid cannot be in strict thermodynamic equilibrium, and thermodynamic variables require special interpretation. For a fluid in nonuniform motion it is customary to define the density ρ and internal energy e in the usual way, that is, such that ρ and ρe are the mass and internal energy per unit volume. The pressure and all other thermodynamic variables are then defined by means of the same functions of ρ and e as would be the case in thermal equilibrium. The "thermodynamic pressure" $p = p(\rho, e)$ so defined is no longer the sole source of normal stress on a surface drawn in the fluid. The mean normal stress is obtained by averaging $p_{ij}n_i n_j$ (see (1.2.5)) over all possible directions of the unit vector **n** defining the normal to a surface element. The average is obtained by evaluating $(1/4\pi)\oint n_i n_j \, dS$ over the surface of a sphere of unit radius. The integral is equal to $\frac{4}{3}\pi\delta_{ij}$, and the mean normal stress is therefore equal to $\frac{1}{3}p_{ii} = p - \eta' \operatorname{div} \mathbf{v}$, which differs from the thermodynamic pressure p if the bulk coefficient of viscosity η' is nonzero. This is the case for a fluid whose molecules possess rotational degrees of freedom, whose relaxation time (during which thermal equilibrium is reestablished after, say, an expansion of the fluid) is large relative to the equilibration time of the translational degrees of freedom. For example, in an expanding diatomic gas ($\operatorname{div} \mathbf{v} > 0$) the temperature must decrease, but the reduction in the rotational energy lags slightly behind that of the translational energy; the thermodynamic pressure $p = (\gamma - 1)\rho e$ accordingly exceeds the actual pressure $\frac{1}{3}p_{ii}$ by an amount equal to $\eta' \operatorname{div} \mathbf{v}$. The rate of adjustment of the rotational degrees of freedom is such that in practice η' is of the same order as η (for dry air $\eta' \sim 0.6\eta$).

It may be shown [2] that whereas the thermodynamic pressure differs from the mean normal stress by a term that is linear in the velocity gradient, the corresponding departure of the thermodynamic specific entropy s from the true entropy is proportional at least to the square of such gradients, and the difference is usually small in practice.

1.2.5 Crocco's Equation and Bernoulli's Equation

The momentum equation (1.2.4) can also be expressed in *Crocco's* form

$$\partial v_i/\partial t + \partial B/\partial x_i = -(\boldsymbol{\omega} \wedge \mathbf{v})_i + T\partial s/\partial x_i + (1/\rho)\partial \sigma_{ij}/\partial x_j + F_i/\rho, \tag{1.2.18}$$

where B is the *total enthalpy*,

$$B = w + \frac{1}{2}v^2, \tag{1.2.19}$$

and $\boldsymbol{\omega} = \operatorname{curl} \mathbf{v}$ is the *vorticity*. It is derived from (1.2.4) by making use of (1.2.12) and the identity $D\mathbf{v}/Dt \equiv \partial \mathbf{v}/\partial t + (\mathbf{v} \cdot \nabla)\mathbf{v} = \partial v/\partial t + \nabla(\frac{1}{2}v^2) + \boldsymbol{\omega} \wedge \mathbf{v}$.

In an *ideal fluid* viscosity and thermal conduction are negligible. If, in addition, the motion is *homentropic* ($s =$ constant throughout the flow), irrotational ($\boldsymbol{\omega} = \mathbf{0}$), and the body force $\mathbf{F}$ is conservative, that is, there is a scalar function of position Φ such that $\mathbf{F} = \rho\nabla\Phi$, Crocco's equation implies the existence of a velocity potential ϕ, such that

$$\mathbf{v} = \nabla\phi. \tag{1.2.20}$$

A first integral of the momentum equation then yields *Bernoulli's equation*, which may be expressed in either of the forms:

$$\partial\phi/\partial t + B - \Phi = f(t), \quad \partial\phi/\partial t + w + \frac{1}{2}v^2 - \Phi = f(t). \tag{1.2.21}$$

The velocity potential ϕ is undefined to within an arbitrary function of time so that the function $f(t)$ may be set equal to a constant or zero. For steady flow $\partial\phi/\partial t = 0$, $f(t) =$ constant, and Bernoulli's equation reduces to $w + \frac{1}{2}v^2 - \Phi =$ constant. It is clear from (1.2.21), and from Crocco's equation (1.2.18), that the potential Φ of the body force may be absorbed into the total enthalpy B.

A flow is said to be *isentropic* when the specific entropy of each fluid particle is constant, that is, when $Ds/Dt = 0$. For steady flow of an ideal fluid subject to conservative body forces, Crocco's equation, $\nabla B = -\boldsymbol{\omega}\wedge\mathbf{v} + T\nabla s$, implies that $DB/Dt \equiv (\mathbf{v}\cdot\nabla)B = -\mathbf{v}\cdot\boldsymbol{\omega}\wedge\mathbf{v} + (\mathbf{v}.\nabla)s = 0$ so that B is constant along a streamline. If the flow is homentropic, B is constant over the fixed fluid surfaces spanned by intersecting streamlines and vortex lines (because $\mathbf{v}\cdot\boldsymbol{\omega}\wedge\mathbf{v} \equiv \mathbf{0}$).

1.2.6 Reynolds Number

An equation for the rate of change of the momentum density $\rho\mathbf{v}$ is obtained by adding the continuity equation (1.2.1) multiplied by $\mathbf{v}$ to the momentum equation (1.2.4) and writing the result in the form

$$\frac{\partial(\rho v_i)}{\partial t} = -\frac{\partial}{\partial x_j}(p\delta_{ij} + \rho v_i v_j - \sigma_{ij}) + F_i. \tag{1.2.22}$$

The quantity $\rho v_i v_j$ is called the *Reynolds stress* and accounts for changes in the momentum ρv_i by convection at velocity v_j. The order of magnitude of the Reynolds stress relative to the viscous stress σ_{ij} at any given point in the flow is determined by the value of the *Reynolds number*

$$\mathrm{Re} = u\ell/\nu, \tag{1.2.23}$$

in which u and ℓ are the respective characteristic values of the velocity and length scale of variation of the flow in the neighborhood of the point in question.

In regions where $\mathrm{Re} \gg 1$ viscosity plays a minor role in the local development of the flow. This is the case, for example, in high-speed jets, where $\ell \sim$ jet diameter and $u \approx$ mean jet velocity. In turbulent flow over a wall, the motion in a thin layer of fluid adjacent to the wall (the "viscous sublayer," where u and ℓ are small and $\mathrm{Re} \to 0$) is dominated by the viscous diffusion of momentum across the layer. Further away from the wall, in the region of fully developed turbulence, $\ell \approx$ boundary layer thickness and $u \approx$ velocity of the main stream, and Re typically exceeds 10^5 or more. The viscous (i.e., molecular) transport of momentum is then negligible compared with that due to turbulence convection.

1.2.7 Vorticity

The motion of a spherical particle of fluid may be decomposed into a uniform translation at the velocity $\mathbf{v}$, say, of its center, a straining motion determined by the rate of strain tensor e_{ij} and a rotation at angular velocity $\frac{1}{2}\omega = \frac{1}{2}\,\mathrm{curl}\,\mathbf{v}$ [3]. The vorticity is therefore a measure of the angular momentum of a fluid particle. Conservation of angular momentum suggests that vorticity may be regarded as a quantity characterizing the *intrinsic* kinetic energy of a flow, inasmuch as it determines the motion that would persist in an incompressible fluid at rest at infinity if all the boundaries were brought to rest. A vortex line in the fluid is tangential to the vorticity vector at all points along its length. Because $\mathrm{div}\,\omega = \mathrm{div}(\mathrm{curl}\,\mathbf{v}) \equiv 0$, a vortex line cannot begin or end within the fluid. The "no-slip" condition requires the fluid velocity at the boundary to be the same as that of the boundary. A vortex line must therefore form a closed loop or end on a rotating surface bounding the flow [3].

Vorticity is transported by convection and molecular diffusion, and for this reason it is often permissible to assume that an initially confined region of vorticity remains localized. The vorticity equation is obtained from the curl of Crocco's equation (1.2.18) and by using the vector identity $\mathrm{curl}(\omega \wedge \mathbf{v}) = (\mathbf{v}\cdot\nabla)\omega + \omega\,\mathrm{div}\,\mathbf{v} - (\omega\cdot\nabla)\mathbf{v}$ and the continuity equation (1.2.1):

$$\begin{aligned} \rho D(\omega/\rho)/Dt = {} & (\omega\cdot\nabla)\mathbf{v} + \nabla T \wedge \nabla s \\ & + \mathrm{curl}((1/\rho)\partial p_{ij}/\partial x_j) + \mathrm{curl}(\mathbf{F}/\rho). \end{aligned} \tag{1.2.24}$$

The terms on the right-hand side account for changes in $\boldsymbol{\omega}/\rho$ following the motion of a fluid particle. They are interpreted as follows:

1. $(\boldsymbol{\omega} \cdot \nabla)\mathbf{v}$: rotation and stretching of vortex lines by the velocity field. The magnitude of $\boldsymbol{\omega}/\rho$ increases in direct proportion to the stretching of vortex lines by the flow. When the remaining terms on the right of (1.2.24) are absent, the motion satisfies *Kelvin's circulation theorem* [2–4]. This asserts that, in an inviscid, homentropic fluid subject only to conservative body forces, the *circulation* C, equal to the line integral $\oint \mathbf{v} \cdot d\mathbf{x}$ taken around a closed material contour in the fluid, does not change with time ($DC/Dt = 0$). It follows that vortex lines move with the fluid, and in particular that the vorticity of a material element of the fluid remains zero if it is initially zero. In other words, vorticity cannot be created in a body of fluid if it is initially irrotational and the motion is inviscid and homentropic.
2. $\nabla T \wedge \nabla s$: production of vorticity when temperature and entropy gradients are not parallel, or equivalently (because, from (1.2.12), $\nabla T \wedge \nabla s = -\nabla(1/\rho) \wedge \nabla p$), when density and pressure gradients are not parallel.
3. $\operatorname{curl}((1/\rho)\partial\sigma_{ij}/\partial x_j)$: molecular diffusion of vorticity. This term is important only in regions of high shear. When variations in the kinematic viscosity ν are negligible, we recover the more familiar expression

$$\operatorname{curl}((1/\rho)\partial\sigma_{ij}/\partial x_j) = \nu\nabla^2\boldsymbol{\omega}. \qquad (1.2.25)$$

If local effects of compressibility are unimportant the vorticity equation (1.2.24) reduces to

$$D\boldsymbol{\omega}/Dt = (\boldsymbol{\omega} \cdot \nabla)\mathbf{v} + \nu\nabla^2\boldsymbol{\omega} + \operatorname{curl}(\mathbf{F}/\rho). \qquad (1.2.26)$$

4. $\operatorname{curl}(\mathbf{F}/\rho)$: production of vorticity by nonconservative body forces.

Vorticity is generated at solid boundaries, and viscosity is responsible for its diffusion into the body of the fluid, where it can subsequently be convected by the flow. When body forces and density variations are neglected, integration of (1.2.26) over a fluid region V contained within a control surface Σ moving with the fluid, and bounded internally by a moving solid surface S, yields

$$\frac{\partial}{\partial t}\int \boldsymbol{\omega}\, d^3\mathbf{x} = -\oint_{S+\Sigma} ((\mathbf{n} \cdot \boldsymbol{\omega})\mathbf{v} + \nu\mathbf{n} \wedge \nabla^2\mathbf{v})\, dS, \qquad (1.2.27)$$

where in the surface integrals over S and Σ, $\mathbf{n}$ is a unit normal directed *into* V. Vorticity changes in V are therefore caused by the rotation and stretching of vortex tubes at the boundaries and diffusion across the boundaries by viscosity.

The momentum equation (1.2.9) and the no-slip condition imply that $(\mathbf{n}\cdot\boldsymbol{\omega})\mathbf{v} = 2(\boldsymbol{\Omega}\cdot\mathbf{n})\mathbf{U}$ and $\nu\mathbf{n}\wedge\nabla^2\mathbf{v} = (1/\rho)\mathbf{n}\wedge\nabla p + D(\mathbf{n}\wedge\mathbf{U})/Dt - (\boldsymbol{\Omega}\wedge n)\wedge\mathbf{U}$ on S, where **U** and $\boldsymbol{\Omega}$ are, respectively, the velocity and angular velocity of the surface element dS. Equation (1.2.27) is therefore equivalent to

$$\frac{\partial}{\partial t}\int \boldsymbol{\omega}\, d^3\mathbf{x} = -\oint_{\mathrm{S}} (2(\boldsymbol{\Omega}\cdot\mathbf{n})\mathbf{U} - (\boldsymbol{\Omega}\wedge\mathbf{n})\wedge\mathbf{U} + (1/\rho)\mathbf{n}\wedge\nabla p + D(\mathbf{n}\wedge\mathbf{U})/Dt)\, d\mathrm{S} - \oint_{\Sigma} ((\mathbf{n}\cdot\boldsymbol{\omega})\mathbf{v} + \nu\mathbf{n}\wedge\nabla^2\mathbf{v})\, d\mathrm{S}. \tag{1.2.28}$$

This shows that the mechanism of vorticity production on S is independent of the value of the viscosity ν. Viscosity merely serves to diffuse the vorticity into the fluid from the surface [3, 5, 6]. The terms in the first integral on the right-hand side may be regarded as sources of vorticity on S. For a rigid solid in translational motion, vorticity production requires that either the tangential surface acceleration, $D(\mathbf{n}\wedge\mathbf{U})/Dt$, or surface pressure gradient, $\mathbf{n}\wedge\nabla p$, should be nonzero.

1.2.8 The Energy Equation

In homentropic flow, the equation $s =$ constant provides a relation between the thermodynamic variables that permits the motion to be determined from the equations of continuity and momentum together with the equation of state. In more general situations, where s is variable, it is necessary to introduce an energy equation to account for coupling between macroscopic motions and the internal energy of the fluid. This equation governs the conservation of the total energy of the system (i.e., the dissipation of mechanical energy) and is derived by analyzing the energy balance for a moving particle of fluid.

Consider a small element of fluid of volume V, bounded by a surface S. The kinetic and internal energies per unit volume are equal respectively to $1/2\rho v^2$ and ρe, and the energy of the fluid in V is $E = \rho\mathrm{V}(\frac{1}{2}v^2 + e)$. Changes in E are produced by the work done by the compressive stress p_{ij} on the boundary S, by the flux of heat energy through S by molecular conduction, and by the work performed by the body force **F** within V. If thermal conduction is expressed in terms of the *heat flux vector* **Q**, energy conservation requires

$$\frac{DE}{Dt} = \oint_{\mathrm{S}} (p_{ij}v_i + Q_j)\, d\mathrm{S}_j + \mathrm{V}v_j F_j, \tag{1.2.29}$$

where the surface element dS is directed *into* V. According to the divergence theorem $\oint_S (p_{ij}v_i + Q_j)\,dS_j = -V(\partial/\partial x_j)(p_{ij}v_i + Q_j)$. Because $DV/Dt = V\,\mathrm{div}\,\mathbf{v}$, it follows that (1.2.29) is equivalent to

$$\frac{\partial}{\partial t}\left(\frac{1}{2}\rho v^2 + \rho e\right) + \mathrm{div}(\rho\mathbf{v}B - \mathbf{v}\cdot\underline{\boldsymbol{\sigma}} + \mathbf{Q}) = \mathbf{v}\cdot\mathbf{F}, \qquad (1.2.30)$$

where $(\mathbf{v}\cdot\underline{\boldsymbol{\sigma}})_i = v_j\sigma_{ij}$.

Equation (1.2.30) balances the rate of working of the body force per unit volume against the rate of increase of the kinetic and internal energies, $\frac{1}{2}\rho v^2 + \rho e$, and the flow of energy at a rate determined by the energy flux vector

$$\mathbf{I} = \rho\mathbf{v}B - \mathbf{v}\cdot\underline{\boldsymbol{\sigma}} + \mathbf{Q}. \qquad (1.2.31)$$

In this expression $\rho\mathbf{v}B \equiv \rho\mathbf{v}w + \frac{1}{2}v^2 = (\rho e + \frac{1}{2}\rho v^2)\mathbf{v} + p\mathbf{v}$ is the energy flux due to convection and by the work done by the pressure; $-\mathbf{v}\cdot\underline{\boldsymbol{\sigma}}$ represents energy transfer by viscous (molecular) diffusion and $\mathbf{Q}$ by thermal conduction.

A more convenient form of the energy equation (1.2.30) is obtained by using Crocco's equation (1.2.18) and the equation of continuity to eliminate the kinetic energy density $\frac{1}{2}\rho v^2$. To do this, multiply Crocco's equation by the momentum density ρv_i and rearrange, using the equation of continuity (1.2.1) and the identity $d(\rho e) = T\rho\,ds + w\,d\rho$ (obtained from (1.2.12) by using $p = \rho(w - e)$), to obtain

$$\begin{aligned}\frac{\partial}{\partial t}\left(\frac{1}{2}\rho v^2 + \rho e\right) &+ \mathrm{div}(\rho\mathbf{v}B - \mathbf{v}\cdot\underline{\boldsymbol{\sigma}}) \\ &= \rho Dw/Dt - Dp/Dt - \sigma_{ij}\partial v_i/\partial x_j + v_i F_i. \qquad (1.2.32)\end{aligned}$$

The desired alternative energy equation is now derived by subtracting this from (1.2.30). The result can be expressed in either of the equivalent forms

$$\rho Dw/Dt - Dp/Dt \equiv \rho T\,Ds/Dt = \sigma_{ij}\partial v_i/\partial x_j - \partial Q_i/\partial x_i. \qquad (1.2.33)$$

The quantity $\rho T\,Ds/Dt$ is the time rate of change following the fluid particles of the heat gained per unit volume of fluid. The term $\sigma_{ij}\partial v_i/\partial x_j \equiv 2\eta(e_{ij} - \frac{1}{3}e_{kk}\delta_{ij})^2 + \eta'(\mathrm{div}\,\mathbf{v})^2 > 0$ represents the production of heat by frictional dissipation of macroscopic motions; the term in $\mathbf{Q}$ gives the diffusion of heat due to molecular conduction.

Heat diffuses from regions of higher to regions of lower temperature, and in most cases temperature gradients are sufficiently small that, to a good approximation, $\mathbf{Q} = -\kappa\nabla T$, where $\kappa > 0$ is the *thermal conductivity*, which usually depends on both temperature and pressure. Equation (1.2.33) then becomes

$$\rho T \frac{Ds}{Dt} = 2\eta \left(e_{ij} - \frac{1}{3} e_{kk} \delta_{ij} \right)^2 + \eta' (\operatorname{div} \mathbf{v})^2 + \frac{\partial}{\partial x_j} \left(\kappa \frac{\partial T}{\partial x_j} \right). \quad (1.2.34)$$

1.3 Equations of Motion of an Elastic Solid

1.3.1 Navier's Equation

Flow induced displacements of solid bodies are generally sufficiently small to be described by the linearized equations of elasticity. It is therefore permissible to assume that the mass density, ρ_s, of the solid does not vary with time. We consider an isotropic solid subject to body forces $\mathbf{F}$ per unit volume and denote by $\mathbf{u}(\mathbf{x}, t)$ the displacement of a material point from its equilibrium position. Small-amplitude motions are governed by the Navier equation, which may be set in either of the forms [7, 8],

$$\rho_s \partial^2 \mathbf{u} / \partial t^2 = (\lambda + \mu) \nabla (\operatorname{div} \mathbf{u}) + \mu \nabla^2 \mathbf{u} + \mathbf{F}, \quad (1.3.1)$$

$$\begin{aligned} \rho_s \partial^2 u_i / \partial t^2 &= \partial \tau_{ij} / \partial x_j + F_i, \\ \tau_{ij} &= 2\mu \varepsilon_{ij} + \lambda \varepsilon_{kk} \delta_{ij}, \\ \varepsilon_{ij} &= \frac{1}{2} (\partial u_i / \partial x_j + \partial u_j / \partial x_i), \end{aligned} \quad (1.3.2)$$

where λ and μ are the Lamé elastic constants, τ_{ij} is the stress tensor, and ε_{ij} is the strain tensor. The second set of equations, (1.3.2), is applicable when ρ_s, λ, and μ vary with position.

Energy transfer within the solid by thermal conduction can usually be ignored in fluid–structure interactions in which heat transfer is not of primary concern. If we temporarily denote by λ_s, μ_s and k_T, μ_T the respective values of the elastic constants for adiabatic and isothermal motions, then [1],

$$\lambda_s - \lambda_T = (\gamma - 1) \left(\lambda_T + \frac{2}{3} \mu_T \right), \quad \mu_s - \mu_T = 0, \quad (1.3.3)$$

where γ is the ratio of the specific heats of the solid. The coefficient μ is associated with shearing motions of the solid, without change in volume or heat generation. At room temperature $\gamma - 1 \approx 0.01$, which usually decreases with temperature. For practical purposes, it may therefore be assumed that the propagation of structural vibrations is governed by the adiabatic Lamé constants.

In engineering applications, it is usual to express the elastic properties in terms of the Young's modulus E and Poisson's ratio σ. Their interrelations with the Lamé constants λ, μ are given by

$$\begin{aligned} \lambda &= \sigma \mathrm{E}/(1+\sigma)(1-2\sigma), \quad \mu = \mathrm{E}/2(1+\sigma), \\ \mathrm{E} &= \mu(3\lambda+2\mu)/(\lambda+\mu), \quad \sigma = \lambda/2(\lambda+\mu). \end{aligned} \tag{1.3.4}$$

Navier's equation then becomes

$$\frac{\partial^2 \mathbf{u}}{\partial t^2} = \frac{\mathrm{E}}{2\rho_s(1+\sigma)}\left(\frac{\nabla(\operatorname{div}\mathbf{u})}{(1-2\sigma)} + \nabla^2\mathbf{u}\right) + \frac{\mathbf{F}}{\rho_s}. \tag{1.3.5}$$

By taking the divergence and curl of Navier's equation, div **u** and curl **u** are seen to satisfy the separate, nondispersive, inhomogeneous wave equations,

$$\begin{aligned} \left(\partial^2/\partial t^2 - c_1^2\nabla^2\right)\operatorname{div}\mathbf{u} &= \operatorname{div}(\mathbf{F}/\rho_s), \\ \left(\partial^2/\partial t^2 - c_2^2\nabla^2\right)\operatorname{curl}\mathbf{u} &= \operatorname{curl}(\mathbf{F}/\rho_s), \end{aligned} \tag{1.3.6}$$

where the wave speeds c_1 and c_2 are given by

$$\begin{aligned} c_1 &= \sqrt{(\lambda+2\mu)/\rho_s} = \left(\frac{\mathrm{E}(1-\sigma)}{\rho_s(1+\sigma)(1-2\sigma)}\right)^{\frac{1}{2}}, \\ c_2 &= \sqrt{\mu/\rho_s} = \left(\frac{\mathrm{E}}{2\rho_s(1+\sigma)}\right)^{\frac{1}{2}}. \end{aligned} \tag{1.3.7}$$

The body force terms on the right-hand sides of equations (1.3.6) represent distributed sources within the solid. Perturbations in div **u** propagate as longitudinal *dilatational* waves, frequently referred to as *P* – *("pressure") waves.* Rotational disturbances, curl **u**, propagate without fluctuations in volume and are called *shear*, or S-, waves.

Typical values of the wave speeds and other mechanical properties are given in Table 1.3.1 for steel, aluminum, and rubber.

Table 1.3.1. *Wave speed and mechanical property values*

Material	c_1 m/s	c_2 m/s	E N/m^2 $\times 10^{-10}$	ρ_s kg/m^3	Poisson's ratio σ
Steel	6,110	3,205	21	7,800	0.31
Aluminum	6,405	3,155	7.2	2,700	0.34
Hard rubber	2,115	865	0.23	1,100	0.4
Soft rubber	–	40	0.0005	950	0.5

The solution of the elastic equations is often facilitated by using the Helmholtz decomposition theorem [9] to express **u** in terms of scalar and vector potentials Φ and **H**, as follows:

$$\mathbf{u} = \nabla\Phi + \operatorname{curl}\mathbf{H}, \quad \operatorname{div}\mathbf{H} = 0. \tag{1.3.8}$$

Partitioning the body force per unit mass, $\mathbf{F}/\rho_s$, in the same way, by writing $\mathbf{F}/\rho_s = \nabla f + \operatorname{curl}\mathsf{F}$, it is found that the Navier equation is satisfied provided Φ and **H** satisfy the following analogs of equations (1.3.6)

$$(\partial^2/\partial t^2 - c_1^2\nabla^2)\Phi = f, \quad (\partial^2/\partial t^2 - c_2^2\nabla^2)\mathbf{H} = \mathsf{F}. \tag{1.3.9}$$

1.3.2 Governing Equations in Two Dimensions

Significant simplifications occur when the motion can be assumed to be two dimensional. If conditions are uniform in the x_3-direction, the equations for **u** in terms of vector and scalar potentials reduce to

$$u_1 = \partial\Phi/\partial x_1 + \partial H_3/\partial x_2, \quad u_2 = \partial\Phi/\partial x_2 - \partial H_3/\partial x_1, \tag{1.3.10}$$

$$u_3 = \partial H_2/\partial x_1 - \partial H_1/\partial x_2, \quad \partial H_1/\partial x_1 + \partial H_2/\partial x_2 = 0. \tag{1.3.11}$$

The components $\tau_{ij} = \tau_{ji}$ of the stress tensor may then be calculated from

$$\begin{aligned}
\tau_{11} &= \rho_s c_1^2(\partial u_1/\partial x_1 + \partial u_2/\partial x_2) - 2\rho_s c_2^2 \partial u_2/\partial x_2,\\
\tau_{12} &= \tau_{21} = \rho_s c_2^2(\partial u_1/\partial x_2 + \partial u_2/\partial x_1),\\
\tau_{13} &= \tau_{31} = \rho_s c_2^2 \partial u_3/\partial x_1,\\
\tau_{22} &= \rho_s c_1^2(\partial u_1/\partial x_1 + \partial u_2/\partial x_2) - 2\rho_s c_2^2 \partial u_1/\partial x_1,\\
\tau_{23} &= \tau_{32} = \rho_s c_2^2 \partial u_3/\partial x_2,\\
\tau_{33} &= \tau_{21} = \rho_s\left(c_1^2 - 2c_2^2\right)(\partial u_1/\partial x_1 + \partial u_2/\partial x_2).
\end{aligned} \tag{1.3.12}$$

It follows from these expressions and equations (1.3.2) that the propagation of the components (u_1, u_2) of the displacement is decoupled from the propagation of u_3. When $u_3 \equiv 0$ the motion is said to be one of *plane strain*, and $\tau_{13} = \tau_{31} = \tau_{23} = \tau_{32} \equiv 0$. When $u_1 = u_2 = 0$, but $u_3 \neq 0$, the disturbances are called *SH-waves*, and $\tau_{13} = \tau_{31}$ and $\tau_{23} = \tau_{32}$ are the only no-zero components of stress.

1.3.3 Elastic Energy Equation

The equation of conservation of energy is derived by multiplying the Navier equation (1.3.2) by $\partial u_i/\partial t$ and rearranging the result in the form

$$\frac{\partial}{\partial t}\left[\frac{1}{2}\rho_s\dot{\mathbf{u}}^2 + \mu\varepsilon_{ij}^2 + \frac{1}{2}\lambda(\operatorname{div}\mathbf{u})^2\right] + \frac{\partial}{\partial x_j}(-\dot{u}_i\tau_{ij}) = \dot{\mathbf{u}}\cdot\mathbf{F}, \qquad (1.3.13)$$

where $\dot{\mathbf{u}} = \partial\mathbf{u}/\partial t$. In this equation $\frac{1}{2}\rho_s\dot{\mathbf{u}}^2$ is the kinetic energy density, and $\mu\varepsilon_{ij}^2 + \frac{1}{2}\lambda(\operatorname{div}\mathbf{u})^2$ is the potential energy density associated with the straining and volumetric expansion of the solid. The rate of change of the energy within a given volume element δV of the solid is balanced by the rate of working of the body force, $\dot{\mathbf{u}}\cdot\mathbf{F}$, in δV, and the flux of elastic energy from the boundaries of δV, which is determined by the energy flux (*Poynting*) vector,

$$I_i = -\dot{u}_j\tau_{ji}. \qquad (1.3.14)$$

1.4 Equations for Vibrating Membranes, Plates, and Shells

The interaction of unsteady flow and sound with thin elastic plates and shells can be treated by use of the general equations of linearized elasticity given in Section 1.3. However in situations in which the characteristic wavelength is large compared to the thickness of the plate or shell, a considerable reduction in complexity is achieved by using specialized equations of motion.

1.4.1 Membrane Equation

A membrane is a homogeneous, perfectly flexible plane sheet of infinitesimal thickness h and mass density $m = \rho_s h$, which is stretched in all directions by a tension τ. Let the sheet lie in the plane $x_2 = 0$ when undisturbed, with fluid in the regions $x_2 \gtrless 0$ (Figure 1.4.1). Let $\zeta(x_1, x_3, t)$ denote the small-amplitude displacement from equilibrium of the point (x_1, x_3) at time t. The membrane has no resistance to shear or bending; the elastic restoring force on a small surface element is supplied by the resolved part of the tension on its periphery in the x_2-direction such that linearized motions of the membrane satisfy [10],

$$m\partial^2\zeta/\partial t^2 - \tau\nabla_2^2\zeta = -[p] + F_2, \qquad (1.4.1)$$

where $\nabla_2^2 = \partial^2/\partial x_1^2 + \partial^2/\partial x_3^2$, F_2 is the normal component of the body force (in the x_2-direction) per unit area of the membrane, and

$$[p] = p(x_1, +0, x_3, t) - p(x_1, -0, x_3, t) \qquad (1.4.2)$$

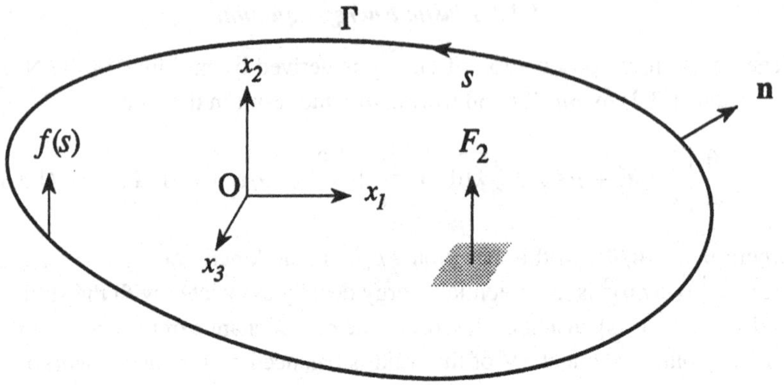

Figure 1.4.1. Membrane with edge force.

is the net normal force per unit area caused by fluid loading when viscous stresses are ignored.

Equation (1.4.1) may also be derived from *Hamilton's principle* [10, 11], which requires the first variation, δJ, of the *action*

$$\mathrm{J}(\zeta) = \int_{t_1}^{t_2} \mathcal{L}_o \, dt, \quad \mathcal{L}_o = \mathrm{T} - \Phi_o \tag{1.4.3}$$

to be zero, where T and Φ_o are the kinetic and potential energies of the membrane, and the variation $\delta\zeta$ vanishes at $t = t_1$ and t_2. The kinetic energy is given by

$$\mathrm{T} = \int \frac{1}{2} m (\partial\zeta/\partial t)^2 \, dx_1 \, dx_3, \tag{1.4.4}$$

where the integration is over the surface of the membrane. The potential energy of a small surface element $dx_1 \, dx_3$ of the deformed membrane is equal to the tension τ multiplied by the increase in area $\frac{1}{2}(\nabla_2\zeta)^2 \, dx_1 \, dx_3$, where $\nabla_2 = (\partial/\partial x_1, \partial/\partial x_3)$. When the virtual work performed by the fluid loading $[p]$ and the force F_2 are also included, we obtain

$$\Phi_o = \int \left(\frac{1}{2} \tau (\nabla_2\zeta)^2 - (-[p] + F_2)\zeta \right) dx_1 \, dx_3. \tag{1.4.5}$$

For a finite membrane, it is necessary to specify boundary conditions at the edges. Let the curvilinear edge of the membrane, Γ (Figure 1.4.1), be subject to a distributed force in the x_2-direction equal to $f(s)$ per unit length s of the edge, where s is measured in the "positive" sense such that the membrane is to the left when Γ is traversed in the direction of increasing s. The boundary condition on Γ is derived by replacing the Lagrangian $\mathcal{L}_o$ in the action (1.4.3) by $\mathcal{L} = \mathcal{L}_o - \Phi_1$, where Φ_1 is the potential energy of the boundary force

$$\Phi_1 = -\oint_\Gamma f(s)\zeta(s,t)\,ds, \tag{1.4.6}$$

where the integration is around the edge Γ. Because ζ satisfies (1.4.1) within Γ, the variation of the action becomes

$$\delta \mathrm{J} = \int_{t_1}^{t_2} \oint_\Gamma (f - \tau\partial\zeta/\partial n)\delta\zeta\,ds\,dt, \tag{1.4.7}$$

where $\delta\zeta$ is the variation of ζ on Γ, and $\partial/\partial n$ denotes differentiation in the direction of the outward normal to Γ in the x_1x_3-plane.

Because $\delta\mathrm{J} = 0$, the membrane displacement at the edge must satisfy $(f - \tau\partial\zeta/\partial n)\,\delta\zeta = 0$, and permissible edge conditions are therefore

$$\zeta = 0 \quad \text{or} \quad \tau\partial\zeta/\partial n = f \quad \text{on } \Gamma. \tag{1.4.8}$$

1.4.2 Bending Wave Equation

Flexural waves on a thin plate are called *bending waves.* Consider an isotropic plate of uniform thickness h, whose midplane coincides with the plane $x_2 = 0$ in the undisturbed state (Figure 1.4.2). Let $\zeta(x_1, x_3, t)$ denote a small perpendicular displacement from the point (x_1, x_3). It is assumed that $\zeta \ll h$ and that the wavelength of the flexural motion is much larger than h. Then ζ satisfies bending wave equation [7, 8, 10],

$$m\partial^2\zeta/\partial t^2 + \mathrm{B}\nabla_2^4\zeta = -[p] + F_2, \tag{1.4.9}$$

where $\nabla_2^4 = \nabla_2^2\nabla_2^2$ is the biharmonic operator, $m = \rho_s h$ is the mass per unit area of the plate, and B is the *bending stiffness*

$$\mathrm{B} = \mathrm{E}h^3/12(1-\sigma^2). \tag{1.4.10}$$

The body force F_2 and the fluid loading pressure force [p] are defined as for the membrane of Section 1.4.1.

The boundary conditions at the edge of a finite plate may be derived from Hamilton's principle by the method of Section 1.4.1. The kinetic energy T is given by (1.4.4). The potential energy of bending of a thin plate is a homogeneous, symmetrical quadratic function of the two principal radii of curvature,

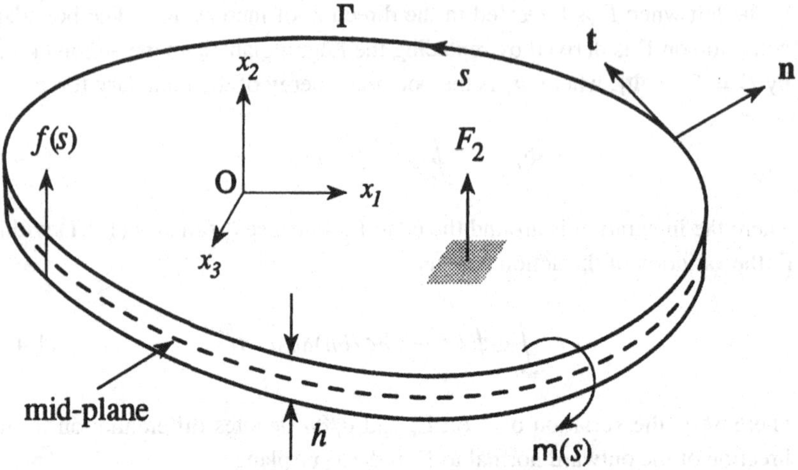

Figure 1.4.2. Forces and moments applied to a thin plate.

that is, of the second derivatives of the displacement ζ. When contributions from the fluid loading $[p]$ and body force potentials are included, Φ_o is given by [10–12],

$$\Phi_o = \int \Big[\frac{1}{2}\mathrm{B}\big((\nabla_2^2\zeta)^2 - 2(1-\sigma)\big[(\partial^2\zeta/\partial x_1^2)(\partial^2\zeta/\partial x_3^2) \\ -(\partial^2\zeta/\partial x_1\partial x_3)^2\big]\big) - (-[p] + F_2)\zeta\Big]\, dx_1\, dx_3. \qquad (1.4.11)$$

Let Γ denote the curvilinear edge of the plate (Figure 1.4.2), and assume the edge is subject to force and moment distributions $f(s)$ and $\mathrm{m}(s)$, where the moment obeys the "right-hand rule" with respect to the direction of increasing s, that is, m is positive in the sense that it tends to stretch the upper surface ($x_2 \approx +0$) of the plate. The additional potential energy Φ_1 for inclusion in the modified Lagrangian $\mathcal{L} = \mathcal{L}_o - \Phi_1$ is then

$$\Phi_1 = -\oint_\Gamma (f(s)\zeta(s,t) - \mathrm{m}(s)\partial\zeta(s,t)/\partial n)\, ds. \qquad (1.4.12)$$

Thus, the contribution of the boundary conditions to $\delta\mathrm{J}$ is

$$\delta\mathrm{J} = -\int_{t_1}^{t_2}\oint_\Gamma ((F - f)\delta\zeta - (M - \mathrm{m})\partial(\delta\zeta)/\partial n)\, ds\, dt, \qquad (1.4.13)$$

where

$$F = -\mathrm{B}\left[\frac{\partial}{\partial n}\nabla_2^2\zeta + (1-\sigma)\frac{\partial}{\partial s}\sum n_i \mathrm{t}_j \frac{\partial^2}{\partial x_i \partial x_j}\zeta\right],$$

$$M = -\mathrm{B}\left[\sigma\nabla_2^2\zeta + (1-\sigma)\sum n_i n_j \frac{\partial^2}{\partial x_i \partial x_j}\zeta\right], \quad (1.4.14)$$

where $\mathbf{t}$ is the unit vector tangential to Γ in the direction of increasing s, and the summations are over $i,\ j = 1$ and 3.

The condition $\delta \mathrm{J} = 0$ supplies the following possible conditions on Γ:

(i) natural edge conditions,

$$F = f, \quad M = \mathrm{m};$$

(ii) simply supported,

$$\zeta = 0, \quad M = 0;$$

(iii) clamped edge,

$$\zeta = 0, \quad \partial\zeta/\partial n = 0. \quad (1.4.15)$$

When the edge coincides with $x_1 =$ constant, and there are no externally applied forces and moments, these conditions become

(i) free edge,

$$\partial^3\zeta/\partial x_1^3 + (2-\sigma)\partial^3\zeta/\partial x_1 \partial x_3^2 = 0, \quad \partial^2\zeta/\partial x_1^2 + \sigma\partial^2\zeta/\partial x_3^2 = 0;$$

(ii) simply supported edge,

$$\zeta = 0, \quad \partial^2\zeta/\partial x_1^2 = 0;$$

(iii) clamped edge,

$$\zeta = 0, \quad \partial\zeta/\partial x_1 = 0. \quad (1.4.16a)$$

For a circular edge of radius R, these conditions are expressed as follows, in terms of polar coordinates (r, θ) with origin at the center of the circle:

(i) free edge,

$$\frac{\partial}{\partial r}\left(\frac{\partial^2}{\partial r^2}+\frac{\partial}{r\partial r}\right)\zeta+\frac{\partial^2}{\partial\theta^2}\left(\frac{(2-\sigma)}{R^2}\frac{\partial}{\partial r}-\frac{(3-\sigma)}{R^3}\right)\zeta=0,$$

$$\frac{\partial^2\zeta}{\partial r^2}+\sigma\left(\frac{1}{R}\frac{\partial}{\partial r}+\frac{1}{R^2}\frac{\partial^2}{\partial\theta^2}\right)\zeta=0;$$

(ii) simply supported edge,

$$\zeta=0,\quad \frac{\partial^2\zeta}{\partial r^2}+\sigma\left(\frac{1}{R}\frac{\partial}{\partial r}+\frac{1}{R^2}\frac{\partial^2}{\partial\theta^2}\right)\zeta=0;$$

(iii) clamped edge,

$$\zeta=0,\quad \partial\zeta/\partial r=0. \tag{1.4.16b}$$

1.4.3 Longitudinal Waves in Plates

The most general deformation of a thin plate is that in which the midplane is stretched and the plate is bent. The kinetic energy of small-amplitude motions is proportional to the plate thickness h. In the absence of bending, the potential energy of deformation (i.e., of stretching) is also proportional to h. This means that when the fluid loading $[p]$ is very small, the equation of motion is independent of plate thickness. However, when bending occurs the potential energy of deformation is proportional to h^3 (cf., (1.4.11), where $\mathrm{B} \propto \mathrm{E}h^3$), and this implies that natural frequencies of vibration are then proportional to h. In practice, the most easily excited plate motions tend to keep the potential energy of deformation as small as possible [10]. For very thin plates this corresponds to bending, or flexural, vibrations.

Long wavelength deformations of a thin plate that are also symmetric with respect to the midplane propagate as quasi-longitudinal waves. They are analogous to P-waves in a solid and have displacement u_α parallel to the direction of propagation α, say. Because the midplane of the plate does not bend, the waves are nondispersive and have wave velocity c_ℓ given by [8, 13]

$$c_\ell=\left(\frac{\mathrm{E}}{\rho_s(1-\sigma^2)}\right)^{\frac{1}{2}}=\frac{c_1\sqrt{1-2\sigma}}{1-\sigma}. \tag{1.4.17}$$

Thus, when $\sigma=0.3$, $c_\ell\approx 0.9c_1$, that is, c_ℓ is about 10% smaller than the purely longitudinal wave speed in the solid.

If the midplane coincides with $x_2 = 0$, the displacement u_2 of the surface of the plate is given by [13],

$$u_2 = \mp \frac{\sigma h}{2(1-\sigma)} \frac{\partial u_\alpha}{\partial x_\alpha}, \tag{1.4.18}$$

where the upper–lower sign is taken on the upper and lower surfaces ($x_2 = \pm\frac{1}{2}h$) of the plate, respectively. For long wavelength disturbances (1.4.18) shows that $u_2 \ll u_\alpha$. The coupling of plate motions to sound waves in the fluid, which occurs only because of *normal* displacements of the plate surface, is therefore likely to be weak.

Similarly, the efficiency with which quasi-longitudinal plate waves are generated by incident sound or other surface pressure fluctuations is small; the most effective fluid driving force is the viscous tangential shear stresses $\sigma_{\pm 2\alpha}$. Because $u_2 \ll u_\alpha$, it follows from (1.2.8) that $\sigma_{\pm 2\alpha} = \pm\eta\partial v_\alpha/\partial x_2$, where v_α is the fluid velocity parallel to the plate. The waves are then governed by the equation [8, 13],

$$\left(\partial^2/\partial t^2 - c_\ell^2 \partial^2/\partial x_\alpha^2\right) u_\alpha = (\eta/\mathrm{m})(\partial v_\alpha/\partial x_2) + F_\alpha/m, \tag{1.4.19}$$

where $\eta(\partial v_\alpha/\partial x_2)$ is the net viscous shear stress due to the fluid on the upper and lower sides of the plate.

1.4.4 Mindlin's Equation

The bending wave equation (1.4.9) ceases to be valid at high frequencies, when the wavelength of flexural motions become comparable with the plate thickness h. Improved accuracy is achieved over the whole frequency range by an approximate equation developed by Mindlin [14], which includes the effects of rotary inertia and shear deformations that are neglected in (1.4.9). Mindlin's equation can be expressed in the form

$$m\frac{\partial^2 \zeta}{\partial t^2} + \mathrm{B}\left(\nabla_2^2 - \frac{1}{\mu_* c_2^2}\frac{\partial^2}{\partial t^2}\right)\left(\nabla_2^2 - \frac{1}{c_\ell^2}\frac{\partial^2}{\partial t^2}\right)\zeta$$
$$= \left[1 - \frac{h^2}{6\mu_*(1-\sigma)}\left(\nabla_2^2 - \frac{1}{c_\ell^2}\frac{\partial^2}{\partial t^2}\right)\right](-[p] + F_2), \tag{1.4.20}$$

where c_2 is the shear wave speed of the material of the plate, defined as in (1.3.7), c_ℓ is the propagation speed of quasi-longitudinal waves (1.4.17), and μ_* is an empirical correction factor. The value of μ_* depends on Poisson's ratio σ and varies in the range 0.76–0.91; $\mu_* = \pi^2/12$ is a good compromise value that optimizes the agreement with the exact theory at lower frequencies.

By comparing Mindlin's equation with (1.4.9) an estimate can be made of the maximum frequency for which the simple plate equation is likely to be valid. To do this, consider time-dependent motions proportional to $e^{-i\omega t}$, of radian frequency ω. The orders of magnitude of the correction terms in (1.4.20) relative to the terms retained in the approximation (1.4.9) can be determined by using the estimates $\partial/\partial t \sim \omega$ and $\nabla_2^2 \sim \omega\sqrt{m/\mathrm{B}}$. The definition (1.4.10) of the bending stiffness B and (1.3.7) then imply that (1.4.9) becomes inapplicable when $\omega h/c_2 \sim 1$.

1.4.5 Wave Propagation in a Cylindrical Shell

The analysis of wave motions on thin shells is much more complicated than for the plane geometry discussed earlier. There exists a multitude of approximate theories [15]; only the very simplest "membrane" theory will be discussed. In this approximation, transverse shear forces and bending and twisting moments are neglected, and the motion is assumed to be governed by the action of normal and shear forces in the midsurface of the shell [7, 16]. The potential energy of deformation is accordingly determined by the stretching of the midsurface and is proportional to the shell thickness h. Small-amplitude vibrations in the *absence* of fluid loading are therefore independent of h.

Consider a circular cylindrical shell of radius R whose axis coincides with the x-axis of the cylindrical coordinate system (x, θ, r). Let the corresponding displacements be (u, v, w) (Figure 1.4.3). When body forces are neglected, the equations of shell membrane theory are [7, 16]

$$\begin{aligned} u_{tt}/c_\ell^2 &= u_{xx} + (\sigma/R)(w_x + v_{x\theta}) + (1-\sigma)(v_{x\theta} + u_{\theta\theta}/R)/2R, \\ v_{tt}/c_\ell^2 &= (\sigma u_{\theta x} + v_{\theta\theta}/R + w_\theta/R)/R + \frac{1}{2}(1-\sigma)(v_{xx} + u_{\theta x}/R), \\ w_{tt}/c_\ell^2 &= -(\sigma u_x + v_\theta/R + w/R)/R - [p]/\left(mc_\ell^2\right), \end{aligned} \tag{1.4.21}$$

where the suffix notation $u_{xx} = \alpha^2 u/\partial x^2$, and so forth has been used, and $-[p]$ is the net pressure force on the shell in the direction of increasing r.

The limiting form of the shell equations should strictly reduce to the bending wave equation (1.4.9) when the radius R is large compared to all other length scales of the motion. Donnell [13, 15] has shown that a consistent approximation that exhibits this behavior is obtained by adding a stiffness term $-(\mathrm{B}/mc_\ell^2)\nabla^4 w$ to the right-hand side of the third of equations (1.4.21), where $\nabla^4 = (\partial^2/\partial x^2 + \partial^2/\partial s^2)^2$, $ds = R\,dh$. Similarly, additional terms are sometimes included that are formally of order $(h/R)^2 \ll 1$ relative to those shown

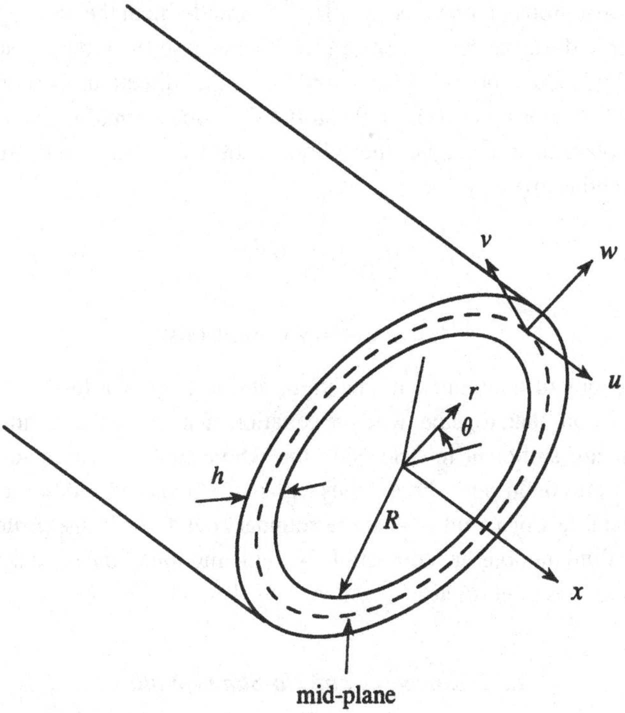

Figure 1.4.3. Coordinates used in the membrane theory of a cylindrical shell.

in equations (1.4.21) [13]; these are important in the modeling of very low-frequency vibrations (well below the "ring frequency" $\omega = c_\ell/R$).

1.4.6 Longitudinal Waves in a Solid Cylinder

Longitudinal waves propagating axially along a uniform, *solid* cylinder are described by an equation analogous to (1.4.19) when the wavelength is much larger than the cylinder diameter. Taking cylindrical coordinates as in Section 1.4.5, u being the displacement in the axial direction x, and assuming the cylinder has circular cross-section of radius R, the lowest order approximate equation of motion (in the absence of body forces) is [16, Section 289]

$$\left(\frac{\partial^2}{\partial t^2} - \frac{\mathrm{E}}{\rho_s}\frac{\partial^2}{\partial x^2}\right)u = \frac{2\eta}{\rho_s R}\frac{\partial v_x}{\partial r} \tag{1.4.22}$$

where v_x is the component of fluid velocity in the x-direction, η is the shear viscosity, and the derivative on the right-hand side is evaluated at $r = R$.

The wave propagation velocity $\sqrt{E/\rho_s}$ is smaller than the wave speed c_ℓ of longitudinal disturbances in a plate (1.4.17) because the effective stiffness is reduced by lateral motions of the cylinder in radial directions, whereas for the plate, such motions can occur only in the direction normal to the plate. The radial displacement w couples the cylinder motions to pressure fluctuations in the fluid and is given by

$$w = -\sigma R \partial u/\partial x. \tag{1.4.23}$$

1.5 Boundary Conditions

The equations of continuity, momentum, and energy of a fluid provide five scalar equations that, together with the equation of state and the thermodynamic identities, are sufficient to determine the whole motion. The solution must satisfy certain conditions at boundaries with other fluids and with solids. Except in the special case of a rigid solid, these boundary conditions must ensure that the predicted fluid motions are compatible with the motion of the solid determined by the equations of elasticity.

1.5.1 Kinematic and No-Slip Conditions

The *no-slip* condition, that the relative velocity of fluid in contact with the surface is zero, is derived from the notion that fluid molecules in the immediate vicinity of the surface are in thermodynamic equilibrium with the surface. At very high Reynolds number there is a thin *boundary layer* between the main fluid motions and the surface, within which viscous stresses are dominant and the fluid velocity rapidly adjusts to that of the surface. In many cases, the presence of the boundary layer can be ignored in a first approximation, and the external flow determined from the Euler equations. The outer edge of the boundary layer becomes the effective boundary of the flow, over which the fluid glides exactly as if the motion is inviscid. It is then unnecessary to impose any condition on the tangential velocity component: The Euler equations are one order less in velocity gradients than the Navier–Stokes equations, and the number of boundary conditions is also reduced by one. The variation of the normal component of velocity across the layer is small, and the inviscid (Euler) boundary condition is equivalent to the kinematic condition that the normal components of velocity of the solid and fluid are equal. However, high Reynolds number flow over a surface often exhibits *separation* [3], where streamlines close to the surface turn abruptly away into the region of high-speed flow, creating a "bubble" of slower moving, turbulent flow near the surface and also

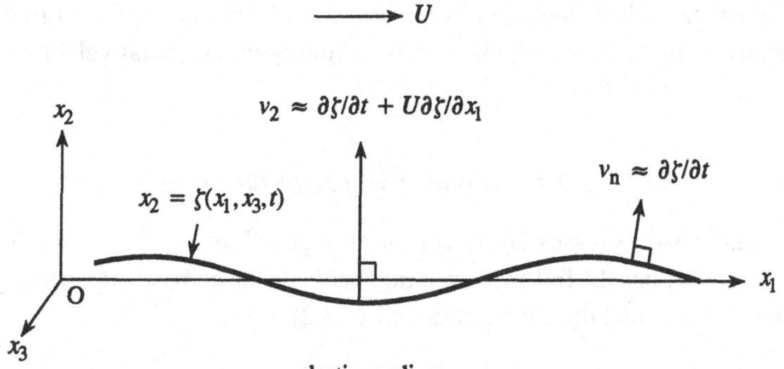

Figure 1.5.1. Kinematic conditions at a moving surface.

possibly a turbulent wake further downstream. Simple approximations based on Euler's equations generally fail in these circumstances.

Let $f(\mathbf{x}, t) = 0$ be the equation of the moving surface of the solid. The normal velocity of the surface is equal to $-(\partial f/\partial t)/|\nabla f|$ (in the direction of increasing f). This equals the normal component of fluid velocity, $\mathbf{v}\cdot\nabla f/|\nabla f|$, provided

$$Df/Dt \equiv \partial f/\partial t + \mathbf{v}\cdot\nabla f = 0. \tag{1.5.1}$$

An important special case occurs when $f \equiv x_2 - \zeta(x_1, x_3, t) = 0$ defines a small deformation of a surface from the plane $x_2 = 0$ in the presence of a uniform mean flow at speed U in the x_1-direction (Figure 1.5.1). If $\mathbf{v}$ is the perturbation fluid velocity relative to its undisturbed value $(U, 0, 0)$, condition (1.5.1) becomes

$$\partial\zeta/\partial t + (U + v_1)\partial\zeta/\partial x_1 - v_2 + v_3\partial\zeta/\partial x_3 = 0, \tag{1.5.2}$$

which can be linearized when $\mathbf{v}/U$ and $|\nabla\zeta|$ are small, yielding

$$v_2 = \partial\zeta/\partial t + U\partial\zeta/\partial x_1. \tag{1.5.3}$$

This expresses the component v_2 of the fluid velocity normal to the *undisturbed* surface $x_2 = 0$ in terms of the motion of the surface. The linearized approximation for the fluid velocity normal to the *moving* surface is

$$v_n = \partial\zeta/\partial t. \tag{1.5.4}$$

The condition $Df/Dt = 0$ of (1.5.1) is a kinematic condition that is satisfied by any surface that moves with the fluid, for example, by the moving interface between two different fluids. The tangential component of velocity is continuous across the interface. However, it is frequently convenient to model regions of high shear, where the velocity varies rapidly across a very thin transition layer, by a *vortex sheet*, which is a discontinuity in tangential velocity whose surface, $f(\mathbf{x}, t) = 0$, also satisfies (1.5.1).

1.5.2 Conditions Satisfied by the Stress

Fluid and elastic stresses are in equilibrium at a fluid-solid interface S. This condition supplies the following relation between the compressive stress p_{ij} of the fluid (1.2.5) and the elastic stress τ_{ij} (1.3.2)

$$n_j(\tau_{ij} + p_{ij}) = 0 \quad \text{on S}, \tag{1.5.5}$$

where n is the unit normal on S. For an immovable, *rigid* body, the stress within the solid cannot be expressed in a useful form, and (1.5.5) reduces to a statement of the surface stress in terms of the fluid motion.

In situations like that illustrated in Figure 1.5.1, involving a small deformation from a planar interface, the undisturbed values of τ_{ij} and p_{ij} may be assumed to vanish. In the linearized approximation the stress condition (1.5.5) is applied at the undisturbed position $x_2 = 0$ of the interface. For a Stokesian fluid (1.2.8) this becomes

$$\tau_{22} = -p + 2\eta\left(\partial v_2/\partial x_2 - \frac{1}{3}\text{div }\mathbf{v}\right),$$
$$\tau_{2\alpha} = 2\eta(\partial v_2/\partial x_\alpha + \partial v_\alpha/\partial x_2), \quad \alpha = 1, 3. \tag{1.5.6}$$

At high Reynolds number viscous stresses are often negligible compared to the normal pressure forces. The relations (1.5.6) are then $\tau_{22} = -p$, $\tau_{2\alpha} = 0$, $\alpha = 1, 3$. If the surface is highly compliant, however, it may be necessary to retain the contribution from the tangential component of the viscous stress.

At an interface between two fluids, $n_j p_{ij}$ is continuous. In the inviscid approximation this reduces to continuity of pressure. This is typically the approximation used for a gas bubble in a liquid, although when the bubble radius is very small (less than about 10^{-2} cm for air bubbles in water) the influence of surface tension should also be included [2, 4]. If the bubble radius is R, surface tension causes the bubble pressure to exceed the ambient pressure by $2\text{T}/R$, where T is the surface tension of the gas–liquid interface.

1.5.3 Thermal Boundary Conditions

The energy equation (1.2.34) describes the transfer of heat by molecular conduction and convection. Convection can be neglected very close to a solid surface so that heat transfer between the solid and fluid is dominated by thermal conduction; this is analogous to the dominance of the viscous diffusion of vorticity near the surface. The condition that the surface layer of fluid be in thermodynamic equilibrium with the solid requires the fluid and solid temperatures to be equal at the surface.

1.6 Sound Propagation in Homentropic, Irrotational Flow

Vibrating mechanical structures, agitated regions of turbulent flow, the mixing of flows of different temperatures, and so forth produce fluctuations in the pressure that are communicated throughout a fluid as sound – a succession of compressions and rarefactions that propagate away from the acoustic source region. The acoustic amplitude is usually small relative to the undisturbed mean pressure, and a first approximation to the wave equation governing propagation may be derived by linearizing the equations of motion. When the propagation distances of interest are not too large it is also permissible to neglect attenuation of the sound by viscosity and thermal conduction.

1.6.1 Acoustic Waves in a Stationary, Uniform Fluid

In the absence of mean flow and dissipation, the linearized energy equation (1.2.34) reduces to $\partial s/\partial t = 0$ so that the compressions and rarefactions of fluid elements take place adiabatically. This approximation tends to be better at lower frequencies because viscous dissipation and heat transfer by thermal conduction are weaker when the acoustic wavelengths (which increase with decreasing frequency) are longer [6]. Consider the propagation of sound in a medium of uniform mean density ρ_o and pressure p_o. The departures of the density and pressure from these mean values will be denoted by p', ρ', where $p'/p_o \ll 1$, $\rho'/\rho_o \ll 1$. The linearized, inviscid momentum equation (1.2.4) is

$$\rho_o \partial \mathbf{v}/\partial t + \nabla p' = \mathbf{F}, \tag{1.6.1}$$

where the body force $\mathbf{F}$ is assumed to vanish in the undisturbed state.

It is often permissible to interpret $\mathbf{F}$ as a localized force applied to the fluid at a solid boundary (e.g., by a vibrating body) rather than a distributed body force within the fluid. Similarly, it is sometimes convenient to consider *volume sources* of fluid by writing the continuity equation (1.2.2) in the form

$$\frac{1}{\rho}\frac{D\rho}{Dt} + \operatorname{div}\mathbf{v} = q, \tag{1.6.2}$$

where the source strength $q \equiv q(\mathbf{x}, t)$ is the rate of increase of fluid volume per unit volume of the fluid. For example, q might represent the influence of volume pulsations of a body immersed in the fluid. Linearizing, (1.6.2) becomes

$$(1/\rho_o)\partial\rho'/\partial t + \operatorname{div}\mathbf{v} = q, \tag{1.6.3}$$

and by eliminating $\mathbf{v}$ between (1.6.1) and (1.6.3) we find

$$\partial^2\rho'/\partial t^2 - \nabla^2 p' = \rho_o \partial q/\partial t - \operatorname{div}\mathbf{F}. \tag{1.6.4}$$

The perturbation density ρ' can be expressed in terms of p' by using the equation of state in the form $p = p(\rho, s)$. For arbitrary and small adiabatic changes, dp and $d\rho$ are related by

$$dp = (\partial p/\partial\rho)_s\, d\rho \equiv d\rho/\rho \mathrm{K}_s \equiv c^2 d\rho, \tag{1.6.5}$$

where the derivative is taken at constant s, $\mathrm{K}_s \equiv 1/\rho c^2$ is the adiabatic compressibility, and c is the *speed of sound*

$$c = \sqrt{(\partial p/\partial\rho)_s}. \tag{1.6.6}$$

When the perturbations are from a uniform state of pressure p_o and density ρ_o, the derivative may be evaluated by setting $p = p_o$, $\rho = \rho_o$. The speed of sound so defined will be denoted by c_o.

Thus, replacing ρ' in (1.6.4) by p'/c_o^2, we find

$$\left(\partial^2/c_o^2\partial t^2 - \nabla^2\right)p = \rho_o\partial q/\partial t - \operatorname{div}\mathbf{F}, \tag{1.6.7}$$

where the prime $'$ on the acoustic pressure has been discarded. Equation (1.6.7) is an inhomogeneous wave equation describing the production of sound waves by the volume source q and the force $\mathbf{F}$. It will often be convenient to replace q and $\mathbf{F}$ by a generalized pressure source $\mathcal{F}(\mathbf{x}, t)$ and to write

$$\left(\partial^2/c_o^2\partial t^2 - \nabla^2\right)p = \mathcal{F}(\mathbf{x}, t). \tag{1.6.8}$$

At large distances (where $\mathcal{F} = 0$) pressure fluctuations propagate away from the source region at speed c_o. In an ideal gas (Section 1.2.3), at temperature T_o,

$$c_o = \sqrt{\gamma p_o/\rho_o} = \sqrt{\gamma \mathrm{R} T_o}. \tag{1.6.9}$$

For dry air at $T_o = 10°\mathrm{C}$, $c_o \approx 337$ m/s. In water at the same temperature,

$$c_o \approx 1447 + 0.16 p_o \text{ m/s}, \tag{1.6.10}$$

provided p_o is expressed in atmospheres; in *sea water* the sound speed given by (1.6.10) should be increased by about 43 m/s.

Example 1. In the absence of mean body forces, show that the sound generated by a volume source q and force $\mathbf{F}$ in a stationary fluid of uniform mean pressure p_o but variable mean density $\rho_o = \rho_o(\mathbf{x})$ is governed by

$$\frac{1}{\rho_o c_o^2}\frac{\partial^2 p}{\partial t^2} - \frac{\partial}{\partial x_j}\left(\frac{1}{\rho_o}\frac{\partial p}{\partial x_j}\right) = \frac{\partial q}{\partial t} - \operatorname{div}(\mathbf{F}/\rho_o). \qquad (1.6.11)$$

Example 2. The body force in (1.6.7) is regarded as a time-dependent *source* of sound. When the influence of gravity is included $\mathbf{F}$ involves a component $\rho_o\mathbf{g}$, where $\mathbf{g}$ is the gravitational acceleration. The mean pressure and density vary with position, such that $\nabla p_o = \rho_o\mathbf{g}$, and in the absence of time-dependent forces (1.6.1) becomes

$$\rho_o \partial\mathbf{v}/\partial t + \nabla p' = \rho'\mathbf{g}.$$

Deduce that gravitational effects are negligible when the acoustic frequency ω satisfies [17]

$$\omega \gg g/c_o. \qquad (1.6.12)$$

For air at sea level $g/c_o \approx 0.03\ \mathrm{s}^{-1}$.

1.6.2 Wave Equation for the Velocity Potential

Small-amplitude acoustic waves in a stationary ideal fluid, with no body forces, can be expressed in terms of the velocity potential of (1.2.20) according to

$$\mathbf{v} = \nabla\varphi, \quad p - p_o = (\rho - \rho_o)c_o^2 = -\rho_o \partial\varphi/\partial t. \qquad (1.6.13)$$

It follows from (1.6.3) that

$$\left(\partial^2/c_o^2\partial t^2 - \nabla^2\right)\varphi = -q. \qquad (1.6.14)$$

Outside the source region (where $q = 0$, $\mathbf{F} = \mathbf{0}$), fluctuations in $\mathbf{v}$ and p, ρ and T (and also e and w, but not the specific entropy s) propagate as sound waves (governed by the homogeneous form of (1.6.14)). The velocity $\mathbf{v}$ associated with the sound is called the *acoustic particle velocity*.

1.6.3 Plane Waves

A plane acoustic wave propagates one dimensionally, say in the x_1-direction, and satisfies

$$(\partial^2/c_o^2\partial t^2 - \partial^2/\partial x_1^2)\varphi = 0. \tag{1.6.15}$$

The general solution of this equation is

$$\varphi = \varphi_+(t - x_1/c_o) + \varphi_-(t + x_1/c_o), \tag{1.6.16}$$

where φ_+, φ_- are arbitrary functions respectively, representing disturbances propagating at speed c_o without change of form in the positive and negative x_1-directions. Equation (1.6.13) shows that the acoustic particle velocity $\mathbf{v}$ is parallel to the propagation direction.

Example 3. For a plane wave propagating parallel to the x_1-axis

$$v_1 = \pm p/\rho_o c_o, \quad \rho' = p/c_o^2, \quad T' = p/\rho_o c_p, \quad w' = p/\rho_o, \tag{1.6.17}$$

where p is the acoustic pressure, and the $\pm$ sign is taken according as the wave propagates in the positive or negative x_1-direction.

1.6.4 Acoustic Waves in an Irrotational Mean Flow

The simplest description of sound propagation in a moving medium occurs when the background flow is steady and *irrotational*. To examine this case we continue to neglect dissipation and to regard the motion as homentropic, but we impose no restriction on the amplitude of the sound. Volume sources and body forces will be ignored so that sound generation occurs only by motion of the boundaries.

Let the undisturbed mean flow be specified by a velocity potential $\varphi_o(\mathbf{x})$, with mean velocity $\mathbf{U} = \nabla\varphi_o$. In an unbounded fluid $\mathbf{U} =$ constant, and $\varphi_o = \mathbf{U}.\mathbf{x}$. $\mathbf{U}$ can vary with position only if the fluid is bounded, either internally by an airfoil, say, or externally by the walls of a duct of variable cross-section; $\varphi_o(\mathbf{x})$ may be multiple valued if the bounding surfaces are multiply connected, but the mean velocity is always single valued.

Consider an irrotational disturbance $\varphi'(\mathbf{x}, t)$, and set

$$\varphi(\mathbf{x}, t) = \varphi_o(\mathbf{x}) + \varphi'(\mathbf{x}, t). \tag{1.6.18}$$

The general equation satisfied by φ is nonlinear [18], but it is considerably simplified when expressed as an equation for $\dot{\varphi}(\mathbf{x}, t) \equiv \partial\varphi'/\partial t$. To do this, substitute $\mathbf{v} = \nabla\varphi$ in the continuity equation (1.2.2), take the partial derivative with respect to time, and use the adiabatic relation (1.6.5) to find

$$\nabla^2\dot{\varphi} + \frac{\partial}{\partial t}\left(\frac{1}{\rho c^2}\frac{Dp}{Dt}\right) = 0, \tag{1.6.19}$$

where $c \equiv c(\mathbf{x}, t)$ is the speed of sound evaluated from (1.6.6) by using the local values of the pressure and density. The latter are given by $p = p_o(\mathbf{x}) + p'(\mathbf{x}, t)$ and $\rho = \rho_o(\mathbf{x}) + \rho'(\mathbf{x}, t)$, where p_o, ρ_o are the undisturbed mean values, and primed quantities represent the perturbations caused by the sound.

In homentropic flow, ρc^2 may be regarded as a function of the pressure alone so that

$$\frac{\partial}{\partial t}\left(\frac{1}{\rho c^2}\frac{Dp}{Dt}\right) \equiv \frac{\partial}{\partial t}\left(\frac{D}{Dt}\int\frac{dp}{\rho c^2}\right) = \frac{D}{Dt}\left(\frac{1}{\rho c^2}\frac{\partial p}{\partial t}\right) + \frac{1}{\rho c^2}\nabla p\cdot\nabla\dot{\varphi}, \tag{1.6.20}$$

where the integral is evaluated at constant specific entropy s. When s is constant, equation (1.2.12) is $dw = dp/\rho$, and the partial time derivative of Bernoulli's equation (1.2.21) (in which $\Phi = 0$ and $f(t) =$ constant) supplies the relation

$$\frac{1}{\rho}\frac{\partial p}{\partial t} = -\frac{D\dot{\varphi}}{Dt}. \tag{1.6.21}$$

This is used to eliminate $\partial p/\partial t$ from the right of (1.6.20), after which substitution into (1.6.19) yields

$$\left(\frac{D}{Dt}\left(\frac{1}{c^2}\frac{D}{Dt}\right) - \frac{\nabla p}{\rho c^2}\cdot\nabla - \nabla^2\right)\dot{\varphi} = 0. \tag{1.6.22}$$

This is the desired propagation equation. It may also be cast in the following equivalent forms (by using the relations $\nabla p = -\rho D\mathbf{v}/Dt = c^2\nabla\rho$)

$$\left(\frac{D}{Dt}\left(\frac{1}{c^2}\frac{D}{Dt}\right) + \frac{1}{c^2}\frac{D\mathbf{v}}{Dt}\cdot\nabla - \nabla^2\right)\dot{\varphi} = 0. \tag{1.6.23}$$

$$\left(\frac{D}{Dt}\left(\frac{1}{c^2}\frac{D}{Dt}\right) - \frac{1}{\rho}\nabla\cdot(\rho\nabla)\right)\dot{\varphi} = 0. \tag{1.6.24}$$

The coefficients of the differential operators in (1.6.22)–(1.6.24) are functions of both mean and perturbation quantities. Small-amplitude acoustic waves satisfy the linearized versions of these equations, obtained by replacing these coefficients by their values in the *absence of the sound.* The mean density and sound speed may be expressed in terms of the variable mean velocity $\mathbf{U}$ (see

Section 1.6.5) and $D/Dt \approx \partial/\partial t + \mathbf{U}\cdot\nabla$. Furthermore, because the mean flow is time independent, we can take the perturbation potential φ', rather than $\dot{\varphi}$, as the acoustic variable. Equation (1.6.24), for example, would then assume the form

$$\left\{\left(\frac{\partial}{\partial t}+\mathbf{U}\cdot\nabla\right)\left[\frac{1}{c^2}\left(\frac{\partial}{\partial t}+\mathbf{U}\cdot\nabla\right)\right]-\frac{1}{\rho}\nabla\cdot(\rho\nabla)\right\}\varphi=\hat{\mathcal{F}}(\mathbf{x},t), \tag{1.6.25}$$

where the notation has been modified so that φ now denotes the perturbation potential, and the equation has been generalized to include a source term $\hat{\mathcal{F}}$, where the hat distinguishes this source from the pressure source $\mathcal{F}$ of equation (1.6.8). It is understood that $c(\mathbf{x})$ and $\rho(\mathbf{x})$ are the local sound speed and density in the steady flow.

1.6.5 Dependence of Sound Speed and Mean Flow Variables on Velocity

According to Bernoulli's equation (1.2.21), in steady irrotational flow at velocity $\mathbf{U}(\mathbf{x})$ the total enthalpy $B = w + \frac{1}{2}U^2$ is constant throughout the fluid. For an ideal gas, (1.2.17) and (1.6.9) give

$$w = \frac{\gamma p}{(\gamma-1)\rho} = \frac{c^2}{\gamma-1}. \tag{1.6.26}$$

This can be used in conjunction with Bernoulli's equation to show that

$$c(\mathbf{x})/c_o = \sqrt{1+\frac{1}{2}(\gamma-1)M_o^2}, \quad M_o = U/c_o, \tag{1.6.27}$$

where c_o is the stagnation, or *reservoir*, sound speed at which $\mathbf{U} = \mathbf{0}$. $M_o = U(\mathbf{x})/c_o$ is the local Mach number referred to the sound speed c_o. The value of c departs significantly from the stagnation sound speed only when M_o becomes large. In dry air, for example, c differs from c_o by less than 1% when $M_o < 0.3$, and the difference does not exceed 5% until M_o is greater than 0.7. The proportional changes in other thermodynamic quantities can be calculated from (1.6.27) and the formulae

$$T/T_o = (c/c_o)^2, \quad \rho/\rho_o = (c/co)^{2/(\gamma-1)}, \quad p/p_o = (c/c_o)^{2\gamma/(\gamma-1)}. \tag{1.6.28}$$

These results are also applicable along individual streamlines of the mean flow when the motion is *isentropic* ($Ds/Dt = 0$, but s may take different values on different streamlines). However, Crocco's equation (1.2.18) shows that such flows are generally rotational.

Mach numbers encountered in underwater applications rarely exceed 0.01 and, apart from exceptions in which, for example, small vapor bubbles are present, it is usually permissible to assume that the mean temperature, density, and sound speed are independent of the mean velocity. Changes in the mean pressure are given by

$$\frac{p - p_o}{\rho_o} + \frac{1}{2}U^2 = 0. \tag{1.6.29}$$

1.6.6 The Convected Wave Equation

At very low Mach numbers ($M^2 \ll 1,\ M = U(\mathbf{x})/c$) variations in the mean density and sound speed can be neglected in equation (1.6.25), which then reduces to the inhomogeneous *convected wave equation*,

$$\left(\frac{1}{c_o^2}\left(\frac{\partial}{\partial t} + \mathbf{U}\cdot\nabla\right)^2 - \nabla^2\right)\varphi = \hat{\mathcal{F}}(\mathbf{x}, t), \tag{1.6.30}$$

where c_o is constant. This equation remains valid when $\mathbf{U}$ is constant for *arbitrary* values of the Mach number because the mean density and sound speed are also constant (although neglected nonlinear terms become important in transonic flows where $M \approx 1$ in the presence of stationary sources or "thin" streamlined bodies; see Section 3.6).

Example 4. Show that the perturbation pressure and the perturbation density are *not* solutions of the convected wave equation, except when $\mathbf{U}$ = constant.

Example 5. In the presence of a uniform mean flow at speed U in the x_1-direction, plane wave solutions of the convected wave equation that propagate parallel to the mean flow are given by

$$\varphi = \varphi_+(t - x_1/c_o(1+M)) + \varphi_-(t + x_1/c_o(1-M)), \quad M = U/c_o,$$

where $\varphi_\pm$ respectively represent waves propagating at speeds $c_o(1 \pm M)$ in the positive and negative x_1-direction.

1.7 Fundamental Solutions of the Acoustic Wave Equations

To investigate the production of sound it is convenient to consider first sources whose strengths vary periodically with time. This is because the wave equation is linear, and solutions corresponding to the different frequency components of the source may be superposed, permitting the full solution to be expressed as a Fourier series or integral.

1.7.1 The Helmholtz Equation

Consider the sound radiated into an unbounded, stationary fluid from a time-harmonic volume source $q(\mathbf{x}, t) = q(\mathbf{x}, \omega)e^{-i\omega t}$ of radian frequency ω. The velocity potential of the unsteady motion is governed by equation (1.6.14) and evidently oscillates at the same frequency. The substitution $\varphi(\mathbf{x}, t) = \varphi(\mathbf{x}, \omega)e^{-i\omega t}$ transforms (1.6.14) into the inhomogeneous Helmholtz equation

$$(\nabla^2 + \kappa_o^2)\varphi = q, \tag{1.7.1}$$

where

$$\kappa_o = \omega/c_o \tag{1.7.2}$$

is the *acoustic wavenumber*. φ and q are understood to represent the *frequency-domain* quantities $\varphi(\mathbf{x}, \omega)$ and $q(\mathbf{x}, \omega)$. This shorthand notation will be adopted henceforth, except when the intended meaning is not clear from the context. In an unbounded fluid, the solution of (1.7.1) must satisfy the *radiation condition* that energy delivered to the fluid by the source radiates away from the source, in other words, the solution must exhibit *outgoing wave behavior.*

The case of a unit point source $q(\mathbf{x}, \omega) = \delta(\mathbf{x} - \mathbf{y}) \equiv \delta(x_1 - y_1)\delta(x_2 - y_2)\delta(x_3 - y_3)$, where δ denotes the Dirac δ-function [19, 20], defines the frequency-domain *Green's function* $G(\mathbf{x}, \mathbf{y}; \omega)$, which satisfies

$$(\nabla^2 + \kappa_o^2)G = \delta(\mathbf{x} - \mathbf{y}). \tag{1.7.3}$$

The identity $q(\mathbf{x}, \omega) = \int_V q(\mathbf{y}, \omega)\delta(\mathbf{x} - \mathbf{y})\, d^3\mathbf{y}$, where the integration is taken over a volume V containing the sources, and the principle of superposition enables the solution of the general problem (1.7.1) to be written

$$\varphi(\mathbf{x}, \omega) = \int_V G(\mathbf{x}, \mathbf{y}; \omega) q(\mathbf{y}, \omega)\, d^3\mathbf{y}. \tag{1.7.4}$$

Equation (1.7.3) will be solved by the method of Fourier transforms [20]. The n-dimensional Fourier space-transform $f(\mathbf{k})$ of a function $f(\mathbf{x})$ of $\mathbf{x} = (x_1, x_2, \ldots, x_n)$ satisfies the reciprocal relations

$$f(\mathbf{k}) = \frac{1}{(2\pi)^n} \int_{-\infty}^{\infty} f(\mathbf{x})e^{-i\mathbf{k}\cdot\mathbf{x}}\, d^n\mathbf{x}, \quad f(\mathbf{x}) = \int_{-\infty}^{\infty} f(\mathbf{k})e^{i\mathbf{k}\cdot\mathbf{x}}\, d^n\mathbf{k}, \tag{1.7.5}$$

where $\mathbf{k}$ is the n-dimensional *wavenumber vector* $\mathbf{k} = (k_1, k_2, \ldots, k_n)$.

Take the Fourier transform of (1.7.3) by applying the integral operator $(1/2\pi)^3 \int_{-\infty}^{\infty}(\cdot)\mathrm{e}^{-i\mathbf{k}\cdot\mathbf{x}}\,d^3\mathbf{x}$. By integration by parts (which is permissible for any generalized function such as $G(\mathbf{x},\mathbf{y};\omega)$), we write $\int_{-\infty}^{\infty}\nabla^2 G(\mathbf{x},\mathbf{y};\omega)\mathrm{e}^{-i\mathbf{k}\cdot\mathbf{x}}\,d^3\mathbf{x}$ $= -k^2\int_{-\infty}^{\infty} G(\mathbf{x},\mathbf{y};\omega) \equiv -(2\pi)^3 k^2 G(\mathbf{k},\mathbf{y};\omega)$, where $k = |\mathbf{k}|$, to obtain

$$G(\mathbf{k},\mathbf{y};\omega) = \frac{-\mathrm{e}^{-i\mathbf{k}\cdot\mathbf{y}}}{(2\pi)^3\left(k^2-\kappa_o^2\right)}. \tag{1.7.6}$$

The inverse transform (the second of equations (1.7.5)) now yields

$$G(\mathbf{x},\mathbf{y};\omega) = \frac{-1}{(2\pi)^3}\int_{-\infty}^{\infty}\frac{\mathrm{e}^{i\mathbf{k}\cdot(\mathbf{x}-\mathbf{y})}}{\left(k^2-\kappa_o^2\right)}\,d^3\mathbf{k}. \tag{1.7.7}$$

To evaluate the integral, introduce spherical polar coordinates (k,θ,ϕ) for $\mathbf{k}$, where the latitude θ is measured from the source–observer direction of $\mathbf{x}-\mathbf{y}$, and $d^3\mathbf{k} = k^2\sin\theta\,d\theta\,d\phi\,dk$. Performing the integrations with respect to θ and ϕ, we find

$$G(\mathbf{x},\mathbf{y};\omega) = \frac{i}{(2\pi)^2|\mathbf{x}-\mathbf{y}|}\int_0^{\infty}\frac{k}{\left(k^2-\kappa_o^2\right)}\left(\mathrm{e}^{ik|\mathbf{x}-\mathbf{y}|}-\mathrm{e}^{-ik|\mathbf{x}-\mathbf{y}|}\right)dk. \tag{1.7.8}$$

The remaining integral is undefined for *real* values of κ_o because of the pole on the real axis at $k = |\kappa_o|$. The pole can be avoided by indenting the path into the complex plane to pass either above or below the singularity. Two alternative solutions of equation (1.7.3) are obtained in this way, but only one, where the path runs below or above the pole according as $\kappa_o \gtrless 0$, exhibits outgoing waves.

To prove this, suppose that $\kappa_o > 0$. Indent the integration path to pass below the pole at $k = \kappa_o$ (Figure 1.7.1), and consider the separate contributions to the integral in (1.7.8) from the two exponential terms in the integrand. For the first, Cauchy's theorem [21] permits the path of integration to be rotated through 90° onto the positive imaginary axis, on which the integrand decays exponentially as $k \to +i\infty$. The pole at $k = \kappa_o$ is crossed during this rotation, giving a residue contribution $i\pi\mathrm{e}^{i\kappa_o|\mathbf{x}-\mathbf{y}|}$ to the value of the integral. The path of integration for the second exponential may be similarly rotated onto the *negative* imaginary axis, this time without encountering any singularities. The two integrals on the positive and negative imaginary axes are equal and opposite, and therefore

$$G(\mathbf{x},\mathbf{y};\omega) = \frac{-\mathrm{e}^{i\kappa_o|\mathbf{x}-\mathbf{y}|}}{4\pi|\mathbf{x}-\mathbf{y}|}. \tag{1.7.9}$$

When the exponential time factor is restored $G(\mathbf{x},\mathbf{y};\omega)\mathrm{e}^{-i\omega t} = -\mathrm{e}^{-i\omega(t-|\mathbf{x}-\mathbf{y}|/c_o)}/4\pi|\mathbf{x}-\mathbf{y}|$, which represents a continuous wave propagating radially outward at

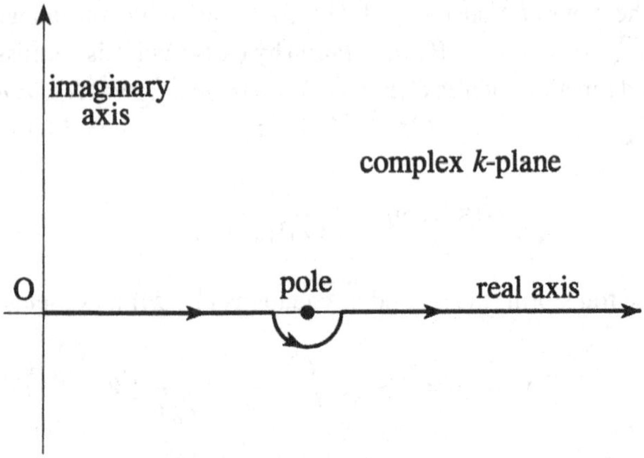

Figure 1.7.1. Path of integration in the wavenumber plane when $\kappa_o > 0$.

speed c_o (with increasing time) from the source at $\mathbf{y}$. The amplitude decreases inversely with distance $|\mathbf{x}-\mathbf{y}|$, in accordance with the inverse square law of energy conservation. When $|\mathbf{x}-\mathbf{y}| \to \infty$, the wave "crests" (surfaces of constant *phase*) become locally plane with wavelength $2\pi/\kappa_o$.

Conversely, when the integration path in (1.7.8) passes *above* the pole at $k = \kappa_o > 0$, the solution is of the form (1.7.9), but with the sign of κ_o reversed. This represents a spherical wave converging toward the source in violation of the radiation condition.

The outgoing wave solution is also obtained from the condition that dissipation within the fluid (which gradually transforms acoustic energy into heat) causes the wave to decay faster than $1/|\mathbf{x}-\mathbf{y}|$ at large distances from the source. This will be the case if, ab initio, κ_o is imagined to be assigned a small positive imaginary part that (for $\omega > 0$) shifts the pole off the real axis into the upper half-plane.

This is effectively identical to a formal method proposed by Landau [22] and Lighthill [23] for obtaining the *causal* solution of time-harmonic wave problems. The frequency ω is temporarily assigned a small positive imaginary part $i\epsilon$, which is equivalent to considering the radiation from a source that is "switched-on" in the infinite past, whose amplitude increases slowly in proportion to $e^{\epsilon t}$. The exponential growth of the solution as $t \to +\infty$ is of no significance because causality ensures that it cannot influence the behavior at finite times. When $\kappa_o > 0$ the pole is displaced to $k = \kappa_o + i\epsilon/c_o$ in the upper half-plane, and the integration is performed as before, leading to (1.7.9). The method is applicable generally to time-harmonic wave problems governed by linear equations. However, misleading predictions can occur for systems that are linearly unstable. For example, the linearized treatment of the motion produced by a source in a steady flow over an elastic surface exhibits linear instabilities that grow by extracting energy from the mean flow. These instabilities are not captured by the Landau–Lighthill procedure because it assumes that all of the perturbation energy is supplied by the source. In these circumstances a causal solution can be derived by considering an initial value problem (Section 1.7.4).

Example 1. Show that solutions of equation (1.7.3) that are radially symmetric with respect to the point source satisfy

$$\frac{1}{r^2}\frac{\partial}{\partial r}\left(r^2\frac{\partial}{\partial r}\right)G + \kappa_o^2 G \equiv \frac{1}{r}\frac{\partial^2}{\partial r^2}(rG) + \kappa_o^2 G = 0, \quad \text{for } r = |\mathbf{x}-\mathbf{y}| > 0$$

and that the solution with outgoing wave behavior is $G = Ae^{i\kappa_o r}/r$. Find the value of the constant A and recover (1.7.9) by integrating (1.7.3) over the interior of a small sphere centered on the source whose radius is subsequently allowed to vanish.

Example 2. Use the procedure of Section 1.7.1 to show that Green's function $G(\mathbf{x}, \mathbf{y}; \omega)$ for the Helmholtz equation in two dimensions,

$$(\nabla_2^2 + \kappa_o^2)G = \delta(\mathbf{x}-\mathbf{y}), \quad \mathbf{x} = (x_1, x_2), \quad \mathbf{y} = (y_1, y_2),$$

is

$$G(\mathbf{x}, \mathbf{y}; \omega) = \frac{-i}{4}\mathrm{H}_0^{(1)}(\kappa_o|\mathbf{x}-\mathbf{y}|), \tag{1.7.10}$$

where $\mathrm{H}_0^{(1)}$ is a Hankel function [24]. This represents a cylindrical disturbance whose behavior at large distances from the source is given by

$$G(\mathbf{x}, \mathbf{y}; \omega) \approx \frac{-e^{i(\kappa_o|\mathbf{x}-\mathbf{y}|+\frac{\pi}{4})}}{\sqrt{8\pi\kappa_o|\mathbf{x}-\mathbf{y}|}}, \quad |\mathbf{x}-\mathbf{y}| \to \infty.$$

Example 3. Derive (1.7.10) by integrating the three-dimensional Green's function (1.7.9) over $-\infty < y_3 < \infty$.

Example 4. Show that Green's function for the one-dimensional Helmholtz equation can be written

$$G(x, y; \omega) = \frac{-ie^{i\kappa_o|x-y|}}{2\kappa_o} \tag{1.7.11}$$

Example 5. The integration in (1.7.7) with respect to the component k_2, say, of $\mathbf{k}$ may be performed by residues. There are simple poles at $k_2 = \pm\sqrt{\kappa_o^2 - k_1^2 - k_3^2}$ that (for $k_1^2 + k_3^2 < \kappa_o^2$) move to opposite sides of the real k_2-axis when κ_o is replaced by $\kappa_o + i\epsilon$ ($\epsilon > 0$). If we define the square root by $\sqrt{\kappa_o^2-k_1^2-k_3^2} = \operatorname{sgn}(\kappa_o)|\sqrt{\kappa_o^2-k_1^2-k_3^2}|$ for $\epsilon = 0$ and $k_1^2 + k_3^2 < \kappa_o^2$, then the poles are displaced respectively into the upper/lower half-planes when $\epsilon > 0$. Analytic continuation for $\epsilon = +0$ then implies that $\sqrt{\kappa_o^2-k_1^2-k_3^2} = +i|\sqrt{\kappa_o^2-k_1^2-k_3^2}|$ for $k_1^2+k_3^2 > \kappa_o^2$. When $x_2 \gtrless y_2$ the integration contour can be shifted from the real k_2-axis to $k_2 = \pm i\infty$, thereby capturing one of these poles and yielding

$$G(\mathbf{x}, \mathbf{y}; \omega) = \frac{-i}{8\pi^2}$$
$$\times \int_{-\infty}^{\infty} \frac{\exp\left(i\left[k_1(x_1-y_1)+k_3(x_3-y_3)+|x_2-y_2|\sqrt{\kappa_o^2-k_1^2-k_3^2}\right]\right)}{\sqrt{\kappa_o^2-k_1^2-k_3^2}} dk_1\, dk_3, \tag{1.7.12}$$

where $\operatorname{sgn}(\sqrt{\kappa_o^2-k_1^2-k_3^2}) = \operatorname{sgn}(\kappa_o)$ for $k_1^2 + k_3^2 < \kappa_o^2$, and $\sqrt{\kappa_o^2-k_1^2-k_3^2} = +i|\sqrt{\kappa_o^2-k_1^2-k_3^2}|$ for $k_1^2 + k_3^2 > \kappa_o^2$.

Contributions to the integral from the high wavenumber region $|k_1^2+k_3^2|^{1/2} > |\kappa_o|$ "outside the acoustic domain" decay exponentially with distance $|x_2 - y_2|$ from the source, that is, the radiation field is dominated by wavenumbers having "supersonic" trace phase velocity $\omega/\sqrt{k_1^2 + k_3^2} > c_o$ in planes of constant x_2. This representation of G is used in problems involving the interaction of sound with a plane boundary $x_2 =$ constant.

1.7.2 Influence of Uniform Mean Flow

In the presence of a uniform flow at speed U in the x_1- direction, $\varphi(\mathbf{x}, \omega)$ satisfies the "time-reduced" form of the convected wave equation (1.6.30), obtained by replacing $\partial/\partial t$ by $-i\omega$. Let us determine the solution G of this equation for *subsonic* mean flow ($M = U/c_o < 1$) when the "source" $\hat{\mathcal{F}}(\mathbf{x}, \omega) = -\delta(\mathbf{x}-\mathbf{y})$, this is, when

$$(\nabla^2 + (\kappa_o + iM\partial/\partial x_1)^2)G = \delta(\mathbf{x} - \mathbf{y}). \tag{1.7.13}$$

We set

$$\begin{aligned} x_1' &= x_1/\sqrt{1-M^2}, \quad x_2' = x_2, \quad x_3' = x_3, \\ \kappa_o' &= \kappa_o/\sqrt{1-M^2}, \quad G = G' \mathrm{e}^{-i\kappa_o' M x_1'}, \end{aligned} \tag{1.7.14}$$

then, because $\delta(x_1 - y_1) \equiv \sqrt{1-M^2}\delta(x_1' - y_1')$, (1.7.13) becomes

$$\left(\nabla'^2 + \kappa_o'^2\right)G' = \frac{\delta(\mathbf{x}' - \mathbf{y}')e^{i\kappa_o' M y_1'}}{\sqrt{1-M^2}}, \tag{1.7.15}$$

where ∇'^2 is the Laplacian with respect to $\mathbf{x}'$. This equation is similar to (1.7.3), and the outgoing wave solution can be written down by inspection from the corresponding solution (1.7.9),

$$G'(\mathbf{x}', \mathbf{y}'; \omega) = \frac{-e^{i\kappa_o'(|\mathbf{x}'-\mathbf{y}'|+My_1')}}{4\pi\sqrt{1-M^2}|\mathbf{x}'-\mathbf{y}'|}. \tag{1.7.16}$$

Reverting to the original coordinates

$$G(\mathbf{x}, \mathbf{y}; \omega) = \frac{-\exp\left(i\kappa_o\left[\left(\frac{|\mathbf{x}-\mathbf{y}|}{1-M^2} + \frac{M^2(x_1-y_1)^2}{(1-M^2)^2}\right)^{\frac{1}{2}} - \frac{M(x_1-y_1)}{1-M^2}\right]\right)}{4\pi\left(|\mathbf{x}-\mathbf{y}|^2 + \frac{M^2(x_1-y_1)^2}{1-M^2}\right)^{\frac{1}{2}}}. \tag{1.7.17}$$

Example 6. The source in equation (1.7.13) is fixed relative to the stationary coordinate axes in a mean flow at velocity $\mathbf{U} = (U, 0, 0)$. Its volume velocity q, say, is therefore different from the value $q = 1$ it would have in the absence of flow by an amount dependent on M^2. By evaluating the integral $\oint \nabla G \cdot d\mathbf{S}$ over the surface of a small sphere enclosing the source, show that $q = (\sqrt{1-M^2}/2M)\ln[(1+M)/(1-M)] \approx 1 - \frac{1}{6}M^2 - 11/120\,M^4 + \cdots$, when $M < 1$.

1.7.3 Time-Domain Green's Function

Green's function for the wave equation (1.6.8) is the solution $G(\mathbf{x}, \mathbf{y}, t-\tau)$ with outgoing wave behavior of

$$\left(\partial^2/c_o^2\partial t^2 - \nabla^2\right)G = \delta(\mathbf{x}-\mathbf{y})\delta(t-\tau). \tag{1.7.18}$$

The right-hand side represents an impulsive point source that vanishes except at $t = \tau$.

The equation is solved by superposition, by observing that $\delta(t-\tau) = (1/2\pi)\int_{-\infty}^{\infty} e^{-i\omega(t-\tau)}\,d\omega$ and, therefore, that (1.7.3) is transformed into (1.7.18) by application of the integral operator $(-1/2\pi)\int_{-\infty}^{\infty}(\cdot)e^{-i\omega(t-\tau)}\,d\omega$. Hence,

$$G(\mathbf{x}, \mathbf{y}, t-\tau) = \frac{-1}{2\pi}\int_{-\infty}^{\infty} G(\mathbf{x}, \mathbf{y}; \omega)e^{-i\omega(t-\tau)}\,d\omega \tag{1.7.19}$$

so that in *three dimensions* the solution (1.7.9) of (1.7.3) supplies

$$G(\mathbf{x}, \mathbf{y}, t-\tau) = \frac{1}{8\pi^2|\mathbf{x}-\mathbf{y}|}\int_{-\infty}^{\infty} e^{-i\omega(t-\tau-|\mathbf{x}-\mathbf{y}|/c_o)}\, d\omega$$

$$= \frac{1}{4\pi|\mathbf{x}-\mathbf{y}|}\delta(t-\tau-|\mathbf{x}-\mathbf{y}|/c_o). \tag{1.7.20}$$

This vanishes for $t < \tau$, in accordance with the causality principle, and represents an impulsive, spherically symmetric wave expanding from the source at $\mathbf{y}$ at the speed of sound.

The causal solution in unbounded fluid of the inhomogeneous wave equation (1.6.8) is obtained from this result by noting that

$$\mathcal{F}(\mathbf{x}, t) = \int\int_{-\infty}^{\infty} \mathcal{F}(\mathbf{y}, \tau)\delta(\mathbf{x}-\mathbf{y})\delta(t-\tau)\, d^3\mathbf{y}\, d\tau$$

and, therefore, that

$$p(\mathbf{x}, t) = \int\int_{-\infty}^{\infty} \mathcal{F}(\mathbf{y}, \tau)G(\mathbf{x}, \mathbf{y}, t-\tau)\, d^3\mathbf{y}\, d\tau \tag{1.7.21}$$

$$= \frac{1}{4\pi}\int_{-\infty}^{\infty} \frac{\mathcal{F}(\mathbf{y}, t-|\mathbf{x}-\mathbf{y}|/c_o)}{|\mathbf{x}-\mathbf{y}|}\, d^3\mathbf{y}. \tag{1.7.22}$$

The integral formula (1.7.22) is called a *retarded potential* and represents the pressure at position $\mathbf{x}$ and time t as a linear superposition of contributions from sources at positions $\mathbf{y}$ that radiated at the earlier times $t - |\mathbf{x}-\mathbf{y}|/c_o$, $|\mathbf{x}-\mathbf{y}|/c_o$ being the time of travel of sound waves from $\mathbf{y}$ to $\mathbf{x}$.

Example 7. Use (1.7.10) and (1.7.11) to show that in

two dimensions,

$$G(\mathbf{x}, \mathbf{y}, t-\tau) = \frac{\mathrm{H}(t-\tau-|\mathbf{x}-\mathbf{y}|/c_o)}{2\pi\sqrt{(t-\tau)^2-(\mathbf{x}-\mathbf{y})^2/c_o^2}}; \tag{1.7.23}$$

one dimension,

$$G(x, y, t-\tau) = \frac{c_o}{2}\mathrm{H}(t-\tau-|x-y|/c_o), \tag{1.7.24}$$

where $\mathrm{H}(x)$ is the *Heaviside unit function* defined by $\mathrm{H}(x)=0$ for $x < 0$, $\mathrm{H}(x)=1$ for $x > 0$.

In three dimensions, Green's function (1.7.20) defines a spherically radiating pulse that is nonzero only at the wavefront, where it has a δ-function singularity.

The singularity in two dimensions is algebraic, and the wavefront is followed by an extensive, slowly decaying "tail." In one dimension the wave consists of a simple discontinuity followed by a tail of constant amplitude.

1.7.4 Initial Value Problems Solved by the Method of Fourier Transforms

The procedure used in Section 1.7.3 to obtain Green's function for the wave equation will now be formalized into a general technique, involving an additional Fourier transform with respect to time, for determining the *causal* solution $\psi(\mathbf{x}, t)$ of the inhomogeneous equation

$$\mathcal{L}(-i\partial/\partial\mathbf{x}, i\partial/\partial t)\psi = \mathcal{F}(\mathbf{x}, t). \tag{1.7.25}$$

$\mathcal{L}$ is a linear differential operator that has no explicit dependence on $\mathbf{x}$ or t. The source distribution $\mathcal{F}(\mathbf{x}, t)$ vanishes prior to some initial time $t = \tau$, and for $|\mathbf{x}| > \ell$ when $t > \tau$. It is also assumed that the temporal growth of $\mathcal{F}(\mathbf{x}, t)$ is *exponentially bounded* as $t \to \infty$, that is, that $|\mathcal{F}| < e^{\alpha t}$ for some fixed number $\alpha > 0$.

Introduce the space–time Fourier transform defined as in (1.7.5), but with the following convention as regards signs, which is usually adopted in the treatment of wave problems,

$$\begin{aligned} f(\mathbf{k}, \omega) &= \frac{1}{(2\pi)^{n+1}} \int_{-\infty}^{\infty} f(\mathbf{x}, t) e^{-i(\mathbf{k}\cdot\mathbf{x} - \omega t)} \, d^n\mathbf{x}\, dt, \\ f(\mathbf{x}, t) &= \int_{-\infty}^{\infty} f(\mathbf{k}, \omega) e^{i(\mathbf{k}\cdot\mathbf{x} - \omega t)} \, d^n\mathbf{k}\, d\omega. \end{aligned} \tag{1.7.26}$$

Because $\mathcal{F}(\mathbf{x}, t) = 0$ for $|\mathbf{x}| > \ell$, its Fourier transform $\mathcal{F}(\mathbf{k}, \omega)$ is a regular function for both real and complex values of the wavenumber $\mathbf{k}$. Also, because $\mathcal{F}(\mathbf{x}, t) = 0$ for $t < \tau$, and it grows no faster than $e^{\alpha t}$ as $t \to \infty$, $\mathcal{F}(\mathbf{k}, \omega)$ is a regular function of ω in the upper half-plane $\mathrm{Im}\, \omega > \alpha$.

Take the space–time Fourier transform of equation (1.7.25) and solve the resulting algebraic equation to obtain $\psi(\mathbf{k}, \omega) = \mathcal{F}(\mathbf{k}, \omega)/\mathcal{L}(\mathbf{k}, \omega)$. The second of equations (1.7.26) then provides a representation of $\psi(\mathbf{x}, t)$ as a fourfold Fourier integral. Now, because of the finite speed of wave propagation in any real system, for any given time $t \geq \tau$, $\psi(\mathbf{x}, t)$ must vanish for sufficiently large $|\mathbf{x}|$, and $\psi(\mathbf{k}, t)$ is therefore a regular function for real and complex values of $\mathbf{k}$. The value of $\psi(\mathbf{x}, t) = \int_{-\infty}^{\infty} \psi(\mathbf{k}, t) e^{i\mathbf{k}\cdot\mathbf{x}} \, d^3\mathbf{k}$ is therefore independent of the route followed by the path of integration between $k = \pm\infty$. As a function of time, however, $\psi(\mathbf{k}, t) = \int_{-\infty}^{\infty} \psi(\mathbf{k}, \omega) e^{-i\omega t} \, d\omega = 0$ for $t < \tau$, and this requires that the path of integration in the complex ω-plane should pass above all of the singularities of $\psi(\mathbf{k}, \omega)$, for in that case Cauchy's theorem permits the contour to be displaced to $+i\infty$ when $t < \tau$ yielding a null result for $\psi(\mathbf{x}, t)$.

Thus, the causal solution of (1.7.25) is

$$\psi(\mathbf{x}, t) = \int_{-\infty}^{\infty} d^3\mathbf{k} \int_{-\infty+i\epsilon}^{\infty+i\epsilon} \frac{\mathcal{F}(\mathbf{k}, \omega)}{\mathcal{L}(\mathbf{k}, \omega)} \mathrm{e}^{i(\mathbf{k}\cdot\mathbf{x}-\omega t)}\, d\omega, \tag{1.7.27}$$

where ϵ ($\geq \alpha$) is sufficiently large and positive that the path of integration in the ω-plane passes above all singularities of the integrand.

Example 8. Express the acoustic Green's function (solution of (1.7.18)) in the form

$$G(\mathbf{x}, \mathbf{y}, t-\tau) = \frac{1}{(2\pi)^4} \int_{-\infty}^{\infty} d^3\mathbf{k} \int_{-\infty+i\epsilon}^{\infty+i\epsilon} \frac{\mathrm{e}^{i[\mathbf{k}\cdot(\mathbf{x}-\mathbf{y})-\omega(t-\tau)]}}{k^2 - \kappa_o^2}\, d\omega. \tag{1.7.28}$$

For real $\mathbf{k}$ the integrand has simple poles on the real ω-axis at $\omega = \pm c_o k$, and ϵ must be positive to ensure that $G(\mathbf{x}, \mathbf{y}, t-\tau) = 0$ for $t < \tau$ (Figure 1.7.2).

When $t > \tau$, we can let $\epsilon \to +0$, that is, take the path in the ω-plane to be just above the real axis. The procedure of Section 1.7.2 could then be reproduced by interchanging the orders of integration, performing that with respect to $\mathbf{k}$ before the ω-integration. Instead of doing this, take the limit $\epsilon \to -\infty$ and show (by evaluating the residues of poles captured during the displacement of

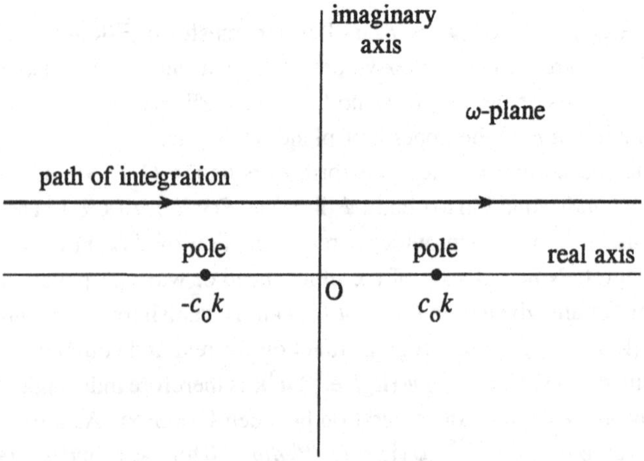

Figure 1.7.2. Integration contour and poles in the ω-plane.

the contour) that for $t > \tau$

$$G(\mathbf{x}, \mathbf{y}, t - \tau) = \frac{ic_o}{16\pi^3} \int_{-\infty}^{\infty} \left[e^{-ic_o k(t-\tau)} - e^{ic_o k(t-\tau)}\right] \frac{e^{i\mathbf{k}\cdot(\mathbf{x}-\mathbf{y})}}{k}\, d^3\mathbf{k}. \tag{1.7.29}$$

The integrand is proportional to $G(\mathbf{k}, \mathbf{y}, t - \tau)$ and is regular for all real and complex values of $\mathbf{k}$ (because $G(\mathbf{x}, \mathbf{y}, t-\tau) \equiv 0$ for $|\mathbf{x}-\mathbf{y}| > c_o(t-\tau)$). Obtain (1.7.20) by evaluating the wavenumber integral by transforming to spherical polar coordinates (k, θ, ϕ), as in Section 1.7.2.

1.8 Multipoles, Radiation Efficiency, and the Influence of Source Motion

1.8.1 Multipole Sources

A pressure source $\mathcal{F}$ in the wave equation (1.6.8) of the form

$$\mathcal{F} = \partial^n F_{ijk\ldots}/\partial x_i \partial x_j \partial x_k \cdots \tag{1.8.1}$$

is called a *multipole of order* 2^n. The acoustic pressure in an unbounded fluid can be found from (1.7.21). Provided $F_{ijk\ldots}$ decreases sufficiently fast as $|\mathbf{y}| \to \infty$, repeated integration by parts furnishes

$$p(\mathbf{x}, t) = (-1)^n \int_{-\infty}^{\infty} F_{ijk\ldots}(\mathbf{y}, \tau) \frac{\partial^n}{\partial y_i \partial y_j \partial y_k \cdots} \times G(\mathbf{x}, \mathbf{y}, t - \tau)\, d^3\mathbf{y}\, d\tau. \tag{1.8.2}$$

Now, $\partial G/\partial y_i = -\partial G/\partial x_i$ in an unbounded medium so that this is equivalent to

$$p(\mathbf{x}, t) = \frac{\partial^n}{\partial x_i \partial x_j \partial x_k \cdots} \int_{-\infty}^{\infty} F_{ijk\ldots}(\mathbf{y}, \tau) G(\mathbf{x}, \mathbf{y}, t - \tau)\, d^3\mathbf{y}\, d\tau. \tag{1.8.3}$$

This formula is applicable for propagation in one, two, or three dimensions. In three dimensions, Green's function (1.7.20) gives

$$p(\mathbf{x}, t) = \frac{\partial^n}{\partial x_i \partial x_j \partial x_k \cdots} \int_{-\infty}^{\infty} \frac{F_{ijk\ldots}(\mathbf{y}, t - \tau - |\mathbf{x} - \mathbf{y}|/c_o)}{4\pi|\mathbf{x} - \mathbf{y}|}\, d^3\mathbf{y}. \tag{1.8.4}$$

1.8.2 Compact Sources

A source of characteristic frequency ω radiates sound of wavelength $\sim 2\pi c_o/\omega$. When the extent of the fluid region occupied by the source $\sim \ell$, the source is *acoustically compact* if $\omega\ell/c_o \equiv \kappa_o \ell \ll 1$, that is, if ℓ is much smaller than the wavelength. Assuming the coordinate origin to be within the source region, a point $\mathbf{x}$ is said to be in the *acoustic far field*, many acoustic wavelengths from the source, when $\kappa_o|\mathbf{x}| \gg 1$. At such points the wave fronts are locally "flat," and the plane wave relations (1.6.17) are approximately satisfied.

For a compact multipole centered on the origin, the representation (1.8.4) may be simplified by neglecting differences in the retarded times $t - |\mathbf{x} - \mathbf{y}\}/c_o \approx t - |\mathbf{x}|/c_o$ for different positions $\mathbf{y}$ within the source region. When $|\mathbf{x}| \to \infty$, $|\mathbf{x} - \mathbf{y}|$ is replaced by $|\mathbf{x}|$ in the denominator of the integrand to obtain

$$p(\mathbf{x}, t) \approx \frac{(-1)^n x_i x_j x_k \cdots}{4\pi c_o^n |\mathbf{x}|^{n+1}} \frac{\partial^n}{\partial t^n} \int_{-\infty}^{\infty} F_{ijk\ldots}(\mathbf{y}, t - |\mathbf{x}|/c_o)\, d^3\mathbf{y},$$

$$|\mathbf{x}| \to \infty, \qquad (1.8.5)$$

where all terms decaying faster than $1/|\mathbf{x}|$ have been discarded because, according to the inverse square law, they can make no contribution to the radiation of energy to infinity. The integral in this expression vanishes only if the source is actually equivalent to a multipole of order higher than 2^n. The acoustic amplitude decreases like $1/|\mathbf{x}|$ as $|\mathbf{x}| \to \infty$; its dependence on the *direction* of propagation (the "field shape") is determined by the product of $F_{ijk\ldots}$ and the direction cosines $(x_i/|\mathbf{x}|)(x_j/|\mathbf{x}|)(x_k/|\mathbf{x}|)\cdots$.

The pressure field (1.8.5) is identical to that generated by the *point* multipole,

$$\mathcal{F}(\mathbf{x}, t) = S_{ijk\ldots}(t) \frac{\partial^n \delta(\mathbf{x})}{\partial x_i \partial x_j \partial x_k \cdots}, \qquad (1.8.6)$$

of strength

$$S_{ijk\ldots}(t) = \int_{-\infty}^{\infty} F_{ijk\ldots}(\mathbf{y}, t)\, d^3\mathbf{y}. \qquad (1.8.7)$$

Example 1. A compact *volume source* concentrated near the origin, $\mathcal{F}(\mathbf{x}, t) = \rho_o \partial q(\mathbf{x}, t)/\partial t$ (Section 1.6.1, equation (1.6.7)), is a multipole of order one (a *monopole*). The source strength $m(t) = \int q(\mathbf{y}, t)\, d^3\mathbf{y}$ is equal to the rate of production of fluid volume, and the pressure waves form the spherically symmetric disturbance

$$p(\mathbf{x}, t) \approx \frac{\rho_o}{4\pi|\mathbf{x}|} \frac{\partial m}{\partial t}(t - |\mathbf{x}|/c_o), \quad |\mathbf{x}| \to \infty,$$

which is the same as the sound produced by a point source $\mathcal{F}(\mathbf{x},t) = \rho_o\delta(\mathbf{x})\partial m(t)/\partial t$.

Example 2. A compact body force $\mathbf{F}(\mathbf{x},t)$ is an acoustic *dipole* (an order two multipole). If the force is centered on $\mathbf{x} = \mathbf{0}$, we find, because $\mathcal{F} = -\operatorname{div}\mathbf{F}$,

$$p(\mathbf{x},t) \approx \frac{x_j}{4\pi c_o|\mathbf{x}|^2}\frac{\partial \mathrm{f}_j}{\partial t}(t - |\mathbf{x}|/c_o), \quad |\mathbf{x}| \to \infty,$$

where $\mathbf{f}(t) = \int \mathbf{F}(\mathbf{y},t)\,d^3\mathbf{y}$. The radiation at large distances is the same as that produced by a point force $\mathbf{f}(t)\delta(\mathbf{x})$. If the direction of $\mathbf{f}$ is constant, the field shape has two lobes, with the peak radiation intensities occurring in the directions of $\pm\mathbf{f}$.

Example 3. *Quadrupole* sources $\mathcal{F}(\mathbf{x},t) = \partial^2 \mathrm{T}_{ij}/\partial x_i \partial x_j$ are important in the Lighthill theory of aerodynamic sound (Section 2.1). They correspond to turbulence stress distributions whose integrated strength vanishes (because the fluid cannot exert a net force on itself) and generate the pressure field

$$p(\mathbf{x},t) = \frac{\partial^2}{\partial x_i \partial x_j}\int_{-\infty}^{\infty}\frac{\mathrm{T}_{ij}(\mathbf{y}, t - \tau - |\mathbf{x}-\mathbf{y}|/c_o)}{4\pi|\mathbf{x}-\mathbf{y}|}\,d^3\mathbf{y}. \tag{1.8.8}$$

It is generally not permissible to assume that the turbulent flow is compact so that retarded time differences across the source region cannot be neglected. If the origin is within the turbulent domain, the far field pressure is therefore

$$p(\mathbf{x},t) \approx \frac{x_i x_j}{4\pi c_o^2|\mathbf{x}|^3}\frac{\partial^2}{\partial t^2}\int_{-\infty}^{\infty}\mathrm{T}_{ij}(\mathbf{y}, t - |\mathbf{x}-\mathbf{y}|/c_o)\,d^3\mathbf{y}, \quad |\mathbf{x}| \to \infty. \tag{1.8.9}$$

1.8.3 Multipole Expansions

The acoustic far field of an arbitrary three-dimensional source $\mathcal{F}(\mathbf{x},t)$ is the same as that generated by the infinite series of point multipoles

$$\sum_{n=0}^{\infty}\frac{\mathrm{S}^n_{ijk\ldots}(t)\partial^n\delta(\mathbf{x})}{\partial x_j \partial x_j \partial x_k \cdots}, \tag{1.8.10}$$

where

$$\mathrm{S}^n_{ijk\ldots}(t) = \frac{(-1)^n}{n!}\int_{-\infty}^{\infty} y_i y_j y_k \cdots \mathcal{F}(\mathbf{y},t)\,d^3\mathbf{y}, \tag{1.8.11}$$

and the integrand contains n factors $y_i, y_j, \ldots$.

To prove this we make the following approximations for $|\mathbf{x}-\mathbf{y}|$ as $|\mathbf{x}| \to \infty$ in the retarded potential formula (1.7.22): (1) in the denominator of the integrand, $|\mathbf{x}-\mathbf{y}|$ is replaced by $|\mathbf{x}|$, and (2) in the retarded time, $t-|\mathbf{x}-\mathbf{y}|/c_o$ is replaced by

$$|\mathbf{x}-\mathbf{y}| \approx |\mathbf{x}| - \mathbf{x}\cdot\mathbf{y}/|\mathbf{x}|, \quad |\mathbf{x}| \gg \ell, \tag{1.8.12}$$

where the second term on the right is of order ℓ, the diameter of the source region, and neglected terms are smaller by a factor of order $\ell/|\mathbf{x}| \ll 1$. This yields the *Fraunhofer approximation*

$$p(\mathbf{x}, t) \approx \frac{1}{4\pi|\mathbf{x}|}\int_{-\infty}^{\infty} \mathcal{F}\left(\mathbf{y}, t - \frac{|\mathbf{x}|}{c_o} + \frac{\mathbf{x}\cdot\mathbf{y}}{c_o|\mathbf{x}|}\right) d^3\mathbf{y}, \quad |\mathbf{x}| \to \infty. \tag{1.8.13}$$

The desired expression in terms of the multipoles (1.8.10) is now obtained by expanding $\mathcal{F}$ in powers of the difference $\mathbf{x}\cdot\mathbf{y}/c_o|\mathbf{x}|$ in the retarded times of different points within the source region:

$$\begin{aligned} p(\mathbf{x}, t) &\approx \frac{1}{4\pi|\mathbf{x}|}\sum_{n=0}^{\infty} \frac{x_i x_j x_k \cdots}{n! c_o^n |\mathbf{x}|^n} \frac{\partial^n}{\partial t^n} \int_{-\infty}^{\infty} y_i y_j y_k \cdots \mathcal{F}(\mathbf{y}, t - |\mathbf{x}|/c_o)\, d^3\mathbf{y} \\ &\approx \sum_{n=0}^{\infty} \frac{\partial^n}{\partial x_i \partial x_j \partial x_k \cdots}\left(\frac{\mathrm{S}^n_{ijk\ldots}(t - |\mathbf{x}|/c_o)}{4\pi|\mathbf{x}|}\right), \quad |\mathbf{x}| \to \infty. \end{aligned} \tag{1.8.14}$$

Each term in the expansion is of order $(\ell/c_o)\partial/\partial t \approx \omega\ell/c_o$ relative to the preceding one. When the source is compact ($\omega\ell/c_o \ll 1$), the far-field behavior is determined by the first nonzero member of the series. If this occurs at $n = m$, the source is equivalent to a multipole of order 2^m. The integral moments (1.8.11) may not converge when n becomes large, in which case the series becomes an asymptotic approximation.

The possibility of approximating a source of sound by an acoustically equivalent set of multipoles is the basis of practical *active control* schemes to suppress undesirable sources of noise [25].

Example 4. Two equal and opposite monopoles separated by a small distance ℓ are together equivalent to a dipole. If the monopoles have source strengths $\pm m(t)$ and are placed respectively at $(\pm\frac{1}{2}\ell, 0, 0)$, show that

$$p(\mathbf{x}, t) \approx \frac{\rho_o \ell \cos\theta}{4\pi c_o|\mathbf{x}|}\frac{\partial^2 m}{\partial t^2}(t - |\mathbf{x}|/c_o), \quad |\mathbf{x}| \to \infty,$$

where θ is the angle between the observer direction $\mathbf{x}/|\mathbf{x}|$ and the x_1-axis.

Example 5.

$$G(\mathbf{x}, \mathbf{y}, t-\tau) \equiv \frac{1}{4\pi|\mathbf{x}-\mathbf{y}|}\delta(t-\tau-|\mathbf{x}-\mathbf{y}|/c_o)$$
$$\approx \frac{1}{4\pi|\mathbf{x}|}\sum_{n=0}^{\infty}\frac{1}{n!}\left(\frac{\mathbf{x}\cdot\mathbf{y}}{c_o|\mathbf{x}|}\right)^n \delta^{(n)}(t-\tau-|\mathbf{x}|/c_o),$$
$$|\mathbf{x}| \to \infty. \qquad (1.8.15)$$

Example 6. For a time-harmonic source $\mathcal{F}(\mathbf{x}, t) \to \mathcal{F}(\mathbf{x})e^{-i\omega t}$,

$$p(\mathbf{x}, t) \approx \frac{\bar{\mathcal{F}}(\theta, \phi)}{4\pi r}e^{i(\kappa_o r-\omega t)}, \quad r = |\mathbf{x}| \to \infty,$$

where (r, θ, ϕ) are spherical polar coordinates, and $\bar{\mathcal{F}}(\theta, \phi) = \int_{-\infty}^{\infty}\mathcal{F}(\mathbf{y})\exp(-i\kappa_o\mathbf{x}\cdot\mathbf{y}/|\mathbf{x}|)\,d^3\mathbf{y}$. Deduce that

$$r\left(\frac{\partial p}{\partial r} - i\kappa_o p\right) \to 0 \quad \text{as } r \to \infty. \qquad (1.8.16)$$

This is the *Sommerfeld radiation condition*, which is always satisfied at sufficiently large distances from the source that the waves may be regarded as locally plane.

1.8.4 Radiation Efficiency

A point whose distance ℓ from a compact source is small compared to the acoustic wavelength is said to lie in the *hydrodynamic domain* of the source. For a multipole $\mathcal{F} = \partial^n F_{ijk\ldots}/\partial x_i \partial x_j \partial x_k \ldots$, the pressure p_h in this region is estimated from (1.8.4) to be of order F/ℓ^{n-2} ($F \sim |F_{ijk\ldots}|$) by neglecting temporal retardations $|\mathbf{x}-\mathbf{y}|/c_o$. On the other hand, in the acoustic far field at a distance $|\mathbf{x}|$, equation (1.8.5) implies that $p \sim F\ell^3\kappa_o^n/|\mathbf{x}|$ for a source of frequency ω ($\kappa_o = \omega/c_o$). The squared ratio$|p/p_h|^2/(\ell/|\mathbf{x}|)^2 \sim (\kappa_o\ell)^{2n}$ is a measure of the *radiation efficiency*, and evidently decreases rapidly with increasing multipole order, because $\kappa_o\ell \ll 1$.

Sources of aerodynamic sound are identified with turbulence of velocity $\sim v$, length scale $\sim\ell$, and frequency $\omega \sim v/\ell$. The efficiencies of low Mach number aerodynamic sources of monopole, dipole, quadrupole, and so forth, type therefore decrease, respectively, in the ratios $1 : M^2 : M^4 : \ldots$, where $M \approx \omega\ell/c_o \sim v/c_o < 1$. Actually the *acoustic efficiency* of an aerodynamic source is usually defined as the ratio of the radiated power to the mechanical power needed to maintain the source (e.g., a turbulent jet), but this modification does not affect the *relative* efficiencies of aerodynamic monopoles, dipoles, and so forth (see Section 2.1.2).

1.8.5 Moving Sources

Consider a pressure point source (equation (1.6.8)) $\mathcal{F}(\mathbf{x}, t) = \mathrm{S}(t)\delta(\mathbf{x} - \mathbf{s}(t))$ of strength $\mathrm{S}(t)$ moving at velocity $\mathbf{U} = \partial \mathbf{s}/\partial t$. The acoustic pressure is given by (1.7.20) and (1.7.21) in the form

$$\begin{aligned} p(\mathbf{x}, t) &= \frac{1}{4\pi}\int_{-\infty}^{\infty} \frac{\mathrm{S}(\tau)\delta(t - \tau - |\mathbf{x} - \mathbf{s}(\tau)|/c_o)}{|\mathbf{x} - \mathbf{s}(\tau)|}\, d\tau \\ &= \frac{\mathrm{S}(\tau_e)}{4\pi|\mathbf{x} - \mathbf{s}(\tau_e)|\left|\frac{\partial}{\partial \tau}(t - \tau - |\mathbf{x} - \mathbf{s}(\tau)|/c_o)\right|_{\tau=\tau_e}}, \end{aligned} \tag{1.8.17}$$

where the retarded time τ_e is the solution of the equation

$$c_o(t - \tau_e) = |\mathbf{x} - \mathbf{s}(\tau_e)|. \tag{1.8.18}$$

This has only one real root for *subsonic* motion, and the following discussion is confined to this case.

To obtain a proper comparison of the solution (1.8.17) with that for a stationary source, it is necessary to introduce the *emission time* coordinate $\mathbf{R}$ of an observer at $\mathbf{x}$, defined by

$$\mathbf{R} \equiv \mathbf{R}(\mathbf{x}, t) = \mathbf{x} - \mathbf{s}(\tau_e), \quad R = c_o(t - \tau_e). \tag{1.8.19}$$

$\mathbf{R}$ is the position of the observer relative to the source at the time of emission of the sound received at $(\mathbf{x}, t)$. Figure 1.8.1 illustrates the relation between $\mathbf{R}$ and $\mathbf{x}$ when the source translates at constant velocity $\mathbf{U}$, for which

$$\mathbf{R} = \mathbf{x} - \mathbf{s}(t) + \mathbf{M}R, \tag{1.8.20}$$

where $\mathbf{s}(t)$ is the source position when the sound arrives at $(\mathbf{x}, t)$, and $\mathbf{M}R \equiv \mathbf{U}R/c_o$ is the vector displacement of the source during the time of travel of the sound from source to observer.

The following differential relations are easily derived:

$$\begin{aligned} \partial \tau_e/\partial t &= 1/(1 - M\cos\Theta), \\ \partial \tau_e/\partial x_i &= -R_i/Rc_o(1 - M\cos\Theta), \\ \partial R/\partial t &= -U\cos\Theta/(1 - M\cos\Theta), \\ \partial R/\partial x_i &= R_i/R(1 - M\cos\Theta), \end{aligned} \tag{1.8.21}$$

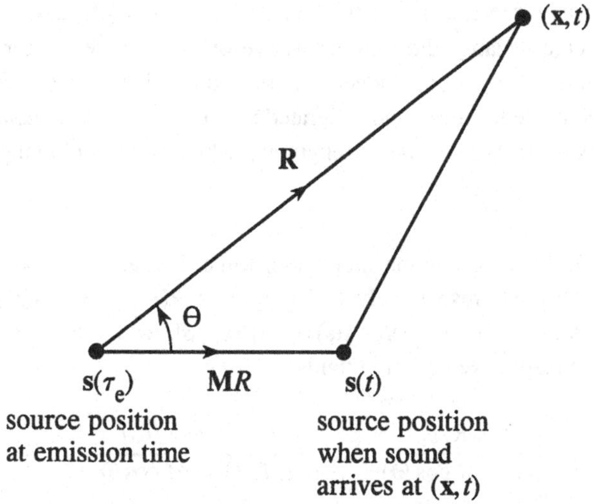

Figure 1.8.1. Emission time coordinates for a uniformly translating source.

where $M = U(\tau_e)/c_o$, and Θ is the angle between **R** and the direction of motion of the source at the emission time.

Thus, for subsonic motion the source field (1.8.17) can be written

$$p(\mathbf{x}, t) = \frac{\mathrm{S}(t - R/c_o)}{4\pi R(1 - M\cos\Theta)}. \tag{1.8.22}$$

In this result $1/(1 - M\cos\Theta)$ is called the *Doppler factor*, which causes sound radiated in the "forward" direction, ahead of the source (where $\cos\Theta > 0$), to be amplified by the motion, whereas sound propagating to the rear is attenuated. When **U** is constant, the characteristic frequency ω of the received sound is modified in a similar manner because

$$\begin{aligned} \omega &= \frac{1}{p}\frac{\partial p}{\partial t} \approx \frac{1}{\mathrm{S}(\tau_e)}\frac{\partial \mathrm{S}(\tau_e)}{\partial t}, \quad \text{as } R \to \infty \\ &= \frac{1}{(1 - M\cos\Theta)}\frac{1}{\mathrm{S}(\tau_e)}\frac{\partial \mathrm{S}(\tau_e)}{\partial \tau_e} \\ &\equiv \frac{\omega_0}{(1 - M\cos\Theta)}, \end{aligned}$$

where ω_0 is the frequency in the absence of motion. The change in frequency is accompanied by a dilation of the acoustic wavelength, the new wavelength being equal to $(1 - M\cos\Theta)$ times the wavelength in the absence of motion.

The frequency dependence of the sound is more complicated for an accelerating source because the time derivative of the Doppler factor in (1.8.22) remains finite as $R \to \infty$. Indeed, the time dependence of this factor is entirely responsible for the sound generated by an accelerating *volume monopole* $q = m_o\delta(\mathbf{x} - \mathbf{s}(t))$ of constant strength m_o, which is "silent" in the absence of motion.

Example 7. Moving volume monopole, force dipole, and quadrupole point sources correspond, respectively, to $\mathcal{F}(\mathbf{x}, t) = \rho_o\partial(m(t)\delta(\mathbf{x} - \mathbf{s}(t)))/\partial t$, $-\text{div}(\mathbf{f}(t)\delta(\mathbf{x} - \mathbf{s}(t)))$, $\partial^2(\mathrm{t}_{ij}(t)\delta(\mathbf{x} - \mathbf{s}(t)))/\partial x_i\partial x_j$. Show that when $\mathbf{U} = \partial\mathbf{s}/\partial t$ is constant the respective radiation fields are

$$p(\mathbf{x}, t) \approx \frac{m'(t - R/c_o)}{4\pi R(1 - M\cos\Theta)^2}, \quad \frac{R_j \mathrm{f}'_j(t - R/c_o)}{4\pi c_o R^2(1 - M\cos\Theta)^2}, \quad \frac{R_i R_j \mathrm{t}'_{ij}(t - R/c_o)}{4\pi c_o^2 R^3(1 - M\cos\Theta)^3}, \quad R \to \infty, \qquad (1.8.23\text{a–c})$$

where a prime$'$ denotes differentiation with respect to the argument.

1.9 Reciprocity

The various Green's functions (1.7.9)–(1.7.11) for the Helmholtz equation in an unbounded fluid satisfy $G(\mathbf{x}, \mathbf{y}; \omega) = G(\mathbf{y}, \mathbf{x}; \omega)$, which implies, rather obviously, that the pressure at $\mathbf{x}$ produced by a point source at $\mathbf{y}$ is equal to the pressure at $\mathbf{y}$ produced when the same source is placed at $\mathbf{x}$. This is a trivial, special case of a vastly more general *reciprocal theorem* due to Rayleigh [10], which has important applications in both theoretical and applied acoustics [13, 26].

1.9.1 Rayleigh's Theorem [10]

Small-amplitude, time-harmonic oscillations of a discrete mechanical system with n degrees of freedom can be represented by a system of generalized coordinates $\psi_i \mathrm{e}^{-i\omega t}$, $i = 1, 2, \ldots, n$ and a corresponding set of generalized forces $\Psi_i \mathrm{e}^{-i\omega t}$. Assume that the undisturbed state ($\psi_i = 0$, $\Psi_i = 0$) is one of stable equilibrium.

The motion is governed by n linear equations

$$C_{ij}\psi_j = \Psi_i, \quad i = 1, 2, \ldots, n, \qquad (1.9.1)$$

where the matrix $C_{ij} \equiv C_{ij}(\omega)$ can be shown to be symmetric ($C_{ij} = C_{ji}$). The natural frequencies of oscillation are the roots of the determinantal equation $\det C_{ij}(\omega) = 0$. The roots are either real or have negative imaginary parts when the vibrations are about a state of stable equilibrium.

Suppose the system is now acted on by a second set of generalized forces Ψ'_i. The resulting displacements ψ'_i are determined by $C_{ij}\psi'_j = \Psi'_i$, $(i = 1, 2, \ldots, n)$, and the symmetry of C_{ij} implies that $\psi'_i C_{ij}\psi_j \equiv \psi_i C_{ij}\psi'_j$ and therefore that ψ_i, ψ'_i, Ψ_i, Ψ'_i satisfy the reciprocal relation

$$\psi_1\Psi'_1 + \psi_2\Psi'_2 + \cdots = \psi'_1\Psi_1 + \psi'_2\Psi_2 + \cdots. \tag{1.9.2}$$

This states that the work performed by the reciprocal forces Ψ'_i subject to the displacements ψ_i is the same as the work performed by forces Ψ_i subject to the reciprocal displacements ψ'_i.

The theorem can also be expressed in terms of the generalized velocities, $\dot{\psi}_i = -i\omega\psi_i$, $\dot{\psi}'_i = -i\omega\psi'_i$:

$$\sum_i \dot{\psi}_i\Psi'_i = \sum_i \dot{\psi}'_i\Psi_i, \tag{1.9.3}$$

which equates the power delivered by the reciprocal forces Ψ'_i subject to the velocities $\dot{\psi}_i$ to the power delivered by the Ψ_i subject to the reciprocal velocities $\dot{\psi}'_i$.

Equations (1.9.2) and (1.9.3) are the most general statements of the reciprocal theorem for systems with a finite number of degrees of freedom. To apply it to small-amplitude oscillations of coupled fluid and continuous elastic media, it is necessary to consider a limit in which the number of degrees of freedom of a discrete representation of the continuum becomes unbounded. However, it is simpler to derive the general acoustic reciprocal theorem directly from the continuum differential equations of motion. The discrete system symmetry condition $C_{ij} = C_{ji}$ is then replaced by the requirement that these equations are *self-adjoint* [9], which is the case for the usual equations describing sound propagation and its interaction with elastic structures *provided there is no mean flow* (see Section 1.9.5).

1.9.2 Reciprocal Theorem for Fluid–Structure Interactions

Consider a stationary (viscous) fluid of uniform mean pressure p_o, but with variable mean density $\rho_o(\mathbf{x})$ and sound speed $c_o(\mathbf{x})$. The fluid is assumed to extend to infinity and to be bounded internally by a system of stationary elastic bodies with surface S (Figure 1.9.1.).

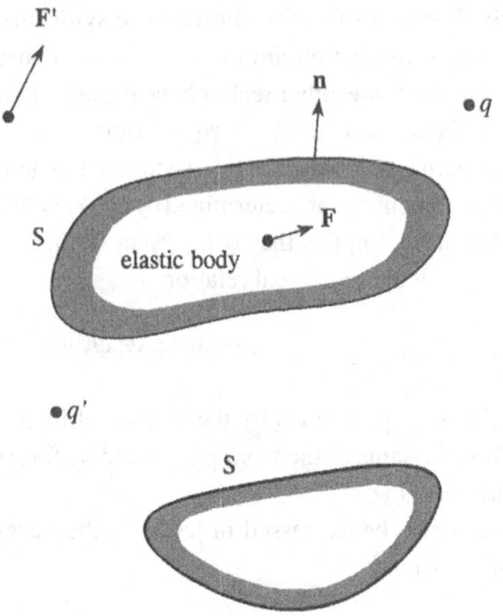

Figure 1.9.1. The reciprocal theorem.

Small-amplitude oscillations in the fluid produced by time-harmonic volume sources $q(\mathbf{x}, \omega)e^{-i\omega t}$ and body forces $\mathbf{F}(\mathbf{x}, \omega)e^{-i\omega t}$ are governed by the linearized forms of the momentum equation (1.2.4) and continuity equation (1.6.2):

$$-i\omega\rho_o v_i + \partial p_{ij}/\partial x_j = F_i, \quad -\left(i\omega/\rho_o c_o^2\right) + \operatorname{div}\mathbf{v} = q, \qquad (1.9.4\text{a, b})$$

where $v_i \equiv v_i(\mathbf{x}, \omega)$ denotes the complex amplitude of the velocity $\mathbf{v}(\mathbf{x}, \omega)e^{-i\omega t}$, and so forth. The simplifying approximation $(1/\rho)D\rho/Dt \equiv (1/\rho c^2)Dp/Dt - (\beta T/c_p)Ds/Dt \approx -(i\omega/\rho_o c_o^2)p$ has been made in the continuity equation. The neglect of the entropy term is equivalent to ignoring dissipation by thermal conduction, which can be just as important in surface thermal boundary layers as the viscous dissipation retained in p_{ij} (see Section 5.1). The reader can verify, however, that the retention of this term does not affect the outcome of the following analysis.

The reciprocal problem involves source and force fields q', $\mathbf{F}'$, and the velocity v_i' and pressure p' are governed by

$$-i\omega\rho_o v_i' + \partial p_{ij}'/\partial x_j = F_i', \quad -\left(i\omega/\rho_o c_o^2\right)p' + \operatorname{div}\mathbf{v}' = q'. \qquad (1.9.5\text{a, b})$$

Add equations (1.9.4a) and (1.9.4b) after multiplying the first by v_i' and the second by $-p'$. Repeat this operation on equations (1.9.5) by using the respective multiplicands v_i and $-p$. Subtract the resulting equations to obtain

$$p'\mathrm{div}\,\mathbf{v} - p\mathrm{div}\,\mathbf{v}' + v_i \partial p'_{ij}\partial x_j - v_i \partial p'_{ij}\partial x_j = (qp' + \mathbf{v}\cdot\mathbf{F}') - (q'p + \mathbf{v}'\cdot\mathbf{F}).$$

Now $p_{ij} = p\delta_{ij} - \sigma_{ij}$, where the viscous stress σ_{ij} is given by (1.2.6). Hence,

$$\frac{\partial}{\partial x_j}\left(v_i p'_{ij} - v'_i p_{ij}\right) = (qp' + \mathbf{v}\cdot\mathbf{F}') - (q'p + \mathbf{v}'\cdot\mathbf{F}), \tag{1.9.6}$$

and integration over the volume V of the fluid, and application of the divergence theorem yields

$$\oint_{\mathrm{S}} \left(v_i p'_{ij} - v'_i p_{ij}\right) n_j dS = \int_{\mathrm{V}} [(qp' + \mathbf{v}\cdot\mathbf{F}'] - (q'p + \mathbf{v}'\cdot\mathbf{F})]\, d^3\mathbf{x} \tag{1.9.7}$$

where the unit normal **n** on S is directed into the fluid.

This equation is also valid for an ideal fluid, where $p_{ij} = p\delta_{ij}$. In that case, however, the absence of dissipation means that sound waves decay like $1/|\mathbf{x}|$ as $|\mathbf{x}| \to \infty$. The surface integral on the left must then be supplemented by a similar integral over a large sphere Σ, say, of radius r that encloses S and all the sources. If ρ_o and the sound speed c_o become constant as $r \to \infty$, the Sommerfeld radiation condition (1.8.16) can be used to show that the integral over Σ vanishes as $r \to \infty$.

The vibrations of the elastic bodies are governed by Navier's equation (1.3.2), for which the above-indicated procedure can be repeated in an obvious manner within the regions V_s occupied by the bodies, yielding

$$\oint_{\mathrm{S}} \left(v'_i \tau_{ij} - v_i \tau'_{ij}\right) n_j dS = \int_{\mathrm{V}_s} (\mathbf{v}\cdot\mathbf{F}' - \mathbf{v}'\cdot\mathbf{F}) d^3\mathbf{x}, \tag{1.9.8}$$

where $\mathbf{v} = -i\omega\mathbf{u}$ is the elastic particle velocity. The density ρ_s and the Lamé constants λ, μ can be functions of position, and structural dissipation can be included by taking λ and μ to be complex functions of ω [8, 13].

Now the no-slip condition requires the velocities of the fluid and solid to be equal on S. Similarly, equilibrium between the fluid and elastic stresses at the interface implies that $(\tau_{ij} + p_{ij})n_j = 0$ on S (Section 1.5.2). Hence, adding equations (1.9.7) and (1.9.8), we obtain the *Reciprocal theorem for fluid-structure interactions*:

$$\int (qp' + \mathbf{v}\cdot\mathbf{F}')\, d^3\mathbf{x} = \int (q'p + \mathbf{v}'.\mathbf{F})\, d^3\mathbf{x}, \tag{1.9.9}$$

where the integrations are over the infinite region occupied by the fluid and solids, and $q \equiv q(\mathbf{x}, \omega)$, $p \equiv p(\mathbf{x}, \omega), \ldots$ are the respective complex amplitudes of the time-harmonic quantities $q(\mathbf{x}, \omega)e^{-i\omega t}$, $p(\mathbf{x}, \omega)e^{-i\omega t}$, and so forth.

This result is equivalent to Rayleigh's theorem (1.9.3), with generalized velocities $\mathbf{v}(\mathbf{x}, \omega)$ and $q(\mathbf{x}, \omega)$ (the "volume velocity") and corresponding generalized forces $\mathbf{F}(\mathbf{x}, \omega)$ and $p(\mathbf{x}, \omega)$. For the discrete point monopoles and body forces

$$\begin{aligned} q &= m(\omega)\delta(\mathbf{x} - \mathbf{x}_q), \qquad & \mathbf{F} &= \mathbf{f}(\omega)\delta(\mathbf{x} - \mathbf{x}_\mathbf{F}), \\ q' &= m'(\omega)\delta(\mathbf{x} - \mathbf{x}_{q'}), \qquad & \mathbf{F}' &= \mathbf{f}'(\omega)\delta(\mathbf{x} - \mathbf{x}_{\mathbf{F}'}), \end{aligned} \tag{1.9.10}$$

where the body force points $\mathbf{x}_\mathbf{F}$, $\mathbf{x}_{\mathbf{F}'}$ can lie in the fluid or within an elastic body (Figure 1.9.1), reciprocity implies

$$mp'(\mathbf{x}_q, \omega) + \mathbf{v}(\mathbf{x}_{\mathbf{F}'}, \omega) \cdot \mathbf{f}' = m'p(\mathbf{x}_{q'}, \omega) + \mathbf{v}'(\mathbf{x}_\mathbf{F}, \omega) \cdot \mathbf{f}. \tag{1.9.11}$$

The reciprocal theorem has been proved for arbitrary inhomogeneous, elastic bodies immersed in an inhomogeneous, compressible, and viscous fluid. Any of these conditions may be relaxed: For example, one or more of the bodies may be rigid and the fluid inviscid and incompressible. The "static" form of the theorem is obtained by dividing (1.9.9) or (1.9.11) by $-i\omega$ and by taking the limit $\omega \to 0$ so that the "velocities" q, q', $\mathbf{v}, \mathbf{v}'$ are replaced by their corresponding time-independent volume and particle displacements.

Example 1. Consider the two reciprocal, time-harmonic problems: (1) a normal force of amplitude F is applied to a thin elastic plate at A, producing a pressure of amplitude p at B in the ambient fluid (Figure 1.9.2); (2) a point

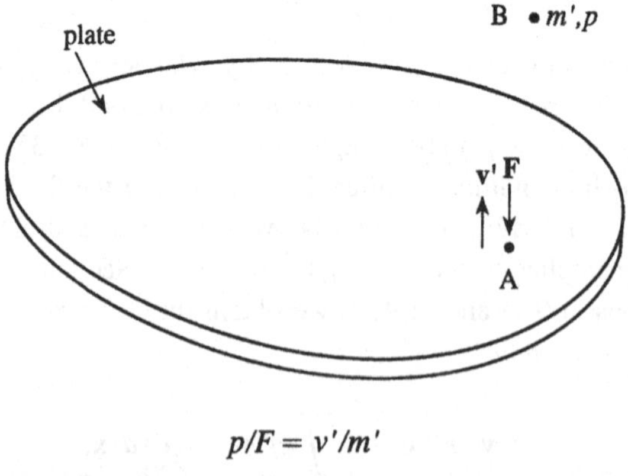

Figure 1.9.2. The reciprocal theorem applied to a thin plate.

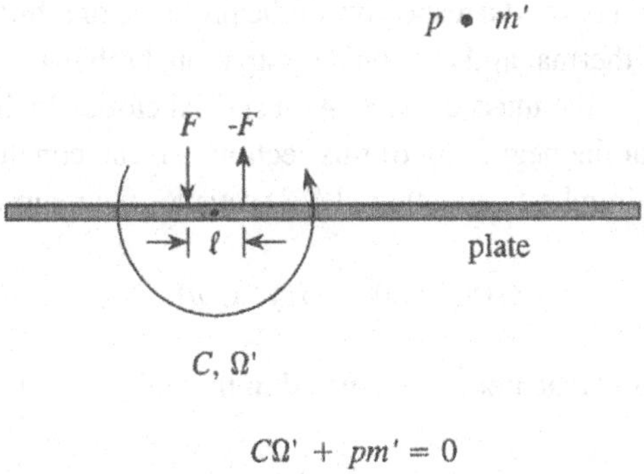

Figure 1.9.3. Couple applied to a plate.

volume source of strength m' at B generates a flexural (normal) velocity v' at A. If the edges of the plate are unrestrained, use the bending wave equation (1.4.9) and edge conditions (1.4.15) to derive the reciprocal relation

$$p/F = v'/m'.$$

A formula of this type is applicable to any elastic structure, and an experiment in which v' is measured at different points on the plate by using a known volume source m' furnishes a "map" of the *transfer function* p/F, which determines the importance of different parts of the structure in generating noise when excited by an external force [26].

Example 2. Two equal and opposite normal forces $\pm F$ applied to a plate at neighboring points distance ℓ apart are equivalent to a *couple* $C = \ell F$ (Figure 1.9.3). Establish the reciprocal relation

$$p/C = -\Omega'/m',$$

where m' is the strength of a reciprocal volume source in the ambient fluid, p is the pressure generated at the source by the couple, and Ω' is the angular velocity of the plate produced by the source m' about the axis of the couple at its point of application.

1.9.3 Reciprocity and Green's Function

The reciprocal formula (1.9.11) implies that $p(\mathbf{x}_{q'}, \omega) = p'(\mathbf{x}_q, \omega)$ when $F = F' = 0$ and $m = m'$, that is, the pressure at $\mathbf{x}_q$ due to a point source at $\mathbf{x}_{q'}$ is equal to the pressure at $\mathbf{x}_{q'}$ when the same source is placed at $\mathbf{x}_q$. This very general conclusion is valid irrespective of the presence of arbitrarily distributed elastic bodies, of thermal and frictional dissipation, both in the fluid and solids, and of variations in the mass density. As a special case, which generalizes the statement made at the beginning of this section, we can conclude that Green's function for the Helmholtz equation (1.7.3) satisfies the reciprocal relation

$$G(\mathbf{x}, \mathbf{y}; \omega) = G(\mathbf{y}, \mathbf{x}; \omega) \qquad (1.9.12)$$

in the presence of elastic bodies immersed in the fluid.

1.9.4 Sound Generated by a Vibrating Surface

Let a closed surface S execute small-amplitude vibrations at frequency ω with velocity $v_n(\mathbf{x}, \omega)$ in the direction of its outward normal (Figure 1.9.4). The motion in the fluid is equivalent to that generated by a distribution of volume sources of strength $v_n(\mathbf{x}, \omega)$ per unit area of S when S is assumed to be *rigid*. To calculate the pressure $p(\mathbf{x}, \omega)$, consider a reciprocal problem in which S is rigid and a unit point volume source is at **x**. The reciprocal theorem asserts that $p(\mathbf{x}, \omega)$ and the pressure $i\omega\rho_o G(\mathbf{y}, \mathbf{x}; \omega) \equiv i\omega\rho_o G(\mathbf{x}, \mathbf{y}; \omega)$ produced by this point source at any point **y** on S satisfy

$$p(\mathbf{x}, \omega) = i\omega\rho_o \oint_S v_n(\mathbf{y}, \omega) G(\mathbf{x}, \mathbf{y}; \omega) dS(\mathbf{y}). \qquad (1.9.13)$$

In this formula, Green's function is the velocity potential of the reciprocal point

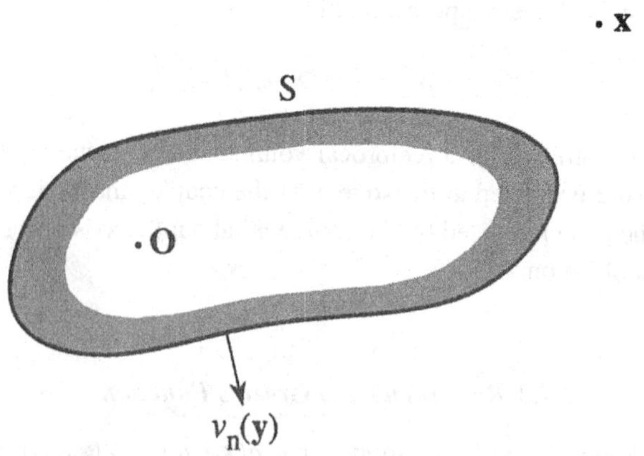

Figure 1.9.4. Sound generation by surface vibration.

source and satisfies, with respect to its dependence on both $\mathbf{x}$ and $\mathbf{y}$, the condition of vanishing normal velocity on S: $\partial G(\mathbf{x}, \mathbf{y}; \omega)/\partial y_n = 0$ for $\mathbf{y}$ on S, $\partial G(\mathbf{x}, \mathbf{y}; \omega)/\partial x_n = 0$ for $\mathbf{x}$ on S.

Example 3. An acoustically compact body of diameter ℓ pulsates with small-amplitude volume velocity $m(t)$. To calculate the sound by using (1.9.13), we write $m(t) = \int_{-\infty}^{\infty} m(\omega)\mathrm{e}^{-i\omega t} d\omega$ and apply the formula to each frequency component of the radiation. If the coordinate origin is within the body and $\mathbf{x}$ is in the acoustic far field,

$$G(\mathbf{x}, \mathbf{y}; \omega) = \frac{-\mathrm{e}^{i\kappa_o|\mathbf{x}|}}{4\pi|\mathbf{x}|}(1 + \mathrm{O}(\kappa_o\ell)) \quad \text{for } |\mathbf{y}| \sim \ell.$$

Because $m(\omega) = \oint_S v_n(\mathbf{y}, \omega)\, dS$,

$$\begin{aligned} p(\mathbf{x}, t) &\approx \frac{\rho_o}{4\pi|\mathbf{x}|}\int_{-\infty}^{\infty} -i\omega m(\omega)\mathrm{e}^{-i\omega(t-|\mathbf{x}|/c_o}\, d\omega \\ &= \frac{\rho_o}{4\pi|\mathbf{x}|}\frac{\partial m}{\partial t}(t - |\mathbf{x}|/c_o), \quad |\mathbf{x}| \to \infty. \end{aligned} \tag{1.9.14}$$

The pressure fluctuations are therefore identical to those produced by the monopole $q = m(t)\delta(\mathbf{x})$.

1.9.5 Reverse Flow Reciprocity [27, 28]

There is no simple relation of the form (1.9.12) when sound propagates in a moving medium because the acoustic equations (Section 1.6.4) are not self-adjoint [9]. Let us consider fluid of mean density $\rho(\mathbf{x})$ and sound speed $c(\mathbf{x})$ in irrotational flow at velocity $\mathbf{U}(\mathbf{x})$ past a system of stationary bodies with surface S. The perturbation velocity potential $\varphi(\mathbf{x}, \omega)$ satisfies the time-reduced form of equation (1.6.25) (obtained by setting $\partial/\partial t = -i\omega$)

$$\left\{(\omega + i\mathbf{U}.\nabla)\left[\frac{1}{c^2}(\omega + i\mathbf{U}.\nabla)\right] - \frac{1}{\rho}\nabla.(\rho\nabla)\right\}\varphi = -\hat{\mathcal{F}}(\mathbf{x}, \omega). \tag{1.9.15}$$

The reciprocal potential will now be defined as the solution φ' of the *reversed flow* equation obtained by replacing $\mathbf{U}$ by $-\mathbf{U}$ in (1.9.15) and the source by $-\hat{\mathcal{F}}'(\mathbf{x}, \omega)$. Multiply this equation by $\rho(\mathbf{x})\varphi$ and subtract from the product of (1.9.15) and $\rho(\mathbf{x})\varphi'$. Integrate the result over the fluid inside a spherical control surface Σ enclosing the bodies and sources, whose radius is sufficiently large that $\mathbf{U} \approx \mathbf{U}_o =$ constant and $c \approx c_o =$ constant on Σ. Because $\operatorname{div}(\rho(\mathbf{x})\mathbf{U}(\mathbf{x})) = 0$ for steady flow, application of the divergence theorem supplies

$$\int_{S+\Sigma} \{(\varphi \partial\varphi' \partial x_i - \varphi' \partial\varphi/\partial x_i)(\delta_{ij} - M_i M_j) - 2i(\omega/c) M_j \varphi\varphi'\} \rho n_j \, dS$$
$$= \int \rho(\varphi \hat{\mathcal{F}}' - \varphi' \hat{\mathcal{F}}) d^3\mathbf{x},$$

where $\mathbf{M} = \mathbf{U}(\mathbf{x})/c(\mathbf{x})$ is the local vector Mach number. The terms involving $\mathbf{U}$ in the surface integral over S vanish because $n_j U_j = 0$ on S. The remaining integral over S is also zero if φ and φ' satisfy

$$\mathcal{Z}(\mathbf{x}) \frac{\partial \varphi}{\partial x_n} = \varphi, \quad \mathcal{Z}(\mathbf{x}) \frac{\partial \varphi'}{\partial x_n} = \varphi' \quad \text{on S}, \tag{1.9.16}$$

where the "impedance" $\mathcal{Z}$ does not depend on $\mathbf{U}$. These conditions include the important special cases in which S is rigid ($\mathcal{Z} = \infty$) or "pressure release" ($\mathcal{Z} = 0$ so that $(-i\omega + \mathbf{U} \cdot \nabla)\varphi = 0$ on S).

On Σ the potential φ resembles an outgoing sound wave convected by a uniform mean flow. By analogy with the solution (1.7.17) for a point source, this is easily seen to imply that

$$\varphi \approx \frac{\chi(\mathbf{x}/|\mathbf{x}|, \mathbf{M}_o) \exp\left(i\kappa_o \left[\left(\frac{|\mathbf{x}|^2}{1-M_o^2} + \frac{(\mathbf{M}_o \cdot \mathbf{x})}{(1-M_o^2)^2}\right)^{\frac{1}{2}} - \frac{\mathbf{M}_o \cdot \mathbf{x}}{1-M_o^2}\right]\right)}{\sqrt{|\mathbf{x}|^2 + \frac{(\mathbf{M}_o \cdot \mathbf{x})^2}{1-M_o^2}}} \quad \text{on } \Sigma,$$

where $\mathbf{M}_o = \mathbf{U}_o/c_o$, and the directivity factor χ is determined by the juxtapositions of the sources $\hat{\mathcal{F}}$ and the surfaces S. By using a similar asymptotic representation for φ' (with different χ, and with $\mathbf{M}_o$ replaced by $-\mathbf{M}_o$), it is concluded that the surface integral over Σ vanishes as the surface recedes to infinity. The argument is the same as that involving the Sommerfeld radiation condition in the absence of flow (cf. the remarks following equation (1.9.7)). This establishes the *reverse flow reciprocal theorem:*

$$\int \rho(\mathbf{x}) \varphi(\mathbf{x}, \omega) \hat{\mathcal{F}}'(\mathbf{x}, \omega) \, d^3\mathbf{x} = \int \rho(\mathbf{x}) \varphi'(\mathbf{x}, \omega) \hat{\mathcal{F}}(\mathbf{x}, \omega) \, d^3\mathbf{x}, \tag{1.9.17}$$

where φ is the solution of (1.9.15) with outgoing wave behavior, and φ' is the corresponding solution of the reverse flow equation with source $-\hat{\mathcal{F}}'$, subject to the boundary conditions (1.9.16) on any surfaces in the flow. The theorem has been proved for irrotational, *subsonic* mean flow, but remains true when the flow is supersonic.

The solution of equation (1.9.15) for the point source $\hat{\mathcal{F}} = -\,\delta(\mathbf{x}-\mathbf{y})$ defines Green's function, which will be denoted by $G_+(\mathbf{x}, \mathbf{y}; \omega)$. The reciprocal theorem yields the following relation between $G_+(\mathbf{x}, \mathbf{y}; \omega)$ and Green's function $G_-(\mathbf{y}, \mathbf{x}; \omega)$ (with the source at $\mathbf{x}$) for the reverse flow equation

$$\rho(\mathbf{x})G_+(\mathbf{x}, \mathbf{y}; \omega) = \rho(\mathbf{y})G_-(\mathbf{y}, \mathbf{x}; \omega). \qquad (1.9.18)$$

To compare this formula with the no-flow case (1.9.12) governed by the Helmholtz equation (1.7.3), note that $G_\pm$ in (1.9.18) are velocity potentials, but G in (1.9.12) may be interpreted as a *pressure*. Thus, Bernoulli's equation (1.2.21) (in which we can take with $\Phi = f = 0$) implies that, in the presence of flow, the perturbed *total pressure* $P = p + \frac{1}{2}\rho v^2$, rather than the pressure p, satisfies the point source reverse flow reciprocal relation $\delta P_+(\mathbf{x}, \mathbf{y}; \omega) = \delta P_-(\mathbf{y}, \mathbf{x}; \omega)$.

1.10 General Solution of the Inhomogeneous Wave Equation

1.10.1 General Solution in the Frequency Domain

The general method of solving wave problems in the presence of arbitrary source distributions and arbitrary surfaces makes use of a Green's function of the *reciprocal* (i.e., *adjoint*) equation. In an ideal, stationary fluid the frequency domain pressure $p(\mathbf{x}, \omega)$ satisfies equation (1.6.8) in which $\partial/\partial t \to -i\omega$, that is, the inhomogeneous Helmholtz equation,

$$\left(\nabla^2 + \kappa_o^2\right) p = -\mathcal{F}(\mathbf{x}, \omega), \qquad (1.10.1)$$

where $\mathcal{F}(\mathbf{x}, \omega) = (1/2\pi) \int_{-\infty}^{\infty} \mathcal{F}(\mathbf{x}, t) e^{i\omega t} dt$ is the Fourier time transform of $\mathcal{F}(\mathbf{x}, t)$.

Green's function $G(\mathbf{y}, \mathbf{x}; \omega) \equiv G(\mathbf{y}, \mathbf{x}; \omega)$ is *any* solution with outgoing wave behavior of the reciprocal problem in which the source is placed at the point $\mathbf{x}$ where the solution $p(\mathbf{x}, \omega)$ is to be found:

$$\left(\partial^2/\partial y_j^2 + \kappa_o^2\right) G(\mathbf{x}, \mathbf{y}; \omega) = \delta(\mathbf{x} - \mathbf{y}). \qquad (1.10.2)$$

The notation implies that $G(\mathbf{x}, \mathbf{y}; \omega)$ is sought as a function of $\mathbf{y}$ for fixed $\mathbf{x}$.

A formal representation of the solution $p(\mathbf{x}, \omega)$ is obtained by combining these equations in the following way: (i) replace $\mathbf{x}$ by $\mathbf{y}$ in (1.10.1) and multiply by $G(\mathbf{x}, \mathbf{y}; \omega)$, (ii) subtract the product of (1.10.2) and $p(\mathbf{y}, \omega)$ and integrate with respect to $\mathbf{y}$ over the whole of the fluid. The divergence theorem permits the result to be cast in the form

$$p(\mathbf{x}, \omega) = \oint_S \left\{ G(\mathbf{x}, \mathbf{y}; \omega) \frac{\partial p}{\partial y_j}(\mathbf{y}, \omega) - p(\mathbf{y}, \omega) \frac{\partial G}{\partial y_j}(\mathbf{x}, \mathbf{y}; \omega) \right\} n_j \, dS(\mathbf{y})$$
$$- \int G(\mathbf{x}, \mathbf{y}; \omega) \mathcal{F}(\mathbf{y}, \omega) \, d^3\mathbf{y}, \qquad (1.10.3)$$

where the unit normal $\mathbf{n}$ on S is directed into the fluid. The Sommerfeld radiation condition (1.8.16) ensures that there are no additional contributions from a surface integral at infinity.

Equation (1.10.3) is applicable for any outgoing solution G of (1.10.2). The surface integral can be evaluated if p and $\partial p/\partial y_n$ are known on S, but these quantities cannot be prescribed independently. Indeed, (1.10.3) is valid for any stationary surface, including a closed control surface, within which the pressure p on the left of (1.10.3) would be replaced by zero because the δ-function in (1.10.2) then makes no contribution to the volume integral outside S. The values of p and $\partial p/\partial y_n$ on the control surface (and on the solid surfaces) are therefore related such that, within S, their contributions exactly cancel sound incident on S from elsewhere in the fluid.

When there are no sources ($\mathcal{F} \equiv 0$), and $G(\mathbf{x}, \mathbf{y}; \omega)$ is chosen to make $\partial G/\partial y_n = 0$ on S, the momentum equation $\partial p/\partial y_j = i\rho_o \omega v_j$ can be used to show that (1.10.3) reduces to the representation (1.9.13) of the sound generated by a vibrating body. In more general cases, involving sound production by sources within the flow, the surface conditions often reduce to $\partial p/\partial x_n = i\omega\rho_o p/\mathcal{Z}$, where $\mathcal{Z}$ is the mechanical impedance of S. The contribution from the surface integral in (1.10.3) can then be eliminated by choosing G to satisfy the same condition on S.

1.10.2 General Solution in the Time Domain

Fourier superposition may now be used to derive an integral formula for the solution $p(\mathbf{x}, t)$ of the wave equation (1.6.8). The application of the *convolution theorem* [9]

$$\int_{-\infty}^{\infty} f_1(\omega) f_2(\omega) \mathrm{e}^{-i\omega t} \, d\omega = \frac{1}{2\pi} \int_{-\infty}^{\infty} f_1(\tau) f_2(t - \tau) \, d\tau \qquad (1.10.4)$$

to (1.10.3) gives *Kirchhoff's formula*

$$p(\mathbf{x}, t) = \oint_S \left[p(\mathbf{y}, \tau) \frac{\partial G}{\partial y_j}(\mathbf{x}, \mathbf{y}, t - \tau) - \frac{\partial p}{\partial y_j}(\mathbf{y}, \tau) G(\mathbf{x}, \mathbf{y}, t - \tau) \right] \times n_j dS(\mathbf{y}) d\tau + \int G(\mathbf{x}, \mathbf{y}, t - \tau) \mathcal{F}(\mathbf{y}, \tau)\, d^3\mathbf{y}\, d\tau, \tag{1.10.5}$$

where the retarded time integration with respect to τ is taken over $(-\infty, \infty)$, and $G(\mathbf{x}, \mathbf{y}, t-\tau)$ is an outgoing solution of (1.7.18) and is related to $G(\mathbf{x}, \mathbf{y}; \omega)$ by (1.7.19).

The linearized momentum equation ((1.6.1) with $\mathbf{F} = \mathbf{0}$) gives the alternative representation

$$p(\mathbf{x}, t) = \oint_S \left[p(\mathbf{y}, \tau) \frac{\partial G}{\partial y_j}(\mathbf{x}, \mathbf{y}, t - \tau) + \rho_o \frac{\partial v_j}{\partial \tau}(\mathbf{y}, \tau) G(\mathbf{x}, \mathbf{y}, t - \tau) \right] \times n_j\, dS(\mathbf{y})\, d\tau + \int G(\mathbf{x}, \mathbf{y}, t - \tau) \mathcal{F}(\mathbf{y}, \tau) d^3\mathbf{y}\, d\tau. \tag{1.10.6}$$

The special case in which G is the free-space Green's function (1.7.20) should be noted:

$$\begin{aligned} p(\mathbf{x}, t) = {} & \frac{\rho_o}{4\pi} \frac{\partial}{\partial t} \oint_S \frac{v_n(\mathbf{y}, t - |\mathbf{x} - \mathbf{y}|/c_o)}{|\mathbf{x} - \mathbf{y}|}\, dS(\mathbf{y}) \\ & - \frac{1}{4\pi} \frac{\partial}{\partial x_j} \oint_S \frac{p(\mathbf{y}, t - |\mathbf{x} - \mathbf{y}|/c_o)}{|\mathbf{x} - \mathbf{y}|} n_j\, dS(\mathbf{y}) \\ & + \frac{1}{4\pi} \int \frac{\mathcal{F}(\mathbf{y}, t - |\mathbf{x} - \mathbf{y}|/c_o)}{|\mathbf{x} - \mathbf{y}|}\, d^3\mathbf{y}. \end{aligned} \tag{1.10.7}$$

According to Section 1.8.1, the surface integrals in this formula represent distributions of monopole and dipole sources on S, respectively, of strengths $v_n(\mathbf{x}, t)$ and $p(\mathbf{x}, t)$ per unit area. These sources cannot be prescribed independently.

1.10.3 Multipole Expansion of the Radiation Field

In the acoustic far field, the method of Section 1.8.3 can be used to expand the right-hand side of (1.10.7) into a series of point multipoles as in the second line of (1.8.14).

1.10.4 Solution of the Wave Equation in Arbitrary, Irrotational Mean Flow

Equation (1.6.25), which is satisfied by the perturbed velocity potential in the presence of steady irrotational flow, is *self-adjoint* when multiplied by the mean density $\rho(\mathbf{x})$. The adjoint of the frequency domain equation (1.9.15) is obtained by reversing the mean flow direction (Section 1.9.5). The reciprocal Green's function $G_-(\mathbf{y}, \mathbf{x}; \omega)$ is the outgoing solution *as a function of* $\mathbf{y}$ of equation (1.9.15) when the sign of $\mathbf{U}$ is reversed and $-\hat{\mathcal{F}}$ is replaced by $\delta(\mathbf{x}-\mathbf{y})$. In order to have a consistent notation, we shall write $G(\mathbf{x}, \mathbf{y}; \omega) = G_-(\mathbf{y}, \mathbf{x}; \omega)$. The method of Sections 1.10.1 and 2 then gives

$$\varphi(\mathbf{x}, \omega)\rho(\mathbf{x}) = \oint_S \left\{ \left(G(\mathbf{x}, \mathbf{y}; \omega)\frac{\partial \varphi}{\partial y_i}(\mathbf{y}, \omega) - \varphi(\mathbf{y}, \omega)\frac{\partial G}{\partial y_i}(\mathbf{x}, \mathbf{y}; \omega) \right) \times (\delta_{ij} - M_i M_j) + \frac{2i\omega M_j}{c(\mathbf{y})}\varphi(\mathbf{y}, \omega) G(\mathbf{x}, \mathbf{y}; \omega) \right\} \rho(\mathbf{y}) n_j \, dS(\mathbf{y}) - \int G(\mathbf{x}, \mathbf{y}; \omega)\rho(\mathbf{y})\hat{\mathcal{F}}(\mathbf{y}, \omega)\, d^3\mathbf{y}, \tag{1.10.8}$$

$$\varphi(\mathbf{x}, t)\rho(\mathbf{x}) = \oint_S \left\{ \left(\varphi(\mathbf{y}, \tau)\frac{\partial G}{\partial y_i}(\mathbf{x}, \mathbf{y}, t-\tau) - \frac{\partial \varphi}{\partial y_i}(\mathbf{y}, \tau) G(\mathbf{x}, \mathbf{y}, t-\tau) \right) \times (\delta_{ij} - M_i M_j) + \frac{2M_j}{c(\mathbf{y})}\varphi(\mathbf{y}, \tau)\frac{\partial G}{\partial t}(\mathbf{x}, \mathbf{y}; t-\tau) \right\} \times \rho(\mathbf{y}) n_j \, dS(\mathbf{y})\, d\tau + \int G(\mathbf{x}, \mathbf{y}, t-\tau)\rho(\mathbf{y})\hat{\mathcal{F}}(\mathbf{y}, \tau)\, d^3\mathbf{y}\, d\tau, \tag{1.10.9}$$

where $\mathbf{n}$ is directed into the fluid, and $G(\mathbf{x}, \mathbf{y}, t-\tau)$ and $G(\mathbf{x}, \mathbf{y}; \omega)$ are related as in (1.7.19). When the stationary boundary S coincides with a solid surface, on which $\mathbf{n} \cdot \mathbf{U} = 0$ (as opposed to a control surface which can be penetrated by the mean flow), the terms involving $\mathbf{M} = \mathbf{U}(\mathbf{x})/c(\mathbf{x})$ are absent.

1.10.5 Spherical and Cylindrical Harmonics

It is frequently useful to expand the solution of a radiation problem in terms of spherical or cylindrical harmonics, which arise naturally when the wave equation is expressed in spherical or cylindrical polar coordinates. The relevant equations and formulae are set down here for future reference.

Spherical harmonics. In spherical polar coordinates (r, θ, ϕ) the wave equation (1.6.8) is

$$\left[\frac{1}{c_o^2}\frac{\partial^2}{\partial t^2} - \frac{1}{r^2}\frac{\partial}{\partial r}\left(r^2\frac{\partial}{\partial r}\right) - \frac{1}{r^2 \sin\theta}\frac{\partial}{\partial \theta}\left(\sin\theta\frac{\partial}{\partial \theta}\right) - \frac{1}{r^2\sin^2\theta}\frac{\partial^2}{\partial \phi^2} \right] p = \hat{\mathcal{F}}(r, \theta, \phi, t). \tag{1.10.10}$$

Solutions of this equation in the frequency-domain ($\partial/\partial t = -i\omega$) are

$$p(r,\theta,\phi,\omega) = \sum_{n=0}^{\infty}\sum_{m=-n}^{n} \alpha_{nm} \mathrm{z}_n(\kappa_o r)\mathrm{P}_n^{|m|}(\cos\theta)\mathrm{e}^{im\phi}, \tag{1.10.11}$$

where α_{nm} are arbitrary constants, z_n is a spherical Bessel function of order n, and $\mathrm{P}_n^{|m|}(\cos\theta)$ is an associated Legendre function of order n, $|m|$ [24]. $\mathrm{P}_n^{|m|}(\cos\theta)\mathrm{e}^{im\phi}$ is called a *spherical surface harmonic*. Any generalized function $f(\theta,\phi)$, $0<\theta<\pi$, $0<\phi<2\pi$ can be expanded in the form

$$f(\theta,\phi) = \sum_{n=0}^{\infty}\sum_{m=-n}^{n} f_{nm}\mathrm{P}_n^{|m|}(\cos\theta)\mathrm{e}^{im\phi},$$

$$f_{nm} = \frac{(2n+1)(n-m)!}{4\pi(n+m)!}\int_0^{\pi} d\theta \int_0^{2\pi} f(\theta,\phi)\mathrm{P}_n^{|m|}(\cos\theta)\mathrm{e}^{-im\phi}\,d\phi, \tag{1.10.12}$$

where the second equation is derived from the orthogonality relation

$$\int_0^{\pi} \mathrm{P}_n^m(\cos\theta)\mathrm{P}_k^m(\cos\theta)\sin\theta\, d\theta = \frac{2\delta_{kn}(n+m)!}{(2n+1)(n-m)!}. \tag{1.10.13}$$

$\mathrm{P}_n^0(\cos\theta)$ is the Legendre polynomial of order n and is usually denoted by $\mathrm{P}_n(\cos\theta)$. It satisfies $\mathrm{P}_n^m(\cos\theta) = \sin^m\theta\,\partial^m\mathrm{P}_n(\cos\theta)/\partial(\cos\theta)^m (m \geq 0)$. In particular,

$$\begin{aligned} &\mathrm{P}_0(\cos\theta) = 1, && \mathrm{P}_1^1(\cos\theta) = \sin\theta \\ &\mathrm{P}_1(\cos\theta) = \cos\theta, && \mathrm{P}_2^1(\cos\theta) = 3\sin\theta\cos\theta \\ &\mathrm{P}_2(\cos\theta) = \tfrac{1}{2}(3\cos^2\theta - 1), && \mathrm{P}_2^2(\cos\theta) = 3\sin^2\theta. \end{aligned}$$

The series (1.10.11) represents an outgoing acoustic wave when $\mathrm{z}_n(x) = \mathrm{h}_n^{(1)}(x) \equiv \sqrt{\pi/2x}\mathrm{H}_{n+\frac{1}{2}}(x)$ is a spherical Hankel function of the first kind [24]. These satisfy

$$\begin{aligned} &\mathrm{h}_0^{(1)}(x) \sim \frac{-i}{x}, \quad x \to 0; \\ &\mathrm{h}_n^{(1)}(x) \sim -i\frac{1.3.5\ldots(2n-1)}{x^{n+1}}, \quad n \geq 1,\ x \to 0; \\ &\mathrm{h}_n^{(1)}(x) \sim \frac{(-i)^{n+1}\mathrm{e}^{ix}}{x}, \quad x \to \infty. \end{aligned} \tag{1.10.14}$$

Hankel functions are singular at the origin. A solution of the wave equation that remains finite at $r=0$ is obtained by using the radial Bessel function $\mathrm{z}_n(x)=\mathrm{j}_n(x)\equiv\sqrt{\pi/2x}\,\mathrm{J}_{n+\frac{1}{2}}(x)$ [24].

For *incompressible*, irrotational motion, the pressure outside any sphere enclosing the sources and solid boundaries can be expanded in the following modified form of (1.10.11):

$$p(r,\theta,\phi,\omega)=\sum_{n=0}^{\infty}\sum_{m=-n}^{n}\frac{\alpha_{nm}}{r^{n+1}}\mathrm{P}_n^{|m|}(\cos\theta)\mathrm{e}^{im\phi}. \tag{1.10.15}$$

Cylindrical harmonics. In cylindrical coordinates (r,θ,x) the wave equation is

$$\left\{\frac{1}{c_o^2}\frac{\partial^2}{\partial t^2}-\frac{1}{r}\frac{\partial}{\partial r}\left(r\frac{\partial}{\partial r}\right)-\frac{1}{r^2}\frac{\partial^2}{\partial\theta^2}-\frac{\partial^2}{\partial x^2}\right\}p=\hat{\mathcal{F}}(r,\theta,x,t). \tag{1.10.16}$$

In the frequency domain

$$p(r,\theta,x,\omega)=\sum_{n=-\infty}^{\infty}\int_{-\infty}^{\infty}\alpha(k)\mathrm{Z}_{|n|}(\gamma r)\mathrm{e}^{i(n\theta+kx)}dk, \tag{1.10.17}$$

where $\gamma=\mathrm{sgn}(\kappa_o)|\kappa_o^2-k^2|^{1/2},\ i|\kappa_o^2-k^2|^{1/2}$ according as $|k|\lessgtr|\kappa_o|$, and $\mathrm{Z}_\nu(x)$ is any linear combination of the cylinder functions $\mathrm{J}_\nu(x)$, $\mathrm{Y}_\nu(x)$, $\mathrm{H}_\nu^{(1)}(x)$, $\mathrm{H}_\nu^{(2)}(x)$ [24]. A solution with outgoing wave behavior is obtained when $\mathrm{Z}_\nu\,(\gamma r)=\mathrm{H}_\nu^{(1)}(\gamma r)$. The behavior of (1.10.17) for small and large values of γr can generally be deduced from the asymptotic formulae:

$$\begin{aligned}
\mathrm{H}_0^{(1)}(x)&=1+\frac{2i}{\pi}[\gamma_\mathrm{E}+\ln(x/2)]+\mathrm{O}(x^2\ln x),\quad x\to 0;\\
\mathrm{H}_1^{(1)}(x)&=\frac{-2i}{\pi x}+\left[\gamma_\mathrm{E}-\frac{i\pi}{2}+\ln(x/2)\right]\frac{ix}{\pi}+\mathrm{O}(x^3\ln x),\quad x\to 0;\\
\mathrm{H}_n^{(1)}(x)&=\frac{-i(n-2)!}{\pi}\left(\frac{2}{x}\right)^n\left(n-1-\frac{1}{4}x^2\right)+\mathrm{O}(x^{4-n}),\\
&\qquad x\to 0,\ n\geq 2;\\
\mathrm{H}_n^{(1)}(x)&\sim\sqrt{\frac{2}{\pi x}}\mathrm{e}^{i[x-(n+\frac{1}{2})\frac{\pi}{2}]},\quad x\to\infty,\ n\ \text{fixed},
\end{aligned} \tag{1.10.18}$$

where $\gamma_\mathrm{E}\approx 0.57722$ is Euler's constant [24]. A solution that remains bounded at $r=0$ is obtained when $\mathrm{Z}_m(\gamma r)=\mathrm{J}_m(\gamma r)$ in (1.10.17).

Example 3: Green's function for the interior of a hard-walled, circular cylindrical duct. In the frequency domain $G(\mathbf{x}, \mathbf{y}; \omega)$ satisfies (1.10.16) for $r < R =$ radius of the duct, with $\hat{\mathcal{F}}(r, \theta, x, \omega) = \frac{-1}{r}\delta(r - r')\delta(\theta - \theta')\delta(x - x')$ and $\partial G/\partial r = 0$ on $r = R$, where (r, θ, x), (r', θ', x') are the respective cylindrical coordinates of $\mathbf{x}$, $\mathbf{y}$. G has the expansion [9]

$$G(\mathbf{x}, \mathbf{y}; \omega) = G(\mathbf{y}, \mathbf{x}; \omega)$$

$$= \frac{-i}{\pi R^2} \sum_{n,m=0}^{\infty} \frac{\sigma_m \mathrm{J}_m(\lambda_{mn} r'/R) \mathrm{J}_m(\lambda_{mn} r/R) \cos[m(\theta - \theta')] \mathrm{e}^{i\gamma_{mn}|x - x'|}}{\gamma_{mn}\left(1 - \frac{m^2}{\lambda_{mn}^2}\right) \mathrm{J}_m^2(\lambda_{mn})}$$

where $\sigma_0 = \frac{1}{2}$, $\sigma_m = 1$ $(m > 0)$, λ_{mn} is the nth nonnegative zero of $\partial \mathrm{J}_m(x)/\partial x$, and

$$\gamma_{mn} = \operatorname{sgn}(\kappa_o)\left|\kappa_o^2 - \lambda_{mn}^2/R^2\right|^{1/2} \quad \text{for } |\kappa_o| > \lambda_{mn}/R,$$

$$= i\left|\kappa_o^2 - \lambda_{mn}^2/R^2\right|^{1/2} \quad \text{for } |\kappa_o| < \lambda_{mn}/R.$$

The nm-mode decays exponentially with axial distance from the source when $|\kappa_o| < \lambda_{mn}/R$, that is, at frequencies ω smaller than $c_o \lambda_{mn}/R$ in absolute value. The smallest positive value of λ_{mn} is $\lambda_{1,0} \approx 1.841$ so that only the axially propagating *plane wave* mode $m = n = 0$ can propagate to $x = \pm\infty$ when the frequency is below the *cutoff frequency* $\sim 0.29 c_o/R$ Hz.

1.11 Compact Green's Functions

The compact Green's function provides a formal and intuitive procedure for calculating the leading order monopole and dipole terms in the multipole expansion of the sound produced by sources near a solid body.

1.11.1 Time-Harmonic Problems

Consider the particular Green's function $G(\mathbf{x}, \mathbf{y}; \omega)$ determined by the Helmholtz equation (1.10.2) that has *vanishing normal derivative* on the surface S of an acoustically compact body of diameter ℓ. This equation is to be solved as a function of $\mathbf{y}$ for fixed $\mathbf{x}$, which we take to be in the *acoustic far field* of S.

The determination of $G(\mathbf{x}, \mathbf{y}; \omega)$ can be posed as a scattering problem in which the spherical wave (1.7.9) generated by a point source at $\mathbf{x}$, which corresponds to the free-space Green's function, is incident on S. If the coordinate origin is within S and $\mathbf{y}$ is close to S, the compactness condition $\kappa_o \ell \ll 1$ permits

$G(\mathbf{x}, \mathbf{y}; \omega)$ to be expanded in the form [29–31]

$$G(\mathbf{x}, \mathbf{y}; \omega) = \frac{-e^{i\kappa_o|\mathbf{x}|}}{4\pi|\mathbf{x}|}\left[1 - \frac{i\kappa_o x_i}{|\mathbf{x}|}(y_i - \varphi_i^*(\mathbf{y})) + \sum_{n\geq 2}(\kappa_o\ell)^n \Phi_n\left(\frac{\mathbf{x}}{|\mathbf{x}|}, \mathbf{y}\right)\right],$$

$$\mathbf{y} \sim O(\ell), \quad |\mathbf{x}| \to \infty. \qquad (1.11.1)$$

The first term in the large braces represents the incident wave (1.7.9) evaluated at $\mathbf{y} = \mathbf{0}$. The next term is $O(\kappa_o\ell)$ and includes a component $-i\kappa_o x_i y_i/|\mathbf{x}|$ from the incident wave plus a correction $-i\kappa_o x_i \varphi_i^*(\mathbf{y})/|\mathbf{x}|$ due to S. To this order of approximation $Y_i(\mathbf{y}) \equiv y_i - \varphi_i^*(\mathbf{y})$ is a solution of Laplace's equation satisfying $\partial Y_i/\partial y_n = 0$ on S. Because $\varphi_i^*(\mathbf{y})$ must decay with distance from S, this implies that φ_i^* is simply the velocity potential of the incompressible motion that would be produced by translational motion of S as a *rigid body* at unit speed in the i-direction. The remaining terms in (1.11.1) are of order $(\kappa_o\ell)^2$ or smaller. When they are neglected, the resulting approximation for G can be used to determine the monopole and dipole terms in the multipole expansion of the solution of a fluid–structure interaction problem.

The potentials φ_i^* are uniquely defined by the shape of the body and satisfy

$$\frac{\partial \varphi_i^*}{\partial y_n}(\mathbf{y}) = n_i \text{ on S.} \qquad (1.11.2)$$

When the body is rigid, they also determine the *added mass* tensor M_{ij} [4], which is symmetric and given by

$$\begin{aligned}\mathrm{M}_{ij} &= -\rho_o \oint_S \varphi_i^*(\mathbf{y}) n_j \, dS(\mathbf{y}) \\ &= -\rho_o \oint_S \varphi_i^* \partial\varphi_j^*/\partial y_n \, dS(\mathbf{y}) \equiv -\rho_o \oint_S \varphi_j^* \partial\varphi_i^*/\partial y_n \, dS(\mathbf{y}). \qquad (1.11.3)\end{aligned}$$

A body in translational accelerated motion at velocity $\mathbf{U}(t)$ in an ideal, incompressible fluid experiences a reaction force from the fluid equal to $-\partial(\mathrm{M}_{ij}\mathrm{U}_j)/\partial t$ in the i-direction ([4, 17]; Section 1.14). An external force F_i acting through the center of mass of a body of mass m therefore produces accelerated motion determined by $\partial(mU_i + \mathrm{M_{ij}}U_j)/\partial t = F_i$.

Example 1. Sound generation by a compact, rigid body executing small-amplitude translational oscillations. Let the translational velocity be $\mathbf{U}(t)$ and consider first the contribution to the sound from the Fourier component $\mathbf{U}(\omega)e^{-i\omega t}$. Use Green's function (1.11.1) in the general solution (1.10.3), with $\mathcal{F} = 0$, and $\partial p/\partial y_n = i\rho_o \mathbf{U}.\mathbf{n}$ on S:

$$p(\mathbf{x}, \omega) = \frac{-i\rho_o \omega U_j(\omega) \mathrm{e}^{i\kappa_o|\mathbf{x}|}}{4\pi|\mathbf{x}|} \times \left\{ \oint_S n_j \, d\mathrm{S}(\mathbf{y}) - \frac{i\kappa_o x_i}{|\mathbf{x}|} \oint_S Y_i n_j \, d\mathrm{S}(\mathbf{y}) + \mathrm{O}[\ell^2(\kappa_o \ell)^2] \right\}.$$

The first integral vanishes identically; the second is equal to $(m_o\delta_{ij} + \mathrm{M}_{ij})/\rho_o$, where m_o is the mass of fluid displaced by the body. When the body is compact $U_j(\omega) \neq 0$ only for $\kappa_o \ell \ll 1$. Hence, multiplying by $\mathrm{e}^{-i\omega t}$ and integrating over $-\infty < \omega < \infty$

$$p(\mathbf{x}, t) \approx \frac{x_i}{4\pi c_o |\mathbf{x}|^2} \frac{\partial^2}{\partial t^2}(m_o U_i + \mathrm{M}_{ij} U_j)(t - |\mathbf{x}|/c_o), \quad |\mathbf{x}| \to \infty. \tag{1.11.4}$$

This is a dipole field of strength $m_o \dot{U}_i + \mathrm{M}_{ij} \dot{U}_j$ (cf. Example 2, Section 1.8). The motion close to the body is effectively incompressible, and the external force that must be applied to the body to generate the motion is $F_i = m\dot{U}_i + \mathrm{M}_{ij}\dot{U}_j$, where m is the mass of the body. Thus, the dipole strength $= \mathbf{F} - (m - m_o)\dot{\mathbf{U}} =$ the applied force less the excess of the inertia of the body over the displaced fluid.

Example 2. In an ideal, incompressible fluid, $\partial(\mathrm{M}_{ij} U_j)/\partial t$ is the force exerted on the fluid by a body translating at velocity $\mathbf{U}$. Use this in the integral representation (1.10.7) (with $\mathcal{F} = 0$) to verify (1.11.4).

Example 3. $\varphi_i^*(\mathbf{y}) = -R^3 y_i / 2|\mathbf{y}|^3$ for a rigid sphere of radius R [4, 17], and the added mass tensor $\mathrm{M}_{ij} = \frac{2}{3}\pi R^3 \rho_o \delta_{ij} \equiv 1/2 m_o \delta_{ij}$. The sound generated by low-frequency, small-amplitude translational oscillations at velocity $\mathbf{U}(t)$ is

$$p(\mathbf{x}, t) \approx \frac{\rho_o R^3 \cos\theta}{2c_o|\mathbf{x}|} \frac{\partial^2 U}{\partial t^2}(t - |\mathbf{x}|/c_o), \quad |\mathbf{x}| \to \infty,$$

where θ is the angle between $\mathbf{U}$ and the observer direction $\mathbf{x}/|\mathbf{x}|$.

Example 4. Calculate the sound generated by translational oscillations of a *small* circular disc in the direction of its normal (Figure 1.11.1). Assume the disc to be rigid, of radius R, and to move at speed $U(t)$ in the direction of the

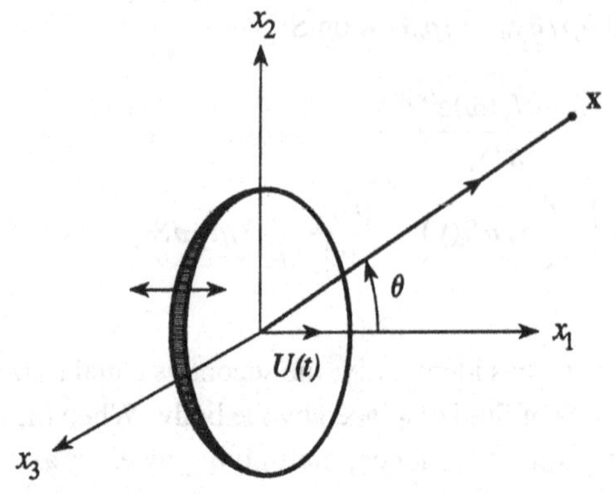

Figure 1.11.1. Radiation from a vibrating disc.

x_1-axis.

$$p(\mathbf{x},t) \approx \frac{-\rho_o \cos\theta}{4\pi c_o|\mathbf{x}|}\frac{\partial^2 U}{\partial t^2}(t-|\mathbf{x}|/c_o)\oint_S \varphi_1^*(\mathbf{y})\,dS(\mathbf{y})$$
$$= \frac{2\rho_o R^3 \cos\theta}{3\pi c_o|\mathbf{x}|}\frac{\partial^2 U}{\partial t^2}(t-|\mathbf{x}|/c_o), \quad |\mathbf{x}|\to\infty,$$

where $\mathrm{M}_{11} \equiv -\rho_o \oint_S \varphi_1^*(\mathbf{y})\,dS(\mathbf{Y}) = 8\rho_o \int_0^R r\sqrt{R^2-r^2}\,dr = \frac{8}{3}\rho_o R^3$ [17], and θ is the angle between the observer direction $\mathbf{x}/|\mathbf{x}|$ and the x_1-axis. In a real fluid, the large acceleration of the flow at the edge of the disc causes flow separation (vortex shedding). This changes the unsteady force on the disc and the amplitude of the radiated sound (Section 1.14).

1.11.2 Time-Domain Problems

Consider the general solution (1.10.6) of the wave equation (1.6.8) when S is *stationary* and *rigid*. Choose Green's function to have vanishing normal derivative on S. Then, using the expansion (1.11.1) to calculate $G(\mathbf{x},\mathbf{y},t-\tau) = \frac{-1}{2\pi}\int_{-\infty}^{\infty} G(\mathbf{x},\mathbf{y};\omega)\mathrm{e}^{-i\omega t}d\omega$, we find

$$p(\mathbf{x},t) \approx \frac{1}{4\pi|\mathbf{x}|}\bigg(\int \mathcal{F}(\mathbf{y},t-|\mathbf{x}|/c_o)\,d^3\mathbf{y}$$
$$+\frac{x_i}{c_o|\mathbf{x}|}\frac{\partial}{\partial t}\int Y_i(\mathbf{y})\mathcal{F}(\mathbf{y},t-|\mathbf{x}|/c_o)\,d^3\mathbf{y} + \sum_{n\geq 2}\frac{(i\ell)^n}{c_o^n}\frac{\partial^n}{\partial t^n}$$
$$\times \int \Phi_n(\mathbf{x}/|\mathbf{x}|,\mathbf{y})\mathcal{F}(\mathbf{y},t-|\mathbf{x}|/c_o)\,d^3\mathbf{y}\bigg), \quad |\mathbf{x}|\to\infty. \qquad (1.11.5)$$

The term of order n in the series $\sim O((\omega\ell/c_o)^n)$. When the body is compact it is usually sufficient to retain only the first two terms in the large brackets, the monopole and dipole. This is equivalent to approximating Green's function by

$$G(\mathbf{x},\mathbf{y},t-\tau) \approx \frac{1}{4\pi|\mathbf{x}|}\left(\delta(t-\tau-|\mathbf{x}|/c_o) + \frac{\mathbf{x}\cdot\mathbf{Y}}{c_o|\mathbf{x}|}\delta'(t-\tau-|\mathbf{x}|/c_o)\right)$$

$$\approx \frac{1}{4\pi|\mathbf{x}|}\delta(t-\tau-(|\mathbf{x}|-\mathbf{x}\cdot\mathbf{Y}/|\mathbf{x}|)/c_o)$$

$$\approx \frac{1}{4\pi\,|\mathbf{x}-\mathbf{Y}|}\delta\big(t-\tau-|\mathbf{x}-\mathbf{Y}|/c_o\big), \quad |\mathbf{x}|\to\infty. \tag{1.11.6}$$

This result can be made symmetric, in accordance with reciprocity, by replacing $\mathbf{x}$ by $\mathbf{X}\equiv\mathbf{x}-\varphi^*(\mathbf{x})$, after which we define the *compact Green's function for a body bounded by a surface S*:

$$G(\mathbf{x},\mathbf{y},t-\tau) = \frac{1}{4\pi|\mathbf{X}-\mathbf{Y}|}\delta(t-\tau-|\mathbf{X}-\mathbf{Y}|/c_o), \tag{1.11.7}$$

where $X_i = x_i - \varphi_i^*(\mathbf{x})$, $Y_i = y_i - \varphi_i^*(\mathbf{y})$, and φ_i^* is the velocity potential of incompressible flow that would be produced by rigid body motion of S at unit speed in the i-direction, so that $\partial G(\mathbf{x},\mathbf{y},t-\tau)/\partial x_n = \partial G(\mathbf{x},\mathbf{y},t-\tau)/\partial y_n = 0$ on S.

In the frequency domain, the corresponding approximation is

$$G(\mathbf{x},\mathbf{y};\omega) = \frac{-e^{i\kappa_o|\mathbf{X}-\mathbf{Y}|}}{4\pi|\mathbf{X}-\mathbf{Y}|}. \tag{1.11.8}$$

Note that because of the symmetrical manner in which $\mathbf{x}$ and $\mathbf{y}$ occur in (1.11.7) and (1.11.8), the location of the coordinate origin is now arbitrary. However, it is still necessary that *either* $\mathbf{x}$ or $\mathbf{y}$ should lie in the far field of the body. When *both* $\mathbf{x}$ and $\mathbf{y}$ are in the far field, predictions made with the compact Green's function will be the same as when the body is *absent*. This is because for distant sources the amplitude of the sound scattered by a compact rigid object is $O[(\kappa_o\ell)^2]$ smaller than the incident sound, that is, is of *quadrupole* intensity [10] (see Example 6).

Example 5. If S is multiply connected, φ_i^* is uniquely defined by the requirement that the circulations about all irreducible circuits are zero [17]. Deduce the necessity for this from the causality principle (let the frequency

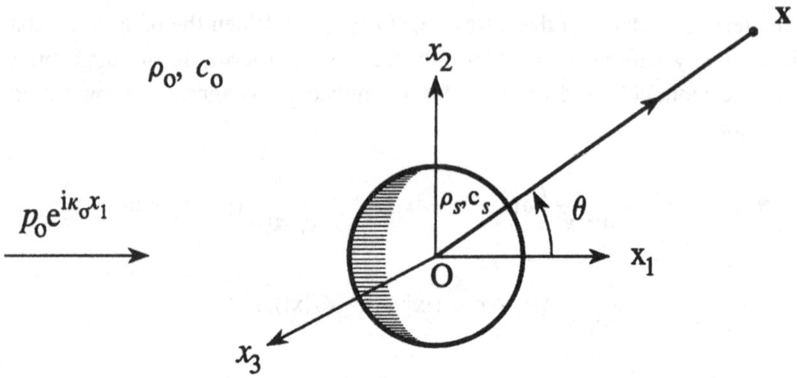

Figure 1.11.2. Rayleigh scattering by a spherical inclusion.

ω have a small positive imaginary part and apply Kelvin's circulation theorem, Section 1.2.7).

Example 6: Rayleigh scattering. A plane, time-harmonic sound wave $p_I = p_0 e^{i\kappa_o x_1}$ is incident on an acoustically compact spherical inclusion of radius R, density ρ_S, and sound speed c_S (Figure 1.11.2). In the ambient fluid (of mean density ρ_o and sound speed c_o), the scattered pressure p_S can be written as the integral (1.9.13) by using the compact Green's function $G(\mathbf{x}, \mathbf{x}; \omega)$ and by taking $v_n = v_{Sn}$, where $\mathbf{v}_S$ is the *scattered* fluid particle velocity. Using (1.11.8) with coordinate origin at the undisturbed center of the sphere, we find

$$p(\mathbf{x}, \omega) = \frac{-i\rho_o \omega e^{i\kappa_o|\mathbf{x}|}}{4\pi|\mathbf{x}|}\left(\oint_S v_{Sn} dS(\mathbf{y}) - \frac{i\kappa_o x_j}{|\mathbf{x}|}\oint_S Y_j v_{Sn} dS(\mathbf{y})\right), \quad |\mathbf{x}| \to \infty. \qquad (1.11.9)$$

With the exception of cases (exemplified by an air bubble in water) in which the inclusion is highly compressible, the pressure in the neighborhood of S is equal to p_0 plus a small correction $\sim O(\kappa_o R)p_0$. This causes the volume V of S to dilate by an amount $\delta V = (-1/i\omega)\oint_S v_n\, dS = (-1/i\omega)\int_V \operatorname{div} \mathbf{v}\, d^3\mathbf{y} \approx -p_0 V/\rho_S c_S^2$, where $\mathbf{v}$ is the total fluid velocity. Thus, the first integral in (1.11.9) is given by

$$\oint_S v_{Sn} dS = \oint_S \left(v_n - \frac{\partial p_I/\partial y_n}{i\rho_o\omega}\right) dS = \int_V \left(\operatorname{div} \mathbf{v} - \frac{\nabla^2 p_I}{i\rho_o\omega}\right) d^3\mathbf{y}$$

$$\approx i\omega p_0 V\left(\frac{1}{\rho_S c_S^2} - \frac{1}{\rho_o c_o^2}\right).$$

The value of the second integral in (1.11.9) depends on the translational velocity U, say, of the sphere parallel to the incident wave direction. U is calculated from the equation of motion $(m + \mathrm{M}_{11})\partial U/\partial t = -\oint_S p' n_1 dS \approx -p_0 \int_S [1 + i\kappa_o Y_1(y)] n_1 dS$, where $m = \mathrm{V}\rho_S$ is the mass of the sphere, $\mathrm{M}_{11} = \frac{1}{2}V\rho_o$ is the added mass, p' the surface pressure due to the incident wave when the motion of the sphere is ignored, and φ_1^* is defined in Example 3. We find $U = (p_0/\rho_o c_o)(m_o + \mathrm{M}_{11})/(m + \mathrm{M}_{11})$, where m_o is the mass of fluid displaced by the sphere. The component of $\mathbf{v}_S$ in the ambient fluid produced by the translational oscillations is equal to $(U - p_0/\rho_o c_o)\nabla\varphi_1^*$, and this is used to evaluate the second integral in (1.11.9). Components of $\mathbf{v}_S$ associated with changes in the *shape* of S are of higher order in $\kappa_o R$ and need not be considered. Collecting together the results of these calculations, we find

$$p_S \approx \frac{p_0(\kappa_o R)^2 R}{3|\mathbf{x}|}\left[\left(\frac{\mathrm{K}_S - \mathrm{K}_o}{\mathrm{K}_o}\right) - 3\left(\frac{\rho_o - \rho_S}{\rho_o + 2\rho_S}\right)\cos\theta\right]\mathrm{e}^{i\kappa_o|\mathbf{x}|}, \quad |\mathbf{x}| \to \infty, \qquad (1.11.10)$$

where $\mathrm{K}_S = 1/\rho_S c_S^2$, $\mathrm{K}_o = 1/\rho_o c_o^2$ are the respective *adiabatic* compressibilities of the sphere and the ambient fluid (see (1.6.5)), and θ is the angle between the radiation direction $\mathbf{x}/|\mathbf{x}|$ and the x_1-axis.

Equation (1.11.10) is the *Rayleigh* approximation to the scattered sound, which can be attributed to monopole and dipole sources whose strengths are respectively proportional to the differences in the compressibilities and densities of the inclusion and its environment. The case of a *rigid* sphere is obtained when $\mathrm{K}_S \to 0$. The limit $\rho_S/\rho_o \to \infty$ corresponds to a *fixed*, rigid sphere, for which

$$p_S \approx \frac{-p_0(\kappa_o R)^2 R}{3|\mathbf{x}|}\left(1 - \frac{3}{2}\cos\theta\right)\mathrm{e}^{i\kappa_o|\mathbf{x}|}, \quad |\mathbf{x}| \to \infty,$$

Example 7. Show that the acoustic pressure p_S scattered when a sound wave $p_I(\mathbf{x}, t)$ is incident on a compact and *stationary* rigid body of volume V is determined at large distances by the outgoing solution of

$$\left(\partial^2/c_o^2\partial t^2 - \nabla^2\right)p_S = \rho_o \partial(m\delta(\mathbf{x}))/\partial t - \mathrm{div}(\mathbf{f}\delta(\mathbf{x})), \qquad (1.11.11)$$

$$m = \frac{\mathrm{V}}{\rho_o c_o^2}\frac{\partial p_I}{\partial t}; \quad \mathrm{f}_i = (m_o\delta_{ij} + \mathrm{M}_{ij})\frac{1}{\rho_o}\frac{\partial p_I}{\partial x_j} \equiv -(m_o\delta_{ij} + \mathrm{M}_{ij})\frac{\partial v_{Ij}}{\partial t}, \qquad (1.11.12)$$

where the coordinate origin is within the body, v_I is the incident wave acoustic particle velocity, and m_o is the mass of fluid displaced by the body. The monopole strength m is equal and opposite to the volume flux into a surface that just encloses the body produced by the incident sound alone; f_i is the net force exerted on the fluid by the body.

If the body executes small, low-frequency translational oscillations at velocity $U_i(t)$, the radiated sound is still given by (1.11.12) but with f_i modified as follows:

$$\mathrm{f}_i = -(m_o\delta_{ij} + \mathrm{M}_{ij})\left(\frac{\partial v_{Ij}}{\partial t} - \frac{\partial U_j}{\partial t}\right). \tag{1.11.13}$$

Except when $m_o = 0$, f_i does not now equal the force on the fluid, but it is the force that *would* be exerted on the fluid if the body were stationary and the incident acoustic particle velocity were reduced to $\mathbf{v}_I - \mathbf{U}$.

Example 8. Show that a multipole of nonzero order located in the neighborhood of a compact solid is equivalent to an acoustic dipole.

Example 9. In incompressible fluid $G(\mathbf{x}, \mathbf{y}, t-\tau) = \delta(t-\tau)/4\pi|\mathbf{X} - \mathbf{Y}|$. Show that this Green's function determines the monopole and dipole components of an arbitrary distribution of sources interacting with a rigid body provided that at least one of the points $\mathbf{x}$, $\mathbf{y}$ is in the *hydrodynamic* far field of the body.

Example 10. A rigid body translates at velocity $\mathbf{U}$ in an incompressible, ideal fluid. Use the Green's function of Example 9 and the incompressible form of equation (1.10.9) ($c \to \infty$) to deduce that the velocity potential φ of the motion satisfies

$$\varphi = U_j\varphi_j^*(\mathbf{x}) \approx \frac{-U_j x_j}{4\pi\rho_o|\mathbf{x}|^3}(m_o\delta_{ij} + \mathrm{M}_{ij}), \quad |\mathbf{x}| \to \infty,$$

where the coordinate origin is within, and translates with the body, m_o is the mass of the displaced fluid, and M_{ij} is the added mass tensor.

Example 11. Show that in two dimensions the *dipole* component of the compact Green's function is given by

$$G(\mathbf{x}, \mathbf{y}, t-\tau) = \frac{X_i Y_i}{4\pi c_o}\frac{\partial}{\partial t}\left(\frac{\mathrm{H}(t-\tau-r/c_o)}{(t-\tau)\sqrt{c_o^2(t-\tau)^2 - r^2}}\right), \quad r \equiv |\mathbf{x} - \mathbf{y}|.$$

1.11.3 Compact Green's Function in Low Mach Number, Irrotational Flow

At sufficiently low Mach number M that terms $\sim O(M^2)$ may be neglected in the convected wave equation (1.6.30) the frequency domain Green's function satisfies the reverse flow equation

$$\left(\frac{\partial^2}{\partial y_j^2} + \kappa_o^2 - 2i\kappa_o M_j \frac{\partial}{\partial y_j}\right) G = \delta(\mathbf{x} - \mathbf{y}), \tag{1.11.14}$$

where $M_i = U_i(\mathbf{y})/c_o$.

Suppose the flow is past a stationary, acoustically compact rigid surface S, with $U_i(\mathbf{y}) = \mathbf{U}_o \cdot \nabla Y_i(\mathbf{y})$, where $\mathbf{U}_o$ is the uniform mean velocity at large distances from S. The substitution $G(\mathbf{x}, \mathbf{y}; \omega) = \bar{G}(\mathbf{x}, \mathbf{y}; \omega)\mathrm{e}^{i\kappa_o \mathbf{M}_o \cdot \mathbf{Y}}$[32], where $\mathbf{M}_o = \mathbf{U}_o/c_o$, transforms (1.11.14) into

$$(\nabla^2 + \kappa_o^2)\bar{G} = \delta(\mathbf{x} - \mathbf{y})\mathrm{e}^{-i\kappa_o \mathbf{M}_o \cdot \mathbf{X}}, \tag{1.11.15}$$

which may be solved in terms of Green's function for the Helmholtz equation. Proceeding as in Section 1.11.1, we can derive the following analogs of (1.11.7) and (1.11.8)

$$G(\mathbf{x}, \mathbf{y}, t - \tau) = \frac{1}{4\pi|\mathbf{X} - \mathbf{Y}|}\delta(t - \tau - |\mathbf{X} - \mathbf{Y}|/c_o + \mathbf{M}_o \cdot (\mathbf{X} - \mathbf{Y})/c_o), \tag{1.11.16}$$

$$G(\mathbf{x}, \mathbf{y}; \omega) = \frac{-\mathrm{e}^{i\kappa_o[|\mathbf{X}-\mathbf{Y}| - \mathbf{M}_o \cdot (\mathbf{X}-\mathbf{Y})]}}{4\pi|\mathbf{X} - \mathbf{Y}|}. \tag{1.11.17}$$

It is sometimes convenient to express these in terms of emission time coordinates (cf. Figure 1.8.1). This would be appropriate when the surface S is regarded as being in uniform motion at velocity $-\mathbf{U}_o$ in stationary fluid, the observer being fixed relative to the fluid. Then

$$|\mathbf{x}| \approx R(1 - M_o \cos\Theta),$$
$$|\mathbf{X} - \mathbf{Y}| - \mathbf{M}_o \cdot (\mathbf{X} - \mathbf{Y}) \approx R - \mathbf{R} \cdot \mathbf{Y}/R(1 - M_o \cos\Theta), \quad R \to \infty,$$

where Θ is the angle between the direction of motion of S and $\mathbf{R}$ at the time of emission of the sound. Then

$$G(\mathbf{x}, \mathbf{y}, t - \tau) = \frac{1}{4\pi R(1 - M_o \cos\Theta)}\delta(t - \tau - R/c_o + \mathbf{R} \cdot \mathbf{Y}/[c_o R(1 - M_o \cos\Theta)]),$$

$$G(\mathbf{x}, \mathbf{y}; \omega) = \frac{-\mathrm{e}^{i\kappa_o(R - \mathbf{R}.\mathbf{Y}/\{c_o R(1 - M_o \cos\Theta)\})}}{4\pi R(1 - M_o \cos\Theta)}, \quad R \to \infty. \tag{1.11.18}$$

These formulae use the emission time coordinate **R** for the observer position, but the source variable $Y_i(\mathbf{y})$ is with respect to a reference frame translating with the body.

Example 12. The influence of low Mach number, translational motion on the sound radiated by an acoustically compact, pulsating sphere [33].

Let the sphere have radius ℓ and translate at velocity U_o in the negative x_1-direction, and let $s(t)$ denote the radial displacement of the surface S of the sphere, with $v_n = \partial s/\partial t$. In a frame of reference translating with the sphere, with origin at the center, the potential of the unsteady motion in the neighborhood of the sphere is $\Phi(\mathbf{x}, t) = -\ell^2 v_n(t)/|\mathbf{x}| + U_o x_1[1 + (\ell + s)^3/2|\mathbf{x}|^3]$, and the perturbation potential is $\varphi(\mathbf{x}, t) \approx -\ell^2 v_n(t)/|\mathbf{x}| + 3s(t)\ell^2 U_o x_1/2|\mathbf{x}|^3$. On the mean position of S, $\partial\varphi/\partial x_n = v_n - 3s\mathbf{U}_o \cdot \mathbf{n}/\ell$. Substituting into (1.10.9) with $\rho = \rho_o$, $\hat{\mathcal{F}} = 0$, and G given by (1.11.18), we find

$$\varphi(\mathbf{R}, t) \approx \frac{-1}{4\pi r(1 - M_o \cos\Theta)} \int_S (v_n - 3U_{oi}n_i s/\ell)\{t - R/c_o + \mathbf{R}\cdot\mathbf{Y}/[c_o R(1 - M_o\cos\Theta)]\}\, dS(\mathbf{y}), \quad R \to \infty,$$

where $\mathbf{Y} = \mathbf{y} - \varphi^*(\mathbf{y})$ and φ_i^* is defined in Example 3. The value of the integral is estimated by expanding the integrand to first order in the retarded time variations $\mathbf{R}\cdot\mathbf{Y}/[c_o R(1 - M_o\cos\Theta)]$ on S. The acoustic pressure is $p(\mathbf{R}, t) = -\rho_o d\varphi(\mathbf{R}(t), t)/dt = -\rho_o[\partial\varphi/\partial t + (\partial\varphi/\partial R)(\partial R/\partial t)]$; using (1.8.21) we find, when terms of order M_o^2 relative to unity are neglected,

$$p(\mathbf{R}, t) \approx \frac{\rho_o}{4\pi r(1 - M_o\cos\Theta)^{7/2}} \frac{\partial m}{\partial t}(t - R/c_o), \quad R \to \infty,$$

where $m(t)$ is the volume flux from the sphere.

Motion of the sphere amplifies the sound by $3\frac{1}{2}$ Doppler factors $1/(1 - M_o\cos\Theta)$ relative to the result (1.9.14) for a stationary sphere. A pulsating sphere is equivalent to a point monopole of strength $m(t)$ when at rest, but the analogy is destroyed during translational motion, because the radiation from an ideal monopole then has two Doppler factors (Section 1.8.5). The difference is caused by the unsteady drag on the sphere (equivalent to an additional dipole) which is equal $2\pi\ell^2\rho_o U_o v_n(t)$.

1.12 The Radiation of Sound from Vibrating Bodies: Matched Expansions

The principal contribution to the sound radiated by a compact body executing torsional oscillations frequently turns out to be of *quadrupole* strength and cannot be evaluated by use of a compact Green's function. We now describe how the radiation from a compact body executing arbitrary, small-amplitude vibrations can be determined correct to quadrupole order by a matching of series expansions [29–31].

1.12.1 Matching in the Absence of Flow

When the frequency of the motion is sufficiently small, the hydrodynamic field of the motion extends out to distances $|\mathbf{x}| \gg \ell \sim$ diameter of the body. Let φ_o denote the *hydrodynamic* velocity potential, valid for *incompressible* motion, and which satisfies Laplace's equation $\nabla^2\varphi_o = 0$. Each component of frequency ω of the actual velocity potential φ is a solution of the Helmholtz equation (1.7.1) with $q = 0$. Thus, $\nabla^2(\varphi_o - \varphi) = \kappa_o^2\varphi$, and near the body, where $\partial/\partial\mathbf{x} \sim 1/\ell$, we must have $\varphi_o - \varphi \approx (\kappa_o\ell)^2\varphi \approx (\kappa_o\ell)^2\varphi_o \ll \varphi_o$.

Take the coordinate origin within the body, and assume that a multipole expansion of φ_o (defined as in the second line of (1.8.14) when $c_o \to \infty$) is known in the form

$$\varphi_o(\mathbf{x}, t) = s^0(t)\frac{1}{4\pi|\mathbf{x}|} + s_i^1(t)\frac{\partial}{\partial x_i}\left(\frac{1}{4\pi|\mathbf{x}|}\right) + s_{ij}^2(t)\frac{\partial^2}{\partial x_i \partial x_j}\left(\frac{1}{4\pi|\mathbf{x}|}\right) + \dots, \quad |\mathbf{x}| \gg \ell, \tag{1.12.1}$$

where the terms on the right are the monopole, dipole, quadrupole, and so forth, components of the potential. We shall attempt to derive from (1.12.1) a second expansion valid in the *acoustic* region $\kappa_o|\mathbf{x}| \gg 1$ by matching terms in the second expansion with those of the "inner"expansion (1.12.1). It will transpire that the information contained in (1.12.1) is generally insufficient to permit matching to be performed correct to quadrupole order, and that additional information about the behavior of φ_o close to the body must also be given.

In the frequency domain the multipole coefficients in (1.12.1) are replaced by their Fourier transforms $s^0(\omega)$, $s_i^1(\omega)$, and so forth. The values of these coefficients can be expressed in terms of φ_o and its normal derivative on an arbitrary surface S that just encloses the body by expanding the integrand of the incompressible version of (1.10.3) (with p replaced by φ_o, $G(\mathbf{x}, \mathbf{y}; \omega) = -1/4\pi|\mathbf{x} - \mathbf{y}|$ and $\mathcal{F} = 0$) in powers of $\mathbf{y}/|\mathbf{x}|$.) Let S be the smallest sphere (of radius a, say) that just encloses the body, and take the coordinate origin at its

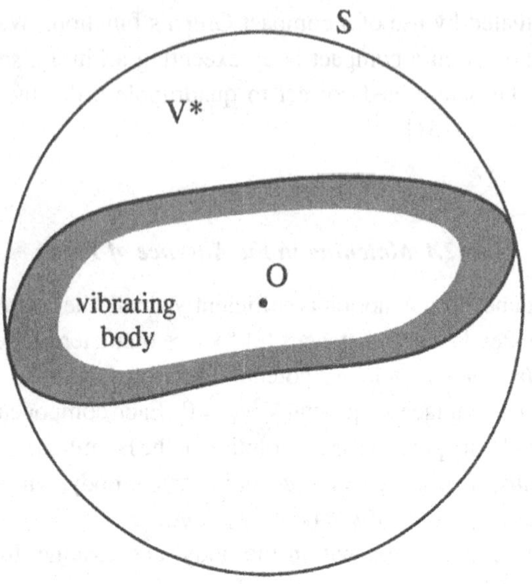

Figure 1.12.1. Calculating the radiation from a small vibrating body.

center (Figure 1.12.1). Then we find

$$s^0(\omega) = -\oint_S \frac{\partial \varphi_o}{\partial y_j} n_j \, dS, \quad s_i^1(\omega) = \oint_S \left(y_i \frac{\partial \varphi_o}{\partial y_j} - \varphi_o \delta_{ij} \right) n_j \, dS,$$

$$s_{ij}^2(\omega) = \oint_S \left(y_i \varphi_o \delta_{jk} - \frac{1}{2} y_i y_j \frac{\partial \varphi_o}{\partial y_k} \right) n_k \, dS + C\delta_{ij}, \tag{1.12.2}$$

where C is an arbitrary function of ω that must be included because $\delta_{ij}\partial^2 (1/|\mathbf{x}|)/\partial x_i \partial x_j \equiv 0$ for $\mathbf{x} \neq \mathbf{0}$.

This procedure is now repeated (by taking $G(\mathbf{x}, \mathbf{y}; \omega) = -\mathrm{e}^{i\kappa_o|\mathbf{x}-\mathbf{y}|}/4\pi|\mathbf{x}-\mathbf{y}|$ in (1.10.3)) to obtain the following expansion in the acoustic domain,

$$\varphi_o(\mathbf{x}, \omega) = S^0(\omega) \frac{\mathrm{e}^{i\kappa_o|\mathbf{x}|}}{4\pi|\mathbf{x}|} + S_i^1(\omega) \frac{\partial}{\partial x_i} \left(\frac{\mathrm{e}^{i\kappa_o|\mathbf{x}|}}{4\pi|\mathbf{x}|} \right)$$
$$+ S_{ij}^2(\omega) \frac{\partial^2}{\partial x_i \partial x_j} \left(\frac{\mathrm{e}^{i\kappa_o|\mathbf{x}|}}{4\pi|\mathbf{x}|} \right) + \dots, \quad \kappa_o|\mathbf{x}| \gg 1. \tag{1.12.3}$$

The coefficients $S^0(\omega)$, $S_i^1(\omega)$, $S_{ij}^2(\omega)$ are again given by (1.12.2), except that φ_o is replaced by the acoustic potential φ and $C \equiv 0$. However, in the acoustic far field each occurrence of a differential operator $\partial/\partial x_i$ in (1.12.3) is equivalent to multiplication by $-i\kappa_o x_i/|\mathbf{x}|$ so that in evaluating the expansion correct to quadrupole order (i.e., to $O((\kappa_o \ell)^2)$), the dipole and quadrupole coefficients $S_i^1(\omega)$, $S_{ij}^2(\omega)$ can be defined *exactly* as in (1.12.2) with $C = 0$.

To match (1.12.1) and (1.12.3), we must express $S^0(\omega)$, $S_i^1(\omega)$, $S_{ij}^2(\omega)$ explicitly in terms of $s^0(\omega)$, $s_i^1(\omega)$, $s_{ij}^2(\omega)$. First, noting that $\partial\varphi/\partial y_n = \partial\varphi_o/\partial y_n$ on the surface of the body, we have

$$S^0(\omega) - s^0(\omega) = -\oint_S \frac{\partial}{\partial y_j}(\varphi - \varphi_o)n_j\, dS$$
$$= -\int_{V*}(\nabla^2\varphi - \nabla^2\varphi_o)d^3\mathbf{y} \approx \kappa_o^2 \int_{V*}\varphi_o d^3\mathbf{y}, \qquad (1.12.4)$$

where V* is the volume of fluid between the body and the sphere S, and the last integral follows from $\nabla^2\varphi = -\kappa_o^2\varphi \approx -\kappa_o^2\varphi_o$. This shows that the"monopole" in (1.12.3) differs from the monopole of (1.12.1) because of the volume flux through S produced by compression of the fluid in V* by the hydrodynamic pressure. The correction is nominally of quadrupole strength.

Next $S_i^1(\omega) = s_i^1(\omega)$, but $S_{ij}^2(\omega) = s_{ij}^2(\omega) + C\delta_{ij}$. To eliminate C, we examine the integral defining $S_{ij}^2(\omega)$. In the region outside S, the potential φ_o can be expanded in spherical harmonics as in (1.10.15). When the final integral in (1.12.2) is evaluated by using this series expansion, only the terms with $n = 0,\ 2$ make nontrivial contributions and yield

$$S_{ij}^2(\omega) = \tfrac{1}{2}s^0(\omega)a^2\delta_{ij} + A_{ij}, \quad s_{ij}^2(\omega) = A_{ij} + C'\delta_{ij}, \ A_{kk} = 0,$$

where A_{ij} is a linear combination of the coefficients α_{2m}, $-2 \le m \le 2$, and C' is arbitrary. These results imply that

$$S_{ij}^2(\omega) = s_{ij}^2(\omega) - \tfrac{1}{3}s_{kk}^2(\omega)\delta_{ij} + \tfrac{1}{2}s^0(\omega)a^2\delta_{ij}.$$

Having established these relations, the expansion (1.12.3) may be reordered so that each term of the series is $O(\kappa_o\ell)$ relative to the preceding one, and the result transformed back to the time domain to give the following truncated multipole expansion of the acoustic field

$$\varphi(\mathbf{x}, t) \approx \sum_{n=0}^{2} \frac{\partial^n}{\partial x_i \partial x_j \ldots}\left(\frac{S_{ij\ldots}^n(t - |\mathbf{x}|/c_o)}{4\pi|\mathbf{x}|}\right), \quad |\mathbf{x}| \to \infty. \qquad (1.12.5a)$$

$$S^0(t) = s^0(t), \quad S_i^1(t) = s_i^1(t),$$
$$S_{ij}^2(t) = s_{ij}^2(t) - \frac{1}{3}s_{kk}^2(t)\delta_{ij} + \delta_{ij}\left(\frac{1}{2}s^0(t)a^2 - \int_{V*}\varphi_o(\mathbf{y}, t)d^3\mathbf{y}\right), \qquad (1.12.5b)$$

where the integral is over the region V* between the mean position of the vibrating surface and the surface of the smallest sphere (of radius a) that just

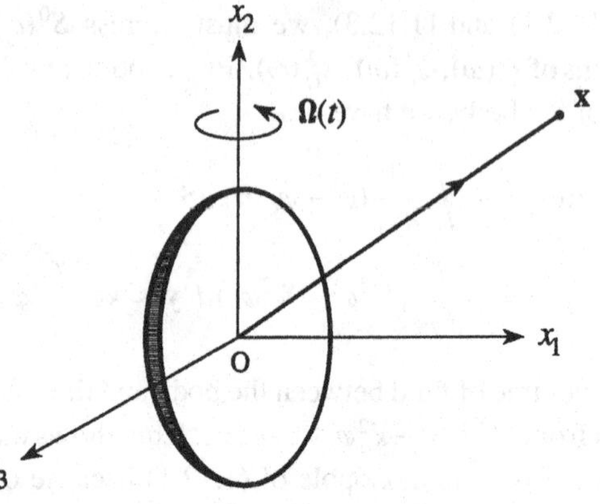

Figure 1.12.2. Sound generated by torsional oscillations.

encloses the body. The presence of this integral implies that, except in the particular case of a spherical body, the hydrodynamic multipole expansion (1.12.1) does not generally supply sufficient data to permit the acoustic field to be calculated to quadrupole order. When the time retardation $|\mathbf{x}|/c_o$ is neglected (1.12.5a) reduces to (1.12.1).

Example 1. A disc of radius R executes low-frequency, small-amplitude torsional oscillations at angular velocity $\Omega(t)$ about a diameter on the x_2-axis (Figure 1.12.2). In the hydrodynamic far field, $\varphi_o \approx -(2\ell^5/15\pi)\Omega(t)\partial^2(1/|\mathbf{x}|)/\partial x_1 \partial x_3$[17] and $\int_{V*} \varphi_o d^3\mathbf{y} = 0$. Show that

$$p(x,t) \approx \frac{-2\rho_o R^5 x_i x_j}{15\pi c_o^2 |x|^3} \frac{d^3\Omega}{dt^2}(t - |x|/c_o), \quad |x| \to \infty.$$

Example 2. A tuning fork consists of two equal, rigid spheres of radius R whose distance apart varies periodically [10]. The mean positions of the centers of the spheres are on the x_1-axis at $\pm s$. When $R/2s \ll 1$, the hydrodynamic velocity potential is given approximately by

$$\varphi_o = U(t)[1 - (R/2s)^3][\varphi_1^*(x_1 - s, x_2, x_3) - \varphi_1^*(x_1 + s, x_2, x_3)],$$

where $2U(t)$ is the relative speed of separation and $\varphi_1^*(\mathbf{x}) = -R^3 x_1/2|\mathbf{x}|^3$. Show that

$$p(\mathbf{x},t) \approx \frac{\rho_o s R^3}{c_o^2|\mathbf{x}|}\left[\left(1 - \frac{r^3}{8s^3}\right)\cos\theta - \frac{R^3}{12s^3}\right]\frac{d^3U}{dt^3}(t - |\mathbf{x}|/c_o), \quad |\mathbf{x}| \to \infty,$$

where θ is the angle between the x_1-axis and the radiation direction $\mathbf{x}/|\mathbf{x}|$.

1.12.2 Matching in Low Mach Number Flow

The procedure of Section 1.12.1 is also applicable in the presence of a low Mach number, irrotational mean flow. The reader can show (by using (1.10.8), with $G(\mathbf{x}, \mathbf{y}; \omega)$ defined as in (1.7.17) in the acoustic domain, and with the neglect of terms $\sim O(M^2)$) that the relations (1.12.5b) remain valid when $M_o^2 \ll 1$, where $\mathbf{M}_o$ is the uniform vector Mach number at large distances from the body, and that the multipole expansion of the acoustic field is

$$\varphi(\mathbf{x}, t) \approx \sum_{n=0}^{2} \frac{\partial^n}{\partial x_i \partial x_j \dots} \left(\frac{\mathrm{S}^n_{ij\dots}(t - |\mathbf{x}|/c_o + \mathbf{M}_o.\mathbf{x}/c_o)}{4\pi |\mathbf{x}|} \right), \quad |\mathbf{x}| \to \infty. \tag{1.12.6}$$

This approximation is correct to orders $\kappa_o \ell M_o$ and $(\kappa_o \ell)^2$ relative to unity, but terms $\sim O((\kappa_o \ell)^2 M_o)$ have been discarded.

Example 4. A rigid, acoustically compact sphere of radius a is in rectilinear motion at speed $U_o + v(t)$ in the x_1-direction in stationary fluid, where U_o is constant and $v(t)$ $(\ll U_o)$ is a small, fluctuating velocity. Show that, in a reference frame moving at the mean velocity U_o, with the origin at the mean location of the center of the sphere, the hydrodynamic velocity potential is

$$\varphi_o(\mathbf{x}, t) = +\frac{1}{2} a^3 v(t) \frac{\partial}{\partial x_1} \left(\frac{1}{|\mathbf{x}|} \right) + \frac{1}{2} a^3 s(t) U_o \frac{\partial^2}{\partial x_1^2} \left(\frac{1}{|\mathbf{x}|} \right),$$

where $s(t) = \int_{-\infty}^{t} v(t)\, dt$ is the displacement of the sphere due to the fluctuating motion. The second term in this expansion is a quadrupole generated by a distribution of forces on the sphere whose net strength is null. Deduce that [33]

$$p(\mathbf{R}, t) \approx \frac{\rho_o a^3}{2Rc_o} \left(\frac{\cos \Theta}{(1 - M_o \cos \Theta)^4} - \frac{1}{3} M_o \right) \frac{d^2 v}{dt^2}(t - R/c_o), \quad R \to \infty,$$

where R is the observer emission-time coordinate (Section 1.8.5), and Θ is the angle between the direction of motion of the sphere and the observer direction at the time of emission of the received sound.

A small, stationary vibrating body is equivalent to an acoustic dipole, but this formula differs from that for an ideal, uniformly translating dipole (Section 1.8.5) in two respects. First the dipole is augmented by a monopole; second, the Doppler amplification involves *four* powers of $1/(1 - M_o \cos \Theta)$ rather than two. These are a result of the quadrupole contribution to the motion, which is important only when $U_o \neq 0$.

1.13 The Acoustic Energy Equation

The transport of mechanical energy by sound waves can be expressed in terms of an acoustic energy conservation equation

$$\frac{\partial E_a}{\partial t} + \text{div}\,\mathbf{I}_a = S_a, \qquad (1.13.1)$$

where E_a is the acoustic energy per unit volume, $\mathbf{I}_a$ the acoustic energy flux vector (the *intensity*), and S_a is a "source" accounting for the generation or absorption of sound. This equation can be derived by rearranging the general energy equation (1.2.30), and it therefore describes, in its most general form, not only the propagation of acoustic energy but also interactions between components of the flow that may only be weakly coupled to the sound. In particular, S_a must implicitly include terms describing alternative mechanisms of energy transport, by convection and by viscous and thermal diffusion [34, 35].

The fraction of the total flow energy contained in the acoustic field is usually very small, and the principal *hydrodynamic* contributions to S_a can individually be very much larger than either of the acoustic terms on the left. This indicates that the evaluation of S_a (from experiments or by numerical simulation of the flow) may be prone to significant errors because the sum of the separate, large contributors to S_a must be very nearly zero to balance the acoustic terms on the left. Thus, the acoustic energy equation is generally ill-conditioned, and its usefulness tends to be limited to situations in which S_a can be determined unambiguously. The simplest case is when there are no interactions of the sound with the background flow through which it propagates, and this occurs when the flow is irrotational and homentropic.

1.13.1 Energy Conservation in Steady, Irrotational Homentropic Flow

In the linearized approximation, the perturbation velocity potential φ in an arbitrary, irrotational steady mean flow satisfies (1.6.25) with $\hat{\mathcal{F}}(\mathbf{x}, t) = 0$. The acoustic energy equation is derived by multiplying this equation by $\rho\partial\varphi/\partial t$ and by rearranging, where $\rho \equiv \rho(\mathbf{x})$ is the density of the undisturbed mean flow. By taking account of the relations

$$p' = -\rho(\partial/\partial t + \mathbf{U}\cdot\nabla)\varphi, \quad \rho' = p'/c^2, \quad \text{div}(\rho\mathbf{U}) = 0,$$

where primed quantities are perturbations from local mean values, equation (1.6.25) may be recast as follows

$$\frac{\partial}{\partial t}\left(\frac{1}{2}\frac{p'^2}{\rho c^2}+\rho'\mathbf{U}\cdot\nabla\varphi+\frac{1}{2}\rho(\nabla\varphi)^2\right)+\operatorname{div}\left[\left(-\frac{\partial\varphi}{\partial t}\right)(\rho\nabla\varphi+\rho'\mathbf{U})\right]=\mathbf{0}. \tag{1.13.2}$$

In homentropic, irrotational flow, the perturbation stagnation enthalpy $B'\equiv -\partial\varphi/\partial t$ (Section 1.2.5), and $\mathbf{v}'=\nabla\varphi$ is the local velocity perturbation from $\mathbf{U}(\mathbf{x})$. Comparing (1.13.2) with (1.13.1), we conclude that, in irrotational, homentropic flow, the acoustic energy density, flux, and source terms are

$$E_a = p'^2/2\rho c^2+\rho'\mathbf{U}\cdot\mathbf{v}'+\frac{1}{2}\rho v'^2,\quad \mathbf{I}_a=(\rho\mathbf{v}'+\rho'\mathbf{U})B',\quad S_a=0. \tag{1.13.3}$$

These expressions are an exact consequence of the linearized wave equation (1.6.25) and are particularly useful because the perturbation quantities that define E_a and $\mathbf{I}_a$ can always be computed from solutions of that equation.

Actually, a formal expansion to *second order* of the full energy equation (1.2.30) might be expected to include second-order terms (in addition to those shown in (1.13.3)) that satisfy a second-order approximation to the *nonlinear* wave equation (1.6.22). We can easily show that the result of such an expansion is consistent with (1.13.3). For example, the expansion of the perturbation internal energy density $(\rho e)'$ is given to second order by

$$(\rho e)'\approx\rho'\frac{\partial(\rho e)}{\partial\rho}+\frac{1}{2}\rho'^2\frac{\partial^2(\rho e)}{\partial\rho^2}=w\rho'+\frac{c^2\rho'^2}{2\rho}, \tag{1.13.4}$$

where the derivatives are evaluated at constant entropy, unprimed quantities are evaluated by using local mean values, and the thermodynamic identities $d(\rho e)=\rho T ds+w d\rho$ and $c^2=\rho(\partial w/\partial\rho)_s$ have been used. To the same order of approximation, the final term on the right of (1.13.4) can be replaced by $p'^2/2\rho c^2$. Similarly, the perturbation kinetic energy density $(\frac{1}{2}\rho v^2)'$ and energy flux vector $\mathbf{I}'=(\rho\mathbf{v}B)'$ are given to second order by

$$\begin{aligned}(\tfrac{1}{2}\rho v^2)' &= \tfrac{1}{2}\rho' U^2+\rho\mathbf{v}'\cdot\mathbf{U}+\rho'\mathbf{v}'\cdot\mathbf{U}+\tfrac{1}{2}\rho v'^2,\\ (\rho\mathbf{v}B)' &= B(\rho\mathbf{v})'+\rho\mathbf{U}B'+\rho'\mathbf{U}B'+\rho\mathbf{v}'B',\end{aligned} \tag{1.13.5}$$

where the undisturbed stagnation enthalpy $B = w + \frac{1}{2}\rho U^2$ is constant throughout the flow. In (1.13.4) and (1.13.5) the contributions from terms that are linear in the perturbation ("primed") quantities include second-order components determined by solutions of the nonlinear equations of motion.

Using the definitions (1.13.3) of E_a and $\mathbf{I}_a$, the complete, second-order, homentropic approximation to the energy equation ((1.2.30) with $\mathcal{F} = 0$) now becomes

$$\frac{\partial}{\partial t}(E_a + \rho' B + \rho \mathbf{U} \cdot \mathbf{v}') + \operatorname{div}(\mathbf{I}_a + B(\rho \mathbf{v})' + \rho \mathbf{U} B') = 0. \tag{1.13.6}$$

But $B =$ constant, and $\operatorname{div}(\rho \mathbf{U}) = 0$. It therefore follows, respectively, (i) from the continuity equation and (ii) from Crocco's equation (1.2.18) (with zero right-hand side) that the contribution from the two terms in B, and from the two terms in $\rho\mathbf{U}$, vanishes. Equation (1.13.1) is accordingly an *exact* second-order acoustic energy equation [36].

Example 1. In the absence of flow in fluid of mean density ρ_o and sound speed c_o, $\mathbf{I}_a = \rho_o \mathbf{v} B' \equiv p'\mathbf{v}$, where $p'\mathbf{v}$ is the rate of working of the pressure fluctuations in the direction of propagation of the sound. For *plane waves* $v = p'/\rho_o c_o$, and

$$E_a = \frac{p'^2}{\rho_o c_o^2}, \quad I_a = \frac{p'^2}{\rho_o c_o}, \tag{1.13.7}$$

so that $\mathbf{I}_a = c_o E_a \mathbf{n}$, where $\mathbf{n}$ is a unit vector parallel to the propagation direction. The energy flux is therefore equal to the energy density propagated at the speed of sound.

1.13.2 Method of Stationary Phase

When the acoustic pressure or velocity potential is represented as a Fourier integral, the method of stationary phase [37, 38] can be used to determine the value of the integral at large distances from the sources and hence, also, the energy radiated by the sound [20, 23, 36, 38]. To illustrate the procedure, consider the radiation of sound into a stationary medium occupying the half-space $x_2 > 0$ from sources distributed over $x_2 = 0$. Then (cf. Section 1.7, Example 5) for each frequency ω the radiation into $x_2 > 0$ can be cast in the form

$$p(\mathbf{x}, \omega) = \iint_{-\infty}^{\infty} \hat{p}(\mathbf{k}, \omega) \mathrm{e}^{i(\mathbf{k}.\mathbf{x} + \gamma(k) x_2)}\, dk_1 dk_3, \tag{1.13.8}$$

where $\hat{p}(\mathbf{k}, \omega) \equiv \hat{p}(k_1, k_3, \omega)$ is the Fourier transform of $p(\mathbf{x}, \omega)$ on $x_2 = 0$ with respect to its dependence on (x_1, x_3), and

$$\begin{aligned} \gamma(k) &= \operatorname{sgn}(\kappa_o)\left|\kappa_o^2 - k^2\right|^{1/2}, \quad \text{for } k < |\kappa_o|, \\ &= i\left|\kappa_o^2 - k^2\right|^{1/2}, \quad \text{for } k > |\kappa_o|, \\ \mathbf{k} &= (k_1, k_3). \end{aligned} \tag{1.13.9}$$

The definition of $\gamma(k)$ ensures that (for the usual implied time dependence $e^{-i\omega t}$) each Fourier component of the integral represents a disturbance that either radiates or decays exponentially *into* the region $x_2 > 0$.

The stationary phase approximation gives the asymptotic behavior of the Fourier integral (1.13.8) for an observer at $\mathbf{x}$ in the far field ($\kappa_o|\mathbf{x}| \to \infty$), provided that $\hat{p}(\mathbf{k}, \omega)$ is a "smooth" function for real values of k_1 and k_3. At large distances, the main contribution to the integral is from the acoustic domain $k_1^2 + k_2^2 < \kappa_o^2$, where $\gamma(k)$ is real. Within this region, however, small changes in $\mathbf{k}$ tend to produce rapid oscillations between ± 1 of the real and imaginary parts of the exponential factor. The net contribution to the integral is therefore very small except when $\mathbf{k}$ varies in the vicinity of a point $\mathbf{k}_s$ where the phase $\Theta = \mathbf{k} \cdot \mathbf{x} + \gamma(k)x_2$ is *stationary*. Near $\mathbf{k}_s$ the fluctuations occur more slowly, and the neighborhoods of such points furnish the dominant contributions to the integral. If the sources are assumed to be concentrated near the coordinate origin (Figure1.13.1), then arriving waves in the far field are travelling parallel to $\mathbf{x}$, and their wavenumber vectors $(k_1, \gamma(k), k_3)$ of length $|\kappa_o|$, which are normal to the wavefronts, must also point in this direction. The stationary phase approximation is therefore dominated by wavenumbers centered around $\mathbf{k}_s = \kappa_o(x_1/|\mathbf{x}|, x_3/|\mathbf{x}|)$. The integral is evaluated by expanding the integrand about this point (Example 2). For future reference, we give the general formula for the stationary phase approximation:

$$\begin{aligned} &\iint_{-\infty}^{\infty} f(k_1, k_3, \omega) e^{i(k_1 x_1 + k_3 x_3 + \gamma(k)|x_2|)} dk_1 dk_3 \\ &\quad \approx \frac{-2\pi i \kappa_o |x_2|}{|\mathbf{x}|^2} f\left(\frac{\kappa_o x_1}{|\mathbf{x}|}, \frac{\kappa_o x_3}{|\mathbf{x}|}, \omega\right) e^{i\kappa_o|\mathbf{x}|}, \quad \kappa_o|\mathbf{x}| \to \infty, \end{aligned} \tag{1.13.10}$$

where $\gamma(k)$ is defined as in (1.13.9).

In particular, the acoustic pressure (1.13.8) becomes

$$p(\mathbf{x}, \omega) \approx \frac{-2\pi i \kappa_o x_2}{|\mathbf{x}|^2} \hat{p}\left(\frac{\kappa_o x_1}{|\mathbf{x}|}, \frac{\kappa_o x_3}{|\mathbf{x}|}, \omega\right) e^{i\kappa_o|\mathbf{x}|}, \quad |\mathbf{x}| \to \infty.$$

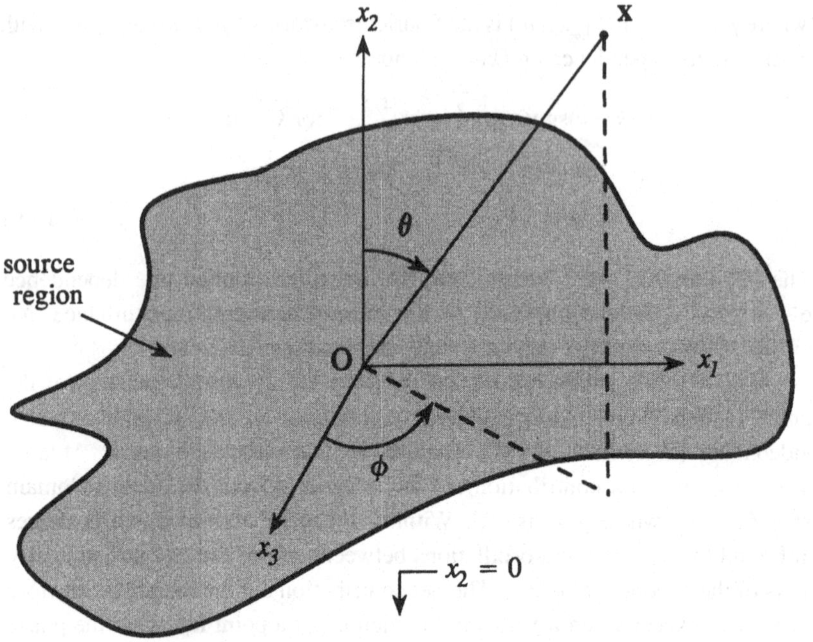

Figure 1.13.1. Acoustic sources radiating into $x_2, > 0$.

The mean acoustic intensity (see (1.13.7)), which determines the power radiated from the sources, is obtained by averaging $I_a = p(\mathbf{x}, t)^2/\rho_o c_o$ over a wave period $2\pi/\omega$. To do this, we take $p(\mathbf{x}, t) = \mathrm{Re}(p(\mathbf{x}, \omega)\mathrm{e}^{-i\omega t})$ to obtain

$$\langle I_a \rangle = \frac{\langle p(\mathbf{x}, t)^2 \rangle}{\rho_o c_o} = \frac{2\pi^2 \kappa_o^2 \cos^2\theta}{\rho_o c_o |\mathbf{x}|^2} \left| \hat{p}\left(\frac{\kappa_o x_1}{|\mathbf{x}|}, \frac{\kappa_o x_3}{|\mathbf{x}|}, \omega\right)\right|^2, \tag{1.13.11}$$

where the angle brackets $\langle\ \rangle$ denote a time average, and $\theta = \arccos(x_2/|\mathbf{x}|)$ is the angle in Figure 1.13.1 between the observer direction and the normal to the source plane.

Example 2. To derive (1.13.10), introduce spherical polar coordinates $x_1 = |\mathbf{x}| \sin\theta \sin\phi$, $|x_2| = |\mathbf{x}| \cos\theta$, $x_3 = |\mathbf{x}| \sin\theta \cos\phi$, $(0 < \theta < \pi/2,\ 0 < \phi < 2\pi)$ The integration may be restricted to the acoustic domain, where we can put $k_1 = \kappa_o \sin\bar{\theta} \sin\bar{\phi}$, $k_3 = \kappa_o \sin\bar{\theta} \cos\bar{\phi}$, $(0 < \bar{\theta}, < \pi/2,\ 0, < \bar{\phi}, < 2\pi)$ so that $\gamma(k) = \kappa_o \cos\bar{\theta}$ and $dk_1 dk_3 = \kappa_o^2 \sin\bar{\theta} \cos\bar{\theta}\, d\bar{\theta}\, d\bar{\phi}$. The phase of the exponential term in the integral becomes

$$\Theta = \kappa_o |\mathbf{x}| (\cos\theta \cos\bar{\theta} + \sin\theta \sin\bar{\theta} \cos(\phi - \bar{\phi})),$$

which is stationary where $\partial\Theta/\partial\bar{\theta} = \partial\Theta/\partial\bar{\phi} = 0$. This occurs only at $\bar{\theta} = \theta$, $\bar{\phi} = \phi$ within the region of integration. The contribution to the integral from the neighborhood of the stationary point is obtained by setting $\xi = \bar{\phi} - \phi$, $\eta = \bar{\theta} - \theta$, by expanding the argument of the exponential to second order in ξ and η, and by putting $k_1 = \kappa_o \sin\theta \sin\phi$, $k_3 = \kappa_o \sin\theta\cos\phi$ elsewhere in the integrand. The integral then reduces to

$$\kappa_o^2 \sin\theta\cos\theta f\left(\frac{\kappa_o x_1}{|\mathbf{x}|}, \frac{\kappa_o x_3}{|\mathbf{x}|}, \omega\right) \mathrm{e}^{i\kappa_o|\mathbf{x}|}$$

$$\times \iint_{-\infty}^{\infty} \exp\left(-\frac{1}{2} i\kappa_o|\mathbf{x}|(\eta^2 + \xi^2\sin^2\theta)\right) d\eta\, d\xi.$$

The remaining integrals are evaluated by using the standard formula

$$\int_{-\infty}^{\infty} \mathrm{e}^{-i\alpha x^2}\, dx = \sqrt{\frac{\pi}{|\alpha|}} \exp\left(-\frac{1}{4}\pi i \,\mathrm{sgn}(\alpha)\right), \tag{1.13.12}$$

which leads directly to (1.13.10).

Example 3: Stationary phase in two space dimensions. When conditions are uniform in the x_3-direction, the Fourier integral for the sound involves an integration over the k_1 wavenumber alone, for which the stationary phase approximation becomes

$$\int_{-\infty}^{\infty} f(k_1, \omega) \mathrm{e}^{i(k_1 x_1 + \gamma(k)|x_2|)} dk_1$$

$$\approx \frac{|x_2|}{|\mathbf{x}|} \sqrt{\frac{2\pi|\kappa_o|}{|\mathbf{x}|}}\, f\left(\frac{\kappa_o x_1}{|\mathbf{x}|}, \omega\right) \mathrm{e}^{i(\kappa_o|\mathbf{x}| - \frac{1}{4}\pi\,\mathrm{sgn}(\kappa_o))},$$

$$\kappa_o|\mathbf{x}| \to \infty. \tag{1.13.13}$$

where $\gamma(k)$ is defined in (1.13.9) with $k = |k_1|$.

1.14 Vorticity

In most problems encountered in aero- and hydroacoustics the ultimate source of acoustic energy is the motion of solid boundaries, such as rotating blades, vibrating electrical power transformers, and so forth. The boundaries can radiate directly or supply energy to the fluid that is transported by convection as the kinetic energy of "shed" vorticity. Vortex shedding always occurs when the Reynolds number is sufficiently large, and it is often very "noisy." The energy radiated as sound is usually a minute fraction of the total energy of the flow, and a substantial amount of the radiation may actually be generated by vorticity after convecting a considerable distance from its source. This is the case, for example, when vorticity subsequently encounters other moving or stationary boundaries (a contraction or bend in a duct, a nozzle exit, blade rows, struts, etc.). When the flow Mach number M is small (more particularly, when $M^2 \ll 1$), vortex–boundary interactions are effectively the same as if the fluid is incompressible. This can greatly simplify their analysis.

Motion generated from rest by a moving boundary S in an ideal, incompressible fluid cannot contain vorticity. It can be described by a velocity potential φ that satisfies Laplace's equation and the condition that $\partial\varphi/\partial x_n$ is the normal velocity on S. When the boundary is brought to rest the fluid motion ceases instantaneously. This unphysical behavior does not occur in a real fluid because (i) no fluid is perfectly incompressible, and "signals" generated by changes in the boundary conditions propagate at the speed of sound, and (ii) diffusion of vorticity from the boundary supplies irrecoverable kinetic energy to the fluid. The crucial difference between rotational and irrotational flows is that, once established, vortical motions proceed irrespective of whether the fluid continues to be driven by moving boundaries or other external agencies.

1.14.1 Motion of an Unbounded Fluid

The velocity $\mathbf{v}$ can always be expressed in terms of scalar and vector potentials φ and $\mathbf{A}$ such that [4, 9, 17]

$$\mathbf{v} = \nabla\varphi + \operatorname{curl}\mathbf{A}, \quad \text{where } \operatorname{div}\mathbf{A} = 0. \tag{1.14.1}$$

In Cartesian coordinates these relations imply

$$\nabla^2\varphi = \operatorname{div}\mathbf{v}, \quad \nabla^2\mathbf{A} = -\operatorname{curl}\mathbf{v} \equiv -\boldsymbol{\omega}. \tag{1.14.2}$$

When the fluid is incompressible, unbounded, and at rest at infinity, we can take $\varphi = 0$ and $\mathbf{A} = \int \boldsymbol{\omega}(\mathbf{y}, t)\, d^3\mathbf{y}/4\pi|\mathbf{x}-\mathbf{y}|$. The velocity is then given by the *Biot–Savart* law [4, 17]

$$\mathbf{v}(\mathbf{x}, t) = \operatorname{curl}\int \frac{\boldsymbol{\omega}(\mathbf{y}, t)d^3\mathbf{y}}{4\pi|\mathbf{x}-\mathbf{y}|}, \tag{1.14.3}$$

which is a purely kinematic relation between a vector $\mathbf{v}$ that vanishes at infinity and $\boldsymbol{\omega} = \operatorname{curl}\mathbf{v}$.

Vorticity is transported by convection and molecular diffusion (Section 1.2.7) It may therefore be assumed that vorticity initially confined to a finite region remains within a bounded domain for any subsequent finite time (i.e., $\boldsymbol{\omega} \to 0$ as $|\mathbf{x}| \to \infty$). Vortex lines are reentrant, and the divergence theorem shows that $\int \omega_i(\mathbf{y}, t) d^3\mathbf{y} = -\oint_\Sigma y_i \omega_j(\mathbf{y}, t) n_j \, dS(\mathbf{y}) \equiv 0$, where the surface Σ (with inward normal $\mathbf{n}$) is large enough to contain all of the vorticity. It follows from this result, by expanding the denominator of the integrand of (1.14.3) in powers of $\mathbf{y}/|\mathbf{x}|$, that $\mathbf{v}(\mathbf{x}, t) \sim O(1/|\mathbf{x}|^3)$ as $|\mathbf{x}| \to \infty$. In two dimensions, when all of the vorticity is parallel to the x_3-direction, say, $\mathbf{v}(\mathbf{x}, t) \sim O(1/|\mathbf{x}|^2)$ as $|\mathbf{x}| \to \infty$, provided that the net vorticity vanishes (i.e., the vortex filaments are reentrant at infinity).

When the density ρ_o is uniform, the kinetic energy of an unbounded incompressible flow is [4, 17]

$$E = \frac{\rho_o}{8\pi} \int\int \frac{\boldsymbol{\omega}(\mathbf{x}, t).\boldsymbol{\omega}(\mathbf{y}, t)}{|\mathbf{x} - \mathbf{y}|} d^3\mathbf{x} d^3\mathbf{y} \equiv \rho_o \int \mathbf{x}.(\boldsymbol{\omega} \wedge \mathbf{v})(\mathbf{x}, t) d^3\mathbf{x}. \tag{1.14.4}$$

Example 1. In two dimensions the kinetic energy per unit length (in the x_3-direction) is

$$E = (-\rho_o/8\pi) \int\int \omega(x_1, x_2, t)\omega(y_1, y_2, t)$$
$$\ln((x_1 - y_1)^2 + (x_2 - y_2)^2) dx_1 dx_2 dy_1 dy_2,$$

provided $\int \omega(x_1, x_2, t) dx_1 dx_2 = 0$.

1.14.2 Incompressible Flow with an Internal Boundary

A rigid body of volume V_o and surface S moves in an incompressible fluid at rest at infinity with velocity

$$\mathbf{U} = \mathbf{U}_o + \boldsymbol{\Omega} \wedge (\mathbf{x} - \mathbf{x}_o(t)), \tag{1.14.5}$$

where $\mathbf{U}_o = d\mathbf{x}_o/dt$ is the velocity of its center of volume $\mathbf{x}_o(t)$, and $\boldsymbol{\Omega}(t)$ is its angular velocity (Figure 1.14.1).

To determine the representation (1.14.1) of the fluid velocity, introduce an indicator function $f(\mathbf{x}, t)$ that vanishes on S, such that $f > 0$ in the fluid, and

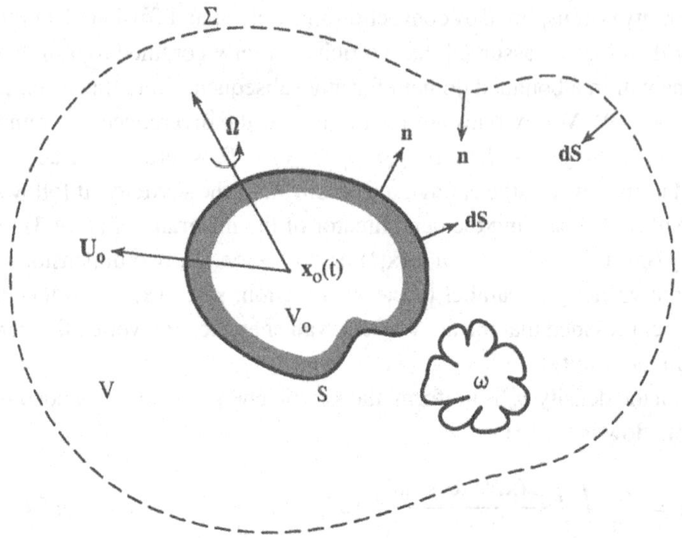

Figure 1.14.1. Motion of a rigid body in fluid at rest at infinity.

write $H(f)\mathbf{v} = \nabla\varphi + \text{curl}\,\mathbf{A}$ ($\text{div}\,\mathbf{A} = 0$), where H is the Heaviside unit function. Then $\mathbf{v} \equiv H(f)\mathbf{v}$ within the fluid, and

$$\nabla^2\varphi = \mathbf{v}.\nabla H(f), \quad \nabla^2\mathbf{A} = \mathbf{v}\wedge\nabla H(f) - H(f)\boldsymbol{\omega}. \tag{1.14.6}$$

$\nabla H(f)$ vanishes except on S where it is equal to $(\nabla f/|\nabla f|)\delta(x_n) \equiv \mathbf{n}\delta(x_n)$, x_n being distance into the fluid measured in the normal direction $\mathbf{n}(\mathbf{x}, t)$ from S. Hence, taking $\mathbf{v} = \mathbf{U}$ on S (no-slip condition), the velocity in the fluid is found to be

$$\mathbf{v}(\mathbf{x}, t) = -\nabla \oint_S \frac{\mathbf{U}(\mathbf{y}, t)\cdot d\mathbf{S}}{4\pi|\mathbf{x}-\mathbf{y}|} + \text{curl} \oint_S \frac{d\mathbf{S}\wedge\mathbf{U}(\mathbf{y}, t)}{4\pi|\mathbf{x}-\mathbf{y}|} + \text{curl}\int_V \frac{\boldsymbol{\omega}(\mathbf{y}, t)d^3\mathbf{y}}{4\pi|\mathbf{x}-\mathbf{y}|}, \tag{1.14.7}$$

where the volume integral is confined to the fluid region V. The velocity predicted by this formula vanishes when $\mathbf{x}$ lies in the interior of the solid.

A representation of $\mathbf{v}$ that is valid everywhere is obtained by writing

$$H(f)\mathbf{v} + H(-f)\mathbf{U} = \nabla\varphi + \text{curl}\,\mathbf{A}, \ (\text{div}\,\mathbf{A} = 0),$$

where the left-hand side defines the velocity in the fluid *and* within the rigid body. The body has constant volume ($\text{div}\,\mathbf{U} = 0$), but $\text{curl}\,\mathbf{U} = 2\boldsymbol{\Omega}$. Now,

$\text{div}(\text{H}(f)\mathbf{v}+\text{H}(-f)\mathbf{U}) = \nabla\text{H}(f)\cdot(\mathbf{v}-\mathbf{U}) \equiv 0$, and the no-slip condition implies

$$\text{curl}(\text{H}(f)\mathbf{v}+\text{H}(-f)\mathbf{U}) = \text{H}(f)\boldsymbol{\omega}+\text{H}(-f)2\boldsymbol{\Omega}. \tag{1.14.8}$$

Hence, $\varphi \equiv 0$, and the velocity *everywhere* is given by

$$\mathbf{v}(\mathbf{x},t) = \text{curl}\int_{\text{V}}\frac{\boldsymbol{\omega}(\mathbf{y},t)d^3\mathbf{y}}{4\pi|\mathbf{x}-\mathbf{y}|}\cdot +\text{curl}\int_{\text{V}_o}\frac{2\boldsymbol{\Omega}(t)d^3\mathbf{y}}{4\pi|\mathbf{x}-\mathbf{y}|}. \tag{1.14.9}$$

Vortex lines may now be imagined to continue into the solid. As for an unbounded flow, the identity $\int \text{curl}\,(\text{H}(f)\mathbf{v}+\text{H}(-f)\mathbf{U})d^3\mathbf{x} = 0$ implies that $\mathbf{v} \sim \text{O}(1/|\mathbf{x}|^3)$ as $|\mathbf{x}| \to \infty$.

1.14.3 Force Exerted on the Fluid by a Rigid Body

In incompressible flow, the force $\mathbf{F}(t)$ exerted on the fluid by a moving body is given by the following general formula in terms of the vorticity and the center of volume velocity $\mathbf{U}_o$ [39]

$$\mathbf{F}(t) = \frac{1}{2}\frac{d}{dt}\int_{\text{V}}\rho_o\mathbf{x}\wedge\boldsymbol{\omega}d^3\mathbf{x} - m_o\frac{d\mathbf{U}_o}{dt}, \tag{1.14.10}$$

where m_o is the mass of the fluid displaced by the body. This equation is applicable for three-dimensional bodies. In two dimensions, m_o is the mass displaced per unit span, and the factor of $\frac{1}{2}$ is omitted.

The integral in (1.14.10) defines the *impulse* of the coupled system [4, 17], which is an invariant of the motion when the body is absent. To evaluate the integral, the vorticity must be defined as in (1.14.8), that is, vortex lines must be continued into the interior of S to form reentrant filaments. For a nonrotating body, vortex lines meeting the surface form reentrant loops by continuation on the surface. The impulse then becomes a function of time because the motion of these vortex lines within and on S is no longer governed by the Navier–Stokes equations. For *irrotational* flow $\boldsymbol{\omega} = \mathbf{0}$ in the fluid, but not on S, where it must be taken as the *bound vorticity* $\boldsymbol{\omega}_o \equiv \nabla\text{H}(f)\wedge(\nabla\varphi-\mathbf{U})$ (φ being the velocity potential of the irrotational flow), nor within S where $\boldsymbol{\omega} = 2\boldsymbol{\Omega}$.

At high Reynolds numbers, involving the interaction of turbulent flows with S, the surface boundary layers are often very thin, and the exterior fluid motion can frequently be approximated by irrotational flow with a superposed distribution of vorticity. In these circumstances it would be convenient to have an expression for $\mathbf{F}$ that minimizes the contribution from the bound vorticity. Such a formula can be derived for a body in *translational* motion.

To do this, suppose the fluid is at rest at infinity and write the force in the form

$$F_i = \frac{d}{dt}\int_V \rho_o v_i d^3\mathbf{x} + \oint_\Sigma p n_i dS, \tag{1.14.11}$$

where V here denotes the fluid between S and a large fixed control surface Σ (Figure 1.14.1) whose normal **n** is directed into V. To express the right-hand side of this formula in terms of the vorticity and the translational velocity of the body, we first cast the momentum equation (1.2.9) in the following form appropriate for an incompressible fluid

$$\frac{\partial(\rho_o \mathbf{v})}{\partial t} + \nabla\left(p + \frac{1}{2}\rho_o v^2\right) = -\rho_o \boldsymbol{\omega} \wedge \mathbf{v} - \eta \,\text{curl}\, \boldsymbol{\omega}. \tag{1.14.12}$$

The scalar product of this equation with $\nabla X_i(\mathbf{x}, t) = \nabla(x_i - \varphi_i^*(\mathbf{x}, t))$ is then integrated over V. X_i is the function used in Section 1.11 to define the compact Green's function; here it is dependent on time because of the translation of S. By applying the divergence theorem, the integrated equation can be manipulated [40–42] to yield an alternative representation of the right-hand side of (1.14.11). Using the definition (1.11.3) of the added mass tensor, the following formula is then obtained for the force exerted on an incompressible fluid by a body in translational motion:

$$F_i = \mathrm{M}_{ij}\frac{dU_j}{dt} - \rho_o \int_V \nabla X_i \cdot \boldsymbol{\omega} \wedge \mathbf{v}_{\text{rel}} d^3\mathbf{x} + \eta \oint_S \boldsymbol{\omega} \wedge \nabla X_i \cdot d\mathbf{S}, \tag{1.14.13}$$

where $\mathbf{v}_{\text{rel}} = \mathbf{v} - \mathbf{U}$ is the fluid velocity relative to S. Bound vorticity makes no contribution to the volume integral because (i) $\mathbf{v}_{\text{rel}} = \mathbf{0}$ on S for viscous flow, or (ii) $\nabla X_i . \boldsymbol{\omega} \wedge \mathbf{v}_{\text{rel}} \equiv 0$ on S for an ideal fluid because ∇X_i, $\boldsymbol{\omega}$ and $\mathbf{v}_{\text{rel}}$ are then all locally parallel to the surface.

This formula expresses the force as a sum of three essentially distinct components: (i) the inertia due to the added mass of the body, (ii) the vector sum of the normal stresses induced on S by the vorticity, and (iii) the viscous skin friction.

Example 5. Calculate the force exerted on a large rigid plate by a small, adjacent region of vorticity (an "eddy"). Let the plate be the strip $|x_1| < a$, $-\infty <$

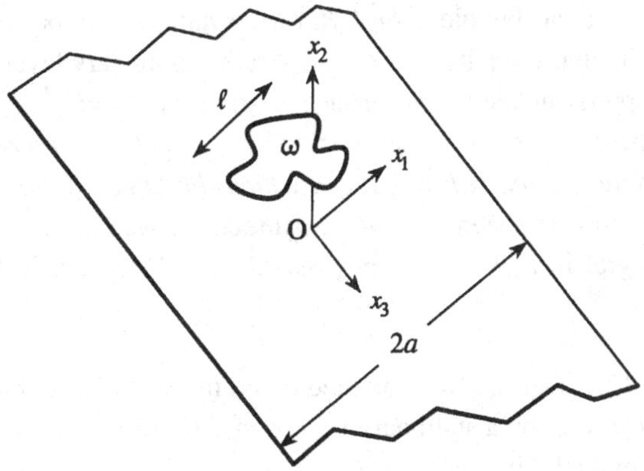

Figure 1.14.2. Vorticity adjacent to a rigid plate.

$x_3 < \infty$ of the plane $x_2 = 0$. Assume the vorticity has length scale $\ell \ll a$ and is located above the plate ($x_2 > 0$) near the coordinate origin O (Figure 1.14.2).

The normal force (in the $-x_2$-direction) is the component F_2 given by equation (1.14.13) when $X_2 = \text{Re}(-i(z^2 - a^2)^{1/2})$, where $z = x_1 + ix_2$ [17]. $\nabla X_2 \approx \mathbf{Z}/a$, where $\mathbf{Z} = (-x_1, x_2, 0)$, when $\ell \ll a$, so that

$$F_2 \approx \frac{-\rho_o}{a} \int \mathbf{Z} \cdot \boldsymbol{\omega} \wedge \mathbf{v} d^3\mathbf{x} + \frac{\eta}{a} \oint_S \mathbf{Z} \wedge \boldsymbol{\omega}\, d\mathbf{S}, \quad a/\ell \to \infty.$$

The vorticity in these integrals must include that generated by the interaction of the eddy with the plate. It may be verified that the additional contribution is finite and therefore that $F_2 \sim (\ell/a)\rho_o v^2 \ell^2 f(t)$ when $\ell \ll a$, where $f(t)$ is a dimensionless function of the time that does not depend on the width a of the plate [44]. The force vanishes for an infinitely large plate.

Example 6: Stokes drag. At very low Reynolds number the vorticity distribution surrounding a spherical inclusion of radius R translating at constant velocity $\mathbf{U}$ is $\boldsymbol{\omega} = CR\mathbf{U} \wedge \mathbf{x}/|\mathbf{x}|^3$, where $\mathbf{x}$ is measured from the center of the moving sphere and C is a constant [4]. Show that in the linearized approximation the sphere experiences a drag force $D = 4\pi C\eta UR$, determined by the final term on the right of equation (1.14.13). For a smooth, rigid sphere ($C = \frac{3}{2}$) this reduces to the Stokes drag $D = 6\pi\eta UR$. For a fluid sphere of interior shear viscosity $\bar{\eta}$, the constant $C = (2\eta + 3\bar{\eta})/2(\eta + \bar{\eta})$. The limit $\bar{\eta} \to 0$ gives the drag $4\pi\eta UR$ experienced by a gas bubble at low Reynolds number.

The drag on a gas bubble at *high Reynolds number* can be calculated similarly. The principal vorticity is confined to a thin boundary layer on the bubble and is well approximated by the surface distribution $\boldsymbol{\omega} = (\frac{-2}{R})\frac{\mathbf{x}}{|\mathbf{x}|} \wedge \nabla\varphi$, where φ is the velocity potential of the exterior irrotational flow *expressed in terms of a coordinate system with respect to which the bubble is at rest* [4]. For a bubble advancing at speed U in the x_i-direction, we can take $\varphi = -UX_i$; the final integral in (1.14.13) then gives $D = 2(U\eta/R)\oint_S(\nabla X_i)^2 dS = 12\pi\,\eta UR$ [43].

Example 7. Consider the force exerted on an incompressible fluid of nonuniform density $\rho(\mathbf{x}, t)$ by a stationary rigid body. If $\rho \to \rho_o =$ constant at large distances from the body, show that

$$F_i = -\rho_o \int \nabla X_i \cdot \boldsymbol{\omega} \wedge \mathbf{v} d^3\mathbf{x} + \eta \oint_S \boldsymbol{\omega} \wedge \nabla X_i \cdot d\mathbf{S} + \int \left(\frac{\rho_o}{\rho} - 1\right) \frac{\partial \varphi_i^*}{\partial x_j} \frac{\partial p'_{jk}}{\partial x_k} d^3\mathbf{x}$$

where $p'_{ij} = p\delta_{ij} - \sigma_{ij}$ is the compressive stress tensor.

Example 8. When a rigid body *rotates* and translates with the velocity $\mathbf{U}$ of (1.14.5), show that the force exerted on the fluid can be written

$$F_i = -\rho_o \int_V \nabla X_i \cdot \boldsymbol{\omega} \wedge \mathbf{v} d^3\mathbf{x} + \eta \oint_S \boldsymbol{\omega} \wedge \nabla X_i \cdot d\mathbf{S} + \rho_o \oint_S X_i \frac{\partial \mathbf{U}}{\partial t} \cdot d\mathbf{S} - m_o \frac{\partial U_{oi}}{\partial t},$$

where m_o is the mass of the fluid displaced by the body.

1.14.4 Vortex Shedding

The motion of a body through a viscous fluid is always accompanied by vortex shedding, and bluff bodies at high Reynolds number have highly energetic turbulent wakes. Even when the fluid is assumed to be inviscid and irrotational, it can be necessary to take account of the residual influences of vorticity shed at some earlier time, but subsequently swept far downstream outside the domain of current interest.

The classic example of an airfoil "starting vortex" is illustrated in Figure 1.14.3. The two-dimensional case is approximately realized by a wing

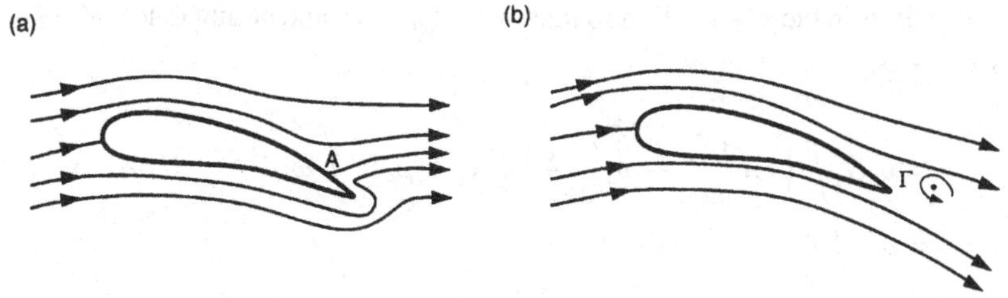

Figure 1.14.3. The starting vortex.

of uniform cross-section spanning a wind tunnel. Vorticity is generated at the surface, but because it diffuses at a finite rate, the motion is initially irrotational, with a high-pressure stagnation point at A on the "suction" side of the airfoil, where the flow separates, and a high-speed, low-pressure region at the trailing edge. Counterclockwise vorticity is produced on the upper surface downstream of A and diffuses into the flow to roll up into a concentrated core. The strength and size of the vortex increase until the combined action of the mean flow close to the separation streamline and the equal and opposite circulation that must now exist about the airfoil are sufficient to eject it into the wake and simultaneously shift the stagnation point toward the trailing edge, where its high pressure is neutralized. A steady state is reached when the starting vortex is far downstream; the mean flow leaves the trailing edge smoothly, and the mean circulation about the airfoil is equal and opposite to that of the starting vortex. At high Reynolds number, new vorticity diffusing from the surfaces of the airfoil is confined to thin boundary layers (when the angle of attack is small enough to avoid "stall"), which may be disregarded in a first approximation. That is, the steady flow in the vicinity of the airfoil may now be assumed to be irrotational, provided circulation is introduced to make the velocity and pressure finite at the edge. This is the Kutta–Joukowski hypothesis [4, 17], and it is usually referred to as the *Kutta condition.*

For a stationary airfoil in a uniform high Reynolds number flow at speed U in the x_1-direction, the lift and drag are dominated by the second integral on the right of (1.14.13). To calculate the familiar Kutta–Joukowski formula for the lift L, for example, take the coordinate origin within the airfoil and assume the motion started from rest at time $t = 0$. If Γ is the mean circulation about the airfoil (in the clockwise sense in Figure 1.14.3) when the starting vortex is far downstream, then

$$\boldsymbol{\omega} \approx \Gamma \mathbf{i}_3 \delta(x_1 - Ut)\delta(x_2),$$

where $\mathbf{i}_3$ is a unit vector in the x_3-direction (out of paper in Figure 1.14.3).

The lift is in the x_2-direction so that, with $\mathbf{v}_{\text{rel}} = \mathbf{U}$, and noting that $\nabla\varphi_2^* \to 0$ as $x_1 \to \infty$,

$$L \approx \rho_o U \int \left(\omega_3 \frac{\partial X_2}{\partial x_2} - \omega_2 \frac{\partial X_2}{\partial x_3} \right) d^3\mathbf{x} \to \rho_o U \int \omega_3 d^3\mathbf{x}, \quad Ut \to \infty$$
$$= \rho_o U \Gamma \ell, \tag{1.14.14}$$

where ℓ is the airfoil span.

1.14.5 Modeling Vortex Shedding in Two Dimensions

Unsteady shedding of vorticity from a straight edge in a two-dimensional incompressible flow (taken to be uniform in the x_3-direction) can sometimes be approximated by the method of Brown and Michael [45–59]. The continuous shedding of vorticity is represented by a sequence of line vortices whose positions and circulation depend on time. Separation is assumed to occur in the form of a thin sheet of vorticity of infinitesimal circulation that rolls up into a concentrated core; the influence of vortex shedding is calculated from a potential flow representation of the interaction of this variable-strength core with the surface and any other vortices in the flow.

Let the flow occur in the neighborhood of a rigid body S (Figure 1.14.4). Suppose high Reynolds number flow induced by a vorticity distribution Σ produces shedding from the edge O. The circulation around a closed contour enclosing S and all of the vorticity is assumed to vanish so that the fluid is at rest at infinity. For simplicity, S is taken to be at rest, although situations in which S is in unsteady translational motion are easily handled. Shedding is modeled by introducing a line vortex of variable circulation $\Gamma(t)$ whose axis is at $\mathbf{x} = \mathbf{x}_\Gamma \equiv (x_{\Gamma 1}, x_{\Gamma 2})$. The vortex is "fed" continuously with additional vorticity that passes along a "connecting sheet" from O, which is assumed to have negligible circulation compared to Γ. It is usual to require the strength Γ to vary monotonically with time, and the vortex to be "released" from the edge when $d\Gamma/dt$ first changes sign, following which it moves away from the edge as a "free" vortex with Γ equal to its value at the time of release. The vortex then becomes a member of the distribution Σ, and a new vortex must be released from O. We are here concerned only with the motion of Γ prior to release.

Let $\mathbf{V}(t) = d\mathbf{x}_\Gamma/dt$ be the translational velocity of Γ. The flow is assumed to be inviscid away from S; the Kutta condition is applied to mimic the generation of vorticity at O by viscous diffusion. Because of its variable strength,

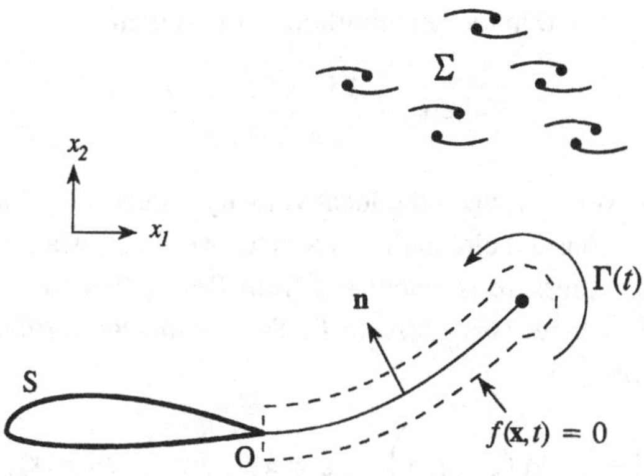

Figure 1.14.4. Modeling vortex shedding in two dimensions.

the motions of Γ and the connecting sheet are not compatible with the equations of motion; they are therefore excluded from the domain of application of these equations by enclosing them within the control surface $f(\mathbf{x}, t) = 0$ of Figure 1.14.4, which moves so as always to enclose Γ and the connecting sheet. Suppose that $f(\mathbf{x}, t) \gtrless 0$ accordingly as $\mathbf{x}$ lies in the exterior/interior of the control surface. Using the procedure of Section 1.14.2, we formally extend the range of validity of the momentum equation (1.2.4) to the *whole* fluid region (including that within $f = 0$) by multiplying by $\mathrm{H}(f)$. Using Crocco's form of the equation for an ideal fluid, we obtain

$$\frac{\partial}{\partial t}(\mathrm{H}(f)\mathbf{v}) + \nabla\left[\left(\frac{p}{\rho_o} + \frac{1}{2}v^2\right)\mathrm{H}(f)\right]$$
$$= -\boldsymbol{\omega}_\Sigma \wedge \mathbf{v}\mathrm{H}(f) + \frac{1}{2}v^2\nabla\mathrm{H}(f) - \mathbf{v}(\mathbf{v}\cdot\nabla)\mathrm{H}(f)$$
$$+ \mathbf{v}\frac{D\mathrm{H}}{Dt}(f) + \frac{p}{\rho_o}\nabla\mathrm{H}(f), \tag{1.14.15}$$

where ρ_o is assumed to be constant, and $\boldsymbol{\omega}_\Sigma$ is the vorticity Σ.

The terms on the right involving $\nabla\mathrm{H} = \nabla f\delta(f)$ and $D\mathrm{H}/Dt = (\mathbf{v} - \bar{\mathbf{V}})\cdot\nabla\mathrm{H}(f)$ represent forces distributed over the control surface, where $\bar{\mathbf{V}}$ is the velocity of the control surface. They are evaluated by first writing the vorticity of the line vortex Γ in the form

$$\boldsymbol{\Omega} = \Gamma\mathbf{i}_3\delta(\mathbf{x} - \mathbf{x}_\Gamma). \tag{1.14.16}$$

The velocity in $f > 0$ in the neighborhood of Γ is then

$$\mathbf{v} = \mathbf{v}_o + \frac{\Gamma \mathbf{i}_3 \wedge (\mathbf{x} - \mathbf{x}_\Gamma)}{2\pi |\mathbf{x} - \mathbf{x}_\Gamma|^2},$$

where $\mathbf{v}_o$ is the velocity when the local velocity induced by Γ is excluded. A "free" vortex of constant circulation at $\mathbf{x} = \mathbf{x}_\Gamma$ would translate at velocity $\mathbf{v}_o$.

The unsteady pressure is calculated from Bernoulli's equation $p = -\rho_o (\partial\varphi/\partial t + \frac{1}{2}v^2)$ where, very close to Γ, the dominant contribution from the velocity potential φ is

$$\varphi \approx \mathrm{Re}\left(\frac{-i\Gamma}{2\pi} \ln (z - z_\Gamma)\right), \quad z = x_1 + ix_2, \quad z_\Gamma = x_{\Gamma 1} + ix_{\Gamma 2}. \tag{1.14.17}$$

We now let the control surface $f = 0$ shrink down to the connecting sheet and vortex Γ and calculate the limiting form of the right-hand side of (1.14.15) by integrating over a neighborhood containing the control surface, and substituting, where necessary, these expressions for $\mathbf{v}$ and p. Noting that the velocity is continuous across the connecting sheet, but the velocity potential is discontinuous because of the time dependence of Γ in the component of φ shown in (1.14.17), we find

$$\frac{\partial \mathbf{v}}{\partial t} + \nabla\left(\frac{p}{\rho_o} + \frac{1}{2}v^2\right) = -\boldsymbol{\omega}_\Sigma \wedge \mathbf{v} - \boldsymbol{\Omega} \wedge \mathbf{v}_o + \frac{\mathbf{F}_\Gamma}{\rho_o}, \tag{1.14.18}$$

where

$$\mathbf{F}_\Gamma = \rho_o \boldsymbol{\Omega} \wedge (\mathbf{v}_o - \mathbf{V}) - \rho_o \mathbf{n} \frac{d\Gamma}{dt} \delta(x_\perp) \mathrm{H}(s) \mathrm{H}(s_\Gamma - s). \tag{1.14.19}$$

In these expressions, $\mathbf{v}_o$ is the velocity at Γ when the local contribution from Γ is excluded. It is *not* the same as the convection velocity $\mathbf{V}$ of Γ unless Γ is constant. With this understanding, the right side of (1.14.18) may be expressed in the more usual form $-\boldsymbol{\omega} \wedge \mathbf{v} + \mathbf{F}_\Gamma/\rho_o$, where $\boldsymbol{\omega}$ now includes $\boldsymbol{\Omega}$. In (1.14.19), $x_\perp$ is the distance from the connecting sheet measured in the direction of the normal $\mathbf{n}$, s_Γ is the total length of the sheet, and s is the distance along the sheet from O. The *force* $\mathbf{F}_\Gamma$ has two components. The first term on the right of (1.14.19) is concentrated at the vortex Γ and is the reaction on the fluid to the Kutta–Joukowski lift (1.14.14) experienced by Γ because of its motion relative to the fluid; the second is a pressure force on the fluid because of the jump in pressure across the connecting sheet.

The equation of motion of Γ is now obtained by requiring that the additional net force on the fluid attributable to $\mathbf{F}_\Gamma$ should vanish. However, it is *not* sufficient to demand that $\int \mathbf{F}_\Gamma \, dx_1 \, dx_2 = \mathbf{0}$ because a distributed system of forces generally involves an unbalanced couple that generates a reaction force $\mathbf{F}_{S\Gamma}$ on S. To completely neutralize the dynamical effect of $\mathbf{F}_\Gamma$, it is necessary that $\mathbf{F}_{S\Gamma} + \int \mathbf{F}_\Gamma \, dx_1 \, dx_2 = \mathbf{0}$.

The method leading from (1.14.12) to the force formula (1.14.13) can be applied to the modified momentum equation (1.14.18) to show that the ith component of the net force $\mathbf{F}$ exerted on the fluid is

$$F_i = -\rho_o \int \boldsymbol{\omega} \wedge \mathbf{v} \cdot \nabla X_i \, dx_1 \, dx_2 + \int \mathbf{F}_\Gamma \cdot \nabla X_i dx_1 dx_2.$$

The first integral is the usual force due to the distributed vorticity (including Γ). Because the fluid is at rest at infinity, and there are no external forces, this force is actually applied at the surface S. The second term is the net force on the fluid produced by the force system $\mathbf{F}_\Gamma$, including the surface reaction force $\mathbf{F}_{S\Gamma}$. If the dynamics of the flow are to be represented entirely in terms of the vortices ω_Σ and Γ, we must have

$$\int \mathbf{F}_\Gamma \cdot \nabla X_i dx_1 dx_2 = \mathbf{0}. \tag{1.14.20}$$

This condition determines the equation of motion of Γ, which can be expressed in differential form by introducing the *stream function* $\Psi_i(\mathbf{x})$ defined such that $w_i = X_i(\mathbf{x}) + i\Psi_i(\mathbf{x})$ is the complex potential of flow past S that has unit speed in the i-direction at large distances from S. They satisfy the Cauchy–Riemann relations [4, 17, 21]

$$\partial X_i/\partial x_1 = \partial \Psi_i/\partial x_2, \quad \partial X_i/\partial x_2 = -\partial \Psi_i/\partial x_1.$$

We are free to require that $\Psi_i = 0$ on S, in which case the reader can evaluate the integral in (1.14.20) to obtain the Brown and Michael equation of motion:

$$\frac{d\mathbf{x}_\Gamma}{dt} \cdot \nabla \Psi_i + \frac{\Psi_i}{\Gamma} \frac{d\Gamma}{dt} = \mathbf{v}_o \cdot \nabla \Psi_i, \quad i = 1, 2, \tag{1.14.21}$$

where $\mathbf{v}_o$ is velocity at Γ when its local self-potential velocity is excluded.

Example 9. Let S be the rigid strip $|x_1| < a,\ x_2 = 0$. Then

$$\Psi_1 = x_2, \quad \Psi_2 = \text{Im}\Big(-i\sqrt{z^2 - a^2}\Big), \quad z = x_1 + ix_2.$$

Deduce the Brown and Michael equation in the complex form

$$\frac{dz^*}{dt} + \left(z - \frac{\text{Re}(a^2\sqrt{z^2 - a^2})}{\text{Re}(z\sqrt{z^2 - a^2})}\right)\frac{1}{\Gamma}\frac{d\Gamma}{dt} = \frac{dw_o}{dz},$$

where z is the complex position of the vortex, the asterisk denotes complex conjugate, and $w_o(z)$ is the complex velocity potential defining $\mathbf{v}_o$ (i.e., $v_{o1} - iv_{o2} = dw_o/dz$).

Example 10. By letting $a \to \infty$ and replacing z by $z + a$ in Example 9, deduce the Brown and Michael equation for the half-plane $x_1 < 0, x_2 = 0$, namely,

$$\frac{dz^*}{dt} + \frac{(|z| + z^*)}{\Gamma}\frac{d\Gamma}{dt} = \frac{dw_o}{dz}. \tag{1.14.22}$$

Example 11: Vortex motion past a half-plane. A line vortex of circulation Γ_o translates around the edge of the half-plane of Example 10. In the absence of vortex shedding, the vortex trajectory is governed by equation (1.14.22) with the second term on the left omitted, and when $\Gamma_o > 0$, the motion is in a clockwise sense along the symmetric path shown in Figure 1.14.5a. Time is measured from the instant at which the vortex crosses the x_1-axis, when its distance from the edge is a minimum and equal to ℓ. Far from the edge the motion is parallel to the plane at speed $U = \Gamma_o/8\pi\ell$, and the points marked on the trajectory indicate the position $\mathbf{x}_o(t)$ of the vortex at different nondimensional times $T = Ut/\ell$:

$$\frac{x_{o1}}{\ell} = \frac{1 - T^2}{\sqrt{1 - T^2}}, \quad \frac{x_{o2}}{\ell} = \frac{-2T}{\sqrt{1 - T^2}}. \tag{1.14.23}$$

The effect on the motion of continuous vortex shedding from the edge can be estimated by modeling the shed vorticity by a concentrated core (line vortex) of circulation Γ. The instantaneous circulation $\Gamma(t)$ is determined by applying the Kutta condition at the edge $z = 0$, where the complex velocity of the whole

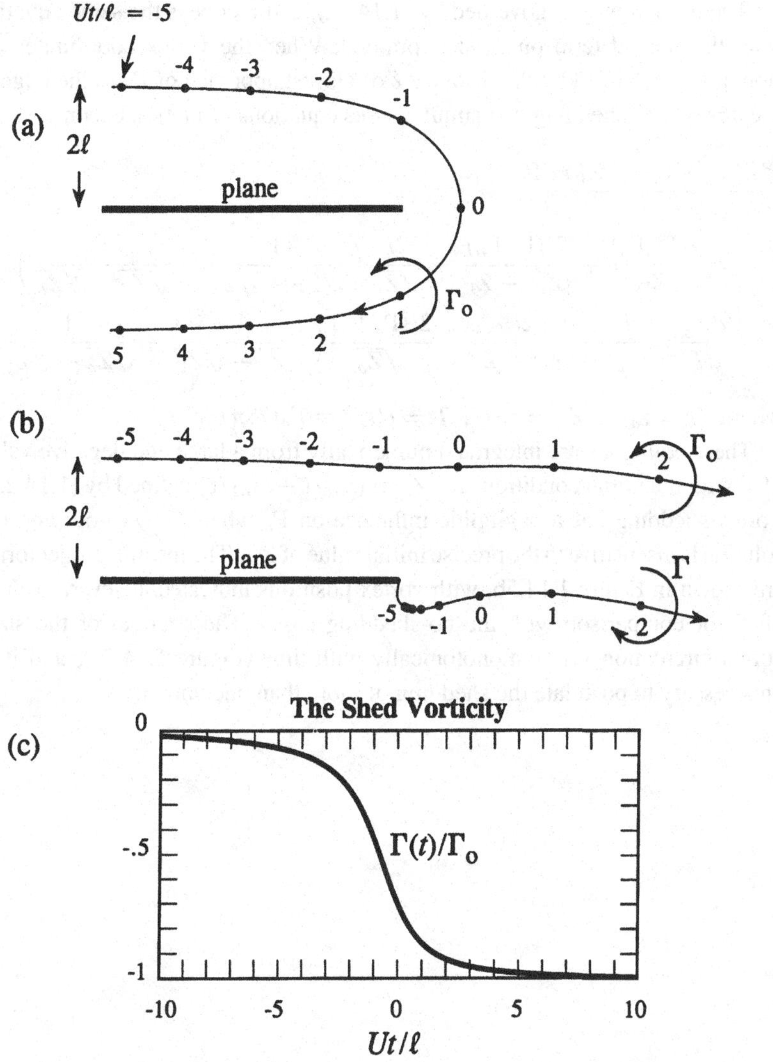

Figure 1.14.5. (a) Vortex motion in the absence of shedding, (b) incident and shed vortex trajectories, and (c) the shed vorticity $\Gamma(t)$.

flow is required to be finite. Then, if $z_o(t)$ and $z_\Gamma(t)$ are the complex positions of Γ_o and Γ,

$$\Gamma = \frac{-\Gamma_o |z_\Gamma|}{|z_o|} \left(\frac{\sqrt{z_o} + \sqrt{z_o}^*}{\sqrt{z_\Gamma} + \sqrt{z_\Gamma}^*} \right),$$

which shows that Γ_o and Γ are opposite in sign.

The motion of Γ is governed by (1.14.22), and Γ_o obeys the same equation with the second term on the left omitted. When the vortex coordinates are nondimensionalized by the distance ℓ of closest approach of Γ_o to the edge in the *absence* of shedding, the simultaneous equations of motion become

$$
\begin{aligned}
&\frac{dZ_\Gamma^*}{dT} + \frac{(|Z_\Gamma| + Z_\Gamma^*)}{\Gamma}\frac{d\Gamma}{dT} \\
&\quad = \frac{i(\Gamma/\Gamma_o)}{Z_\Gamma} + \frac{2i(\Gamma/\Gamma_o)}{|Z_\Gamma| + Z_\Gamma} - \frac{2i}{\sqrt{Z_\Gamma}}\left(\frac{1}{\sqrt{Z_\Gamma} - \sqrt{Z_o}} - \frac{1}{\sqrt{Z_\Gamma} - {}^{*}\!\sqrt{Z_o}}\right) \\
&\frac{dZ_o^*}{dT} = \frac{i}{Z_o} + \frac{2i}{|Z_o| + Z_o} - \frac{2i(\Gamma/\Gamma_o)}{\sqrt{Z_o}}\left(\frac{1}{\sqrt{Z_o} - \sqrt{Z_\Gamma}} - \frac{1}{\sqrt{Z_o} - {}^{*}\!\sqrt{Z_\Gamma}}\right),
\end{aligned}
$$

where $Z_o = z_o/\ell,\ Z_\Gamma = z_\Gamma/\ell,\ T = Ut/\ell \equiv \Gamma_o t/8\pi\ell^2$.

These equations are integrated numerically from a large and negative value of T with the initial conditions for $Z_o \equiv (x_{o1}/\ell + ix_{o2}/\ell)$ defined by (1.14.23). Vortex shedding has a negligible influence on Γ_o when $T \to -\infty$, and the solution is insensitive to the precise initial value of Z_Γ. The resulting trajectories are shown in Figure 1.14.5b, with vortex positions indicated at several values of T for comparison with the no-shedding case. The strength of the shed vortex circulation varies monotonically with time (Figure 1.14.5c), and it is unnecessary to postulate the shedding of more than one vortex.

2

Aerodynamic sound in unbounded flows

2.1 Aerodynamic Sound

The sound generated by vorticity in an unbounded fluid is called *aerodynamic sound* [60, 61]. Most unsteady flows of technological interest are of high Reynolds number and turbulent, and the acoustic radiation is a very small by-product of the motion. The turbulence is usually produced by fluid motion relative to solid boundaries or by the instability of free shear layers separating a high-speed flow (such as a jet) from a stationary environment. In this chapter the influence of boundaries on the production of sound as opposed to the production of vorticity will be ignored. The aerodynamic sound problem then reduces to the study of mechanisms that convert kinetic energy of rotational motions into acoustic waves involving longitudinal vibrations of fluid particles. There are two principal source types in free vortical flows: a quadrupole, whose strength is determined by the unsteady Reynolds stress, and a dipole, which is important when mean mass density variations occur within the source region.

2.1.1 Lighthill's Acoustic Analogy

The theory of aerodynamic sound was developed by Lighthill [60], who reformulated the Navier–Stokes equation into an exact, inhomogeneous wave equation whose source terms are important only within the turbulent (vortical) region. Sound is expected to be such a very small component of the whole motion that, once generated, its back-reaction on the main flow is usually negligible. In a first approximation the motion in the source region may then be determined by neglecting the production and propagation of the sound. This would be inappropriate if the Mach number M is large enough for compressibility to be important in the source flow, when the source flow is coupled to a resonator, such as an organ pipe, or when bubbles are present in the case of liquids. However, there are many technologically important flows where M is sufficiently small that the hypothesis is obviously correct, and where the theory leads to unambiguous predictions of the sound.

Consider the sound generated by a finite region of rotational flow in an unbounded fluid at rest at infinity. Let us compare the equations for the *density* fluctuations in the real flow with those for an ideal, linear acoustic medium that coincides with the real fluid at large distances from the sources. The difference between these equations will be shown to be equivalent to a distribution of sources in the ideal acoustic medium, whose radiation field is the same as that in the real flow and may therefore be calculated by the methods of linear acoustics. To do this, body forces are neglected, and the momentum equation (1.2.22) is written in the form

$$\partial(\rho v_i)/\partial t = -\partial \pi_{ij}/\partial x_j, \tag{2.1.1}$$

where π_{ij} is the *momentum flux* tensor

$$\pi_{ij} = \rho v_i v_j + (p - p_o)\delta_{ij} - \sigma_{ij}, \tag{2.1.2}$$

and p_o is the uniform pressure at infinity.

By integrating over a fixed region V, it can be seen that (2.1.1) equates the rate of change of momentum in V to the action of the pressure and viscous stresses on its boundary and to the convection of momentum across the boundary at a rate determined by the Reynolds stress $\rho v_i v_j$. In an ideal, linear acoustic medium, momentum transfer is produced solely by the pressure, that is, $\pi_{ij} \to \pi_{ij}^o = (p - p_o)\delta_{ij} \equiv c_o^2(\rho - \rho_o)\delta_{ij}$, where ρ_o and c_o are the mean density and sound speed. The equation of linear acoustics for the perturbation density $\rho - \rho_o$,

$$\left(\partial^2/c_o^2\partial t^2 - \nabla^2\right)\left[c_o^2(\rho - \rho_o)\right] = 0, \tag{2.1.3}$$

is obtained by eliminating ρv_i between (2.1.1) (with π_{ij} replaced by π_{ij}^o) and the continuity equation (1.2.1). In the absence of externally applied forces or moving boundaries, this equation has only the trivial solution $\rho - \rho_o = 0$ because the radiation condition ensures that sound waves cannot enter from infinity.

The sound generated in the real fluid may now be seen to be exactly equivalent to that produced in an ideal, stationary acoustic medium that is forced by the stress distribution $T_{ij} = \pi_{ij} - \pi_{ij}^o$. T_{ij} is called the *Lighthill stress tensor*

$$T_{ij} = \rho v_i v_j + \left((p - p_o) - c_o^2(\rho - \rho_o)\right)\delta_{ij} - \sigma_{ij}. \tag{2.1.4}$$

The Reynolds stress $\rho v_i v_j$ is nonlinear and is significant only within the rotational source region. The second term is the *excess* of momentum transfer by the pressure over that in an ideal ("linear") fluid of density ρ_o and sound speed c_o. This is caused by wave amplitude nonlinearity and by mean density variations in the source flow. The viscous stress tensor σ_{ij} is linear in the perturbation quantities and properly accounts for the attenuation of the sound; in most applications, the Reynolds number in the source region is sufficiently large that σ_{ij} can be neglected, and attenuation in the radiation zone is usually ignored in a first approximation.

Lighthill's *acoustic analogy* equation for the production of aerodynamic sound is obtained by first rewriting (2.1.1) as the momentum equation for an ideal, stationary fluid of density ρ_o and sound speed c_o subject to the externally applied stress T_{ij}:

$$\partial(\rho v_i)/\partial t + \partial\left(c_o^2(\rho - \rho_o)\right)/\partial x_i = -\partial T_{ij}/\partial x_j.$$

Elimination of the momentum density ρv_i between this and the continuity equation yields *Lighthill's equation*, which is the exact, nonlinear counterpart of (2.1.3):

$$\left(\frac{1}{c_o^2}\frac{\partial^2}{\partial t^2} - \nabla^2\right)\left[c_o^2(\rho - \rho_o)\right] = \frac{\partial^2 T_{ij}}{\partial x_i \partial x_j}. \qquad (2.1.5)$$

The fluid mechanical problem of calculating the aerodynamic sound is therefore formally equivalent to solving this equation for the radiation into a stationary, ideal fluid produced by a distribution of *quadrupole* sources whose strength per unit volume is the Lighthill stress tensor T_{ij}.

The solution with outgoing wave behavior is given by (1.8.8) with $p(\mathbf{x}, t)$ replaced by $c_o^2(\rho - \rho_o)$, which tends to $p - p_o$ as $|\mathbf{x}| \to \infty$ in the linearly disturbed fluid outside the source flow. This solution should strictly be regarded as an alternative, integral equation formulation of the Navier–Stokes equation, which provides a useful representation of the sound when T_{ij} is known. T_{ij} accounts not only for the generation of sound but also for self-modulation due to acoustic nonlinearity, convection by the flow, refraction due to sound speed variations, and attenuation due to thermal and viscous actions. Nonlinear effects on propagation and dissipation are usually sufficiently weak to be neglected within the source region, although they may affect propagation to a distant observer. Convection and refraction of sound within and near the source flow can be important, for example, when the sources are contained in a turbulent shear layer, or are adjacent to a large, quiescent region of fluid whose mean thermodynamic properties differ from those in the radiation zone. Effects of this kind are accounted for by contributions to T_{ij} that are *linear* in the perturbation quantities relative to a mean background flow.

Thus, with the exception of flows amenable to special treatment, the practical utility of Lighthill's equation rests on the hypothesis that all of these effects, which actually depend on the *compressibility* of the source flow, can be ignored, and that adequate predictions of the aerodynamic sound are obtained by taking for T_{ij} an estimate based on the equations of motion of an incompressible fluid. This approximation is likely to be acceptable when $M^2 \ll 1$ and when the wavelength of the sound is much larger than the dimension of the source region. The remainder of this section is devoted to a consideration of such cases.

2.1.2 Aerodynamic Sound Generated by Low Mach Number Turbulence of Uniform Mean Density

When the mean density and sound speed are uniform, the variations in ρ produced by low Mach number, high Reynolds number velocity fluctuations are of order $\rho_o M^2$, and $\rho v_i v_j \approx \rho_o v_i v_j$ with a relative error $\sim O(M^2) \ll 1$. Similarly, $p - p_o - c_o^2(\rho - \rho_o) \approx (p - p_o)(1 - c_o^2/c^2) \sim O(\rho_o v^2 M^2)$ (Section 1.6.5). Thus, $T_{ij} \approx \rho_o v_i v_j$, when viscous stresses are neglected, and the solution (1.8.8) of Lighthill's equation becomes

$$
\begin{aligned}
p(\mathbf{x}, t) &\approx \frac{\partial^2}{\partial x_i \partial x_j} \int \frac{\rho_o v_i v_j(\mathbf{y}, t - |\mathbf{x} - \mathbf{y}|/c_o)}{4\pi |\mathbf{x} - \mathbf{y}|} d^3\mathbf{y} \\
&\approx \frac{x_i x_j}{4\pi c_o^2 |\mathbf{x}|^3} \frac{\partial^2}{\partial t^2} \int \rho_o v_i v_j(\mathbf{y}, t - |\mathbf{x} - \mathbf{y}|/c_o)\, d^3\mathbf{y}, \quad |\mathbf{x}| \to \infty,
\end{aligned}
\tag{2.1.6}
$$

where $p(\mathbf{x}, t) = c_o^2(\rho - \rho_o)$ is the perturbation pressure in the far field. Quantitative predictions can be made from this equation provided the behavior of the Reynolds stress is known.

The order of magnitude of p can be estimated in terms of the characteristic velocity v and length scale ℓ (of the energy containing eddies) in the source region. ℓ is determined by the scale of the mechanism responsible for turbulence production, such as the width of a jet mixing layer. Fluctuations in $v_i v_j$ occurring in different regions of the turbulent flow separated by distances $> O(\ell)$ will tend to be statistically independent, and the sound may therefore be considered to be generated by a collection of V_o/ℓ^3 independent eddies, where V_o is the volume occupied by the turbulence. The dominant frequency of the motion $\sim v/\ell$ so that the wavelength of the radiated sound $\sim \ell/M \gg \ell\, (M \sim v/c_o)$, and each eddy is therefore acoustically *compact*. Hence, the acoustic pressure generated by a single eddy is of order $p \sim (\ell/|\mathbf{x}|)\rho_o v^2 M^2$, and the acoustic power it radiates $\sim 4\pi |\mathbf{x}|^2 p^2/\rho_o c_o \approx \ell^2 \rho_o v^8/c_o^5 = \ell^2 \rho_o v^3 M^5$. This is

Lighthill's "eighth power" law. The total acoustic power is $\Pi_a \approx (\mathrm{V}_o/\ell^3)(\ell^2\rho_o v^3 M^5) = \rho_o v^3 M^5 \mathrm{V}_o/\ell$. Dimensional arguments and experiment [62, 63] indicate that the rate of decay of the turbulence kinetic energy $\sim O(\mathrm{V}_o \rho_o v^3/\ell)$. In a statistically steady state this must equal the rate Π_o, say, at which energy is supplied to the flow by the action of external forces. The *efficiency* Π_a/Π_o with which this energy is converted into sound is therefore proportional to M^5, confirming Lighthill's hypothesis that the flow-generated sound is an infinitesimal by-product of the motion.

2.1.3 Aerodynamic Sound Generated by Low Mach Number Turbulence of Variable Mean Density

When the mean density in the source region is not constant, the Reynolds stress quadrupoles are augmented by a distribution of *dipoles*, the dipole strength being equal to the hydrodynamic force experienced by a fluid particle of density ρ relative to the force the same particle would experience had its density been equal to ρ_o. This dipole is "hidden" within the Lighthill quadrupole T_{ij} and might therefore be expected to have the same radiation efficiency as the Reynolds stresses, in spite of our conclusion in Section 1.8.4 that the acoustic pressure radiated by a dipole is an order of magnitude larger than that from a quadrupole of comparable source strength. However, there is no contradiction of the rank ordering of Section 1.8.4 because the component of T_{ij} responsible for the dipole is $c_o^2(\rho - \rho_o)$, which is larger than the Reynolds stress $\rho v_i v_j$ by a factor of order $O(1/M^2)$, provided the mean value of $\rho - \rho_o \neq 0$ in the source region.

To obtain a precise estimate of the dipole radiation, the solution of Lighthill's equation (2.1.5) is written

$$p(\mathbf{x}, t) \approx \frac{1}{4\pi c_o^2}\frac{\partial^2}{\partial t^2}\int [\rho v_i v_j]\frac{(x_i - y_i)(x_j - y_j)\, d^3\mathbf{y}}{|\mathbf{x}-\mathbf{y}|^3} + \frac{1}{4\pi}\int \frac{\partial^2}{\partial t^2}\left[\frac{p - p_o}{c_o^2} - (\rho - \rho_o)\right]\frac{d^3\mathbf{y}}{|\mathbf{x}-\mathbf{y}|}, \quad |\mathbf{x}-\mathbf{y}| \to \infty, \tag{2.1.7}$$

where the square brackets [ ] indicate evaluation at the retarded time $t - |\mathbf{x}-\mathbf{y}|/c_o$.

The density of a fluid particle changes by $(p - p_o)/c^2$ when isentropically compressed by departures of p from the ambient pressure p_o, where c is the local speed of sound. Let us assume that the motion of each fluid particle is isentropic, but that the mean density, which will be denoted by $\bar{\rho}(\mathbf{x}, t)$, varies from point to point in the source region, where $(\bar{\rho}-\rho_o)/\rho_o$ need not necessarily be small. This means that $(p-p_o)/c^2 = \rho - \bar{\rho}$ and $D\bar{\rho}/Dt = 0$. The Reynolds stress on the right (2.1.7) must be replaced by $\bar{\rho}v_i v_j$, and the integrand of the second term in (2.1.7) becomes

$$\frac{\partial^2}{\partial t^2}\left\{\frac{p-p_o}{c_o^2} - (\rho - \rho_o)\right\} = \frac{\partial^2}{\partial t^2}\left\{(p-p_o)\left(\frac{1}{c_o^2} - \frac{1}{c^2}\right)\right\} - \frac{\partial^2}{\partial t^2}(\bar{\rho} - \rho_o).$$

The first term on the right is already in a form suitable for estimating the acoustic radiation. However, the second term would yield zero when retarded time variations are neglected over a coherent source region of constant mass. To deal with this, the relation $D(\bar{\rho}-\rho_o)/Dt = 0$ is used to show that

$$\frac{\partial^2}{\partial t^2}(\bar{\rho}-\rho_o) = -\frac{\partial}{\partial x_j}\left((\bar{\rho}-\rho_o)\frac{Dv_j}{Dt}\right) + \frac{\partial^2}{\partial x_i \partial x_j}((\bar{\rho}-\rho_o)v_i v_j) + \frac{\partial}{\partial t}((\bar{\rho}-\rho_o)\operatorname{div}\mathbf{v}) - \frac{\partial}{\partial x_j}(v_j(\bar{\rho}-\rho_o)\operatorname{div}\mathbf{v}). \tag{2.1.8}$$

This is further simplified by substitution from the momentum equation $Dv_j/Dt = (-1/\rho)\partial p/\partial x_j$ and from the continuity equation $\operatorname{div}\mathbf{v} = (-1/\rho c^2)Dp/Dt \approx -D[(p-p_o)/\rho c^2]/Dt$. The amplitude of sound produced by the final term on the right of (2.1.8) is $O(M)$ relative to the preceding one and is neglected.

Inserting these results into (2.1.7), and discarding the overbar on $\bar{\rho}$, we obtain

$$p(\mathbf{x},t) \approx \frac{x_i x_j}{4\pi c_o^2|\mathbf{x}|^3}\frac{\partial^2}{\partial t^2}\int [\rho_o v_i v_j]\, d^3\mathbf{y} + \frac{\rho_o x_j}{4\pi c_o|\mathbf{x}|^2}\frac{\partial}{\partial t} \times \int \left[\left(\frac{1}{\rho_o} - \frac{1}{\rho}\right)\frac{\partial p}{\partial y_j}\right] d^3\mathbf{y} + \frac{\rho_o}{4\pi|\mathbf{x}|}\frac{\partial^2}{\partial t^2} \times \int \left[\left(\frac{1}{\rho_o c_o^2} - \frac{1}{\rho c^2}\right)(p-p_o)\right] d^3\mathbf{y}, \quad |\mathbf{x}| \to \infty, \tag{2.1.9}$$

where quantities in square brackets are evaluated at time $t - |\mathbf{x}-\mathbf{y}|/c_o$. The new terms in this formula, in addition to the Reynolds stress radiation of (2.1.6), are respectively nonzero when the mean density in the source region $\rho \neq \rho_o$, and when $\rho_o c_o^2 \neq \rho c^2$, and are the sound fields of dipole and monopole sources.

All of these source types occur in the turbulent mixing region of a hot gas jet exhausting into cold air: "Hot spots" or "entropy inhomogeneities" behave as scattering centers at which dynamic pressure fluctuations are converted directly into sound. The dipole source strength is proportional to $(1/\rho - 1/\rho_o)\nabla p$, that is, to the difference between the acceleration of fluid of density ρ in the jet and that which fluid of ambient mean density ρ_o would experience in the same pressure gradient. For an ideal gas with $p - p_o \sim O(\rho_o v^2)$ in the jet, the order of magnitude of the dipole sound from an eddy of scale ℓ is

$$p \sim (\ell/|\mathbf{x}|)(1 - \rho_o/\rho)\rho_o v^2 M \approx (\ell/|\mathbf{x}|)(\Delta T/T)\rho_o v^2 M,$$

where $\Delta T/T \approx (\rho - \rho_o)/\rho$ is the fractional temperature difference between the hot spot and its environment, which can be large. The sound power $\sim\ell^2(\Delta T/T)^2\rho_o v^3 M^3$ exceeds that from a Reynolds stress quadrupole by a factor of order $(\Delta T/T)^2/M^2$. Thus, "entropy noise" may be an important component of the noise of a hot gas jet at very low Mach numbers [64].

The final term on the right of (2.1.9) is a monopole whose strength is determined by the difference between the adiabatic compressibilities ($K_s = 1/\rho c^2$, Section 1.6.1.) in the source region and in the ambient medium. In an ideal gas, $K_s = 1/\gamma p$ so that this source is generally small, although it may be significant during the turbulent mixing of gases with different values of the specific heat ratios γ. It can be very important in multiphase flows where, for example, the presence of small air bubbles in water often leads to an immense increase in the turbulence generated noise [65, 66].

Example 1: Aerodynamic sound in two dimensions [66]. Let T_{ij} be independent of x_3 so that i and j vary over one and two. Show, using Green's function (1.7.23), that the sound generated by low Mach number turbulence of uniform mean density is given by

$$p(\mathbf{x}, t) = \frac{\partial^2}{\partial x_i \partial x_j} \int d^2\mathbf{y} \int_{-\infty}^{t-|\mathbf{x}-\mathbf{y}|/c_o} \frac{(\rho_o v_i v_j)(\mathbf{y}, \tau)\, d\tau}{2\pi\sqrt{(t-\tau)^2 - (\mathbf{x}-\mathbf{y})^2/c_o^2}}.$$

The sound received at $\mathbf{x}$ depends on the source strength at all times prior to the retarded time $t - |\mathbf{x} - \mathbf{y}|/c_o$. The square root in the denominator of the integrand weights the contributions from each of these times differently, but the singularity at $\tau = t - |\mathbf{x} - \mathbf{y}|/c_o$ indicates that the principal contributions are from the neighborhood of this time. For a compact source, this implies

$$p(\mathbf{x}, t) \approx \frac{x_i x_j}{4\pi c_o^{3/2}|\mathbf{x}|^{5/2}} \frac{\partial^2}{\partial t^2}$$
$$\times \int d^2\mathbf{y} \int_{-\infty}^{0} \rho_o(v_i v_j)(\mathbf{y}, \tau' + t - |\mathbf{x}|/c_o) \frac{d\tau'}{\sqrt{-\tau'}}, \quad |\mathbf{x}| \to \infty.$$

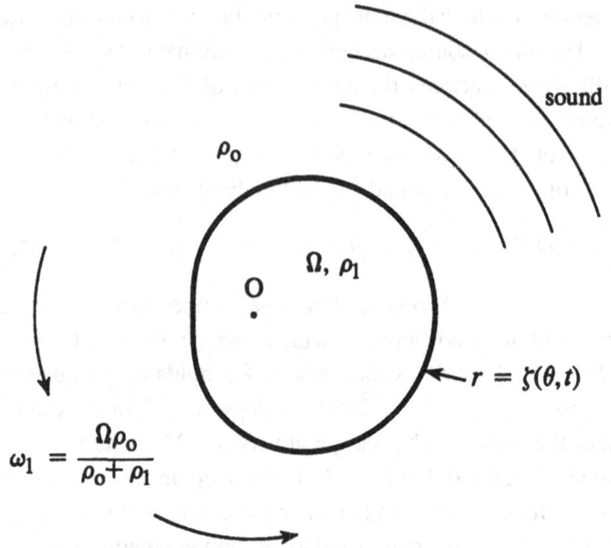

Figure 2.1.1. Spinning vortex.

The radiation from an eddy of dimension ℓ is coherent over times $\sim\ell/v$, and $p \sim \rho_o v^2 M^{3/2}(\ell/|\mathbf{x}|)^{1/2}$. The radiated power (per unit length in the x_3-direction) is proportional to $\ell\rho_o v^3 M^4$, which exceeds that from a three-dimensional turbulent eddy by a factor $\sim 1/M \gg 1$.

Example 2. Show that the acoustic power generated by low Mach number turbulence confined to a volume V in an infinite pipe of diameter $<\ell/M$ is of order $(\mathrm{V}/\ell)\rho_o v^3 M^3$ and that the acoustic efficiency $\sim O(M^3)$ [67].

Example 3: Sound radiation by a spinning vortex. Calculate the sound generated by a columnar vortex parallel to the x_3-axis, whose core is bounded by the curve

$$r = \zeta(\theta, t) \equiv a(1 + \epsilon\cos(n\theta - \omega_n t)), \quad \epsilon \ll 1,$$

where (r, θ) are polar coordinates in the x_1x_2-plane, a is the nominal core radius, and n is a positive integer (Figure 2.1.1). The fluid in the core has mean density ρ_1 and uniform vorticity $\boldsymbol{\omega} = (0, 0, \Omega)$, and the core rotates by self-induction about the x_3-axis with angular velocity ω_n/n, radiating sound of frequency ω_n into the ambient fluid where $\rho = \rho_o$, $c = c_o$.

Assume the core to be acoustically compact ($\Omega a/c_o \ll 1$) and that the tangential velocity is continuous at $r = a$ for $\epsilon = 0$. When $\epsilon \neq 0$, the interaction between the pressure gradient and the infinite density gradient at $r = \zeta(\theta, t)$ results in the formation of a vortex sheet at the interface (Section 1.2.7). Elsewhere, the unsteady motion is irrotational [68] and can be expressed in terms of the vector potential $\mathbf{A} = (0, 0, \psi)$ (Section 1.14.1), such that the velocity $(v_r, v_\theta) = (\partial\psi/r\partial\theta, -\partial\psi/\partial r)$ and $\nabla^2\psi = -\Omega, 0$ according as $r \lessgtr \zeta(\theta, t)$.

For $n > 0$, the boundary conditions are satisfied in the incompressible core region by

$$\psi = -\frac{1}{4}\Omega r^2 + A(r/a)^n \cos(n\theta - \omega_n t),$$
$$-\frac{1}{2}a^2\Omega \ln r + B(a/r)^n \cos(n\theta - \omega_n t), \quad r \lessgtr \zeta(\theta, t),$$

where the terms involving Ω describe the mean flow. To first order in ϵ, the condition (Section 1.5.1) $D(r - \zeta(\theta, t))/Dt = 0$ at the interface implies that $v_r = a\epsilon(\omega_n - n\Omega/2)\sin(n\theta - \omega_n t)$ and therefore that $A = B = -(a^2\epsilon/n)(\omega_n - n\Omega/2)$. Admissible values of the frequency ω_n are found from the condition of continuity of pressure at $r = \zeta(\theta, t)$. The pressure can be expressed in terms of the velocity potential φ of the unsteady motion

$$\varphi = -A(r/a)^n \sin(n\theta - \omega_n t), \quad B(a/r)^n \sin(n\theta - \omega_n t), \quad r \lessgtr \zeta(\theta, t).$$

In the exterior potential flow, Bernoulli's equation gives

$$p = \epsilon\rho_o a^2\left(\frac{\Omega^2}{4} - \frac{(\omega_n - n\Omega/2)^2}{n}\right)\cos(n\theta - \omega_n t), \quad r = \zeta(\theta, t) + 0.$$

The pressure within the core is obtained from the momentum equation $\nabla(\partial\varphi/\partial t + p/\rho_1 + \frac{1}{2}v^2) + \boldsymbol{\omega} \wedge \mathbf{v} = 0$, where $\boldsymbol{\omega} = (0, 0, \Omega)$, which leads to

$$p = \epsilon\rho_1 a^2\left(\frac{\Omega^2}{4} - \frac{\left(\omega_n - \frac{1}{2}n\Omega\right)\left(\omega_n - \frac{1}{2}n\Omega + \Omega\right)}{n}\right)\cos(n\theta - \omega_n t),$$
$$r = \zeta(\theta, t) - 0.$$

Equating these two expressions for the pressure, and solving for ω_n, we find

$$\omega_n = \frac{\Omega}{2}\left(n - \frac{\rho_1 \pm \sqrt{n\rho_o^2 - (n-1)\rho_1^2}}{\rho_o + \rho_1}\right).$$

This result confirms the compactness hypothesis that $\omega_n a/c_o \ll 1$ when $\Omega a/c_o < 1$.

When the mean density is uniform $n = 2$ is the minimum value of n for which nontrivial solutions exist, that is, for which $A = B \neq 0$ when $\omega_n \neq 0$. The core then has elliptic cross-section, and the sound has quadrupole intensity (see Section 2.3, Example 2). If $\rho_1 \neq \rho_o$, a nontrivial solution exists for $n = 1$, with $\omega_1 = \Omega\rho_o/(\rho_o + \rho_1)$ and $A = B = \frac{\epsilon}{2}a^2\Omega(\rho_1 - \rho_o)/(\rho_1 + \rho_o)$, the core having circular cross-section. The vortex now generates dipole sound that the reader can determine by evaluating the second integral on the right of (2.1.9), where $\partial p/\partial y_j$ is the unsteady pressure gradient within the core. The same result can be obtained by *matching* the velocity potential of the hydrodynamic motion in $r > \zeta(\theta, t)$ to the near field limit of the acoustic potential. As $|\mathbf{x}| \to \infty$, the dipole velocity potential can be written (Section 1.10.5)

$$\varphi \approx \alpha \mathrm{H}_1^{(1)}\left(\frac{\omega_1 r}{c_o}\right)\mathrm{e}^{i(\theta-\omega_1 t)} + \text{c.c.}, \quad \alpha = \text{constant}.$$

From (1.10.18) $\mathrm{H}_1^{(1)}(\omega_1 r/c_o) \approx -2ic_o/\pi\omega_1 r$, as $\omega_1 r/c_o \to 0$. By equating the resulting near-field approximation for φ to the incompressible potential calculated above for $n = 1$, we find $\alpha = (\pi\epsilon a^3\Omega\omega_1/8c_o)(\rho_1 - \rho_o)/(\rho_1 + \rho_o)$ and therefore that the acoustic pressure is the dipole field

$$p \approx 2\epsilon\rho_o U^2\left(\frac{\pi a M}{r}\right)^{\frac{1}{2}}\left(\frac{\rho_o}{\rho_1 + \rho_o}\right)^{\frac{3}{2}}\left(\frac{\rho_1 - \rho_o}{\rho_1 + \rho_o}\right) \times \cos\left[\theta - \frac{\pi}{4} - \omega_1(t - r/c_o)\right], \quad r \to \infty,$$

where $U = a\Omega/2$ is the mean velocity at the edge of the core, and $M = U/c_o$.

2.2 The Ffowcs Williams–Hawkings Equation

2.2.1 General Solution of Lighthill's Equation

Most practical problems of sound generation by flow involve moving boundaries, moving sources interacting with such boundaries, or turbulence in shear layers separating a quiescent medium from a high-speed flow. To apply Lighthill's equation in these circumstances, *control surfaces*, S, are introduced. These may coincide with the surface of a moving solid or mark a convenient interface between fluid regions of widely differing mean properties. A solution is then sought by imposing boundary conditions on S, either by first performing subsidiary calculations to determine the pressure or velocity on S or, when S coincides with the surface of a solid, by application of suitable impedance conditions.

Let $f(\mathbf{x}, t)$ be an indicator function that vanishes on S and satisfies $f(\mathbf{x}, t) > 0$ in the fluid where Lighthill's equation is to be solved, and $f(\mathbf{x}, t) < 0$ elsewhere (cf. Sections 1.14.2 and 5), and set

$$p'_{ij} = (p - p_o)\delta_{ij} - \sigma_{ij}. \tag{2.2.1}$$

Multiply the momentum equation (2.1.1) by $\mathrm{H}(f)$ and rearrange into the form

$$\frac{\partial}{\partial t}(\rho v_i \mathrm{H}(f)) + \frac{\partial}{\partial x_i}\left(\rho v_i \mathrm{H}(f) c_o^2(\rho - \rho_o)\right)$$
$$= -\frac{\partial}{\partial x_j}(\mathrm{H}(f) T_{ij}) + (\rho v_i(v_j - \bar{v}_j) + p'_{ij})\frac{\partial \mathrm{H}}{\partial x_j}(f),$$

where $\bar{\mathbf{v}}$ is the velocity of S (so that $\partial \mathrm{H}(f)/\partial t = -\bar{v}_j \partial \mathrm{H}(f)/\partial x_j$), and ρ_o, c_o are the mean density and sound speed in $f > 0$. To simplify the notation, we shall henceforth write H instead of $\mathrm{H}(f)$ when there is no danger of confusion. The same procedure applied to the equation of continuity supplies

$$\frac{\partial}{\partial t}(\mathrm{H}(\rho - \rho_o)) + \frac{\partial}{\partial x_i}(\mathrm{H}\rho v_i) = (\rho(v_i - \bar{v}_i) + \rho_o \bar{v}_i)\frac{\partial \mathrm{H}}{\partial x_i}.$$

The elimination of $\mathrm{H}\rho v_i$ between these two equations yields the differential form of the *Ffowcs Williams–Hawkings equation*

$$\left(\frac{1}{c_o^2}\frac{\partial^2}{\partial t^2} - \nabla^2\right)[\mathrm{H}c_o^2(\rho - \rho_o)]$$
$$= \frac{\partial^2(\mathrm{H}T_{ij})}{\partial x_i \partial x_j} - \frac{\partial}{\partial x_i}\left([\rho v_i(v_j - \bar{v}_j) + p'_{ij}]\frac{\partial \mathrm{H}}{\partial x_j}\right)$$
$$+ \frac{\partial}{\partial t}\left([\rho(v_j - \bar{v}_j) + \rho_o \bar{v}_j]\frac{\partial \mathrm{H}}{\partial x_j}\right). \tag{2.2.2}$$

This equation is valid throughout the whole of space. By using Green's function (1.7.20) to write down the formal, outgoing wave solution, it is transformed into an integral equation known as the Ffowcs Williams–Hawkings equation [69]

$$\mathrm{H}c_o^2(\rho - \rho_o) = \frac{\partial^2}{\partial x_i \partial x_j}\int_{\mathrm{V}(\tau)} [T_{ij}]\frac{d^3\mathbf{y}}{4\pi|\mathbf{x} - \mathbf{y}|}$$
$$- \frac{\partial}{\partial x_i}\oint_{\mathrm{S}(\tau)} [\rho v_i(v_j - \bar{v}_j) + p'_{ij}]\frac{dS_j(\mathbf{y})}{4\pi|\mathbf{x} - \mathbf{y}|}$$
$$+ \frac{\partial}{\partial t}\oint_{\mathrm{S}(\tau)} [\rho(v_j - \bar{v}_j) + \rho_o \bar{v}_j]\frac{dS_j(\mathbf{y})}{4\pi|\mathbf{x} - \mathbf{y}|}, \tag{2.2.3}$$

where quantities in the square brackets [ ] are evaluated at the retarded time $\tau = t - |\mathbf{x} - \mathbf{y}|/c_o$, the surface integrals are over the retarded surface $\mathrm{S}(\tau)$ defined by $f(\mathbf{y}, \tau) = 0$, with the surface element $d\mathbf{S}$ directed into the region $\mathrm{V}(\tau)$ where $f > 0$.

If the control surface is stationary ($\bar{\mathbf{v}} = \mathbf{0}$), S is defined by $f(\mathbf{y}, t) \equiv f(\mathbf{y}) = 0$, and (2.2.3) reduces to Curle's equation [70]

$$\begin{aligned} \mathrm{H}c_o^2(\rho - \rho_o) = {} & \frac{\partial^2}{\partial x_i \partial x_j} \int_{\mathrm{V}} [T_{ij}] \frac{d^3\mathbf{y}}{4\pi|\mathbf{x} - \mathbf{y}|} \\ & - \frac{\partial}{\partial x_i} \oint_{\mathrm{S}} [\rho v_i v_j + p'_{ij}] \frac{d\mathrm{S}_j(\mathbf{y})}{4\pi|\mathbf{x} - \mathbf{y}|} \\ & + \frac{\partial}{\partial t} \oint_{\mathrm{S}} [\rho v_j] \frac{d\mathrm{S}_j(\mathbf{y})}{4\pi|\mathbf{x} - \mathbf{y}|}, \end{aligned} \tag{2.2.4}$$

which is a generalization of Kirchhoff's formula (1.10.7) of linear acoustics. When S is stationary and *rigid*, this simplifies to

$$\mathrm{H}c_o^2(\rho - \rho_o) = \frac{\partial^2}{\partial x_i \partial x_j} \int_{\mathrm{V}} [T_{ij}] \frac{d^3\mathbf{y}}{4\pi|\mathbf{x} - \mathbf{y}|} - \frac{\partial}{\partial x_i} \oint_{\mathrm{S}} [p'_{ij}] \frac{d\mathrm{S}_j(\mathbf{y})}{4\pi|\mathbf{x} - \mathbf{y}|}. \tag{2.2.5}$$

The surface integrals in these formulae may be interpreted as distributions of monopole or dipole sources (Section 1.10.2). They are associated with mass and momentum transfer across S, but their values cannot normally be specified independently. Indeed, because the power in the radiated sound is usually a mere fraction of the total power, small errors made in prescribing the surface terms can seriously impair the accuracy of predictions of the flow-generated sound.

2.2.2 Source-Fixed Coordinates

Most applications of the Ffowcs Williams–Hawkings equation are made to high-speed flows involving rapidly moving control surfaces and source distributions. Both the directivity and amplitude of the sound are strongly affected by high-speed source motion (Section 1.8.5). Source strengths are generally more easily calculated in a reference frame translating with the source, and it is sometimes useful to rewrite the integrals in the Ffowcs Williams–Hawkings equation in terms of a moving coordinate system.

Consider a generalized aerodynamic source $\mathcal{F}(\mathbf{x}, t)$ of arbitrary multipole order, whose radiation is given explicitly in terms of Green's function (1.7.20) by

$$\mathrm{H}c_o^2(\rho - \rho_o) = \iint \mathcal{F}(\mathbf{y}, \tau)\delta(t - \tau - |\mathbf{x} - \mathbf{y}|/c_o)\frac{d^3\mathbf{y}\, d\tau}{4\pi|\mathbf{x} - \mathbf{y}|}. \qquad (2.2.6)$$

Introduce a Lagrangian coordinate $\boldsymbol{\eta}(\mathbf{y}, \tau)$ with respect to which each element of the source distribution is at rest. If $\mathbf{u}(\boldsymbol{\eta}, \tau)$ is the velocity (with respect to the fixed Eulerian frame $\mathbf{y}$) of the source element labeled $\boldsymbol{\eta}$, and $\boldsymbol{\eta} = \mathbf{y}$ at some initial instant $\tau = t_o$, then

$$\mathbf{y}(\boldsymbol{\eta}, \tau) = \boldsymbol{\eta} + \int_{t_o}^{\tau} \mathbf{u}(\boldsymbol{\eta}, \tau')\, d\tau', \quad \tau > t_o. \qquad (2.2.7)$$

The mass of a source element is conserved during arbitrary motion so that the volume element $d^3\mathbf{y}$ in (2.2.6) is equal to $(\rho^*/\rho)d^3\boldsymbol{\eta}$, where ρ^*/ρ is the Jacobian of the transformation, $\rho^* \equiv \rho^*(\boldsymbol{\eta})$ is the mass density of the source element at $\tau' = t_o$, and ρ is the density at time τ. For surface source distributions that are fixed relative to a *rigid* body moving in the fluid, the $\boldsymbol{\eta}$ coordinate is a simple translation and rotation of the $\mathbf{y}$ system, and $d^3\boldsymbol{\eta} = d^3\mathbf{y}$.

The integral (2.2.6) now becomes

$$\begin{aligned}
&\iint \mathcal{F}(\mathbf{y}, \tau)\delta(t - \tau - |\mathbf{x} - \mathbf{y}|/c_o)\frac{d^3\mathbf{y}\, d\tau}{4\pi|\mathbf{x} - \mathbf{y}|} \\
&= \iint \mathcal{F}(\boldsymbol{\eta}, \tau)(\rho^*/\rho)\delta(t - \tau - R/c_o)\frac{d^3\boldsymbol{\eta}\, d\tau}{4\pi R} \\
&= \int \left[\frac{\mathcal{F}(\boldsymbol{\eta}, \tau)(\rho^*/\rho)}{4\pi R\left|\frac{\partial}{\partial\tau}(t - \tau - R(\mathbf{x}, \boldsymbol{\eta}, \tau)/c_o)\right|}\right] d^3\boldsymbol{\eta} \\
&= \int \left[\frac{\mathcal{F}(\boldsymbol{\eta}, \tau)(\rho^*/\rho)}{4\pi R|1 - M\cos\Theta|}\right] d^3\boldsymbol{\eta}, \qquad (2.2.8)
\end{aligned}$$

where $\mathbf{R} = \mathbf{x} - \mathbf{y}(\boldsymbol{\eta}, \tau)$, Θ is the angle between $\mathbf{R}$ and the velocity $\mathbf{u}$ of the source, $M = |\mathbf{u}|/c_o$, and the terms in square brackets [ ] are evaluated at the retarded time $\tau_e = \tau_e(\boldsymbol{\eta}, \mathbf{x})$ determined from the equation

$$c_o(t - \tau_e) = |\mathbf{x} - \mathbf{y}(\boldsymbol{\eta}, \tau_e)|. \qquad (2.2.9)$$

Applying the transformation (2.2.8) to each term on the right of (2.2.3), we obtain the Ffowcs Williams–Hawkings equation in source-fixed coordinates:

$$\begin{aligned}
\mathrm{H}c_o^2(\rho-\rho_o) &= \frac{\partial^2}{\partial x_i \partial x_j}\int_{\mathrm{V}_e}\left[\frac{T_{ij}(\rho^*/\rho)}{4\pi R|1-M\cos\Theta|}\right]d^3\eta \\
&\quad -\frac{\partial}{\partial x_i}\int_{\mathrm{S}_e}\left[\frac{(\rho v_i(v_j-\bar{v}_j)+p'_{ij})}{4\pi R|1-M\cos\Theta|}\left(\frac{\rho^*|\partial f/\partial \mathbf{y}|}{\rho|\partial f/\partial \eta|}\right)n_j\right]d\mathrm{S}(\eta) \\
&\quad +\frac{\partial}{\partial t}\int_{\mathrm{S}_e}\left[\frac{(\rho(v_j-\bar{v}_j)+\rho_o\bar{v}_j)}{4\pi R|1-M\cos\Theta|}\left(\frac{\rho^*|\partial f/\partial \mathbf{y}|}{\rho|\partial f/\partial \eta|}\right)n_j\right]d\mathrm{S}(\eta),
\end{aligned} \tag{2.2.10}$$

where the terms in [ ] are evaluated at the retarded time τ_e of (2.2.9), S_e is the surface $f(\eta,\tau_e)=0$ with normal n_j directed into the region V_e where $f(\eta,\tau_e)>0$, and we have used the formula [20]

$$\frac{\partial \mathrm{H}(f)}{\partial y_j} = n_j\delta(\eta_n)\frac{|\partial f/\partial \mathbf{y}|}{|\partial f/\partial \eta|},$$

where η_n is a local coordinate measured in the normal direction from S_e into the region $f>0$.

The Ffowcs Williams–Hawkings equation simplifies considerably when the control surface S moves with the fluid particles ($\bar{\mathbf{v}}=\mathbf{v}$), when $f(\mathbf{y},\tau)\to f(\eta)=0$ becomes independent of τ. One of the most important cases occurs when S is the surface of a *rigid* body whose motion is defined by the velocity $\mathbf{U}_o(\tau)$ of its center of volume η_o and its angular velocity $\boldsymbol{\Omega}(\tau)$. Then $\bar{\mathbf{v}}\equiv\mathbf{u}=\mathbf{U}_o+\boldsymbol{\Omega}\wedge(\eta-\eta_o)$ on S, and $\rho^*|\partial f/\partial\mathbf{y}|/\rho|\partial f/\partial\eta|\equiv 1$ in the surface integrals, and (2.2.10) becomes

$$\begin{aligned}
\mathrm{H}c_o^2(\rho-\rho_o) &= \frac{\partial^2}{\partial x_i \partial x_j}\int_{\mathrm{V}}\left[\frac{T_{ij}(\rho^*/\rho)}{4\pi R|1-M\cos\Theta|}\right]d^3\eta \\
&\quad -\frac{\partial}{\partial x_i}\int_{\mathrm{S}}\left[\frac{p'_{ij}n_j}{4\pi R|1-M\cos\Theta|}\right]d\mathrm{S}(\eta) \\
&\quad +\frac{\partial}{\partial t}\int_{\mathrm{S}}\left[\frac{\rho v_n}{4\pi R|1-M\cos\Theta|}\right]d\mathrm{S}(\eta),
\end{aligned} \tag{2.2.11a}$$

where S and V are now fixed relative to the moving coordinate η. Because there can be no net outflow of volume from a rigid surface, the final "monopole" integral must actually be equivalent to a higher order multipole. The integral can be replaced by two volume integrals taken over the interior $\bar{\mathrm{V}}$, say, of the rigid body (Example 1) that are equivalent to dipole and quadrupole sources, yielding the alternative representation

$$\mathrm{H}c_o^2(\rho - \rho_o) = \frac{\partial^2}{\partial x_i \partial x_j} \int_{\mathrm{V}} \left[\frac{T_{ij}(\rho^*/\rho)}{4\pi R|1 - M\cos\Theta|} \right] d^3\eta$$

$$- \frac{\partial}{\partial x_i} \int_{\mathrm{S}} \left[\frac{p'_{ij} n_j}{4\pi R|1 - M\cos\Theta|} \right] d\mathrm{S}(\eta)$$

$$- \frac{\partial}{\partial x_i} \int_{\bar{\mathrm{V}}} \left[\frac{\rho \dot{u}_i}{4\pi R|1 - M\cos\Theta|} \right] d^3\eta$$

$$+ \frac{\partial^2}{\partial x_i \partial x_j} \int_{\bar{\mathrm{V}}} \left[\frac{\rho u_i u_j}{4\pi R|1 - M\cos\Theta|} \right] d^3\eta \qquad (2.2.11\mathrm{b})$$

where $\dot{u}_i = Du_i/Dt$. Applications of this equation are discussed Section 3.6.

Example 1. The rigid body velocity $\mathbf{u} = \mathbf{U}_o + \boldsymbol{\Omega} \wedge (\eta - \eta_o)$ is *solenoidal* (i.e., div $\mathbf{u} = 0$). If $f = 0$ on the surface of the rigid body bounded by S, with $f < 0$ within S, then

$$\frac{\partial}{\partial t}\left(u_i \frac{\partial}{\partial x_i} \mathrm{H}(f) \right) = -\frac{\partial}{\partial x_i}\left(\frac{Du_i}{Dt} \mathrm{H}(-f) \right) + \frac{\partial^2}{\partial x_1 \partial x_j}(u_i u_j \mathrm{H}(-f)).$$

Use this result in (2.2.2) when $\bar{\mathbf{v}} = \mathbf{v}$ to deduce the particular form (2.2.11b) of the Ffowcs Williams–Hawkings equation.

2.2.3 Sound Generated by Moving Aerodynamic Quadrupoles [71]

The first term on the right of the Ffowcs Williams–Hawkings equation (2.2.3) describes the direct radiation from quadrupole sources. The transformation (2.2.7) with $\mathbf{u} = \mathbf{v}$ permits the integral to be expressed as follows, in terms of the Lagrangian coordinate η moving with the fluid particles,

$$\mathrm{H}c_o^2(\rho - \rho_o) = \frac{\partial^2}{\partial x_1 \partial x_j} \int\!\!\int T_{ij}(\eta, \tau)(\rho^*/\rho)\delta(t - \tau - R/c_o) \frac{d^3\eta\, d\tau}{4\pi R}. \qquad (2.2.12)$$

In the acoustic far field, the operator $\partial^2/\partial x_i \partial x_j$ need be applied only to the δ-function, and

$$\frac{\partial}{\partial x_i}\delta(t-\tau-R/c_o) = \frac{-R_i}{c_o R}\frac{\partial}{\partial t}\delta(t-\tau-R/c_o)$$
$$= \frac{R_i}{c_o R(1-M\cos\Theta)}\frac{\partial}{\partial \tau}\delta(t-\tau-R/c_o),$$

where the final term is obtained from the relation

$$\partial R/\partial\tau \equiv \partial|\mathbf{x}-\mathbf{y}(\boldsymbol{\eta},\tau)|/\partial\tau = -\mathbf{v}\cdot(\mathbf{x}-\mathbf{y})/|\mathbf{x}-\mathbf{y}| = -c_o M\cos\Theta.$$

When this formula is applied to the derivative $\partial/\partial x_j$ in (2.2.12), the result may be integrated by parts with respect to τ, which has the effect of transferring the differential operator $\partial/\partial\tau$ from the δ-function to the other terms in the integrand. Repeating this procedure for the derivative $\partial/\partial x_i$, we find, for $R\to\infty$,

$$\mathrm{H}c_o^2(\rho-\rho_o) = \int\!\!\int \rho^* \frac{\partial}{\partial\tau}\left\{\frac{R_i}{R(1-M\cos\Theta)}\frac{\partial}{\partial\tau}\left(\frac{R_j T_{ij}}{\rho R(1-M\cos\Theta)}\right)\right\}$$
$$\times\frac{\delta(t-\tau-R/c_o)}{4\pi c_o^2 R}d^3\eta\, d\tau.$$

Because R is large, this is equivalent to

$$\mathrm{H}c_o^2(\rho-\rho_o) = \frac{1}{4\pi c_o^2}\int\!\!\int\left[\rho^*\frac{R_i R_j}{R^3}\frac{D}{Dt}\left\{\frac{1}{(1-M\cos\Theta)}\right.\right.$$
$$\left.\left.\times\frac{D}{Dt}\left(\frac{T_{ij}}{\rho(1-M\cos\Theta)}\right)\right\}\frac{d^3\eta}{(1-M\cos\Theta)}\right],$$
$$= \frac{1}{4\pi c_o^2}\int\!\!\int\left[\rho\frac{R_i R_j}{R^3}\frac{D}{Dt}\left\{\frac{1}{(1-M\cos\Theta)}\right.\right.$$
$$\left.\left.\times\frac{D}{Dt}\left(\frac{T_{ij}}{\rho(1-M\cos\Theta)}\right)\right\}\right]d^3\mathbf{y},\quad |\mathbf{x}|\to\infty,$$

(2.2.13)

where in the first line the square brackets [ ] denote evaluation at the retarded time τ_e defined by (2.2.8), and in the second at $\tau = t-|\mathbf{x}-\mathbf{y}|/c_o$. These formulae are an exact representation of the radiation into an ideal, stationary fluid from the quadrupole distribution T_{ij} moving with the flow. The presence of the material derivatives can make it easier to obtain reliable estimates of the sound because the time dependence of the sources is usually well defined in a reference frame convecting with the sources.

As an illustration, let the convection velocity $\mathbf{u}$ of turbulence quadrupoles in a jet be approximated by a uniform, mean velocity $\mathbf{U}$. The Mach number M is then constant and $D^2T_{ij}/Dt^2 \sim \rho_o U^4/\ell^2$, where ℓ is the length scale of the "eddies." The first line of (2.2.13) then implies that the mean square acoustic pressure radiated by an eddy $\sim(\ell/R)^2\rho_o^2U^4M^4/(1-M\cos\Theta)^6$ (cf. Section 1.8 Example 7). Integration of the acoustic intensity over the surface of a large sphere centered on the retarded position of the quadrupole reveals that the source motion increases the net radiated power of a quadrupole whose strength is independent of the convection velocity. For a region of turbulence occupying the interval $0 < y_1 < L$ in the mean flow direction (corresponding, for example, to turbulence in a jet that decays over a distance L from the jet nozzle), the integration with respect to η_1 may be taken to extend over the range $-U(t-|\mathbf{x}|/c_o) < \eta_1 < L - U[t - |\mathbf{x}|/c_o + (L/c_o)\cos\theta]$ when $|\mathbf{x}| \to \infty$ so that the effective length of the source region is decreased for radiation in "forward" directions ($0 < \Theta < 90°$) and increased to the rear ($\Theta > 90°$). If V_o is the volume of the turbulent flow, the number of independent eddies contributing to the acoustic intensity is therefore equal to $V_o(1-M\cos\Theta)/\ell^3$ and not V_o/ℓ^3, and the net acoustic intensity varies as $(V_o/\ell R^2)\rho_o^2U^4M^4/(1-M\cos\Theta)^5$ and involves five rather than six powers of the Doppler factor [72].

These elementary estimates do not exhaust the full range of possible effects of source motion on aerodynamic sound. For example, when accounting for unsteady convection of the quadrupoles, the material derivatives D/Dt in (2.2.13) should also be applied to the Doppler factors. This can lead to predictions of acoustic intensity varying as $1/(1-M\cos\Theta)^{10}$ for a single eddy or as $1/(1-M\cos\Theta)^9$ for an extended region of turbulence. It is unlikely, therefore, that the influence of source convection on aerodynamic sound generated by a turbulent jet can easily be embodied in a simple power law dependence on Doppler factor. Furthermore, experiments [73–76] indicate that explicit account should be taken of the "refraction" of sound generated within a high-speed jet whenever the propagation path through the nonuniform jet shear layer exceeds an acoustic wavelength, that is, whenever the source region is not "compact."

Example 2: Influence of source motion on entropy noise. In the notation of Section 2.1.3, the component of T_{ij} responsible for entropy noise is

$$T_{ij} = c_o^2\left\{(p-p_o)\left(\frac{1}{c_o^2}-\frac{1}{c^2}\right) - (\bar{\rho}-\rho_o)\right\}\delta_{ij}.$$

By transforming to Lagrangian coordinates convecting with the fluid particles,

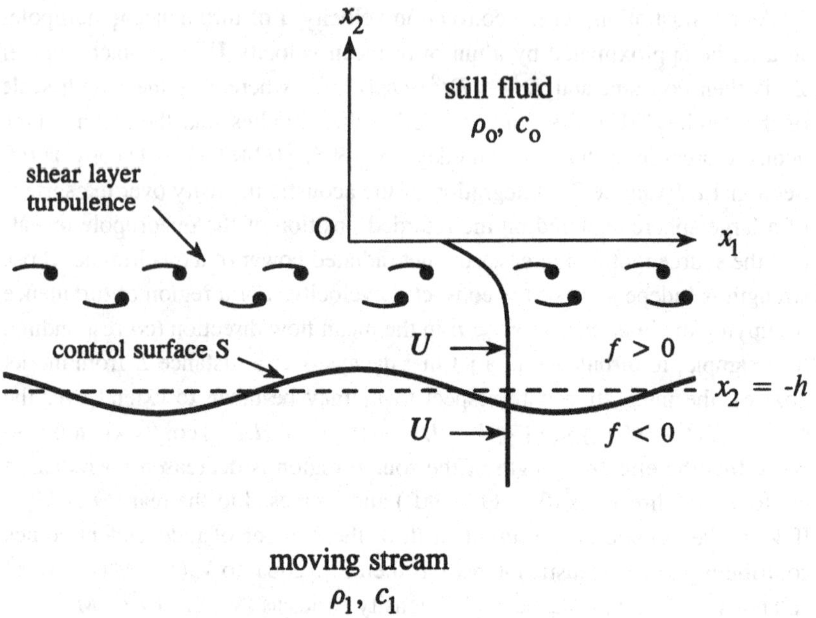

Figure 2.2.1. Plane turbulent shear layer.

show that the leading order entropy radiation at finite Mach numbers is given for each source type by

$$c_o^2(\rho - \rho_o) \approx \frac{\rho_o}{4\pi}\int\left[\frac{1}{R(1 - M\cos\Theta)^3}\frac{D^2}{Dt^2}\left\{(p - p_o)\left(\frac{1}{\rho_o c_o^2} - \frac{1}{\rho c^2}\right)\right\}\right.$$
$$\left. + \left(\frac{1}{\rho_o} - \frac{1}{\rho}\right)\frac{R_j}{c_o R^2(1 - M\cos\Theta)^4}\frac{D}{Dt}\left(\frac{\partial p}{\partial y_j}\right)\right] d^3\eta,$$
$$R \to \infty.$$

Example 3: Shear layer quadrupole noise [71]. Turbulence quadrupoles lie in a plane, mean shear layer separating a quiescent half-space of mean density ρ_o and sound speed c_o from fluid of density ρ_1 and sound speed c_1 moving uniformly at speed U in the x_1-direction (Figure 2.2.1). The x_1x_3-plane is parallel to the mean flow and lies within the shear layer, with the quiescent fluid in $x_2 > 0$. We apply the Ffowcs Williams–Hawkings equation (2.2.3) to estimate the sound radiated into the quiescent region. In the undisturbed state, the control surface S is taken to be the plane $y_2 = -h =$ constant in the region where the mean flow speed is uniform and equal to U. The surface is assumed to move with the fluid particles ($\bar{\mathbf{v}} = \mathbf{v}$), but h is sufficiently large that S is only *linearly* displaced by an amount ζ in the x_2-direction from its undisturbed position.

As $|\mathbf{x}| \to \infty$ in the quiescent region

$$c_o^2(\rho - \rho_o) \approx p_+(\mathbf{x}, t) + \frac{x_2}{4\pi c_o|\mathbf{x}|^2}\frac{\partial}{\partial t}\int_S [p']\,dy_1\,dy_3$$
$$+ \frac{\rho_o}{4\pi|\mathbf{x}|}\frac{\partial}{\partial t}\int_S \left[\frac{\partial \zeta}{\partial t}\right] dy_1\,dy_3, \tag{2.2.14}$$

where p_+ is the acoustic pressure that would be produced in $x_2 > 0$ when the quadrupoles radiate into an unbounded, stationary medium, which is given by the first line of (2.2.3). Viscous stresses have been ignored on S where $p'_{ij} = p'\delta_{ij}(p' \approx p - p_o)$, and $\partial\zeta/\partial t \approx v_n$ is the normal velocity of S (see Section 1.5.1).

To determine the contribution from the surface integrals, take the Fourier transform with respect to time. As $|\mathbf{x}| \to \infty$

$$\frac{1}{2\pi}\int_{-\infty}^{\infty} [p']\mathrm{e}^{i\omega t}\,dt \to p'(\mathbf{y}, \omega)\exp(i(\kappa_o|\mathbf{x}| - \mathbf{k}\cdot\mathbf{y})),$$

where $\mathbf{k} = \kappa_o\mathbf{x}/|\mathbf{x}|$. Equation (2.2.14) becomes

$$c_o^2\rho'(\mathbf{x}, \omega) = p_+(\mathbf{x}, \omega) - \left(\frac{\pi}{|\mathbf{x}|}\right)(ik_2\hat{p} + \rho_o\omega^2\hat{\zeta})\mathrm{e}^{i(\kappa_o|\mathbf{x}| - k_2 h)}, \tag{2.2.15}$$

where $\rho' = \rho - \rho_o$ and $\hat{p} = (1/2\pi)^2\int_{-\infty}^{\infty} p'(y_1, -h, y_3, \omega)\exp(-i(k_1 y_1 + k_3 y_3))\,dy_1\,dy_3$ is the Fourier transform in the y_1y_3-plane, with a similar definition for $\hat{\zeta}$. The radiation condition implies that each Fourier component $\hat{p}$ of the pressure on $y_2 = -h$ represents the boundary value of a plane wave $\bar{p}$, say, propagating *into* the region $y_2 < -h$ of uniform mean flow. $\bar{p}$ is a solution of the convected wave equation (1.6.30) (with $\hat{\mathcal{F}} = 0$) and is given by

$$\bar{p} = \hat{p}\mathrm{e}^{i[k_1 y_1 + k_3 y_3 - \Gamma(y_2 + h)]}, \quad y_2 < -h,$$

where $\Gamma = ((\kappa_o - Mk_1)^2 - (k_1^2 + k_3^2))^{1/2}$ and $M = U/c_1$; Γ is positive imaginary when the argument of the square root is negative, corresponding to a disturbance that decays with increasing distance from the shear layer in $y_2 < -h$; otherwise $\operatorname{sgn}(\Gamma) = \operatorname{sgn}(\kappa_o - Mk_1)$.

The pressure p and the displacement ζ are related by the y_2-component of the linearized momentum equation: $\rho_1 Dv_2/Dt \equiv \rho_1 D^2\zeta/Dt^2 = -\partial p/\partial y_2$ on $y_2 = -h$, which implies that $\hat{\zeta} = [-i\Gamma/\rho_1(\omega - Uk_1)^2]\hat{p}$ and therefore that (2.2.15) becomes

$$c_o^2\rho'(\mathbf{x}, \omega) = p_+(\mathbf{x}, \omega) - \frac{i\pi}{|\mathbf{x}|}\left(k_2 - \frac{\rho_o\omega^2\Gamma}{\rho_1(\omega - Uk_1)^2}\right)\hat{p}\mathrm{e}^{i(\kappa_o|\mathbf{x}| - k_2 h)}. \tag{2.2.16}$$

The remaining Fourier coefficient $\hat{p}$ can be calculated by applying the representation (2.2.3) at the image $(x_1, -x_2, x_3)$ of (x_1, x_2, x_3) in $x_2 = 0$, where $\mathrm{H}c_o^2(\rho - \rho_o) \equiv 0$, leading to

$$0 = p_-(\mathbf{x}, \omega) + \frac{i\pi}{|\mathbf{x}|}\left(k_2 + \frac{\rho_o\omega^2\Gamma}{\rho_1(\omega - Uk_1)^2}\right)\hat{p}\mathrm{e}^{i(\kappa_o|\mathbf{x}|+k_2h)},$$

where p_- is the pressure that would be received at $(x_1, -x_2, x_3)$ if the medium were homogeneous with no mean flow.

Eliminating $\hat{p}$ between this result and (2.2.16), and taking the inverse Fourier transform, we obtain

$$p(\mathbf{x}, t) \approx p_+(\mathbf{x}, t) + \int_{-\infty}^{\infty} \mathcal{R}(\mathbf{x}/|\mathbf{x}|, \omega)p_-(\mathbf{x}, \omega)\mathrm{e}^{-i\omega(t-2x_2h/c_o|\mathbf{x}|)}\,d\omega \tag{2.2.17}$$

where, for $\omega > 0$

$$\mathcal{R}\left(\frac{\mathbf{x}}{|\mathbf{x}|}, \omega\right) = \frac{\frac{x_2}{|\mathbf{x}|} - \frac{\rho_o}{\rho_1}\sqrt{\frac{c_o^2}{c_1^2} - \frac{(x_1^2 + x_2^3)}{(|\mathbf{x}| - Mx_1)^2}}}{\frac{x_2}{|\mathbf{x}|} + \frac{\rho_o}{\rho_1}\sqrt{\frac{c_o^2}{c_1^2} - \frac{(x_1^2 + x_2^3)}{(|\mathbf{x}| - Mx_1)^2}}},$$

and the square roots are defined to be either positive or positive imaginary. When $\omega(1 - Mx_1/|\mathbf{x}|) < 0$, the signs of the square roots are reversed when real.

The quadrupole sources T_{ij} that define $p_\pm(\mathbf{x}, t)$ contain terms proportional to $Uv_j\delta_{i1}$ *linear* in the perturbation velocity and describe the *refraction* of sound in the vicinity of the shear layer rather than its generation. In order to minimize the contribution to the sources from such terms, it is necessary to choose the smallest possible value for the offset h defining the position of the control surface. When $M \ll 1$, the dominant wavelengths of the sound will be much larger than the scale of the shear layer motions, and we can position S within a linearly disturbed region of the flow where $\kappa_o h \ll 1$. In that case, we can set $h = 0$ in (2.2.17).

The function $\mathcal{R}$ is equivalent to a reflection coefficient and represents the effect of refraction and reflection of the shear layer generated sound by the moving fluid. When $\mathcal{R}$ is real, part of the "incident" sound p_- is transmitted into the moving medium and part is reflected; for complex values of $\mathcal{R}$ it is clear $|\mathcal{R}| = 1$, and the incident sound is totally reflected by the moving stream. It may be verified [71] that $\mathcal{R}$ is the same as the reflection coefficient for sound impinging on a plane *vortex sheet* separating stagnant fluid from a uniform flow at speed U. However, whereas the linear theory of sound interacting with a vortex sheet involves the excitation of unstable *Kelvin–Helmholtz* waves at the interface [4], which grow exponentially with time, no such instabilities arise in the present case because the motion of the real shear layer is always bounded.

2.3 Vorticity and Entropy Fluctuations as Sources of Sound

2.3.1 The Rôle of Vorticity in Lighthill's Theory

At low Mach numbers, the velocity defining the Reynolds stress quadrupole in Lighthill's equation can be determined by regarding the source flow as incompressible and by using the Biot–Savart law (1.14.3) to express $\mathbf{v}$ in terms of the vorticity. This suggests the existence of a fundamental relation between the sound waves, which diverge from the source region and transport energy to the whole of space, and the vorticity, which is confined by Kelvin's theorem (Section 1.2.7) to a finite domain where most of its kinetic energy is ultimately dissipated by the action of viscosity.

Consider an acoustically compact, homentropic vorticity distribution $\boldsymbol{\omega}$ of scale ℓ centered on the coordinate origin in a medium of density ρ_o and sound speed c_o. The velocity $\mathbf{v} = \mathbf{u} + \nabla\varphi$, where $\mathbf{u}$ is defined in terms of $\boldsymbol{\omega}$ by the Biot–Savart formula (1.14.3) so that $u \sim O(1/|\mathbf{x}|^3)$ as $|\mathbf{x}| \to \infty$. Because $\operatorname{div} \mathbf{u} = 0$, the value of φ is determined by the compressibility of the fluid. In the source region $p - p_o \sim \rho_o u^2$, and the characteristic frequency of the source flow $\sim u/\ell$. Thus, $Dp/Dt \sim \rho_o u^3/\ell$, and the continuity equation $\nabla^2\varphi + D\rho/\rho Dt \approx \nabla^2\varphi + (1/\rho_o c_o^2) Dp/Dt = 0$ implies that

$$\nabla\varphi = O(uM^2), \quad M = u/c_o \quad \text{for} \quad |\mathbf{x}| \sim \ell. \tag{2.3.1}$$

Let us now write

$$\partial^2(u_i u_j)/\partial x_i \partial x_j = \operatorname{div}(\boldsymbol{\omega} \wedge \mathbf{u}) + \nabla^2(u^2/2) \tag{2.3.2}$$

and express the solution (2.1.6) of Lighthill's equation in the form

$$\begin{aligned} p(\mathbf{x}, t) &= p_1(\mathbf{x}, t) + p_2(\mathbf{x}, t), \\ p_1(\mathbf{x}, t) &= \frac{-\rho_o x_i}{4\pi c_o |\mathbf{x}|^2} \frac{\partial}{\partial t} \int [(\boldsymbol{\omega} \wedge \mathbf{u})_i] \, d^3\mathbf{y}, \\ p_2(\mathbf{x}, t) &= \frac{\rho_o}{4\pi c_o^2 |\mathbf{x}|} \frac{\partial^2}{\partial t^2} \int \left[\frac{1}{2} u^2\right] d^3\mathbf{y}, \quad |\mathbf{x}| \to \infty, \end{aligned} \tag{2.3.3}$$

where [ ] denotes evaluation at time $t - |\mathbf{x} - \mathbf{y}|/c_o$. When retarded time variations across the source region are neglected, the identity (2.3.2) and the divergence theorem imply that $\int [\boldsymbol{\omega} \wedge \mathbf{u}]\, d^3\mathbf{y} \equiv \mathbf{0}$ because $u \sim O(1/|\mathbf{y}|^3)$ as $|\mathbf{y}| \to \infty$. To estimate the value of the first integral in (2.3.3), it is therefore necessary to expand $(\boldsymbol{\omega} \wedge \mathbf{u})(t - |\mathbf{x} - \mathbf{y}|/c_o)$ in powers of the retarded time element $\mathbf{x} \cdot \mathbf{y}/c_o|\mathbf{x}|$. The first term in the expansion yields

$$p_1(\mathbf{x}, t) \approx \frac{-\rho_o x_i x_j}{4\pi c_o^2 |\mathbf{x}|^3} \frac{\partial^2}{\partial t^2} \int y_i (\boldsymbol{\omega} \wedge \mathbf{u})_j (t - |\mathbf{x}|/c_o)\, d^3\mathbf{y}$$
$$\sim O(\ell/|\mathbf{x}|) \rho_o u^2 M^2, \tag{2.3.4}$$

where $\partial/\partial t \sim u/\ell$.

The order of magnitude of $p_2(\mathbf{x}, t)$ is estimated by first writing the momentum equation in the form

$$\partial \mathbf{u}/\partial t + \nabla\left(\int dp/\rho + v^2/2 + \partial\varphi/\partial t\right) = -\boldsymbol{\omega} \wedge \mathbf{u} - \boldsymbol{\omega} \wedge \nabla\varphi.$$

Take the scalar product with $\mathbf{u}$

$$\partial\left(\frac{1}{2}u^2\right)\Big/\partial t + \operatorname{div}\left(\mathbf{u}\left(\int dp/\rho + \frac{1}{2}v^2 + \partial\varphi/\partial t\right)\right) = -\mathbf{u} \cdot \boldsymbol{\omega} \wedge \nabla\varphi$$

and integrate over the whole of space. The contribution from the divergence vanishes because $\mathbf{u}(\int dp/\rho + \frac{1}{2}v^2 + \partial\varphi/\partial t)$ tends to zero at least as fast as $1/|\mathbf{y}|^3$ as $|\mathbf{y}| \to \infty$. Hence, using (2.3.1)

$$\frac{\partial}{\partial t} \int \frac{1}{2} u^2(\mathbf{y}, t)\, d^3\mathbf{y} = -\int (\mathbf{u} \cdot \boldsymbol{\omega} \wedge \nabla\varphi)(\mathbf{y}, t)\, d^3\mathbf{y} \sim \ell^2 u^3 M^2, \tag{2.3.5}$$

where the final estimate is really a crude upper bound that takes no account of the details of the interactions between the vorticity and irrotational velocity.

Thus,

$$p_2(\mathbf{x}, t) \sim O((\ell/|\mathbf{x}|)\rho_o u^2 M^4),$$

and by comparison with (2.3.4), we see that $p_2 \sim O(M^2) p_1 \ll p_1$ when $M \ll 1$. The component $\operatorname{div}(\rho_o \boldsymbol{\omega} \wedge \mathbf{v})$ of $\partial^2(\rho_o v_i v_j)/\partial x_i \partial x_j$ is therefore the principal source of sound at low Mach numbers [77]. This is consistent with Lighthill's hypothesis that the back-reaction of the sound on the flow is negligible because (2.3.5) states that the rate of change of the kinetic energy of the source flow due to compressible effects is at most of order $\rho_o u^3 M^2/\ell$ per unit volume, which is much smaller than the power $\rho_o u^3/\ell$ needed to maintain the flow.

2.3.2 Acoustic Analogy in Terms of the Total Enthalpy

This analysis and the discussion of Section 2.1.3 show that the dominant aeroacoustic sources at low Mach numbers are vorticity and entropy fluctuations. For homentropic, irrotational flow, sound propagation is governed by a nonlinear equation in any of the forms (1.6.22)–(1.6.24), where the acoustic variable is $\partial\varphi/\partial t$. When Lighthill's equation is cast in a form that emphasizes vorticity and entropy as the sound sources, it will therefore be necessary to select an independent acoustic variable that reduces to $\partial\varphi/\partial t$ in irrotational regions. Bernoulli's equation (1.2.21) suggests that the total enthalpy $B = w + \frac{1}{2}v^2$ is the appropriate choice. B is constant in steady irrotational flow, and far from the acoustic sources perturbations in $B = -\partial\varphi/\partial t$ represent acoustic waves and propagate according to equation (1.6.25) (with $\hat{\mathcal{F}} = 0$), where the coefficients in the wave operator assume values determined by the local mean flow.

To reformulate Lighthill's equation in terms of B for a *homogeneous* fluid, whose chemical composition is the same everywhere, we start from Crocco's equation (1.2.18) (with no body forces)

$$\partial \mathbf{v}\, \partial t + \nabla B = -\boldsymbol{\omega} \wedge \mathbf{v} + T\nabla_s + \boldsymbol{\sigma} \tag{2.3.6}$$

$$\sigma_i = (1/\rho)\partial\sigma_{ij}/\partial x_j \tag{2.3.7}$$

and rearrange the continuity equation (1.2.2) to read

$$\frac{1}{\rho c^2}\frac{Dp}{Dt} + \operatorname{div}\mathbf{v} = \frac{\beta T}{c_p}\frac{Ds}{Dt}, \tag{2.3.8}$$

where $D\rho/Dt$ has been eliminated by means of the relations

$$d\rho = \frac{dp}{c^2} + \left(\frac{\partial\rho}{\partial s}\right)_p ds \quad \text{and} \quad \left(\frac{\partial\rho}{\partial s}\right)_p = \left(\frac{\partial\rho}{\partial T}\right)_p\left(\frac{\partial T}{\partial s}\right)_p \equiv \frac{-\beta\rho T}{c_p},$$

β and c_p being, respectively, the coefficient of expansion and the specific heat at constant pressure (Section 1.2.3). For an ideal gas, $\beta = 1/T$.

Subtract the divergence of (2.3.6) from the time derivative of (2.3.8):

$$\frac{\partial}{\partial t}\left(\frac{1}{\rho c^2}\frac{Dp}{Dt}\right) - \nabla^2 B = \operatorname{div}(\boldsymbol{\omega} \wedge \mathbf{v} - T\nabla s - \boldsymbol{\sigma}) + \frac{\partial}{\partial t}\left(\frac{\beta T}{c_p}\frac{Ds}{Dt}\right). \tag{2.3.9}$$

This equation is already in a form that strongly suggests that the terms on the right-hand side constitute the most important sources of flow-generated sound because they are unchanged when the compressibility $K_s = 1/\rho c^2 \to 0$ and accordingly dominate the hydrodynamic far field of B, which must ultimately match the outgoing sound waves.

The first term on the left of (2.3.9) is expanded as follows:

$$\frac{\partial}{\partial t}\left(\frac{1}{\rho c^2}\frac{Dp}{Dt}\right) = \frac{D}{Dt}\left(\frac{1}{\rho c^2}\frac{\partial p}{\partial t}\right) + \frac{1}{\rho c^2}\frac{\partial v_j}{\partial t}\frac{\partial p}{\partial x_j}$$
$$+ v_j\left\{\frac{\partial}{\partial t}\left(\frac{1}{\rho c^2}\right)\frac{\partial p}{\partial x_j} - \frac{\partial}{\partial x_j}\left(\frac{1}{\rho c^2}\right)\frac{\partial p}{\partial t}\right\}. \quad (2.3.10)$$

The last term on the right will be discarded. It vanishes when the adiabatic compressibility $K_s = 1/\rho c^2$ is a function of the pressure alone, that is, when the motion is homentropic, or in the particular case of an ideal gas, where $K_s = 1/\gamma p$ and $\gamma = c_p/c_v =$ constant. Variations of γ in a real gas are significant only in the presence of large temperature gradients; problems of this kind are best treated separately. In liquids the compressibility may be regarded as constant; acoustic sources involving variations of K_s are important only when gas bubbles are present. The other terms on the right of (2.3.10) are transformed as follows. First,

$$\frac{1}{\rho}\frac{\partial p}{\partial t} = \frac{1}{\rho}\frac{Dp}{Dt} - \frac{1}{\rho}\mathbf{v}\cdot\nabla p = \frac{DB}{Dt} - T\frac{Ds}{Dt} - \mathbf{v}\cdot\boldsymbol{\sigma}, \quad (2.3.11)$$

where the identity (1.2.12) and the momentum equation have been used, and second, from (2.3.6),

$$\frac{\partial v_j}{\partial t}\frac{\partial p}{\partial x_j} = -\nabla p\cdot(\nabla B + \boldsymbol{\omega}\wedge\mathbf{v} - T\nabla s - \boldsymbol{\sigma}). \quad (2.3.12)$$

Thus, substituting from (2.3.10) into (2.3.9), we arrive at the acoustic analogy equation for the total enthalpy

$$\left(\frac{D}{Dt}\left(\frac{1}{c^2}\frac{D}{Dt}\right) - \frac{\nabla p\cdot\nabla}{\rho c^2} - \nabla^2\right)B$$
$$= \left(\frac{\partial}{\partial \mathbf{x}} + \frac{\nabla p}{\rho c^2}\right)\cdot(\boldsymbol{\omega}\wedge\mathbf{v} - T\nabla s - \boldsymbol{\sigma}) \frac{\partial}{\partial t}\left(\frac{\beta T}{c_p}\frac{Ds}{Dt}\right)$$
$$+ \frac{D}{Dt}\left\{\frac{1}{c^2}\left(T\frac{Ds}{Dt} + \mathbf{v}\cdot\boldsymbol{\sigma}\right)\right\}. \quad (2.3.13)$$

2.3.3 Vorticity and Entropy Sources

The nonlinear operator on the left of this equation is identical with that in (1.6.22) governing the propagation of sound in irrotational, homentropic flow.

All terms on the right-hand side vanish in irrotational regions, and in the absence of such terms and of boundary motions, B = constant. The radiation condition ensures that the terms on the right may be identified as acoustic sources. They are confined to the region in which $\boldsymbol{\omega} \neq \mathbf{0}$ and in which $\nabla s \neq \mathbf{0}$. The wave operator on the left describes propagation of the sound through the nonuniform flow although, as for Lighthill's equation, it will not usually be permissible to neglect the interaction of the aerodynamically generated sound with the vorticity and entropy gradients when the source flow is very extensive. The following special cases should be noted.

High Reynolds Number, Homentropic Flow

When dissipation is neglected and s = constant, (2.3.13) becomes

$$\left(\frac{D}{Dt}\left(\frac{1}{c^2}\frac{D}{Dt}\right) - \frac{1}{\rho}\nabla\cdot(\rho\nabla)\right)B = \frac{1}{\rho}\operatorname{div}(\rho\boldsymbol{\omega}\wedge\mathbf{v}). \tag{2.3.14}$$

At low Mach numbers when the flow is at rest at infinity (where $\rho = \rho_o$ and $c = c_o$) further simplification is possible (a) by neglecting nonlinear effects of propagation and the scattering of sound by the vorticity and (b) by taking $c = c_o$, and $\rho = \rho_o$. Then

$$\left(\frac{1}{c_o^2}\frac{\partial^2}{\partial t^2} - \nabla^2\right)B = \operatorname{div}(\boldsymbol{\omega}\wedge\mathbf{v}), \tag{2.3.15}$$

and in the far field the acoustic pressure is given by the linearized approximation

$$p(\mathbf{x}, t) \approx \rho_o B(\mathbf{x}, t). \tag{2.3.16}$$

Nonhomentropic Source Flow

When the source region is not homentropic, it is important to distinguish between terms on both sides of (2.3.13) that account principally for *scattering* of the sound and those (on the right) that can unequivocally be recognized as sources. For compact sources we can argue that the correct source terms are those remaining when the fluid is temporarily taken to be *incompressible*. These are the sources on the right of (2.3.9). When dissipation is ignored, (2.3.13) then reduces to

$$\left\{\left(\frac{\partial}{\partial t} + \mathbf{U}\cdot\nabla\right)\left[\frac{1}{c^2}\left(\frac{\partial}{\partial t} + \mathbf{U}\cdot\nabla\right)\right] - \frac{1}{\rho}\nabla\cdot(\rho\nabla)\right\}B$$

$$= \operatorname{div}(\boldsymbol{\omega}\wedge\mathbf{v} - T\nabla s) + \frac{\partial}{\partial t}\left(\frac{\beta T}{c_p}\frac{Ds}{Dt}\right). \tag{2.3.17}$$

This approximation is useful when the background mean flow may be regarded as irrotational (e.g., steady flow past a streamlined body) at mean velocity $\mathbf{U}(\mathbf{x})$ and density and sound speed $\rho(\mathbf{x})$ and $c(\mathbf{x})$. At very small Mach numbers it simplifies further to

$$\left(\frac{1}{c_o^2}\frac{\partial^2}{\partial t^2} - \nabla^2\right)B = \operatorname{div}(\boldsymbol{\omega}\wedge\mathbf{v} - T\nabla s) + \frac{\partial}{\partial t}\left(\frac{\beta T}{c_p}\frac{Ds}{Dt}\right). \qquad (2.3.18)$$

The final entropy source on the right of (2.3.17) and (2.3.18) represents the production of sound by unsteady *heating* of the fluid. According to the equation of continuity (2.3.8), this source is equivalent to a volume monopole of strength,

$$q(\mathbf{x}, t) = \frac{\beta T}{c_p}\frac{Ds}{Dt}. \qquad (2.3.19)$$

Example 1. The sound produced by a compact distribution of vorticity of scale ℓ in a stationary, uniform fluid is given by the solution of (2.3.15). The source term may be approximated by its value for incompressible flow, for which $\int(\boldsymbol{\omega}\wedge\mathbf{v})(\mathbf{x}, t)\,d^3\mathbf{x} = \mathbf{0}$. The acoustic pressure is given by (2.3.4).

Example 2: Spinning vortex pair [78]. Two parallel line vortices each of circulation Γ and distance 2ℓ apart rotate about the x_3-axis midway between them at angular velocity $\Omega = \Gamma/4\pi\ell^2$ in fluid of mean density ρ_o (Figure 2.3.1). At time t their positions in the x_1x_2-plane are $\mathbf{x} = \pm\mathbf{s}(t) \equiv \pm\ell(\cos\Omega t, \sin\Omega t)$. Show that

$$\operatorname{div}(\boldsymbol{\omega}\wedge\mathbf{v}) = \frac{\Gamma^2}{2\pi\ell^2}\frac{\partial^2}{\partial x_i \partial x_j}(s_i(t)s_j(t)\delta(\mathbf{x})) + \cdots, \qquad \mathbf{x} = (x_1, x_2)$$

where the terms omitted are higher order multipoles. If $M = \Gamma/c_o\ell \ll 1$, then use the method of Section 2.1, Example 1, to obtain the acoustic pressure in the form

$$p \approx -4\sqrt{\frac{\pi\ell}{r}}\rho_o U^2 M^{3/2}\cos\left(2\theta - 2\Omega(t - r/c_o) + \frac{\pi}{4}\right), \qquad \Omega r/c_o \to \infty,$$

where $U = \Gamma/4\pi\ell$, $M = U/c_o$, and (r, θ) are polar coordinates of (x_1, x_2). The power radiated per unit length of vortex $\sim\ell\rho_o U^3 M^4$, which agrees with the estimate given in Section 2.1 for two-dimensional turbulence.

Example 3: Kirchhoff's spinning vortex. Repeat the analysis of Example 2 for a spinning, columnar vortex of elliptic cross-section defined by the polar

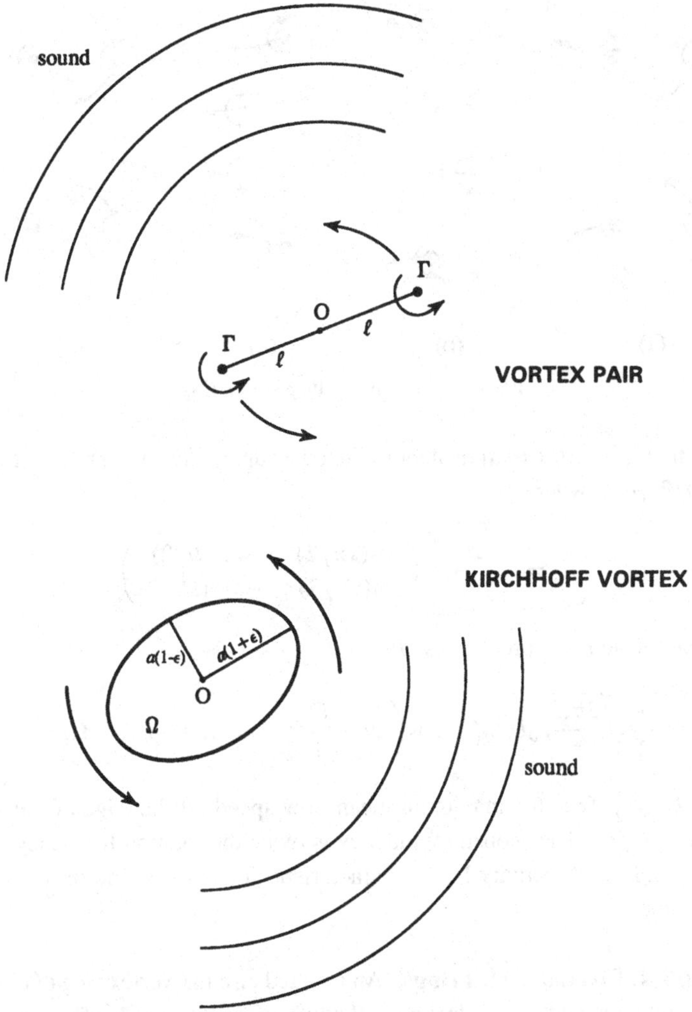

Figure 2.3.1. Spinning vortices and Kirchhoff's vortex.

equation $r = a(1 + \epsilon \cos(2\theta - \Omega t/2))$, where $\epsilon \ll 1$, and Ω is the uniform vorticity in the core (Figure 2.3.1). The ellipse rotates at angular velocity $\frac{1}{4}\Omega$ (Section 2.1, Example 3), and the velocity distribution within the core is given by [17],

$$\mathbf{v} = (v_1, v_2) = -\frac{1}{2}\Omega r(\sin\theta + \epsilon \sin(\theta - \Omega t/2), -\cos\theta + \epsilon \cos(\theta - \Omega t/2)).$$

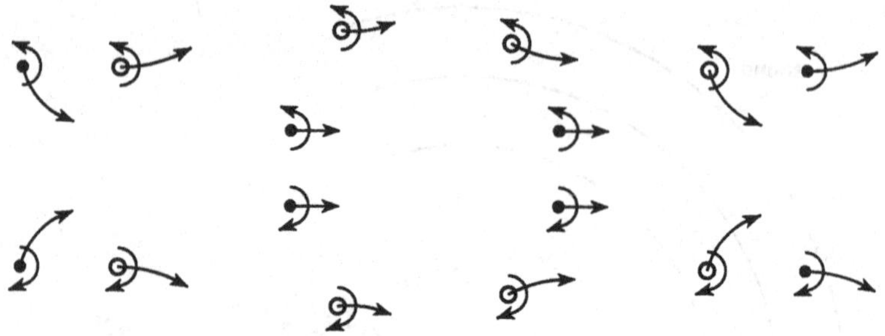

Figure 2.3.2. Interacting vortex rings.

Show that the vortex is equivalent to the quadrupole div$(\boldsymbol{\omega} \wedge \mathbf{v}) \approx \partial^2(\mathrm{t}_{ij}\delta(x_1)\delta(x_2))/\partial x_i \partial x_j$, where

$$\mathrm{t}_{ij} = \frac{\epsilon\pi\Omega^2 a^4}{8}\begin{pmatrix}\cos(\Omega t/2) & \sin(\Omega t/2)\\ \sin(\Omega t/2) & -\cos(\Omega t/2)\end{pmatrix}$$

and that the acoustic pressure is [79],

$$p \approx -\frac{\epsilon}{8}\sqrt{\frac{2\pi a}{r}}\rho_o U^2 M^{3/2}\cos\left(2\theta - \frac{\Omega}{2}(t - r/c_o) + \frac{\pi}{4}\right), \quad \Omega r/c_o \to \infty,$$

where $U = \frac{1}{2}a\Omega$ is the maximum mean flow speed (at the edge of the core), and $M = U/c_o$. The acoustic frequency is twice the rotation frequency of the ellipse, and the directivity has a characteristic four-lobe shape of a "lateral" quadrupole.

Example 4: Coaxial vortex rings. An isolated circular vortex ring of circulation Γ, radius R, and core radius $a \ll R$ advances along its axis of symmetry at a self-induced velocity $V_\Gamma \sim (\Gamma/4\pi R)(\ln[8R/a] - \frac{1}{4})$ [17]. When two or more coaxial rings are present, the motion of a ring is modified by the velocity field of its neighbors. Figure 2.3.2 illustrates the interaction of two rings of equal circulations when viscous dissipation is ignored. In (a) the radii and translational velocities of the vortices are the same; the motion of each vortex under the action of the other vortex causes the radius of the right-most vortex to increase and that of the following vortex to decrease (b); the self-induced velocity of the larger vortex decreases while that of the second increases, and the smaller vortex can pass through the larger (c); the smaller vortex now starts to expand and the larger one to contract, and the process is ready to repeat itself at stage (d).

Assume the vortex system is acoustically compact and that the motion occurs in a stationary, uniform fluid. Use equation (2.3.15) to calculate the sound produced by the interacting rings. In the source term take the vorticity of the nth vortex to be $\omega_n = \Gamma_n \delta(x - X_n(t))\delta(r - R_n(t))\mathbf{i}_\theta$, where (r, θ, x) are cylindrical polar coordinates with x measured along the common axis of symmetry, $R_n(t)$ is the vortex radius, $X_n(t)$ its location in the x-direction, and $\mathbf{i}_\theta$ is a unit vector in the azimuthal direction. Show that

$$p \approx \frac{\rho_o}{4c_o^2|\mathbf{x}|} \mathrm{P}_2(\cos\Theta) \frac{\partial^2}{\partial t^2}\left[\sum_n \Gamma_n X_n(\tau) \frac{dR_n^2(\tau)}{d\tau}\right], \quad |\mathbf{x}| \to \infty, \qquad (2.3.20)$$

where Θ is the angle between the observer direction and the positive x-axis, and the term in square brackets is evaluated at $\tau = t - |\mathbf{x}|/c_o$.

The conservation of energy and momentum for the ideal, incompressible motion of the vortices implies that [17],

$$\sum_n \Gamma_n R_n (R_n\, dX_n/dt - X_n\, dR_n/dt) = \text{constant}, \quad \sum_n \Gamma_n R_n^2 = \text{constant}.$$

If $S(t) = \sum_n \Gamma_n R_n^2 X_n \equiv \text{constant} \times \langle X \rangle$, where $\langle X \rangle = \sum_n \Gamma_n R_n^2 X_n / \sum_n \Gamma_n R_n^2$ is the centroid of the vortex system, then

$$p \approx \frac{\rho_o}{6c_o^2|\mathbf{x}|} \mathrm{P}_2(\cos\Theta)\left[\frac{d^3 S}{dt^3}\right], \quad |\mathbf{x}| \to \infty.$$

Example 5: Entropy noise. When the source region has nonuniform temperature but the motion is isentropic ($Ds/Dt = 0$), the enthalpy of a fluid particle is $w = \int_{p_o}^{p} dp'/\rho(p', s(\eta))$, where η is the Lagrangian coordinate of the fluid particle. Then

$$-\mathrm{div}(T\nabla s) = -\nabla^2\left\{\int_{p_o}^{p}\left(\frac{1}{\rho(p', s(\eta))} - \frac{1}{\rho_o}\right)dp'\right\}$$
$$- \mathrm{div}\left\{\left(\frac{1}{\rho_o} - \frac{1}{\rho}\right)\nabla p\right\}. \qquad (2.3.21)$$

The second term is a dipole corresponding to the dipole in Lighthill's equation (2.1.9) (second term on the right). The first term is a quadrupole associated with the scattering of the hydrodynamic pressure fluctuations by the density variations. It should be discarded unless an account is also taken of scattering by sound speed variations. This can be done by adding

$$\frac{\partial}{\partial t}\left(\frac{1}{\rho_o c_o^2}\frac{\partial p}{\partial t}\right) \approx \frac{1}{c_o^2}\frac{\partial^2 B}{\partial t^2}$$

to both sides of the acoustic analogy equation, which is then rearranged so that the left-hand side becomes $(\partial^2/c_o^2\partial t^2 - \nabla^2)B$. The final, monopole term on the right of (2.1.9) is now recovered when the radiation from the additional source is combined with that from the first source on the right of (2.3.21).

Example 6: Fluctuating source of heat. Calculate the sound generated in an ideal gas by a small, rigid sphere of radius R and temperature $T_o + \Delta T e^{-i\omega t}$, $\Delta T/T_o \ll 1$, where T_o is the mean fluid temperature.

The temperature fluctuations produce periodic heating and cooling of the gas, accompanied by expansions and contractions in a thin thermal boundary layer of thickness $\sqrt{\chi/\omega}$ on the surface of the sphere ($\chi = \kappa/\rho_o c_p$ is the thermometric conductivity, and κ is the coefficient of thermal conductivity, Section 1.2.8). The sphere behaves as a volume source whose strength is determined by the entropy sources in the acoustic analogy equation. Write $T\,ds \approx T_o\,ds'$, where s' is the fluctuating specific entropy. The acoustic analogy equation (2.3.13) becomes

$$\left(\frac{D}{Dt}\left(\frac{1}{c^2}\frac{D}{Dt}\right) - \frac{\nabla p.\nabla}{\rho c^2} - \nabla^2\right)(B - T_o s') = \frac{\partial}{\partial t}\left(\frac{\beta T}{c_p}\frac{Ds}{Dt}\right), \qquad (2.3.22)$$

where $s' \equiv 0$ in the far field, and $\beta \approx 1/T_o$ for an ideal gas. The linearized Crocco's equation (1.2.18) implies that $\partial(B - T_o s')/\partial x_n = 0$ on the rigid sphere, and in the absence of mean flow, $D/Dt \approx \partial/\partial t$ so that the left-hand side of (2.3.23) becomes $(\partial^2/c_o^2\partial t^2 - \nabla^2)(B - T_o s')$.

When the acoustic wavelength greatly exceeds the boundary layer thickness, the motion in the neighborhood of the sphere may be assumed to be incompressible, any variations in density being attributed to volume changes caused by the periodic heating and cooling essentially at constant pressure [2]. The temperature perturbation T' is calculated from the linearized energy equation (1.2.34), where $\rho T\,Ds/Dt \approx \rho_o T_o(\partial s/\partial T)_p DT/Dt \equiv \rho_o c_p \partial T'/\partial t$ so that (taking the origin at the center of the sphere)

$$\nabla^2 T' + i(\omega/\chi)T' = 0, \quad |\mathbf{x}| > R,$$

where $T' = \Delta T$ for $|\mathbf{x}| = R$, and $T' \to 0$ as $|\mathbf{x}| \to \infty$. Hence,

$$s' = c_p(\Delta T/T_o)(R/|\mathbf{x}|)\exp((1 - i)\sqrt{\omega/2\chi}(R - |\mathbf{x}|)), \quad |\mathbf{x}| > R.$$

The solution of the acoustic analogy equation can be expressed in the form (1.10.3), in which p is replaced by $B - T_o s'$, and $\mathcal{F}(\mathbf{y}, \omega) = -(\omega^2/c_p)s'(\mathbf{y}, \omega)$. For an acoustically compact sphere, we take $G(\mathbf{x}, \mathbf{y}; \omega)$ to be the compact Green's function (1.11.8), with $Y_i = y_i(1 + R^3/2|\mathbf{y}|^3)$. As $\kappa_o|\mathbf{x}| \to \infty$,

$$\frac{p}{\rho_o} \approx -\left(\frac{\Delta T}{T_o}\right)\left(\frac{R}{|\mathbf{x}|}\right)\frac{\omega R\sqrt{2\omega\chi}}{(1-i)}\left(1 + \frac{1}{(1-i)R\sqrt{\omega/2\chi}}\right)\mathrm{e}^{i\kappa_o|\mathbf{x}|}.$$

In the time domain, this can also be written

$$\frac{p}{\rho_o} \approx \frac{1}{4\pi c_p T_o|\mathbf{x}|}\frac{\partial}{\partial t}\oint_S \left[\kappa\frac{\partial T}{\partial y_n}\right]\partial S(\mathbf{y}),$$

where [ ] denotes evaluation at the retarded time $t - |\mathbf{x}|/c_o$, and the integration is over the surface of the sphere. This expresses the monopole radiation in terms of the heat flux through the surface of the sphere.

Example 7. Determine the sound produced at low Mach numbers by monopole sources associated with localized heating due to the dissipation of mechanical energy and during the mixing of gases of different temperatures [80].

We have to estimate the source term on the right of (2.3.22) when entropy generation is by dissipation, governed by equation (1.2.34). Viscous dissipation in a turbulent flow is dominated by small-scale fluctuations of characteristic velocity $v_\lambda \sim u(\lambda/\ell)^{1/3}$, where $\mathbf{u}$ is a velocity of the energy containing eddies of length scale ℓ, $\lambda = \ell/R_\ell^{3/4}$ is the Kolmogorov dissipation length [62, 63], and $R_\ell = u\ell/\nu \gg 1$ is the Reynolds number. The motion is "self-similar" at these very small scales, and the power dissipated depends only on the rate at which energy is supplied by the large-scale velocity fluctuations, which is of order $\rho_o u^3/\ell$ per unit volume. For an ideal gas ($\beta = 1/T$), we then have

$$\frac{1}{c_p}\frac{Ds}{Dt} = \frac{(\gamma-1)}{c_o^2}\frac{u^3}{\ell},$$

where γ is the ratio of the specific heats. The corresponding acoustic pressure is

$$p \sim \frac{\rho_o}{4\pi c_o^2|\mathbf{x}|}\frac{\partial}{\partial t}\int\left[(\gamma-1)\frac{u^3}{\ell}\right]d^3\mathbf{y}, \quad |\mathbf{x}| \to \infty.$$

The remaining time derivative is the rate of change following the motion of a dissipation eddy and may be approximated in order of magnitude by $\partial/\partial t \sim (u/\ell)R_\ell^{1/2}$ [63, 64]. These eddies are statistically independent so that there are V/λ^3 independently radiating monopoles in a region of volume V, and the total sound power is

$$\Pi_\nu \sim \rho_o \frac{V}{\lambda^3 c_o^4}\left(\frac{u R_\ell^{1/2}}{\ell}\right)^2 \frac{(\gamma-1)u^6\lambda^6}{\ell^2} = \frac{V}{\ell^3}(\gamma-1)^2 \rho_o \frac{u^3}{\ell} M^5 R_\ell^{-5/4}.$$

The efficiency of sound generation by viscous dissipation is therefore $O(1/R_\ell^{5/4}) \ll 1$ relative to that of the turbulence quadrupoles. A similar argument indicates that the sound produced by the entropy monopoles (the final term on the right of (2.1.9)) generated during the mixing of gases of different temperatures is also negligible.

Example 8: Laser-excited broadside array. A time-modulated laser beam supplies heat to a stationary acoustic medium in the region $x_1 > 0$ at a rate per unit volume given by [81]

$$Q = \mathrm{Re}\left(\frac{\Pi}{\ell}\delta(x_2)\delta(x_3)(1 - \mathrm{e}^{-i\omega t})\mathrm{e}^{-x_1/\ell}\right), \quad x_1 > 0.$$

ℓ is the attenuation length due to thermal losses to the fluid, and Π is the power output of the laser.

The source term is

$$\frac{\partial}{\partial t}\left(\frac{\beta T}{c_p}\frac{Ds}{Dt}\right) = \frac{\beta}{\rho_o c_p}\frac{\partial Q}{\partial t}.$$

Show that

$$p \approx \mathrm{Re}\left(\frac{-i\omega\beta\Pi\mathrm{e}^{-i\omega(t-|\mathbf{x}|/c_o}}{4\pi c_p|\mathbf{x}|(1+i\kappa_o\ell\cos\theta)}\right), \quad \kappa_o|\mathbf{x}| \to \infty,$$

where $\theta = \arccos(x_1/|\mathbf{x}|)$.

Example 9: Radiation damping of turbulence. Let $\mathbf{v} = \mathbf{u} + \nabla\varphi$ for a low Mach number, compact region of turbulence of uniform mean density ρ_o, where $\mathbf{u}$ is the rotational velocity determined by the Biot–Savart law (Section 1.14.1). Define $S_{ij} = u_i u_j - 1/2 u^2 \delta_{ij}$.

The "compressible" component $\nabla\varphi$ is given by

$$\frac{-\partial\varphi}{\partial t} = \frac{\partial^2}{\partial x_i \partial x_j}\int (S_{ij}(\mathbf{y}, t - |\mathbf{x}-\mathbf{y}|/c_o) - S_{ij}(\mathbf{y}, t))\frac{d^3\mathbf{y}}{4\pi|\mathbf{x}-\mathbf{y}|}. \tag{2.3.23}$$

By expanding in powers of the retarded time $|\mathbf{x}-\mathbf{y}|/c_o$, use (2.3.5) to deduce that the *mean* rate of change of the rotational kinetic energy $E = \int \frac{1}{2}\rho_o u^2\, d^3\mathbf{x}$ is given by

$$\frac{\partial E}{\partial t} \approx \frac{-\rho_o}{60\pi c_o^5} \int\int (\delta_{ij}\delta_{kl} + \delta_{ik}\delta_{jl} + \delta_{il}\delta_{jk})$$
$$\times \frac{\partial^2 S_{ij}}{\partial t^2}(\mathbf{x}, t) \frac{\partial^2 S_{kl}}{\partial t^2}(\mathbf{y}, t)\, d^3\mathbf{x}\, d^3\mathbf{y} \sim \ell^2 \rho_o u^3 M^5.$$

This is smaller than the crude estimate (2.3.5) by a factor of $M^3 \ll 1$. Show that it is the same as the acoustic power radiated by the quadrupole Reynolds stress pressure field

$$p \approx \frac{\rho_o x_i x_j}{4\pi c_o^2 |\mathbf{x}|^3} \frac{\partial^2}{\partial t^2} \int S_{ij}(\mathbf{y}, t - |\mathbf{x}|/c_o)\, d^3\mathbf{y}, \quad |\mathbf{x}| \to \infty.$$

2.4 Two-Phase Flow

The efficiency with which sound is generated by turbulence in water is so small that the presence of any kind of inhomogeneity usually causes a large increase in the radiation. Bubbles formed by gaseous entrainment or by cavitation in regions of high acceleration are particularly important; they behave as monopole sources that are easily excited by turbulence pressures and changes in the mean pressure caused by convection through a variable geometry duct, say. The musical sound of a shallow stream is attributed to entrained air bubbles pulsating at resonance [82], and the excitation and formation of bubbles by breaking surface waves is an important source of background noise in the ocean.

2.4.1 Bubble Excited by an Applied Pressure

Consider a small spherical gas bubble in liquid of density ρ_o and sound speed c_o. Let the undisturbed bubble radius be R, and let the density and speed of sound of the gas be ρ_b and c_b, respectively. We first determine the sound radiated from the bubble when a uniform, time harmonic pressure $p_\mathrm{I} \mathrm{e}^{-i\omega t}$ is applied in its vicinity.

The volume pulsations of the bubble produce monopole radiation determined by the final term on the right of (2.1.9), wherein $\rho \to \rho_b$, $c \to c_b$. However, because of the large difference between the compressibilities of the gas and liquid, the bubble internal pressure $p_b = p - p_o$ can differ significantly from p_I. To calculate p_b it will be assumed that the pulsations are adequately described by linear perturbation theory. Take the origin of coordinates at the center of the bubble and suppose the monopole pressure produced by the pulsations is

$$p'(\mathbf{x}, \omega) = \frac{-i\omega\rho_o v_b R^2}{|\mathbf{x}|(1 - i\kappa_o R)} \mathrm{e}^{i\kappa_o(|\mathbf{x}|-R)}, \quad |\mathbf{x}| > R. \tag{2.4.1}$$

The effective volume velocity of the monopole is $4\pi R^2 v_b(\omega)$, where v_b is the normal velocity at the surface of the bubble *less* the normal velocity that would exist if the compressibility of the bubble were the same as that of the liquid. Spherically symmetric pressure fluctuations within the bubble are proportional to the zeroth order ($m = n = 0$) term in the expansion (1.10.11), with

$$\mathrm{z}_0 = \mathrm{j}_0(\omega|\mathbf{x}|/c_b) \equiv \frac{\sin(\omega|\mathbf{x}|/c_b)}{\omega|\mathbf{x}|/c_b}$$

so that the pressure is uniform in the bubble provided $(\omega R/c_b)^2 \ll 1$. This condition is assumed to apply. The normal velocities can be expressed in terms of this pressure by integrating the linearized continuity equation over $|\mathbf{x}| < R$, first when the bubble is present, and second when the region is imagined to be filled with liquid. If the motion is adiabatic, we have to evaluate

$$\int_{|\mathbf{x}|<R} \left(\frac{1}{\rho c^2} \frac{\partial p}{\partial t} + \operatorname{div} \mathbf{v} \right) d^3\mathbf{x},$$

first for $p = p_b,\ c = c_b,\ \rho = \rho_b$ and then for $p = p_\mathrm{I},\ c = c_o,\ \rho = \rho_o$. The divergence theorem gives $\int \operatorname{div} \mathbf{v}\, d^3\mathbf{x} = 4\pi R^2 v_n$, where v_n is the radial velocity at $|\mathbf{x}| = R$. Hence,

$$v_b = \frac{-i\omega R}{3} \left(\frac{p_\mathrm{I}}{\rho_o c_o^2} - \frac{p_b}{\rho_b c_b^2} \right). \tag{2.4.2}$$

The condition of continuity of pressure at the surface of the bubble ($p_\mathrm{I} + p' = p_b$) supplies a second relation involving p_b and v_b. This is written by using (2.4.1)

$$p_b = p_\mathrm{I} - i\omega\rho_o v_b R/(1 - i\kappa_o R),$$

and together with (2.4.2) supplies

$$v_b = \frac{-i\omega R(1 - i\kappa_o R) p_\mathrm{I}}{3\left(1 - i\kappa_o R - \omega^2/\omega_b^2\right)} \left(\frac{1}{\rho_o c_o^2} - \frac{1}{\rho_b c_b^2} \right), \tag{2.4.3}$$

where

$$\omega_b = \sqrt{3\rho_b c_b^2/\rho_o R^2} = \sqrt{3\gamma_b p_o/\rho_o R^2} \tag{2.4.4}$$

is the *adiabatic resonance frequency* of the bubble, and γ_b is the specific heat ratio of the gas at the mean pressure p_o. Equation (2.4.1) now gives for the monopole radiation

$$p'(\mathbf{x}, \omega) = \frac{-\rho_o \omega^2 \mathrm{V}_b p_\mathrm{I}}{4\pi|\mathbf{x}|\left(1 - i\kappa_o R - \omega^2/\omega_b^2\right)} \left(\frac{1}{\rho_o c_o^2} - \frac{1}{\rho_b c_b^2} \right) \mathrm{e}^{i\kappa_o(|\mathbf{x}|-R)}, \tag{2.4.5}$$

Table 2.4.1. *Resonance frequencies and loss factors for bubbles in water*

R cm	f_b Hz	η
0.4	820	0.014
0.1	3,300	0.04
0.01	33,000	0.07

where V_b is the mean bubble volume. This is the frequency-domain representation of the final term on the right of (2.1.9), in which $p - p_o = p_I/(1 - i\kappa_o R - \omega^2/\omega_b^2)$.

The bubble radiates sound because of the difference between the compressibilities $1/\rho_b c_b^2$ and $1/\rho_o c_o^2$ of the gas and liquid. The amplitude of the radiation becomes very large when the frequency approaches ω_b and is then controlled by the damping. The influence of radiation damping is included in the above-indicated calculation and is represented by the term $i\kappa_o R$ in the denominators of (2.4.3) and (2.4.5). Viscous dissipation in the liquid near the surface of the bubble, and thermal losses involving an irreversible transfer of heat between the liquid and gas, are also important. For air bubbles in water, the damping is dominated by thermal dissipation when the radius is smaller than about 0.1 cm [83, 84]. These effects are usually accounted for by replacing $\kappa_o R$ in (2.4.5) by a *loss factor* η, defined such that the energy of a transient resonant oscillation of the bubble decays like $e^{-\omega\eta t}$.

The bubble constitutes a forced, simple harmonic oscillator whose inertia and stiffness are determined respectively by the mass density of the liquid and the compressibility of the gas. Typical values of the adiabatic resonance frequency $f_b = \omega_b/2\pi$ (Hz) and the loss factor η at resonance for air bubbles in water at a mean pressure p_o of one atmosphere are given in Table 2.4.1 [83].

It follows from (2.4.4) that $(\omega_b R/c_b)^2 = 3\rho_b/\rho_o \approx 0.04$ at one atmosphere. Thus, the condition $(\omega R/c_b)^2 \ll 1$, that must be satisfied if the pressure in the bubble is to be uniform, is valid for $\omega < 3\omega_b$.

Example 1. Pressure fluctuations in the bubble occur isothermally when $\sqrt{\omega R^2/\chi} \ll 1$, where $\chi = \kappa/\rho_b c_p$ is the thermometric conductivity, and κ is the thermal conductivity (Section 1.2.8; $\chi \approx 2 \times 10^{-5}$ m^2/s for air at 300°K and one atmosphere). The resonance frequency is then equal to $\sqrt{3p_o/\rho_o R^2}$.

Example 2. Surface tension influences bubble motions when R is very small (<0.001 cm for air bubbles in water) [2, 4]. The adiabatic resonance frequency is

$$\omega_b = \frac{1}{R\sqrt{\rho_o}}\left(3\gamma_b p_o + \frac{2(3\gamma_b - 1)T}{R}\right)^{\frac{1}{2}},$$

where T is the surface tension of the liquid–gas interface (Section 1.5.3). $T \approx 0.07$ N/m for an air–water interface at 300°K.

2.4.2 Bubble in a Turbulent Flow [65]

Bubble oscillations are driven by the dynamic pressure fluctuations in a turbulent flow. When this occurs it is usually permissible to assume that the bubble radius $R < \ell$, the scale of the turbulent eddies, because a larger bubble would tend to be distorted by shearing motions and breakup into smaller ones. The length scale of the dominant turbulence pressures experienced by the bubble are then much larger than R so that the sound radiated by the bubble can be estimated by the method of Section 2.4.1. For a single bubble in the flow, the applied pressure $p_I(\mathbf{x}, t)$ is the pressure that would be generated by the turbulence velocity fluctuations if the bubble were absent, which may be taken to be given by the *first* line of (2.1.6). Retarded times can be neglected in this formula because only scattering of local hydrodynamic pressures are important for the *production* of sound by the flow. Thus $p_I \sim \rho_o v^2$, where v is a turbulence velocity. A small bubble tends to be convected with the turbulence in any large-scale mean motion so that the appropriate characteristic frequency is $\omega \sim v/\ell$.

To use these estimates in equation (2.4.5), we first introduce the following simplifications. The ratio $\omega^2/\omega_b^2 \sim (v/c_b)^2(R/\ell)^2(\rho_o/\rho_b)$, and although $\rho_o/\rho_b \sim 10^3$ for air bubbles in water, the Mach number v/c_b is unlikely to exceed about 10^{-2} so that resonant oscillations of the bubble are not important at the dominant frequencies of the turbulent motions. Similarly, the compressibility of the gas is so much larger than that of the liquid that only the second term in the large brackets on the right of (2.4.5) need be retained. Hence the bubble radiated sound pressure is estimated by

$$p \sim \frac{\rho_o\omega^2 \mathrm{V}_b}{|\mathbf{x}|\rho_b c_b^2}\rho_o v^2 \sim \frac{\rho_o v^2 M^2}{|\mathbf{x}|}\frac{\mathrm{V}_b}{\ell^2}\frac{\rho_o c_o^2}{\rho_b c_b^2},$$

where $M \sim v/c_o$. The acoustic power is

$$\Pi_b \sim \left(\frac{\mathrm{V}_b}{\ell^3}\right)^2\left(\frac{\rho_o c_o^2}{\rho_b c_b^2}\right)^2 \Pi_a,$$

where $\Pi_a \sim (\rho_o v^3/\ell)M^5\ell^3$ is the sound power generated by a single turbulent eddy of scale ℓ when the bubble is absent. The increase in the efficiency is $10 \times \log_{10}(\Pi_b/\Pi_a)$ dB. For an air bubble in water, at a mean pressure of one atmosphere, the increase is about 24 dB when $V_b/\ell^3 = 0.001$.

Example 3: Sound waves in a bubbly medium [85–87]. When long wavelength sound propagates through a bubbly medium, the bubbles may be treated as a fine-scale substructure whose principal effect is to modify the compressibility. The modified wave equation is obtained by averaging the equations of motion over regions of volume v containing many bubbles but of dimension $v^{1/3}$ small compared to the wavelength. Let $\bar{p}$, $\bar{\mathbf{v}}$ be the averaged acoustic pressure and velocity, and let there be $\mathcal{N}$ bubbles per unit volume. In the frequency domain, the averaged continuity and momentum equations are

$$\operatorname{div}\bar{\mathbf{v}} - i\omega(1-\alpha)\bar{p}/\rho_o c_o^2 = \frac{1}{v}\sum_n 4\pi R_n^2 v_n,$$

$$-i\omega\bar{p} + \nabla\bar{p}/\rho_o(1-\alpha) = 0,$$

where $\alpha \ll 1$ is the volume fraction of the bubbles, and the summation is over the bubbles within v, where the nth bubble, of radius R_n, has normal surface velocity v_n, which can be expressed in terms of $\bar{p}$ as in (2.4.3).

Deduce that

$$\nabla^2\bar{p} + \left((1-\alpha)^2\kappa_o^2 + \frac{\rho_o(1-\alpha)\omega^2\mathcal{N}}{\rho_b c_b^2} \times \int_0^\infty \frac{P(V)V\,dV}{[1 - i\eta(V) - \omega^2/\omega_b^2(V)]}\right)\bar{p} = 0, \tag{2.4.6}$$

where $P(V)\,dV$ is the probability that the volume of a bubble lies between V and $V + dV$, and summations over n have been replaced by integrals. The substitution $\bar{p}(\mathbf{x},\omega) = \mathrm{e}^{i\mathbf{k}.\mathbf{x}}$ yields the *dispersion equation*

$$k^2 = (1-\alpha)^2\kappa_o^2 + \frac{\rho_o(1-\alpha)\omega^2\mathcal{N}}{\rho_b c_b^2}\int_0^\infty \frac{P(V)V\,dV}{[1 - i\eta(V) - \omega^2/\omega_b^2(V)]}. \tag{2.4.7}$$

At very low frequencies $\omega \ll \omega_b$, and for $\alpha \ll 1$, this gives

$$\frac{1}{c_m^2} = \frac{1}{c_o^2} + \frac{\alpha\rho_o}{\rho_b c_b^2}, \tag{2.4.8}$$

where $c_m = \omega/k$ is the phase speed of sound waves in the bubbly medium. c_m is typically very much smaller than both c_o and c_b. It is about 40 m/s when $\alpha = 0.1$, and the bubbles contain air in water at a mean pressure of one atmosphere; c_m increases as the bubble concentration decreases, but α must be smaller than 10^{-3} before c_m exceeds the sound speed in air ($\approx 0.23c_o$).

2.5 Absorption of Aerodynamic Sound

Acoustic waves are attenuated by turbulence. Part of the loss is caused by *scattering* [88–90], where energy extracted from an incident wave reappears as sound propagating in other directions. When the time scale ℓ/v of the turbulence is comparable to the frequency of the sound, a fraction of the acoustic energy is also *absorbed* by the turbulence. The absorption is generally small, however, and is of no importance in, say, estimating the noise generated by a turbulent jet [91].

2.5.1 Absorption and Scattering by Turbulence

Let us consider the propagation of sound through a homogeneous turbulent flow of uniform mean density ρ_o and infinitesimal Mach number. Although viscosity is ultimately responsible for the conversion of mechanical energy into heat, it has a negligible influence on the transfer of energy between the sound and the turbulence. We can therefore use the homentropic form of Lighthill's equation (2.1.5) to study sound propagation:

$$(\partial^2/c_o^2\partial t^2 - \nabla^2)p = \rho_o \partial^2 v_i v_j/\partial x_i \partial x_j, \tag{2.5.1}$$

where the perturbation pressure $p \equiv c_o^2(\rho - \rho_o)$.

Set

$$v_i = U_i + V_i, \tag{2.5.2}$$

where $U_i(\mathbf{x}, t)$ denotes the turbulence velocity in the *absence* of the incident sound, which may be assumed to have zero mean value, and $V_i(\mathbf{x}, t)$ is the total perturbation produced by the sound. Substitute into the right of (2.5.1) and discard the nonlinear terms $U_i U_j$, $V_i V_j$ that account for the production of aerodynamic sound by the turbulence independently of the incident wave and for nonlinear self-modulation of the sound. Hence, (2.5.1) becomes

$$(\partial^2/c_o^2\partial t^2 - \nabla^2)p = 2\rho_o \partial^2 U_i V_j/\partial x_i \partial x_j. \tag{2.5.3}$$

Let us now specialize to the case in which the wavelength of the incident sound is much longer than the turbulence correlation scale ℓ. Because the wavelength of aerodynamic sound $\sim\ell/M$ ($M \sim U/c_o$), this covers the range of acoustic frequencies corresponding to those of the dominant turbulent motions, and for which strong coupling between sound and turbulence would be expected. The separation of length scales permits a *coherent* component of the sound to be introduced by averaging over regions of the turbulence that are large relative to ℓ, but small compared to the acoustic wavelength. We therefore set

$$p = \bar{p} + p', \quad V_i = \bar{V}_i + V_i', \quad \text{and so forth},$$

where $\bar{p}$, $\bar{V}_i$, and so forth, are the coherent acoustic pressure, velocity, and so forth, and p', $V_i', \ldots$ are the corresponding fluctuating components, which have zero mean values, and are statistically independent of the incident sound. In this long-wave approximation, the sound scattered out of the incident wave direction tends to be distributed evenly over all solid angles [89], and the coherent wave can therefore be identified with the incident sound, whose amplitude and phase are progressively modified by scattering and absorption.

The coherent wave satisfies the averaged equation (2.5.3)

$$(\partial^2/c_o^2\partial t^2 - \nabla^2)\bar{p} = 2\rho_o\partial^2\overline{U_i V_j'}/\partial x_i \partial x_j, \tag{2.5.4}$$

and by subtraction from (2.5.3)

$$\begin{aligned}(\partial^2/c_o^2\partial t^2 - \nabla^2)p' &= 2\rho_o\partial^2(U_i \bar{V}_j)/\partial x_i \partial x_j \\ &\quad + 2\rho_o\partial^2(U_i V_j' - \overline{U_i V_j'})/\partial x_i \partial x_j. \end{aligned}\tag{2.5.5}$$

This equation is used in conjunction with the perturbation momentum equation

$$\partial V_i/\partial t + (V_j \partial U_i/\partial x_j + U_j \partial V_i/\partial x_j) = (-1/\rho_o)\partial p/\partial x_i \tag{2.5.6}$$

to evaluate the interaction Reynolds stress $\overline{U_i V_j'}$ on the right of (2.5.4) in terms of $\bar{p}$, yielding a "renormalized" wave equation for $\bar{p}$ alone. To do this, we use a "closure approximation" based on the observation that significant contributions to the product $\overline{U_i(\mathbf{x}, t)V_j'(\mathbf{x}, t)}$ arise from those components of $V_j'(\mathbf{x}, t)$ that are correlated with the turbulent velocity $U_i(\mathbf{x}, t)$ at $\mathbf{x}$ at time t. Such components of V_j' must have experienced an interaction with the turbulence at distances $\leq O(\ell)$ from $\mathbf{x}$ and possibly multiple interactions with the turbulence within a distance $\sim\ell/M$ from $\mathbf{x}$ (the distance traveled by a sound wave during the time scale of the turbulence $\approx$ the acoustic wavelength). Provided that the scattering is sufficiently weak over distances of the order of the acoustic wavelength, we can neglect terms describing multiple interactions in equations (2.5.5) and (2.5.6) when the solutions of these equations are to be used to evaluate $\overline{U_i V_j'}$.

In other words, in calculating p' and V_i' in terms of $\bar{p}$ and $\bar{V}_i$, it is sufficient to approximate these equations by

$$(\partial^2/c_o^2\partial t^2 - \nabla^2)p' = 2\rho_o\partial^2 U_i\bar{V}_j/\partial x_i\partial x_j, \tag{2.5.7}$$

$$\partial V_i'/\partial t = -(\bar{V}_j\partial U_i/\partial x_j + U_j\partial\bar{V}_i/\partial x_j) - (1/\rho_o)\partial p'/\partial x_i. \tag{2.5.8}$$

The solution p' of the first of these equations represents the pressure generated by a quadrupole of strength $2\rho_o U_i\bar{V}_j$, formed by the interaction of the coherent acoustic particle velocity and the turbulence velocity, and may be expressed as a retarded potential integral as in (1.8.8). When this is substituted into the final term on the right of (2.5.8), that equation determines V_i' in terms of $\bar{V}_i$ and U_i. To the same order of approximation, V_i' can be calculated from (2.5.8) in terms of $\bar{p}$ and U_i alone by using the linearized coherent momentum equation $\partial\bar{V}_i/\partial t = -\partial\bar{p}/\rho_o\partial x_i$ to replace $\bar{V}_i$ in (2.5.8) by $(-1/\rho_o)\int_{-\infty}^{t}\{\partial\bar{p}(\mathbf{x},\tau)/\partial x_i\}\,d\tau$. V_i' thus obtained is now substituted into (2.5.4) to yield an integro-differential equation for $\bar{p}$ that describes propagation of the coherent sound through the turbulence.

In making these various substitutions, we encounter mean values of the form $\overline{U_i(\mathbf{x},t)U_j(\mathbf{X},T)}$. These *correlation products* are simplified by assuming that, during the time in which the interaction with the sound occurs, the turbulence velocity fluctuations may be regarded as stationary random functions of position and time (i.e., as functions of $\mathbf{x}-\mathbf{X}$ and $t-T$ alone). This permits us to introduce the turbulence velocity correlation tensor [62, 63]

$$\mathcal{R}_{ij}(\mathbf{x},t) = \overline{U_i(\mathbf{x},t)U_j(\mathbf{X},T)}. \tag{2.5.9}$$

The equation for $\bar{p}$ then assumes the form [92]

$$(\partial^2/c_o^2\partial t^2 - \nabla^2)\bar{p} = \iint_{-\infty}^{\infty} F(\mathbf{X}-\mathbf{x}, T-t)\bar{p}(\mathbf{X},T)\,d^3\mathbf{X}\,dT,$$

where $F(\mathbf{x},t)$ is proportional to $\mathcal{R}_{ij}(\mathbf{x},t)$. The integral on the right-hand side is a *convolution product*, and an equation of this type always admits plane wave solutions $\bar{p} = \mathrm{e}^{i(\mathbf{k}\cdot\mathbf{x}-\omega t)}$, where $\mathbf{k}$ and ω are related by the dispersion relation obtained by substitution into the equation. In the present case we find

$$k^2 = \kappa_o^2 + \frac{2}{\omega}\iint_{-\infty}^{\infty}\frac{k_i k_j\Psi_{ij}(\mathbf{k}-\mathbf{K},\omega-\Omega)}{(\Omega+i0)} \times\left(\frac{2(\mathbf{K}\cdot\mathbf{k})^2}{K^2-(\Omega+i0)^2/c_o^2} - k^2\right)d^3\mathbf{K}\,d\Omega, \tag{2.5.10}$$

where Ψ_{ij} is the *wavenumber-frequency spectrum* of the turbulence velocity, which is the space–time Fourier transform of $\mathcal{R}_{ij}(\mathbf{x}, t)$ defined as in (1.7.26) (for $n = 3$), and the notation $\Omega + i0$ in (2.5.10) implies that the integration contour along the real Ω-axis passes above the poles at $\Omega = 0, \pm c_o K$. The latter condition is a consequence of the requirement that the solutions of equations (2.5.7) and (2.5.8) should be *causally* related to the interactions between the turbulence and the coherent wave (Section 1.7.4). Thus, in the integral on the right of (2.5.10), $\mathbf{K}$ and Ω may be interpreted as the wavenumber vector and frequency of a disturbance generated by an interaction of the coherent wave with the turbulence. Those disturbances that satisfy $\Omega = \pm c_o K$ correspond to scattered sound waves.

Equation (2.5.10) can be solved for k by successive approximations, in which the zeroth order approximation $k = \kappa_o$ corresponds to propagation through a turbulent-free medium. The integrated term is $O(M^2)$ smaller than κ_o^2. To determine the correction, set

$$\mathbf{k} = (\kappa_o + \delta k)\mathbf{n}, \tag{2.5.11}$$

where $\mathbf{n}$ is a unit vector in the propagation direction. To first order

$$\delta k \approx \frac{\kappa_o^3}{\omega} \iint_{-\infty}^{\infty} \frac{n_i n_j \Psi_{ij}(\kappa_o \mathbf{n} - \mathbf{K}, \omega - \Omega)}{(\Omega + i0)} \times \left(\frac{2(\mathbf{K}\cdot\mathbf{n})^2}{K^2 - (\Omega + i0)^2/c_o^2} - 1 \right) d^3\mathbf{K}\, d\Omega. \tag{2.5.12}$$

The attenuation is determined by the value of $\text{Im}(\delta k) \equiv \mu$, say. It can be shown that $n_i n_j \Psi_{ij}$ is real and nonnegative for stationary random fluctuations [63, 93], so that μ is given by the contributions to the integral from the contour indentations around the poles at $\Omega = \pm c_o K$ and 0. These correspond respectively to attenuation due to scattering and absorption by the turbulence, and we denote their respective contributions to μ by μ_S and μ_A.

Attenuation by Scattering

The imaginary part of the integral when K varies in the neighborhood of $|\Omega|/c_o$ is obtained by setting $d^3\mathbf{K} = K^2\, dK\, d\boldsymbol{\sigma}$, where $d\boldsymbol{\sigma}$ is the element of solid angle occupied by (K, dK). Then $\text{Im}(1/[K^2 - (\Omega + i0)^2]) = (\pi/2K)\text{sgn}(\Omega)$ $\delta(K - |\Omega|/c_o)$ [20], and the damping due to scattering is determined by

$$\mu_S = \frac{\pi\kappa_o^2}{c_o^4} \iint \Omega^2 (\boldsymbol{\sigma}\cdot\mathbf{n})^2 n_i n_j \Psi_{ij}\left(\kappa_o\mathbf{n} - \frac{|\Omega|}{c_o}\boldsymbol{\sigma}, \omega - \Omega\right) d\boldsymbol{\sigma}\, d\Omega, \tag{2.5.13}$$

the integrations being over all directions of the unit vector $\boldsymbol{\sigma}$ and frequencies $-\infty < \Omega < \infty$ [88–90].

The intensity of the scattered sound vanishes in directions at right angles to the coherent wave propagation direction $\mathbf{n}$ (Example 2). Similarly, when the Mach number of the velocities U_i is small enough that the turbulence may be regarded as incompressible, the condition $\operatorname{div}\mathbf{U} = 0$ implies that $\kappa_i\kappa_j\Psi_{ij}(\boldsymbol{\kappa}, \omega) \equiv 0$ [62, 63], and therefore that the scattered acoustic intensity is null in the back-scattering direction $\boldsymbol{\sigma} = -\mathbf{n}$, and in the forward direction $\boldsymbol{\sigma} = \mathbf{n}$ provided $\Omega \neq \omega$. For aerodynamically generated sound, whose wavelength $\lambda \sim \ell/M$ is much greater than the dimension ℓ of the energy containing turbulence eddies, $\mu_S \sim M^4/\ell$ [89] that is, damping by scattering is very weak.

Attenuation by Absorption

The component μ_A of μ accounts for acoustic energy conversion into kinetic energy of the turbulent motions and is determined by the δ-function contribution $\operatorname{Im}[1/(\Omega + i0)] = -\pi\delta(\Omega)$ to the integral (2.5.12):

$$\mu_A \approx \frac{\pi\kappa_o^2}{c_o}\int n_i n_j \Psi_{ij}(\kappa_o\mathbf{n} - \mathbf{K}, \omega)\left(1 - \frac{2(\mathbf{K}\cdot\mathbf{n})^2}{K^2}\right)d^3\mathbf{K}. \tag{2.5.14}$$

Thus, the spectral components of the turbulence responsible for absorption have the same frequency ω as the coherent wave. The interactions producing this absorption have frequency $\Omega = 0$, which means that the corresponding components of the local perturbation pressure p' are effectively *hydrodynamic* and governed by Laplace's equation. The first term in the brace brackets of (2.5.14) arises from local straining of the vorticity by the incident sound (first term on the right of (2.5.8)), and the second accounts for a longer range tendency of the incompressible components of p' to restore the strained turbulence to its undistorted state and comes from the pressure term on the right of (2.5.8) [94].

For aerodynamic sound, where $\omega \sim U/\ell$, we have the order of magnitude estimate $\mu_A \sim M^3/\ell$. Aerodynamically generated sound of wavelength $\lambda \sim \ell/M$ would therefore have to propagate a distance $\sim\lambda/M^2$ before the influence of absorption becomes important. This attenuation is $O(1/M)$ stronger than that attributable to scattering, but both mechanisms have a negligible influence on the aerodynamic sound generated by extensive regions of turbulence. It should be noted, however, that the predicted absorption is sufficient to ensure that the acoustic pressure fluctuations $p_a(\mathbf{x}, t)$, say, within an unbounded domain of low Mach number turbulence, are always small compared to the hydrodynamic pressure. Indeed, the acoustic pressure at $\mathbf{x}$ is then generated by turbulence quadrupoles located within a sphere of radius $\sim 1/\mu_A$ centered on $\mathbf{x}$. The acoustic power generated per unit volume of turbulence $\sim \rho_o v^3 M^5/\ell$ (Section 2.1.2). The mean square acoustic pressure at $\mathbf{x}$ is therefore $\overline{p_a^2} \sim (\rho_o v^2)^2 M$.

Example 1. The damping coefficient μ_D for a plane wave propagating in a stationary fluid subject to viscous and thermal dissipation is given by [2]

$$\mu_D = \frac{\kappa_o^2}{2\rho_o c_o}\left\{\left(\frac{4}{3}\eta + \eta'\right) + \kappa\left(\frac{1}{c_v} - \frac{1}{c_p}\right)\right\},$$

where η, η' are the shear and bulk viscosities, κ is the thermal conductivity, and c_v, c_p are the specific heats at constant volume and constant pressure. Deduce that for aerodynamically generated sound propagating through turbulence of characteristic velocity U and length scale ℓ,

$$\frac{\mu_A}{\mu_D} \sim \frac{U\ell}{\nu} \gg 1.$$

Example 2: Scattering of sound by a vortex ring [95]. Determine the scattered sound when a plane acoustic wave $p(\mathbf{x}, \omega) = p_{\mathrm{I}} \mathrm{e}^{i\kappa_o \mathbf{n}\cdot\mathbf{x}}$ interacts with an acoustically compact vortex ring.

Let the vortex ring have radius R, circulation Γ, and core radius $a \ll R$ and neglect any effects associated with self-induced motion of the vortex (Section 2.3, Example 4). Take the origin at the center of the ring with the x_1-axis parallel to the direction of its mean motion (Figure 2.5.1). The scattered pressure is the solution of (2.5.7), in which $\bar{\mathbf{V}} = (\mathbf{n} p_{\mathrm{I}}/\rho_o c_o)\mathrm{e}^{i\kappa_o \mathbf{n}\cdot\mathbf{x}}$:

$$p' \approx \frac{-4\pi^2\kappa_o^2 p_{\mathrm{I}}}{c_o|\mathbf{x}|^3} n_i x_i x_j U_j(\boldsymbol{\kappa})\mathrm{e}^{i\kappa_o|\mathbf{x}|}, \quad |\mathbf{x}| \to \infty,$$

where $\boldsymbol{\kappa} = \kappa_o(\mathbf{x}/|\mathbf{x}| - \mathbf{n})$, and $U_j(\boldsymbol{\kappa}) = (1/2\pi)^3 \int_{-\infty}^{\infty} U_j(\mathbf{x}) \exp(-i\boldsymbol{\kappa}\cdot\mathbf{x})\, d^3\mathbf{x}$.

$\mathbf{U}(\boldsymbol{\kappa})$ is determined from the vorticity $\boldsymbol{\omega} = \Gamma \mathbf{i}_\theta \delta(x_1)\delta(r - R)$, where $\mathbf{i}_\theta$ is a unit vector in the azimuthal direction, parallel to $\boldsymbol{\omega}$, and $r = \sqrt{x_2^2 + x_3^2}$. Show that

$$\mathbf{U}(\boldsymbol{\kappa}) = i(\boldsymbol{\kappa} \wedge \boldsymbol{\omega})/\kappa^2, \quad \boldsymbol{\omega}(\boldsymbol{\kappa}) = (-iR\Gamma/4\pi^2\kappa_\perp)(\mathbf{i}_1 \wedge \boldsymbol{\kappa})\mathrm{J}_1(\kappa_\perp R),$$

where $\kappa_\perp = |\mathbf{i}_1 \wedge \boldsymbol{\kappa}|$ ($\mathbf{i}_1$ being a unit vector in the x_1-direction), and that

$$p'(\mathbf{x}, \omega) \approx -(p_{\mathrm{I}}(\kappa_o R)^2\Gamma/4c_o|\mathbf{x}|)\cos\Theta(\cos\Theta_{\mathrm{I}} + \cos\Theta_{\mathrm{S}})\mathrm{e}^{i\kappa_o|\mathbf{x}|}, \quad |\mathbf{x}| \to \infty.$$

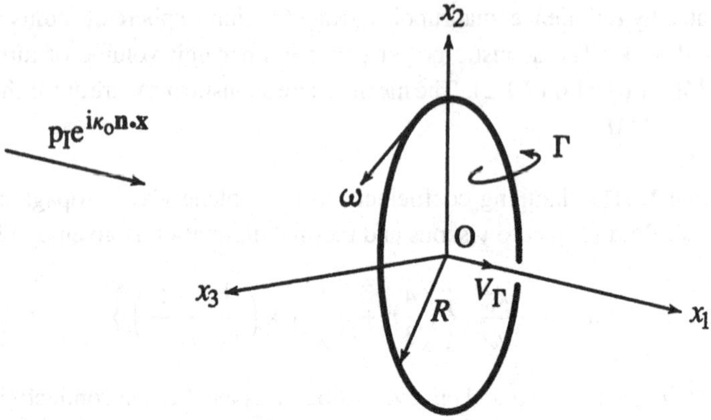

Figure 2.5.1. Scattering by a vortex ring.

where $\Theta = \arccos(\mathbf{n} \cdot \mathbf{x}/|\mathbf{x}|)$ = angle between the incident wave direction and the radiation direction; $\Theta_I = \arccos(\mathbf{n} \cdot \mathbf{i}_1)$ = angle between the incident wave direction and the direction of self-induced motion of the vortex ring; and $\Theta_S = \arccos(\mathbf{i}_1 \cdot \mathbf{x}/|\mathbf{x}|)$ = angle between the radiation direction and the direction of motion of the ring.

The directivity of the scattered sound is illustrated in Figure 2.5.2 when the propagation direction of the incident wave is (a) normal to the plane of the ring and (b) parallel to the plane of the ring.

Example 3. Show that the first approximation to the sound scattered when a plane wave $p(\mathbf{x}, \omega) = p_I e^{i\kappa_o x_1}$ impinges on a line vortex of strength Γ whose axis coincides with the x_3-axis is

$$p' \approx \frac{ip_I\Gamma}{2}\sqrt{\frac{i\kappa_o}{2\pi|\mathbf{x}|}}\left(\sin\theta - \cot\frac{\theta}{2}\right)e^{i\kappa_o|\mathbf{x}|}, \quad |\mathbf{x}| \to \infty,$$

where θ is the angle between the scattering direction and the x_1-axis, and $|\mathbf{x}| = \sqrt{x_1^2 + x_2^2}$. The scattering is singular in the "forward" direction $\theta = 0$. This breakdown in first-order scattering theory is a result of the long range "refraction" of sound by the velocity field of the vortex, which decays slowly with distance like $\Gamma/|\mathbf{x}|$ [96].

2.5.2 Attenuation by Particulate Inhomogeneities

There is a practical need to suppress sound of aerodynamic origin (usually generated by fluid–structure interactions rather than directly by relatively weak-volume quadrupole sources). It is found that significant damping occurs in a

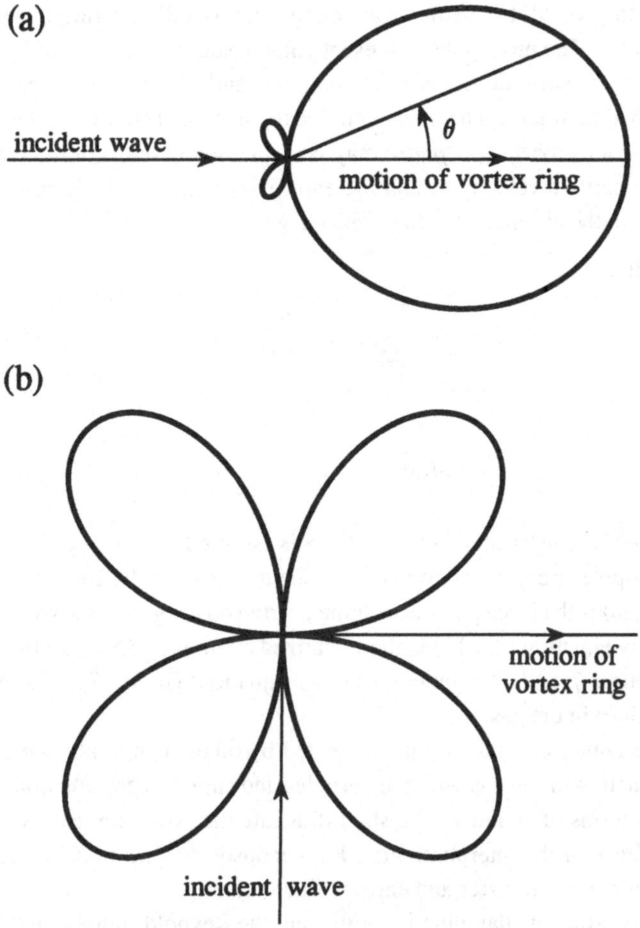

Figure 2.5.2. Linear field shape $|p'(\Theta)/p_{\rm I}|$.

fluid containing small particles or droplets, for example, in the gaseous combustion products of a solid-fueled rocket motor, where their presence tends to inhibit the excitation of acoustic instabilities. Similarly, water droplets injected into the exhaust or inlet flows of jet engines and rocket motors tend to promote acoustic damping [97–100]. Particulate attenuation is caused by relaxation processes associated with viscous drag and with heat and mass transfer between the particles and gas. In the following discussion we shall consider sound propagation through a gas containing a homogeneous distribution of small spherical particles.

When the acoustic wavelength is much larger than the average distance between the particles, the equations governing the propagation of sound may be derived by the procedure suggested in Section 2.4.3, Example 3 for a bubbly liquid. In general, the particles are equivalent to a distribution of monopoles and dipoles. The monopoles represent volume sources determined by heat and mass transfer between the particles and gas, and the dipoles are produced by unsteady drag forces. The linearized, locally averaged continuity and momentum equations for the *gas phase* may be expressed in terms of the pressure $\bar{p}$, acoustic particle velocity $\bar{\mathbf{v}}$, and the entropy perturbation $\bar{s}$. In the frequency domain, in the absence of mean flow, we write

continuity,

$$\operatorname{div} \bar{\mathbf{v}} - \frac{i\omega}{\rho_o c_o^2}\bar{p} = \frac{-i\omega\beta T_o}{c_p}\bar{s} + \bar{q}, \tag{2.5.15}$$

momentum,

$$-i\omega\bar{\mathbf{v}} + \frac{1}{\rho_o}\nabla\bar{p} = \frac{1}{\rho_o}\bar{\mathbf{F}}, \tag{2.5.16}$$

where the volume fraction of the particles is assumed to be negligible, $\bar{q}(\mathbf{x}, \omega)$ is the monopole strength per unit volume due to mass transfer from the particles, and $\bar{\mathbf{F}}(\mathbf{x}, \omega)$ is the force per unit volume exerted on the gas by the particles. The continuity equation (2.5.15) is the linearized form of (2.3.8) with the addition of the source $\bar{q}$, and other quantities in the equations (ρ_o, c_o, T_o, etc.) represent mean values in the gas.

These equations must be supplemented by others that model the details of the interaction of the gas with the particles and supply representations of $\bar{s}$, $\bar{q}$, and $\bar{\mathbf{F}}$ in terms of $\bar{p}$ and $\bar{\mathbf{v}}$. We shall illustrate the procedure by investigating the influence of the interphase drag $\bar{\mathbf{F}}$ on acoustic propagation, neglecting for the moment mass transfer and entropy fluctuations.

When particulate damping is significant, the Reynolds number uR/ν based on particle translational velocity $\mathbf{u}$ and radius R (typically $<10^{-5}$ m) is sufficiently small that the drag is dominated by viscosity. Also $\omega R^2/\nu \ll 1$ for all audible frequencies, which implies [17] that the viscous force on a particle is the Stokes drag $6\pi\eta R(\bar{\mathbf{v}} - \mathbf{u})$, where $\eta = \rho_o\nu$ is the shear viscosity. Particulate motion is therefore governed by

$$m\partial\mathbf{u}/\partial t = 6\pi\eta R(\bar{\mathbf{v}} - \mathbf{u}), \tag{2.5.17}$$

where m is the particle mass. Hence, for harmonic time dependence

$$\bar{\mathbf{F}} = \sum -m\frac{\partial\mathbf{u}}{\partial t} = \frac{i\omega\mathcal{N}m\bar{\mathbf{v}}}{(1 - i\omega\tau_\eta)}, \tag{2.5.18}$$

where the summation is over all particles in unit volume ($=\mathcal{N}$), and τ_η =

$m/6\pi\eta R$ is the *Stokes relaxation time*. The equation for the acoustic pressure $\bar{p}$ is now formed by substituting this result into (2.5.16) and eliminating $\bar{\mathbf{v}}$ by use of the continuity equation (with $\bar{s} \equiv 0$ and $\bar{q} \equiv 0$):

$$\nabla^2 \bar{p} + \kappa_o^2 \left(1 + \frac{\alpha_p}{(1 - i\omega\tau_\eta)}\right) \bar{p} = 0, \qquad (2.5.19)$$

where $\alpha_p = \mathcal{N} m/\rho_o$ is the particle mass fraction. This equation has the same structure as an equation describing the propagation of sound in a medium where chemical or molecular relaxation processes occur, in which *equilibrium* and *frozen* sound speeds, denoted, respectively, by c_e, c_f can be defined [1, 6, 101, 102]. The motion is frozen for $\omega\tau_\eta \gg 1$, when $c_f = c_o$ and the particles remain fixed, whereas equilibrium conditions prevail for $\omega\tau_\eta \ll 1$, when the particles move at the same speed as the gas and $c_e = c_o/\sqrt{1 + \alpha_p}$. When the radius R varies from particle to particle, the summation in equation (2.5.18) must be replaced by an integral involving a suitable radius or volume distribution function. This integral will then appear in the second term in the brace brackets of (2.5.19), as in the analogous problem of propagation through a bubbly liquid.

For a plane wave $\bar{p}(\mathbf{x}, \omega) = e^{i\mathbf{k}\cdot\mathbf{x}}$ equation (2.5.19) yields

$$k^2 = \kappa_o^2 \left(1 + \frac{\alpha_p}{(1 - i\omega\tau_\eta)}\right).$$

Wave dispersion is determined by the variation with frequency of the phase speed $c = \mathrm{Re}(\omega/k)$, which increases monotonically from c_e to c_f as ω increases, the rate of change being greatest when $\omega \approx 1/\tau_\eta$, the inverse Stokes relaxation time (Figure 2.5.3). Setting $\mathbf{k} = (\kappa_o + \delta k)\mathbf{n}$, where $\mathbf{n}$ is a unit vector in the direction of propagation, the attenuation per wavelength of propagation is then proportional to $\mathrm{Im}(\delta k)/\kappa_o \equiv \mu_A/\kappa_o$, where for small values of the particle mass fraction α_p

$$\frac{\mu_A}{\kappa_o} \approx \frac{\alpha_p}{2} \frac{\omega\tau_\eta}{[1 + (\omega\tau_\eta)^2]}. \qquad (2.5.20)$$

This vanishes at very low and very high frequencies (for equilibrium and frozen flows) and attains its maximum at $\omega = 1/\tau_\eta$, as illustrated in the figure. In the equilibrium limit ($\omega\tau_\eta \to 0$), $\mu_A \propto \omega^2$, which is the same dependence on frequency as classical viscous and thermal damping in the gas.

The influence of heat transfer between the particles and the gas is governed by the entropy term $\bar{s}$ on the right of (2.5.15). This is readily calculated for an ideal gas ($\beta = 1/T_o$) in the absence of mass transfer, and when the temperature within a particle is assumed to be uniform. A plane sound wave produces a

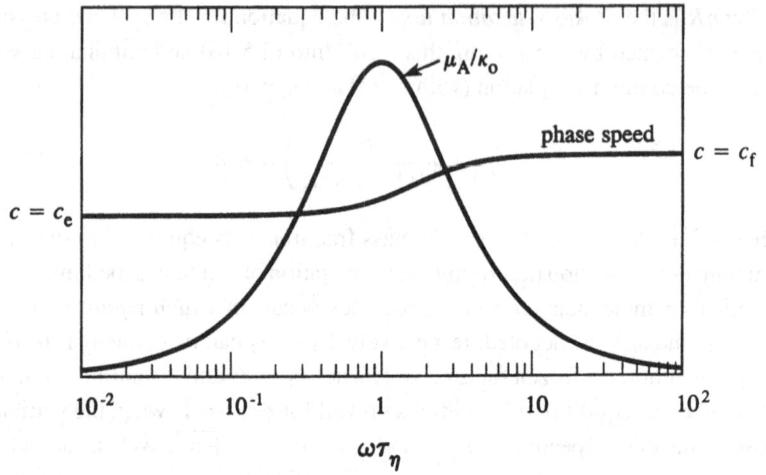

Figure 2.5.3. Stokes drag dispersion and absorption curves.

temperature fluctuation $\bar{T} \approx \bar{p}/\rho_o c_p$ in the bulk of the gas. The actual temperature distribution in the neighborhood of a spherical particle can be found by the method of Example 6, Section 2.3. Separate diffusion equations are applicable within the gas and sphere, and their solutions must satisfy continuity of temperature and heat flux $\kappa \partial T/\partial r$ at the surface $r = R$ of the sphere. The entropy perturbation $\bar{s}$ is calculated by equating the net heat flux from the spheres into unit volume of gas to $\rho_o T_o \partial \bar{s}/\partial t$. This yields

$$\frac{-i\omega\beta T_o}{c_p}\bar{s} = \frac{i\omega\alpha_p(\gamma - 1)(c_s/c_o)\bar{p}}{c_o^2(1 - i\omega\tau_T)},$$

where γ is the ratio of specific heats of the gas, c_s the specific heat of the particles (at constant pressure),

$$\tau_T = mc_s/4\pi\kappa R,$$

is the thermal relaxation time, κ being the thermal conductivity of the gas, and it is assumed that the particle radius $R \ll \sqrt{\kappa/\rho_o c_p \omega}$ = thermal length scale in the gas. When this result is substituted into the continuity equation (2.5.15) (with $\bar{q} = 0$), the dispersion equation that incorporates thermal dissipation and the Stokes drag (2.5.18) can be derived as before. For small values of the particle mass fraction α_p, the attenuation is given by

$$\frac{\mu_A}{\kappa_o} \approx \frac{\alpha_p}{2}\left(\frac{\omega\tau_\eta}{[1 + (\omega\tau_\eta)^2]} + \frac{c_s(\gamma - 1)}{c_p}\frac{\omega\tau_T}{[1 + (\omega\tau_T)^2]}\right). \tag{2.5.21}$$

There are now two frequency absorption bands of the type shown in Figure 2.5.3, centered on $\omega = 1/\tau_\eta$ and $1/\tau_T$, where the attenuation for each mechanism is optimal. In practice, however, these bands will be close together because $\tau_\eta/\tau_T = \frac{3}{2}(c_s/c_p)\mathrm{Pr}$, where both the *Prandtl number* $\mathrm{Pr} = \eta c_p/\kappa$ and the ratio c_s/c_p are of order unity ($\mathrm{Pr} \approx 0.7$ for air; $c_s/c_p \approx 4.2$ for water droplets in air) [2, 4].

Damping by Stokes drag and heat transfer can be large in solid rocket combustion chambers, where the particulate mass fraction α_p often exceeds 0.2. When the particles consist of very small water droplets, occurring naturally in fog, for example, or sprayed into an industrial flow to attenuate aerodynamic noise, α_p is usually small, and viscous and thermal dissipation are important only at higher frequencies (usually much greater than 1,000 Hz). However, there is now an additional damping mechanism that is particularly effective at low frequencies. Sound can cause condensation and vaporization of the droplets, involving the transfer of both heat and mass between the bulk gas and the liquid phase. The droplets adjust to the temperature of the bulk gas on a time scale $\sim O(\tau_T)$, following which evaporation or condensation occurs essentially at constant temperature in an effort to restore equilibrium between the saturation vapor pressure and the liquid. Latent heat of vaporization is extracted or delivered to the gas, producing a volume dilation governed by the entropy term in the continuity equation (2.5.15). Vapor diffuses to or from the droplets at a rate proportional to the volume source $\bar{q}$, whose magnitude, however, is usually smaller than the volume source determined by the entropy fluctuation. The time scale for mass diffusion sources is $\tau_D \sim m/4\pi D\rho_o R$, where D is the mass transfer coefficient [2]. The *Schmidt number* $\mathrm{Sc} = \nu/D \sim O(1)$ for a gas, and this would normally imply that the frequency absorption band for diffusive attenuation occurs in the neighborhood of the viscous and thermal bands. In the present case, however, the mass of vapor that must be diffused is proportional to the vapor present in the fluid, but it is *independent* of the mass fraction α_p of the droplets. The time during which condensation or evaporation occurs is inversely proportional to the liquid surface area that, for droplets of given size, is proportional to α_p. Thus, the effective relaxation time for phase exchange is actually of order τ_D/α_p, which can be as much as 10 to 100 times longer than the viscous and thermal relaxation times. The peak attenuation therefore occurs at progressively smaller frequencies as the droplet mass fraction decreases, and it is usually optimal at frequencies $\leq$1,000 Hz. In principle, the magnitude of the absorption band center frequency can be controlled by suitably adjusting the rate of injection of water into a flow [99, 100].

2.6 Jet Noise

The aerodynamic sound generated by a turbulent jet can usually be attributed to several different sources. The principal ones for a high-speed air jet (when the exhaust velocity exceeds about 100 m/s) are *jet mixing noise* and, for supersonic flow, *shock associated noise*. Acoustic sources located within the "jet pipe" also contribute to the noise, especially at lower jet speeds. These include *combustion noise* and sound produced by interactions of a turbulent stream with fans, compressors, and turbine systems. The properties of sources in real jets generally differ considerably from those of the idealized models considered in Sections 2.1–2.3, and the following discussion is based on accumulated experimental data [103].

2.6.1 Jet Mixing Noise

This is the sound produced by the turbulent mixing of a jet with the ambient fluid. The simplest free jet is an air stream issuing from a large reservoir through a circular convergent nozzle (Figure 2.6.1). The gas accelerates from near zero velocity in the reservoir to a peak velocity in the narrowest cross section (the "throat") of the nozzle. The jet Mach number $M_J = U/c_J$, where U is the jet speed, and c_J the local speed of sound. Using relations given in Section 1.6.5, and assuming the flow to be adiabatic, we can write

$$M_J = \sqrt{\frac{2}{\gamma - 1}\left(\left(\frac{p_R}{p}\right)^{\frac{(\gamma-1)}{\gamma}} - 1\right)}, \tag{2.6.1}$$

where p_R is the reservoir pressure, and p is the local pressure in the jet. If p is identified with the mean ambient pressure p_o, sonic flow occurs in the nozzle when $p_R/p_o = 1.89$ for air ($\gamma = 1.4$). An increase above this critical pressure ratio leads to the appearance of *shock cells* downstream of the nozzle and "choking" of the flow unless the throat is followed by a divergent section in which the flow "expands" and the pressure decreases smoothly to p_o [104].

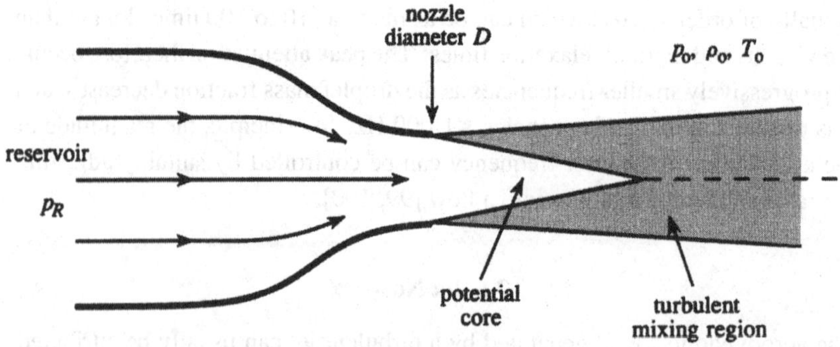

Figure 2.6.1. Subsonic, turbulence-free jet.

In an ideal jet, in the absence of shock waves, the flow upstream of the nozzle is assumed to be free from turbulence so that interactions between the flow and solid boundaries may be neglected. Noise is produced entirely by turbulent mixing of the jet with the ambient fluid. Mixing first occurs in an annular shear layer, which grows with distance from the nozzle; the irrotational, inner region of the jet is called the *potential core*, which disappears when the mixing region fills the whole jet, usually at four or five diameters from the nozzle. The sources extend over a considerable distance. The high-frequency components of the sound are generated predominantly close to the nozzle, where the dimensions of the turbulence eddies are small (of the order of the thickness of the mixing layer); lower frequency sound emanates from sources further downstream where the eddy size is comparable to the jet diameter. The sources are usually regarded as quadrupoles, whose strength and directivity are modified both by nonuniform density (temperature) and convection[105].

The total radiated sound power Π_a of the mixing noise is usually expressed in terms of the mechanical stream power of the jet, $\frac{1}{2}m_J U^2$, where $m_J = \rho_J U A$ is the mass flow from the nozzle, and ρ_J, U, and A henceforth denote the jet density, mean velocity, and cross-sectional area in the "fully expanded" region downstream of the nozzle [104]. The efficiency of sound generation $\Pi_a/\frac{1}{2}m_J U^2$ can be expressed in terms of the Mach number $M = U/c_o$ and density ratio ρ_J/ρ_o, where c_o and ρ_o are the sound speed and mean density in the ambient medium. It is given by [103]

$$\Pi_a \Big/ \frac{1}{2}m_J U^2 \approx 1.3 \times 10^{-4}(\rho_J/\rho_o)^{w-1}M^n, \tag{2.6.2}$$

where w is called the *jet density exponent* [106], which is defined for arbitrary values of $M \geq 0.3$ by

$$w = 3M^{3.5}/(0.6 + M^{3.5}) - 1. \tag{2.6.3}$$

The exponent n in (2.6.2) ≈ 5 for $M \leq 1.8$ but becomes progressively smaller at higher Mach numbers, such that for $M \geq 3$ the efficiency apparently attains a maximum that probably does not exceed about 1% of the total jet power. To put this in perspective, note that, at least for subsonic jets, most of the sound is radiated by sources within a distance of about five or so diameters from the jet nozzle and that over this distance, the kinetic energy of the jet suffers a reduction of about 25% due to viscous dissipation. For very large jet Mach numbers, it has been speculated that n becomes negative [107]. When M is less than about 1.05, the sound power increases as ρ_J/ρ_o decreases (as the jet temperature increases), but it decreases for larger values of M.

The behavior at subsonic Mach numbers is not satisfactorily explained by Lighthill's theory, according to which the efficiency should increase more rapidly than M^5 because of convection of the quadrupoles (Doppler amplification, Section 2.2.3). Model experiments [73] suggest that the discrepancy arises because convection has little or no effect on acoustic intensity when the depth of a source within a shear layer exceeds an acoustic wavelength, whereas at lower frequencies ($\omega D/c_o \leq 1$, where D is the nozzle exit diameter), predictions of Doppler amplification by Lighthill's theory are correct.

Source convection and refraction in the jet shear layers cause the sound field to be directive, the maximum noise being radiated in directions inclined at angles θ to the direction of the jet flow of about 30°–45°. At $\theta = 90°$, the mean square acoustic pressure $\langle p^2 \rangle$ for an axisymmetric gas jet exhausting into air at Mach number $M \leq 3$ is given by [106]

$$\frac{\langle p^2(r, 90°)\rangle}{p_{\mathrm{ref}}^2} \approx \frac{9 \times 10^{13} A}{r^2} \frac{(\rho_J/\rho_o)^w M^{7.5}}{[1 - 0.1M^{2.5} + 0.015M^{4.5}]}, \quad r \to \infty, \tag{2.6.4}$$

where r is the observer distance from the nozzle exit, and $p_{\mathrm{ref}} = 2 \times 10^{-5}$ N/m^2 is a standard reference pressure used for airborne sound. This formula shows that when $\rho_J = \rho_o$ and $M \ll 1$, the dependence on Mach number is close to the U^8-law for quadrupoles predicted by Lighthill's theory (Section 2.1). The behavior at other Mach numbers and density ratios is illustrated in Figure 2.6.2.

Typical field shapes for an air jet at a fixed observer distance from the nozzle are shown in Figure 2.6.3 for three different jet Mach numbers $M = U/c_o$

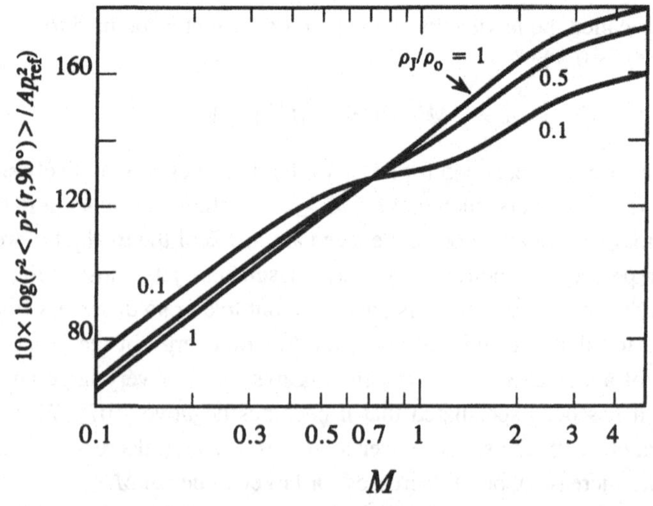

Figure 2.6.2. Mach number dependence of mixing noise ($M = U/c_o, \theta = 90°$).

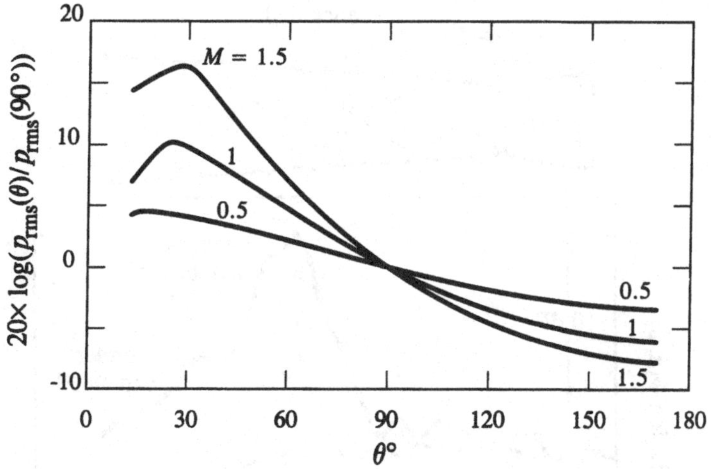

Figure 2.6.3. Directivity of mixing noise.

[103]. The curves give the variation of $20 \times \log[p_{\text{rms}}(\theta)/p_{\text{rms}}(90^\circ)]$ (dB), where p_{rms} is the root mean square acoustic pressure, and $p_{\text{rms}}(90^\circ)$ is given by (2.6.4).

The mean square acoustic pressure $\langle p^2(\mathbf{x}, t)\rangle$ can be expressed in terms of the *acoustic pressure frequency spectrum* $\Phi(\mathbf{x}, \omega)$ by $\langle p^2(\mathbf{x}, t)\rangle = \int_0^\infty \Phi(\mathbf{x}, \omega)\, d\omega$. The spectrum is broadband and exhibits a broad peak centered on a frequency ω_p that depends on radiation direction θ, Mach number M, and density ratio ρ_J/ρ_o. The *Strouhal number* $\omega_p D/U$, based on this frequency and the jet exit diameter D, varies from about 2 when $\rho_J/\rho_o \approx 0.3$ to about 5 when $\rho_J = \rho_o$. For subsonic jets, $\Phi(\mathbf{x}, \omega)$ is roughly proportional to ω^2 at low frequencies and decays like $1/\omega^2$ for $\omega > \omega_p$.

Real jets exhausting from pipes and engine nozzles are not turbulence free but are generally disturbed or *spoiled* before leaving the nozzle. Unless the jet velocity U exceeds about 100 m/s, the intensity of jet mixing noise may then be smaller than that produced by sources in the flow upstream of the nozzle exit.

2.6.2 Noise of Imperfectly Expanded Supersonic Jets

Supersonic, underexpanded, or "choked" jets contain shock cells through which the flow repeatedly expands and contracts (Figure 2.6.4) [104]. Seven or more distinct cells are often visible extending up to ten jet diameters from the nozzle. They are responsible for two additional components of jet noise: *screech* tones and broadband *shock-associated noise* [108, 109]. Screech is produced by a feedback mechanism in which a disturbance convecting in the shear layer generates sound as it traverses the standing system of shock waves in the jet. The

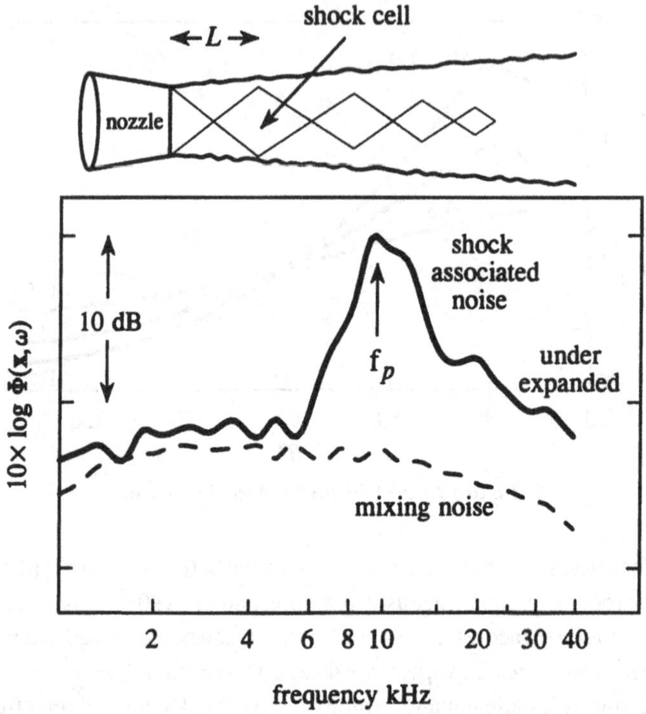

Figure 2.6.4. Jet noise spectra for shock-free and underexpanded jets at $\theta = 90°, \beta = 1$. (After [109].)

sound propagates upstream through the ambient medium and causes the release of a new flow disturbance at the nozzle. This is amplified as it convects downstream, and the feedback loop is completed when it encounters the shocks. A notable feature of screech tones is that the frequencies are independent of the radiation direction, the fundamental frequency occurring at $\omega \approx 2\pi U_c/L(1 + M_c)$, where $M_c = U_c/c_o$; $U_c \approx 0.7U$ is the convection velocity of the disturbance in the shear layer, and L is the axial length of the first shock cell.

Screech is readily suppressed by appropriate nozzle design. If shocks are still present in the flow, it is found [103] that a good estimate of the net radiation from the jet is obtained by adding the separate mean square pressures that can separately be attributed to mixing noise for a shock-free jet and the residual broadband, shock associated noise.

The intensity of the shock-associated noise appears to be independent of the radiation direction and is given by [109]

$$\frac{\langle p^2(r)\rangle}{p_{\text{ref}}^2} \approx \frac{5 \times 10^{15}\beta^n A}{r^2}, \quad r \to \infty, \tag{2.6.5}$$

where $\beta = \sqrt{U^2/c_J^2 - 1}$, and $n = 4$ when $\beta < 1$. For larger values of β, the exponent n increases from about 1 when the stagnation temperature of the jet T_J (Section 1.6.5) equals the ambient temperature T_o to about 2 when $T_J/T_o \geq 1.1$. The frequency spectrum $\Phi(\mathbf{x}, \omega)$ has a well-defined peak at

$$\omega_p \approx \frac{5.7U_c}{D\beta|1 - M_c \cos\theta|}, \tag{2.6.6}$$

where the convection velocity $U_c \approx 0.7U$, and θ is the angle made between the observer direction and the direction of the mean jet flow. This frequency is determined by constructive interference between the sound waves generated when a large-scale disturbance (a "coherent structure" [110]) convecting at speed U_c interacts with successive shock cells.

The contribution from shock-associated noise is illustrated in Figure 2.6.4, in which the spectrum of the jet mixing noise of a shock-free supersonic jet (formed by fully expanding the jet to the ambient pressure p_o by means of a convergent–divergent nozzle [104]) is compared with that for an underexpanded jet at the same pressure ratio p_R/p_o when $\theta = 90°$. The spectral peak occurs at $\mathrm{f}_p \equiv \omega_p/2\pi$ defined by (2.6.6).

2.6.3 *Effect of Flight*

The noise generated by a jet engine in flight is modified because of the joint effects of Doppler amplification (Section 2.2.3) and the reduced turbulence levels caused by a decrease in mean shear between the jet flow and the stationary atmosphere. Empirical formulae have been developed [103] that enable the noise levels in flight to be estimated from corresponding static measurements.

For mixing noise, $\langle p^2_{\text{flight}}\rangle$ is related to $\langle p^2_{\text{static}}\rangle$ by

$$\frac{\langle p^2_{\text{flight}}\rangle}{\langle p^2_{\text{static}}\rangle} \approx \frac{1}{(1 - M_\mathrm{f}\cos\Theta)}\left(\frac{U - V}{U}\right)^{\mathrm{m}(\theta)}, \tag{2.6.7}$$

where V is the flight speed, $M_\mathrm{f} = V/c_o$, and θ, Θ are respectively the angles between the observer direction and the jet axis and flight direction at the time of emission of the sound. Equation (2.6.7) contains only one Doppler factor $1/(1 - M_\mathrm{f}\cos\Theta)$, which is very different from predictions that would be made by a naive application of the theoretical results of Section 2.2.3 for idealized quadrupole and dipole sources. This is indicative of the profound effect on sound generation of the turbulent shear layers, which contain the aerodynamic sources, and through which the sound must propagate. The exponent $\mathrm{m}(\theta)$ has a complicated dependence on θ and on jet Mach number

$M = U/c_o$. When $1.1 \le M \le 1.95$ and θ is expressed in degrees in the range $20^\circ < \theta < 160^\circ$, $\mathrm{m}(\theta)$ is well approximated by

$$\mathrm{m}(\theta) = \left\{ \left(\frac{6{,}959}{|\theta - 125|^{2.5}} \right)^7 + \left(\frac{1}{31 - 18.5M - (0.41 - 0.37M)\theta} \right)^7 \right\}^{-\frac{1}{7}};$$

for $M < 1.1$ and $M > 1.95$, $\mathrm{m}(\theta)$ is given by this formula at $\mathrm{M} = 1.1$ and 1.95, respectively.

In the case of shock-associated noise, the influence of flight is given simply by

$$\frac{\langle p_{\text{flight}}^2 \rangle}{\langle p_{\text{static}}^2 \rangle} \approx \frac{1}{(1 - M_{\mathrm{f}} \cos \Theta)^4}. \tag{2.6.8}$$

3

Sound generation in a fluid with rigid boundaries

3.1 Influence of Rigid Boundaries on the Generation of Aerodynamic Sound

The Ffowcs Williams–Hawkings equation (2.2.3) enables aerodynamic sound to be represented as the sum of the sound produced by the aerodynamic sources in unbounded flow together with contributions from monopole and dipole sources distributed on boundaries. For turbulent flow near a fixed rigid surface, the direct sound from the quadrupoles T_{ij} is augmented by radiation from surface dipoles whose strength is the force per unit surface area exerted on the fluid. If the surface is in accelerated motion, there are additional dipoles and quadrupoles, and neighboring surfaces in relative motion also experience "potential flow" interactions that generate sound. At low Mach numbers, M, the acoustic efficiency of the surface dipoles exceeds the efficiency of the volume quadrupoles by a large factor $\sim O(1/M^2)$ (Sections 1.8 and 2.1). Thus, the presence of solid surfaces within low Mach number turbulence can lead to substantial increases in aerodynamic sound levels. Many of these interactions are amenable to precise analytical modeling and will occupy much of the discussion in this chapter.

3.1.1 Acoustically Compact Bodies [70]

Consider the production of sound by turbulence near a compact, stationary rigid body. Let the fluid have uniform mean density ρ_o and sound speed c_o, and assume the Mach number is sufficiently small that convection of the sound by the flow may be neglected. This particular situation arises frequently in applications. In particular, M rarely exceeds about 0.01 in water, and sound generation by turbulence is usually negligible except where the flow interacts with a solid boundary [111].

The acoustic pressure in the far field is given by Curle's equation (2.2.4) in the form

$$p(\mathbf{x},t) \approx \frac{x_i x_j}{4\pi c_o^2|\mathbf{x}|^3}\frac{\partial^2}{\partial t^2}\int [\rho_o v_i v_j]\, d^3\mathbf{y} + \frac{x_i}{4\pi c_o|\mathbf{x}|^2}\frac{\partial}{\partial t}\oint_S [n_j p'_{ij}]\, dS(\mathbf{y}),$$
$$|\mathbf{x}| \to \infty, \qquad (3.1.1)$$

where, in the usual notation, the terms in square brackets are evaluated at the retarded time $t - |\mathbf{x}-\mathbf{y}|/c_o$, $p'_{ij} = (p - p_o)\delta_{ij} - \sigma_{ij}$ is the compressive stress tensor, and the unit normal $\mathbf{n}$ on S is directed into the fluid. The first term on the right is the quadrupole noise, which has the same formal structure as when the body is absent. The surface integral is the dipole contribution.

In practice, the length scale ℓ of the turbulence is comparable to the dimensions of the body. The frequency of the hydrodynamic fluctuations $\sim v/\ell$, and the whole motion in the neighborhood of the body is therefore acoustically compact if $(v/\ell)\ell/c_o = v/c_o \ll 1$ (i.e., provided the Mach number is small). The retarded time variations on S, which are of order $\mathbf{x}\cdot\mathbf{y}/c_o|\mathbf{x}|$ when $|\mathbf{x}| \to \infty$ may then be neglected so that, with the origin in the neighborhood of the body, the first approximation to the surface integral yields the dipole sound field

$$\frac{x_i}{4\pi c_o|\mathbf{x}|^2}\frac{\partial F_i}{\partial t}(t - |\mathbf{x}|/c_o),$$

where $F_i(t) = \oint_S p'_{ij}(\mathbf{y},t) n_j\, dS(\mathbf{y})$ is the net force exerted on the fluid by the rigid body. In this compact approximation the force can be calculated in terms of the vorticity $\boldsymbol{\omega}$ and the velocity $\mathbf{v}$ from equation (1.14.13) (in which $dU_j/dt \equiv 0$ for a stationary body), and the acoustic pressure cast in the form

$$p(\mathbf{x},t) \approx \frac{-\rho_o x_i}{4\pi c_o|\mathbf{x}|^2}\frac{\partial}{\partial t}\left\{\int [(\boldsymbol{\omega}\wedge\mathbf{v})\cdot\nabla Y_i]\, d^3\mathbf{y} - \nu\oint_S [\boldsymbol{\omega}\wedge\nabla Y_i\cdot\mathbf{n}]\, dS(\mathbf{y})\right\}$$
$$\approx \frac{-\rho_o x_i}{4\pi c_o|\mathbf{x}|^2}\frac{\partial}{\partial t}\int (\boldsymbol{\omega}\wedge\mathbf{v})(\mathbf{y}, t - |\mathbf{x}|/c_o)\cdot\nabla Y_i(\mathbf{y})\, d^3\mathbf{y}, \quad |\mathbf{x}| \to \infty,$$
$$(3.1.2)$$

where [ ] denotes evaluation at $t - |\mathbf{x}|/c_o$, and $Y_i = y_i - \varphi_i^*(\mathbf{y})$ is the velocity potential of incompressible flow past the body having unit speed in the i-direction at large distances from S (the corresponding contribution to the radiation being equivalent to a dipole orientated in the i-direction). The approximation in the second line is applicable at high Reynolds numbers, $v\ell/\nu \gg 1$, when the "skin friction" contribution to the surface force is negligible.

The order of magnitude of $F \sim \rho_o v^2 \ell^2$, and $\partial/\partial t \sim v/\ell$. The dipole generated acoustic pressure is therefore of order $(\ell/|\mathbf{x}|)\rho_o v^2 M$ $(M = v/c_o)$, with corresponding sound power $\Pi_a \sim \ell^2 \rho_o v^3 M^3$ and acoustic efficiency (Section 2.1.2) $\Pi_a/(\ell^2 \rho_o v^3) \sim O(M^3)$. For the direct quadrupole radiation, the acoustic pressure $\sim(\ell/|\mathbf{x}|)\rho_o v^2 M^2$, the same as in the absence of the body, and the efficiency is $\sim O(M^5)$. The radiation is accordingly dominated by the dipole when M is small, and as $M \to 0$ the acoustic power exceeds the quadrupole power by a factor $\sim 1/M^2 \gg 1$. Precisely how small M should be for this to be true depends on the details of the flow.

This increase in acoustic efficiency brought about by surface dipoles on an acoustically compact body occurs also for arbitrary, noncompact bodies when vorticity interacts with compact structural elements, such as edges, corners, and protuberances. In the extreme case of a *flat*, rigid surface, whose dimension greatly exceeds the acoustic wavelength, the normal force exerted on the flow vanishes identically (Section 1.14, Example 5). The integrated contribution from the dipoles then vanishes unless account is taken of retarded time differences, when their aggregate effect becomes the same as a higher order quadrupole that coincides with the *image* of the turbulence volume quadrupoles in the wall [112]; the overall efficiency is therefore $O(M^5)$, the same as for quadrupoles in free space (see Section 2.1).

3.1.2 Influence of Low Mach Number Flow

Consider vorticity convected past the body in low Mach number, homentropic mean flow (Figure 3.1.1). Let us determine the generated sound by solving equation (2.3.17). The outgoing wave solution is given by equation (1.10.9), where $\varphi(\mathbf{x}, t)$ and $\hat{\mathcal{F}}(\mathbf{x}, t)$ are replaced by B and $\operatorname{div}(\boldsymbol{\omega} \wedge \mathbf{v})$, respectively. The mean flow velocity $\mathbf{U}(\mathbf{x})$ is irrotational and assumed to be of sufficiently low Mach number M that M^2 is small relative to unity (say, $M < 0.3$), so that variations in the mean density and sound speed can be ignored. Equation (2.3.17) may be approximated by

$$\left\{\frac{1}{c_o^2}\left(\frac{\partial}{\partial t} + \mathbf{U}\cdot\nabla\right)^2 - \nabla^2\right\} B = \operatorname{div}(\boldsymbol{\omega} \wedge \mathbf{v}), \qquad (3.1.3)$$

where $\mathbf{U} \to \mathbf{U}_o =$ constant at large distances from the body.

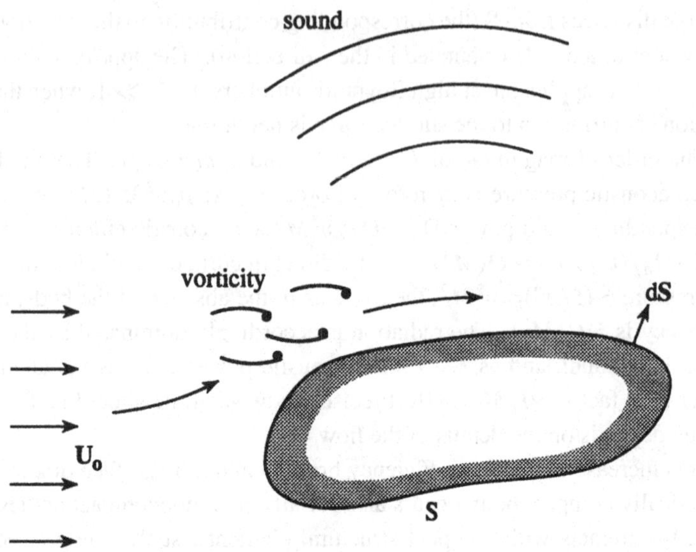

Figure 3.1.1. Turbulent flow past a stationary, rigid body.

When Green's function is chosen to have vanishing normal derivative on S, and $M^2 \ll 1$, (1.10.9) supplies

$$B(\mathbf{x},t) \approx -\oint_S G(\mathbf{x},\mathbf{y},t-\tau)\frac{\partial B}{\partial y_j}(\mathbf{y},\tau)n_j\, dS(\mathbf{y})\, d\tau$$
$$+\int G(\mathbf{x},\mathbf{y},t-\tau)\frac{\partial}{\partial y_j}(\omega\wedge\mathbf{v})_j(\mathbf{y},\tau)\, d^3\mathbf{y}\, d\tau, \quad |\mathbf{x}|\to\infty. \tag{3.1.4}$$

The volume integral is transformed by use of the divergence theorem: $\int G\,\mathrm{div}(\omega\wedge\mathbf{v})\,d^3\mathbf{y} = -\oint_S G(\omega\wedge\mathbf{v})_j n_j\, dS - \int(\omega\wedge\mathbf{v})\cdot\nabla G\, d^3\mathbf{y}$. When the motion near S is regarded as incompressible, Crocco's equation (1.2.18) may be approximated by $\partial\mathbf{v}/\partial t + \nabla B = -\,\omega\wedge\mathbf{v} - \nu\,\mathrm{curl}\,\omega$, and (3.1.4) then becomes

$$B(\mathbf{x},t) \approx -\int(\omega\wedge\mathbf{v})_j(\mathbf{y},\tau)\frac{\partial G}{\partial y_j}(\mathbf{x},\mathbf{y},t-\tau)\, d^3\mathbf{y}\, d\tau$$
$$+\nu\oint_S \omega(\mathbf{y},\tau)\wedge\frac{\partial G}{\partial\mathbf{y}}(\mathbf{x},\mathbf{y},t-\tau)\cdot d\mathbf{S}(\mathbf{y})\, d\tau$$
$$+\oint_S G(\mathbf{x},\mathbf{y},t-\tau)\frac{\partial v_j}{\partial\tau}(\mathbf{y},\tau)n_j\, dS(\mathbf{y})\, d\tau. \tag{3.1.5}$$

The last term vanishes when S is rigid. Thus, when the compact Green's function (1.11.16) is used, we find, correct to dipole order,

$$\begin{aligned}B(\mathbf{x},t) &\approx \frac{-1}{4\pi c_o|\mathbf{x}|}\left(\frac{x_i}{|\mathbf{x}|}-M_{oi}\right)\\&\quad\times\frac{\partial}{\partial t}\left\{\int[(\boldsymbol{\omega}\wedge\mathbf{v})\cdot\nabla Y_i]\,d^3\mathbf{y}-\nu\oint_S[\boldsymbol{\omega}\wedge\nabla Y_i\cdot\mathbf{n}]\,dS(\mathbf{y})\right\}\\&\approx\frac{-1}{4\pi c_o|\mathbf{x}|}\left(\frac{x_i}{|\mathbf{x}|}-M_{oi}\right)\frac{\partial}{\partial t}\int[(\boldsymbol{\omega}\wedge\mathbf{v}).\nabla Y_i]\,d^3\mathbf{y},\\&\qquad\qquad |\mathbf{x}|\to\infty,\quad \frac{v\ell}{\nu}\gg 1,\end{aligned} \tag{3.1.6}$$

where $\mathbf{M}_o = \mathbf{U}_o/c_o$, and [ ] now denotes evaluation at $t-(|\mathbf{x}|-\mathbf{M}_o\cdot\mathbf{x})/c_o$; in the second line the skin friction contribution has been discarded for high Reynolds number flow.

In the absence of mean flow, $B\equiv p/\rho_o$ in the acoustic far field and (3.1.6) then reduces to Curle's result (3.1.2). In the general case, the pressure as $|\mathbf{x}|\to\infty$ is calculated from the relation

$$\frac{\partial p}{\partial t}=\rho_o\left(\frac{\partial}{\partial t}+\mathbf{U}_o\cdot\nabla\right)B \Rightarrow p(\mathbf{x},t)\approx\frac{\rho_o B(\mathbf{x},t)}{\left(1+\frac{\mathbf{M}_o.\mathbf{x}}{|\mathbf{x}|}\right)},\quad M_o^2\ll 1. \tag{3.1.7}$$

When the solution is expressed in terms of the *emission time* coordinate ($\mathbf{x}=\mathbf{R}+\mathbf{M}_o R$) introduced in Section 1.8.5, the acoustic pressure becomes

$$\begin{aligned}p(\mathbf{R},t)&\approx\frac{-\rho_o R_i}{4\pi c_o R^2(1-M_o\cos\Theta)^3}\frac{\partial}{\partial t}\int(\nabla Y_i\cdot\boldsymbol{\omega}\wedge\mathbf{v})(\mathbf{y},t-R/c_o)\,d^3\mathbf{y}\\&=\frac{R_i}{4\pi c_o R^2(1-M_o\cos\Theta)^3}\frac{\partial F_i}{\partial t}(t-R/c_o),\\&\qquad\qquad M^2\ll 1,\quad R\to\infty.\end{aligned} \tag{3.1.8}$$

This formula is appropriate for a fixed observer in stationary fluid, through which the body translates at constant velocity $-\mathbf{U}_o$; $\mathbf{R}$ is the observer position relative to the body at the time of emission of the sound, and Θ is the angle between $\mathbf{R}$ and the direction of motion of the body at emission time. $\mathbf{F}$ is the force exerted on the fluid, and the time derivative in the first integral is taken at constant R. This result can also be derived from the Ffowcs Williams–Hawkings equation (2.2.10), although the analysis in terms of pressure rather than total enthalpy is more complicated because the radiation from the surface pressure dipoles must be augmented by a contribution from the interaction of the volume quadrupoles with the moving surface S [113].

According to (3.1.8) the acoustic pressure is modified by three Doppler factors $1/(1 - M_o \cos\Theta)$ as a result of the uniform motion of the body. This should be contrasted with (1.8.23b) for a point dipole, which only involves two Doppler factors. Other differences of this kind have already been noted in Sections 1.11.3 and 1.12.2 [28, 33, 113, 114].

By writing $\mathbf{v} = \mathbf{U} + \mathbf{u}$, the velocity $\mathbf{u}$ can be identified with the vorticity $\boldsymbol{\omega} = \operatorname{curl} \mathbf{u}$ impinging on the body. For streamlined bodies it frequently happens that $u \ll U$. No sound is generated when $\mathbf{u} = \mathbf{0}$ but, in contrast to sound production by the volume quadrupoles or when $\mathbf{U} = \mathbf{0}$, mean flow convection of vorticity past the body increases the amplitude of the sound from being proportional to u^2 to uU. This *linear* component of the radiation is furnished by the source $\operatorname{div}(\boldsymbol{\omega} \wedge \mathbf{U})$. Furthermore, $R_i \nabla Y_i . \boldsymbol{\omega} \wedge \mathbf{U} \equiv 0$ when $\mathbf{R}$ is parallel to $\mathbf{U}_o$ because $\mathbf{U} = U_{oi} \nabla Y_i$, so that the linear surface dipoles have no net component parallel to the mean velocity $\mathbf{U}_o$ at large distances from the body.

For a bluff body, the magnitude of the rotational velocity $\mathbf{u}$ in the wake is comparable to the mean flow speed, although for moderate Reynolds numbers the translational velocity of a typical vortical structure remains small until it is released into the flow. When this happens, the core accelerates to a streamwise velocity close to that of the mean flow, and a sound pulse is radiated. The periodic nature of these events is responsible for the Æolian tones produced by wires in a wind [78, 115, 116]. The conclusion above – that the intensity of the sound is strongest in directions at right angles to the mean flow – corresponds to dipole radiation produced by a fluctuating lift force, which reverses its direction as successive vortices are shed into the wake (Example 3).

3.1.3 Convected Entropy Inhomogeneities

In isentropic motion $(Ds/Dt = 0)$ of fluid of nonuniform mean density past a solid, it is necessary to retain the entropy source $-\operatorname{div}(T\nabla s)$ in equation (2.3.17) for B. At low Mach numbers, and when the acoustic field is required only to dipole order, this source can be approximated by the second term on the right of (2.3.21), $-\operatorname{div}(T\nabla s) \approx -\operatorname{div}((1/\rho_o - 1/\rho)\nabla p)$. Proceeding as in Section 3.1.2, the sound generated by the combined interactions of the vorticity and entropy inhomogeneities with the body is found to be

$$p(\mathbf{R}, t) \approx \frac{-\rho_o R_i}{4\pi c_o R^2(1 - M_o \cos\Theta)^3} \frac{\partial}{\partial t} \int \nabla Y_i \cdot \left(\omega \wedge \mathbf{v} - \left(\frac{1}{\rho_o} - \frac{1}{\rho}\right)\nabla p\right)$$

$$\times (\mathbf{y}, t - R/c_o)\, d^3\mathbf{y} = \frac{R_i}{4\pi c_o R^2(1 - M_o \cos\Theta)^3}$$

$$\times \frac{\partial}{\partial t}\left\{F_i(t - R/c_o) + \rho_o \int \left(\frac{1}{\rho_o} - \frac{1}{\rho}\right)\frac{\partial p}{\partial y_i}(\mathbf{y}, t - R/c_o)\, d^3\mathbf{y}\right\},$$

$$M^2 \ll 1, \quad R \to \infty, \qquad (3.1.9)$$

where $\mathbf{F}$ is the force exerted on the fluid by the body when compressibility and viscous stresses are neglected (Section 1.14.3, Example 7), and the integrated term in the second line is the radiation caused by the acceleration of the density inhomogeneities relative to their environment and corresponds to the identical term in (2.1.9) in the absence of the body.

3.1.4 Aerodynamic Sound Generated by Sources Near a Compact, Rigid Body at Low Mach Numbers

For future reference, we present here the general formula for calculating aerodynamic sound at low Mach numbers ($M^2 \ll 1$):

$$B(\mathbf{x}, t) \approx -\int \left((\omega \wedge \mathbf{v})_j - \left(\frac{1}{\rho_o} - \frac{1}{\rho}\right)\frac{\partial p}{\partial y_j}\right)(\mathbf{y}, \tau)$$

$$\times \frac{\partial G}{\partial y_j}(\mathbf{x}, \mathbf{y}, t - \tau)\, d^3\mathbf{y}\, d\tau,$$

$$p(\mathbf{x}, t) \approx \frac{\rho_o B(\mathbf{x}, t)}{\left(1 + \frac{\mathbf{M}_o \cdot \mathbf{x}}{|\mathbf{x}|}\right)}, \qquad (3.1.10)$$

where $\mathbf{U}_o$ is the uniform mean flow velocity at large distances from the body. G is given by the compact approximation (1.11.7) in the absence of mean flow ($\mathbf{U}_o = \mathbf{0}$), or by (1.11.16) in the presence of low Mach number irrotational flow past the body.

Example 1. A circular vortex ring of circulation Γ and small core radius translates by self-induction in stationary, ideal fluid directly toward a plane rigid wall. If the motion occurs at infinitesimal Mach number M, show that the efficiency of sound production is $\sim M^5$ and that the acoustic pressure can be expressed as in (2.3.20), in terms of Γ and an equal and opposite image vortex in the wall.

Example 2. Show from (3.1.5), using the compact Green's function and emission time coordinates, that the sound produced when the body executes small amplitude rotational and translational oscillations is given by

$$p(\mathbf{x}, t) \approx \frac{-R_i}{4\pi c_o R^2(1 - M_o \cos\Theta)^3} \frac{\partial}{\partial t}\left[\rho_o \int \nabla Y_i \cdot \boldsymbol{\omega} \wedge \mathbf{v}\, d^3\mathbf{y} \right.$$
$$\left. - \eta \oint_S \boldsymbol{\omega} \wedge \nabla Y_i \cdot \mathbf{n}\, dS - (\mathrm{M}_{ij} + m_o \delta_{ij}) \frac{dV_j}{dt}\right]_{t-R/c_o}$$
$$= \frac{R_i}{4\pi c_o R^2(1 - M_o \cos\Theta)^3} \frac{\partial}{\partial t}\left[F_i + m_o \frac{dV_i}{dt}\right]_{t-R/c_o},$$
$$M^2 \ll 1, \quad R \to \infty,$$

where **V** is the velocity of the center of volume of the body, **F** is the force on the fluid, and M_{ij} and m_o are the added mass tensor and the mass of fluid displaced by the body (cf. Section 14, Example 8).

Example 3: Æolian tones [116]. A vortex wake is formed in steady flow past a circular cylinder. Two symmetrically placed standing vortices appear behind the cylinder when the Reynolds number $\mathrm{Re} = U_o d/\nu > 5 - 15$ ($U_o =$ mean stream velocity, $d =$ cylinder diameter). The vortices are stretched downstream and break down when Re > 50, leading to a régime in which vortices are shed alternately and periodically from opposite sides of the cylinder. Further downstream the vortices arrange themselves into a double row, a "Kármán vortex street" (Figure 3.1.2). The distance between neighboring vortices in the same row is about $3.5d$, and each vortex is roughly opposite the midpoint of the interval between two vortices in the other row. For $40 < \mathrm{Re} < 5 \times 10^4$ the motion contains an identifiably periodic component, whose amplitude relative to the background, increasingly random flow decreases with increasing Reynolds number until the motion is fully turbulent when Re exceeds about 10^5. Before this occurs, however, the wake exhibits turbulent characteristics for Re > 300 and is diffuse by Re $\sim 2{,}500$. The frequency f (Hz) of the periodic component is given approximately by [117],

$$\mathrm{S_t} \equiv f d/U_o = 0.198(1 - 19.7/\mathrm{Re}), \quad \mathrm{Re} < 5 \times 10^5, \qquad (3.1.11)$$

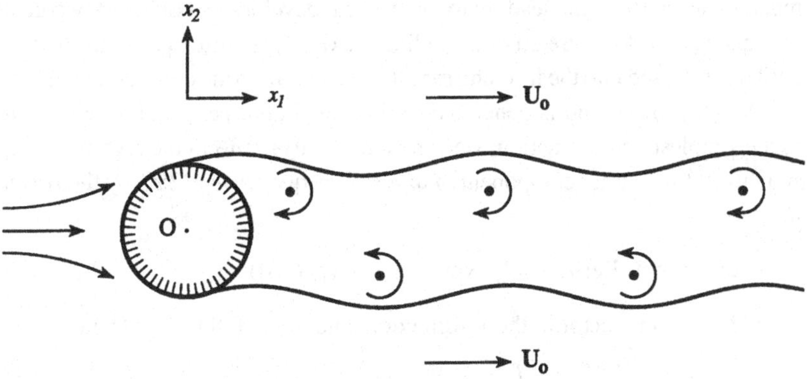

Figure 3.1.2. Generation of æolian tones.

where S_t is the *Strouhal number*. When vortices are present, they translate downstream at velocity $U_c \approx 3.5 S_t U_o$. For $\text{Re} \geq 3 \times 10^6$ a turbulent vortex street appears to be established with a variable Strouhal number exceeding 0.2 [118].

Vortex shedding is accompanied by periodic fluctuations in the lift leading to the production of sound of the same frequency. Fluctuations in the drag also occur, but they are typically an order of magnitude smaller [116, 117]. To estimate the amplitude of the radiation, assume that the cylinder is rigid and fixed and that the mean flow Mach number is negligible. Take the x_1-axis in the mean flow direction, and let the x_3-axis coincide with that of the cylinder.

The radiation is calculated from (3.1.2) with

$$Y_i = y_i\left(1 + \frac{d^2}{4\sqrt{y_1^2 + y_2^2}}\right), \quad i = 1,\ 2; \quad Y_3 = y_3. \qquad (3.1.12)$$

The cylinder is not necessarily compact in the axial direction, but this is accounted for by retaining retarded time differences between different sections of the cylinder in the argument of $\boldsymbol{\omega} \wedge \mathbf{v}$. This will not affect the following analysis, because the axial coherence length of the shed vortices normally does not exceed 5–10 diameters when the cylinder is fixed (and is less than about $3d$ when $\text{Re} > 300$), so that each correlated element of the wake near the cylinder can be assumed to be acoustically compact.

The convection velocity of the vortices is predominantly in the direction of the mean flow, and $\boldsymbol{\omega}$ is parallel to the cylinder, so that only the component $\partial Y_i/\partial y_2$ of ∇Y_i is significant. Now although vorticity streams out from the boundary layer at points of separation on the cylinder, the process may be regarded as quasi-static, leading to the gradual development of a vortex core at distance $\ell_o > d$ downstream of the cylinder axis. This vortex grows in strength until it is released into the flow and rapidly accelerates to the convection velocity U_c. Most of the sound is generated during this initial period of acceleration. In the simplest approximation, we can take $\partial Y_i/\partial y_2 \approx \delta_{i2}$ in the region $y_1 > \ell_o$ and model the Fourier component of $\boldsymbol{\omega} \wedge \mathbf{v}$ of frequency $\omega = 2\pi f$ by the *vortex sheet* source

$$\boldsymbol{\omega} \wedge \mathbf{v} \approx \mathbf{i}_2 \operatorname{Re}(\alpha(y_3) U_o^2 \exp[-i\omega(t - y_1/U_c)])\delta(y_2), \quad y_1 \geq \ell_o,$$

where $\mathbf{i}_2$ is a unit vector in the x_2-direction, and $|\alpha| \approx 0.8$ [117]. Then

$$F_2 \sim -\operatorname{Re}\left(\int dy_3 \int_{\ell_o}^{\infty} \alpha(y_3)\rho_o U_o^2 \exp[-i\omega(t - y_1/U_c)]\, dy_1\right).$$

The integration with respect to y_1 converges in the sense of a generalized function [19, 20], or by appealing to *causality* by first considering the case Im $(\omega) > 0$ (Section 1.7.4), which requires the vortices to decay gradually as they convect downstream. Thus,

$$F_2 \sim -\operatorname{Re}(0.56 i \rho_o U_o^2 d \int \alpha(y_3) \exp[-i\omega(t - \ell_o/U_c)]\, dy_3),$$

where the empirical relation $U_c \approx 3.5 S_t U_o$ has been used. Hence, if the correlation scale of the vortices in the axial direction is ℓ (defined by $\langle|\alpha|^2\rangle\ell = \int_{-\infty}^{\infty}\langle\alpha(0)\alpha(y_3)\rangle dy_3$, where $\langle\,\rangle$ denotes an ensemble or time average), the mean acoustic intensity $\mathrm{I}_a = \langle p^2/\rho_o c_o\rangle$ generated by a section of the cylinder of length L is

$$\mathrm{I}_a \approx 0.026 \rho_o \ell L U_o^3 M^3 S_t^2 \frac{\cos^2\theta}{|\mathbf{x}|^2}, \quad |\mathbf{x}| \to \infty, \tag{3.1.13}$$

where $\theta = \arccos(x_2/|\mathbf{x}|)$. A formula of this type was first given by Phillips [116], who estimated the lift from an analysis of experimental data and obtained the value of 0.016 instead of 0.026 for the numerical coefficient when $100 < \mathrm{Re} < 160$.

The intensity of the sound is significantly increased when the fluctuating lift causes the cylinder to vibrate. The motion correlates vortex shedding along the length of the cylinder to such an extent that ℓ may actually equal the total wetted length L. This is probably characteristic of sound production by wind blowing across telegraph wires, exposed tree branches, twigs, and so on. The effect is particularly noticeable if the natural frequency of vibration of a wire is close to the vortex shedding frequency, when the intensity can increase by a factor of several hundred.

3.2 Special Cases Treated by the Method of Compact Green's Functions

A deeper understanding of the mechanisms of sound production by fluid–structure interaction can be obtained from the study of model problems involving idealized distributions of vorticity and simplified surface geometries. When the source region is acoustically compact, the radiation can frequently be evaluated in analytical closed form. Typical cases at low Mach are discussed in this section.

3.2.1 Convection of Turbulence Through a Contraction

A hard-walled duct of infinite length contains fluid of mean density ρ_o and sound speed c_o in steady, irrotational flow at velocity $\mathbf{U}(\mathbf{x})$ and small Mach number M. The duct has uniform cross-sectional areas A_1 and A_2 upstream and downstream of a contraction of length L in which the mean velocity increases from U_1 to U_2 (where $A_1 U_1 = A_2 U_2$ for $M^2 \ll 1$; Figure 3.2.1). The unsteady drag exerted on the flow when turbulence convects through the contraction generates sound of dipole intensity and wavelength $\sim L/M \gg L$, which radiates as plane waves in the duct and augments the direct radiation from the quadrupoles.

The dipole radiation is determined by the vorticity term in the integral of (3.1.10), where $G(\mathbf{x}, \mathbf{y}, t - \tau)$ is the compact Green's function for the duct. Take the origin near the contraction, with x_1 measured in the axial direction of the mean flow. For observer positions at $x_1 \gg L$, *downstream* of the contraction [119]

$$G(\mathbf{x}, \mathbf{y}, t - \tau) = \frac{c_o}{A_1 + A_2} \mathrm{H}\left(t - \tau - \frac{x_1}{c_o(1 + M_2)} + \frac{(A_1 + A_2)Y(\mathbf{y})}{2A_2 c_o(1 + M_1)}\right), \quad x_1 \to +\infty, \qquad (3.2.1)$$

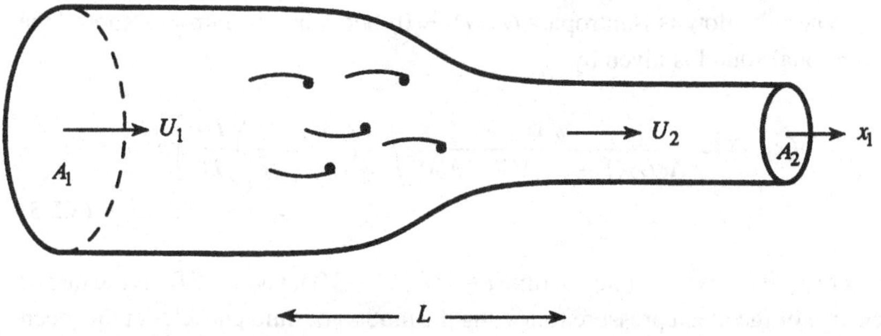

Figure 3.2.1. Turbulence convected through a contraction.

where $M_1 = U_1/c_o$, $M_2 = U_2/c_o$, and $Y(\mathbf{y})$ is the velocity potential of incompressible flow through the duct, normalized such that $Y \to 2A_1y_1/(A_1 + A_2)$, $2A_2y_1/(A_1 + A_2)$ respectively as $y_1 \to \pm\infty$.

According to (3.1.7), the acoustic pressure $p(x_1, t) \approx \rho_o B(x_1, t)/(1 + M_2)$ as $x_1 \to +\infty$. Hence,

$$p(x_1, t) \approx \frac{-\rho_o}{2A_2(1 + M_1)(1 + M_2)} \times \int (\nabla Y \cdot \boldsymbol{\omega} \wedge \mathbf{v})(\mathbf{y}, t - x_1/c_o[1 + M_2])\, d^3\mathbf{y}, \quad x_1 \to +\infty.$$

By writing $\mathbf{v} = \mathbf{U} + \mathbf{u}$, where $\mathbf{U}(\mathbf{y}) \equiv \{(A_1+A_2)/2A_1\}U_1\nabla Y(\mathbf{y})$ and $\boldsymbol{\omega} = \operatorname{curl} \mathbf{u}$ we see that *no* dipole sound is generated by the mean flow source term $\operatorname{div}(\boldsymbol{\omega} \wedge \mathbf{U})$ and that the amplitude of the sound, $\propto u^2$, is quadratic in the turbulence velocity. Using the identity $\boldsymbol{\omega} \wedge \mathbf{u} = (\mathbf{u} \cdot \nabla)\mathbf{u} - \nabla(\frac{1}{2}u^2)$ and the low Mach number approximation $\operatorname{div} \mathbf{u} = 0$, we find

$$p(x_1, t) \approx \frac{1}{2A_2(1 + M_1)(1 + M_2)} \times \int (\rho_o u_i u_j \bar{\epsilon}_{ij})(\mathbf{y}, t - x_1/c_o[1 + M_2])\, d^3\mathbf{y}, \quad x_1 \to +\infty, \tag{3.2.2}$$

where $\bar{\epsilon}_{ij} = \partial^2 Y/\partial y_i\, \partial y_j$ is proportional to strain tensor of the mean flow in the contraction. The integral is equal to the drag on the duct walls at the contraction and is nonzero even if Reynolds stress $\rho_o u_i u_j$ following a fluid particle is effectively *frozen* during convection through the contraction (although this is unlikely to be the case when A_1/A_2 is large [120, 121]). The acoustic efficiency $\sim O(M)$, which is a factor of order $1/M^2$ larger than the efficiency of the direct production of sound by turbulence quadrupoles in a duct (Section 2.1, Example 2).

When the flow is isentropic ($Ds/Dt = 0$) but contains entropy "spots," the additional sound is given by

$$p(x_1, t) \approx \frac{\rho_o A_1}{A_2 U_2(1 + M_1)(1 + M_2)} \int \left[\left(\frac{1}{\rho_o} - \frac{1}{\rho}\right)\frac{Dp}{Dt}\right] d^3\mathbf{y}, \quad x_1 \to +\infty, \tag{3.2.3}$$

where [ ] denotes evaluation at time $t - x_1/c_o(1+M_2)$, and Dp/Dt is the rate of change of the mean pressure following the motion of fluid particles at the mean velocity $\mathbf{U}$. The amplitude of the sound is determined by the rate of convection of the entropy inhomogeneities through the contraction, which occurs over time $\sim L/U$, and the characteristic acoustic wavelength $\sim L/M$.

For "hot spots" we can write $(\rho - \rho_o)/\rho \sim \Delta T/T_o$, where T_o is the mean temperature and ΔT a characteristic temperature variation. A comparison of the acoustic pressures produced by the Reynolds stress and entropic sources indicates that their respective amplitudes scale as $\rho_o u^2$ and $(\Delta T/T_o)\rho_o U^2$, and therefore that the entropy-generated sound can be dominant for quite moderate values of $\Delta T/T_o$ [122, 123].

Example 1. For source points $\mathbf{y}$ in the neighborhood of the contraction and $x_1 \to -\infty$ (upstream of the contraction), the compact Green's function (3.2.1) becomes

$$G(\mathbf{x}, \mathbf{y}, t - \tau) = \frac{c_o}{A_1 + A_2} \mathrm{H}\left(t - \tau + \frac{x_1}{c_o(1 - M_1)} - \frac{(A_1 + A_2)Y(\mathbf{y})}{2A_1 c_o(1 - M_2)}\right), \quad x_1 \to -\infty.$$

3.2.2 Aerodynamic Sources Near a Nozzle Exit

A semi-infinite, circular cylindrical duct terminates in an axisymmetric nozzle (Figure 3.2.2). Take the origin on the axis of symmetry in the nozzle exit plane, with x_1 directed axially out of the nozzle. The compact Green's function governing the generation of low-frequency sound in the *exterior* region by

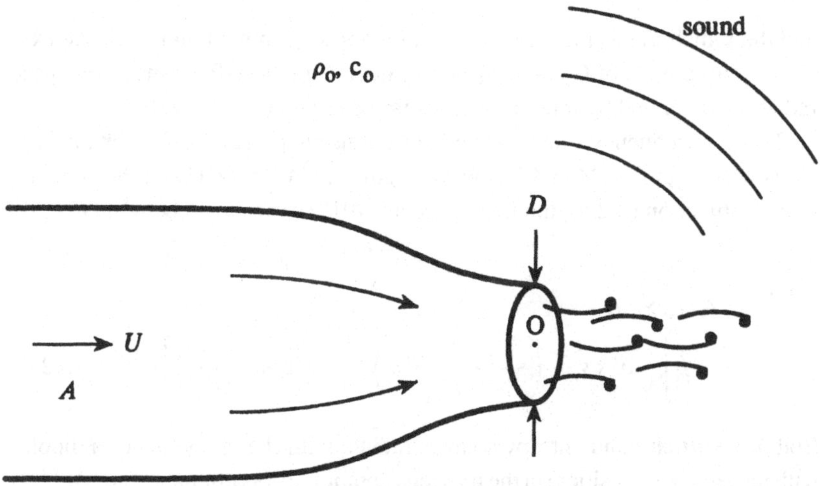

Figure 3.2.2. Unsteady flow at a nozzle exit.

sources close to the nozzle is (Example 2):

$$G(\mathbf{x}, \mathbf{y}, t-\tau) = \frac{1}{4\pi|\mathbf{x}|}\delta\left(t-\tau-\frac{(|\mathbf{x}-\mathbf{Y}(\mathbf{y})|-\bar{Y}(\mathbf{y}))}{c_o}\right), \quad |\mathbf{x}|\to\infty. \tag{3.2.4}$$

In this formula, $\bar{Y}(\mathbf{y})$ is the potential of incompressible flow from the nozzle normalized such that

$$\begin{aligned} \bar{Y} &\approx y_1 - \ell_1, \quad \text{for } |y_1| \gg D \quad \text{within the nozzle} \\ &\approx -A/4\pi|\mathbf{y}| \quad \text{as } |\mathbf{y}|\to\infty \quad \text{in free space,} \end{aligned}$$

where D is the nozzle exit diameter, A the cross-sectional area of the uniform duct upstream of the contraction, and $\ell_1 \sim O(D)$ is an "end correction" determined by the nozzle geometry [10, 124]. The potential $\mathbf{Y}(\mathbf{y}) \equiv (y_1 - \bar{Y}, Y_2, Y_3)$, where $Y_i, i = 2, 3$, is the potential of ideal flow in free space having unit speed in the i-direction at large distances from the nozzle, and such that $Y_2, Y_3 \to 0$ as $y_1 \to -\infty$ within the nozzle.

The function $\bar{Y}$ determines the monopole component of sound generated when an aerodynamic source near the nozzle causes the volume flux from the nozzle to fluctuate; $Y_{2,3}$ determine the sound generated by unsteady surface forces. When $y_1 \to -\infty$ within the nozzle,

$$G(\mathbf{x}, \mathbf{y}, t-\tau) = \frac{1}{4\pi|\mathbf{x}|}\delta(t-\tau-(|\mathbf{x}|-y_1)/c_o), \quad |\mathbf{x}|\to\infty, \tag{3.2.5}$$

and the radiation has the character of a monopole centered on the nozzle exit. This limiting form of G can be used to calculate the low-frequency free-space radiation generated by internal sources far upstream of the nozzle.

The low-frequency sound generated by a turbulent jet issuing from the nozzle at Mach number $M \ll 1$ has wavelength $\sim D/M \gg D$. Using the compact Green's function (3.2.4), the free-space acoustic pressure can be written

$$p(\mathbf{x}, t) \approx \frac{-\rho_o}{4\pi c_o|\mathbf{x}|}\frac{\partial}{\partial t} \times \int \left[\left(\omega\wedge\mathbf{v} + \left(\frac{1}{\rho_o}-\frac{1}{\rho}\right)\nabla p\right)\cdot\frac{\partial}{\partial \mathbf{y}}\left(\bar{Y} + \frac{\mathbf{x}\cdot\mathbf{Y}}{|\mathbf{x}|}\right)\right] d^3\mathbf{y}. \tag{3.2.6}$$

Both the vorticity and entropy sources radiate with the efficiency of a dipole, with respective intensities (in the previous notation) proportional to $\rho_o v^3 M^3/D$,

$\rho_o(\Delta T/T_o)^2 v^3 M^3/D$. The radiation actually consists of monopole and dipole components of comparable strengths (corresponding respectively to $\bar{Y}$ and $\mathbf{Y}$ in (3.2.6)). These fields of different multipole orders have the same radiation efficiencies because the monopole strength $q \sim O(M)$ relative to the dipole strength $\rho_o v^2$, as is shown by the following argument.

q is the volume flux from the nozzle (Section 1.6.1), and it is determined by the flow velocity averaged over the nozzle exit. The unsteady force exerted on the fluid upstream of the nozzle exit by the exterior turbulence fluctuations $\sim \rho_o v^2 A$. If the fluid were incompressible, this force would be incapable of accelerating the infinite mass of fluid within the nozzle, and the unsteady nozzle flux would vanish. For a compressible fluid, the effective length of the slug of fluid upstream of the nozzle that responds to this force is of the order of the acoustic wavelength $\lambda \sim D/M$, and its equation of motion is

$$\lambda \rho_o \partial q/\partial t \sim \rho_o v^2 A.$$

Thus, $q \sim v D^2 M$, and the corresponding sound pressure $\sim (D/|\mathbf{x}|)\rho_o v^2 M$, the same as for an aerodynamic dipole.

Real jets are unstable and become fully turbulent at some distance downstream of the nozzle. This occurs even under carefully controlled experimental conditions, when the flow in the nozzle is very smooth [125]. A prominent large-scale, axisymmetric instability of the jet, analogous to the periodic wake of a cylinder, is preferentially excited at a Strouhal number $fD/U \sim 0.5$. In terms of a system of cylindrical polar coordinates (r, ϑ, x_1), only the radial and axial components of $\boldsymbol{\omega} \wedge \mathbf{v}$ are non-zero for this mode. If the mean density is uniform, equation (3.2.6) and the condition $\int \boldsymbol{\omega} \wedge \mathbf{v}.\nabla Y_i d^3\mathbf{x} \equiv \mathbf{0},\ i = 2, 3$ imply that the sound produced by this mode is [126]

$$p(\mathbf{x}, t) \approx \alpha \frac{D}{|\mathbf{x}|} \rho_o U^2 M (1 - \cos\theta), \quad |\mathbf{x}| \to \infty, \quad M = U/c_o,$$

where U is the mean jet speed, α is proportional to the amplitude of the instability wave, and θ is the angle between the jet axis and the observer direction. The angular dependence shows that the peak acoustic pressures radiate in the "forward" direction, opposite to that of the jet velocity. A jet can also support azimuthal ("spinning") instability modes, proportional to $e^{in\vartheta}$. However, because the potentials Y_2, Y_3 in (3.2.4) must themselves be proportional to $e^{\pm i\vartheta}$, only the spinning modes $n = \pm 1$ can generate sound with the efficiency of a dipole. In that case, the acoustic pressure $\propto (D/|\mathbf{x}|)\rho_o U^2 M \sin\theta$, with the peak radiation occurring in directions at right angles to the jet.

Example 2. Compact Green's function for a nozzle. An approximate representation of $G(\mathbf{x}, \mathbf{y}; \omega)$ when $\kappa_o D \ll 1$ is obtained by application of the reciprocal theorem, with the reciprocal source at $\mathbf{x}$ in free space, as described in Section 1.11. In the neighborhood of the nozzle, as $|\mathbf{x}| \to \infty$,

$$\begin{aligned} G(\mathbf{x}, \mathbf{y}; \omega) &\approx G_0(1 - i\kappa_o x_j Y_j/|\mathbf{x}| + \alpha \bar{Y} + \ldots) \quad (\alpha = \text{constant}) \\ &\to G_0(1 + i\kappa_o x_1 \ell_1/|\mathbf{x}| + \alpha(y_1 - \ell_1) + \ldots), \\ &\qquad\qquad \text{as } y_1/D \to -\infty \text{ in the nozzle,} \end{aligned}$$

where $G_0 = -\mathrm{e}^{i\kappa_o|\mathbf{x}|}/4\pi|\mathbf{x}|$. In the duct, G consists of a plane wave propagating away from the nozzle exit

$$\begin{aligned} G(\mathbf{x}, \mathbf{y}; \omega) &\approx \beta G_0 \mathrm{e}^{-i\kappa_o y_1} (\beta = \text{constant}), \\ &\to \beta G_0(1 - i\kappa_o y_1 + \ldots) \text{ as } \kappa_o y_1 \to 0. \end{aligned}$$

The constants α and β are determined by equating the coefficients of y_1 in these asymptotic formulas, which are both applicable in the interval $1/\kappa_o \ll -y_1 \ll D$, yielding $\alpha \approx -i\kappa_o$ to leading order. $G(\mathbf{x}, \mathbf{y}, t - \tau)$ is now found by taking the inverse Fourier transform, as in Section 1.11.

Example 3. Show that for an observer at $x_1 \to -\infty$ within the nozzle of Figure 3.2.2 and in the limit of infinitesimal mean flow Mach number, the compact Green's function for aerodynamic sources at $\mathbf{y}$ in the vicinity of the nozzle is

$$G(\mathbf{x}, \mathbf{y}, t - \tau) \approx -\frac{\bar{Y}(\mathbf{y})}{A}\delta(t - \tau + x_1/c_o),$$

where $\bar{Y}$ is defined as in equation (3.2.4). Deduce that acoustic waves are generated within the duct by turbulence near the nozzle with an efficiency $\sim$O(M) and that the acoustic pressure is

$$p(x_1, t) \approx -(1/A)\int [\rho_o v_i v_j \partial^2 \bar{Y}/\partial y_i \partial y_j]_{t+x_1/c_o} d^3\mathbf{y}, \quad x_1 \to -\infty.$$

Example 4. If the uniform speed of sound in the duct of Figure 3.2.2 is $c_1 \neq c_o$, show that the nozzle Green's function (3.2.4) becomes

$$G(\mathbf{x}, \mathbf{y}, t - \tau) = \frac{1}{4\pi|\mathbf{x}|}\delta\left(t - \tau - \frac{|\mathbf{x} - \mathbf{Y}(\mathbf{y})|}{c_o} + \frac{\bar{Y}(\mathbf{y})}{c_1}\right), \quad |\mathbf{x}| \to \infty,$$

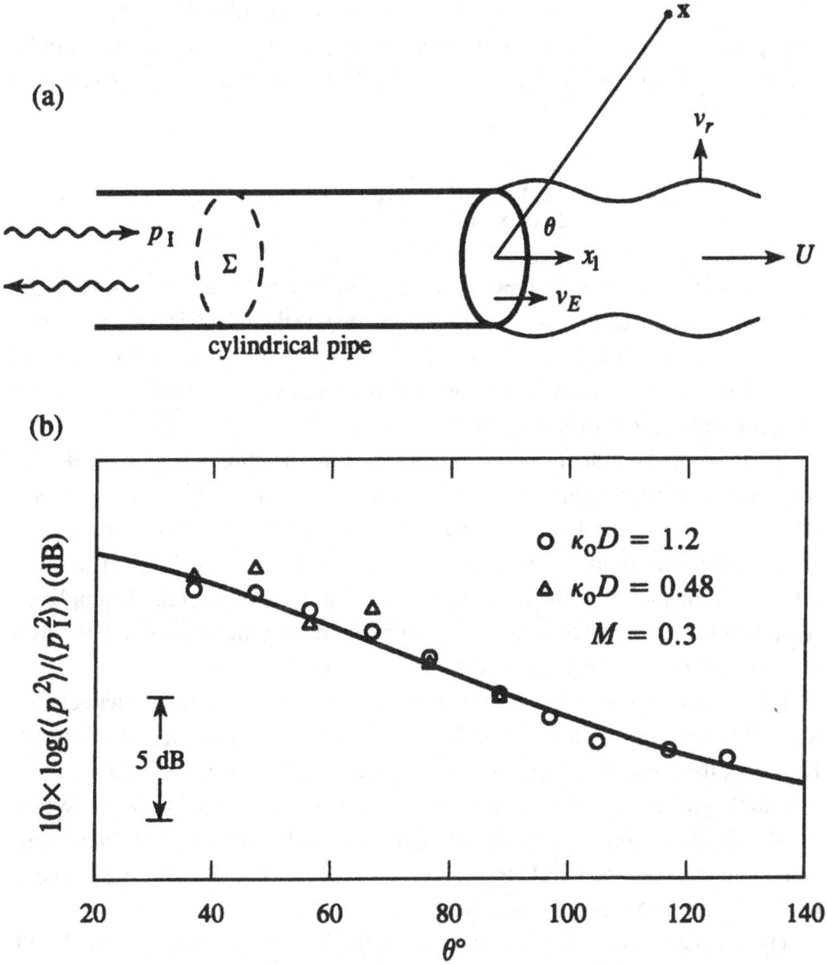

Figure 3.2.3. (a) Radiation from a jet pipe and (b) comparison of (3.2.11) with experiment [128].

3.2.3 Emission of Internally Generated Sound from Low Mach Number Nozzle Flow

Let us investigate the influence of flow from the nozzle of Figure 3.2.3 on the radiation of low-frequency sound generated upstream of the nozzle exit [127].

Consider first the case of no mean flow, when a plane wave $p_I(t - x_1/c_o)$ is incident on the nozzle exit from $x_1 = -\infty$ within the pipe. The wave may be assumed to be generated by a low-frequency volume source of strength $q(t)$ situated at distance s far upstream of the nozzle. Using the infinite duct Green's function (3.2.1), with $A_1 = A_2 = A$ we have $p_I = (c_o\rho_o/2A)q(t - (x_1 + s)/c_o)$. The pressure waves radiated out of the nozzle can be calculated in a similar manner by using the asymptotic form (3.2.5) of the nozzle Green's function, which yields

$$p(\mathbf{x}, t) \approx \frac{A}{2\pi c_o|\mathbf{x}|}\frac{\partial p_I}{\partial t}(t - |\mathbf{x}|/c_o), \quad |\mathbf{x}| \to \infty. \tag{3.2.7}$$

Because $\omega D/c_o \ll 1$, this shows that only a very small fraction $\sim\mathrm{O}(\omega D/c_o)^2$ of the incident acoustic power escapes from the nozzle, which is consistent with the approximation of classical acoustics [10], where the perturbation pressure is assumed to vanish at the open end, and the incident wave is reflected with a pressure reflection coefficient $\mathcal{R} = -1$.

To examine the influence of mean nozzle flow, the motion within the duct is assumed to be irrotational, with uniform mean velocity U and Mach number $M = U/c_o$ ($M^2 \ll 1$). The incident wave becomes $p(x_1, t) = p_I(t - x_1/c_o(1 + M))$. The sound produces vorticity fluctuations in the shear layer of the free jet, which interact with the nozzle to produce additional sound. A simplified treatment of the interaction can be given by modeling the free shear layer as a cylindrical vortex sheet lying in the continuation of the pipe.

If the mean flow were irrotational everywhere, it would have the character of a steady source flow from the nozzle, instead of the jet that is actually formed. However, the mean flow can always be represented as the sum of such an ideal potential flow and the flow generated by the mean vorticity in the jet. When the density is uniform, the equation governing both the mean and fluctuating components of the total enthalpy can then be taken in the form (3.1.3), where $\mathbf{U}$ denotes the irrotational source flow velocity.

The free-space radiation can be cast in the form (1.10.9) with φ replaced by B and $\hat{\mathcal{F}} = \mathrm{div}(\boldsymbol{\omega} \wedge \mathbf{v})$, and terms $\sim\mathrm{O}(M^2)$ are discarded. For a rigid walled duct, the surface integration may be confined to a plane control surface Σ (Figure 3.2.3a) conveniently positioned several nozzle diameters upstream of the exit, where the flow is parallel to the duct; at low frequencies this distance is much less than the acoustic wavelength. Because $\boldsymbol{\omega} \wedge \mathbf{v} = \mathbf{0}$ within the nozzle, the radiation condition ensures that only the component $B_I \approx (1 + M)p_I/\rho_o$ due to the incident wave makes a finite contribution to this integral. To evaluate it correct to $\mathrm{O}(M)$, the compact Green's function (3.2.4) must be modified to account for convection of sound by the flow in the duct. The reader may easily show that this is accomplished by replacing the function $\bar{Y}(\mathbf{y})$, where it appears explicitly in (3.2.4), by $\bar{Y}(\mathbf{y})/(1 + M)$. We then find that

$$p(\mathbf{x},t) \approx \frac{A(1+M)}{2\pi c_o|\mathbf{x}|}\frac{\partial p_{\mathrm{I}}}{\partial t}(t-|\mathbf{x}|/c_o) - \rho_o \int (\boldsymbol{\omega}\wedge\mathbf{v})(\mathbf{y},\tau)\frac{\partial G}{\partial \mathbf{y}}(\mathbf{x},\mathbf{y},t-\tau) \times d^3\mathbf{y}\,d\tau, \quad |\mathbf{x}|\to\infty. \tag{3.2.8}$$

Because the motion is axisymmetric, only the terms in Y_1 and $\bar{Y}$ in (3.2.4) contribute to the remaining integral. It vanishes in the linearized approximation when there is no mean flow. The leading order correction is therefore at least $O(M)$ and can be written

$$\frac{\rho_o(\cos\theta-1)}{4\pi c_o|\mathbf{x}|}\frac{\partial}{\partial t}\int\left[(\boldsymbol{\omega}\wedge\mathbf{v})_r\frac{\partial \bar{Y}}{\partial r}\right]d^3\mathbf{y} - \frac{\rho_o\cos\theta}{4\pi c_o|\mathbf{x}|}\frac{\partial}{\partial t}\int[(\boldsymbol{\omega}\wedge\mathbf{v})_1]\,d^3\mathbf{y}, \tag{3.2.9}$$

where $\theta=\arccos(x_1/|\mathbf{x}|)$, $r=\sqrt{y_2^2+y_3^2}$ is the radial distance from the jet axis in the cylindrical coordinate system (r,ϑ,y_1), and the square brackets are evaluated at time $t-|\mathbf{x}|/c_o$.

At very low frequencies, the fluctuating volume flux from the nozzle causes the strength of the cylindrical vortex sheet just downstream of the nozzle to vary quasi-statically over distances $\sim U/\omega$ large compared to D. The vortex lines are circular, and near the nozzle

$$\boldsymbol{\omega} = (U+v_E)\delta(r-D/2)\mathbf{i}_\vartheta,$$

where $\mathbf{i}_\vartheta$ is the unit vector in the azimuthal (ϑ) direction, and v_E is the perturbation velocity in the nozzle exit. The convection velocity of this vorticity is the average of the velocities on either side of the vortex sheet [4] and is therefore equal to $\frac{1}{2}(U+v_E)$ in the axial direction, so that for the unsteady components,

$$(\boldsymbol{\omega}\wedge\mathbf{v})_r \approx Uv_E\delta(r-D/2), \quad (\boldsymbol{\omega}\wedge\mathbf{v})_1 \approx -Uv_r\delta(r-D/2),$$

where $v_r(x_1,t)$ is the radial velocity of the vortex sheet. By noting that $\oint \partial\bar{Y}/\partial r\, dS = A$, where the integration is over the cylindrical vortex sheet, and similarly, when compressibility is neglected near the nozzle, that $\oint v_r dS \approx$ volume flux from the nozzle $\equiv Av_E$, the two terms (3.2.9) evaluate to

$$\frac{\rho_o(2\cos\theta-1)}{4\pi c_o|\mathbf{x}|}AU\frac{\partial v_E}{\partial t}(t-|\mathbf{x}|/c_o).$$

The velocity v_E is calculated in terms of p_I by noting that the pressure reflection coefficient of the sound incident on the nozzle exit can only differ by terms ~O(M) from its value $\mathcal{R} \approx -1$ in the absence of flow. The reflected acoustic wave is therefore $p \approx -p_I(t + x_1/c_o[1 - M])$, which, combined with the incident wave at $x_1 = 0$ yields $v_E \approx 2p_I(t)/\rho_o c_o$ (Section 1.6.3).

Collecting together these various results and substituting into (3.2.8), we find

$$p(\mathbf{x}, t) \approx \frac{A(1 + 2M\cos\theta)}{2\pi c_o|\mathbf{x}|} \frac{\partial p_I}{\partial t}(t - |\mathbf{x}|/c_o), \quad |\mathbf{x}| \to \infty. \tag{3.2.10}$$

Comparison with the no-flow result (3.2.7) shows that the interaction of the sound with the jet produces an acoustic dipole. Because $M^2 \ll 1$, the effect of flow can be expressed in terms of a Doppler factor correction by noting that $1 + 2M\cos\theta \approx 1/(1 - M\cos\theta)^2$. For sound of frequency ω, the mean square acoustic pressure is therefore

$$\langle p^2 \rangle \approx \frac{(\kappa_o A)^2}{4\pi^2|\mathbf{x}|^2} \frac{\langle p_I^2 \rangle}{(1 - M\cos\theta)^4}, \quad |\mathbf{x}| \to \infty. \tag{3.2.11}$$

This formula is compared with directivity measurements [128] in Figure 3.2.3b, for $\kappa_o D = 0.48$ and 1.2 when M = 0.3. The experiments used a convergent nozzle in which the upstream, uniform area of the duct was about 3.7 times the nozzle exit area A. More detailed calculations [129, 130] suggest that the field shape is not significantly dependent on the nozzle area ratio at low Mach numbers.

Example 5. Derive equation (3.2.10) from Curle's equation (2.2.4). Take S to be the exterior surface S_N of the nozzle and the nozzle cross section in the nozzle-exit plane, S_E. The contribution from S_N vanishes to leading order. On S_E: $\rho v \approx \rho_o v_E$, $\rho v_i v_j + p'_{ij} \approx 2\rho_o U v_E \delta_{i1}\delta_{j1}$ [130].

Example 6. The efficiency with which sound is generated by turbulence in an infinite, uniform duct ~O(M^3) (Section 2.1, Example 2). Deduce from equation (3.2.7) that turbulence far upstream of the open end of a uniform duct generates sound radiated into free space with the efficiency ~O(M^5) of *free-field* quadrupoles. This conclusion is not necessarily correct for ducts of finite length, when the excitation of resonant duct modes can produce large amplitude radiation from the open end.

3.2.4 Radiation from the Edge of a Large, Rigid Plate

A thin, semiinfinite rigid plate furnishes a tractable analytical model for investigating the production of sound by turbulence near the edge of a surface whose size is much larger than the acoustic wavelength [131, 132]. The compact Green's function for a half-plane does not exist in the form discussed previously because there is no surface dimension that determines the scale of the motion when vorticity interacts with the edge. However, the efficiency of sound generation by an eddy adjacent to an infinite, plane wall is uninfluenced by the wall (Section 3.1.1), so that any augmentation of the acoustic field must be the result of an interaction with the edge. The frequency of the interaction $\sim v/\ell_0$, where v is the velocity of the eddy and ℓ_0 is its distance from the edge. The wavelength of the edge generated sound is therefore of order ℓ_0/M $(M \sim v/c_o)$, so that the source region, which includes the edge and the eddy, is acoustically compact provided M is small.

Let us determine the influence of the edge on the sound of frequency ω generated by a time-harmonic aerodynamic source when $\kappa_o\ell_0$ and $M \ll 1$ [131]. In an ideal fluid, the velocity becomes unbounded like $1/\sqrt{r}$ as the distance r from the edge decreases [4]. The actual motion near the edge is controlled by viscosity, which must have some bearing on the production of sound, but for the moment this will be ignored. The radiation is given by the time-harmonic form of equation (3.1.10), where only the vorticity term is retained. Green's function has vanishing normal derivatives on S, and in the absence of mean flow

$$p(\mathbf{x}, \omega) \approx \rho_o \int (\boldsymbol{\omega} \wedge \mathbf{v})(\mathbf{y}, \omega) \cdot \frac{\partial G}{\partial \mathbf{y}}(\mathbf{x}, \mathbf{y}; \omega)\, d^3\mathbf{y}. \tag{3.2.12}$$

An expression for $G(\mathbf{x}, \mathbf{y}; \omega)$ can be written down in terms of Fresnel integrals [133, 134] or as an infinite series of cylindrical harmonics (Example 8), but we need only the approximate form of G for $\mathbf{y}$ close to the edge and $\mathbf{x}$ in the acoustic far field. Let the half-plane occupy the region $x_1 < 0,\ x_2 = 0$ (Figure 3.2.4 with $\chi = 0$), then [79, 131, 135]

$$G(\mathbf{x}, \mathbf{y}; \omega) = G_0(\mathbf{x}, \mathbf{y}; \omega) + G_1(\mathbf{x}, \mathbf{y}; \omega) + \cdots, \tag{3.2.13}$$

where, for $|\mathbf{x} - y_3\mathbf{i}_3| \to \infty$ and $\kappa_o\sqrt{y_1^2 + y_2^2} \ll 1$,

$$\begin{aligned} G_0(\mathbf{x}, \mathbf{y}; \omega) &= \frac{-1}{4\pi|\mathbf{x} - y_3\mathbf{i}_3|} e^{i\kappa_o|\mathbf{x} - y_3\mathbf{i}_3|}, \\ G_1(\mathbf{x}, \mathbf{y}; \omega) &= \frac{-1}{\pi\sqrt{2\pi i}} \frac{\sqrt{\kappa_o}\varphi^*(\mathbf{x})\varphi^*(\mathbf{y})}{|\mathbf{x} - y_3\mathbf{i}_3|^{3/2}} e^{i\kappa_o|\mathbf{x} - y_3\mathbf{i}_3|}. \end{aligned} \tag{3.2.14}$$

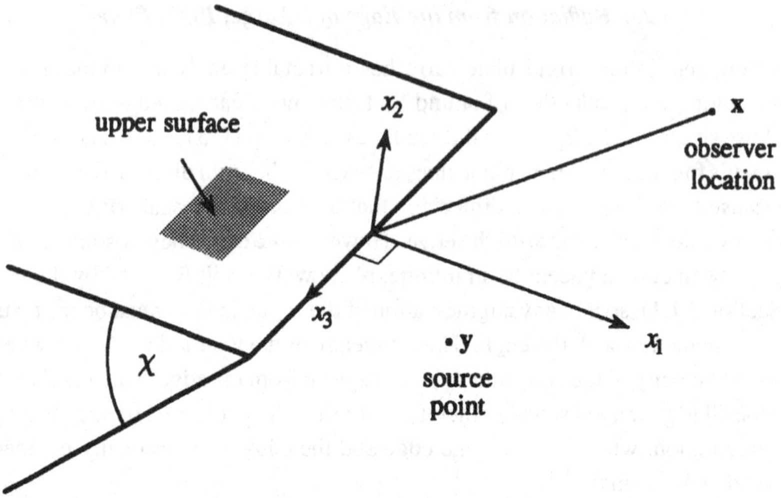

Figure 3.2.4. Coordinate system for the wedge.

$\mathbf{i}_3$ is a unit vector parallel to the x_3-axis (the edge), and

$$\varphi^*(\mathbf{x}) = \sqrt{r}\sin(\theta/2) \tag{3.2.15}$$

is a velocity potential of incompressible flow around the edge expressed in terms of polar coordinates $(x_1, x_2) = r(\cos\theta, \sin\theta)$. The component G_0 of G is the radiation from a point source at $\mathbf{y}$ when scattering is neglected; G_1 is the first correction due to presence of the half-plane.

Thus, the first approximation to integral in (3.2.12) is

$$p(\mathbf{x}, \omega) \approx \frac{\sqrt{\kappa_o}}{\pi\sqrt{2\pi i}} \int (\rho_o v_i v_j)(\mathbf{y}, \omega) \frac{\partial^2 \varphi^*(\mathbf{y})}{\partial y_i \partial y_j} \frac{\varphi^*(\mathbf{x})}{|\mathbf{x} - y_3\mathbf{i}_3|^{3/2}}$$
$$\times \exp(i\kappa_o|\mathbf{x} - y_3\mathbf{i}_3|)\, d^3\mathbf{y}, \quad |\mathbf{x} - y_3\mathbf{i}_3| \to \infty.$$

In the time domain $p(\mathbf{x}, t) = \int_{-\infty}^{\infty} p(\mathbf{x}, \omega)\mathrm{e}^{-i\omega t}\, d\omega$, but it is not necessary to evaluate the integral to estimate the acoustic efficiency. For an eddy of dimension ℓ and frequency v/ℓ, distance ℓ_0 from the edge, we find

$$p(\mathbf{x}, t) \sim \frac{\ell}{|\mathbf{x}|}\left(\frac{\ell}{\ell_0}\right)^{\frac{3}{2}} \rho_o v^2 \sqrt{M} \sin^{\frac{1}{2}}\psi \sin\left(\frac{1}{2}\theta\right),$$
$$M \sim v/c_o \ll 1, \quad |\mathbf{x}| \to \infty,$$

where ψ is the angle between the observer direction and the edge of the plate. The acoustic pressure is a maximum in the directions $\theta \approx \pm\pi$.

Thus, the acoustic efficiency $\sim(\ell/\ell_0)^3 M^2$, which is larger than the efficiency of turbulence near a compact solid or nozzle by a factor $1/M$. For the case of turbulence in a boundary layer flow over the edge, for example, the edge-generated sound power is the same as the direct acoustic power produced by turbulence quadrupoles (of efficiency $\sim M^5$) in a region of the boundary layer extending a distance $\ell_0/M^3 \gg \ell_0$ upstream of the edge. In water, this distance can be so large that *all* of the noise can be considered to be generated at the edge. The frequency of the edge noise will also tend to be higher by a factor $\sim U/v$, where U and v are the mean and fluctuating velocity components, because the frequency of the quadrupole sound $\sim v/\delta$, where δ is of the order of the boundary layer thickness (Section 3.4), whereas for edge noise the frequency $\sim U/\delta$ is set by the rate at which the turbulence is swept past the edge, where $\partial^2\varphi^*/\partial y_i \partial y_j$ is large [136].

The increased efficiency with which sound is generated is not directly related to the sharp edge, but rather to the presence of the adjacent rigid surface whose dimension greatly exceeds the acoustic wavelength. The same efficiency obtains for aerodynamic sources near the edge of any sufficiently large surface whose thickness is much smaller than the acoustic wavelength [132], irrespective of the precise geometrical form of the edge (wavy, variable thickness, etc.). This follows from the reciprocal theorem, which permits equations (3.2.13) and (3.2.14) to be interpreted as a representation of the pressure at $\mathbf{y}$ close to the edge produced by a distant point source at $\mathbf{x}$. Evidently, those formulae remain applicable at a distance $r_o = \sqrt{y_1^2 + y_2^2}$ from the mean position of an irregular edge provided $\kappa_o \ell \ll \kappa_o r_o \ll 1$, where ℓ is the length scale of the irregularities. The actual behavior of $G_1(\mathbf{x}, \mathbf{y}; \omega)$ at the edge would be obtained by replacing $\varphi^*(\mathbf{y})$ in (3.2.14) by the potential function $\Phi^*(\mathbf{y})$, say, that describes potential flow about the irregular edge, where $\Phi^*(\mathbf{y}) \to \varphi^*(\mathbf{y})$ when $r_o \gg \ell$.

Example 7. Determine the compact Green's function for two-dimensional sources near the edge of the half-plane. When the source has no dependence on the coordinate x_3 parallel to the edge, we can use the compact Green's function obtained by integrating (3.2.14) over $-\infty < y_3 < \infty$. This is done by the method of stationary phase (Section 1.13.2) for $\kappa_o\sqrt{x_1^2 + x_2^2} \to \infty$. Multiplying the result by $(-1/2\pi)e^{-i\omega(t-\tau)}$ and integrating over all frequencies, we find for the component G_1 of G

$$G_1(\mathbf{x}, \mathbf{y}, t - \tau) \approx \frac{\varphi^*(\mathbf{x})\varphi^*(\mathbf{y})}{\pi|\mathbf{x}|}\delta(t - \tau - |\mathbf{x}|/c_o), \quad |\mathbf{x}| \to \infty, \tag{3.2.16}$$

where $\mathbf{x} = (x_1, x_2)$, $\mathbf{y} = (y_1, y_2)$ in two dimensions. The two-dimensional counterpart of G_0 is independent of y and is therefore of use only in applications involving monopole sources, which are uninfluenced by the edge to first order.

Example 8: Green's function for a wedge. Verify that Green's function for the rigid wedge of Figure 3.2.4 (which is infinite in the x_3-direction) can be expanded in the form [133, 134]

$$G(\mathbf{x},\mathbf{y};\omega)=\frac{-i}{4}\int_{-\infty}^{\infty}\sum_{n=0}^{\infty}\sigma_n\big(\mathrm{H}_\nu^{(1)}(\gamma r_o)\mathrm{J}_\nu(\gamma r)\mathrm{H}(r_o-r)+\mathrm{H}_\nu^{(1)}(\gamma r o)\mathrm{J}_\nu(\gamma r_o)$$
$$\times\,\mathrm{H}(r-r_o)\big)\cos\nu(\theta-\pi)\cos\nu(\theta_o-\pi)\mathrm{e}^{ik_3(x_3-y_3)}\,dy_3,$$

in terms of the cylindrical polar coordinates

$$\mathbf{x}=(r\cos\theta, r\sin\theta, x_3),\quad \mathbf{y}=(r_o\cos\theta_o, r_o\sin\theta_o, y_3),$$

where $\mathrm{H}_\nu^{(1)}$, J_ν are Hankel and Bessel functions (Section 1.10.5), $\sigma_0=1/(2\pi-\chi)$, $\sigma_n=2\sigma_0\ (n\geq 1)$; $\nu=n\pi/(2\pi-\chi)$; $\gamma=\sqrt{\kappa_o^2-k_3^2}$, and $\mathrm{sgn}(\gamma)=\mathrm{sgn}(\omega)$ when γ is real, or γ is positive imaginary.

Show that

$$G(\mathbf{x},\mathbf{y};\omega)\approx\frac{-1}{2|\mathbf{x}-y_3\mathbf{i}_3|}\sum_{n=0}^{\infty}\sigma_n\mathrm{J}_\nu(\kappa_o r_o\sin\psi)\cos\nu(\theta-\pi)$$
$$\times\cos\nu(\theta_o-\pi)\mathrm{e}^{i(\kappa_o|\mathbf{x}-y_3\mathbf{i}_3|-\nu\pi/2)},\quad |\mathbf{x}-y_3\mathbf{i}_3|\to\infty \tag{3.2.17}$$

where $\mathbf{i}_3$ is a unit vector in the x_3-direction and ψ is the angle between $\mathbf{x}$ and the edge of the wedge. Use the formula $\mathrm{J}_\nu(z)\approx(\frac{1}{2}z)^\nu/\nu!$ for $|z|\ll 1$ to derive equations (3.2.14) for the half-plane.

Example 9. Show that the acoustic efficiency of low Mach number turbulence near the apex of a large, rigid wedge of interior angle $\chi\,(<\pi)\sim M^{(4\pi-\chi)/(2\pi-\chi)}$.

3.2.5 Influence of Vortex Shedding

The neglect of viscous boundary forces is a satisfactory approximation when the Reynolds number (Re) is large, provided $\rho_o v_i v_j$ (or the vorticity) is known everywhere. Indeed, for turbulence of length scale ℓ and frequency v/ℓ, the depth of viscous penetration from a stationary boundary $\sim O(\sqrt{\nu\ell/v})$, which is always much smaller than the dimension ℓ of the turbulent flow. However, an impinging turbulence eddy, or the mean flow, can induce "separation" at a sharp edge, where a substantial quantity of new vorticity appears in the flow. Accurate estimates of the sound produced in these circumstances must include contributions from the shed vorticity. It is still not usually necessary to include the viscous surface stresses, provided $\mathrm{Re} \gg 1$.

The importance of sound generation by shed vorticity can be established by considering the idealized flow depicted in Figure 3.2.5a [79]. A line vortex of circulation Γ is parallel to the edge ($x_3 = 0$) of a semi-infinite, rigid plate ($x_1 < 0,\ x_2 = 0$). If $\mathbf{x}_o(t) = (x_{o1}, x_{o2})$ is the position of the vortex at time t, then in the absence of vortex shedding $\boldsymbol{\omega} = \Gamma \mathbf{i}_3 \delta(\mathbf{x} - \mathbf{x}_o(t))$ and the sound generated by the vortex–edge interaction is given by the vorticity term in (3.1.10) and Green's function (3.2.16). The result can be written

$$p(\mathbf{x}, t) \approx \frac{\rho_o \Gamma \sin(\theta/2)}{\pi\sqrt{|\mathbf{x}|}} \left[\frac{\partial \mathbf{x}_o}{\partial t} \cdot \nabla \psi^*\right] \equiv \frac{\rho_o \Gamma \sin(\theta/2)}{\pi\sqrt{|\mathbf{x}|}} \left[\frac{D\psi^*}{Dt}\right],$$

$$|\mathbf{x}| \equiv \sqrt{x_1^2 + x_2^2} \to \infty, \qquad (3.2.18)$$

where $[D\psi^*/Dt]$ is evaluated at the retarded position of the vortex and $\mathbf{x} = r(\cos\theta, \sin\theta)$. The function $\psi^*(\mathbf{x}) = -\sqrt{r}\cos(\theta/2)$ is the imaginary part of the complex potential $w = \varphi^* + i\psi^* \equiv -i\sqrt{z}$ of ideal flow around the edge, where $z = x_1 + ix_2$.

$\psi^* =$ constant on each of the parabolic streamlines of the potential flow φ^* around the edge. A vortex that translates along one of these streamlines would be "silent," and the actual edge-noise depends on the rate at which the trajectory of the vortex cuts across the streamlines of this ideal edge-flow.

We can now form the following qualitative picture of the influence of vortex shedding on sound generation. Let the circulation Γ be in the sense indicated Figure 3.2.5, so that fluid near the plate is induced to flow in a clockwise direction around the edge, as indicated by the dashed curve in Figure 3.2.5a, and is impeded by viscous stresses. When the Reynolds number is large, separation occurs near the edge and vorticity of opposite sign is shed into the flow. Let us assume for simplicity that the shed vorticity rolls up into a concentrated core of strength Γ_s. Equation (3.2.18) gives the total radiated sound as

$$p(\mathbf{x}, t) \approx \frac{\rho_o \sin(\theta/2)}{\pi\sqrt{|\mathbf{x}|}} \left(\Gamma \left[\frac{D\psi^*}{Dt}\right]_\Gamma + \Gamma_s \left[\frac{D\psi^*}{Dt}\right]_{\Gamma_s}\right), \quad |\mathbf{x}| \to \infty,$$

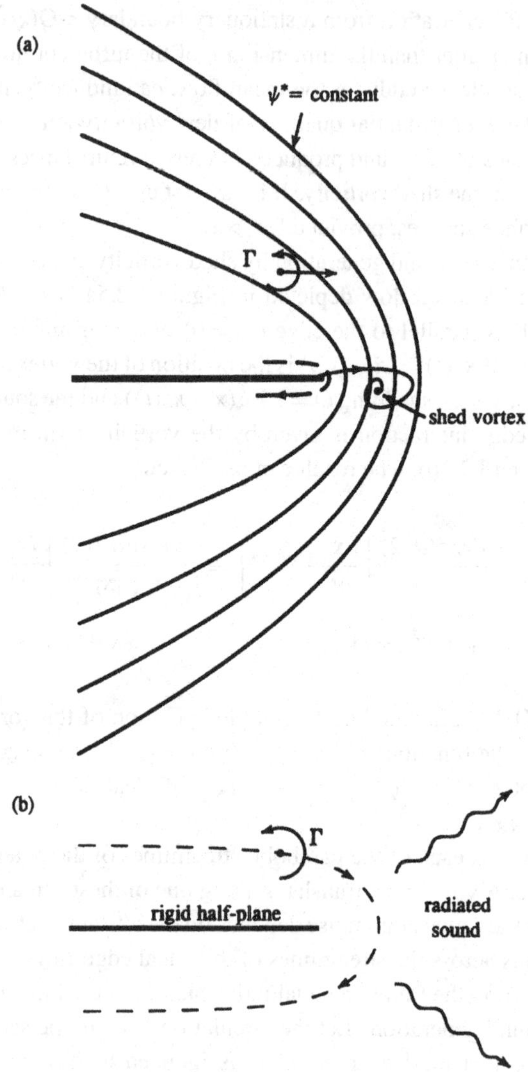

Figure 3.2.5. Vortex motion at an edge.

where the derivatives are evaluated at the retarded positions of Γ and Γ_s, respectively. Both vortices translate across the curves $\psi^* =$ constant in the direction of decreasing ψ^*, and the derivatives therefore have the same sign. Hence, because Γ and Γ_s have opposite signs, sound produced by the shed vortex will tend to cancel the edge-generated sound attributable to the incident vortex Γ

alone. A conclusion of this kind is valid for a wide range of fluid–structure interactions, although not necessarily without qualification in cases where the back-reaction of the shed vorticity on the flow produces an overall increase in the amplitude of the motion (e.g., in wind instruments and other resonant systems, Section 6.3).

Example 10. Use the method of conformal transformation [4] to show that in an ideal fluid, at rest at infinity, the vortex Γ in Figure 3.2.5b traverses a path given by the polar equation

$$r_o = \ell \sec\left(\frac{1}{2}\theta_o\right) \quad (-\pi < \theta_o < \pi),$$

where ℓ is the distance of closest approach to the edge of the plate, provided the Mach number $M = \Gamma/8\pi\ell c_o \ll 1$ (cf. Section 1.14, Example 10). If the vortex is at $\mathbf{x}_o = (-\infty, 2\ell)$ at $t = -\infty$, and crosses the x_1-axis at $t = 0$, show further that

$$r_o = \ell\sqrt{1 + (\Gamma t/8\pi\ell^2)^2}, \quad \theta_o = 2\arctan(-\Gamma t/8\pi\ell^2)$$

and that, when vorticity production at the edge is ignored, the acoustic pressure is [137]

$$p(r,\theta,t) \approx \frac{\rho_o\Gamma^2}{(4\pi\ell)^2}\left(\frac{\ell}{|\mathbf{x}|}\right)^{\frac{1}{2}} \sin\left(\frac{1}{2}\theta\right)\left[\frac{\Gamma t/8\pi\ell^2}{[1 + (\Gamma t/8\pi\ell^2)^2]^{5/4}}\right]_{t-|\mathbf{x}|/c_o}, \quad r \equiv |\mathbf{x}| \to \infty.$$

When r is large the total radiated energy per unit length is

$$E_o = \int_{-\infty}^{\infty} dt \int_0^{2\pi} (p^2(r,\theta,t)/\rho_o c_o) r\, d\theta = \frac{32\pi}{3}\rho_o U^2 M\ell^2, \tag{3.2.19}$$

where $M = U/c_o$, $U = \Gamma/8\pi\ell$.

Example 11. The method of Section 1.14.5 can be used to illustrate the effect of vortex shedding on the solution of Example 10. The shed vorticity is modeled by a concentrated core (line vortex) as in Figure 1.14.5b. Equation (3.2.18) is applied to each vortex by replacing $D\psi^*/Dt$ by $\mathbf{v}.\nabla\psi^*$, where $\mathbf{v}$ is the *fluid velocity* evaluated at the vortex axis (the singular, rotational component of velocity being discarded). As explained in Section 1.14.5, this velocity is the same as the convection velocity of the vortex only when the circulation is constant. The incident vortex is labeled Γ_o in Figure 3.2.6a; $\Gamma \equiv \Gamma(t)$ is the shed

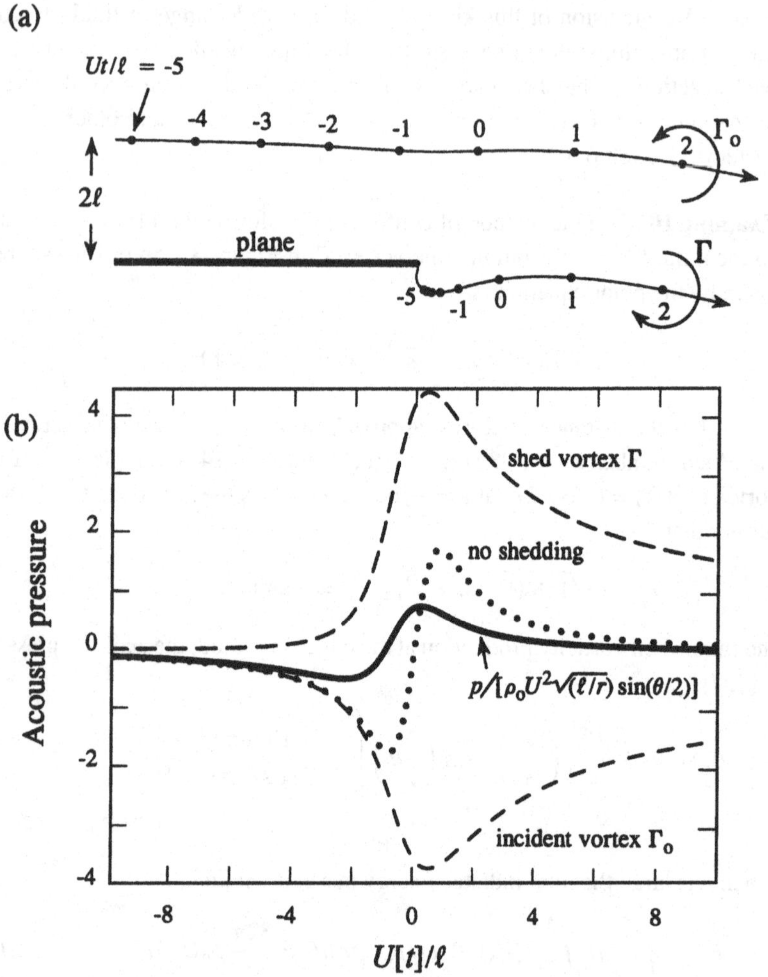

Figure 3.2.6. (a) Vortex trajectories and (b) acoustic pressures.

vortex, and the vortex locations are indicated at different times. The solid curve in Figure 3.2.6b is the net acoustic pressure signature $p/[\rho_o U^2 \sqrt{\ell/r}\sin(\theta/2)]$, where $U = \Gamma_o/8\pi\ell$; the broken curves are the separate contributions from the two vortices, which are effectively equal and opposite at positive retarded times, when the amplitude of the sound becomes very much less than that calculated in Example 10 in the absence of shedding (shown dotted in the figure). Very little sound is generated by the shed vorticity prior to the arrival of the incident vortex at the edge, and the acoustic pressure then coincides with that in the absence of shedding.

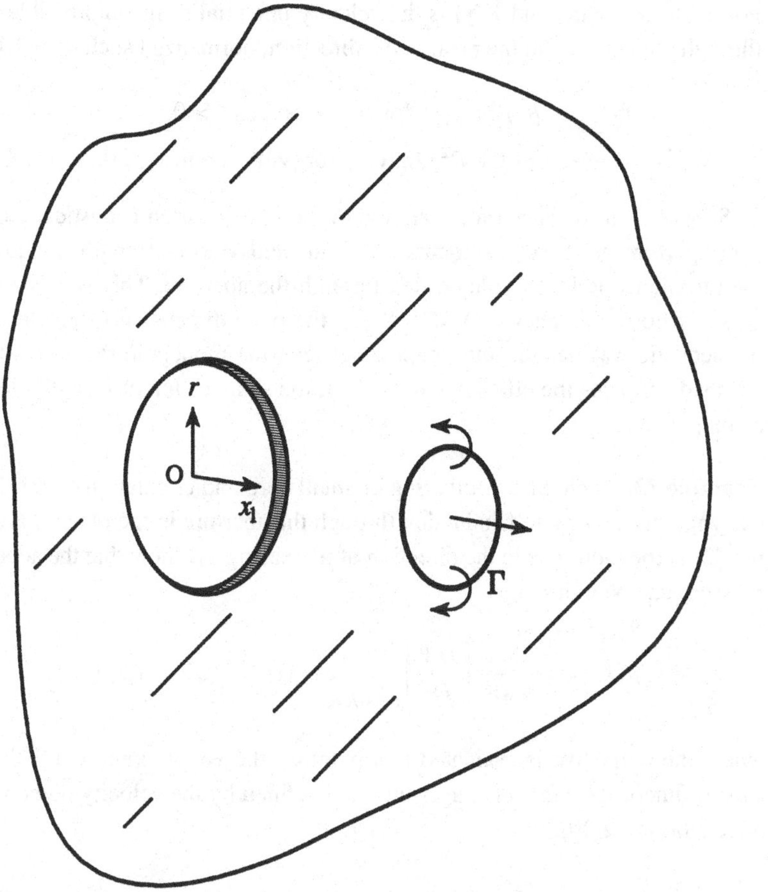

Figure 3.2.7. Vortex ring and circular aperture.

Numerical evaluation of the integral (3.2.19) for the radiated energy E with shedding reveals that $E/E_o \approx 0.18$ (~ -7.5dB), where E_o is the value in the absence of shedding.

Example 12. Derive the compact Green's function for aerodynamic sources in the neighborhood of a circular aperture of radius R in an infinite, thin rigid plate (Figure 3.2.7):

$$G(\mathbf{x}, \mathbf{y}, t-\tau) \approx \frac{1}{2\pi|\mathbf{x}|}\left(1+\frac{2\bar{Y}(\mathbf{y})}{R}\right)\delta(t-\tau-|\mathbf{x}|/c_o),$$

$$|\mathbf{x}| \to \infty, \quad x_1 > 0, \qquad (3.2.20)$$

where the coordinate origin is taken in the center of the aperture, the x_1-axis is normal to the plate, and $\bar{Y}(\mathbf{y})$ is the velocity potential of incompressible flow through the aperture in the positive y_1-direction, normalized such that [10]

$$\begin{aligned} \bar{Y}(\mathbf{y}) &\sim -R^2/2\pi|\mathbf{y}| \quad \text{for } |\mathbf{y}| \to \infty \text{ in } y_1 > 0, \\ &\sim -R/2 + R^2/2\pi|\mathbf{y}| \quad \text{for } |\mathbf{y}| \to \infty \text{ in } y_1 < 0. \end{aligned}$$

Show that turbulence on either side of the plate near an acoustically compact aperture generates a monopole sound field whose strength is equal to the turbulence-induced volume flux through the aperture. This is a powerful source whose efficiency $\sim O(M)$. When the plate dimension is smaller than the acoustic wavelength, interference between the monopole radiations from each side reduces the efficiency to $O(M^3)$, the same as for surface dipoles on a compact body.

Example 13. A circular vortex ring of small core and circulation Γ translates axisymmetrically by self-induction through the aperture in the plate of Example 12. If the motion is in the direction of increasing x_1, show that the acoustic pressure can be written

$$p(\mathbf{x}, t) \approx \frac{2\rho_o \Gamma}{|\mathbf{x}|} \left[\frac{D\Psi}{Dt} \right]_{t - |\mathbf{x}|/c_o}, \quad |\mathbf{x}| \to \infty, \quad x_1 > 0,$$

where the derivative is evaluated at a point on the vortex core, and Ψ is the stream function for the ideal aperture flow defined by the velocity potential $\bar{Y}$ of equation (3.2.20).

3.3 Vortex–Airfoil Interaction Noise

The generation of sound by turbulence or discrete vortices impinging on airfoils and control surfaces is of great practical importance. It occurs, for example, when helicopter main-rotor blade-tip vortices are cut by following blades or ingested by the helicopter tail rotor [138–144]. With few exceptions [145], the sound produced by these and related interactions involving high-speed flow and complex geometries must be determined by numerical integration of the acoustic analogy equation or the full Navier–Stokes equations [139, 146–149]. A limited number of problems can be treated analytically, especially at low Mach numbers, and these are discussed in this section. Their solutions provide much insight and can be used to validate computer predictions; they are also relevant to many industrial and marine applications where the characteristic Mach number is small.

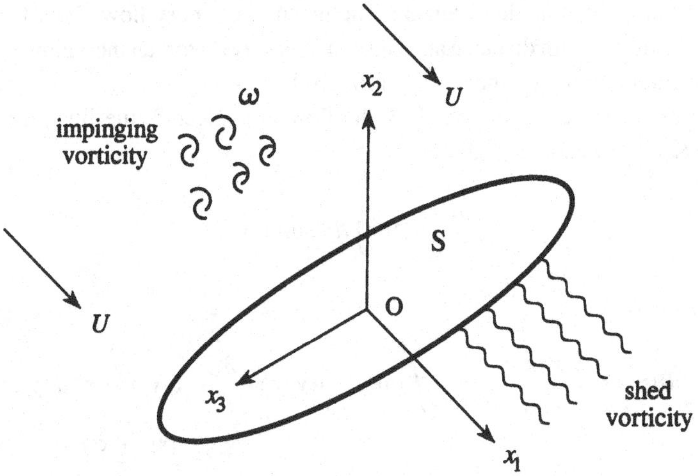

Figure 3.3.1. Vorticity in mean flow past an airfoil.

Consider an inhomogeneous field of vorticity, a "gust," convected in high Reynolds number, homentropic flow past a stationary rigid airfoil (Figure 3.3.1). The undisturbed flow has speed U in the x_1-direction, where the origin is at a convenient point within the airfoil, with x_3 along the span and x_2 vertically "upward." The Mach number $M = U/c_o$ is taken to be sufficiently small that convection of sound by the flow can be ignored. Doppler factor corrections for finite values of M (when $M^2 \ll 1$) may be introduced when appropriate, as in equation (3.1.8).

The aerodynamic sound problem can be formulated in terms of the exact equation (2.3.14) for homentropic flow. The sources include vorticity in the gust together with any shed from the airfoil, either in response to excitation by the gust or as "tip" vortices responsible for the mean lift. The equation is linearized by assuming that $u \ll U$, where curl $\mathbf{u} = \boldsymbol{\omega}$, that is, by requiring the gust velocity, and the perturbation velocities caused by airfoil thickness, twist, camber, and the angle of attack, to be small. When the source term $\operatorname{div}(\boldsymbol{\omega} \wedge \mathbf{v})$ is expanded about the undisturbed mean flow, only the gust vorticity and additional vorticity shed when the gust encounters the airfoil contribute to the acoustic radiation to first order. In other words, thickness, twist, camber, and angle of attack may be ignored, and the airfoil may be regarded as a rigid lamina in the plane $x_2 = 0$. In this approximation, quadrupoles are neglected, and the vorticity convects as a *frozen* pattern of vortex filaments at the undisturbed mean stream velocity $\mathbf{U} = (U, 0, 0)$. In particular, the wake vorticity is confined to a vortex sheet downstream of the trailing edge. Linearization about the undisturbed mean flow is not always satisfactory, however, especially in two dimensions for an airfoil at a mean angle of attack: The mean circulatory flow about the airfoil decays slowly with distance and can considerably distort an incoming gust prior to its interaction with the airfoil [150, 151].

When convection of sound by the flow is neglected, the linearized equation (2.3.14) becomes

$$\left(\frac{1}{c_o^2}\frac{\partial^2}{\partial t^2} - \nabla^2\right) B = \operatorname{div}(\boldsymbol{\omega} \wedge \mathbf{U}), \tag{3.3.1}$$

with solution

$$B(\mathbf{x}, t) = \frac{p(\mathbf{x}, t)}{\rho_o} = -\int (\boldsymbol{\omega} \wedge \mathbf{U})(\mathbf{y}, \tau) \cdot \frac{\partial G}{\partial \mathbf{y}}(\mathbf{x}, \mathbf{y}, t - \tau)\, d^3\mathbf{y}\, d\tau, \quad |\mathbf{x}| \to \infty, \tag{3.3.2}$$

where $\partial G/\partial y_2 = 0$ on both sides ($y_2 = \pm 0$) of the projection of the airfoil planform onto the $y_1 y_3$-plane. At sufficiently small Mach numbers, the airfoil chord will be acoustically compact, and G may be approximated by the compact Green's function. The solution obtained by this means for small but finite Mach numbers (when $M^2 \ll 1$, but $M < 1$) can be corrected by incorporating Doppler factors as in (3.1.8).

3.3.1 Blade–Vortex Interactions in Two Dimensions [152]

Consider a two dimensional airfoil of chord $2a$, and suppose the flow is independent of the spanwise coordinate x_3. The airfoil is replaced by a flat, rigid strip at zero angle of attack (Figure 3.3.2), which occupies $-a < x_1 < a$, $x_2 = 0$.

When $M \ll 1$, the wavelength of the sound is much larger than the chord, and the radiation integral (3.3.2) may be evaluated by using the compact Green's function (1.11.7), where

$$Y_1 = y_1, \quad Y_2 = \operatorname{Re}\left(-i\sqrt{z^2 - a^2}\right), \quad Y_3 = y_3; \quad z = y_1 + iy_2. \tag{3.3.3}$$

In two dimensions, $\boldsymbol{\omega} = (0, 0, \omega_3(x_1, x_2, t))$, $\boldsymbol{\omega} \wedge \mathbf{U} = (0, \omega_3 U, 0)$, and equation (3.3.2) yields

$$p(\mathbf{x}, t) \approx \frac{-\rho_o x_2}{4\pi c_o}\frac{\partial}{\partial t}\int (\nabla Y_2 \cdot \boldsymbol{\omega} \wedge \mathbf{U})(y_1, y_2, t - |\mathbf{x} - y_3\mathbf{i}_3|/c_o)\frac{d^3\mathbf{y}}{|\mathbf{x} - y_3\mathbf{i}_3|^2}, \quad r \equiv \sqrt{x_1^2 + x_2^2} \to \infty, \tag{3.3.4}$$

where the dependence of the retarded time on y_3 has been retained because the source extends over the infinite interval $-\infty < y_3 < \infty$. This formula simplifies

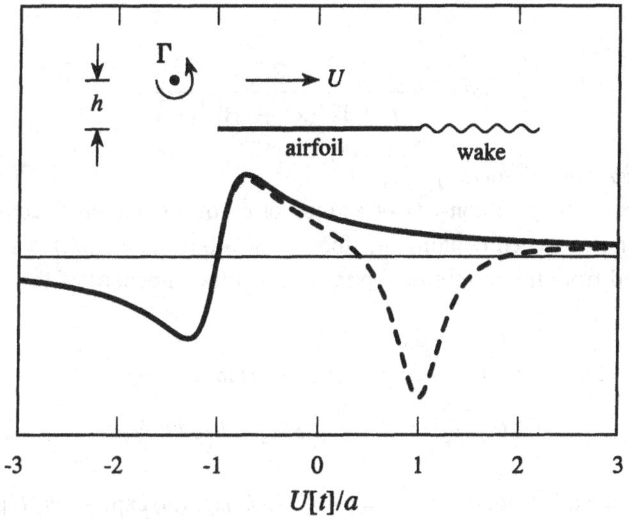

Figure 3.3.2. Blade–vortex interaction in two dimensions.

in the frequency domain. When the y_3-integration is performed by the method of stationary phase (cf. Section 3.2, Example 7), we find

$$p(r, \Theta, \omega) \approx \frac{\cos\Theta}{2}\sqrt{\frac{\kappa_o}{2\pi i r}}\,F_2(\omega)\mathrm{e}^{i\kappa_o r}, \quad \kappa_o r \to \infty,$$

where $\Theta = \arccos(x_2/r)$ and $F_2 \equiv -\rho_o \int \nabla Y_2 \cdot \boldsymbol{\omega} \wedge \mathbf{U}\, dy_1\, dy_2$ is the force exerted on the fluid per unit span of the airfoil when the motion is regarded as incompressible (Section 1.14.3). This is a dipole pressure field, with peak radiation in directions normal to the airfoil. Transforming back to the time domain,

$$p(r, \Theta, t) \approx \frac{\cos\Theta}{2\sqrt{2\pi i c_o r}}\int_{-\infty}^{\infty} \sqrt{\omega}\,F_2(\omega)\mathrm{e}^{-i\omega(t-r/c_o)}d\omega, \quad r \to \infty. \tag{3.3.5}$$

The path of integration passes just above the real axis (Section 1.7), so that $\sqrt{\omega} = i\,|\omega|^{1/2}$ when $\omega < 0$.

Let $\Omega(x_1, x_2, t)\mathbf{i}_3$ denote the vorticity in the incident gust. It convects with the mean flow and is actually a function of $x_1 - Ut$ and x_2. In the frequency domain, the x_2-component of velocity induced by the gust in the plane $x_2 = 0$ of the airfoil (the "upwash") can therefore be written $u_2(\omega)\mathrm{e}^{i\omega x_1/U}$. According to *thin airfoil theory* [152, 153], the force F_2 is given in terms of u_2 by

$$F_2(\omega) = -2\pi a \rho_o u_2(\omega) U S(\omega a/U), \tag{3.3.6}$$

where

$$\mathcal{S}(x) = \frac{2}{\pi x \left(\mathrm{H}_0^{(1)}(x) + i\mathrm{H}_1^{(1)}(x) \right)} \tag{3.3.7}$$

is the *Sears function* [154].

Suppose the gust consists of a line vortex of circulation Γ convecting at the mean flow speed U along the line $x_2 = h > 0$ (Figure 3.3.2). When time is measured from the instant the vortex crosses the midchord of the airfoil, we have

$$\Omega(x_1, x_2, t) = \Gamma\delta(x_1 - Ut)\delta(x_2 - h),$$

$$\Omega(x_1, x_2, \omega) = \frac{\Gamma}{2\pi U}\delta(x_2 - h)\mathrm{e}^{i\omega x_1/U},$$

and the upwash velocity $u_2(\omega) = (-i\Gamma/4\pi U)\mathrm{sgn}(\omega)\exp(-|\omega h/U|)$ (Example 1). The sound generated by the vortex–airfoil interaction can now be calculated from (3.3.5) and (3.3.6). The integral in (3.3.5) cannot be evaluated in closed form, but the strong interaction case $h \ll a$, where the vortex passes close to the airfoil, is approximated asymptotically by [155]

$$p(r, \Theta, t) \approx \frac{\rho_o \Gamma U \cos\Theta}{4\pi}\sqrt{\frac{M}{ra}}\left[\frac{(Ut + a)a/h^2}{1 + (Ut + a)^2/h^2}\right]_{t - r/c_o}, \quad r \to \infty.$$

The pressure signature (illustrated for $h/a = 0.2$ by the solid curve in Figure 3.2.2) is generated predominantly as the vortex passes the leading edge of the airfoil at $t = -a/U$ and has characteristic frequency $\omega \sim U/h$. The dominance of sound generation at the leading edge is typical of all interactions occurring at high *reduced frequencies* $\omega a/U (\sim a/h \gg 1)$ and occurs for the reason discussed in Section 3.2.5: At high Reynolds numbers, the induced velocity near the trailing edge is very large, and the inertia of fluid approaching the edge causes the flow to separate and shed new vorticity into the wake of the airfoil. When the reduced frequency is large, the sound generated by the wake is just equal and opposite to that produced by the incident vortex interacting with the edge. This conclusion can also be interpreted in terms of the surface pressure fluctuations. According to linear theory, a gust convecting at the mean stream velocity U produces no fluctuations in pressure except when the upwash interacts with the edges of the airfoil. At high reduced frequencies, the surface pressure generated when the vortex is upstream of the trailing edge is very small and remains small as it convects over the edge because the shed vorticity cancels the upwash that would otherwise be generated. The linear theory wake consists of a vortex sheet downstream of the edge, whose elements convect at the mean velocity U. Thin airfoil theory determines the wake vorticity by imposing the Kutta condition (Section 1.14.4) that the pressure (and velocity) should be finite at the edge. Thus, surface forces responsible for the sound tend to be concentrated near the leading edge. Observe that linear theory does not permit the corresponding singular pressure at the leading edge to be removed in this way because vorticity shed there is swept over the rigid surface of the airfoil, where its effects are cancelled by image vortices in the airfoil [4].

The component p_w of the sound pressure attributable to the wake can be found indirectly by first calculating the surface force

$$F_2^{\Omega} = -\rho_o \int \nabla Y_2 \cdot \Omega \mathbf{i}_3 \wedge \mathbf{U}\, dy_1\, dy_2,$$

produced by the incident vortex Γ alone. The wake-induced force $F_2^w = F_2 - F_2^{\Omega}$, where $F_2(\omega)$ is given by Sears formulas (3.3.6) and (3.3.7), and when this is substituted for F_2 in (3.3.5) we find

$$p_w(r,\Theta,t) \approx \frac{\rho_o \Gamma U \cos\Theta}{4\pi}\sqrt{\frac{M}{ra}}\left[\frac{a/h}{1+(Ut-a)^2/h^2}\right]_{t-r/c_o}, \quad r\to\infty.$$

Most of the shedding occurs when the vortex is close to the trailing edge when $h \ll a$, and the sound produced by the wake is radiated when the vortex is close to the edge ($t \approx a/U$). If the contribution from the wake is excluded, the pressure signature (caused by a *potential flow* interaction of Γ with the airfoil) assumes the form indicated by the broken curve in Figure 3.3.2.

Example 1. In an ideal, incompressible fluid, the linearized motion induced by vorticity $\Omega(x_1, x_2, \omega)\mathbf{i}_3$ convecting in a uniform flow at speed U in the x_1-direction satisfies $\nabla^2 B = -\operatorname{div}(U\Omega \mathbf{i}_2)$. Show that when $\Omega(x_1, x_2, \omega) = (\Gamma/2\pi U)\delta(x_2 - h)e^{i\omega x_1/U}$, and the fluid is unbounded, the vorticity generates the *evanescent* traveling wave

$$B = (-\Gamma/4\pi)\operatorname{sgn}(x_2 - h)\exp(-|\omega(x_2 - h)/U| + i\omega x_1/U).$$

Hence determine the upwash velocity at $x_2 = 0$.

3.3.2 Three-Dimensional Interactions

When a *frozen* three-dimensional gust $\boldsymbol{\Omega}(\mathbf{x} - \mathbf{U}t) \equiv \boldsymbol{\Omega}(x_1 - Ut, x_2, x_3)$ is swept past the airfoil of Figure 3.3.1 at low Mach number, the acoustic pressure is given by

$$p(\mathbf{x},t) \approx \frac{-\rho_o U \cos\Theta}{4\pi c_o|\mathbf{x}|}\frac{\partial}{\partial t}\int\left[\omega_3\frac{\partial Y_2}{\partial y_2} - \omega_2\frac{\partial Y_2}{\partial y_3}\right]_{t-|\mathbf{x}|/c_o} d^3\mathbf{y}, \quad |\mathbf{x}| \to \infty, \tag{3.3.8}$$

provided the chord is also compact, where $\Theta = \arccos(x_2/|\mathbf{x}|)$ is the angle between the radiation direction and the normal to the airfoil, and the origin is taken in the airfoil within the interaction region. If the aspect ratio of the airfoil is large, then $\partial Y_2/\partial y_3 \ll \partial Y_2/\partial y_2$, except in the tip regions of the airfoil.

Let the interaction occur at an inboard location where the chord may be regarded as constant, with both the leading and trailing edges at right angles to the mean flow. The planform in the interaction region is then locally the same as the two-dimensional airfoil of Figure 3.2.2, and Y_2 can be approximated by (3.3.3) where $2a$ is the local chord, and (3.3.8) becomes

$$\begin{aligned} p(\mathbf{x},t) &\approx \frac{\cos\Theta}{4\pi c_o|\mathbf{x}|}\frac{dF_2}{dt}(t - |\mathbf{x}|/c_o), \quad |\mathbf{x}| \to \infty, \\ F_2(t) &= -\rho_o U\int \omega_3(\mathbf{y},t)\frac{\partial Y_2}{\partial y_2}(\mathbf{y})\,d^3\mathbf{y}, \end{aligned} \tag{3.3.9}$$

where $-F_2$ is the unsteady lift force on the airfoil during the interaction when the motion is regarded as incompressible, and the vorticity ω_3 includes contributions from the impinging gust together with any shed into the vortex sheet wake. This formula is applicable only when M is sufficiently small that the airfoil chord is compact. Amiet [156, 157] has devised approximate formulae for a noncompact airfoil at moderate mean flow Mach numbers. At very high frequencies, however, the scale of the unsteady motion ultimately becomes comparable to the radius of curvature of the nose of the airfoil, and more detailed calculation is necessary [158, 159]. A limited number of results are available for supersonic flows [160].

To calculate F_2 by the method of thin airfoil theory, the incident gust is first represented by the Fourier integral

$$\boldsymbol{\Omega}(\mathbf{x} - \mathbf{U}t) = \frac{1}{U}\int_{-\infty}^{\infty}\boldsymbol{\Omega}(\omega/U, k_2, k_3)\mathrm{e}^{i(\mathbf{k}.\mathbf{x}-\omega t)}\,d\omega\,dk_2\,dk_3, \quad k_1 = \omega/U, \tag{3.3.10}$$

where $\boldsymbol{\Omega}(\mathbf{k})$ is the spatial Fourier transform of the gust vorticity at $t = 0$ (equations (1.7.5) with $n = 3$). A similar representation is easily obtained for the solenoidal induced velocity $\mathbf{u}(\mathbf{x} - \mathbf{U}t)$ in the absence of the airfoil, by using the relation $\mathbf{u}(\mathbf{k}) = i\mathbf{k} \wedge \boldsymbol{\Omega}(\mathbf{k})/k^2$. The upwash on the airfoil is then

$$u_2(x_3, \omega)e^{i\omega x_1/U} = \frac{ie^{i\omega x_1/U}}{U} \int_{-\infty}^{\infty} \frac{(\mathbf{k} \wedge \boldsymbol{\Omega}(\mathbf{k}))_2 e^{ik_3x_3}}{k_1^2 + k_2^2 + k_3^2} \, dk_2 \, dk_3. \qquad (3.3.11)$$

By analogy with (3.3.6), the force per unit span due to each upwash velocity component $u_2(k_3, \omega) \exp[i(\omega x_1/U + k_3x_3)]$ of $u_2(x_3, \omega)e^{i\omega x_1/U}$ can be written

$$-2\pi a\rho_o u_2(k_3, \omega) U \hat{\mathcal{S}}(\omega a/U, k_3) e^{ik_3x_3},$$

where $\hat{\mathcal{S}}(\omega a/U, k_3)$ is the generalization of the Sears function (3.3.7) to oblique gusts, such that $\mathcal{S}(\omega a/U) = \hat{\mathcal{S}}(\omega a/U, 0)$. This force can be evaluated for each Fourier component of (3.3.11), and the net force can be obtained by integration with respect to x_3 over the span. The integration yields $\delta(k_3)$ when the length scale of the gust is small compared to the airfoil span, so that

$$F_2(\omega) = 4\pi^2 i\rho_o aU\omega\mathcal{S}(\omega a/U) \int_{-\infty}^{\infty} \frac{\Omega_3(\omega/U, k_2, 0)}{\omega^2 + U^2k_2^2} \, dk_2. \qquad (3.3.12)$$

Equation (3.3.9) accordingly gives the radiation in the form

$$p(\mathbf{x}, t) \approx \frac{2i\rho_o U^2 \cos\Theta}{c_o|\mathbf{x}|} \times \frac{\partial}{\partial t} \int\!\!\int_{-\infty}^{\infty} \frac{\Omega_3(\omega/U, k_2, 0)e^{-i\omega[t]} \, dk_2 \, d\omega}{(\omega^2 + U^2k_2^2)\left(\mathrm{H}_0^{(1)}(\omega a/U) + i\mathrm{H}_1^{(1)}(\omega a/U)\right)}, \quad |\mathbf{x}| \to \infty, \qquad (3.3.13)$$

where $[t] = t - |\mathbf{x}|/c_o$.

This result is based on thin airfoil theory and thus represents the radiation produced by the vorticities in both the gust and wake. Similar formulae can be derived for their separate contributions. The component of F_2 produced by the gust alone is obtained by substituting the representation (3.3.10) of the gust vorticity into the second of equations (3.3.9). The first equation then yields the following formula for the sound generated by the potential interaction of the gust and airfoil:

$$p_\Omega(\mathbf{x}, t) \approx \frac{\pi\rho_o aU \cos\Theta}{c_o|\mathbf{x}|} \times \frac{\partial}{\partial t} \int\!\!\int_{-\infty}^{\infty} \frac{\omega\Omega_3(\omega/U, k_2, 0)\mathrm{J}_1(\omega a/U)e^{-i\omega[t]} \, dk_2 \, d\omega}{\omega^2 + U^2k_2^2}, \quad |\mathbf{x}| \to \infty. \qquad (3.3.14)$$

The wake generated acoustic pressure is $p(\mathbf{x}, t) - p_\Omega(\mathbf{x}, t)$.

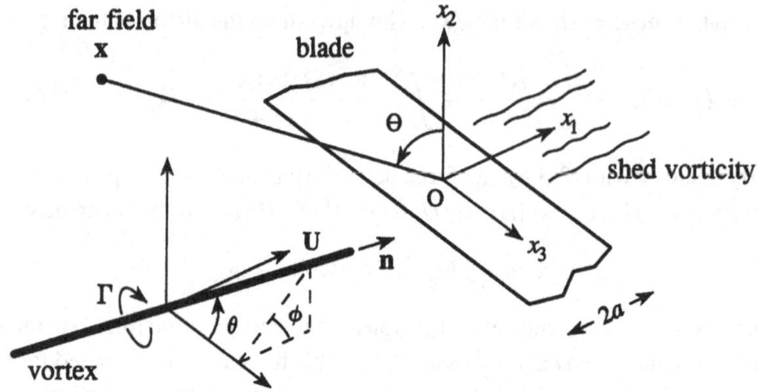

Figure 3.3.3. Blade–vortex interaction.

3.3.3 Blade–Vortex Interactions in Three Dimensions

Let the incident gust be a rectilinear vortex of circulation Γ orientated with its axis in the direction of the unit vector **n**. Choose the origin on the midchord of the airfoil such that the axis of the vortex passes through the origin at time $t = 0$ (Figure 3.3.3.). Let the distribution of axial vorticity be

$$\Omega(\mathbf{x}) = \frac{\Gamma \mathbf{n} e^{-(r_\perp/R)^2}}{\pi R^2}, \qquad (3.3.15)$$

where $r_\perp$ denotes perpendicular distance from the axis, and R is the nominal radius of the vortex core, which is assumed to be small compared to the chord $2a$. Then

$$\Omega(\mathbf{k}) = \frac{\Gamma \mathbf{n}}{4\pi^2}\delta(\mathbf{n}\cdot\mathbf{k})e^{-(kR/2)^2},$$

and (3.3.13) yields the acoustic pressure in the form

$$p(\mathbf{x}, t) \approx \frac{\rho_o U^2 \Gamma |\sin\phi| \cos\Theta}{2\pi^2 c_o \tan\theta |\mathbf{x}|} \int_{-\infty}^{\infty} \frac{\exp\{-(\omega R/2U \sin\phi)^2 - i\omega[t]\}\, d\omega}{\omega\left(\mathrm{H}_0^{(1)}(\omega a/U) + i\mathrm{H}_1^{(1)}(\omega a/U)\right)},$$
$$|\mathbf{x}| \to \infty, \qquad (3.3.16)$$

where the polar angles θ, ϕ define the unit vector **n** by

$$\mathbf{n} = (\sin\theta\cos\phi, \sin\theta\sin\phi, \cos\theta), \quad 0 < \theta < \pi,\ 0 < \phi < 2\pi.$$

$2R/|\sin\phi|$ is the length in the mean flow direction of the projected area of the circular cross section of the core onto the airfoil. Most of the sound is produced when this area is swept over the leading edge of the airfoil and is characterized by frequencies $\omega \sim U|\sin\phi|/R$, so that the acoustic signature is more sharply peaked when $\phi \approx \pi/2$ or $3\pi/2$. Thus, when $|\sin\phi| \sim \mathrm{O}(1)$ and $R \ll a$, the dominant frequencies satisfy $\omega a/U \gg 1$, and the value of the integral in (3.3.16) can be approximated by replacing the Hankel functions by their large argument asymptotic approximations (1.10.18). This procedure supplies [161]

$$p(\mathbf{x}, t) \approx \frac{\rho_o U^2 \Gamma \cos\Theta}{8c_o|\mathbf{x}|} \frac{|\sin\phi|^{\frac{3}{2}} (a/\pi R)^{\frac{1}{2}}}{\tan\theta} \Im(\alpha_+), \quad |\mathbf{x}| \to \infty, \tag{3.3.17}$$

where $\Im$ and arguments $\alpha_\pm$ are defined by

$$\begin{aligned} \Im(\alpha) &= |\alpha|^{\frac{1}{2}} \left(\mathrm{I}_{-\frac{1}{4}}(\alpha^2/8) + \mathrm{sgn}(\alpha) \mathrm{I}_{\frac{1}{4}}(\alpha^2/8) \right) \mathrm{e}^{-\alpha^2/8}, \\ \alpha_\pm &= (2a|\sin\phi|/R)\{1 \pm U[t]/a\}, \end{aligned} \tag{3.3.18}$$

and $\mathrm{I}_{\pm\frac{1}{4}}$ are modified Bessel functions of the first kind [24]. The potential flow interaction acoustic pressure p_Ω is similarly determined from equation (3.3.14):

$$p_\Omega(\mathbf{x}, t) \approx \frac{\rho_o U^2 \Gamma \cos\Theta}{8c_o|\mathbf{x}|} \frac{|\sin\phi|^{\frac{3}{2}} (a/\pi R)^{\frac{1}{2}}}{\tan\theta} (\Im(\alpha_+) - \Im(\alpha_-)), \quad |\mathbf{x}| \to \infty. \tag{3.3.19}$$

$\Im(\alpha)$ is nonnegative and exhibits a unique maximum (≈ 1.938) at $\alpha = 0$. Equation (3.3.17) therefore describes the generation of a single acoustic pulse when the vortex is severed by the leading edge of the blade, as illustrated by the solid curve in Figure 3.3.4a for $R/a = 0.1$. The sign of the pulse is determined by the angle θ between the vortex axis and the span; no sound is generated when $\theta = \pi/2$, when the upwash vanishes. The broken curve is the potential flow interaction pressure p_Ω, which is an odd function of the retarded time [t]. The large negative peak in p_Ω, from the potential flow interaction with the trailing edge, is canceled by an equal and opposite contribution generated by the wake.

In practice, the axial velocity in the core of the vortex differs from that in the ambient, irrotational flow, giving rise to an axial velocity defect. This occurs, for example, in tip vortices shed from helicopter rotor blades and is associated with an azimuthal component of vorticity, so that the vortex lines spiral around the vortex axis. The jetlike flow in the core produces a localized upwash when the vortex is cut by the airfoil. Let the maximum velocity defect

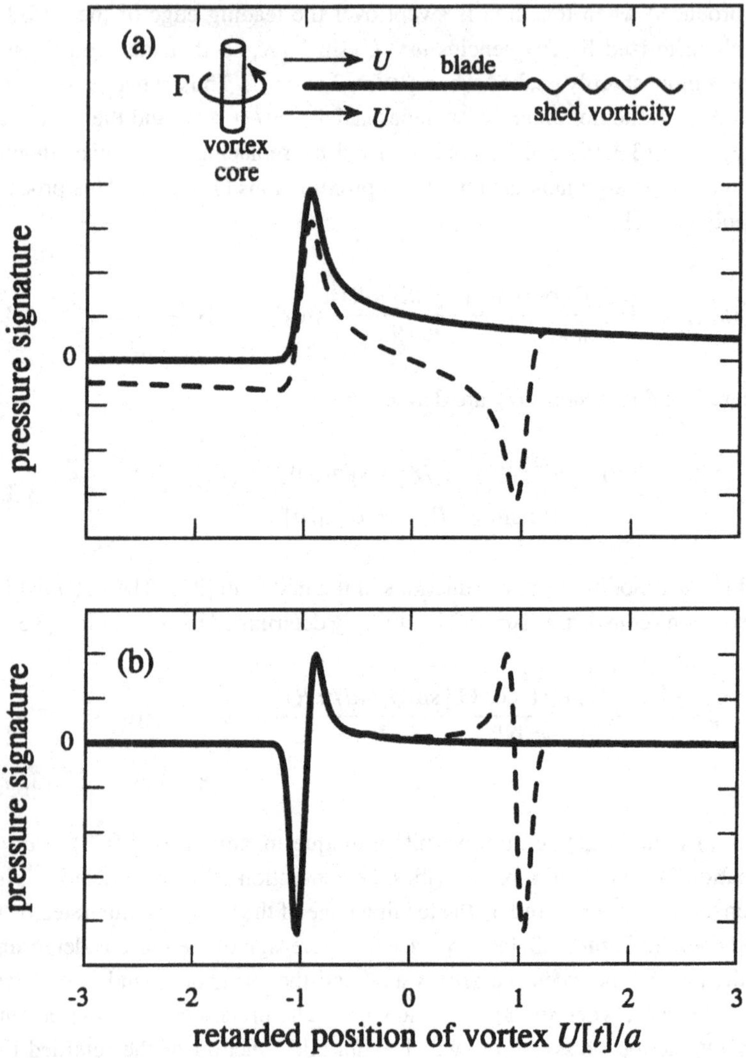

Figure 3.3.4. Blade–vortex interaction acoustic pressure: (a) axial vorticity and (b) azimuthal vorticity.

V occur on the vortex axis, and let the variation across the core be modeled by $\mathbf{v} = V\mathbf{n}\exp[-(r_\perp/R)^2]$, so that the axial volume flux is $\pi R^2 V$. The Fourier transform of the azimuthal vorticity is

$$\boldsymbol{\Omega}(\mathbf{k}) = (iVR^2/4\pi)(\mathbf{k}\wedge\mathbf{n})\delta(\mathbf{n}\cdot\mathbf{k})\exp[-(kR/2)^2].$$

When $R \ll a$, the corresponding acoustic pressure $p(\mathbf{x}, t)$ and the potential flow interaction acoustic pressure are given by

$$p(\mathbf{x},t) \approx \frac{-\rho_o U^2 V\sqrt{\pi a R}\cos\Theta}{4c_o|\mathbf{x}|}\sin\phi|\sin\phi|^{\frac{1}{2}}\Im'(\alpha_+),$$

$$p_\Omega(\mathbf{x},t) \approx \frac{-\rho_o U^2 V\sqrt{\pi a R}\cos\Theta}{4c_o|\mathbf{x}|}\sin\phi|\sin\phi|^{\frac{1}{2}}(\Im'(\alpha_+)+\Im'(\alpha_-)),$$

$$|\mathbf{x}|\to\infty, \qquad (3.3.20)$$

where $\Im'(\alpha) = d\Im(\alpha)/d\alpha$ (which assumes the maximum and minimum values 3.585, −1.647 respectively at $\alpha = -0.78, 2.47$). These pressures do not depend on θ, the angle between the spanwise direction and the vortex axis, because the acoustic source is equivalent to two equal and opposite monopoles on the upper and lower surfaces of the airfoil whose strengths are proportional to the volume flux in the core, which is independent of the vortex orientation (except for the cases $\theta = 0,\ \pi$, which are not covered by this approximation). The shapes of the pressure signatures are shown in Figure 3.3.4b.

Experimental arrangements of the kind shown in Figure 3.3.5 have been used to investigate blade–vortex interaction noise [162, 163]. An airfoil A is set at a finite angle of attack to a low Mach number mean stream at velocity U_o in an open jet wind tunnel. The tip vortex from the airfoil is periodically "chopped" by a rotor at a downstream station B whose plane of rotation is perpendicular to the mean flow. Observations are made of the blade–vortex interaction noise by filtering from the measured acoustic pressure the (typically much stronger) dipole radiation generated by the accelerating mean pressure loading of the rotor (Section 3.6).

Figure 3.3.6a shows details of a blade–vortex interaction acoustic pressure signature measured by Schlinker and Amiet [162] downstream of the rotor ($\Theta = 60°$ in the x_1x_2-plane of Figure 3.3.3). The axial vorticity $\Gamma = 4\pi a U_o \beta_\Gamma$ and the mean velocity defect $V = -\beta_V U_o$, where the parameters β_Γ and β_V and the other conditions of the experiment are given by

$$R/a = 0.15, \quad \theta = 104.5°, \quad \phi = 90°, \quad \beta_\Gamma = 0.07, \quad \beta_V = 0.1,$$
$$U = 155\ \mathrm{ms}^{-1},\ U_o = 9\ \mathrm{ms}^{-1}. \qquad (3.3.21)$$

The vortex is intercepted by a rotor blade approximately 1.25 blade chords from the tip. The chord $2a \approx 5.7$ cm, so that the duration of each interaction is about 0.3 ms. The pressure $p(\mathbf{x},t)$ determined by (3.3.17) and (3.3.20) is plotted in Figure 3.3.6b as a function of $U[t]/a$, where $[t] = t - R_o/c_o$ and R_o is the observer distance from the point of interaction of the vortex axis with the plane of the rotor. Because the blade Mach number $M = U/c_o \approx 0.45$ is not small, it

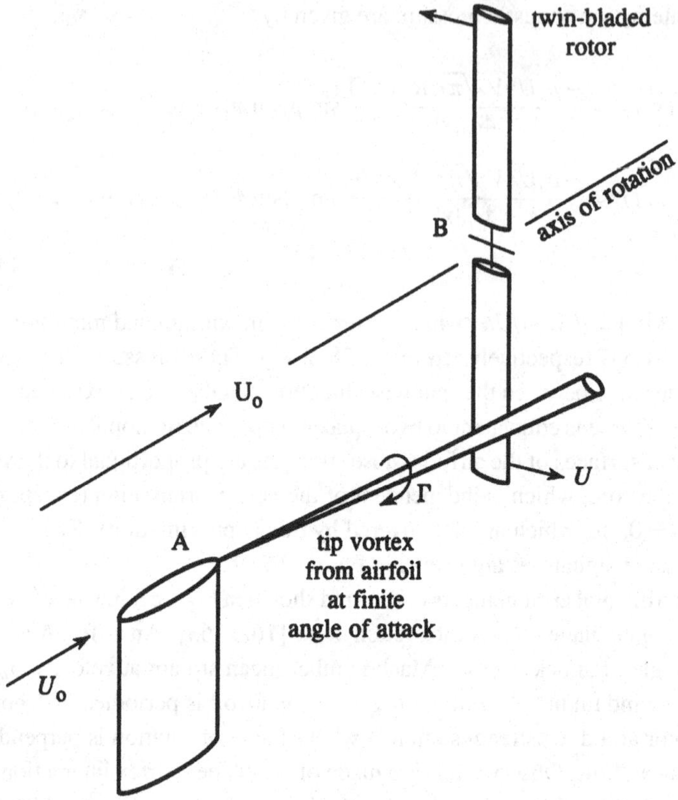

Figure 3.3.5. Experimental arrangement for studying blade–vortex interaction noise.

is necessary to account for the Doppler shift in the frequencies of the radiated sound, and this has been done by replacing [t] in the definition (3.3.18) of $\alpha_\pm$ by $[t]/(1 - M\sin\Theta)$ (cf. Section 1.8.5).

The predicted pressure does not exhibit the observed positive peak near $U[t]/a \approx -1$. However, in the experiment [162], the principal velocity defect impinging on the rotor was actually caused by the wake of the upstream airfoil A, the magnitude of the defect being about the same as in the vortex core. The prediction can be corrected for this by assuming that the section of a rotor blade between the vortex and blade tip passes through the wake. If the wake thickness is taken to be the same as the core diameter, the additional volume flux impinging on a blade is approximately the same as that obtained by increasing β_V to 0.6. This modification yields pressure signature in Figure 3.3.6c. The qualitative agreement is now greatly improved, although the widths of the predicted pressure peaks are probably too narrow; a numerical treatment of the

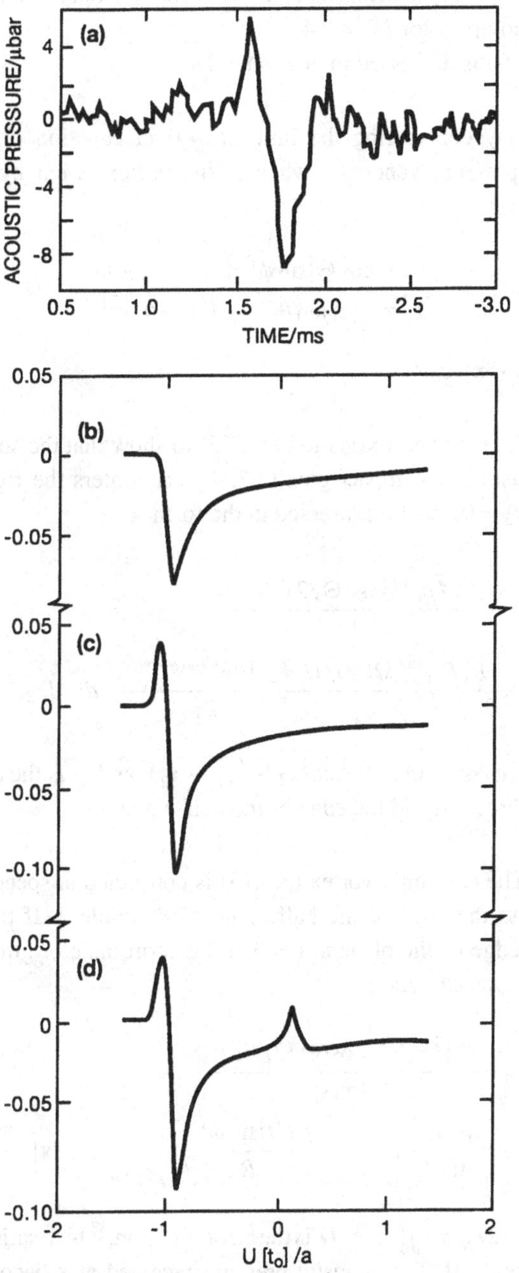

Figure 3.3.6. (a) Typical blade–vortex interaction acoustic pressure [162]; predictions of $p(\mathbf{x}, t)|\mathbf{x}|/a\rho_o U^2 M_o \cos\Theta$: (b) equations (3.3.17) and (3.3.20) for conditions (3.3.21); (c) $\beta_V = 0.6$; and (d) quasi-linear approximation when $U_c/U = 0.8$.

interaction [157, 164] suggests that this is because the blade chord is not strictly acoustically compact for $M \approx 0.45$.

Figure 3.3.6d is discussed in Section 3.3.4.

Example 2. By considering the limit $R \to 0$ in equation (3.3.16), deduce the acoustic pressure generated when a line vortex is cut by the airfoil of Figure 3.3.3:

$$p(\mathbf{x}, t) \approx \frac{\rho_o U^2 \Gamma \cos\Theta |\sin\phi|}{2^{\frac{3}{2}} \pi c_o |\mathbf{x}| \tan\theta} \frac{\mathrm{H}(\mathrm{U}[\mathrm{t}]/\mathrm{a}+1)}{\sqrt{U[t]/a+1}}, \quad |\mathbf{x}| \to \infty,$$

where $[t] = t - |\mathbf{x}|/c_o$.

Example 3. Use the approximation (3.2.13) to show that the sound generated when an acoustically compact gust (3.3.10) encounters the rigid half-plane, $0 < x_1 < \infty$, $x_2 = 0$, can be expressed in the form

$$p(\mathbf{x}, t) \approx \frac{-\sqrt{2M}\rho_o U \cos(\Theta/2) \sin^{\frac{1}{2}} \psi}{|\mathbf{x}|} \times \frac{\partial}{\partial t} \int\int_{-\infty}^{\infty} \frac{\Omega_3(\omega/U, k_2, 0) \mathrm{e}^{-i\omega(t-|\mathbf{x}|/c_o)}}{\omega^2 + U^2 k_2^2} \, d\omega \, dk_2, \quad |\mathbf{x}| \to \infty,$$

where $M = U/c_o \ll 1$, $\Theta = \arcsin(x_2/\sqrt{x_1^2 + x_2^2})$, and ψ is the angle between the radiation direction and the edge of the half-plane.

Example 4. The columnar vortex (3.3.15) is convected at speed U in the x_1-direction across the edge of the half-plane of Example 3. If the vortex axis intersects the edge of the plane at $t = 0$ at the coordinate origin, show that in the linearized approximation

$$p(\mathbf{x}, t) \approx \frac{\sqrt{2M}\rho_o U \Gamma \cos(\Theta/2) \sin^{\frac{1}{2}} \psi}{4\pi |\mathbf{x}|} \times \frac{|\sin\phi|}{\tan\theta} \left[1 + \mathrm{erf}\left(\frac{Ut|\sin\phi|}{R}\right)\right]_{t-|\mathbf{x}|/c_o}, \quad |\mathbf{x}| \to \infty,$$

where $\mathrm{erf}(x) = (2/\sqrt{\pi}) \int_0^x \mathrm{e}^{-\lambda^2} d\lambda$ is the error function, which satisfies $\mathrm{erf}(x) \to \pm 1$ as $x \to \pm\infty$ [24]. The acoustic pressure received at $\mathbf{x}$ becomes *constant* when $U[t]|\sin\phi|/R \geq 1$. For an airfoil of large but finite chord $2a$, the pressure remains constant until $[t] > 2a/c_o(1 + M)$, which is the time required for the sound to propagate to the trailing edge of the airfoil. This is the explanation of the broadening to be expected of the pressure signatures in Figure 3.3.6 when account is taken of the noncompactness of the airfoil chord [157].

Example 5. Consider the *spanwise* component $\gamma_o(k_3, \omega)e^{i\omega x_1/U_c}$ of the circulation per unit length of the vortex sheet wake of Figure 3.3.3 produced by the component $\int_{-\infty}^{\infty} \boldsymbol{\Omega}(\omega/U, k_2, k_3)e^{i\omega x_1/U+k_2x_2}\,dk_2$ of the gust (3.3.10), where U_c is the convection velocity in the wake. Thin airfoil theory requires $U_c = U$.

Use the method of conformal transformation for ideal, incompressible flow and the Kutta condition to show that [161]

$$\gamma_o(0, \omega) = \frac{-4\omega[\mathrm{J}_0(\omega a/U) + i\mathrm{J}_1(\omega a/U)]}{\left[\mathrm{H}_0^{(1)}(\omega a/U_c) + i\mathrm{H}_1^{(1)}(\omega a/U_c)\right]} \int_{-\infty}^{\infty} \frac{\Omega_3(\omega/U, k_2, 0)\,dk_2}{\omega^2 + U^2k_2^2}. \tag{3.3.22}$$

3.3.4 Quasi-Linear Approximation

Thin airfoil theory is a linearized approximation that requires vorticity to convect at the undisturbed mean stream velocity U. The sound generated by new vorticity shed into the wake has a profound effect on the blade–vortex interaction noise, by effectively canceling the sound generated by the potential flow interaction of the incident vortex with the edge. Thus, at the trailing edge two equal and opposite sources are involved, each of which would separately generate sound comparable in intensity to that produced at the leading edge.

It might be expected, therefore, that a modest departure of the trailing edge conditions from those postulated by thin airfoil theory would be sufficient to prevent perfect cancellation. For an airfoil of finite thickness, for example, when separation occurs upstream of the edge, the wake has small but finite thickness, and wake vorticity travels downstream at a mean velocity $U_c < U$. A satisfactory analysis of these complex events must be undertaken numerically [165–167]; however, the likely acoustic consequences can be examined simply by taking a *reduced* value for the wake convection velocity in the formulae of Section 3.3.3. If the mean flow were known exactly, the radiation would still be given to first order by equation (3.3.2), in which $\boldsymbol{\omega} \wedge \mathbf{U}$ would be replaced in the wake by the unsteady value of $\boldsymbol{\omega} \wedge \mathbf{v}$, and $Y_2(y_1, y_2)$ in the compact Green's function would be determined by airfoil geometry. The principal effect of these modifications is dominated not by the altered form of Y_2 but rather by the changed convection speed in the wake because this causes phase differences between the gust and wake radiations; any additional radiation will be the result of an unsteady lift force near the trailing edge caused by these differences [168]. A correction of this kind must be applicable only at high reduced frequencies $\omega a/U$, when the scale of the motions is comparable to the thickness, and is actually a particular instance of a nonlinear trailing edge noise source discussed in Section 3.5.

In this quasi-linear modification of the theory, the effective value of the wake source term at frequency ω becomes

$$(\boldsymbol{\omega} \wedge \mathbf{v})_2 = \gamma_o(\omega, 0) U_c \delta(x_2) \mathrm{e}^{i\omega x_1/U_c}, \quad x_1 > a,$$

where the circulation per unit length of the wake $\gamma_o(\omega, 0)$ is determined in Example 5. When this correction is introduced into the theory of the blade–vortex interaction studied by Schlinker and Amiet [162] (Figure 3.3.5), the respective contributions to the acoustic pressures from the axial and azimuthal components of vorticity, which we now denote by p_Γ, p_A, become

$$p_\Gamma(\mathbf{x}, t) \approx \frac{\rho_o U^2 \Gamma \cos\Theta}{8c_o|\mathbf{x}|} \frac{|\sin\phi|^{\frac{3}{2}} (a/\pi R)^{\frac{1}{2}}}{\tan\theta} \Big(\Im(\alpha_+) - (1 - U_c/U)\Im(\alpha_-)\Big),$$

$$p_A(\mathbf{x}, t) \approx \frac{-\rho_o U^2 V \sqrt{\pi a R} \cos\Theta}{4c_o|\mathbf{x}|} \times \sin\phi |\sin\phi|^{\frac{1}{2}} \Big(\Im'(\alpha_+) + (1 - U_c/U)\Im'(\alpha_-)\Big). \tag{3.3.23}$$

The second term in each of the large brackets vanishes when $U_c = U$ and represents the trailing edge correction for finite thickness of the airfoil and wake. Figure 3.3.6d illustrates how this correction improves the comparison with experiment. In this final figure, conditions are the same as for Figure 3.3.6c except that $U_c = 0.8U$.

3.3.5 Influence of Finite Span

The calculations described above have ignored the influence of planform curvature and the finite span of the airfoil. The simplest analytical model of these effects is obtained by considering an airfoil of *circular* planform. The blade–vortex interaction sound calculated for this case exhibits only minor differences with those of Figure 3.3.4 for the locally two-dimensional airfoil [169].

Example 6. Derive the compact Green's function for the rigid, circular disc $x_1^2 + x_3^2 < a^2$, $x_2 = 0$ in the form (1.11.7) where [170]

$$Y_1 = y_1, \quad Y_2 = \frac{2a\,\mathrm{sgn}(y_2)}{\pi} \int_0^\infty \frac{1}{\lambda} \left(\frac{\sin(\lambda a)}{\lambda a} - \cos(\lambda a) \right) \mathrm{e}^{-\lambda|y_2|} \mathrm{J}_0(\lambda r)\, d\lambda,$$

$$r = \sqrt{y_1^2 + y_3^2}, \quad Y_3 = y_3.$$

Example 7. Investigate the influence on blade–vortex interaction noise of a weak dependence of airfoil chord on x_3, that is, take $a = a(x_3)$ and $Y_2 = \mathrm{Re}(-i\sqrt{z^2 - a^2(x_3)})$ in (3.3.8).

Example 8: Airfoil in a turbulent stream. Use the general solution (3.3.13) to obtain the *acoustic pressure frequency spectrum* $\Phi(\mathbf{x}, \omega)$ of the sound generated when a turbulent stream interacts with a thin airfoil of uniform, compact chord.

The mean square acoustic pressure $\langle p^2(\mathbf{x}, t)\rangle = \int_0^\infty \Phi(\mathbf{x}, \omega)\, d\omega$, where the angle brackets represent an ensemble average. Assume the vorticity correlation function $\langle \Omega_i(\mathbf{x})\Omega_j(\mathbf{X})\rangle$ is homogeneous and isotropic [62, 63], that is, a function $\bar{\Omega}_{ij}(|\mathbf{x} - \mathbf{X}|)$ of $|\mathbf{x} - \mathbf{X}|$, whose spatial Fourier transform (defined as in (1.7.5) with $n = 3$) is the *vorticity spectrum tensor*

$$\bar{\Omega}_{ij}(\mathbf{k}) = \frac{E(k)}{4\pi k^2}(k^2\delta_{ij} - k_i k_j), \tag{3.3.24}$$

where $E(k)$ is the turbulence kinetic energy spectrum, which satisfies $\langle \frac{1}{2}u^2\rangle = \int_0^\infty E(k)\, dk$.

Then

$$\Phi(\mathbf{x}, \omega) \approx \frac{2\rho_o^2 U^5 L \cos^2\Theta}{\pi^2 c_o^2 |\mathbf{x}|^2} \frac{\omega^2}{\left|\mathrm{H}_0^{(1)}(\omega a/U) + i\mathrm{H}_1^{(1)}(\omega a/U)\right|^2}, \\ \times \int_0^\infty \frac{E\left(\sqrt{\omega^2/U^2 + k_2^2}\right) dk_2}{\left(\omega^2 + U^2 k_2^2\right)^2}, \quad |\mathbf{x}| \to \infty, \tag{3.3.25}$$

where L is the span of the airfoil "wetted" by the turbulent flow.

Deduce that

$$\frac{\Phi(\mathbf{x}, \omega)}{\rho_o^2 U \langle u^2\rangle a} \approx \frac{16\cos^2\Theta}{27\pi^2}\left(\frac{a}{|\mathbf{x}|}\right)^2 \\ \times \frac{M^2(L/a)(\ell/a)^2(\omega\ell/U)^2}{(1 + (\omega\ell/U)^2)^{7/3}\left|\mathrm{H}_0^{(1)}(\omega a/U) + i\mathrm{H}_1^{(1)}(\omega a/U)\right|^2},$$

when $E(k)$ is given by the von Kármán approximation [62]

$$E(k) = \frac{55\Gamma\left(\frac{5}{6}\right)\langle u^2\rangle \ell (k\ell)^4}{27\sqrt{\pi}\,\Gamma\left(\frac{1}{3}\right)(1 + (k\ell)^2)^{17/6}},$$

where ℓ is the turbulence correlation scale ($\Gamma(z)$ is the Euler Gamma function [24]).

3.4 Boundary Layers and Boundary Layer Noise

Intense surface pressure fluctuations beneath a turbulent boundary layer generate sound and structural vibrations. Sound is produced directly by aerodynamic sources within the fluid and indirectly by the interaction of hydrodynamic pressures and the surface vibrations with discontinuities of the wall (steps, ribs, support struts, etc.) [171–175]. At very high supersonic speeds, thermal gradients produced by frictional heating can be important, and when $M > 6$ ("hypersonic flow") chemical dissociation and ionization can occur. In this section, we review the principal concepts and terminology of boundary layer theory needed in applications and discuss boundary layer noise in flows over smooth and rough walls.

3.4.1 The Wall Pressure Wavenumber–Frequency Spectrum

The pressure developed beneath a boundary layer on a hard wall is called the *blocked pressure*, and it is twice the pressure that a nominally identical flow would produce if the wall were absent. The response of a flexible wall to boundary layer forcing depends on both the temporal and spatial characteristics of the pressure. When the wall is locally plane and the fluctuations are regarded as statistically stationary in time, these characteristics are usually expressed in terms of the wall-pressure wavenumber–frequency spectrum $\mathrm{P}(\mathbf{k}, \omega)$, which is the Fourier transform (Section 1.7.4) of the space–time correlation function of the wall pressure. By convention, coordinate axes (x_1, x_2, x_3) are taken with x_1 and x_3 respectively parallel and transverse to the mean flow and with x_2 measured outward from the wall (Figure 3.4.1a). $\mathrm{P}(\mathbf{k}, \omega)$ is then given by

$$\mathrm{P}(\mathbf{k}, \omega) = \frac{1}{(2\pi)^3} \int_{-\infty}^{\infty} \mathcal{R}(y_1, y_3, \tau) \mathrm{e}^{-i(\mathbf{k}.\mathbf{y}-\omega\tau)}\, dy_1\, dy_3\, d\tau, \qquad \mathbf{k} = (k_1, 0, k_3), \tag{3.4.1}$$

where

$$\mathcal{R}(y_1, y_3, \tau) = \langle p_s(x_1, x_3, t) p_s(x_1 + y_1, x_3 + y_3, t + \tau) \rangle \tag{3.4.2}$$

is the wall pressure correlation function, $p_s(x_1, x_3, t)$ is the *fluctuating* component of the wall pressure, and the angle brackets $\langle\ \rangle$ denote an ensemble average. The wavenumber $\mathbf{k}$ is parallel to the wall, and (because $\mathcal{R}$ is real) $\mathrm{P}(-\mathbf{k}, -\omega) \equiv \mathrm{P}(\mathbf{k}, \omega)$.

In the simplest applications, it is usually assumed that variations of $\mathcal{R}(y_1, y_3, \tau)$ and $\mathrm{P}(\mathbf{k}, \omega)$ with x_1, x_3 are significant only over distances that are much

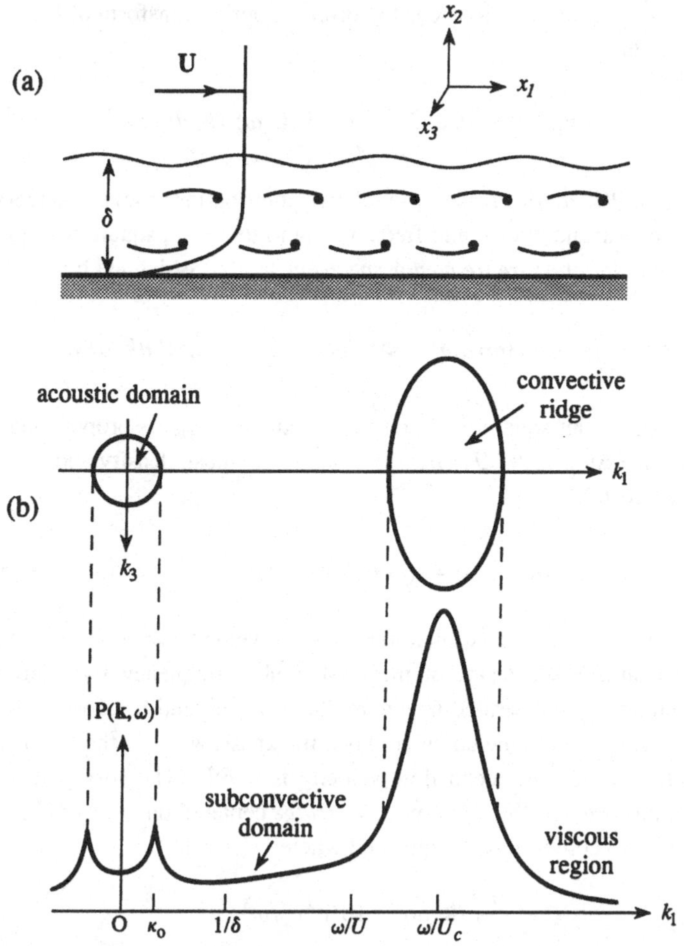

Figure 3.4.1. Turbulent boundary layer and wall pressure spectrum at low Mach number when $\omega\delta/U \gg 1$.

larger than the length scales of the dominant turbulent motions, although these distances may not be large compared to the wavelengths of the boundary layer generated sound. In the ideal case in which $\mathcal{R}(y_1, y_3, \tau)$ is independent of (x_1, x_3) and t, the wall pressure is said to be stationary random in position and time, and the following important relation is then satisfied by the Fourier transform $p_s(\mathbf{k}, \omega)$ of the wall pressure

$$\langle p_s(\mathbf{k}, \omega) p_s^*(\mathbf{k}', \omega') \rangle = \delta(\mathbf{k} - \mathbf{k}')\delta(\omega - \omega')\mathrm{P}(\mathbf{k}, \omega), \qquad (3.4.3)$$

where the asterisk denotes complex conjugate.

The correlation function $\mathcal{R}$ is the inverse Fourier transform of P($\mathbf{k}$, ω), and the formula

$$\langle p_s^2 \rangle \equiv \mathcal{R}(0, 0, 0) = \int_{-\infty}^{\infty} \mathrm{P}(\mathbf{k}, \omega)\, dk_1\, dk_3\, d\omega \tag{3.4.4}$$

shows that P($\mathbf{k}$, ω) determines the relative contribution from a boundary layer "eddy" of wavenumber $\mathbf{k}$ and frequency ω to the mean square wall pressure. The wall *point pressure frequency spectrum* $\Phi_{pp}(\omega)$ is defined by

$$\langle p_s^2 \rangle = \int_{-\infty}^{\infty} \Phi_{pp}(\omega)\, d\omega, \quad \Phi_{pp}(\omega) = \int_{-\infty}^{\infty} \mathrm{P}(\mathbf{k}, \omega)\, dk_1\, dk_3. \tag{3.4.5}$$

The root-mean-square wall pressure fluctuation p_{rms} is proportional to the *dynamic pressure* $1/2\rho_o U^2$ (ρ_o and U being the mean density and velocity of the free stream),

$$p_{rms} = \sqrt{\langle p_s^2 \rangle} = \sigma \frac{1}{2} \rho_o U^2, \tag{3.4.6}$$

where for a smooth, hard wall at low subsonic velocities $\sigma \approx 0.01$ [176, 177].

The characteristic shape of the wavenumber–frequency spectrum at low Mach numbers is illustrated in Figure 3.4.1b for a fixed frequency satisfying $\omega\delta/U > 1$, where δ is the boundary layer thickness, which is the distance from the wall at which the mean flow velocity is $0.99U$. The strongest pressure fluctuations occur within the *convective ridge* centered on $k_1 = \omega/U_c,\ k_3 = 0$, where U_c is a convection velocity, and where

$$\mathrm{P}(\mathbf{k}, \omega) \sim 10^3 \rho_o^2 v_*^6/\omega^3. \tag{3.4.7}$$

In this formula, v_* is the *friction velocity*, which varies with the Reynolds number $U\delta/\nu$, but is typically of order $0.03U - 0.04U$ and is defined such that $\rho_o v_*^2$ is the mean wall shear stress or drag per unit area [62, 178]. The velocity U_c can be interpreted as the translational speed of the principal boundary layer eddies, and both experiment and numerical simulations indicate that $U_c \approx 0.5U - 0.7U$, with only a weak dependence on frequency ω [179, 180].

At low subsonic Mach numbers, the phase velocity ω/k of surface pressures in the convective domain is subsonic, and these pressures therefore decay rapidly with distance from the wall and do not correspond to sound waves. However, convective pressures can generate sound (and structural vibrations) when the wall is rough or has other discontinuities (ribs, rivets, joints, etc.) at which convective energy is *scattered*. In the case of a highly compliant elastomeric wall, in which the shear wave speed is comparable to U, the unsteady surface *shear stresses* that accompany the convective pressures may also cause significant structural excitation. For a typical homogeneous surface, the strongest coupling between the wall and flow is usually attributed to longer wavelength, "subconvective" pressures in the low wavenumber region $|\kappa_o| < k \ll |\omega|/U_c$, where "resonant" forcing of the wall occurs because the phase velocity ω/k coincides with that of a vibrational mode of the wall. The spectral levels here are usually 30 to 60 dB below the convective peak. In the acoustic domain $k < |\kappa_o|$, the phase velocity is supersonic; here wall pressure fluctuations are actually sound waves produced directly by boundary layer quadrupoles or by the scattering of convective pressures and structural motions. The secondary circular ridge in Figure 3.4.1b at $k = |\kappa_o|$ is discussed in Section 3.4.2.

Beyond the convective ridge, say for $k > v_*/30\nu$, the scale of the motion is small enough to be controlled by viscosity, which governs the ultimate decay of the spectrum [181].

3.4.2 Smooth-Wall Spectra at Low Mach Numbers

Theoretical models of $\mathrm{P}(\mathbf{k}, \omega)$ generally assume the flow to be homentropic, of low Mach number over a flat, rigid wall with no mean pressure gradient. The steady growth of the boundary layer in the streamwise direction is usually ignored, and the mean flow velocity is taken to be parallel to the wall and dependent only on distance x_2 from the wall. A sufficient body of experimental data is available, derived principally from wind-tunnel experiments, to permit the formulation of empirical models of the convective domain, but reported properties of the subconvective and acoustic domains often vary significantly between different experiments.

The Hydrodynamic Domain

Chase [179] has devised the following model for $\mathrm{P}(\mathbf{k}, \omega)$ at low Mach numbers in the hydrodynamic domain $k \gg |\kappa_o|$:

$$\frac{\mathrm{P}(\mathbf{k},\omega)}{\rho_o^2 v_*^3 \delta^3} = \frac{1}{[(k_+\delta)^2 + 1/b^2]^{\frac{5}{2}}}\left(\mathrm{C_M}(k_1\delta)^2 + \mathrm{C_T}(k\delta)^2 \frac{(k_+\delta)^2 + 1/b^2}{(k\delta)^2 + 1/b^2}\right)$$

$$k_+^2 = (\omega - U_c k_1)^2/(h v_*)^2 + k^2,$$

$$M = U/c_o \ll 1, \quad k \gg |\omega|/c_o, \quad \omega\delta/U > 1. \qquad (3.4.8)$$

The adjustable coefficients in these expressions are fixed by comparison with experiment, and recommended values are

$$b \approx 0.75, \quad \mathrm{C_M} \approx 0.1553, \quad \mathrm{C_T} \approx 0.0047, \quad h \approx 3. \tag{3.4.9}$$

The coefficients $\mathrm{C_M}$ and $\mathrm{C_T}$ correspond respectively to contributions to the wall pressure from interactions between the turbulence velocity and the mean shear and between turbulent eddies [172, 179]. The former controls the behavior of $\mathrm{P}(\mathbf{k}, \omega)$ in the convective domain, whereas the term in $\mathrm{C_T}$ is dominant at sub-convective wavenumbers ($|\kappa_o| \ll k \ll |\omega|/U_c$), where conditions are isotropic in k.

A simpler approximation, which is applicable for $\omega\delta/U > 1$ in the immediate neighborhood of the convective ridge, is due to Corcos [182]:

$$\begin{aligned} \mathrm{P}(\mathbf{k}, \omega) &= \Phi_{pp}(\omega) \frac{\ell_1}{\pi\left[1 + \ell_1^2(k_1 - \omega/U_c)^2\right]} \frac{\ell_3}{\pi\left[1 + \ell_3^2 k_3^2\right]} \\ \ell_1 &\approx 9U_c/\omega, \quad \ell_3 \approx 1.4U_c/\omega, \end{aligned} \tag{3.4.10}$$

where ℓ_1 and ℓ_3 are turbulence correlation lengths in the x_1- and x_3-directions. The point pressure spectrum Φ_{pp} is based on data collated by Chase [179] and may be approximated by

$$\frac{(U/\delta_*)\Phi_{pp}(\omega)}{(\rho_o v_*^2)^2} = \frac{(\omega\delta_*/U)^2}{\left[(\omega\delta_*/U)^2 + \alpha_p^2\right]^{\frac{3}{2}}}, \quad \alpha_p = 0.12. \tag{3.4.11}$$

In this formula, $\delta_* = \int_0^\infty \left(1 - \bar{v}_1(x_2)/U\right) dx_2$ is the boundary layer displacement thickness, where $\bar{v}_1(x_2)$ is the mean velocity in the streamwise direction at distance x_2 from the wall [62, 178]. It is a measure of the outward displacement of the mean flow caused by the reduction in velocity near the wall due to surface friction. For practical purposes it may be assumed that $\delta_* \approx \delta/8$.

Equations (3.4.8) imply that $\mathrm{P}(\mathbf{k}, \omega)/\rho_o^2 v_*^3 \delta^3 \sim (k\delta)^2$ when $k\delta \ll 1$. This is sometimes referred to as the *Kraichnan–Phillips* theorem [183, 184], which states that the net normal force on the wall produced by the turbulence vanishes in incompressible flow (Section 1.14.3, Example 5). In the interval $1 < k\delta \ll |\omega|\delta/U_c$ (where the term in $\mathrm{C_T}$ is still dominant in (3.4.8)), the spectrum is *wavenumber-white*, that is,

$$\mathrm{P}(\mathbf{k}, \omega)/\rho_o^2 v_*^3 \delta^3 \approx \mathrm{C_T} h^3 (v_*/U)^3/(\omega\delta/U)^3, \quad 1 < k\delta \ll |\omega|\delta/U_c. \tag{3.4.12}$$

This may be contrasted with a low wavenumber model due to Sevik [185], based on measurements of $\mathrm{P}(\mathbf{k}, \omega)$ at $k = 0$ for both smooth and rough walls on buoyant bodies in water and in quiet wind tunnels:

$$\begin{aligned} &\mathrm{P}(\mathbf{0}, \omega)/\rho_o^2 v_*^3 \delta^3 \approx 127 M^2 (v_*/U)/(\omega\delta/U)^{4.5}, \\ &24 \le |\omega|\delta/U \le 240, \quad 0.01 \le M \le 0.15, \quad M = U/c_o. \end{aligned} \tag{3.4.13}$$

Sevik's formula explicitly involves fluid compressibility because of its dependence on Mach number M and was originally proposed as a model for the acoustic domain. However, practically all published data suggest that the wall pressure spectrum is ultimately wavenumber white as $k\delta \to 0$, starting at values of k that are much greater than the acoustic wavenumber κ_o. If equation (3.4.13) is applied at subconvective wavenumbers ($1 \le k\delta \ll |\omega|\delta/U_c$), predictions of $\mathrm{P}(\mathbf{0}, \omega)$ agree in magnitude with (3.4.12) at $\omega\delta/U \approx 24$ for $M = 0.01$. At higher Mach numbers, estimates based on Sevik's formula can be larger by up to 10 dB, although this is offset by a more rapid decrease with increasing frequency.

The Acoustic Domain

Consider the solution of Lighthill's equation (2.1.5) expressed in Curle's form (2.2.5) for a boundary layer of infinitesimal Mach number, when S coincides with the plane of the wall. The surface integral in (2.2.5) represents the *reflection* of the quadrupole generated sound by the wall [112]. When viscous shear stresses on the wall are ignored (Section 3.4.6), the reflection coefficient is unity, and the pressure on the wall $x_2 = 0$ becomes

$$p_s(\mathbf{x}, t) = \int \mathcal{F}(\mathbf{y}, t - |\mathbf{x} - \mathbf{y}|/c_o) \frac{d^3\mathbf{y}}{2\pi|\mathbf{x} - \mathbf{y}|}, \quad \mathbf{x} = (x_1, 0, x_3), \tag{3.4.14}$$

where $\mathcal{F}(\mathbf{x}, t) \approx \partial^2(\rho_o v_i v_j)/\partial x_i \partial x_j$ for homentropic flow, and the integration is over the unsteady sources in the boundary layer.

When this expression is used in (3.4.1) to evaluate $\mathrm{P}(\mathbf{k}, \omega)$, the resulting multiple integral diverges for $k = |\kappa_o|$, at the edge of the acoustic domain, if the boundary layer is assumed to be of unlimited extent. The singularity is caused by the collective action of "grazing" acoustic waves, whose mean square decay like $1/|\mathbf{x}|^2$ with propagation distance over the wall is countered by the growth in the surface area of the boundary layer as $|\mathbf{x}| \to \infty$. If L is characteristic of the size of the boundary layer, we find [186]

$$\mathrm{P}(\mathbf{k}, \omega) \sim \frac{\int\int_0^\delta \Psi(y_2, y_2', \mathbf{k}, \omega) \exp[i\gamma(k) y_2 - \gamma^*(k) y_2']\, dy_2\, dy_2'}{|k^2 - \kappa_o^2| + \epsilon^2 \kappa_o^2} \tag{3.4.15}$$

where $\epsilon = (\pi/2\kappa_o L)^{\frac{1}{2}} \ll 1, \gamma(k) = \sqrt{\kappa_o^2 - k^2}$ is defined as in (1.13.9), and $\Psi(y_2, y_2', \mathbf{k}, \omega)$ is the Fourier transform of $\langle \mathcal{F}(x_1, y_2, x_3, t)\mathcal{F}(x_1 + y_1, y_2', x_3 + y_3, t + \tau)\rangle$ defined as in (3.4.1). The magnitude of $\mathrm{P}(\mathbf{k}, \omega)$ at $k = |\kappa_o|$ is determined by the value of ϵ, which depends on the size of the boundary layer relative to the acoustic wavelength at frequency ω.

This suggests that on a smooth wall $\mathrm{P}(\mathbf{k}, \omega)$ should have the general form [187],

$$\frac{\mathrm{P}(\mathbf{k}, \omega)}{\rho_o^2 v_*^3 \delta^3} = \frac{\mathcal{P}(\mathbf{k}U/\omega, \omega\delta/U)}{(\omega\delta/U)^2}\left((k\delta)^2 + \frac{\alpha\kappa_o^4\delta^2}{\left|k^2 - \kappa_o^2\right| + \epsilon^2\kappa_o^2}\right), \quad (3.4.16)$$

where $\mathcal{P}$ ($\approx$ constant as $kU/\omega \to 0$) is a nonnegative, dimensionless function governing the behavior in the hydrodynamic region, $\epsilon = \epsilon(\omega)$, and α is a constant.

The value of ϵ governs the magnitude of the spectral peak in Figure 3.4.1b at $k = |\kappa_o|$ when $M \ll 1$, although the existence of this theoretical maximum has yet to be confirmed experimentally. In addition to finite boundary layer size, ϵ also depends on the attenuation and refraction of surface propagating sound waves by the boundary layer (Section 5.1, [188]), on diffraction by a wall of finite curvature [186], and on scattering at surface discontinuities and edges.

Chase [172] has extrapolated (3.4.8) into the acoustic domain and has obtained an approximation similar to (3.4.16):

$$\frac{\mathrm{P}(\mathbf{k}, \omega)}{\rho_o^2 v_*^3 \delta^3} = \frac{1}{[(k_+\delta)^2 + 1/b^2]^{\frac{5}{2}}}\left[\frac{\mathrm{C_M}(k_1\delta)^2 k^2}{\left|k^2 - \kappa_o^2\right| + \epsilon^2\kappa_o^2} + \mathrm{C_T}(k\delta)^2\frac{(k_+\delta)^2 + 1/b^2}{(k\delta)^2 + 1/b^2}\right.$$
$$\left.\times\left(c_1 + \frac{c_2\left|k^2 - \kappa_o^2\right|}{k^2} + \frac{c_3 k^2}{\left|k^2 - \kappa_o^2\right| + \epsilon^2\kappa_o^2}\right)\right], \quad (3.4.17)$$

where $c_1 = \frac{2}{3}, c_2 = c_3 = \frac{1}{6}$, and the values of the other coefficients are given in (3.4.9). The validity of this formula has not been confirmed by experiment; predictions in the acoustic domain are typically 20 dB smaller than those given by Sevik's formula (3.4.13) when $\omega\delta/U \approx 20 - 30$.

3.4.3 High Mach Numbers

Wall pressure fluctuations at high Mach numbers are generally accompanied by significant variations in temperature and density. The root mean square wall pressure is often approximated by the following modification of (3.4.6) [189, 190]

$$p_{rms} = \frac{\sigma \frac{1}{2}\rho_o U^2}{\frac{1}{2}\left(1 + \frac{T_w}{T}\right) + 0.1(\gamma - 1)M^2},$$

where $\sigma \approx 0.01$, γ is the ratio of specific heats, T is temperature, and all quantities are evaluated outside the boundary layer except the wall temperature T_w.

There is no clear separation of the wavenumber–frequency plane into acoustic and convective ("hydrodynamic") domains when the Mach number is large, and the properties of $P(\mathbf{k}, \omega)$ are not well understood. For practical purposes, the point wall pressure spectrum can be approximated by the Laganelli and Wolfe formula [189–191]

$$\frac{(U/\delta_*)\Phi_{pp}(\omega)}{(\rho_o U^2)^2} \approx \frac{\sigma^2 \lambda^{2(1+\mu)}}{4\pi[1 + \lambda^{4\mu}(\omega\delta_*/U)^2]}, \quad (3.4.18)$$

where $\lambda = 1/(\frac{1}{2}(1 + T_w/T) + 0.1(\gamma - 1)M^2)$ and $\mu \approx -0.717$. This is frequently applied at low Mach numbers and is consistent with (3.4.6) for any M.

3.4.4 Boundary Layer Generated Sound

At low Mach numbers, the wavelength of sound generated by the boundary layer greatly exceeds its thickness δ, and there is then a simple relation between the acoustic domain of $P(\mathbf{k}, \omega)$ and the frequency spectrum of the sound. When M is not small, this relation is complicated by refraction, scattering, and by sound generation by turbulence in the outer regions of the boundary layer at distances from the wall comparable to the acoustic wavelength. When the flow is supersonic, a substantial part of the sound consists of *Mach waves* generated by the deflection of the high speed main stream by slower moving eddies in the boundary layer. The following discussion is limited to low Mach numbers.

Consider the sound radiated from a large region of the boundary layer of area $\mathcal{A}$, which may be assumed to be centered on the origin. Let

$$p_\delta(\mathbf{k}, \omega) = \int\int_{\mathcal{A}} dx_1\, dx_3 \int_{-\infty}^{\infty} p_\delta(x_1, x_3, t) e^{-i(\mathbf{k}\cdot\mathbf{x} - \omega t)}\, dt$$

denote the Fourier transform of the fluctuating pressure p_δ on the plane $x_2 = \delta$ just outside the boundary layer. At very low Mach number, when the convection of sound can be ignored, the pressure in $x_2 > \delta$ can be written $p(\mathbf{x}, t) =$

$\int\int_{-\infty}^{\infty} p_\delta(\mathbf{k}, \omega) \exp\{i(\mathbf{k}\cdot\mathbf{x} + \gamma(k)(x_2 - \delta) - \omega t)\} d^2\mathbf{k}\, d\omega$, where $\gamma(k)$ is defined as in (1.13.9). At large distances in the exterior flow, the wavenumber integral can be evaluated using the stationary phase formula (1.13.10), which yields

$$\langle p^2(\mathbf{x}, t)\rangle \approx \frac{4\pi^2 \cos^2\Theta}{|\mathbf{x}|^2} \int\int_{-\infty}^{\infty} \kappa_o \kappa_o' \langle p_\delta(\kappa_o \mathbf{x}/|\mathbf{x}|, \omega) p_\delta^*(\kappa_o' \mathbf{x}/|\mathbf{x}|, \omega')\rangle \times e^{i(\omega'-\omega)t}\, d\omega\, d\omega', \quad |\mathbf{x}| \to \infty,$$

where $\Theta = \arccos(x_2/|\mathbf{x}|)$. The reader will readily verify that stationarity implies that

$$\langle p_\delta(\mathbf{k}, \omega) p_\delta^*(\mathbf{k}, \omega')\rangle = \frac{\mathcal{A}}{(2\pi)^2} \delta(\omega - \omega') \mathrm{P}_\delta(\mathbf{k}, \omega), \tag{3.4.19}$$

provided the dimensions of $\mathcal{A}$ are much larger than the correlation scale of the boundary layer turbulence, where $\mathrm{P}_\delta(\mathbf{k}, \omega)$ is defined as in (3.4.1) in terms of p_δ.

At low Mach numbers, the acoustic wavelengths are very much larger than the boundary layer thickness ($\kappa_o \delta \ll 1$), and surface pressures at these wavenumbers are constant across the boundary layer, so that $\mathrm{P}_\delta(\mathbf{k}, \omega) = \mathrm{P}(\mathbf{k}, \omega)$ for $k < |\kappa_o|$. Hence, we obtain the acoustic pressure frequency spectrum $\Phi(\mathbf{x}, \omega)$ of the boundary layer noise in the form

$$\Phi(\mathbf{x}, \omega) = \frac{2\mathcal{A}\kappa_o^2 \cos^2\Theta}{|\mathbf{x}|^2} \mathrm{P}(\kappa_o \mathbf{x}/|\mathbf{x}|, \omega), \quad |\mathbf{x}| \to \infty, \quad \text{where}$$

$$\langle p^2(\mathbf{x}, t)\rangle = \int_0^\infty \Phi(\mathbf{x}, \omega)\, d\omega. \tag{3.4.20}$$

There are no universally accepted measurements of the wall pressure spectrum in the acoustic domain for use in (3.4.20). Sevik's wavenumber-white approximation (3.4.13) yields

$$\frac{\Phi(\mathbf{x}, \omega)}{\rho_o^2 v_*^3 \delta} \approx \frac{254\mathcal{A}M^4(v_*/U)\cos^2\Theta}{|\mathbf{x}|^2(\omega\delta/U)^{\frac{5}{2}}}, \tag{3.4.21}$$

according to which the peak radiation direction is normal to the wall. A more complicated dependence on radiation direction is predicted by the Chase formula (3.4.17). Unless $\omega\delta/U$ is very large, predictions made with Sevik's formula are larger (by about 20 dB at $\omega\delta/U \approx 20$ for $M = 0.01$). Thus, equation (3.4.21) should probably be regarded as an upper bound for the quadrupole noise.

3.4.5 Rough Walls

The wall pressure wavenumber–frequency spectrum for flow over a rough wall whose characteristic roughness height R is much smaller than δ can be approximated by a smooth wall empirical formula provided the friction velocity is increased to compensate for the increased surface drag and turbulence production [171, 178]. This procedure is satisfactory in the convective domain but does not account for changes at low wavenumbers, which are produced by scattering of convective pressures by roughness elements. In particular, it seems that when $\omega R/v_* > 5$, the aerodynamic sound generated by the boundary layer is dominated by scattering.

To investigate the influence of surface roughness, let us assume that the motion is at low Mach number and of sufficiently high Reynolds number that viscous stresses can be disregarded; this will be the case if the roughness elements protrude beyond the viscous sublayer, when $Rv_*/\nu > 10$.

The roughness elements behave like point *dipoles*, the dipole strength being the unsteady drag on an element. If we regard the drag as due primarily to interactions with larger scale boundary layer structures, the dipole strength can be estimated from the solution of the corresponding Rayleigh scattering problem. This problem was studied in Section 1.11.2 (Example 6) when the incident disturbance was a sound wave of wavenumber κ_o. The calculation must be modified when the incident wave is a boundary layer wall pressure $p_s(\mathbf{k}, \omega)$ where $k \sim \omega\delta/U \gg \kappa_o$. In this case the monopole component of the scattered field is negligible, so that if roughness elements are modeled by rigid hemispheres distributed randomly at the points $\mathbf{x}_n$ in $\mathcal{A}$ (with $\mathcal{N}$ per unit area), the scattered acoustic pressure is found to be

$$p_{\mathrm{R}}(\mathbf{x}, t) \approx \frac{R^3 x_j}{2|\mathbf{x}|^2} \sum_n \int\!\!\int_{-\infty}^{\infty} \kappa_o k_j p_s(\mathbf{k}, \omega) \mathrm{e}^{-i\omega(t-|\mathbf{x}-\mathbf{x}_n|/c_o)}\, d^2\mathbf{k}\, d\omega, \quad |\mathbf{x}| \to \infty.$$

If the hemispheres are independently distributed in $\mathcal{A}$, this yields the roughness component of the acoustic frequency spectrum

$$\Phi_R(\mathbf{x}, \omega) \approx \frac{\alpha \mathcal{A} R^4 \kappa_o^2}{2\pi |\mathbf{x}|^4} \int_{-\infty}^{\infty} (\mathbf{k}\cdot\mathbf{x})^2 \mathrm{P}(\mathbf{k}, \omega)\, d^2\mathbf{k}, \quad |\mathbf{x}| \to \infty, \qquad (3.4.22)$$

where $\alpha = \mathcal{N}\pi R^2$ is the "roughness density," and $\mathrm{P}(\mathbf{k}, \omega)$ is the wall pressure wavenumber–frequency spectrum that the same turbulent flow would produce on a smooth wall at $x_2 = 0$.

This approximate result assumes that the turbulence scales contributing to the integral exceed the roughness height R, which implies that the typical wavenumber contributing to the integral satisfies $kR < 1$. Because the dominant contributions are from $k \sim \omega/U_c$, this requires $\omega R/U_c < 1$, or roughly $\omega R/v_* < 20$. To approximate the remaining integral, $\mathbf{k} \cdot \mathbf{x}$ in the integrand is replaced by its value $\omega x_1/U_c$ at the convective peak, so that

$$\Phi_R(\mathbf{x}, \omega) \approx \frac{\alpha A \cos^2\theta}{2\pi |\mathbf{x}|^2} (\kappa_o R)^2 (\omega R/U_c)^2 \Phi_{pp}(\omega), \tag{3.4.23}$$

where θ is the angle between the radiation direction and the mean flow. For typical small-scale roughness of height $R \ll \delta$, the nondimensional frequency $\omega\delta_*/U$ will be large, and, because roughness noise is important for $\omega R/v_* > 5$, $\Phi_{pp}(\omega)$ in this equation can be approximated by setting $\alpha_p = 0$ in (3.4.11). This implies that $\Phi_R \sim \omega^3$ in the range $5 < \omega R/v_* < 20$ for which (3.4.23) is applicable. This conclusion is consistent with measurements by Hersh [192] of the noise generated by low Mach number flow of air from a sand roughened pipe (Figure 3.4.2b). If R is identified with the mean roughness height, the data show that $\Phi_R(\omega)$ peaks near $\omega R/v_* \approx 15$ and decays rapidly at higher frequencies.

An empirical spectrum for the whole frequency range is furnished by

$$\frac{\Phi_R(\mathbf{x}, \omega)}{\rho_o^2 v_*^3 \delta} \approx \tau_o \frac{A \cos^2\theta}{|\mathbf{x}|^2} \frac{R}{\delta} \frac{v_*^2}{c_o^2} \frac{(\omega R/v_*)^3}{[1 + \beta(\omega R/v_*)^2]^{5.5}}, \tag{3.4.24}$$

where τ_o and β are empirical constants. The best fit to the *shape* of the experimental spectrum is obtained for $\beta = 0.0025$, as illustrated by the solid curve in Figure 3.4.2b, whose absolute position has been adjusted to overlie the data, because no directivity information is available from the Hersh experiments. This value of β and the exponent 5.5 of the denominator in (3.4.24) should be regarded as tentative. The "roughness" parameter τ_o depends on the spacing of the roughness elements, determined by α. When $\alpha \ll 1$, the approximation (3.4.23) implies that $\tau_o \approx (v_*/U_c)^2(\alpha/\pi)$.

3.4.6 Surface Shear Stress Fluctuations

Curle's equation (2.2.5) suggests that the unsteady wall shear stress contributes a dipole source to the boundary layer noise. This dipole is very weak (smaller than the quadrupole radiation at low Mach numbers) and does not actually lead to the generation of additional acoustic energy over and above that already produced by the quadrupoles. This is because very close to the wall the equations of motion are linear, and any motion in the immediate neighborhood of the wall is forced by turbulent sources in the outer flow. Evidently, when the outer flow

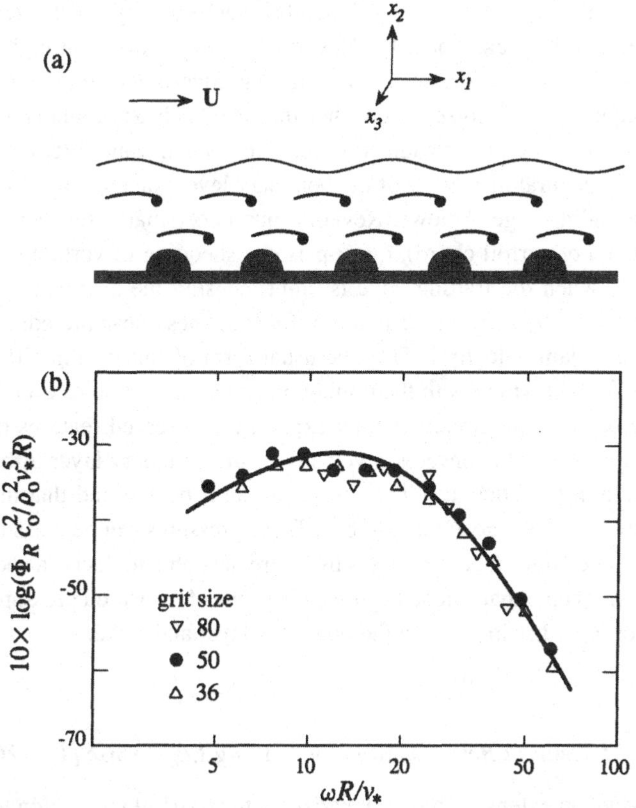

Figure 3.4.2. (a) Wall roughness modeled by rigid hemispheres and (b) roughness noise measured by Hersh [192].

disturbances are resolved into wavenumber–frequency components $(\mathbf{k}, \omega)$, a linear interaction with a smooth *plane*, the wall cannot change the value of $(\mathbf{k}, \omega)$. In particular there can be no transfer of energy at the wall from the energetic convective region ($k \gg |\kappa_o|$) to acoustic wavenumbers; any acoustic disturbance at the wall must be already in the acoustic domain prior to its incidence on the wall. Accordingly, a plane wall merely *reflects* sound waves already generated by the boundary layer quadrupoles. When shear stresses are ignored, the reflection coefficient is unity, but it is smaller when viscosity is taken into account during the wall interaction (Section 5.1)

3.5 Trailing Edge Noise

Vortex–airfoil interaction noise is generated principally when an impinging gust is close to the leading edge of an airfoil. The contribution from the trailing edge is small because the gust-generated upwash velocity is canceled there by that produced by wake vorticity. Nevertheless, experiments at high Reynolds numbers [193–195] indicate that the trailing edge is an important source of high-frequency "self-noise," even when the airfoil is in a nominally turbulence-free flow. The sound is attributed to relatively small-scale turbulence, generated by the natural instability of the boundary layers on the airfoil, interacting with the trailing edge. At lower Reynolds numbers (smaller than about 2×10^5 when based on airfoil chord), quasi-periodic shedding of vorticity can occur, especially when the trailing edge is blunt, causing the airfoil to "sing" at a frequency $\omega \approx U/h$, where h is the airfoil thickness near the edge and U is the mean stream velocity [171]. The usual form of linear, thin airfoil theory (Section 3.3), in which both the impinging and shed vorticities translate at the velocity of the main stream, cannot explain the observed features of trailing edge noise: A single convection velocity of the boundary layer turbulence or large-scale instabilities does not exist, nor can it be asserted that the surface pressures near the edge are negligible. These pressures can be measured using flush-mounted transducers, and it is therefore desirable to devise an edge-noise prediction scheme that can be expressed in terms of measured pressure, instead of the vorticity distributions in the boundary layer and wake.

3.5.1 General Representation of Trailing Edge Noise [196–200]

We consider an extension of the linearized vortex-airfoil interaction theory for an airfoil with a sharp trailing edge. Because we are dealing with sound at high frequencies, whose wavelength is generally small relative to the chord, the airfoil will be modeled by the semi-infinite, rigid plate: $-\infty < x_1 < 0,\ x_2 = 0$. The main stream outside the boundary layers has subsonic speed U in the x_1-direction (Figure 3.5.1). The details of the boundary layer turbulence are unknown, but it will be assumed that during the interaction with the edge the statistical properties of the impinging turbulence are the same as those several boundary layer thicknesses upstream. This will permit the edge noise to be expressed in terms of the boundary layer wall pressure in the upstream region, which may be regarded as uninfluenced by the proximity of the edge.

Let the motion be homentropic, with the boundary layer turbulence confined to the "upper" surface ($x_2 > 0$) of the airfoil. The calculations can be formulated as a scattering problem, in which the pressure p_I, say, that would be produced by the same turbulent flow if the airfoil were removed, is scattered by the edge. The scattered pressure p' includes both acoustic and hydrodynamic components, the latter accounting for the modification of the near field pressure by the airfoil. The condition that the normal velocity vanishes on the airfoil will be taken in

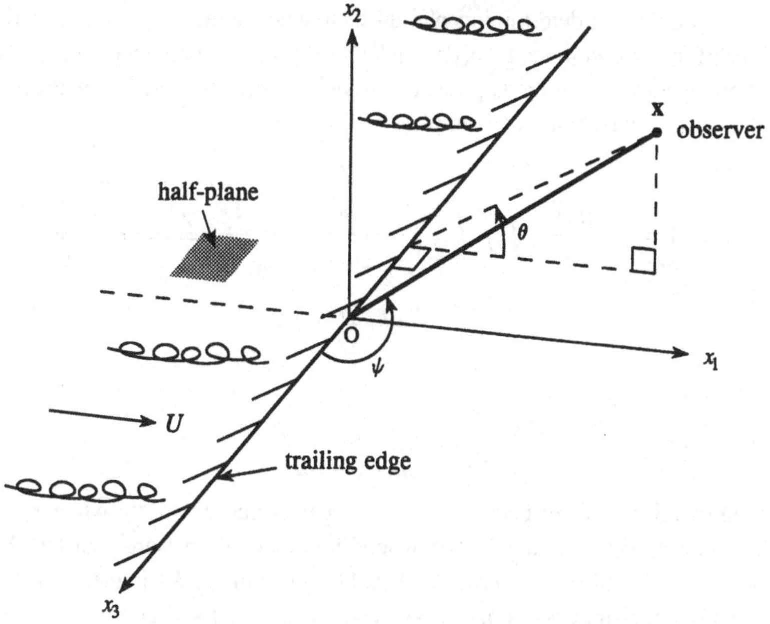

Figure 3.5.1. Turbulent flow over a trailing edge.

the high Reynolds number approximation

$$\partial p_{\mathrm{I}}/\partial x_2 + \partial p'/\partial x_2 = 0, \quad \text{for } x_2 = 0,\ x_1 < 0. \tag{3.5.1}$$

In practice, trailing edge noise is important only at low Mach numbers, and we shall therefore assume that in turbulence-free regions pressure fluctuations of frequency ω satisfy the Helmholtz equation (1.10.1) with $\mathcal{F} = 0$. No explicit account will be taken of the boundary layer turbulence, except insofar as it is responsible for the pressure p_{I}; in particular $p'(\mathbf{x}, \omega)$ is assumed to be a solution of the homogeneous Helmholtz equation everywhere except in the wake, which is modeled by a vortex sheet. Thus, scattering of sound by the shear flow is ignored, although this approximation cannot be valid at very high frequencies when the acoustic wavelength is comparable to the thickness of the boundary layer [201–203].

Because of its definition, the pressure $p_{\mathrm{I}}(\mathbf{x}, \omega)$ must be an outgoing solution of Helmholtz's equation in $x_2 \leq 0$, in the region "below" the boundary layer sources, and equal to *half* the boundary layer blocked pressure p_s on $x_2 = +0$ (see (3.4.14)). Hence, in the notation of (1.13.8),

$$p_{\mathrm{I}}(\mathbf{x}, \omega) = \frac{1}{2} \int\int_{-\infty}^{\infty} p_s(\mathbf{k}, \omega) e^{i(\mathbf{k}\cdot\mathbf{x} - \gamma(k)x_2)}\, dk_1\, dk_3, \quad x_2 \leq 0, \tag{3.5.2}$$

and the problem of determining p' reduces to a consideration of the scattering of each Fourier component $\frac{1}{2}p_s(\mathbf{k},\omega)e^{i(\mathbf{k}\cdot\mathbf{x}-\gamma(k)x_2)}$ of p_{I} by the half-plane. The scattered pressure p' is an outgoing disturbance both above and below the plate and has the representation (Example 1)

$$p'(\mathbf{x},\omega) = \frac{-\mathrm{sgn}(x_2)}{4\pi i}\int\!\!\int\!\!\int_{-\infty}^{\infty} \frac{p_s(\mathbf{k},\omega)\sqrt{\left(\kappa_o^2-k_3^2\right)^{1/2}+k_1}}{(K_1-k_1+i0)\sqrt{\left(\kappa_o^2-k_3^2\right)^{1/2}+K_1}} \times \mathrm{e}^{i\{K_1x_1+k_3x_3+\gamma(K)|x_2|\}}\,dk_1\,dk_3\,dK_1, \quad (3.5.3)$$

where $K=\sqrt{K_1^2+k_3^2}$.

The Surface Pressure

The integration with respect to K_1 can be performed explicitly when $x_2\to\pm0$. It is zero in the wake ($x_1\geq 0$), where the scattered pressure vanishes. For $x_1<0$ the integration contour is displaced to $-i\infty$ in the K_1-plane, capturing the residue contribution from the pole at $K_1=k_1-i0$ and an integral along the branch cut of $\sqrt{(\kappa_o^2-k_3^2)^{1/2}+K_1}$, which extends from $-(\kappa_o^2-k_3^2)^{1/2}$, just below the real axis, to $-i\infty$. The branch-cut integral can be expressed in terms of the error function $\mathrm{erf}(x)=(2/\sqrt{\pi})\int_0^x \mathrm{e}^{-\lambda^2}d\lambda$ [24]. When the result is combined with the pressure $p_{\mathrm{I}}=\frac{1}{2}p_s$, the total surface pressure is found to be

$$p(x_1,\pm0,x_3,\omega) = \frac{1}{2}\int\!\!\int_{-\infty}^{\infty} p_s(\mathbf{k},\omega) \times\left[1\pm\mathrm{erf}\left(\mathrm{e}^{-\frac{i\pi}{4}}|x_1|^{\frac{1}{2}}\sqrt{\left(\kappa_o^2-k_3^2\right)^{1/2}+k_1}\,\right)\right] \times \mathrm{e}^{i\mathbf{k}\cdot\mathbf{x}}\,d^2\mathbf{k}, \quad x_1<0. \quad (3.5.4)$$

The argument of the error function has positive real part for all real values of k_1, so that the error function ≈ 1 as $x_1\to-\infty$. Thus, $p\to 0$ on the lower surface ($x_2=-0$) far upstream of the edge, whereas $p\to p_s\equiv 2p_{\mathrm{I}}$ on the surface $x_2=+0$ exposed to the turbulent stream. This occurs at distances upstream of the edge exceeding the characteristic eddy dimension and represents a pressure doubling by "specular reflection" of the incident pressure p_{I} by the plane, rigid surface. Thus, if the impinging boundary layer turbulence is assumed to convect as a frozen pattern over the edge, measurements of the upstream wall pressure p_s (or, rather, of its wavenumber–frequency spectrum) can be used in (3.5.3) to predict the acoustic spectrum.

The Acoustic Pressure

At large distances from the edge the acoustic pressure is obtained from (3.5.3) by evaluating the double integral with respect to K_1 and k_3 by the method of stationary phase, using the formula (1.13.10). Transforming the result to the time domain, we find

$$p(\mathbf{x}, t) \approx \frac{x_2}{2|\mathbf{x}|^2\sqrt{(1-(x_3/|\mathbf{x}|)^2)^{1/2} + x_1/|\mathbf{x}|}}$$
$$\times \int\!\!\int_{-\infty}^{\infty} \frac{\kappa_o^{\frac{1}{2}}\sqrt{\kappa_o(1-(x_3/|\mathbf{x}|)^2)^{1/2} + k_1}}{(\kappa_o x_1/|\mathbf{x}| - k_1)} p_s(k_1, \kappa_o x_3/|\mathbf{x}|, \omega)$$
$$\times \, \mathrm{e}^{-i\omega(t-|\mathbf{x}|/c_o)}\, dk_1\, d\omega, \quad |\mathbf{x}| \to \infty. \tag{3.5.5}$$

The integrand in this formula is singular at $k_1 = \kappa_o x_1/|\mathbf{x}|$, where the stationary phase approximation breaks down. However, this occurs when k_1 lies in the *acoustic domain*, that is, for a blocked pressure Fourier component $p_s(k_1, \kappa_o x_3/|\mathbf{x}|, \omega)$ that actually represents a sound wave generated by the boundary layer quadrupoles in the absence of the edge. Such contributions can be neglected compared with the edge-generated sound unless the Mach number is large, when it would also be necessary to add the direct quadrupole radiation from the boundary layer to that generated at the edge.

Example 1: Solution of the edge scattering problem. The condition that $\partial p'(\mathbf{x}, \omega)/\partial x_2$ is continuous across $x_2 = 0$, including the region ($x_1 > 0$) occupied by the vortex sheet wake, implies that we can write

$$\bar{p}(x_1, x_2; \mathbf{k}, \omega) = \operatorname{sgn}(x_2) \int_{-\infty}^{\infty} \mathcal{A}(K_1) \mathrm{e}^{i(K_1 x_1 + \gamma(K)|x_2|)}\, dK_1,$$
$$K = \left(K_1^2 + k_3^2\right)^{1/2}, \quad x_2 \neq 0, \tag{3.5.6}$$

where the function $\mathcal{A}(K_1)$ remains to be determined,

$$p'(\mathbf{x}, \omega) = \int\!\!\int_{-\infty}^{\infty} \bar{p}(x_1, x_2; \mathbf{k}, \omega) \mathrm{e}^{ik_3 x_3}\, dk_1\, dk_3, \tag{3.5.7}$$

and $\bar{p}(x_1, x_2; \mathbf{k}, \omega)\mathrm{e}^{ik_3x_3}$ is the pressure produced by scattering of the Fourier component

$$p_{\mathrm{I}}(\mathbf{k}, \omega)\mathrm{e}^{i(k_1x_1 - \gamma(k)x_2 + k_3x_3)} \equiv \frac{1}{2} p_s(\mathbf{k}, \omega) \mathrm{e}^{i(k_1x_1 - \gamma(k)x_2 + k_3x_3)}$$

by the edge.

Equation (3.5.1) and the condition that the pressure is continuous across the vortex sheet respectively yield the following *dual integral equations* for $\mathcal{A}(K_1)$:

$$\int_{-\infty}^{\infty}\left(\gamma(K)\mathcal{A}(K_1)+\frac{\gamma(k)p_{\mathrm{I}}(\mathbf{k},\omega)}{2\pi i(K_1-k_1+i0)}\right)\mathrm{e}^{iK_1x_1}\,dK_1=0,\quad x_1<0; \tag{3.5.8a}$$

$$\int_{-\infty}^{\infty}\mathcal{A}(K_1)\mathrm{e}^{iK_1x_1}dK_1=0,\quad x_1>0. \tag{3.5.8b}$$

We shall solve these by the *Wiener–Hopf* procedure [204–206].

Equation (3.5.8b) is satisfied provided

$$\mathcal{A}(K_1)=U(K_1),$$

where $U(K_1)$ is an "upper function," which is regular and of algebraic growth in $\mathrm{Im}(K_1)>0$ (i.e., $\sim\mathrm{O}(K_1^{\nu})$ as $|K_1|\to\infty$ for some fixed ν). Indeed, for any generalized function [19, 20]

$$\int_{-\infty}^{\infty}U(K_1)\mathrm{e}^{iK_1x_1}\,dK_1\equiv\left(-\frac{i\partial}{\partial x_1}\right)^n\int_{-\infty}^{\infty}\frac{U(K_1)\mathrm{e}^{iK_1x_1}\,dK_1}{(K_1+i0)^n}=0 \quad \text{for}\quad x_1>0$$

where the integer $n>0$ can be chosen to make $U(K_1)/(K_1+i0)^n\to0$ as $\mathrm{Im}(K_1)\to+\infty$.

Similarly, (3.5.8a) is satisfied if

$$\gamma(K)\mathcal{A}(K_1)+\frac{\gamma(k)p_{\mathrm{I}}(\mathbf{k},\omega)}{2\pi i(K_1-k_1+i0)}=L(K_1), \tag{3.5.9}$$

where $L(K_1)$ is a "lower function."

Equation (3.5.9), with $\mathcal{A}$ replaced by U for real K_1, is now rearranged so that all terms on the left-hand side are upper functions and all those on the right are lower functions with respect to their dependence on K_1. To do this, we observe that

$$\begin{aligned}\gamma(K)&\equiv\left(\kappa_o^2-K^2\right)^{1/2}\\&=\left\{\left(\kappa_o^2-k_3^2\right)^{1/2}-K_1\right\}^{1/2}\left\{\left(\kappa_o^2-k_3^2\right)^{1/2}+K_1\right\}^{1/2}\\&=\text{“lower function”}\times\text{“upper function,”}\end{aligned}$$

provided $(\kappa_o^2-k_3^2)^{1/2}$ always has a positive imaginary part (which is the case when Im $\omega > 0$), and the branch cuts for $\{(\kappa_o^2 - k_3^2)^{1/2} \pm K_1\}^{1/2}$ are taken from $\mp(\kappa_o^2 - k_3^2)^{1/2}$ to $\mp i\infty$ respectively in the lower and upper halves of the K_1-plane. After dividing by $\{(\kappa_o^2 - k_3^2)^{1/2} - K_1\}^{1/2}$, equation (3.5.9) may be cast in the form

$$\begin{aligned}
U(K_1)\left\{\left(\kappa_o^2 - k_3^2\right)^{1/2} + K_1\right\}^{1/2} &+ \frac{\gamma(k)p_{\mathrm{I}}(\mathbf{k},\omega)}{2\pi i(K_1 - k_1 + i0)\left\{\left(\kappa_o^2 - k_3^2\right)^{1/2} - k_1\right\}^{1/2}} \\
&= \frac{L(K_1)}{\left\{\left(\kappa_o^2 - k_3^2\right)^{1/2} - K_1\right\}^{1/2}} + \frac{\gamma(k)p_{\mathrm{I}}(\mathbf{k},\omega)}{2\pi i(K_1 - k_1 + i0)} \\
&\quad \times \left(\frac{1}{\left\{\left(\kappa_o^2 - k_3^2\right)^{1/2} - k_1\right\}^{1/2}} - \frac{1}{\left\{\left(\kappa_o^2 - k_3^2\right)^{1/2} - K_1\right\}^{1/2}}\right),
\end{aligned}$$

where the left-hand side is regular in $\mathrm{Im}(K_1) > 0$ and coincides on $\mathrm{Im}(K_1) = 0$ with the right-hand side, which is regular in $\mathrm{Im}(K_1) < 0$. The equation therefore defines a function that is regular in the whole of the K_1-plane and has algebraic growth as $|K_1| \to \infty$. According to *Liouville's* theorem [21], this function must be a polynomial $\mathcal{P}(K_1)$, and therefore

$$\begin{aligned}
\mathcal{A}(K_1) &\equiv U(K_1) \\
&= \frac{-\gamma(k)p_{\mathrm{I}}(\mathbf{k},\omega)}{2\pi i(K_1 - k_1 + i0)\left\{\left(\kappa_o^2 - k_3^2\right)^{1/2} + K_1\right\}^{1/2}\left\{\left(\kappa_o^2 - k_3^2\right)^{1/2} - k_1\right\}^{1/2}} \\
&\quad + \frac{\mathcal{P}(K_1)}{\left\{(\kappa_o^2 - k_3^2)^{1/2} + K_1\right\}^{1/2}}. \qquad (3.5.10)
\end{aligned}$$

The reader should verify from the integral in (3.5.6) that the Kutta condition (that the pressure is finite at the trailing edge) is satisfied only if $\mathcal{P}(K_1) = 0$. This observation leads directly to the representation (3.5.3).

3.5.2 The Acoustic Spectrum

To calculate the mean square pressure in the acoustic far field from the asymptotic approximation (3.5.5), we need to express the ensemble average $\langle p_s(k_1, \kappa_o x_3/|\mathbf{x}|, \omega) p_s^*(k_1', \kappa_o' x_3/|\mathbf{x}|, \omega')\rangle$ in terms of the turbulence wall pressure wavenumber–frequency spectrum $\mathrm{P}(\mathbf{k}, \omega)$ (Section 3.4.1). In practice, however, only a finite section $-\frac{1}{2}L < x_3 < \frac{1}{2}L$, say, of the trailing edge is actually wetted by the turbulent flow. Provided L greatly exceeds the boundary layer thickness δ, the wall pressure fluctuations may be assumed to be stationary random in both position and time, and the reader may easily discern from the definition of $p_s(\mathbf{k}, \omega)$ that the following modified form of equation (3.4.3) must now be used:

$$\langle p_s(k_1, k_3, \omega) p_s^*(k_1', k_3, \omega')\rangle \approx \frac{L}{2\pi}\delta(\omega - \omega')\delta(k_1 - k_1')P(k_1, k_3, \omega),$$

$$L \gg \delta. \qquad (3.5.11)$$

This, together with equation (3.5.5), determines the acoustic pressure frequency spectrum $\Phi(\mathbf{x}, \omega)$ of the edge noise (defined such that $\langle p^2(\mathbf{x}, t)\rangle = \int_0^\infty \Phi(\mathbf{x}, \omega)\, d\omega$, as in (3.4.20)):

$$\Phi(\mathbf{x}, \omega) \approx \frac{\omega L x_2^2}{4\pi c_o |\mathbf{x}|^4 \left|\left(1 - x_3^2/|\mathbf{x}|^2\right)^{\frac{1}{2}} + x_1/|\mathbf{x}|\right|} \times \int_{-\infty}^{\infty} \frac{\left|\kappa_o\left(1 - x_3^2/|\mathbf{x}|^2\right)^{\frac{1}{2}} + k_1\right|}{(\kappa_o x_1/|\mathbf{x}| - k_1)^2} \mathrm{P}\left(k_1, \frac{\kappa_o x_3}{|\mathbf{x}|}, \omega\right) dk_1,$$

$$|\mathbf{x}| \to \infty. \qquad (3.5.12)$$

This is applicable for turbulence on one side of the airfoil. When the reduced frequency $\omega a/U$ is large ($2a$ being the airfoil chord), the turbulent flows on opposite surfaces are statistically independent, and the net acoustic spectrum is obtained by adding the separate contributions from each side.

For applications to aircraft trailing edge noise, the spectrum is expressed in terms of the emission time coordinates (R, Θ) of Section 1.8.5 for a ground-based observer at rest relative to the fluid. In the present case, the trailing edge must be imagined to be in "flight" at Mach number $M = U/c_o$ in the *negative* x_1-direction, so that $\mathbf{x} = \mathbf{R} + MR\mathbf{i}_1$. For an overflying aircraft ($x_3 = R_3 = 0$), we find that

$$\begin{aligned}\Phi(\mathbf{R}, \omega) &\approx \frac{L \sin^2(\Theta/2)}{2\pi R^2 c_o(1 + M)(1 - M\cos\Theta)^2} \int_{-\infty}^{\infty} \frac{|1 + k_1/\kappa_o|}{(\cos\Theta + k_1/\kappa_o)^2} \\ &\quad \times \mathrm{P}(k_1, 0, \omega(1 - M\cos\Theta))\, dk_1 \\ &\approx \frac{M_c(1 + M_c - M\cos\Theta)L\, \sin^2(\Theta/2)}{2\pi R^2 c_o(1 + M)(1 - M\cos\Theta)^2\{1 + (M_c - M)\cos\Theta\}^2} \\ &\quad \times \int_{-\infty}^{\infty} \mathrm{P}(k_1, 0, \omega(1 - M\cos\Theta))\, dk_1,\end{aligned}$$

$$M^2 \ll 1, \; R_3 = 0, \; R \to \infty. \qquad (3.5.13)$$

In this formula, $M_c = U_c/c_o$ is the convection Mach number of the principal turbulence eddies (relative to the wing), and in the second line we have set $k_1 = (\omega/U_c)(1 - M\cos\Theta)$ in the integrand except in the argument of the wall pressure spectrum, where $\omega(1 - M\cos\Theta)$ corresponds to the source frequency when the observed frequency is ω.

When $M \to 0$, the difference between the fixed and emission time coordinates can be neglected, and (3.5.12) becomes

$$\Phi(\mathbf{x}, \omega) \approx \frac{\omega L \sin^2(\theta/2)\sin\psi}{2\pi c_o|\mathbf{x}|^2}\int_{-\infty}^{\infty}\frac{\mathrm{P}(k_1, \kappa_o\cos\psi, \omega)\,dk_1}{|k_1|},$$
$$M \ll 1, \quad |\mathbf{x}| \to \infty, \qquad (3.5.14)$$

where $\theta = \arcsin(x_2/\sqrt{x_1^2 + x_2^2})$, and ψ is the angle between the observer direction and the edge of the half-plane. The peak acoustic pressures are radiated in the "forward" directions $\theta = \pm\pi$, in contrast to the directivity of an airfoil of compact chord, for which the peaks are in directions normal to the wing.

In principle these formulae are to be applied using measured values of the wall pressure spectrum $\mathrm{P}(\mathbf{k}, \omega)$. In practice it is difficult and time-consuming to make sufficiently accurate measurements (which involve instrumenting the airfoil with an array of small transducers) and most estimates and comparisons with experiment are based on a measurement of the surface pressure frequency spectrum $\Phi_{pp}(\omega)$ at a single point. This is used in an empirical model of $\mathrm{P}(\mathbf{k}, \omega)$, which depends linearly on $\Phi_{pp}(\omega)$. The simplest is the Corcos model (3.4.10), which represents the behavior of $\mathrm{P}(\mathbf{k}, \omega)$ at low Mach numbers when $\omega\delta_*/U > 0.1$ ($\delta_* \approx \delta/8$) in the immediate vicinity of the convective ridge (Figure 3.4.1), where the spectrum has a large maximum and where the motion has the characteristics of an incompressible flow.

The Corcos model is used in (3.5.14) by first replacing $|k_1|$ in the denominator of the integrand by ω/U_c, to obtain

$$\Phi(\mathbf{x}, \omega) \approx \frac{U_c L \sin^2(\theta)\sin\psi}{2\pi^2 c_o|\mathbf{x}|^2}\ell_3\Phi_{pp}(\omega)$$
$$= a_o M\left(\frac{\delta_* L}{|\mathbf{x}|^2}\right)\sin^2(\theta/2)\sin\psi\,\frac{\Phi_{pp}(\omega)}{\omega\delta_*/U}, \quad a_o = 0.035.$$
$$(3.5.15)$$

The spectrum is illustrated by the dashed curve in Figure 3.5.2 (labeled Corcos (i)), using the Chase approximation (3.4.11) for $\Phi_{pp}(\omega)$, which is also plotted in the figure. This prediction agrees well with measurements of the self-noise of airfoils with sharp trailing edges [171, 193] when $\omega\delta_*/U < 1$. At higher frequencies, however, Chase's $\Phi_{pp}(\omega)$ decays too slowly (like $1/\omega$), and the actual edge noise tends to be much smaller than the prediction. This is probably because separation just upstream of the edge of a real airfoil of finite thickness prevents smaller scale boundary layer eddies from flowing along

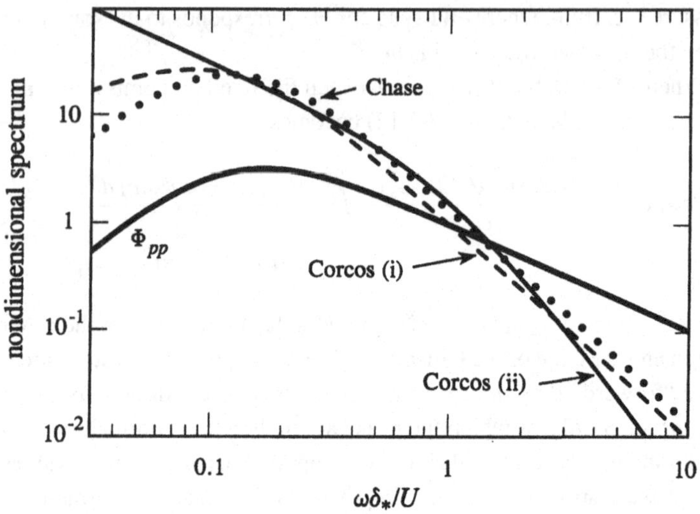

Figure 3.5.2. The Chase approximation for $(U/\delta_*)\Phi_{pp}(\omega)/(\rho_o v_*^2)^2$, and edge noise spectra $(U/\delta_*)\Phi(\mathbf{x},\omega)/[a_o(\rho_o v_*^2)^2 M(\delta_* L/|\mathbf{x}|^2)\sin^2(\theta/2)\sin\psi]$.

paths that would take them close enough to the edge to interact effectively (Example 10).

Alternative and more complicated empirical representations of $\mathrm{P}(\mathbf{k},\omega)$ are available, which are applicable over larger ranges of wavenumber (Section 3.4.2). The following formula was also proposed by Chase [179] (cf. (3.4.8)) for $\omega\delta_*/U > 0.1$:

$$\mathrm{P}(\mathbf{k},\omega) = \frac{\mathrm{C_M}\rho_o^2 v_*^3 k_1^2 \delta^5}{[(\omega - U_c k_1)^2(\delta/hv_*)^2 + (k\delta)^2 + 1/b^2]^{\frac{5}{2}}}, \tag{3.5.16}$$

where $\mathrm{C_M}$, h, and b are defined as in (3.4.8). It can be used to evaluate the integral in equation (3.5.14), and it supplies the following approximation for the trailing edge noise spectrum:

$$\Phi(\mathbf{x},\omega) \approx 1.38 a_o M\left(\frac{\delta_* L}{|\mathbf{x}|^2}\right)\sin^2(\theta/2)\sin\psi \frac{(\delta_*/U)(\rho_o v_*^2)^2(\omega\delta_*/U)^2}{\left[(\omega\delta_*/U)^2 + \alpha_p^2\right]^2}, \tag{3.5.17}$$

where it is assumed that $U_c \approx 0.7U$, $\delta_* = \delta/8$, and α_p and a_o have the values given in equations (3.4.11) and (3.5.15). The spectrum is shown dotted in Figure 3.5.2; the behavior at higher frequencies ($\sim 1/\omega^2$) is the same as for the Corcos model. A more rapid decay at high frequencies (like $1/\omega^3$) is obtained by using the Laganelli and Wolfe approximation (3.4.18) for $\Phi_{pp}(\omega)$ in the Corcos formula (3.5.15). This is the "Corcos (ii)" curve in Figure 3.5.2 obtained by taking $\lambda = 1$ and $v_* = 0.04U$ in (3.4.18).

Example 2. Use the representation (1.10.3) of the solution of Helmholtz's equation to express the trailing edge scattered pressure (Example 1) in the form

$$p(\mathbf{x}, t) = -\oint_S \frac{\partial p_I}{\partial y_n}(y_1, 0, y_3, \omega) G(\mathbf{x}, \mathbf{y}; \omega)\, dS(\mathbf{y}), \qquad (3.5.18)$$

where the integration is over the upper and lower surfaces S of the half-plane ($y_1 < 0$, $y_2 = \pm 0$), y_n being normal to S and directed into the fluid, and Green's function has vanishing normal derivative on S.

By using the compact Green's function (3.2.14), show that, in the notation of Figure 3.5.1,

$$p(\mathbf{x}, \omega) \approx \frac{-\sin(\theta/2)\sin^{\frac{1}{2}}\psi}{\sqrt{2}|\mathbf{x}|} \iint_{-\infty}^{\infty} \frac{\sqrt{\kappa_o}\, p_s(k_1, \kappa_o \sin\psi, \omega) \mathrm{e}^{-i\omega(t-|\mathbf{x}|/c_o)}}{\sqrt{k_1 + i0}} \times dk_1\, d\omega, \quad M \ll 1, \quad |\mathbf{x}| \to \infty. \qquad (3.5.19)$$

Deduce the approximation (3.5.14) for the acoustic pressure frequency spectrum.

Example 3. When the mean flow is inclined at an angle $\beta (< 90°)$ to the x_1-axis, show that the low Mach number, Corcos approximation (3.5.15) for the frequency spectrum of trailing edge noise becomes

$$\Phi(\mathbf{x}, \omega) \approx a_o M \left(\frac{\delta_* L}{|\mathbf{x}|^2}\right) \frac{\sin^2(\theta/2)\sin\psi}{\cos\beta[1 + (1.4)^2\tan^2\beta]} \frac{\Phi_{pp}(\omega)}{\omega\delta_*/U}, \qquad (3.5.20)$$

where L denotes the transverse width of the turbulent stream (so that the wetted edge of the airfoil has length $L/\cos\beta$).

Example 4. In analyzing the noise generated when a low Mach number turbulent jet impinges on a large screen and flows over its edge, Chase [198] modeled $\mathrm{P}(\mathbf{k}, \omega)$ by the following modified form of (3.5.16):

$$\mathrm{P}(\mathbf{k}, \omega) = \frac{\mathrm{C}\rho_o^2 v_*^3 k_1^2 \Delta^5}{[(\omega - U_c k_1)^2 (\Delta/3v_*)^2 + (k\Delta)^2 + 1/b^2]^{\nu + \frac{1}{2}}},$$

where C is a constant, Δ is proportional to the jet diameter, and $\nu = \frac{7}{3}$. Show that the edge noise acoustic pressure spectrum is

$$\Phi(\mathbf{x}, \omega) \approx \frac{3\mathrm{C}\Gamma(\nu)}{2\sqrt{\pi}\Gamma\left(\nu + \frac{1}{2}\right)c_o} \frac{\Delta^2 L}{|\mathbf{x}|^2} \sin^2(\theta/2)\sin\psi \frac{(\rho_o v_*^2)^2(\omega\Delta/U_c)^2}{[(\omega\Delta/U_c)^2 + 1/b^2]^{\nu}}, \quad |\mathbf{x}| \to \infty.$$

Example 5: Right-angled wedge. Consider the edge noise generated by turbulent boundary layer flow off the upper surface $x_2 = +0$ of the rigid, right-angled wedge bounded by the planes $x_1 < 0,\ x_2 = 0$ and $x_1 = 0,\ x_2 < 0$. Assume the boundary layer turbulence translates in the x_1-direction and may be regarded as frozen during its interaction with the edge.

Express the acoustic pressure in the form (3.5.18), where G is given by equation (3.2.17) when $M \ll 1$, and deduce that

$$p(\mathbf{x}, t) \approx \frac{-2^{\frac{1}{3}}}{\sqrt{3} c_o^{\frac{2}{3}} |\mathbf{x}|} \cos\left(\frac{2}{3}(\theta - \pi)\right) \sin^{\frac{2}{3}} \psi$$
$$\times \int_{-\infty}^{\infty} \frac{\omega^{\frac{2}{3}} p_s(k_1, \kappa_o \cos\psi, \omega) \mathrm{e}^{-i\omega(t - |\mathbf{x}|/c_o)}}{(k_1 + i0)^{\frac{2}{3}}} dk_1\, d\omega \quad |\mathbf{x}| \to \infty.$$

Using the Corcos model (3.4.10) show that

$$\Phi(\mathbf{x}, \omega) \approx 0.033 M^{\frac{4}{3}} \left(\frac{\delta_* L}{|\mathbf{x}|^2}\right) \cos^2\left(\frac{2}{3}(\theta - \pi)\right) \sin^{\frac{4}{3}} \psi \frac{\Phi_{pp}(\omega)}{\omega \delta_* / U},$$
$$|\mathbf{x}| \to \infty.$$

This approximation is strictly applicable when both faces of the wedge are large compared to the acoustic wavelength. For a thin airfoil with a wedge-shaped (or beveled) trailing edge, the Mach number dependence and directivity are the same as for the thin airfoil discussed in Section 3.5.1, although the frequency dependence is different [207].

Example 6: Airfoil of compact chord. Consider the airfoil modeled by the rigid strip $|x_1| < a,\ x_2 = 0$, and assume there exists a mean flow in the x_1-direction with $M \ll 1$. Use the compact Green's function (1.11.7), with $\mathbf{Y}(\mathbf{y})$ defined as in (3.3.3), to deduce that the acoustic pressure frequency spectrum of the trailing edge noise generated by turbulent flow on one side of the airfoil is given by

$$\Phi(\mathbf{x}, \omega) \approx \frac{aL}{8|\mathbf{x}|^2} \frac{\omega^2 \cos^2\Theta}{c_o^2} \int_{-\infty}^{\infty} \frac{\mathrm{P}(k_1, \kappa_o \cos\psi, \omega)\, dk_1}{|k_1|}$$
$$\approx 0.027 \frac{aLM^2 \cos^2\Theta \Phi_{pp}(\omega)}{|\mathbf{x}|^2}, \quad M \ll 1, \quad \omega a/U \gg 1,$$
$$|\mathbf{x}| \to \infty,$$

where L is the wetted span of the trailing edge, Θ is the angle between the radiation direction and the normal to the airfoil, and the Corcos formula (3.4.10) has been used.

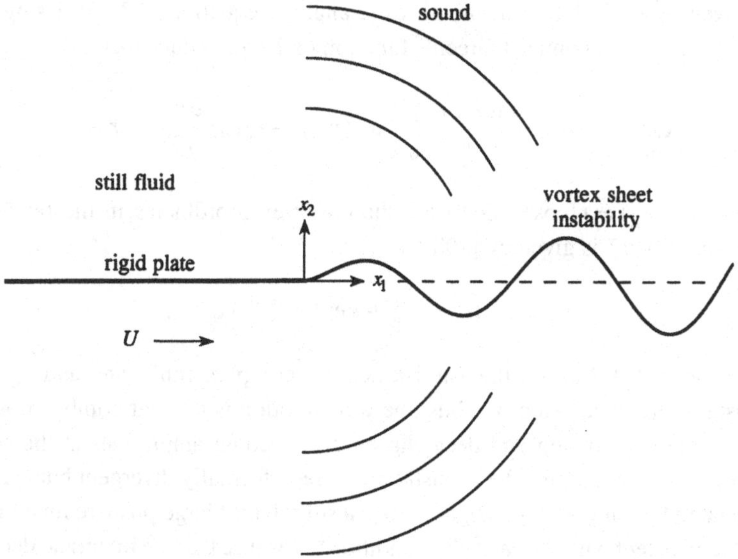

Figure 3.5.3. Kelvin–Helmholtz instability waves.

Example 7. An ideal fluid is in low subsonic uniform motion at speed U in the x_1-direction on the lower side ($x_2 < 0$) of the semi-infinite rigid plate $x_1 < 0$, $x_2 = 0$. The fluid in $x_2 > 0$ is in a mean state of rest (Figure 3.5.3). The vortex sheet in the wake of the plate is unstable, and for time-harmonic disturbances of frequency ω, the linearized equations of motion predict the existence of two-dimensional Kelvin-Helmholtz instability waves proportional to e^{ikx_1}, where $k = (1 \pm i)\omega/U$ (for $M \ll 1$) [2, 4, 10]. One of these waves grows exponentially with distance from the edge of the plate, but both must be included in a solution of the equations of motion that is required to have vanishing normal velocity on the plate. In the absence of external excitation (e.g., by an incident sound wave or gust), the only such solution is one that does not satisfy the Kutta condition at the edge [208]; experiments [171, 193] suggest that this may model quasi-periodic flow at moderate Reynolds numbers just downstream of a plate of finite thickness.

To determine the edge noise generated by these waves, let $\mathbf{u}^{\pm}$ denote the perturbation velocity just above and below the sheet. In the linearized approximation,

$$\omega \wedge \mathbf{v} = [U\mathbf{i}_1 + \mathbf{u}^- - \mathbf{u}^+] \wedge \tfrac{1}{2}[U\mathbf{i}_1 + \mathbf{u}^+ + \mathbf{u}^+]\delta(x_2) \approx U u_1^- \mathbf{i}_2 \delta(x_2).$$

When the convection of sound by the mean flow is neglected, the edge noise is given by the solution of the acoustic analogy equation (2.3.15). Using the two-dimensional compact Green's function (3.2.16), deduce that

$$p(\mathbf{x},t) \approx -\rho_o U \frac{\sin(\theta/2)}{2\sqrt{r}} \int_0^\infty u_1^-(y_1, t - r/c_o) \frac{dy_1}{\sqrt{y_1}}, \quad r \to \infty,$$

where $(x_1, x_2) = r(\cos\theta, \sin\theta)$ are the observer coordinates in the far field. The velocity u_1^- is given by [208]

$$u_1^- = u_o\left(\kappa^* \mathrm{e}^{i\kappa^* x_1} + \kappa \mathrm{e}^{i(\kappa x_1 - \pi/4)}\right)\mathrm{e}^{-i\omega t},$$

where $\kappa = (1+i)\omega/U$, the asterisk denotes complex conjugate, and u_o is a constant reference velocity. Thus, the wake motion is a linear combination of exponentially growing and decaying modes of equal amplitudes at the edge $x_1 = 0$. The integral for the acoustic pressure is formally divergent but can be evaluated by temporarily assigning to ω a sufficiently large positive imaginary part (consistent with the causality condition) to ensure that the integrand decays as $x_1 \to \infty$.

Deduce that the exponentially growing and decaying waves make *equal* contributions to the pressure and that

$$p(\mathbf{x},t) \approx -\rho_o u_o \left(\frac{\sqrt{2i\omega U}}{\pi}\right)^{\frac{1}{2}} \frac{\sin(\theta/2)}{\sqrt{r}} \mathrm{e}^{-i\omega(t - r/c_o)}, \quad r \to \infty.$$

Example 8: Influence of finite thickness on edge noise. Consider the trailing edge noise generated by the low Mach number convection of a frozen turbulent field over an edge of acoustically compact thickness h (Figure 3.5.4a, b). Let the flow be in the x_1-direction over the upper surface $x_2 = +0$ and represent the sound by (3.5.18). Take the compact Green's function in the form (3.2.14) with $\varphi^*(\mathbf{y})$ replaced by the potential function $\Phi^*(\mathbf{y})$, which satisfies $\partial\Phi^*(\mathbf{y})/\partial y_n = 0$ on the surface S of the thick edge, and

$$\Phi^*(y_1, y_2) \to \varphi^*(y_1, y_2), \quad \text{as } \sqrt{y_1^2 + y_2^2} \to \infty.$$

Show that

$$p(\mathbf{x},t) \approx \frac{\sin^{\frac{1}{2}}\psi \sin\left(\frac{1}{2}\theta\right)}{|\mathbf{x}|\sqrt{2\pi i c_o}} \int\int_{-\infty}^{\infty} \sqrt{\omega}\mathcal{I}(k_1) p_s\left(k_1, \frac{\kappa_o x_3}{|\mathbf{x}|}, \omega\right)$$
$$\times\, \mathrm{e}^{-i\omega(t - |\mathbf{x}|/c_o)}\, dk_1\, d\omega, \quad |\mathbf{x}| \to \infty,$$

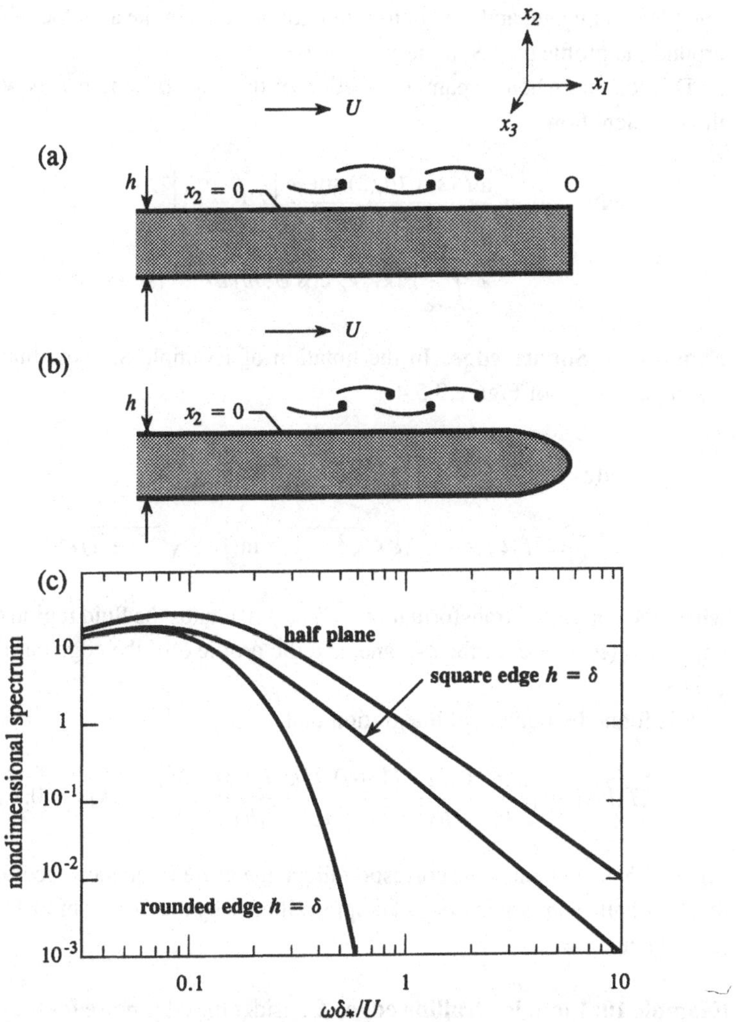

Figure 3.5.4. Predicted edge noise spectrum $(U/\delta_*)\Phi(\mathbf{x},\omega)/[a_o(\rho_o v_*^2)^2 M(L\delta_*/|\mathbf{x}|^2)\sin\psi\sin^2(\theta/2)]$ for the half-plane; for a square edge of thickness $h=\delta$ (a); and for a rounded edge of thickness $h=\delta$ (b).

where

$$\mathcal{I}(k_1) = \left(i \oint_C \frac{dw(z)}{dz} e^{-ik_1 z} dz \right)^*, \quad k_1 > 0, \quad z = y_1 + iy_2;$$

$$\mathcal{I}(k_1) = \mathcal{I}^*(-k_1), \quad k_1 < 0,$$

$w(z)$ is the analytic function whose real part is $\Phi^*(\mathbf{y})$, the asterisk denotes complex conjugate, and the contour integral is taken in the anticlockwise sense around the profile $\mathcal{C}$ of S in the y_1y_2-plane.

Deduce that when a spanwise section of the edge of length L is wetted by the turbulent flow,

$$\Phi(\mathbf{x},\omega) \approx \frac{\omega L \sin^2(\theta/2)\sin\psi}{2\pi^2 c_o|\mathbf{x}|^2}\left|\mathcal{I}\left(\frac{\omega}{U_c}\right)\right|^2 \times \int_{-\infty}^{\infty} \mathrm{P}(k_1, \kappa_o\cos\psi, \omega)\,dk_1, \quad |\mathbf{x}|\to\infty.$$

Example 9: Square edge. In the notation of Example 8, show that for the rectangular edge of Figure 3.5.4a

$$w(z) = -\sqrt{\frac{h}{\pi}}\zeta,$$

$$\frac{z}{h} = f(\zeta) \equiv -\frac{1}{\pi}\{\zeta\sqrt{\zeta^2-1} - \ln(\zeta+\sqrt{\zeta^2-1})\} - i,$$

where the conformal transformation $z/h = f(\zeta)$ maps the fluid region onto the upper half ($\mathrm{Im}\,\zeta > 0$) of the ζ-plane, and the profile $\mathcal{C}$ of the edge onto the real ζ-axis.

It is found by numerical integration that

$$|\mathcal{I}(k_1)| \approx \sqrt{\frac{\pi}{k_1}}\frac{0.71}{(k_1h)^{\frac{1}{6}}}\frac{\{1+0.25k_1h+(k_1h)^2\}}{1+(k_1h)^2}, \quad k_1h > 0.1.$$

Figure 3.5.4c compares the corresponding edge noise frequency spectrum with that for a half-plane when $\mathrm{P}(\mathbf{k},\omega)$ is approximated by the Corcos model (3.4.10), (3.4.11), and $h=\delta$.

Example 10: Rounded trailing edge. Consider the edge noise for the rounded trailing edge of Figure 3.5.4b, defined by the conformal transformation

$$\frac{z}{h} = \frac{f(\zeta)+\alpha f(\zeta/\beta)}{1+\alpha},$$

where $f(\zeta)$ is given in Example 9, and α, β are numerical coefficients. The real ζ-axis maps onto the edge profile $\mathcal{C}$ in the z-plane, the region $\mathrm{Im}\,\zeta > 0$ into the fluid region, and

$$w(z) = -\sqrt{\frac{h(1+\alpha/\beta^2)}{\pi(1+\alpha)}}\zeta.$$

For the case shown in Figure 3.5.4b, $\alpha = 0.0013$, $\beta = 0.02$, the mean radius of curvature of the edge $\sim 1.14h$, and

$$|\mathcal{I}(k_1)| \approx \sqrt{\frac{\pi}{k_1}} \exp\left(-\frac{k_1 h}{4} - \frac{(k_1 h)^2}{20.25}\right), \quad k_1 h > 0.$$

The edge noise frequency spectrum obtained using the Corcos model (3.4.10), (3.4.11) for $\mathrm{P}(\mathbf{k}, \omega)$ is depicted in Figure 3.5.4c for the case $h = \delta$. Note the rapid fall off with increasing frequency relative to the spectra for the half-plane and square trailing edge. This is because, for the range of high frequencies covered by these calculations, the length scale (radius of curvature) of the rounded edge is always large compared to the turbulence length scale.

3.5.3 Serrated Trailing Edge

The spanwise turbulence wall pressure correlation length $\ell_3 \sim U/\omega$ in the Corcos model (3.4.10), so that the order of magnitude of trailing edge noise furnished by (3.5.15) can be written

$$\Phi(\mathbf{x}, \omega) \sim M \frac{\ell_3 L}{|\mathbf{x}|^2} \Phi_{pp}(\omega), \quad |\mathbf{x}| \to \infty.$$

Thus, the edge-generated sound should be of reduced intensity if the effective values of either of the lengths ℓ_3, L can be reduced. To do this, one might attempt to decrease spanwise coherence by serrating the trailing edge, and sawtooth serrations of the type illustrated in Figure 3.5.5 have yielded reductions of 4 dB or more in wind tunnel tests [209].

Let us investigate the influence of serrations by considering turbulent flow over the upper surface of the airfoil: $x_1 < \zeta(x_3)$, $x_2 = 0$, where the $\zeta(x_3)$ is a periodic function of amplitude h and "wavelength" λ. Equation (3.5.18) supplies the following representation of the edge noise

$$p(\mathbf{x}, \omega) = \frac{i}{2} \int_{-\infty}^{\infty} dy_3 \int_{-\infty}^{0} dy_1' \int\int_{-\infty}^{\infty} \gamma(k) p_s(\mathbf{k}, \omega) \times [G(\mathbf{x}, y_1' + \zeta(y_3), y_3; \omega)] \mathrm{e}^{i(k_1[y_1 + \zeta(y_3)] + k_3 y_3)} d^2\mathbf{k}, \qquad (3.5.21)$$

where

$$[G(\mathbf{x}, y_1, y_3; \omega)] = G(\mathbf{x}, y_1, +0, y_3; \omega) - G(\mathbf{x}, y_1, -0, y_3; \omega),$$

and G is a suitable Green's function whose normal derivative vanishes on the airfoil.

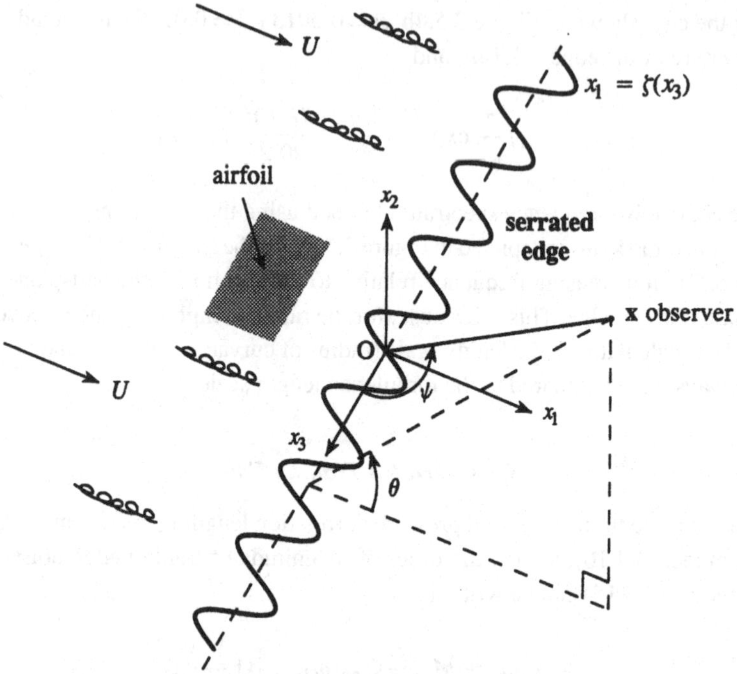

Figure 3.5.5. Serrated trailing edge.

Boundary layer surface pressures of wavenumber $\mathbf{k} = (k_1, 0, k_3)$ generate sound of frequency $\omega \sim U_c k_1$ by interaction with the edge. When the acoustic wavelength is much larger than h, and when $|\mathbf{x}| \to \infty$, the scale of variation of $G(\mathbf{x}, \mathbf{y}; \omega)$ near the edge is of the same order as the smaller of h and λ. The amplitude of the sound is therefore controlled by the oscillations of the exponential in (3.5.21).

The noise generated by turbulence whose length scale $1/k_1$ is much larger than h is approximately the same as when the serrations are absent. When $k_1 \zeta(y_3) \gg 1$, however, the integrand of (3.5.21) oscillates rapidly as y_3 varies, and this will tend to reduce the amplitude of higher frequency edge noise. The y_3-integral can be estimated by the method of stationary phase (Section 1.13.2) when the serrations vary smoothly, according to which, for each fixed $\mathbf{k}$, the principal contributions to the integral are from the neighborhoods of those parts of the edge where $\mathbf{k}$ *is normal to the edge*, that is, where

$$\frac{d\zeta}{dy_3} = \frac{-k_3}{k_1} \approx \frac{-U_c k_3}{\omega}. \tag{3.5.22}$$

However, the dominant wall pressures occur within $|k_3| \leq 1/\ell_3 \sim \omega/U$. Hence, the principal sources of high-frequency edge noise are from those sections of the trailing edge where $|d\zeta/dy_3| < U_c/U < 1$, where the angle between the mean flow direction and the local tangent to the edge exceeds about 45°. The serrations therefore tend to reduce the effective wetted span of the trailing edge responsible for high-frequency sound from the actual value L to an aggregate length L_ζ of those sections where $|d\zeta/dy_3| < 1$, and in order of magnitude, the corresponding acoustic spectrum $\Phi_\zeta(\mathbf{x}, \omega) \sim (L_\zeta/L)\Phi(\mathbf{x}, \omega)$, where $\Phi(\mathbf{x}, \omega)$ is the spectrum for the straight edge.

For sinusoidal serrations, $\zeta = h\sin(2\pi y_3/\lambda)$ the total spanwise length per wavelength λ within which $|d\zeta/dy_3| < 1$ is

$$(2\lambda/\pi)\arcsin(\lambda/2\pi h) \approx \lambda^2/\pi^2 h, \quad \text{when } \lambda \leq 4h.$$

Thus, $L_\zeta \sim (L/\lambda)(\lambda^2/\pi^2 h) = \lambda L/\pi^2 h$, provided λ/h is not too large, and

$$\Phi_\zeta(\mathbf{x}, \omega) \approx \frac{\lambda}{\pi^2 h}\Phi(\mathbf{x}, \omega), \quad \omega h/U_c \gg 1. \tag{3.5.23}$$

It appears that, provided edge geometry does not affect the impinging turbulent boundary layers on the airfoil, the high-frequency components of the edge noise will be reduced to a minimum by straight-sided *sawtooth* serrations whose edges are inclined to the mean flow direction at angles smaller than 45°. Then condition (3.5.22) cannot be satisfied, and the main contributions to the y_3-integral are an order of magnitude smaller than the stationary phase estimate and come from the vicinities of the tips and roots of the sawtooth profile, where $d\zeta/dy_3$ is discontinuous [19].

More detailed calculations are described in [210, 211], where it is shown that

$$\begin{aligned}\Phi_\zeta(\mathbf{x}, \omega) &\leq \frac{\lambda}{2\pi h}\Phi(\mathbf{x}, \omega), \quad \text{for sinusoidal serrations } \zeta = h\sin\left(\frac{2\pi x_3}{\lambda}\right),\\ &\leq \frac{\lambda^2}{\lambda^2 + 16h^2}\Phi(\mathbf{x}, \omega), \quad \text{for sawtooth serrations,}\end{aligned}$$

where for the sawtooth serrations $2h$ corresponds to the streamwise length of the teeth. The smaller attenuation for the sinusoidal serrations is consistent with the stationary phase approximation (3.5.23).

The curves in Figure 3.5.6 (plotted using formulae from [211]) illustrate the predicted edge noise for a sawtooth trailing edge, when the streamwise root-to-tip length of the teeth $2h = 10\delta$, δ being the boundary layer thickness at the edge. The attenuation increases rapidly when $\lambda/h < 5$, when the angle between the edges of the teeth and the mean flow direction ($= \arctan(\lambda/4h)$) is

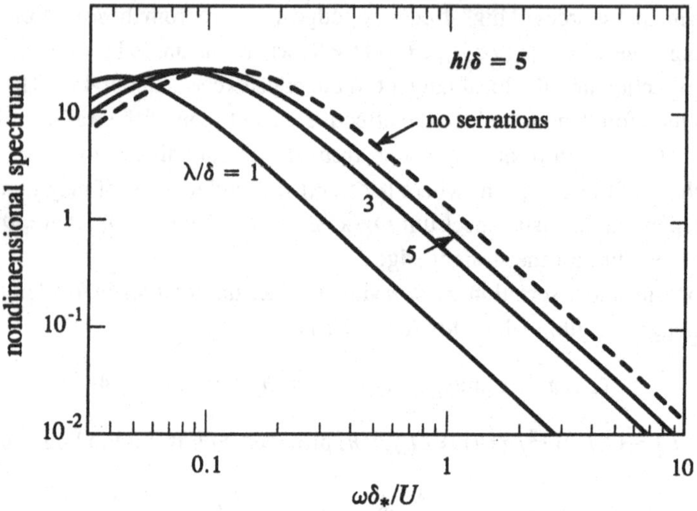

Figure 3.5.6. Predicted acoustic spectra $(U/\delta_*)\Phi_\zeta(\mathbf{x},\omega)/[a_o(\rho_o v_*^2)^2 M(\delta_* L/|\mathbf{x}|^2)\sin^2(\theta/2)\sin\psi]$ for a sawtooth trailing edge [211].

less than 45°. These results make use of the Chase approximation (3.5.16) for the wall pressure wavenumber–frequency spectrum, and the spectrum shown for the unserrated edge corresponds to the dotted curve in Figure 3.5.2.

3.5.4 Porous Trailing Edge

Attempts have been made to attenuate trailing edge-noise by modifying the mechanical properties of the edge. The intensity of edge scattering should be smaller if the "impedance discontinuity" experienced by an eddy in convecting over the edge can be reduced. This might be possible by fabricating the trailing edge region from material of variable porosity, so that the surface becomes progressively "softer" as the edge is approached. Experiments and theory [212, 213] suggest that reductions of 10 dB or more are possible.

3.6 Sound Generation by Moving Surfaces

A rigid body in accelerated motion, or in steady motion relative to another body, generates sound even in the absence of vortex shedding. The sound produced by a propeller whose blades rotate at constant angular velocity in a nominally irrotational incident mean stream is called *Gutin noise* [214]. Vorticity shed from the blade tips produces thrust ("lift") and drag forces that are constant in a frame rotating with the blades. At low Mach numbers the sound is generated principally by the accelerated motions of these forces in a circle and by the (irrotational) displacement of fluid by the advancing blades, both of which cause time-varying fluctuations in pressure and velocity with respect to a stationary observer. Thus, when broadband turbulence sources are ignored, and for a rotor with equal, evenly spaced blades, the acoustic frequency spectrum consists of a series of discrete spikes occurring at multiples of the *blade passage frequency*, Doppler shifted to account for forward motion of the rotor. Similarly, the potential flow interaction between a train traveling at constant speed and a tunnel entrance generates a compression wave that propagates into the tunnel ahead of the train at the speed of sound. Part of the wave energy is radiated from the far end of the tunnel as a spherically spreading pulse, known as a *micro-pressure wave* [215, 216]. For high-speed trains, nonlinear steepening of the wave in a long tunnel can produce micro-pressure wave amplitudes in excess of 1 lb/ft^2 near the tunnel exit, which is comparable to the sonic boom from a supersonic aircraft.

3.6.1 Propeller and Rotor Noise: The Sonic Cylinder

The rotor of blade radius r_t depicted in plan view in Figure 3.6.1 rotates at constant angular velocity Ω with subsonic blade tip Mach number $M_t = \Omega r_t / c_o$. At a radial distance r from the rotor axis, just outboard of the blade tips (on the

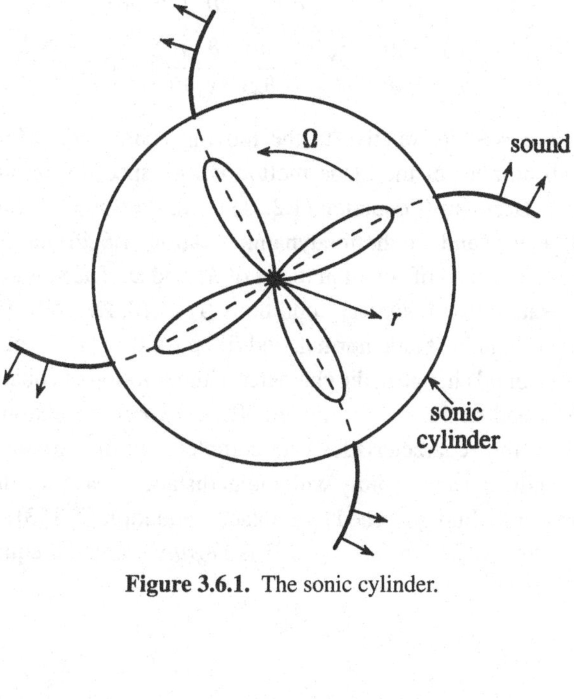

Figure 3.6.1. The sonic cylinder.

dashed lines in the figure) the perturbation velocity and pressure fields rotate with the blades as frozen patterns at the local "phase" velocity Ωr, just as if the fluid were incompressible. However, the velocity becomes supersonic when r exceeds a critical value r_s, which defines the *sonic cylinder* where $\Omega r_s = c_o$. Small amplitude disturbances cannot propagate faster than the speed of sound, and the physical surfaces of constant phase that link the motion in $r > r_s$ to that of the rotor must therefore curve backwards outside the sonic cylinder to form spiral-like surfaces whose normal velocities of advance are just equal to the speed of sound and which actually constitute sound waves propagating away from the rotor.

The motion may be assumed to be irrotational in a first approximation except at certain points on the surface of the cylindrical "jet" (coaxial with the sonic cylinder) projected from the high-pressure side of the rotor, on which the blade-tip vortices form helical filaments. The velocity potential φ in the irrotational region satisfies equation (1.6.24). The flow is steady with respect to a cylindrical coordinate system (r, θ, x_3) rotating with the blades (x_3 being measured along the rotor axis in the direction of mean thrust) in terms of which equation (1.6.24) becomes [217],

$$\begin{aligned}&\left(1+\frac{M(\gamma-1)}{c_o r}\frac{\partial\varphi}{\partial r}\right)\left(\frac{\partial^2\varphi}{\partial x_3^2}+\frac{\partial^2\varphi}{\partial r^2}+\frac{1}{r}\frac{\partial\varphi}{\partial r}\right)\\&\quad+\left(1-M^2+\frac{M(\gamma+1)}{c_o r}\frac{\partial\varphi}{\partial r}\right)\frac{1}{r^2}\frac{\partial^2\varphi}{\partial\theta^2}\\&\quad+\frac{2M}{c_o r}\left(\frac{\partial\varphi}{\partial r}\frac{\partial^2\varphi}{\partial r\partial\theta}+\frac{\partial\varphi}{\partial x_3}\frac{\partial^2\varphi}{\partial x_3\partial\theta}\right)=0,\end{aligned}\qquad(3.6.1)$$

where $\nabla\varphi$ is the velocity relative to the moving frame, and $M = \Omega r/c_o$ is the local Mach number of the blade motion with respect to the undisturbed sound speed c_o. Bernoulli's equation (1.2.21) (with $\partial\varphi/\partial t = -\Omega\partial\varphi/\partial\theta$, and $\mathbf{v} = (0, \Omega r, 0) + \nabla\varphi$) and the thermodynamic relation (1.6.26) have been used to express the local speed of sound in terms of M and φ. The solution must be constructed to satisfy the boundary conditions (i) $\mathbf{n}.((0, \Omega r, 0) + \nabla\varphi) = 0$ on the blades, where $\mathbf{n}$ is the blade normal, and (ii) $\varphi \to 0$ as $|\mathbf{x}| \to \infty$.

There is a profound change in the character of the solution of equation (3.6.1) in the neighborhood of the sonic cylinder. The *linearized* equation is *elliptic* for $r < r_s$, where the "characteristics" are complex and the motion generated by the blades tends to decay rapidly with radial distance r (as for a disturbance in an incompressible fluid governed by Laplace's equation [2, 153]). In $r > r_s$, where $M^2 > 1$, the coefficient of $\partial^2\varphi/\partial\theta^2$ is *negative*, and the equation is of hyperbolic type. The solution in this region exhibits wavelike behavior with real characteristics (acoustic "rays"), and distortions of the sonic cylinder produced by the passage of a blade (whose influence must first be transmitted through the elliptic domain) cause sound waves to be "launched" into the far field. The behavior is similar for the full, nonlinear equation, although the boundary between the elliptic and hyperbolic domains is now dependent on the solution φ (Section 3.6.4). It may be concluded that the most important blade force and displacement sources are those near the blade tips, which produce the greatest distortion of the sonic cylinder.

Gutin noise is important when the blade-tip speed approaches the speed of sound. For noncavitating marine propellers and other flow-control devices operating at very low Mach numbers, it is usually negligible compared to the sound produced by impinging turbulence, or to trailing edge self-noise, and also occurs at very much lower frequencies (typically smaller by a factor $\sim\ell/D$, where ℓ is the turbulence length scale and D the diameter of the rotor). Compare Section 1.12, where the efficiency of sound generation by an acoustically compact body, executing small amplitude torsional oscillations about its center of volume, was shown to equal that of a volume quadrupole.

3.6.2 Loading and Thickness Noise

Consider sound production by steady rotation of a rigid propeller with subsonic blade-tip Mach number. Although it is possible in principle to investigate the radiation by direct numerical integration of equation (3.6.1), solutions derived by this means tend to be unreliable at large distances from the blades, and it is more convenient to proceed via the Ffowcs Williams-Hawkings equation (2.2.11), which relates the acoustic pressure $p(\mathbf{x}, t) = c_o^2(\rho - \rho_o)$ to the blade forces and motion. Provided the blade tip Mach number M_t does not exceed 0.7–0.8, and in the absence of inflow turbulence, volume quadrupoles may be neglected in (2.2.11a) [139, 218–222]. The remaining integrals are over the blade surfaces S, which are fixed relative to the coordinate system η rotating with the blades and respectively represent dipole and monopole source fields.

The dipole term represents sound production by blade forces (the thrust and drag distributions) and is called the *loading noise*. This dominates the radiation ahead of and to the rear of the rotor plane, although it is obvious that for identical, evenly spaced blades, the integral is independent of t when $\mathbf{x}$ lies on the rotor axis, where the loading noise must vanish. To make quantitative predictions, $p'_{ij}n_j$ must first be calculated by solving the equations of compressible

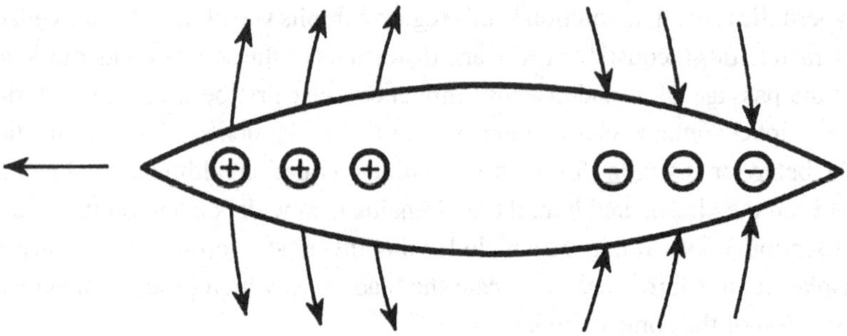

Figure 3.6.2. Thickness noise sources and sinks.

aerodynamics [153]; this is usually done by neglecting surface shear stresses, by taking $p'_{ij} = (p - p_o)\delta_{ij}$, where p_o is the uniform mean pressure.

The final, monopole integral of (2.2.11a) gives the *thickness noise*, generated by the displacement of fluid by the advancing blades. Because ρ_o is the fluid density in the far field, thickness noise is determined by the fluid *volume* displaced by the blades. The volume flux from the surface of a typical blade section is positive and negative respectively on the forward and aft halves of the blade (Figure 3.6.2). This characterizes an acoustic source of dipole or higher multipole order, and we have already seen in equation (2.2.11b) that the thickness noise integral is equivalent to dipole and quadrupole sources distributed within the blade. The axes of the dipoles and quadrupoles are parallel to the plane of rotation, so that the peak thickness noise should be in directions close to the rotor plane, but are modified by the Doppler factor $1/|1 - M \cos \Theta|$ where the source Mach number M is determined both by blade rotation and forward motion of the rotor.

The acoustic pressure signature produced by the loading and thickness sources is generally smoothly varying unless the blade tip Mach number exceeds about 0.7 [139, 219], when Doppler amplification becomes important and the pressure becomes impulsive in character. Impulsive pressure profiles produced by a single blade of a stationary rotor (a fixed propeller or the main rotor of a hovering helicopter) are illustrated schematically in Figure 3.6.3 for an observer in the plane of rotation. The peak pressures are generated at the retarded time at which a blade is moving directly toward the observer.

3.6.3 Linear Theory

Let ϵ be a nondimensional measure of both the blade thickness/chord ratio and the angle of attack to the relative mean flow. Both of the surface integrals in

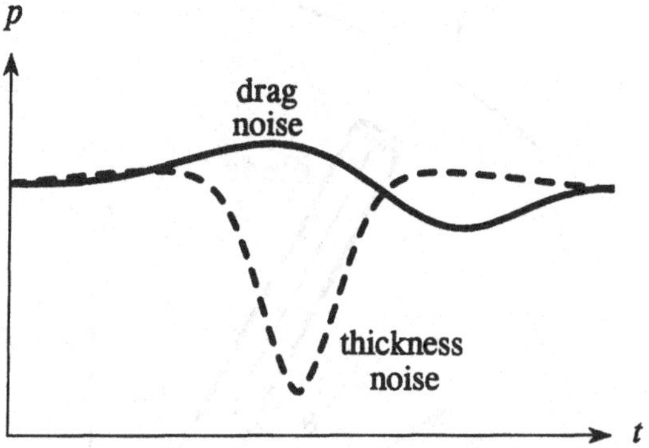

Figure 3.6.3. Impulsive pressures in plane of rotation ($M \geq 0.7$).

(2.2.11a) are zero when $\epsilon = 0$, and linear theory evaluates them correctly to $O(\epsilon)$. The surface of integration S may therefore be collapsed to a mean surface S_m that is locally aligned with the relative mean flow and has relative mean velocity of magnitude $U = \sqrt{\Omega^2 r^2 + V^2}$, where r is distance from the rotor axis and V is the forward velocity of the rotor. The sources are now distributed on the helicoidal surface swept out by S_m as the rotor advances. It is implicit in this approximation that the acoustic wavelength is much larger than the airfoil thickness, so that retarded time differences between the radiation from sources on the actual surface S of the airfoil and the mean surface S_m are negligible. The approximation must therefore break down at high Mach numbers, when the Doppler contracted wavelength (Section 1.8.5) is comparable to the blade thickness.

The source integrals are usually evaluated with the help of a computer, both to determine the aerodynamic surface loading (by numerical integration of the linearized form of equation (3.6.1) in the immediate vicinity of the blades) and to account effectively for contributions to $p(\mathbf{x}, t)$ from the different blade sections at different retarded times [139, 218–225]. Let us examine the case of a fixed rotor with B identical, evenly spaced blades; in the linear approximation the noise sources are distributed on the surface S_m rotating in the fixed plane of the rotor. The acoustic pressure may be written [226, 227] as

$$p(\mathbf{x}, t) = \frac{1}{4\pi c_o} \frac{\partial}{\partial t} \oint_{S_m} \left[\frac{F_R}{R|1 - M\cos\Theta|} \right] dS(\eta)$$

$$+ \frac{\rho_o}{4\pi} \frac{\partial^2}{\partial t^2} \oint_{S_m} \left[\frac{h}{R|1 - M\cos\Theta|} \right] dS(\eta), \quad |\mathbf{x}| \to \infty, \tag{3.6.2}$$

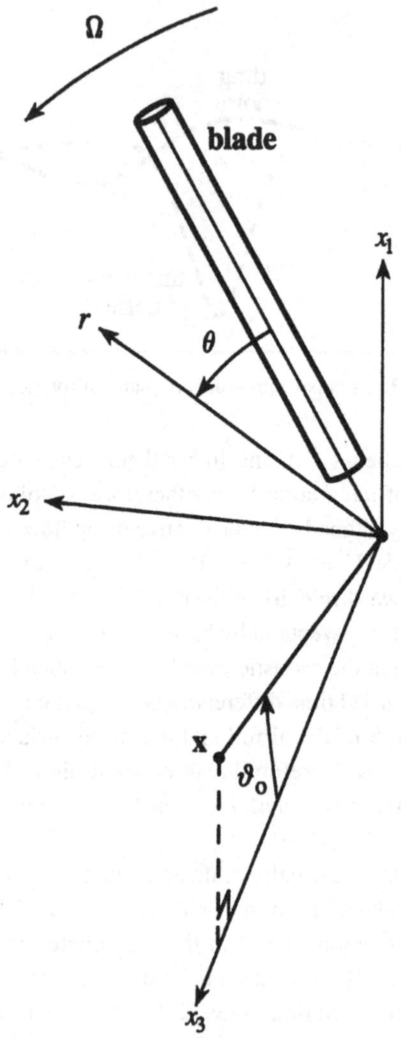

Figure 3.6.4. Coordinates used in the calculation of rotor noise.

where $F_R = p'_{ij} n_j R_i / R \approx (p - p_o) \cos \vartheta$ is the surface force applied to the fluid per unit area of S_m in the radiation direction, and ϑ is the angle between a blade normal and **R**, and h is the blade thickness.

Choose fixed coordinate axes (x_1, x_2, x_3) with S_m in the x_1x_2-plane and the origin at the center of the rotor disc, and let the observer be in the x_1x_3-plane at distance $|\mathbf{x}|$ from the origin, where **x** makes an angle ϑ_o with the x_3-axis (Figure 3.6.4). The motion is periodic with fundamental frequency $B\Omega$, and the pressure can be expanded in the Fourier series

$$p(\mathbf{x}, t) = \sum_{n=-\infty}^{\infty} p_{n\mathrm{B}}(\mathbf{x}) \mathrm{e}^{-in\mathrm{B}\Omega t}. \tag{3.6.3}$$

Introduce polar coordinates (r, θ) fixed with respect to the rotating blades, where $\theta\,(0 < \theta < 2\pi)$ is the azimuthal angle in the rotor plane, $0 < r < r_t(\theta)$, and $d\mathrm{S}(\boldsymbol{\eta}) = r d\theta dr$. Then, if $T(r, \theta)$, $D(r, \theta)$ are the thrust and drag per unit area of S_m, (T being positive in the x_3-direction) and $h(r, \theta)$ is the blade thickness distribution [227]

$$\begin{aligned} p_{n\mathrm{B}}(\mathbf{x}) \approx {} & \frac{in\mathrm{B}^2\Omega}{4\pi c_o|\mathbf{x}|} \exp\left\{ in\mathrm{B}\left(\frac{\Omega|\mathbf{x}|}{c_o} - \frac{\pi}{2}\right)\right\} \\ & \times \oint_{\mathrm{S}_m} \left(T\cos\vartheta_o - \frac{D}{M} + i\rho_o n\mathrm{B}\Omega c_o h\right)(r, \theta) \\ & \times \mathrm{J}_{n\mathrm{B}}(n\mathrm{B}M\sin\vartheta_o)\mathrm{e}^{-in\mathrm{B}\theta} r dr d\theta, \quad |\mathbf{x}| \to \infty, \end{aligned} \tag{3.6.4}$$

where the integration is taken over the mean surface S_m of a *single* blade. This integral must normally be evaluated numerically, using values of the thrust and drag distributions T and D determined by experiment or by numerical integration of equation (3.6.1).

The Bessel function $\mathrm{J}_{n\mathrm{B}}(n\mathrm{B}x)$ is positive and increases monotonically in the range $0 < x < 1$ [228]. Equation (3.6.4) accordingly states that the sound of a subsonic rotor is generated predominantly at the blade tips where $x = M\sin\vartheta_o \equiv \Omega r/c_o \sin\vartheta_o$ assumes its largest absolute value. Furthermore, when nB is large [24],

$$\mathrm{J}_{n\mathrm{B}}(n\mathrm{B}x) \approx \frac{1}{\sqrt{2\pi n\mathrm{B}}(1 - x^2)^{1/2}} \left(\frac{x\mathrm{e}^{\sqrt{1-x^2}}}{1 + \sqrt{1 - x^2}}\right)^{n\mathrm{B}}, \quad \text{for } |x| < 1, \quad n\mathrm{B} \gg 1,$$

where the quantity in the brace brackets increases monotonically from zero to unity as x increases over the range $0 < x < 1$. This formula (with $x = M\sin\vartheta_o$) implies that the radiation is dominated by the low-order harmonics when M_t is small and in particular by the fundamental at the *blade passage frequency* $\mathrm{B}\Omega$. Increasing the number of blades increases the fundamental frequency but also leads to an increased rate of decay of contributions from all harmonics, indicating that steady blade sources become progressively weaker as the number of blades increases. In practice, however, it is found that periodic rotor noise continues to be important at higher frequencies [139, 220, 229–231], and this is usually attributed to the ingestion of nonuniform flow by the rotor.

The Fourier coefficients $p_{nB}(\mathbf{x})$ determine the acoustic pressure frequency spectrum $\Phi(\mathbf{x}, \omega)$ from the formulae

$$\langle p^2(\mathbf{x}, t)\rangle = \sum_{n=1}^{\infty} 2|p_{nB}(\mathbf{x})|^2,$$
$$\Phi(\mathbf{x}, \omega) = \sum_{n=1}^{\infty} 2|p_{nB}(\mathbf{x})|^2 \delta(\omega - nB\Omega). \tag{3.6.5}$$

The spectrum is discrete, with "spikes" at multiples of $B\Omega$.

When thickness noise is neglected (which is a reasonable approximation at small blade-tip Mach number, and therefore when the blade chord is acoustically compact), a useful approximation to $\Phi(\mathbf{x}, \omega)$ is obtained by assuming that the net thrust $\bar{T}$ and drag $\bar{D}$ on a blade act at a fixed radial distance $r = \bar{r}$ from the rotor axis. We then obtain *Gutin's* formula [214]:

$$\Phi(\mathbf{x}, \omega) \approx \frac{B^4\Omega^2}{8\pi^2|\mathbf{x}|^2 c_o^2}\left|\bar{T}\cos\vartheta_o - \frac{\bar{D}}{M_{\bar{r}}}\right|^2 \times \sum_{n=1}^{\infty} n^2 |J_{nB}(nBM_{\bar{r}}\sin\vartheta_o)|^2 \delta(\omega - nB\Omega), \tag{3.6.6}$$

where $M_{\bar{r}} = \Omega\bar{r}/c_o$ is the Mach number of the effective source velocity. The principal characteristics of the acoustic field are exhibited by this formula. The radiation is null in the directions $\vartheta_o = 0, \pi$ (straight ahead and to the rear of the rotor plane, respectively) and where $\cos\vartheta_o = \bar{D}/M_{\bar{r}}\bar{T}$ (provided $\bar{D}/M_{\bar{r}}\bar{T} < 1$, which is usually the case) and has a large maximum in $\vartheta_o > \pi/2$, to the "rear" of the rotor plane. Figure 3.6.5 illustrates this for a four-bladed propeller when $M_{\bar{r}} = \frac{1}{2}$ and $\bar{D} = 0.1\bar{T}$. Equation (3.6.6) was originally derived by Gutin [214], who obtained good agreement with experiment for $nB \leq 8$ and $\bar{r} = 0.8r_t$.

The theory can be adapted to cases in which a rotor or fan is enclosed in a duct. Complications frequently arise because of variable duct geometry and the presence of neighboring control surfaces, for example, when stators and multiple fan configurations are involved. For an isolated fan in a hard-walled duct, the rotor noise can be estimated in terms of a Green's function whose normal derivative vanishes on the duct walls [232]. In practice, however, the walls are often "acoustically transparent" at low frequencies (small values of $nB\Omega$). The rotor noise then radiates into the ambient medium as if the duct walls

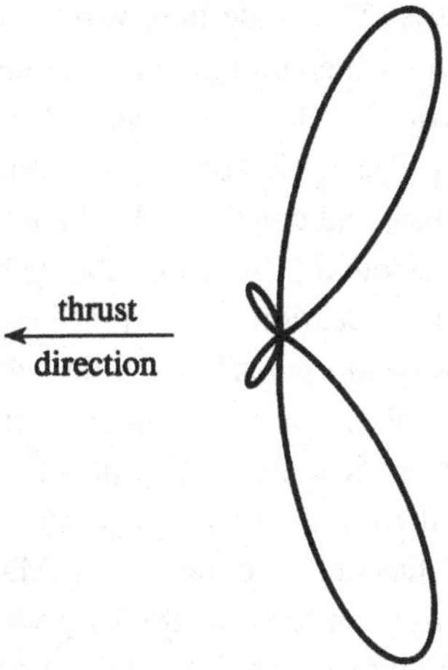

Figure 3.6.5. Angular distribution of Gutin noise $\langle p^2(\mathbf{x}, t)\rangle$ for a four-bladed propeller.

were absent, and acoustic levels within the duct are not predicted correctly by use of the hard wall Green's function.

Example 1. The thickness noise term in (3.6.2) is formally the solution with outgoing wave behavior of

$$\left(\partial^2/c_o^2\partial t^2 - \nabla^2\right)p = \rho_o\partial(\mathbf{v}\cdot\nabla \mathrm{H}(f))/\partial t,$$

where $f(\mathbf{x}, t) \gtrless 0$ respectively inside and outside of the blade surface S. Use the relation $\mathbf{v}\cdot\nabla\mathrm{H} = \partial[1 - \mathrm{H}(f)]/\partial t$ and equation (2.2.8) to show that

$$\begin{aligned} p(\mathbf{x}, t) &= \frac{\rho_o}{4\pi}\frac{\partial^2}{\partial t^2}\oint_{\mathrm{S}(\tau)}[1 - \mathrm{H}(f)]\frac{d^3\mathbf{y}}{|\mathbf{x}-\mathbf{y}|} \\ &\approx \frac{\rho_o}{4\pi|\mathbf{x}|}\frac{\partial^2}{\partial t^2}\int_{\mathrm{S}_m}\left[\frac{h(r,\theta)}{|1 - M\cos\Theta|}\right] r\,dr\,d\theta, \end{aligned}$$

where in the first integral the square brackets denote evaluation at the retarded $\tau = t - |\mathbf{x}-\mathbf{y}|/c_o$, and $\mathrm{S}(\tau)$ is the surface $f(\mathbf{y}, \tau) = 0$. Hence, deduce the thickness noise contribution to (3.6.4).

3.6.4 Nonlinearity and Shock Waves

The Mach number $\Omega r/c_o$ at which the relative velocity of the flow over a blade section at radius r first becomes equal to the local speed of sound is called the *critical Mach number*. The blade flow varies rapidly with the azimuthal angle θ of equation (3.6.1), which implies that the linearized approximation becomes inadequate at transonic velocities, where $M \approx 1$, because the coefficient of $\partial^2\varphi/\partial\theta^2$ is then small. The motion becomes essentially nonlinear in character, shock waves may form, and considerable changes can occur in the nature and levels of the radiated sound [139, 217–222, 233]. The linearized loading and thickness noise theory of Section 3.6.3 must be corrected and account taken of contributions from the quadrupoles in the fluid, which may actually become the dominant source of noise. The quadrupole strengths and surface loading must be calculated by numerical integration of the nonlinear equation (3.6.1) in the neighborhood of the blades. Only quadrupoles in the fluid adjacent to blade sections where M lies between the critical Mach number and unity are significant, and the intensity of the sound they produce can exceed that predicted by linear theory by a factor of 4 (6 dB) or more. The importance of the quadrupoles can be reduced by incorporating blade sweep, which effectively ensures that all blade sections operate below their critical Mach numbers.

The dominant radiation from propellers with supersonic blade tips occurs in directions that satisfy the Mach wave condition $M\cos\Theta = 1$ at some point on a blade, where the integrands of (3.6.2) have (integrable) singularities, and the surface sources approach the observer at precisely the sound speed c_o [226, 234–236]. In other directions, the acoustic field is again determined principally by sources near the blade tips, as for a subsonic rotor. Blade sources on a transonic rotor must be calculated by numerical integration of (3.6.1). However, with the exception of radiation directions close to the rotor plane, it is generally not necessary to do this when $M_t > 1$, because the radiation is from supersonic blade sections where $1/\cos\Theta > 1$, and where linear aerodynamics remains valid [153].

Helicopter main rotor blades usually operate at subsonic tip Mach numbers, at least when the aircraft is hovering. The relative velocity is often transonic, however, and weak shock waves may appear in the flow over the upper blade surfaces near the tips when M_t exceeds about 0.8–0.9. At larger values of M_t, the shocks are said to become delocalized and are directly "connected" to the acoustic far field, resulting in the generation of intense, high-frequency noise in directions close to the plane of the rotor.

The phenomenon of delocalization occurs as follows. A localized region of low pressure, supersonic flow (within which equation (3.6.1) is *hyperbolic*) appears near the tips on the suction side of a blade, typically when $M_t = \Omega r_t/c_o \geq 0.8$; pressure recovery occurs through a standing shock wave to the rear of the blade (Figure 3.6.6a). This region, being inboard of the tip, moves subsonically relative to a distant, stationary observer. Although the rotor noise is

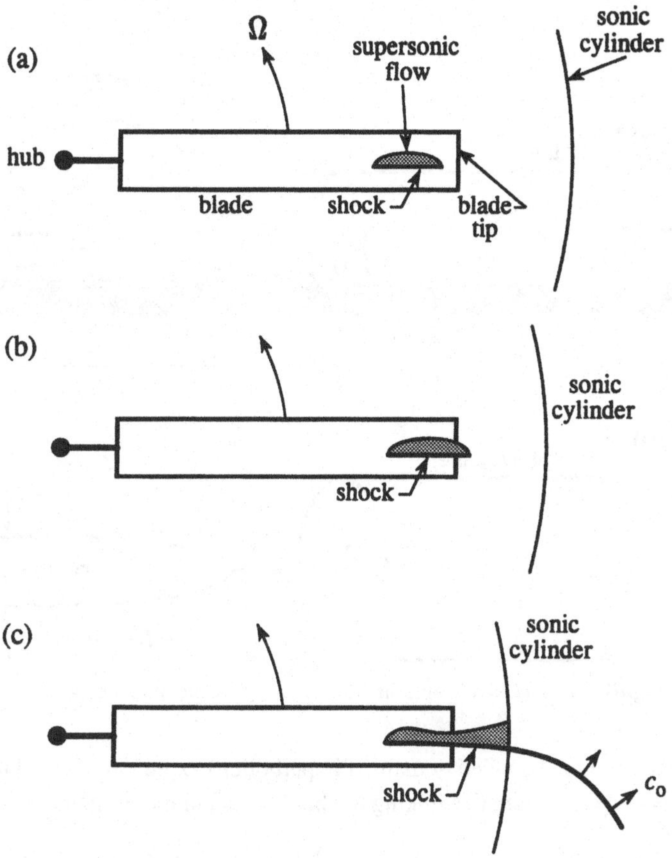

Figure 3.6.6. Shock delocalization on a transonic rotor (after [139]).

affected by the transonic volume quadrupoles, whose influence extends through the elliptic domain between the blade tip and the sonic cylinder causing the latter to be distorted, there is initially no dramatic change in the character of the radiation. At a somewhat larger but still subsonic value of M_t, the transonic zone and shock wave expand radially and eventually extend beyond the blade tip toward the sonic cylinder (Figure 3.6.6b). At the same time, the radius of the sonic cylinder is reduced and is ultimately pierced by the shock after a further increase of M_t (say to 0.9). The flow discontinuity on the sonic cylinder is now launched directly into the far field as a weak acoustic shock, producing a dramatic increase in sound levels and a steepening of the acoustic waveform (Figure 3.6.6c). At this stage, there exists a continuous region of relative supersonic flow that extends inward from the sonic cylinder toward the blade tips, like thickened spokes of a wheel: The shock waves have been delocalized by

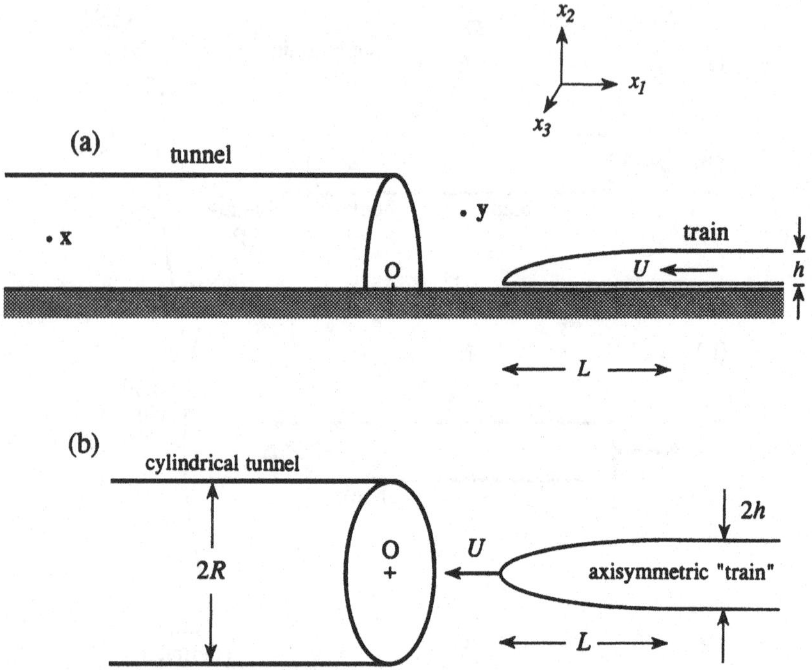

Figure 3.6.7. (a) Train entering tunnel and (b) scale model experiment.

the creation of an extended domain of hyperbolicity of equation (3.6.1), which now has real characteristics linking the blade tips and the far field.

3.6.5 High-Speed Train Compression Wave in a Tunnel

A body in steady, subsonic translational motion is "silent" unless its hydrodynamic near field is "scattered" by other bodies or structures close to its path. When a high-speed train enters a tunnel at constant speed U (Figure 3.6.7a), the accompanying near field potential flow (which is steady with respect to the train prior to arrival at the tunnel entrance) interacts with the tunnel portal and generates a compression wave in the tunnel that propagates at about the speed of sound. The pressure rise across the wave, which can be as much as 1% of atmospheric pressure p_o, is approximately $\rho_o U^2 \mathcal{A}_o/\mathcal{A} \sim M^2(\mathcal{A}_o/\mathcal{A})p_o$, where $\mathcal{A}_o$ and $\mathcal{A}$ are the cross-sectional areas of the train and tunnel. Part of the wave energy radiates from the far end of the tunnel as a spherically spreading pulse (the *micro-pressure wave*). Nonlinear steepening of the compression wave in a very long tunnel (3 km or more in length) can increase the amplitude of the micro-pressure wave, which can cause considerable structural damage and annoyance. This is less important for a tunnel with *ballasted* track, where viscous damping in ballast interstices usually attenuates the wave, but the environmental problems are often severe for modern tunnels with concrete slab track and train speeds approaching $M = 0.4$ [237, 238].

Laboratory tests are conducted [239] by projecting axisymmetric model "trains" along the axis of a long, rigid circular cylinder (Figure 3.6.7b). An analytical model for this case is formulated by considering a semi-infinite, *unflanged* circular cylindrical tunnel of radius R ($\mathcal{A} = \pi R^2$) extending along the negative x_1-axis with the origin on the axis of symmetry in the plane of the tunnel mouth. Let the train enter the tunnel at constant speed U, and suppose its cross section becomes constant and of area $\mathcal{A}_o \equiv \pi h^2$ at a distance L from the nose, where h is its maximum radius. In typical applications the ratio $\mathcal{A}_o/\mathcal{A}$ is small.

The compression wave is calculated with the help of the differential form (2.2.2) of the Ffowcs Williams–Hawkings equation, where S is the moving boundary of the train (so that $\bar{\mathbf{v}} = \mathbf{v}$). In practice, $\mathcal{A}_o/\mathcal{A} < 1/4$ and in a first approximation the flow around the train may be assumed to be the same as when the tunnel is absent. The nose profile is assumed to be streamlined and the aspect ratio h/L sufficiently small to avoid separation. The flow over the train is then effectively *irrotational*, and Bernoulli's equation implies that the dipole source strength (the second term on the right of (2.2.2)) is proportional to $(h/L)^2 \ll 1$, whereas the monopole strength varies linearly with h/L. Hence, when the irrotational volume quadrupoles are ignored at low Mach numbers, only the final, monopole source in (2.2.2) is significant. This accounts for the displacement of air by the moving train, whose effect is equivalent to a distribution of constant strength volume sources $q(x_1 + Ut, x_2, x_3)$ translating with the train, where $q(\mathbf{x})$ is the source distribution at $t = 0$, when the nose of the train may be supposed to enter the tunnel.

The monopole strength q is non-zero only near the front and rear of the train, where the cross-sectional area varies. When the train travels along the axis of symmetry of the tunnel, the unsteady pressure is the solution of equation (1.6.8), where

$$\mathcal{F}(\mathbf{x}, t) = \rho_o \frac{\partial q}{\partial t}(x_1 + Ut, x_2, x_3), \quad q(\mathbf{x}) = -U \frac{\partial}{\partial x_1} \mathrm{H}(r - f(x_1)),$$
$$r = \sqrt{x_2^2 + x_3^2}, \tag{3.6.7}$$

and $f(x_1)$ is the radius of the train cross-section at distance x_1 from the nose, so that the train cross-sectional area $\mathcal{A}_\mathrm{T}(x_1) = \pi f^2(x_1)$, and $\mathcal{A}_\mathrm{T}(L) \equiv \mathcal{A}_o$. Equation (1.6.8) is applicable within the tunnel only when nonlinear steepening of the compression wave front is ignored. Although this will be important in long tunnels, there is an extensive region, many tunnel diameters ahead of the train, where the unsteady motion is one-dimensional and of small amplitude, and where linear theory supplies a good approximation to the perturbation pressure $p(x,t)$. The solution is given by the last term on the right of Kirchhoff's formula (1.10.5), provided G is chosen to have vanishing normal derivative on the interior and exterior walls of the tunnel (Example 2), on which $\partial p/\partial y_n$ also vanishes.

The *micro-pressure wave* $p'(\mathbf{x},t)$, radiated from the far end of the tunnel (located at $x = x_{\mathrm{E}}$) when the compression wave arrives, is given approximately by (3.2.7)

$$p'(\mathbf{x},t) \approx \frac{\mathcal{A}}{4\pi c_o \wp} \frac{\partial p}{\partial t}(x_{\mathrm{E}}, t - \wp/c_o), \quad \wp \gg R, \tag{3.6.8}$$

at distance $\wp$ from the tunnel exit. Because of nonlinear steepening, the value of the derivative on the right-hand side cannot normally be evaluated from the linear theory pressure. However, the linear theory solution determines the initial waveform that can subsequently be used in a nonlinear propagation model to calculate the pressure at the far end of the tunnel. Note that for a real tunnel, the factor of 4π in the denominator of (3.6.8) must be replaced by the effective solid angle Ω of the "open air" into which the spherical wave spreads when account is taken of environmental factors such as the presence of embankments, near the tunnel portal.

For trains with "slender" nose and tail profiles,

$$q(\mathbf{x}) \equiv U \frac{\partial}{\partial x_1}(\pi f^2)\frac{1}{2\pi r}\delta(r-f) \approx U \frac{\partial \mathcal{A}_{\mathrm{T}}}{\partial x_1}\delta(x_2)\delta(x_3). \tag{3.6.9}$$

In this approximation, the source distribution is concentrated on the axis of symmetry of the model train.

A first estimate of the compression wave generated by a long train is obtained by making the further approximation $\partial \mathcal{A}_{\mathrm{T}}/\partial x \approx \mathcal{A}_o \delta(x_1)$ at the nose, so that the equivalent source distribution for the train nose is a single point source $q(\mathbf{x}) = \mathcal{A}_o U \delta(\mathbf{x})$ traveling along the axis of symmetry of the tunnel. The compression wave $p(x_1,t)$ propagates one-dimensionally ahead of the train, and its profile and that of the pressure gradient $\partial p(x_1,t)/\partial t$ (calculated using the Green's function $G(\mathbf{x},\mathbf{y},t-\tau)$ derived in Example 2) are plotted in Figure 3.6.8 (where $[t] = t + x_1/c_o$ for a wave travelling to the left) for $M = U/c_o$ between 0.1 and 0.4. The compression wave *rise time* is essentially independent of Mach number, being equal to the effective transit time $\sim 2R/U$ of the nose across the entrance plane of the tunnel.

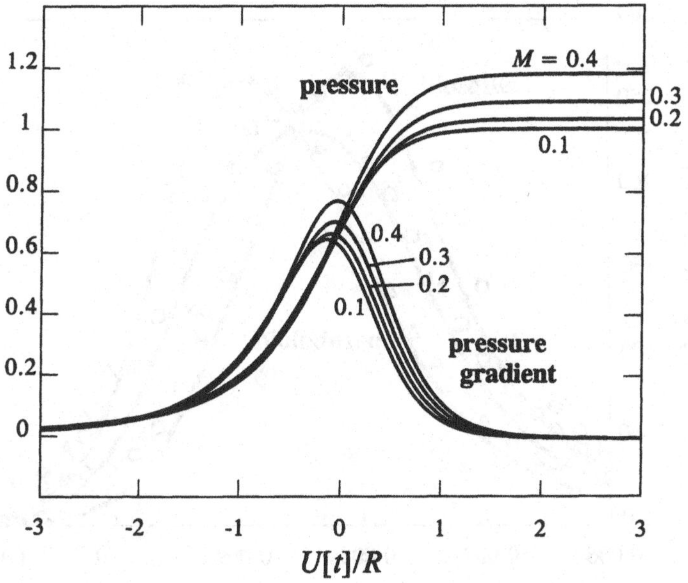

Figure 3.6.8. Compression wave $p/(\rho_o U^2 \mathcal{A}_o/\mathcal{A})$ and pressure gradient $(\partial p/\partial t)/(\rho_o U^3 \mathcal{A}_o/R\mathcal{A})$ when the train nose is modeled by a point source.

In the experiments of Maeda *et al.* [239], model "trains" with nose profiles in the shape of a cone and of a paraboloid and ellipsoid of revolution were used, with respective cross-sectional areas

$$\frac{\mathcal{A}_T(x_1)}{\mathcal{A}_o} = \frac{x_1^2}{L^2}, \quad \frac{x_1}{L}, \quad \frac{x_1}{L}\left(2 - \frac{x_1}{L}\right), \qquad 0 < x < L.$$
$$= 1, \quad x \geq L.$$

Figure 3.6.9 shows the pressure gradient $\partial p/\partial t$ measured about 1 m from the entrance of a circular cylindrical tunnel 7 m long and radius $R = 0.0735$ m. Model trains of aspect ratio $h/L = 0.2$ and area ratio $\mathcal{A}_o/\mathcal{A} = 0.116$ were projected into the tunnel along the axis at $U = 230$ km/h ($M = 0.188$). The solid curves are the corresponding theoretical predictions using the monopole source (3.6.7).

Example 2: Calculation of the compression wave. Green's function for source positions $\mathbf{y}$ near the tunnel entrance of the circular cylinder of Figure 3.6.7b and observation points $\mathbf{x}$ deep within the tunnel is given by [205]

$$G(\mathbf{x}, \mathbf{y}; t - \tau) = G_o(x_1, y_1; t - \tau) + G_s(\mathbf{x}, \mathbf{y}; t - \tau),$$

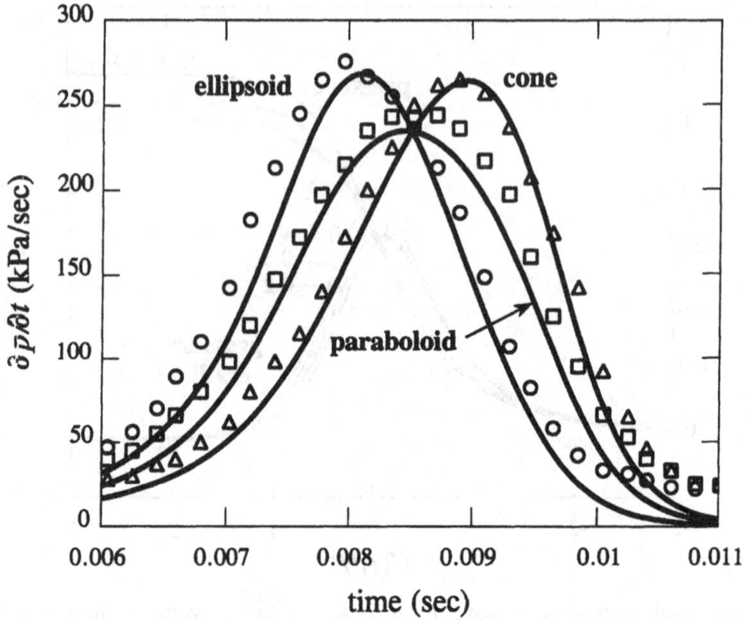

Figure 3.6.9. Comparison of measured [239] and predicted (——) pressure gradients $\partial p/\partial t$ for model conical, paraboloidal, and ellipsoidal nose profiles when $\mathcal{A}_o/\mathcal{A} = 0.116$, $R = 0.0735\ m$, $M = 0.188$, and $h/L = 0.2$.

where

$$G_o(x_1, y_1; t-\tau) = \frac{c_o}{2\mathcal{A}}\mathrm{H}\left(t - \tau + \frac{(x_1 - y_1)}{c_o}\right),$$

$$G_s(\mathbf{x}, \mathbf{y}; t-\tau) = \frac{-R}{8\pi^2\mathcal{A}}\int\int_{-\infty}^{\infty} \frac{\mathrm{I}_0(\bar{\gamma}r')\mathcal{K}_+(\kappa_o)\mathcal{K}_-(k)}{\bar{\gamma}\mathrm{I}_1(\bar{\gamma}R)}$$
$$\times\, \mathrm{e}^{-i\omega(t-\tau+x_1/c_o)+iky_1}\, dk\, d\omega, \quad r' = \sqrt{y_2^2 + y_3^2},$$

and the coordinate origin is on the axis of symmetry in the tunnel entrance plane. The functions $\mathcal{K}_\pm(k)$ are regular and nonzero in Im $k \gtrless 0$, and satisfy for real k: $\mathcal{K}_+(k)\mathcal{K}_-(k) = 2\mathrm{I}_1(\bar{\gamma}R)\mathrm{K}_1(\bar{\gamma}R)$, where I_n and K_n are modified Bessel functions [24]. $\bar{\gamma} = \sqrt{(k^2 - \kappa_o^2)}$ has branch cuts extending from $k = \pm\kappa_o$ to $\pm i\infty$ respectively in the upper and lower halves of the complex k-plane, such that on the real axis, $\bar{\gamma}$ is real and positive when $|k| > |\kappa_o|$, and $\bar{\gamma} = -i\,\mathrm{sgn}(\kappa_o)\sqrt{(\kappa_o^2 - k^2)}$ for $|k| < |\kappa_o|$.

The integration with respect to k in the definition of $G_s(\mathbf{x}, \mathbf{y}; t-\tau)$ must be evaluated numerically. The computations are simplified by first calculating the pressure *gradient* $\partial p(x_1,t)/\partial t$, and then using $p(x_1,t) = \int_{-\infty}^{t} \frac{\partial p}{\partial t'}(x_1,t')dt'$ to find the pressure. When the various integrations are cast in nondimensional forms, we find

$$\left(\frac{R\mathcal{A}}{\rho_o U^3 \mathcal{A}_o}\right)\frac{\partial p}{\partial t}(x_1,t) = \frac{1}{2\pi\sqrt{(1-M^2)}}\int_0^\infty \frac{\lambda}{\mathrm{I}_1(\lambda\sqrt{(1-M^2)})} \times \mathrm{Re}\left\{\mathcal{Q}(\lambda)\mathcal{K}_+\left(\frac{\lambda M}{R}\right)\mathcal{K}_-\left(\frac{\lambda}{R}\right)\mathrm{e}^{-i\lambda U[t]/R}\right\}d\lambda,$$

where $[t] = t + x_1/c_o$ is the retarded time, and

$$\mathcal{Q}(\lambda) = \int_{-\infty}^{\infty}\frac{1}{\mathcal{A}_o}\frac{\partial \mathcal{A}_T}{\partial x'}(x')\mathrm{I}_0\left(\lambda\sqrt{\frac{\mathcal{A}_T}{\mathcal{A}}(x')(1-M^2)}\right)\mathrm{e}^{i\lambda x'/R}\,dx'.$$

The functions $\mathcal{K}_+(\lambda M/R)$ and $\mathcal{K}_-(\lambda/R)$ are evaluated by the Cauchy integral method [205] from the formulae

$$\mathcal{K}_+(\lambda M/R) = \exp\left\{\frac{-i\lambda M}{\pi}\int_0^\infty \frac{\ln[2\mathrm{I}_1(\sqrt{(\mu^2-\lambda^2M^2)})\mathrm{K}_1(\sqrt{(\mu^2-\lambda^2M^2)})]}{\mu^2-\lambda^2M^2}\,d\mu\right\},$$

$$\mathcal{K}_-(\lambda/R) = \sqrt{2\mathrm{I}_1(\lambda\sqrt{(1-M^2)})\mathrm{K}_1(\lambda\sqrt{(1-M^2)})} \times \exp\left\{\frac{i\lambda}{\pi}\int_0^\infty \ln\left[\frac{\mathrm{I}_1(\sqrt{(\mu^2-\lambda^2M^2)})\mathrm{K}_1(\sqrt{(\mu^2-\lambda^2M^2)})}{\mathrm{I}_1(\lambda\sqrt{(1-M^2)})\mathrm{K}_1(\lambda\sqrt{(1-M^2)})}\right]\frac{d\mu}{\mu^2-\lambda^2}\right\}$$

where in the interval $0 < \mu < M\lambda$

$$\mathrm{I}_1(\Gamma)\mathrm{K}_1(\Gamma) = -\frac{\pi}{2}\left(\mathrm{J}_1(|\Gamma|)\mathrm{Y}_1(|\Gamma|) - i\mathrm{J}_1^2(|\Gamma|)\right), \quad \Gamma = \sqrt{(\mu^2-\lambda^2M^2)}.$$

Example 3. When $M \ll 1$, the compression wave thickness is very much larger than the tunnel diameter. Show that the compression wave can be calculated using the following compact Green's function [240]

$$G(\mathbf{x},\mathbf{y};t-\tau) \approx \frac{c_o}{2\mathcal{A}}\{\mathrm{H}(t-\tau-|\bar{Y}(\mathbf{x})-\bar{Y}(\mathbf{y})|/c_o) - \mathrm{H}(t-\tau+(\bar{Y}(\mathbf{x})+\bar{Y}(\mathbf{y}))/c_o)\}, \tag{3.6.10}$$

where $\bar{Y}(\mathbf{x})$ is the solution of Laplace's equation that describes irrotational flow *out* of the tunnel mouth, normalized such that $\bar{Y}(\mathbf{x}) \approx x_1 - \ell$ deep inside the tunnel, and $\bar{Y} \sim O(1/|\mathbf{x}|)$ as $|\mathbf{x}| \to \infty$ outside the tunnel. The tunnel-mouth end-correction $\ell \approx 0.61R$ for the circular cylindrical tunnel of Figure 3.6.7b (cf. Section 3.2, Example 3), but (3.6.10) is applicable for *any* tunnel, the shape of which determines the function $\bar{Y}(\mathbf{x})$. Deduce the compression wave pressure gradient in the form

$$\frac{\partial p}{\partial t}(x_1, t) \approx -\frac{\rho_o U^2}{\mathcal{A}} \int q(y_1 + U[t'], y_2, y_3) \frac{\partial^2 \bar{Y}}{\partial y_1^2}(\mathbf{y})\, d^3\mathbf{y},$$
$$\text{where} \quad [t'] = t + (x_1 - \ell)/c_o.$$

This approximation is adequate for practical purposes when $\mathcal{A}_o/\mathcal{A}$ is small and the train Mach number $M < 0.2$.

4

Sound generation in a fluid with flexible boundaries

To determine the sound produced by turbulence near an elastic boundary, it is necessary to know the response of the boundary to the turbulence stresses. These stresses not only generate sound but also excite structural vibrations that can store a significant amount of flow energy. The vibrations are ultimately dissipated by frictional forces, but they can contribute substantially to the radiated noise because elastic waves are "scattered" at structural discontinuities, and some of their energy is transformed into sound. Thus, flow-generated sound reaches the far field via two paths: directly from the turbulence sources and indirectly from possibly remote locations where the scattering occurs. The result is that the effective acoustic efficiency of the flow can be very much larger than for a geometrically similar rigid surface, even when only a small fraction of the structural energy is scattered into sound. Typical examples include the cabin noise produced by turbulent flow over an aircraft fuselage and the noise radiated from ship and submarine hulls, from duct flows, piping systems, and turbomachines [26]. Interactions of this kind are discussed in this chapter.

4.1 Sources Near an Elastic Plate

The simplest flexible boundary is the homogeneous, nominally flat, thin elastic plate, which supports structural modes in the form of bending waves. The effects of *fluid loading* are usually important in liquids, where the Mach number M is small, and in this section, it will be assumed that $M \ll 1$ and, therefore, that mean flow has a negligible effect on the propagation of sound and plate vibrations.

4.1.1 Fluid Loading and the Coincidence Frequency

Flexural waves and sound are strongly coupled when the bending wave phase speed is comparable to the speed of sound in the fluid. The phase speed increases with frequency, and the *coincidence* or *critical* frequency ω_c is the frequency at which the phase speed *in the absence of fluid loading* (when the plate is imagined to be placed in a vacuum) would just equal the speed of sound c_o.

Let the plate have bending stiffness B and mass m per unit area and occupy the plane $x_2 = 0$ in the undisturbed state. A small-amplitude flexural displacement $\zeta(x_1, x_3, t)$ in the x_2-direction satisfies the bending wave equation (1.4.9). The coincidence frequency is calculated by neglecting the pressure force $[p]$ and considering a bending wave (in the absence of other applied forces)

$$\zeta = \zeta_o \mathrm{e}^{i(kx_1-\omega t)}, \quad \omega > 0, \tag{4.1.1}$$

of constant amplitude ζ_o. This satisfies equation (1.4.9), with $[p] = F_2 \equiv 0$, provided $k = K_o$, where

$$K_o = (m\omega^2/\mathrm{B})^{\frac{1}{4}} > 0 \tag{4.1.2}$$

is the *vacuum bending wavenumber*. Because K_o is real, the wave propagates without attenuation at phase speed ω/K_o that exceeds c_o when ω is greater than the coincidence frequency:

$$\omega_c = c_o^2\sqrt{\frac{m}{\mathrm{B}}}. \tag{4.1.3}$$

Bending waves can also propagate without attenuation when the plate is immersed in a uniform fluid, but the phase velocity cannot exceed c_o. The pressure wave produced in the fluid must be "outgoing" and proportional to $e^{i(kx_1-\omega t)}$. In an ideal fluid, the pressure satisfies equation (1.6.8) with $\mathcal{F} \equiv 0$ in $x_2 \gtrless 0$ and is expressed in terms of ζ_o by observing that $v_2 = \partial\zeta/\partial t$ at both surfaces of the plate and therefore that the x_2-component of the linearized momentum equation (1.6.1) requires that

$$\frac{\partial^2\zeta}{\partial t^2} = \frac{-1}{\rho_o}\frac{\partial p}{\partial x_2}, \quad x_2 \to \pm 0, \tag{4.1.4}$$

where ρ_o is the mean fluid density. This implies that $\partial p/\partial x_2$ is continuous across the plate and that

$$p = -\mathrm{sgn}(x_2)\frac{i\rho_o\omega^2\zeta_o}{\gamma(k)}\mathrm{e}^{i(kx_1+\gamma(k)|x_2|-\omega t)}, \tag{4.1.5}$$

where $\gamma(k) = \sqrt{\kappa_o^2 - k^2}$ is defined as in (1.13.9).

Expressions (4.1.1) and (4.1.5) are compatible with the bending wave equation (1.4.9) (with $F_2 = 0$), provided k satisfies the dispersion equation

$$\mathcal{D}(k, \omega) \equiv \mathrm{B}k^4 - m\omega^2 - \frac{2i\rho_o\omega^2}{\gamma(k)} = 0. \tag{4.1.6}$$

This equation has precisely one real and positive root, which will be denoted in this chapter by κ; it exceeds both κ_o and K_o. Thus, $\gamma(\kappa)$ is positive imaginary, which means that the pressure waves (4.1.5) are *evanescent* (decay exponentially with distance $|x_2|$ from the plate). The phase velocity ω/κ is subsonic but approaches c_o asymptotically as $\omega \to \infty$, when $\kappa \to \kappa_o$.

Further discussion is simplified by introducing the dimensionless

Fluid loading parameter,

$$\epsilon = \rho_o\kappa_o/mK_o^2;$$

and the *Vacuum* bending wave Mach number,

$$\mu = \kappa_o/K_o. \tag{4.1.7}$$

Because $K_o^2 \propto \omega$, the fluid loading parameter ϵ depends only on the material properties of the fluid and plate and is also given by

$$\epsilon = \frac{\rho_o c_1}{\rho_s c_o}\left(\frac{1-2\sigma}{12(1-\sigma)^2}\right)^{\frac{1}{2}} \equiv \frac{\rho_o}{\rho_s}\left(\frac{\mathrm{E}}{12\rho_s c_o^2(1-\sigma^2)}\right)^{\frac{1}{2}}, \tag{4.1.8}$$

where σ, ρ_s, c_1, and E are respectively the Poisson's ratio, mass density, P-wave speed, and Young's modulus of the material of the plate (Section 1.3.1). Similarly,

$$\mu = \sqrt{\omega/\omega_c},$$

$$\frac{\omega_c h}{c_o} = \frac{c_o}{c_1}\left(\frac{12(1-\sigma)^2}{1-2\sigma}\right)^{\frac{1}{2}} \equiv \frac{\rho_o}{\epsilon\rho_s}, \tag{4.1.9}$$

where h is the plate thickness.

Typical values of ϵ and $\omega_c h/c_o$ are given Table 4.1.1 for plates in water and in air.

The coincidence frequencies for steel and aluminium plates are approximately equal. For 1-cm thick plates, $\omega_c/2\pi \sim 22.7$ kHz and 1.2 kHz, respectively, in water and air. For a plywood plate of the same thickness, the corresponding frequencies are about 40 kHz and 2.5 kHz. The ratio κ/κ_o becomes

Table 4.1.1. *Plate values in water and in air*

	Water		Air	
Plate	ϵ	$\omega_c h/c_o$	ϵ	$\omega_c h/c_o$
Steel	0.135	0.95	0.00073	0.22
Aluminium	0.391	0.95	0.0021	0.22
Plywood~	1	1.70	0.005	0.40

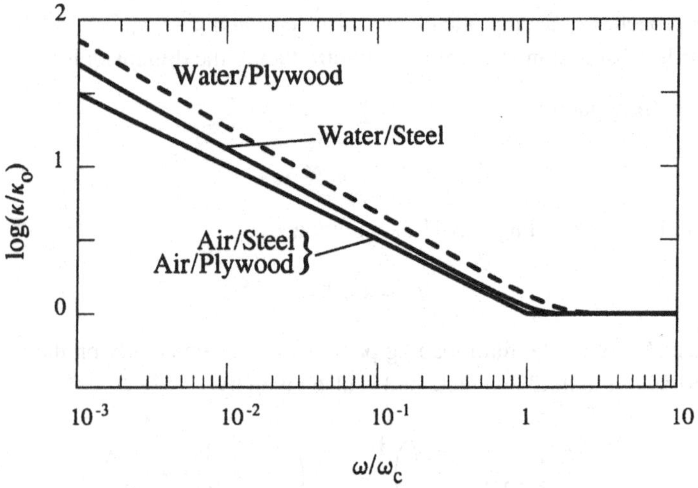

Figure 4.1.1. Dependence of bending wavenumber on frequency.

large when $\omega \ll \omega_c$; the bending wavelength is then much shorter than the corresponding acoustic wavelength. This behavior is illustrated in Figure 4.1.1 for steel and plywood plates.

A convenient nondimensional form of equation (4.1.6) is obtained by dividing by $m\omega^2$:

$$\sigma_2 - (K^4 - 1)\sqrt{K^2 - \mu^2} = 0, \tag{4.1.10}$$

where

$$K = \kappa/K_o, \quad \sigma_2 = 2\epsilon/\mu. \tag{4.1.11}$$

The nondimensional wavenumber $K \to \mu$ for $\omega > \omega_c$. In the absence of fluid loading ($\sigma_2 = 0$), bending waves correspond to the roots $K = \pm 1$ of (4.1.10). When ϵ is fixed, the influence of fluid loading on the mechanical properties of the plate increases as μ decreases, and the plate is said to be subject to *heavy fluid loading* when

$$\omega/\omega_c < \epsilon^2. \tag{4.1.12}$$

Example 1. In the notation of Section 1.4.4, show that the coincidence frequency $\bar{\omega}_c$ for flexural waves governed by the Mindlin plate equation (1.4.20) satisfies

$$\bar{\omega}_c = \frac{\omega_c}{\sqrt{\left(1 - c_o^2/\mu_* c_2^2\right)\left(1 - c_o^2/c_\ell^2\right)}}, \tag{4.1.13}$$

where ω_c is the thin-plate coincidence frequency. Thus, a consequence of Mindlin's corrections to thin-plate theory, which incorporate rotary inertia and shear deformations, is that $\bar{\omega}_c > \omega_c$. For a steel plate in water, $\bar{\omega}_c$ is about 22% larger than ω_c.

4.1.2 Bending Wave Power

The energy of a subsonic bending wave (whose wavenumber satisfies (4.1.6)) is shared between the evanescent pressure field in the fluid and the elastic plate. The bending wave power flux Π (per unit span of wave front parallel to the plate) for the plane wave (4.1.1) can be found by calculating the separate contributions from the energy flux Π_p in the plate and the flux Π_{f} in the fluid.

Π_p is found by considering the energy equation formed by multiplying the bending wave equation (1.4.9) by $\dot{\zeta} = \partial\zeta/\partial t$ and by rearranging to obtain

$$\frac{\partial}{\partial t}\left(\frac{1}{2}m\dot{\zeta}^2 + \frac{1}{2}\mathrm{B}\left(\nabla_2^2\zeta\right)^2\right)$$
$$+ \operatorname{div}\left(\mathrm{B}\dot{\zeta}\nabla\left(\nabla_2^2\zeta\right) - \mathrm{B}\left(\nabla_2^2\zeta\right)\nabla\dot{\zeta}\right) = \dot{\zeta}(-[p] + F_2). \tag{4.1.14}$$

The first term on the left is the rate of increase of mechanical energy of the plate per unit area; the terms on the right represent the rate at which energy is supplied to the plate by surface forces. The argument of the divergence is the instantaneous energy flux vector within the plate. Thus, for a plane wave whose displacement is the *real part* of (4.1.1), the time averaged flux of elastic energy is

$$\Pi_p = \omega\kappa^3 \mathrm{B}|\zeta_o|^2. \tag{4.1.15}$$

In the fluid $\Pi_{\mathrm{f}} = \int_{-\infty}^{\infty}\langle p v_1\rangle\, dx_2$, where p is the real part of the expression on the right of (4.1.5), the velocity v_1 is expressed in terms of p from the x_1-momentum equation $\rho_o \partial v_1/\partial t = -\partial p/\partial x_1$, and $\langle\ \rangle$ denotes a time average.

Performing the integration, we find

$$\Pi_{\mathrm{f}} = \frac{\rho_o \omega^3 \kappa |\zeta_o|^2}{2\left(\kappa^2 - \kappa_o^2\right)^{3/2}}. \tag{4.1.16}$$

Because $\kappa \to \kappa_o$ for $\omega > \omega_c$, these formulae imply that the flexural wave power is carried primarily by the evanescent, marginally subsonic surface-acoustic waves on either side of the plate when ω is large.

The net bending wave power $\Pi = \Pi_p + \Pi_{\mathrm{f}}$ may be written

$$\Pi = c_o m \omega^2 |\zeta_o|^2 \frac{\mu K(5K^4 - 4K^2\mu^2 - 1)}{4(K^2 - \mu^2)}, \quad K = \kappa/K_o. \tag{4.1.17}$$

4.1.3 Surface Wave Energy Flux

This formula for the flexural wave power can be obtained directly from the definition (4.1.6) of the dispersion function $\mathcal{D}(k, \omega)$ as a special case of a general energy flux formula satisfied by arbitrary, *lightly damped* linear systems [241, 242]. It is applicable to surface waves coupled to plane layered media, such as a plate coated with several layers of different viscoelastic materials, which dissipate energy during the extensional–compressional motions produced by the passage of the wave. The damping is usually expressed in terms of a frequency-dependent loss factor [26, 242] $\eta(\omega) \ll 1$.

To fix ideas, consider wave propagation at frequency ω in the x_1-direction along the "coated" plate of Figure 4.1.2. The motion can be expressed in terms of the displacement of a convenient element of the system, which we take to be the flexural displacement ζ of the plate, whose equation of motion is

$$\mathcal{L}(-i\partial/\partial x_1, \omega, \eta_0)\zeta = p_- - p_+, \tag{4.1.18}$$

where $p_\pm$ are the normal stresses applied on the upper and lower surfaces of the plate. The linear operator $\mathcal{L}$ (which may be more general than the bending wave operator of (1.4.9)) determines the response of the plate to forcing, and η_0 is the plate loss factor. In the absence of externally applied forces, $p_\pm$ and ζ also satisfy a second relation

$$p_- - p_+ = \mathcal{H}(-i\partial/\partial x_1, \omega, \eta')\zeta, \tag{4.1.19}$$

which describes the response of the layered structure and the fluid to plate displacements, where η' is a loss factor for the coating. $\mathcal{H}$ will usually involve integral operators that can be interpreted formally as inverse (and possibly fractional) powers of $\partial/\partial x_1$. It corresponds to equation (4.1.5) at $x_2 = \pm 0$

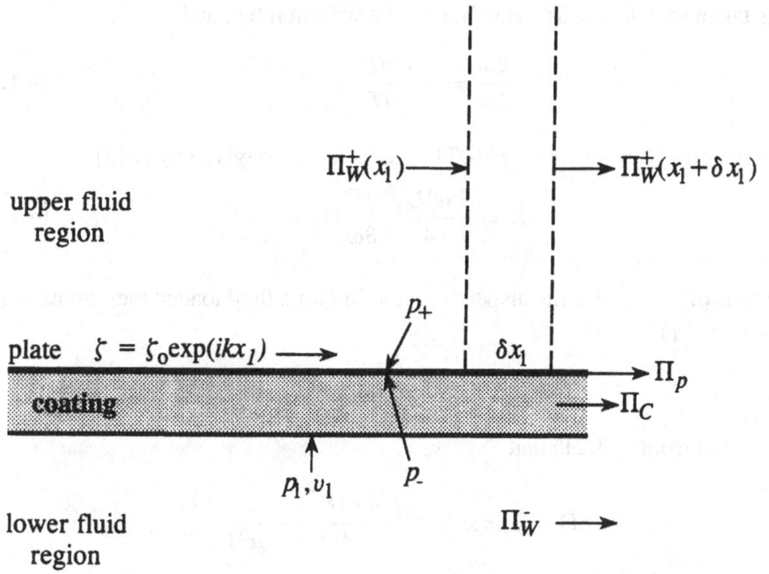

Figure 4.1.2. Flexural wave power flow.

for an *uncoated* plate, whose inverse Fourier transform yields $p_- - p_+ = i\rho_o\omega^2 \int_{-\infty}^{\infty} \mathrm{H}_0^{(1)}(\kappa_o|x_1 - y_1|)\zeta(y_1, \omega)\, dy_1$.

The dispersion equation is obtained by substituting from (4.1.19) into the right of (4.1.18) and taking ζ in the form (4.1.1):

$$\mathcal{D}(k, \omega, \boldsymbol{\eta}) \equiv \mathcal{L}(k, \omega, \eta_0) - \mathcal{H}(k, \omega, \eta') = 0, \quad \boldsymbol{\eta} = (\eta_0, \eta'). \tag{4.1.20}$$

Then the total flux of wave energy Π in the x_1-direction (per unit length out of the plane of the paper in Figure 4.1.2), both within the structure and in the fluid, and the rate Δ at which wave energy is dissipated per unit length in the x_1-direction are given by

$$\Pi = \frac{\omega|\zeta_o|^2}{4}\frac{\partial\mathcal{D}}{\partial k}(k, \omega, \mathbf{0}),$$
$$\Delta = \frac{-\omega|\zeta_o|^2}{2}\boldsymbol{\eta}\cdot\left(\frac{\partial\mathcal{D}_i}{\partial\boldsymbol{\eta}}(k, \omega, \boldsymbol{\eta})\right)_{\boldsymbol{\eta}=\mathbf{0}}, \tag{4.1.21}$$

where $\mathcal{D}(k, \omega, \mathbf{0}) = 0$, and $\mathcal{D}_i(k, \omega, \boldsymbol{\eta}) = \mathrm{Im}(\mathcal{D}(k, \omega, \boldsymbol{\eta}))$ (Example 3).

We can also write $\Pi = E\partial\omega/\partial k$, where E is the wave energy density per unit distance in the x_1-direction. The dispersion equation (4.1.20) with $\boldsymbol{\eta} = \mathbf{0}$

is taken to define ω as a function of the wavenumber, and

$$\frac{\partial\omega}{\partial k} = -\left(\frac{\partial\mathcal{D}/\partial k}{\partial\mathcal{D}/\partial\omega}\right)_{\eta=0} \tag{4.1.22}$$

is the *group velocity* [38, 241–244]. E is therefore given by [245]

$$E = \frac{-\omega|\zeta_o|^2}{4}\frac{\partial\mathcal{D}}{\partial\omega}(k,\omega,0). \tag{4.1.23}$$

Example 2. Derive the dispersion function for a fluid loaded membrane (equation (1.4.1)):

$$\mathcal{D}(k,\omega) = \tau k^2 - m\omega^2 - 2i\rho_o\omega^2/\gamma(k).$$

Deduce from (4.1.21) that

$$\Pi = c_o m\omega^2|\zeta_o|^2\frac{\mu K(3K^2 - 2\mu^2 - 1)}{4(K^2 - \mu^2)},$$

where $K = k/K_o$, $\mu = \kappa_o/K_o$, and $K_o = \sqrt{m\omega^2/\tau}$ is the vacuum wavenumber.

Example 3. Proof of (4.1.21) for the coated plate of Figure 4.1.2.

Let $\Pi_W^+(x_1)$ be the power flux in the fluid in the "upper region" and neglect dissipation in the fluid. Because k is real and the damping is small, disturbances in the fluid are evanescent, and the average power entering the fluid within the interval δx_1 is

$$\left\langle p_+\frac{\partial\zeta}{\partial t}\right\rangle\delta x_1 = \frac{\partial\Pi_W^+}{\partial x_1}(x_1)\delta x_1.$$

Similarly, the power flux Π_W^- in the fluid below the coated plate is related to the pressure p_1 and normal velocity v_1 of the coating at the interface with the fluid, by $\langle v_1 p_1\rangle = -\partial\Pi_W^-/\partial x_1$.

If Π_C and Δ_C are respectively, the power flux and dissipation in the coating, the same argument implies

$$\frac{\partial\Pi_C}{\partial x_1} + \Delta_C = \langle v_1 p_1\rangle - \left\langle p_-\frac{\partial\zeta}{\partial t}\right\rangle$$

and therefore that

$$\left\langle (p_- - p_+)\frac{\partial\zeta}{\partial t}\right\rangle \equiv \left\langle\frac{\partial\zeta}{\partial t}\mathcal{H}\left(-\frac{i\partial}{\partial x_1},\omega,\eta'\right)\zeta\right\rangle = -\frac{\partial\Pi_W}{\partial x_1} - \frac{\partial\Pi_C}{\partial x_1} - \Delta_C,$$

where $\Pi_W = \Pi_W^- + \Pi_W^+$ is the total power flux in the fluid.

The energy equation for the whole system is formed by averaging the product of equation (4.1.18) and $\partial\zeta/\partial t$. By observing that

$$\left\langle \frac{\partial \zeta}{\partial t}\mathcal{L}\left(-\frac{i\partial}{\partial x_1}, \omega, \eta_0\right)\zeta \right\rangle = \frac{\partial \Pi_p}{\partial x_1} + \Delta_p,$$

where Π_p and Δ_p are the power flux and dissipation within the plate, and making use of the above relations, we find

$$\frac{\partial \Pi}{\partial x_1} + \Delta \equiv \left\langle \frac{\partial \zeta}{\partial t}\mathcal{L}\left(-\frac{i\partial}{\partial x_1}, \omega, \eta_0\right)\zeta \right\rangle - \left\langle \frac{\partial \zeta}{\partial t}\mathcal{H}\left(-\frac{i\partial}{\partial x_1}, \omega, \eta'\right)\zeta \right\rangle = 0, \tag{4.1.24}$$

where $\Pi = \Pi_p + \Pi_C + \Pi_W$ is the total wave power flux, and $\Delta = \Delta_p + \Delta_C$ is the total rate of dissipation.

Equations (4.1.21) are now obtained by evaluating the time averages in the identity (4.1.24). Set $\zeta = \mathrm{Re}(\zeta_o(x_1)\mathrm{e}^{i(kx_1-\omega t)})$, where $\partial\zeta_o\partial x_1 \ll k\zeta_o$, and deduce that

$$\left\langle \frac{\partial \zeta}{\partial t}\mathcal{L}\left(-\frac{i\partial}{\partial x_1}, \omega, \eta_0\right)\zeta \right\rangle - \left\langle \frac{\partial \zeta}{\partial t}\mathcal{H}\left(-\frac{i\partial}{\partial x_1}, \omega, \eta'\right)\zeta \right\rangle$$
$$= \frac{i\omega\zeta_o^*}{4}\mathcal{D}\left(k - \frac{i\partial}{\partial x_1}, \omega, \boldsymbol{\eta}\right)\zeta_o + \text{c.c.},$$

where ζ_o^* is the complex conjugate of ζ_o. Now $\mathcal{D}(k - i\partial/\partial x_1, \omega, \boldsymbol{\eta}) \approx \mathcal{D}(k, \omega, \mathbf{0}) - i[\partial\mathcal{D}(k, \omega, \mathbf{0})/\partial k]\partial/\partial x_1 + \boldsymbol{\eta}\cdot[\partial\mathcal{D}(k, \omega, \boldsymbol{\eta})/\partial\boldsymbol{\eta}]_{\boldsymbol{\eta}=\mathbf{0}}$ and $\mathcal{D}(k, \omega, \mathbf{0}) \equiv 0$, so that to first order (4.1.24) becomes

$$\frac{\partial \Pi}{\partial x_1} + \Delta \equiv \frac{\partial}{\partial x_1}\left(\frac{\omega|\zeta_o|^2}{4}\frac{\partial \mathcal{D}}{\partial k}(k, \omega, 0)\right) - \frac{\omega|\zeta_o|^2}{2}\boldsymbol{\eta}\cdot\left(\frac{\partial \mathcal{D}_i}{\partial \boldsymbol{\eta}}(k, \omega, \boldsymbol{\eta})\right)_{\boldsymbol{\eta}=\mathbf{0}}.$$

Equations (4.1.21) are now obtained by equating corresponding terms in this identity.

4.1.4 Green's Function

In the absence of mean flow, Green's function satisfies (1.6.8) with $\mathcal{F} = \delta(\mathbf{x} - \mathbf{y})\delta(t - \tau)$, where it will be assumed that $\mathbf{y}$ lies in the upper region ($y_2 > 0$) (Figure 4.1.3).

In the frequency domain, equation (4.1.4) implies that $\zeta = (1/\rho_o\omega^2)\partial G(\mathbf{x}, \mathbf{y}; \omega)/\partial x_2$ at $x_2 = \pm 0$. This and the bending wave equation (1.4.9) (with $F_2 = 0$)

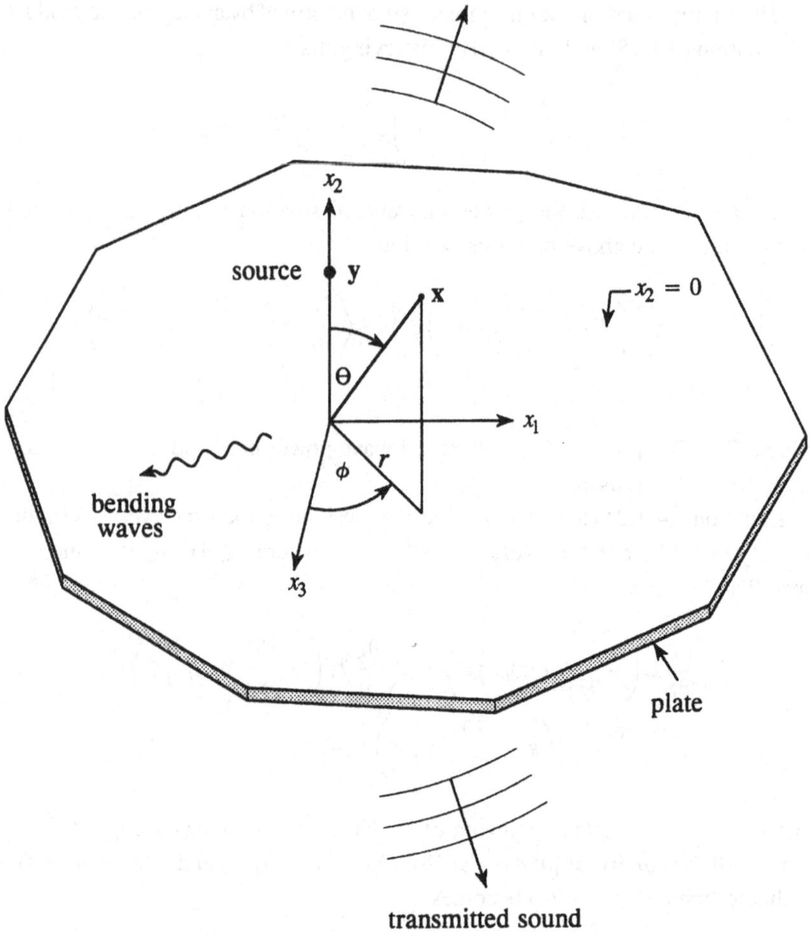

Figure 4.1.3. Point source near a thin elastic plate.

therefore require that

$$\frac{\partial G}{\partial x_2}(x_1, +0, x_3, \mathbf{y}; \omega) = \frac{\partial G}{\partial x_2}(x_1, -0, x_3, \mathbf{y}; \omega);$$
$$\left(\mathrm{B}\nabla_2^4 - m\omega^2\right)\frac{\partial G}{\partial x_2} + \rho_o\omega^2[G] = 0, \quad x_2 = 0. \tag{4.1.25}$$

$G(\mathbf{x}, \mathbf{y}; \omega)$ consists of the spherical wave (1.7.9) incident on the plate from $x_2 > 0$, together with reflected and transmitted waves. These are determined from the two boundary conditions (4.1.25) by first writing the incident wave as a superposition of plane waves by means of the Fourier integral (1.7.12). For each plane wave of wavenumber $\mathbf{k}$ $(=(k_1, 0, k_3))$, there exist reflection and transmission coefficients $\mathcal{R}(k, \omega)$, $\mathcal{T}(k, \omega)$, which depend only on frequency and wavenumber magnitude k when the plate is isotropic, and we therefore have

$$
\begin{aligned}
G(\mathbf{x}, \mathbf{y}; \omega) \\
&= \frac{-i}{8\pi^2} \int_{-\infty}^{\infty} \frac{\left(e^{i\gamma(k)|x_2-y_2|} + \mathcal{R}(k,\omega) e^{i\gamma(k)(x_2+y_2)}\right)}{\gamma(k)} e^{i\mathbf{k}\cdot(\mathbf{x}-\mathbf{y})}\, d^2\mathbf{k}, \quad x_2 > 0, \\
&= \frac{-i}{8\pi^2} \int_{-\infty}^{\infty} \frac{\mathcal{T}(k,\omega) e^{i\gamma(k)(y_2-x_2)}}{\gamma(k)} e^{i\mathbf{k}\cdot(\mathbf{x}-\mathbf{y})}\, d^2\mathbf{k}, \quad x_2 < 0, \qquad (4.1.26)
\end{aligned}
$$

where $\gamma(k) = \sqrt{\kappa_o^2 - k^2}$.

Conditions (4.1.25) must be satisfied separately for each wavenumber $\mathbf{k}$, and supply the following equations for the determination of $\mathcal{R}$ and $\mathcal{T}$:

$$
\mathcal{T} + \mathcal{R} = 1, \qquad \left(1 + \frac{i\gamma(k)[\mathrm{B}k^4 - m\omega^2]}{\rho_o \omega^2}\right)\mathcal{T} - \mathcal{R} = 1, \qquad (4.1.27)
$$

which yield

$$
\mathcal{R} = 1 - \mathcal{T} = \frac{\mathrm{B}k^4 - m\omega^2}{\mathcal{D}(k,\omega)}, \qquad (4.1.28)
$$

where $\mathcal{D}(k, \omega)$ is the dispersion function (4.1.6).

The Acoustic Far Field

Take the origin on the plate, beneath the source (which now lies at $(0, y_2, 0)$). The acoustic field as $\kappa_o|\mathbf{x}| \to \infty$ is found by using the stationary phase formula (1.13.10) to evaluate the Fourier integrals for the reflected and transmitted waves. Hence,

$$
G(\mathbf{x}, \mathbf{y}; \omega) \approx \frac{-e^{i\kappa_o|\mathbf{x}-\mathbf{y}|}}{4\pi|\mathbf{x}-\mathbf{y}|} - \frac{\mathcal{R}(\kappa_o \sin\Theta, \omega) e^{i\kappa_o|\mathbf{x}-\bar{\mathbf{y}}|}}{4\pi|\mathbf{x}-\bar{\mathbf{y}}|}, \qquad |\mathbf{x}| \to \infty,\ x_2 > 0,
$$

$$
\mathcal{R}(\kappa_o \sin\Theta, \omega) = \frac{\mu^2 \cos\Theta(\mu^4 \sin^4\Theta - 1)}{\mu^2\cos\Theta(\mu^4\sin^4\Theta - 1) - 2i\epsilon}, \qquad (4.1.29)
$$

where $\bar{\mathbf{y}} = (0, -y_2, 0)$ is the image of the source point in the plate, and Θ is the angle between the radiation direction and the plate normal (x_2-axis). For a rigid plate, $\mu \to \infty$, $\mathcal{R} \to 1$, and the reflected sound may be attributed to an identical image source. For an elastic plate, however, the reflected field is more complicated because of the dependence of $\mathcal{R}$ on frequency and radiation angle Θ.

Similarly, below the plate in $x_2 < 0$

$$G(\mathbf{x}, \mathbf{y}; \omega) \approx -\frac{\mathcal{T}(\kappa_o \sin\Theta', \omega)\mathrm{e}^{i\kappa_o|\mathbf{x}-\mathbf{y}|}}{4\pi|\mathbf{x}-\mathbf{y}|}, \quad |\mathbf{x}| \to \infty, \ x_2 < 0,$$

$$\mathcal{T}(\kappa_o \sin\Theta', \omega) = \frac{-2i\epsilon}{\mu^2 \cos\Theta'(\mu^4 \sin^4\Theta' - 1) - 2i\epsilon}, \tag{4.1.30}$$

where $\Theta' = \pi - \Theta$ is the radiation angle measured from the negative x_2-direction. The fact that $|\mathcal{R}(\kappa_o \sin\Theta,\omega)|^2 + |\mathcal{T}(\kappa_o \sin\Theta', \omega)|^2 = 1$ shows that acoustic energy is conserved during interaction with the plate: There is no exchange of energy between the acoustic and bending wave modes of the motion.

These results can be expressed in an alternative form, in which we compare the mean acoustic intensities $I_a = \langle p^2/\rho_o c_o\rangle$ generated by the source with and without the plate. To do this, the pressure is identified with $\mathrm{Re}(G(\mathbf{x}, \mathbf{y}; \omega)\mathrm{e}^{-i\omega t})$ for a given frequency ω. The formulae simplify considerably if $\kappa_o y_2 \ll 1$, when the distance of the source from the plate is much smaller than the acoustic wavelength. Then, if I and I_o are the intensities with and without the plate, we find as $|\mathbf{x}| \to \infty$

$$\frac{I}{I_o} = \frac{4[\mu^4 \cos^2\Theta(\mu^4 \sin^4\Theta - 1)^2 + \epsilon^2]}{\mu^4 \cos^2\Theta(\mu^4 \sin^4\Theta - 1)^2 + 4\epsilon^2}, \quad x_2 > 0,$$

$$= \frac{4\epsilon^2}{\mu^4 \cos^2\Theta'(\mu^4 \sin^4\Theta' - 1)^2 + 4\epsilon^2}, \quad x_2 < 0.$$

These ratios are plotted as a function of radiation direction in Figure 4.1.4 for a steel plate in water, for different values of $\mu^2 \equiv \omega/\omega_c$. Figure 4.1.4a indicates that the plate is acoustically transparent when $\omega/\omega_c < \epsilon \sim 0.135$. At higher frequencies the plate becomes "hard" and the transmitted power is generally small, except for $\Theta' \approx 90°$, and in the vicinity of the *leaky wave angle* $\Theta' = \arcsin\sqrt{\omega_c/\omega}$, where $\mu^4 \sin^4\Theta' = 1$ when $\omega/\omega_c > 1$. At this angle, a complex pole of $\mathcal{R}(k, \omega)$ and $\mathcal{T}(k, \omega)$ approaches the real k-axis within the acoustic domain $k < \kappa_o$ as $\sigma_2 \equiv 2\epsilon\sqrt{\omega_c/\omega} \to 0$. This pole occurs approximately at the vacuum wavenumber $k = K_o$ of bending waves of frequency ω, which has supersonic phase velocity when $\omega/\omega_c > 1$ and may be regarded as displaced from the real axis as a consequence of fluid loading. The flexural displacement of the plate produced by this leaky wave decays exponentially with distance from the source because of the radiation of leaky wave energy into the fluid in the stationary phase direction Θ' where $K_o \approx \kappa_o \sin\Theta'$. The stationary phase approximation (1.13.10) is still applicable when the pole approaches the real axis, but careful analysis [246, 247] shows that in directions close to the

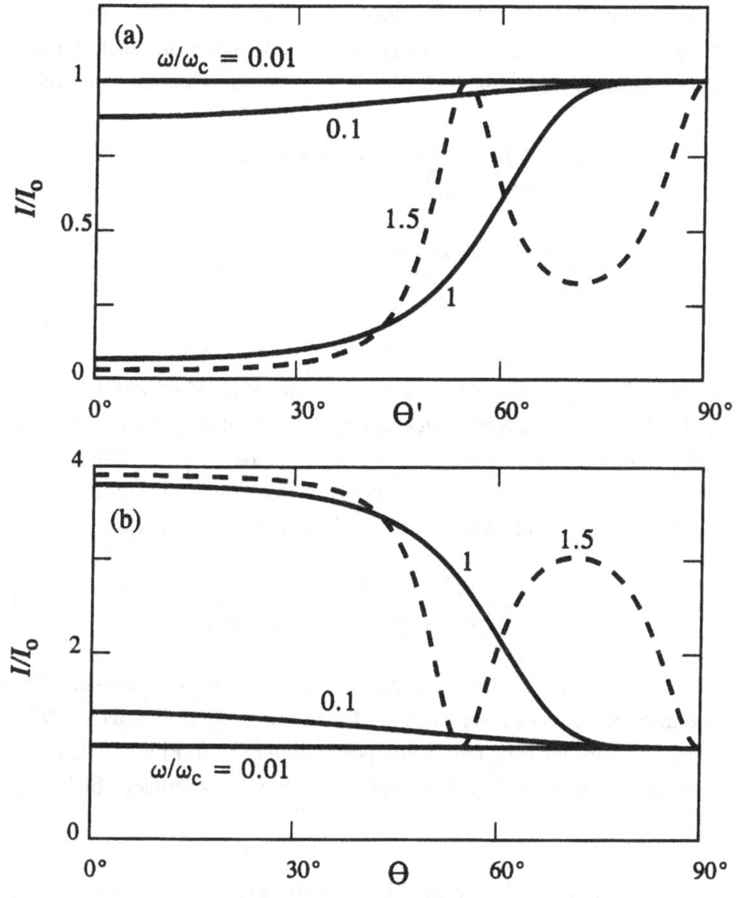

Figure 4.1.4. Sound generated by point source next to a steel plate in water: (a) transmitted intensity and (b) direct plus reflected intensity.

leaky wave angle the usual far-field condition $\kappa_o|\mathbf{x}| \gg 1$ for its validity must be replaced by $\kappa_o|\mathbf{x}| \gg 1/\epsilon^2$.

Actually, these conclusions are only qualitatively correct when $\omega/\omega_c > 1$ because the bending wave equation (1.4.9) strictly becomes invalid when $\omega h/c_o \sim 1$. Table 4.1.1 shows that this occurs at $\omega \approx \omega_c$ for a steel plate in water, but at a somewhat higher relative frequency in air.

Far Field on the Plate

The acoustic component of $G(\mathbf{x}, \mathbf{y}; \omega)$ decays like $1/|\mathbf{x}|$ at large distances from the source. However, as $|\mathbf{x}| \approx r \equiv \sqrt{x_1^2 + x_3^2} \to \infty$ near the plate, account must also be taken of the bending wave excited by the source. This dominates the far field near the plate, but because its influence decreases exponentially with distance from the plate, its energy diffuses into a two-dimensional domain, where its amplitude decreases only like $1/\sqrt{r}$ (in the absence of structural or viscous losses).

The displacement is calculated from the relation $\zeta = (1/\rho_o\omega^2)\partial G(x_1, 0, x_3, \mathbf{y}; \omega)/\partial x_2$. Using the formula $\int_0^{2\pi} e^{ikr\cos\phi}\, d\phi = 2\pi J_0(kr)$ [24], we find

$$\zeta(r, y_2; \omega) = \frac{i}{2\pi}\int_0^\infty \frac{kJ_0(kr)e^{i\gamma(k)y_2}}{\gamma(k)\mathcal{D}(k,\omega)}\,dk, \quad r \equiv \sqrt{x_1^2 + x_3^2} \to \infty. \tag{4.1.31}$$

The path of integration along the real axis is indented to avoid the bending wave pole at $k = \kappa > \kappa_o$ where $\mathcal{D}(k, \omega) = 0$. In the usual way (Section 1.7.4), this is done by temporarily replacing ω by $\omega + i\varepsilon$, $\varepsilon > 0$. $\mathcal{D}(k, \omega)$ and its derivatives are real for real k near the pole, so that when ε is small, the pole at $k = \kappa$ is shifted to $\kappa + i\delta$, where $\mathcal{D}(\kappa + i\delta, \omega + i\varepsilon) = 0$, that is,

$$\delta \approx -\varepsilon\frac{\partial\mathcal{D}(\kappa,\omega)/\partial\omega}{\partial\mathcal{D}(\kappa,\omega)/\partial\kappa} = \frac{\varepsilon}{\partial\omega/\partial\kappa}. \tag{4.1.32}$$

The group velocity $\partial\omega/\partial\kappa$ and ω have the same sign when $\kappa > 0$ so that the pole is shifted into the upper or lower halves of the k-plane according as $\omega \gtrless 0$. Thus, the integration contour runs below the pole when $\omega > 0$ (Figure 4.1.5).

To estimate the integral (4.1.31), the Bessel function is replaced by its asymptotic formula [24]

$$J_0(kr) \approx \frac{1}{\sqrt{2\pi kr}}\left(e^{i(kr-\frac{\pi}{4})} + e^{-i(kr-\frac{\pi}{4})}\right), \quad kr \to \infty.$$

The two exponentials decrease rapidly, respectively, as $\operatorname{Im} k \to \pm i\infty$, and their respective contributions to the integral (4.1.31) are determined as follows. For the first term, the integration contour is deformed onto the positive imaginary axis, a large quarter circle in the first quadrant (whose contribution ultimately vanishes) and, for $\omega > 0$, the branch cut of $\gamma(k) = \sqrt{\kappa_o^2 - k^2}$; for the second term, the contour is deformed onto the negative imaginary k-axis, a large quarter circle in the fourth quadrant and, when $\omega < 0$, the branch cut of $\gamma(k)$ from $-\kappa_o$ to $-i\infty$. Only one of these operations captures a residue from the pole at the bending wavenumber, which yields an outgoing bending wave whose amplitude decreases like $1/\sqrt{r}$. Complex poles may be captured in either half-plane, but their contributions are exponentially small when r is large. The integrations along the imaginary axis and branch cut give contributions that decay more

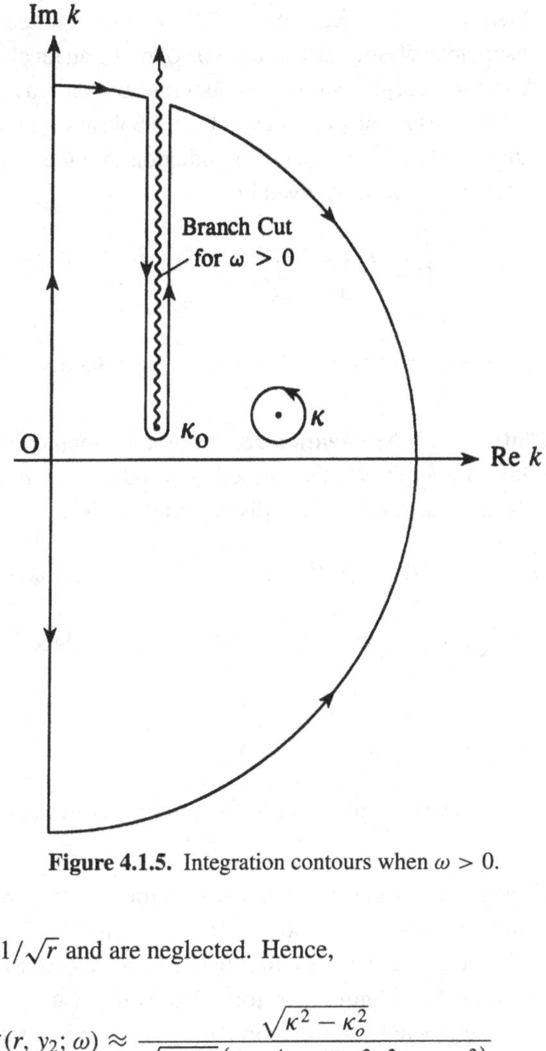

Figure 4.1.5. Integration contours when $\omega > 0$.

rapidly than $1/\sqrt{r}$ and are neglected. Hence,

$$\zeta(r, y_2; \omega) \approx \frac{\sqrt{\kappa^2 - \kappa_o^2}}{\sqrt{2\pi\kappa r}\left(5B\kappa^4 - 4B\kappa^2\kappa_o^2 - m\omega^2\right)} \times e^{i\,\mathrm{sgn}(\omega)(\kappa r + \frac{\pi}{4}) - y_2\sqrt{\kappa^2 - \kappa_o^2}}, \quad r \to \infty. \qquad (4.1.33)$$

This shows that the efficiency with which the bending wave is excited decreases exponentially with source distance y_2 on a length scale approximately equal to the bending wavelength. In other words, only the *hydrodynamic* component of the source field can excite the structural modes of the plate. These modes decay relatively slowly with distance from the source and include both flexural motions of the plate and subsonically traveling, evanescent waves in the fluid.

Example 5. A *leaky wave* $\zeta = \text{Re}(\zeta_o e^{i(\kappa x_1 - \omega t)})$, $\omega > 0$ is excited by a localized source on a conservative elastic surface $x_2 = 0$ bounding an ideal, compressible fluid in $x_2 > 0$. The wave represents the residue contribution to a Fourier integral representation of the surface displacement ζ from a pole at $\kappa = \kappa_R + i\kappa_I$, where $0 < \kappa_R < \kappa_o$ and $\kappa_I \ll \kappa_R$ and decays by radiation into the fluid. Show that the power radiated by the wave is given by

$$\Pi = \frac{\omega|\zeta_o|^2}{4}\left|\frac{\partial \mathcal{D}}{\partial k}(k,\omega)\right|_{k=\kappa_R}, \tag{4.1.34}$$

where $\mathcal{D}(k,\omega)$ is the dispersion function defined as in Section 4.1.3.

Example 6: Plate excited by a point force. A time-harmonic force $F_2(x_1, x_3, \omega) = -F\delta(x_1)\delta(x_3)$ is applied to the surface $x_2 = +0$ of a *vacuum backed* thin elastic plate. The pressure and flexural displacement satisfy

$$(\nabla^2 + \kappa_o^2)p = 0, \quad x_2 > 0, \quad (\text{B}\nabla_2^4 - m\omega^2)\zeta = -p - F\delta(x_1)\delta(x_3), \quad x_2 = 0.$$

By taking $p = \int_{-\infty}^{\infty} \mathcal{A}(k_1, k_3) e^{i(\mathbf{k}\cdot\mathbf{x} + \gamma(k)x_2)} d^2\mathbf{k}$ $(x_2 \geq 0)$, deduce that

$$|p^2| \approx \frac{\epsilon^2\kappa_o^2|F|^2\cos^2\Theta}{8\pi^2|\mathbf{x}|^2[\mu^4\cos^2\Theta(\mu^4\sin^4\Theta - 1)^2 + \epsilon^2]}, \quad |\mathbf{x}| \to \infty,$$

where Θ is the angle between the radiation direction and the x_2-axis.

Example 7. Repeat the calculation of Example 6 for a plate governed by the Mindlin equation (1.4.20) with coincidence frequency $\bar{\omega}_c$ [248].

The acoustic directivity $|p^2(|\mathbf{x}|, \Theta)|$ (normalized with respect to the intensity at $\Theta = 0$) is illustrated in Figure 4.1.6 for $\omega/\bar{\omega}_c = 0.6, 1.4$ for a steel plate and water. The curve labeled (c) is the directivity predicted by the formula of Example 6 for the thin plate when $\omega/\omega_c = 1.4$. The peak relative intensity in the latter case is about ten times greater than for the Mindlin plate at the same frequency ratio; the peaks occur at the corresponding leaky wave angles, which increase when account is taken of rotary inertia and shear deformations of the plate.

Example 8. When the force in Example 6 is replaced by theline force $F_2(x_1, \omega) = -F\delta(x_1)$, show that the mean square acoustic pressure is given

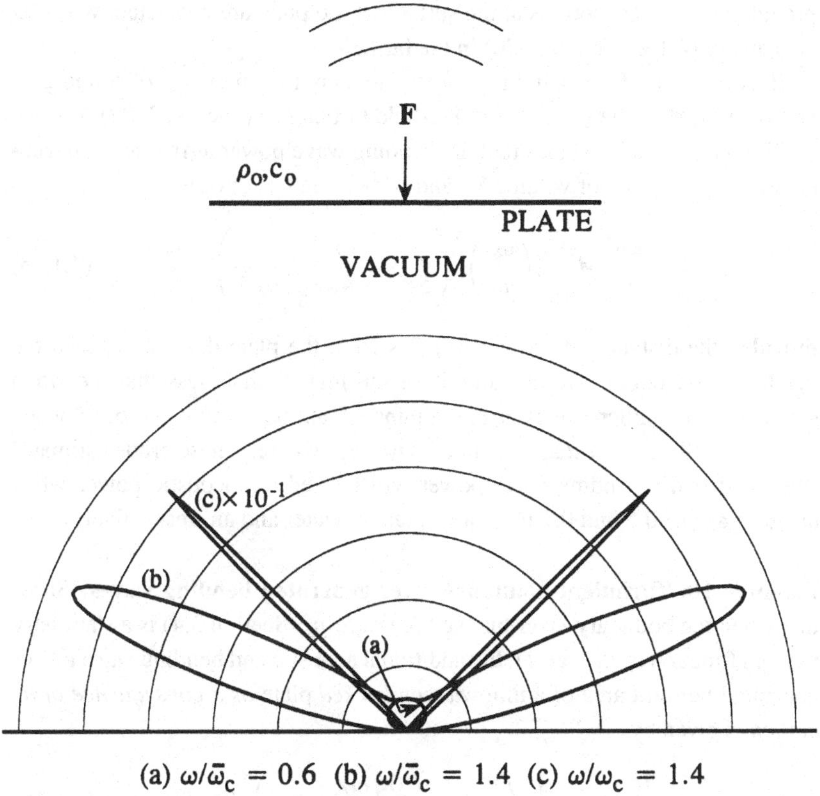

Figure 4.1.6. Directivity of sound produced by a point force applied to a vacuum backed steel plate in water: (a) and (b) Mindlin plate, (c) thin plate.

by [246, 249, 250],

$$|p^2| \approx \frac{\epsilon^2 |\kappa_o| |F|^2 \cos^2\theta}{2\pi |\mathbf{x}| [\mu^4 \cos^2\theta (\mu^4 \sin^4\theta - 1)^2 + \epsilon^2]}, \quad |\mathbf{x}| = \sqrt{x_1^2 + x_2^2} \to \infty,$$

where $(x_1, x_2) = |\mathbf{x}|(\sin\theta, \cos\theta)$, $|\theta| < \pi/2$.

Example 9: Quadrupoles near a thin plate. The sound generated in low Mach number flow by a Reynolds stress quadrupole $\rho_o v_i v_j$ adjacent to a thin plate is given by

$$p(\mathbf{x}, \omega) \approx -\int (\rho_o v_i v_j)(\mathbf{y}, \omega) \frac{\partial^2 G}{\partial y_i \partial y_j}(\mathbf{x}, \mathbf{y}, \omega)\, d^3\mathbf{y},$$

provided nonlinear motions at the surface of the plate are neglected, where G is given by (4.1.29) and (4.1.30) in the far field.

Because $|\mathcal{R}|, |\mathcal{T}| < 1$ in the acoustic domain, the efficiency of sound generation $\sim O(M^5)$, the same as for free-field turbulence (Section 2.1.2).

Similarly, (4.1.33) shows that the bending wave power at frequency ω generated by turbulence of volume V_o and eddy scale ℓ is of order

$$\frac{\rho_o v^3}{\ell} M^5 V_o \left(\frac{\omega_c}{\omega}\right)^3 \left(\frac{\epsilon K^4}{5K^4 - 4K^2\mu^2 - 1}\right), \tag{4.1.35}$$

provided the distance of the quadrupoles from the plate does not exceed the bending wavelength. The magnitude of the final term is less than 2ϵ when $\epsilon > 0.1$. For smaller values of ϵ, it attains a maximum value $\sim 1/6\epsilon^{1/3}$ when ω/ω_c is a little greater than one and is $O(\epsilon)$ elsewhere. These crude estimates suggest that the bending wave power will exceed the acoustic power when $\omega/\omega_c < \epsilon^{1/3}$ (~ 0.5 and 0.1 for a steel plate in water and air, respectively).

Example 10: Turbulent boundary layer generated bending waves. Show that when the boundary layer blocked pressure p_s (Section 3.4) is a stationary random function of (x_1, x_3, t), the additional acoustic and bending wave power generated per unit area of a thin *vacuum backed* plate *as a consequence of its motion* is given by

$$\left\langle p_s \frac{\partial \zeta}{\partial t} \right\rangle = -\frac{1}{2}\int_{-\infty}^{\infty} \left(\frac{i\omega P(\mathbf{k}, \omega)}{\mathcal{D}(k, \omega)} + \text{c.c.}\right) d^2\mathbf{k}\, d\omega, \tag{4.1.36}$$

where $P(\mathbf{k}, \omega)$ is the blocked pressure wavenumber-frequency spectrum, and

$$\mathcal{D}(k, \omega) = \mathrm{B}k^4 - m\omega^2 - \frac{i\rho_o\omega^2}{\gamma(k)}. \tag{4.1.37}$$

The contribution to the integral from the acoustic domain $k < |\kappa_o|$ is the acoustic power generated as a consequence of plate motion. This is in addition to that generated by the boundary layer when wall motion is ignored (Section 3.4.4). The integrand vanishes in the hydrodynamic region $k > |\kappa_o|$ except at the bending wavenumber $k = \kappa$, where $\mathcal{D}(k, \omega) = 0$. By assigning to ω a small positive imaginary part, deduce that the flexural wave power generated per unit area of the plate is

$$\Pi = \pi \int_{-\infty}^{\infty} |\omega| P(\mathbf{k}, \omega) \delta(\mathcal{D}(k, \omega))\, d^2\mathbf{k}\, d\omega, \tag{4.1.38}$$

where the argument of the δ-function vanishes at $k = \kappa$.

4.2 Scattering of Bending Waves by Cracks, Joints, and Ribs

In this section, we investigate the sound produced when a bending wave encounters rectilinear or point inhomogeneities, such as a crack or joint between identical plates, or isolated points where the plate is riveted or perforated [249, 251–262]. Elsewhere the plate is homogeneous, with uniform bending stiffness, mass density, and so forth. For brevity, a rectilinear inhomogeneity will be referred to as a *straight joint*, except when the discussion refers to a specific configuration.

The width of the joint is assumed to be small compared to the wavelength of the plate motions. A "closed" joint may be regarded as connecting abutting edges of adjacent plates; the edges may be clamped, free to vibrate independently (without contact), or individually or jointly attached to resilient structural supports, but there is no transfer of fluid through the joint from one side of the plate to the other. In the absence of a structural path between the edges, bending wave energy is transmitted across the joint only via the coupling provided by the fluid. At high frequencies ($\omega \geq \omega_c$), when practically all of the bending wave energy is in the evanescent waves on either side of the plate, a wave is transmitted across the joint with little or no attenuation. The sound produced during such interactions rises to a maximum in the immediate vicinity of the coincidence frequency. At an "open" joint fluid is "pumped" through the opening between the edges during the interaction; the opening is equivalent to a monopole that can significantly modify the sound scattered from the joint.

4.2.1 Closed Cracks and Joints

An infinite thin plate (at $x_2 = 0$, Figure 4.2.1a) is immersed in stationary fluid of mean density ρ_o and sound speed c_o (the case of a "vacuum backed" plate is treated similarly). A straight joint centerd on the x_3-axis has negligible width and divides the plate into two halves; there may or may not be direct physical contact between the edges at $x_1 = \pm 0$.

Various simple and familiar edge conditions may be imposed at the edges [7, 10] involving linear relations between the edge displacement and its derivatives in the plane of the plate, corresponding to mass, spring, torsional, dashpot, and so forth, loadings and possibly including terms representing mechanical coupling between the edges (Figure 4.2.1b).

The pressure p must satisfy equation (1.6.8) in the fluid, and the plate displacement ζ and pressure are related by the bending wave equation (1.4.9) (with $F_2 = 0$) at $x_2 = 0$, *except* at the joint where possibly ζ and one or more of its derivatives are discontinuous.

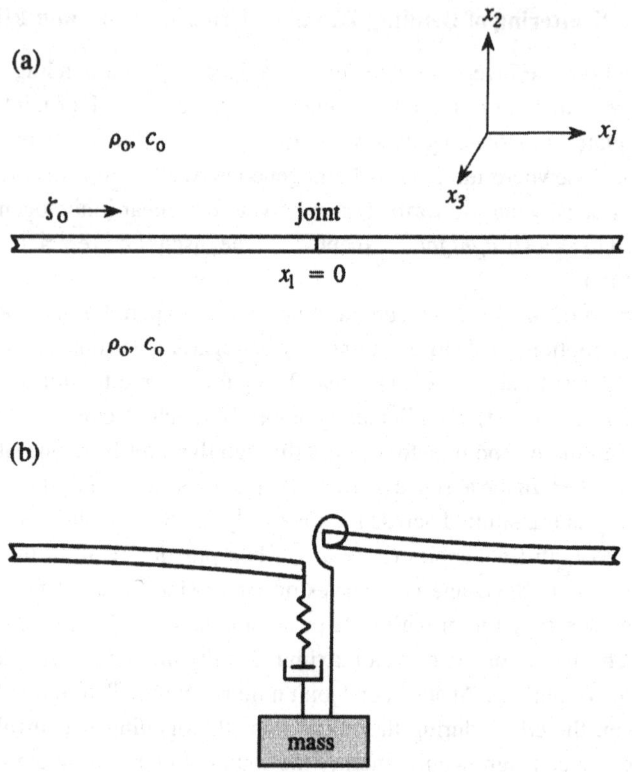

Figure 4.2.1. (a) Scattering by a closed straight crack or joint. (b) Schematic spring, dashpot, torsion, and mass loading edge restraints.

The domain of validity of the bending wave equation can be extended to include the infinitesimal interval occupied by the joint by observing that

$$\zeta(x_1, x_3, t) = \zeta(x_1, x_3, t)\mathrm{H}(|x_1| - \varepsilon), \quad \varepsilon \to +0.$$

When equation (1.4.9) is multiplied by $\mathrm{H}(|x_1| - \varepsilon)$, and repeated use is made of the identity,

$$\mathrm{H}(|x_1| - \varepsilon)\frac{\partial f}{\partial x_1} = \frac{\partial}{\partial x_1}(f\mathrm{H}(|x_1| - \varepsilon)) - [f(+0, x_3, t) - f(-0, x_3, t)]\delta(x_1), \quad \varepsilon \to +0,$$

where the coefficient of the δ-function is the jump in $f(x_1, x_2, t)$ across the joint, we find

$$\left(m\frac{\partial^2}{\partial t^2} + \mathrm{B}\nabla_2^4\right)\zeta + [p] = \sum_{n=0}^{3} a_n \delta^{(n)}(x_1), \quad -\infty < x_1 < \infty. \tag{4.2.1}$$

The function $a_n(x_3, t)$ is the jump in f_n, where

$$\begin{aligned} f_0 &= \mathrm{B}\left(\frac{\partial^3}{\partial x_1^3} + \frac{2\partial^3}{\partial x_1 \partial x_3^2}\right)\zeta, \quad f_1 = \mathrm{B}\left(\frac{\partial^2}{\partial x_1^2} + \frac{2\partial^2}{\partial x_3^2}\right)\zeta, \\ f_2 &= \mathrm{B}\frac{\partial \zeta}{\partial x_1}, \quad f_3 = \mathrm{B}\zeta. \end{aligned} \tag{4.2.2}$$

The Heaviside function has been omitted on the left of (4.2.1) on the understanding that ζ and its derivatives are to remain undefined at $x_1 = 0$.

Equation (4.2.1) is applicable everywhere on $x_2 = 0$. The presence of the joint is represented by the four multipole line sources, whose strengths depend on the mechanical conditions imposed on the abutting edges. Four conditions are required for a unique solution. The equation can be generalized further in an obvious manner to deal with a multiplicity of parallel joints at $x_1 = x_A, x_B, \ldots$.

4.2.2 Scattering of a Time-Harmonic Bending Wave

Let us consider the interaction of the bending wave

$$\zeta(x_1, \omega) = \zeta_o e^{i\kappa x_1}, \quad \zeta_o = \text{constant}, \quad \omega > 0, \tag{4.2.3}$$

normally incident on the joint at $x_1 = 0$ from $x_1 = -\infty$, where κ is the positive real root of the dispersion equation (4.1.6).

Scattered flexural waves $\zeta'(x_1, \omega)$ and outgoing pressure fluctuations $p'(x_1, x_2, \omega)$ are generated at the joint. They are related by equation (4.1.4), which is applicable both at the surfaces of the plates and at the joint and can therefore be expressed as the Fourier integrals

$$\begin{aligned} \zeta'(x_1, \omega) &= \int_{-\infty}^{\infty} \zeta'(k, \omega) e^{ikx_1}\, dk, \\ p'(x_1, x_2, \omega) &= -i\rho_o\omega^2 \operatorname{sgn}(x_2) \int_{-\infty}^{\infty} \frac{\zeta'(k, \omega)}{\gamma(k)} e^{i(kx_1 + \gamma(k)|x_2|)}\, dk. \end{aligned} \tag{4.2.4}$$

If $\zeta'(x_1, \omega)$ is discontinuous at the joint, then $\zeta'(k, \omega) \sim \mathrm{O}(1/k)$ as $k \to \infty$ [19]. The integrand of the second integral is then $\mathrm{O}(1/k^2)$ as $k \to \infty$, and this implies that the pressure is continuous in x_1 at the joint.

Inserting these integral representations into the left-hand side of (4.2.1), we find

$$\zeta'(k,\omega) = \sum_{n=0}^{3} \frac{(ik)^n a_n}{2\pi \mathcal{D}(k,\omega)}, \tag{4.2.5}$$

where $\mathcal{D}(k,\omega)$ is the dispersion function (4.1.6), and the coefficients a_n are *constants* that remain to be determined. The total displacement of the plate, including that of the incident wave, may now be cast in the form

$$\frac{\zeta(X,\omega)}{\zeta_o} = \sum_{n=0}^{3} \int_{-\infty}^{\infty} \frac{A_n \lambda^n \sqrt{\lambda^2-\mu^2} e^{i\lambda X}\, d\lambda}{(\lambda^4-1)\sqrt{\lambda^2-\mu^2}-\sigma_2} + e^{iKX}, \quad -\infty < X < \infty,$$

$$A_n = \frac{a_n K_o (iK_o)^n}{2\pi \zeta_o m \omega^2}, \quad X = K_o x_1, \tag{4.2.6}$$

where the A_n are dimensionless source strengths, and $\mu = \sqrt{\omega/\omega_c}$, $\sigma_2 = 2\epsilon/\mu$, $K = \kappa/K_o$ are defined as in Section 4.1.1.

The causal path of integration is the contour Γ (Figure 4.2.2), which passes just above singularities of the integrand on the negative real axis and below those on the positive real axis. Branch cuts for $\sqrt{\lambda^2-\mu^2} \equiv -i\gamma(k)/K_o$ run from $\lambda = \pm\mu$ to $\pm i\infty$, such that on the real axis $\sqrt{\lambda^2-\mu^2}$ is real and positive for $|\lambda| > \mu$ and is otherwise negative imaginary. There are either four or six simple poles in the cut plane, depending on the magnitude of μ. Two of these poles are at $\lambda = \pm K$ on the real axis, whose residue contributions yield the bending waves transmitted and reflected at the joint. Two more, labeled P_1 in the figure, occur near $\lambda = \pm i$ and correspond to localized motions near the joint. The final pair are labeled P_2 and represent leaky waves, which have small imaginary parts when $\omega/\omega_c \equiv \mu^2 > 1$. When μ decreases through unity these poles cross the branch cuts onto the "nonphysical" sheet of the Riemann surface but reappear as rapidly decaying modes when $\mu \ll 1$.

The Transmitted and Reflected Bending Waves

The transmission coefficient T determines the amplitude of the bending wave $T\zeta_o e^{iKX}$ transmitted across the joint into $x_1 > 0$. This wave consists of the incident wave plus the residue contribution from the pole at $\lambda = +K$ (which lies above Γ) so that

$$T = 1 + \frac{2\pi i (K^2-\mu^2)}{K(5K^4-4K^2\mu^2-1)} \sum_{n=0}^{3} A_n K^n. \tag{4.2.7}$$

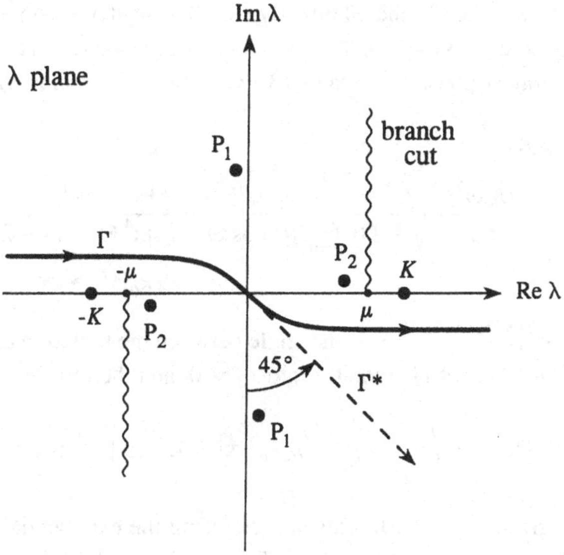

Figure 4.2.2. The complex λ-plane.

Because $K > \mu$, and $K \approx \mu$ when ω exceeds the coincidence frequency (Figure 4.1.1), it follows that $\mathrm{T} \approx 1$ for $\omega > \omega_c$, and the bending wave is then transmitted across the joint without attenuation. This is because bending wave power is carried primarily by the evanescent wave in the fluid when $\omega > \omega_c$, with negligible surface displacement (Section 4.1.2).

Similarly, the reflection coefficient R of the reflected wave $\mathrm{R}\zeta_o \mathrm{e}^{-iKX}$ is determined by the pole at $\lambda = -K$ (just below Γ), which yields

$$\mathrm{R} = \frac{2\pi i(K^2 - \mu^2)}{K(5K^4 - 4K^2\mu^2 - 1)} \sum_{n=0}^{3} A_n(-K)^n, \tag{4.2.8}$$

where $\mathrm{R} \to 0$ for $\omega > \omega_c$.

The Scattered Sound

The acoustic power Π_a scattered by unit length of the joint may be calculated indirectly by subtracting the transmitted and reflected bending wave powers from the incident power. If Π is the incident power per unit length of the joint (including that carried by the evanescent waves on either side of the plate), then the ratio Π_a/Π is the radiation efficiency of the joint, and

$$\Pi_a/\Pi = 1 - |\mathrm{T}|^2 - |\mathrm{R}|^2. \tag{4.2.9}$$

Π_a/Π may also be calculated directly from the scattered pressure given by the second of equations (4.2.4). The integral is first evaluated in the far field by using the stationary phase formula (1.13.13), which supplies, in $x_2 > 0$,

$$p'(x_1, x_2; \omega) = \frac{-i\rho_o\omega^3\zeta_o}{K_o}\sqrt{\frac{2\pi}{\kappa_o|\mathbf{x}|}}\sum_{n=0}^{3}\frac{A_n\mu^{n+2}\cos\Theta\sin^n\Theta e^{i(\kappa_o|\mathbf{x}|+\frac{\pi}{4})}}{\mu^2\cos\Theta(\mu^4\sin^4\Theta-1)-2i\epsilon}, \quad \kappa_o|\mathbf{x}|\to\infty, \tag{4.2.10}$$

where $|\mathbf{x}| = \sqrt{x_1^2+x_2^2}$, and Θ is the angle between the radiation direction and the x_2-axis. One-half of Π_a radiates into $x_2 > 0$, and therefore

$$\Pi_a = 2\int_{-\pi/2}^{\pi/2}(\langle p'^2\rangle/\rho_o c_o)|\mathbf{x}|\,d\Theta, \quad |\mathbf{x}|\to\infty,$$

where p'^2 is given by (4.2.10) after first restoring the exponential time factor $e^{-i\omega t}$ and taking the real part. Because Π is given by (4.1.17), we find

$$\frac{\Pi_a}{\Pi} = \frac{4\pi\sigma_2\mu^4(K^2-\mu^2)}{K(5K^4-4K^2\mu^2-1)} \times \int_{-\frac{\pi}{2}}^{\frac{\pi}{2}}\frac{|\sum_{n=0}^{3}A_n\mu^n\sin^n\Theta|^2\cos^2\Theta\,d\Theta}{\mu^4\cos^2\Theta(\mu^4\sin^4\Theta-1)^2+4\epsilon^2}. \tag{4.2.11}$$

This result is equivalent to (4.2.9), but it also gives the angular distribution of the acoustic power because the fraction of the acoustic energy radiating into the angular element $d\Theta$ in direction Θ is proportional to the integrand of (4.2.11).

These formulae give some insight into the radiation efficiencies of different joints. When $\omega < \omega_c$, the radiation is generally dominated by the term in A_n, with the smallest index n corresponding to the lowest order multipole source in (4.2.1). This is analogous to the low Mach number ranking of compact aerodynamic sound sources (Section 2.1). The effective size of the source region in the present case is the extent of the near-field flexural motions at the joint, which is determined by the bending wavelength and is therefore acoustically compact when $\omega < \omega_c$. At higher frequencies the acoustic and bending wavelengths are approximately equal, and all multipoles radiate with equal efficiencies.

Example 1: Scattering by a closed crack. Let the edges at $X = \pm 0$ be free to vibrate independently. The coupling between the sections $X < 0$ and $X > 0$ of the plate is via the fluid; total reflection of a bending wave would occur in the absence of fluid.

The edge conditions are given by (1.4.16a), (i), with no dependence on x_3:

$$\partial^2\zeta/\partial X^2 = 0, \quad \partial^3\zeta/\partial X^3 = 0, \quad X \to \pm 0. \tag{4.2.12}$$

It follows from (4.2.2) that $A_0 = A_1 = 0$ so that only quadrupole and octupole sources occur in (4.2.1). The coefficients A_2 and A_3 are evaluated by substituting from the general solution (4.2.6) into conditions (4.2.12). It is not permissible to do this by differentiating under the integral sign in (4.2.6) because ζ is discontinuous at the crack and the integral is only conditionally convergent [19]. We therefore recast (4.2.6) in the form

$$\frac{\zeta(X,\omega)}{\zeta_o} = \sum_{n=0}^{3}\int_{-\infty}^{\infty}\frac{\sigma_2 A_n\lambda^n \mathrm{e}^{i\lambda X}\,d\lambda}{[(\lambda^4-1)\sqrt{\lambda^2-\mu^2}-\sigma_2](\lambda^4-1)} + \sum_{n=0}^{3}\int_{-\infty}^{\infty}\frac{A_n\lambda^n \mathrm{e}^{i\lambda X}\,d\lambda}{(\lambda^4-1)} + \mathrm{e}^{iKX}, \tag{4.2.13}$$

where the paths of integration run below singularities on the positive real axis and above those on the negative axis. The second integral, which is conditionally convergent for $n = 3$, is evaluated by residues prior to the application of the boundary conditions. The first integrand $\leq \mathrm{O}(1/\lambda^6)$ as $|\lambda| \to \infty$, and differentiation under the integral sign is permissible, after which we can set $X = 0$.

Conditions (4.2.12) now yield

$$A_2 = -K^2\Big/\left[\frac{\pi}{2}(i-1)+\sigma_2\mathcal{I}_2\right], \quad A_3 = -K^3\Big/\left[\frac{\pi}{2}(i+1)+\sigma_2\mathcal{I}_3\right],$$

where

$$\mathcal{I}_n = 2\int_0^{\infty}\frac{\lambda^{2n}\,d\lambda}{[(\lambda^4-1)\sqrt{\lambda^2-\mu^2}-\sigma_2](\lambda^4-1)}, \tag{4.2.14}$$

and the integration contour runs below singularities on the real axis. This integral is readily evaluated numerically by first rotating the path of integration through 45° onto the contour Γ^* (Figure 4.2.2) in the lower half-plane, along which the integrand decreases smoothly and rapidly. No singularities of the integrand are encountered during the rotation.

The acoustic efficiency $10 \times \log(\Pi_a/\Pi)$ (dB) calculated from (4.2.9) for a closed crack in a steel plate in water is plotted against frequency (dashed curve) in Figure 4.2.3a. The sound power rises to a maximum just above the coincidence frequency, when about 50% of the incident bending wave power is converted into sound. The fraction of the incident power transmitted across the

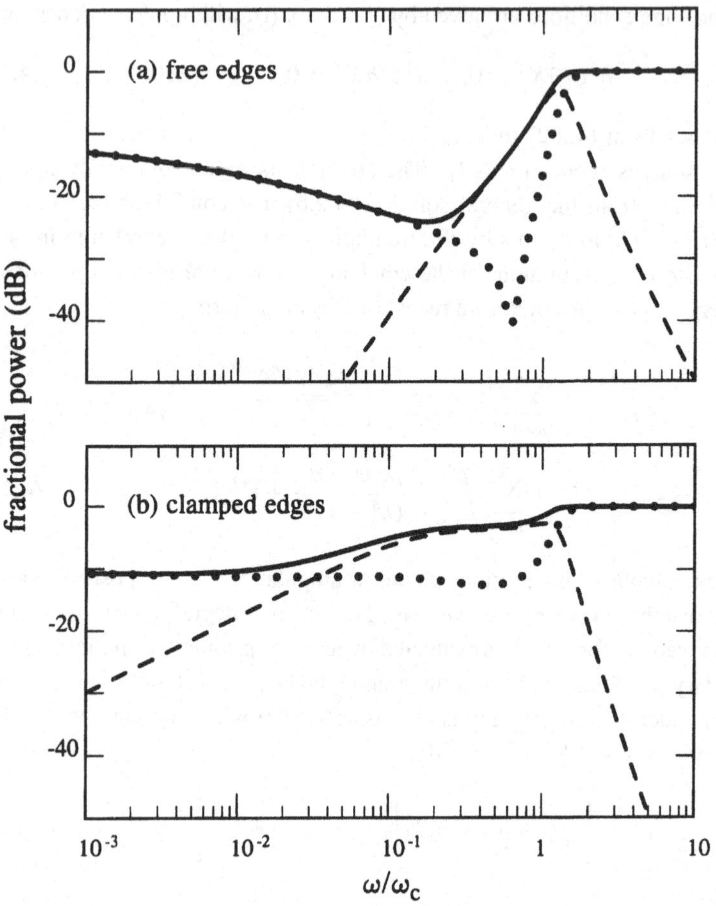

Figure 4.2.3. Scattering by (a) a closed crack and (b) a clamped joint for a steel plate in water: •••• transmitted power; – – – – sound power; and —— total bending wave power loss.

crack is $10 \times \log(|T|^2)$ (dB), shown dotted in the figure. Bending wave energy is transmitted without loss when $\omega > \omega_c$, when the scattered sound falls rapidly to zero.

The solid curve is the total loss of power $10 \times \log(1 - |R|^2)$ (dB) by transmission across the crack and the generation of sound. At very low and high frequencies, the lost power goes into the transmitted bending wave; the losses are dominated by radiation into the fluid over an intermediate range of frequencies extending up to the coincidence frequency.

Example 2: Scattering by a clamped joint. When the edges are clamped, conditions (1.4.16a), (iii), require

$$\zeta = 0, \quad \partial\zeta/\partial X = 0, \quad X \to \pm 0.$$

Deduce by the method of Example 1 that $A_2 = A_3 = 0$,

$$A_0 = -1 \Big/ \left[\frac{\pi}{2}(i-1) + \sigma_2 \mathcal{I}_0\right],$$

$$A_1 = -K \Big/ \left[\frac{\pi}{2}(i+1) + \sigma_2 \mathcal{I}_1\right]. \qquad (4.2.15)$$

The corresponding acoustic and transmitted powers and the total incident wave power loss are shown in Figure 4.2.3b for a steel plate in water. The joint is now equivalent to monopole and dipole sources, whose radiation efficiencies at low frequencies are considerably greater than the quadrupole and octupole sources of the closed crack.

For both the closed crack and clamped joint, the acoustic intensity peaks at the coincidence frequency, when roughly half of the incident power is scattered into sound. The acoustic directivities at $\omega = \omega_c$ are shown in Figure 4.2.4. These curves are drawn to the same scale and are plots of the integrand of (4.2.11). The radiation normal to the plate vanishes unless the monopole coefficient $A_0 \neq 0$.

Example 3. Show that in the absence of fluid loading ($\sigma_2 = 0$, $K = 1$), T = 0, |R| = 1 for a closed crack and a clamped joint, thereby confirming that the incident wave is totally reflected.

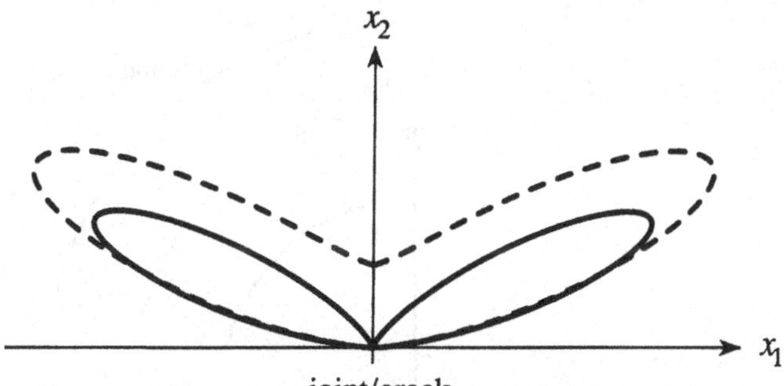

Figure 4.2.4. Directivity of the scattered sound when $\omega = \omega_c$ for a steel plate in water. ———, closed crack; – – – –, clamped joint.

Example 4. Consider the scattering of a bending wave

$$\zeta = \zeta_o e^{i(\kappa_1 x_1 + \kappa_3 x_3 - \omega t)}, \qquad \kappa = \sqrt{\kappa_1^2 + \kappa_3^2}, \qquad \omega > 0,$$

obliquely incident on a straight joint at $x_1 = 0$. Show that no sound is generated unless $|\kappa_3| < \kappa_o$. Deduce that at low frequencies the flexural waves that are scattered into sound tend to be incident normally on the joint.

Example 5. A time-harmonic point force $F_2(\omega) = -F(\omega)\delta(x_1 + d)\delta(x_3)$ is applied to a thin elastic plate at distance d from a joint along the x_3-axis. Show that in the acoustic far field the sound produced by the interaction of the force with the joint has the general form

$$p'(\mathbf{x}, t) \approx \frac{F(\omega)\rho_o m\omega^4 \operatorname{sgn}(x_2) e^{-i\omega[t]}}{2\pi K_o |\mathbf{x}|}$$

$$\times \int_{-\infty}^{\infty} \frac{\sum_{n=0}^{3} \left(\frac{\mu x_1}{|\mathbf{x}|}\right)^n A_n e^{ikd}\, dk}{\mathcal{D}\left(\frac{\kappa_o\sqrt{x_1^2+x_3^2}}{|\mathbf{x}|}, \omega\right)\mathcal{D}\left(\sqrt{k^2 + \frac{k^2 x_3^2}{|\mathbf{x}|}}, \omega\right)}, \qquad |\mathbf{x}| \to \infty.$$

The coefficients $A_n \equiv A_n(k, \kappa_o x_3/|\mathbf{x}|)$ may be determined by the method of Example 1 with the help of conditions (1.4.15).

4.2.3 Line Force and Moment Loads

When a continuous plate is subjected to force $F(x_3, t)$ and moment $\mathrm{M}(x_3, t)$ distributions concentrated along the x_3-axis (Figure 4.2.5), ζ and $\nabla\zeta$ are continuous, and the definitions (4.2.2) imply that the bending wave equation

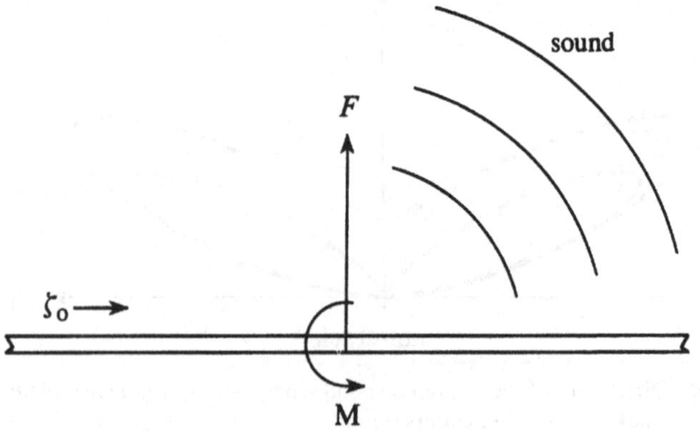

Figure 4.2.5. Line force and moment loading.

(4.2.1) includes only monopole and dipole sources. If F is measured in the x_2-direction, and M is reckoned positive with respect to the positive direction of the x_3-axis, it follows from (1.4.14) that

$$a_0 = F, \quad a_1 = \mathrm{M}. \tag{4.2.16}$$

For example, we can take $F = -\mathrm{m}\partial^2\zeta(0,t)/\partial t^2$ and $\mathrm{M} = 0$ when the load consists of an inelastic rib of mass m per unit length with no rotational inertia, rigidly attached to the plate.

Suppose the plate is restrained along $x_1 = 0$ by a restoring force $F = -\Lambda_f\,\zeta(0,t)$ per unit length, and a couple $\mathrm{M} = -\Lambda_m\partial\zeta(0,t)/\partial x_1 \equiv -\Lambda_m\zeta'(0, t)$, where Λ_f, Λ_m are known spring constants. For a time-harmonic incident wave (4.2.3) we have, in (4.2.1),

$$a_0 = -\Lambda_f\zeta(0,\omega), \quad a_1 = -\Lambda_m\zeta'(0,\omega), \quad a_2 = a_3 = 0,$$

and the flexural displacement is given by

$$\zeta(x_1,\omega) = \frac{-1}{2\pi}\int_{-\infty}^{\infty}\frac{[\Lambda_f\zeta(0,\omega) + ik\Lambda_m\zeta'(0,\omega)]\mathrm{e}^{ikx_1}}{\mathcal{D}(k,\omega)}\,dk + \zeta_o\mathrm{e}^{i\kappa x_1}. \tag{4.2.17}$$

The values of $\zeta(0,\omega)$ and $\zeta'(0,\omega)$ are determined by setting $x_1 = 0$ in this equation and in the equation obtained by differentiating with respect to x_1. When the results are expressed in terms of the source amplitudes A_n, we find $A_2 = A_3 = 0$ and

$$A_0 = \frac{-K_o\Lambda_f/2\pi m\omega^2}{\left[1 + \frac{K_o\Lambda_f}{2\pi m\omega^2}\left(\frac{\pi}{2}(i-1) + \sigma_2\mathcal{I}_0\right)\right]},$$

$$A_1 = \frac{KK_o^3\Lambda_m/2\pi m\omega^2}{\left[1 - \frac{K_o^3\Lambda_f}{2\pi m\omega^2}\left(\frac{\pi}{2}(i+1) + \sigma_2\mathcal{I}_1\right)\right]},$$

where $\mathcal{I}_0$ and $\mathcal{I}_1$ are defined in (4.2.14). These expressions can be used to calculate the scattered sound and bending waves. In the limits in which the spring constants Λ_f and Λ_m are large, displacement and rotational motions of the plate cannot occur at $x_1 = 0$, and A_0 and A_1 reduce to their respective values (4.2.15) for a clamped joint.

Example 6. Determine the sound generated when the bending wave (4.2.3) is incident on the spring-mass-dashpot line load at $x_1 = 0$ shown in Figure 4.2.6.

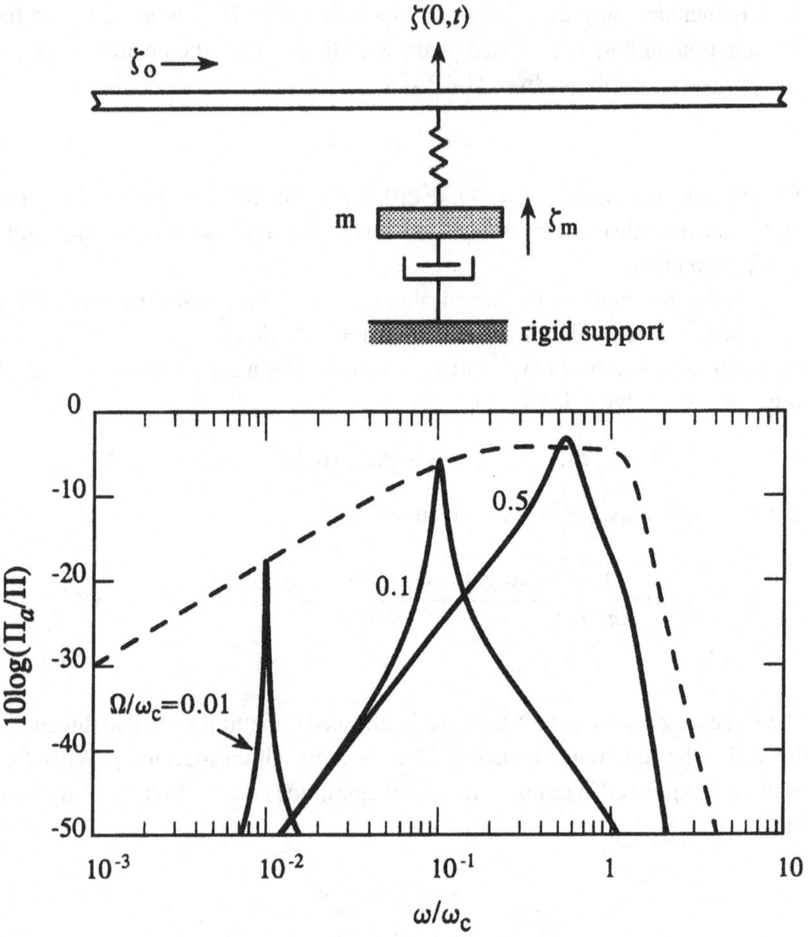

Figure 4.2.6. Resonant line load.

Assume the displacement ζ_m of the mass m (per unit length) satisfies

$$\frac{d^2\zeta_m}{dt^2} + 2\chi\frac{d\zeta_m}{dt} = \Omega^2(\zeta(0,t) - \zeta_m),$$

where χ is a damping coefficient, Ω is the natural frequency of the mass when the plate is clamped at $x_1 = 0$, and $\zeta(0,t) - \zeta_m$ is the extension of the spring. A force $F = -\mathrm{m}\Omega^2(\zeta(0,t) - \zeta_m)$ is exerted on the plate, and the moment M may be assumed to vanish.

When $\chi \ll \Omega$, the multipole coefficients are

$$A_0 = \frac{\frac{-\mu}{2\pi}\left(\frac{\rho_o \mathrm{m}}{\epsilon\rho_s mh}\right)\Omega^2}{(\omega + i\chi)^2 - \Omega^2\left[1 - \frac{\mu}{2\pi}\left(\frac{\rho_o \mathrm{m}}{\epsilon\rho_s mh}\right)\left[\frac{\pi}{2}(i-1) + \sigma_2\mathcal{I}_0\right]\right]},$$

$$A_1 = A_2 = A_3 = 0,$$

where h is the plate thickness, ϵ is the fluid loading parameter, and ρ_s is the mass density of the material of the plate.

When the resonance frequency Ω is fixed and the mass loading becomes large ($\mathrm{m}/mh \to \infty$)

$$A_0 \approx -1\bigg/\left[\frac{\pi}{2}(i-1) + \sigma_2\mathcal{I}_0\right].$$

In this limit, the displacement ζ vanishes along the load line $x_1 = 0$, but the plate can still rotate about the x_3-axis (where $\partial\zeta/\partial x_1 \neq 0$); the corresponding acoustic power $10 \times \log(\Pi_a/\Pi)$ (dB) calculated from (4.2.9) is illustrated by the dashed curve in Figure 4.2.6 for a steel plate in water. The solid curves represent the acoustic spectra for the finite mass ratio $\mathrm{m}/mh = 2$ for different values of Ω/ω_c and $\chi = 0$. The peaks occur at the resonance frequencies of the line load, where the sound power is approximately the same as for the infinite mass load.

4.2.4 Open Cracks, Apertures, and Point Loads

When a bending wave impinges on an open joint, fluid is pumped through the gap between the edges, which then behaves as a monopole sound source. The scattering problem for small gaps can be formulated as an extension of the method of Section 4.2.1 [263]. The details are too involved to be given here, and we shall be content to discuss briefly two examples where the inhomogeneity is localized to a small region of the plate.

Example 7: Scattering by a point mass load. If a concentrated mass m is attached to the plate at $x_1 = x_3 = 0$, the bending wave equation can be taken in the form (4.2.1) with the sources replaced by

$$-\mathrm{m}\frac{\partial^2\zeta}{\partial t^2}(\mathbf{0}, t)\delta(x_1)\delta(x_3).$$

When the bending wave (4.2.3) is incident on the mass, the scattered pressure is

$$p'(\mathbf{x},\omega) = \frac{-i\,\mathrm{sgn}(x_2)\rho_o \mathrm{m}\omega^4}{(2\pi)^2}\int_{-\infty}^{\infty}\frac{\zeta(\mathbf{0},\omega)\mathrm{e}^{i(\mathbf{k}\cdot\mathbf{x}+\gamma(k)|x_2|)}}{\gamma(k)\mathcal{D}(k,\omega)}\,d^2\mathbf{k},$$

where, from the solution of the bending wave equation (4.2.1) applied at $x_1 = x_3 = 0$,

$$\zeta(\mathbf{0},\omega) = \frac{\zeta_o}{1+\frac{\mathrm{m}K_o^2}{4\pi m}\left[\frac{i\pi}{2}+\sigma_2\mathcal{I}_{1/2}\right]},$$

and $\mathcal{I}_{1/2}$ is defined as in (4.2.14). Deduce that

$$|p'^2| \approx \frac{(\rho_o\omega^2\mathrm{m}/m)^2|\zeta(\mathbf{0},\omega)|^2\mu^4\cos^2\Theta}{8\pi^2|\mathbf{x}|^2[\mu^4\cos^2\Theta(\mu^4\sin^4\Theta-1)^2+4\epsilon^2]}, \quad |\mathbf{x}|\to\infty, \tag{4.2.18}$$

where Θ is the angle between the radiation direction and the normal to the plate. At very low frequencies, $\omega/\omega_c \ll \epsilon$, the directivity is that of a dipole normal to the plate. When $\mathrm{m}K_o^2 \gg m$, the strength of the scattered sound and bending wavefields become independent of the loading.

Example 8: Circular aperture. Determine the sound generated when a bending wave impinges on a small circular aperture.

Let the incident bending wave be given by (4.2.3), and let the aperture radius be R where $\kappa R \ll 1$. When viewed from the side $x_2 > 0$ of the plate, the flow through the aperture into the upper region may be attributed to a monopole adjacent to a uniform continuous plate. Let us write this source in the form $\alpha(\omega)\delta(x_1)\delta(x_2-d)\delta(x_3)$, $d\to 0$, where $\alpha(\omega)$ is the unknown source strength.

As $d\to 0$, the first of equations (4.1.26) gives the pressure p'_α generated by this source in the form

$$p'_\alpha = \frac{-i\alpha(\omega)}{8\pi^2}\int_{-\infty}^{\infty}\frac{(1+\mathrm{R}(k,\omega))}{\gamma(k)}\mathrm{e}^{i(\mathbf{k}\cdot\mathbf{x}+\gamma(k)x_2)}\,d^2\mathbf{k}, \quad x_2 > 0. \tag{4.2.19}$$

With respect to the fluid in $x_2 < 0$, the aperture is point *sink* $-\alpha(\omega)\delta(x_1)\delta(x_2+d)\delta(x_3)$ $(d\to +0)$, which produces a pressure field $p'_{-\alpha}$, say, in the upper region, given by the second of (4.1.26) after first reversing the direction of the x_2-axis. Thus, the total scattered pressure in $x_2 > 0$ is $p_\alpha \equiv p'_\alpha + p'_{-\alpha}$. Repeating the argument for points in $x_2 < 0$, and combining with the incident evanescent pressure field, the net pressure in the vicinity of the aperture, where $|\mathbf{x}| \gg R$, becomes

$$p(\mathbf{x}, \omega) = -\mathrm{sgn}(x_2)\frac{i\rho_o\omega^2\zeta_o}{\gamma(k)}\mathrm{e}^{i(kx_1+\gamma(k)|x_2|)}$$
$$-\frac{i\,\mathrm{sgn}(x_2)\alpha(\omega)}{4\pi^2}\int_{-\infty}^{\infty}\frac{(\mathrm{B}k^4 - m\omega^2)\mathrm{e}^{i(\mathbf{k}\cdot\mathbf{x}+\gamma(k)|x_2|)}}{\gamma(k)\mathcal{D}(k,\omega)}\,d^2\mathbf{k}, \quad |\mathbf{x}| \gg R.$$

Near the aperture the pressure can be written

$$p(\mathbf{x}, \omega) = \beta(\omega)\left(\frac{R}{4} + \bar{Y}(\mathbf{x})\right),$$

where $\beta(\omega)$ is to be determined, and $\bar{Y}$ is the velocity potential of *incompressible* flow through the aperture introduced in Section 3.2, Example 13, which in the present notation satisfies

$$\bar{Y} \to \frac{-R^2}{2\pi|\mathbf{x}|}, \quad |\mathbf{x}| \gg R \text{ in } x_2 > 0;$$
$$\bar{Y} \to -\frac{R}{2} + \frac{R^2}{2\pi|\mathbf{x}|}, \quad |\mathbf{x}| \gg R \text{ in } x_2 < 0.$$

These two expressions for $p(\mathbf{x}, \omega)$ must agree in a common region of validity where $R \ll |\mathbf{x}| \ll 1/K_o$. The coefficients α and β are found by matching corresponding terms from the two representations in this overlap region. This gives, in particular, $\alpha = -4\rho_o\omega^2 R\zeta_o/\sqrt{\kappa^2 - \kappa_o^2}$, and the mean square acoustic pressure can then be obtained in the form

$$|p_\alpha^2| \approx \frac{2(\rho_o\omega^2|\zeta_o|R)^2\mu^4\cos^2\Theta(\mu^4\sin^4\Theta - 1)^2}{(\pi K_o|\mathbf{x}|)^2(K^2-\mu^2)[\mu^4\cos^2\Theta(\mu^4\sin^4\Theta - 1)^2 + 4\epsilon^2]}, \quad |\mathbf{x}| \to \infty, \qquad (4.2.20)$$

where Θ is the angle between the radiation direction and the plate normal. At low frequencies, this is a dipole; when $\omega/\omega_c > \epsilon$, it exhibits monopole directivity.

By considering the limiting form of the pressure (4.2.18) scattered by a point mass load as $\mathrm{m} \to \infty$, it is seen that the mean square acoustic pressure is of order $1/(K_o R)^2$ ($\gg$1) larger than the field scattered by the aperture.

4.3 Sources Adjacent to an Inhomogeneous Elastic Wall

The generation of sound by turbulent flow over an inhomogeneous elastic surface involves the excitation of structural vibrations and their subsequent scattering. The response of finite elastic panels to turbulent boundary layer forcing is dominated by resonance modes of the panel, the amplitude of the motion being largely controlled by structural damping, usually including damping in viscoelastic coatings or other attached "trim" [264–266]. It is therefore difficult to make general pronouncements regarding the production of sound, except to say that it tends to be dominated by scattering. This will be illustrated by considering the simplest possible configuration consisting of an elastic plate with a single rib stiffener. The generation of sound by scattering at the junction of different structural elements (e.g., at a change in wall thickness, stiffness, and so forth) can be treated similarly.

4.3.1 Boundary Layer Noise Determined by Reciprocity

The general problem is amenable to simple analytical treatment only when convection by the mean flow is ignored, and our discussion is limited to this case, which has important applications in underwater acoustics. For boundary layer generated sound, it is convenient to express the radiation in terms of the boundary layer blocked pressure (Section 3.4). To do this, we use the reciprocal theorem (Section 1.9.2) in the form

$$p(\mathbf{x},\omega)q^R(\omega) = -i\omega\zeta^R(\mathbf{y},\mathbf{x};\omega)F(\mathbf{y},\omega), \tag{4.3.1}$$

where $p(\mathbf{x},\omega)$ is the perturbation pressure at a point $\mathbf{x}$ in the fluid generated by a force $F(\mathbf{y},\omega)$ applied to the upper surface of the plate of Figure 4.3.1 (in the $-x_2$-direction) at $\mathbf{y}$, and $\zeta^R(\mathbf{y},\mathbf{x};\omega)$ is the flexural displacement of the plate at $\mathbf{y}$ produced by a point volume source of strength $q^R(\omega)$ at $\mathbf{x}$.

This formula is applied to the problem envisaged in the figure, in which a turbulent "gust" interacts with the plate in the vicinity of a structural inhomogeneity. By taking the volume source to have unit strength and the force to be the gust-generated blocked surface pressure p_s, the acoustic pressure is determined by the equation

$$p(\mathbf{x},\omega) = -i\omega\int_S p_s(y_1,y_3,\omega)\zeta^R(y_1,y_3,\mathbf{x};\omega)\,dy_1\,dy_3, \tag{4.3.2}$$

where $\zeta^R(y_1,y_3,\omega)$ is the flexural displacement at (y_1,y_3) produced by a unit volume source at $\mathbf{x}$.

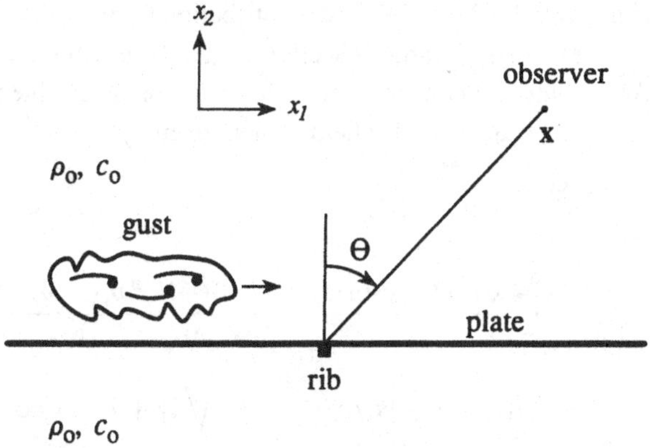

Figure 4.3.1. Gust adjacent to a rib-stiffened plate.

4.3.2 Rib-Stiffened Plate

This formula will be applied to the rib-stiffened plate of Figure 4.3.1. The plate is clamped to a rib along the x_3-axis so that the surface wetted by the turbulent flow may be assumed to be aerodynamically smooth. We make the further simplification that the blocked pressure is *independent* of x_3, (i.e., of position parallel to the rib). Because the plate is uniform in the x_3-direction, the reader will have no difficulty in extending the calculations to deal with more general surface force distributions. To determine ζ^R, we now need to consider a *line* volume source at $\mathbf{x} = (x_1, x_2)$, which may be assumed to be in the acoustic far field. The velocity potential of a unit line source is given by (1.7.10). As $\kappa_o|\mathbf{x}| \to \infty$, this can be approximated by

$$\frac{-\mathrm{e}^{i(\kappa_o|\mathbf{x}|+\frac{\pi}{4})}}{\sqrt{8\pi\kappa_o|\mathbf{x}|}}\mathrm{e}^{-i\kappa_o\mathbf{x}\cdot\mathbf{y}/|\mathbf{x}|}.$$

When the rib is ignored, this produces a flexural displacement (as a function of y_1)

$$\zeta_o^R = \zeta_o \mathrm{e}^{-i\kappa_o y_1 \sin\Theta}, \qquad \zeta_o = \frac{-\cos\Theta\mathcal{T}(\kappa_o\sin\Theta,\omega)\mathrm{e}^{i(\kappa_o|\mathbf{x}|+\frac{\pi}{4})}}{c_o\sqrt{8\pi\kappa_o|\mathbf{x}|}}, \tag{4.3.3}$$

where $\mathcal{T}(\kappa_o \sin\Theta, \omega)$ is the homogeneous plate transmission coefficient, defined as in (4.1.30), and $\mathbf{x} = |\mathbf{x}|(\cos\Theta, \sin\Theta)$.

The influence of the rib is now determined by the method of Section 4.2, by expressing the Fourier transform of scattered displacement field in the form (4.2.5). The total displacement ζ^R is given by a formula of the type (4.2.6) with K replaced by the nondimensional wavenumber $-(\kappa_o/K_o)\sin\Theta = -\mu\sin\Theta$

of the incident wave (4.3.3). If the weak contribution from the acoustic domain of the blocked pressure is discarded (Section 3.4), only the rib-generated part of ζ^R will make a nontrivial contribution to the radiation determined by (4.3.2), and the pressure in the acoustic far field is then given by

$$p(\mathbf{x},t) \approx \frac{i\cos\Theta}{2\sqrt{2\pi}\, c_o|\mathbf{x}|} \times \iint_{-\infty}^{\infty} \sum_{n=0}^{3} \frac{\sqrt{i\omega}\mathcal{T}(\kappa_o\sin\Theta,\omega) a_n(\omega)(-ik)^n p_s(k,\omega)\mathrm{e}^{-i\omega[t]}}{\mathcal{D}(k,\omega)}\, dk\, d\omega,$$

$$[t] = t - |\mathbf{x}|/c_o, \quad |\mathbf{x}| \equiv \sqrt{x_1^2 + x_2^2} \to \infty, \qquad (4.3.4)$$

where $\mathcal{T}(\kappa_o \sin\Theta,\omega)$ is defined as in (4.1.30), $\mathcal{D}(k,\omega)$ is the dispersion function (4.1.6), and $p_s(k,\omega) = (1/2\pi)^2 \iint_{-\infty}^{\infty} p_s(x_1,t)\mathrm{e}^{-i(kx_1-\omega t)}\, dx_1\, dt$ is the Fourier transform of the blocked pressure. In performing the integration, ω is taken to have a small positive imaginary part, and for $\omega < 0$, $a_n(\omega)$ is calculated from the relation (which follows from (4.2.5)) $a_n(-\omega) = a_n^*(\omega)$.

4.3.3 Time-Harmonic gust

The blocked pressure generated by a uniformly convecting quadrupole, which is independent of x_3, is equivalent to a superposition of elementary gusts of the form $p_s = p_o(\omega)\mathrm{e}^{-i\omega(t-x_1/v)}$, where v is the constant convection velocity, which we take to be subsonic. The sound is produced by the scattering by the rib of the flexural wave excited by the gust in a homogeneous plate. However, because the plate is infinite, its response to forcing by the gust will also be infinite when the gust wavenumber ω/v coincides with the real root of the bending wave dispersion equation (4.1.6). To avoid this resonance, the influence of structural damping must be included. This is conveniently done by introducing a loss factor $\eta(\omega)(>0$ for $\omega > 0)$ by requiring the bending stiffness of the plate to have the frequency-dependent form [26]

$$\mathrm{B} = \mathrm{B}_o[1 - i\eta(\omega)], \qquad (4.3.5)$$

where B_o is real, and $\eta(-\omega) = -\eta(\omega)$.

It now follows from (4.3.4) that the gust-generated sound is given by [267]

$$\frac{p(\mathbf{x},t)}{p_o\sqrt{h/|\mathbf{x}|}} \approx \frac{-\sqrt{\frac{\pi}{2}}\cos\Theta\mathcal{T}(\kappa_o\sin\Theta,\omega)M^4\left(\sum_{n=0}^{3} A_n(-\mu/M)^n\right)\mu^2\mathrm{e}^{-i(\omega[t]+\frac{\pi}{4})}}{\sqrt{\omega_c h/c_o}\left([(1-i\eta)\mu^4 - M^4]\mu^2 - \frac{2\epsilon M^5}{\sqrt{1-M^2}}\right)},$$

$$|\mathbf{x}| \to \infty, \qquad (4.3.6)$$

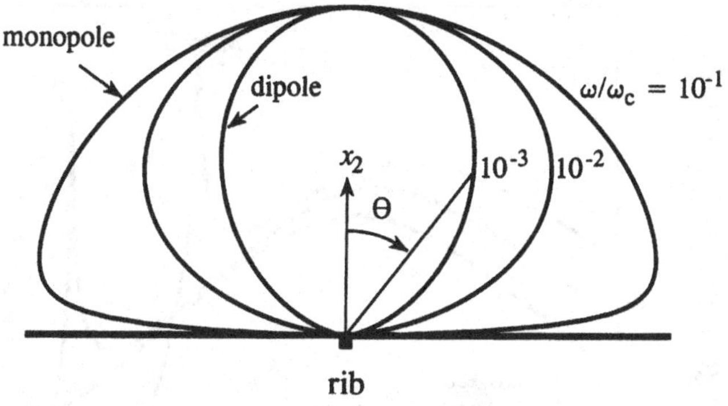

Figure 4.3.2. Directivity of sound scattered from a rib.

where h is the plate thickness, $M = v/c_o$ is the convection Mach number, and the coefficients A_n are defined as in (4.2.6).

The dependence of $p(\mathbf{x}, t)$ on the radiation direction Θ is governed by $\mathcal{T}(\kappa_o \sin\Theta, \omega)$ and the Θ-dependence of the scattering coefficients A_n. The influence of the latter is small, and it follows from the definition (4.1.30) of $\mathcal{T}$ that the radiation has a dipole directivity ($\propto \cos\Theta$) when $\omega/\omega_c < \epsilon$, whereas at higher frequencies ($<\omega_c$) it is more like a monopole. This is illustrated in Figure 4.3.2 for an aluminium plate in air (cf. Table 4.1.1) for clamped conditions at the rib (Section 4.2.2, Example 2). $|p(\mathbf{x}, t)|^2$ is plotted for $\mathrm{M} = 0.5$ (although the dependence on Mach number is small) and is scaled to a constant level in the normal direction $\Theta = 0°$. At very high frequencies (not shown) the directivity is ultimately dominated by the radiation from "leaky" flexural waves generated at the rib.

The efficiency of sound production at different frequencies is shown in Figure 4.3.3, where $10 \times \log(|p(\mathbf{x}, t)/p_o|^2/(h/|\mathbf{x}|))$ (dB) is plotted against ω/ω_c at $\Theta = 0°$. The resonance peaks occur near $\omega/\omega_c \approx M^2$, where the gust convection velocity is the same as the bending wave phase speed. The magnitudes of the peaks depend on the loss factor η, which has been assigned a nominal value of 0.01.

Example 1: Vortex interacting with a rib [267]. Calculate the sound produced when a line vortex of circulation $\Gamma > 0$ translates by self-induction over the rib in the infinite, thin elastic plate of Figure 4.3.1.

For a rigid plate, the translational velocity $v = \Gamma/4\pi d$, where d is the height of the vortex above the plate. Let us suppose that the fluid loading is sufficiently small ($\epsilon \ll 1$) such that the compliance of the plate may be neglected

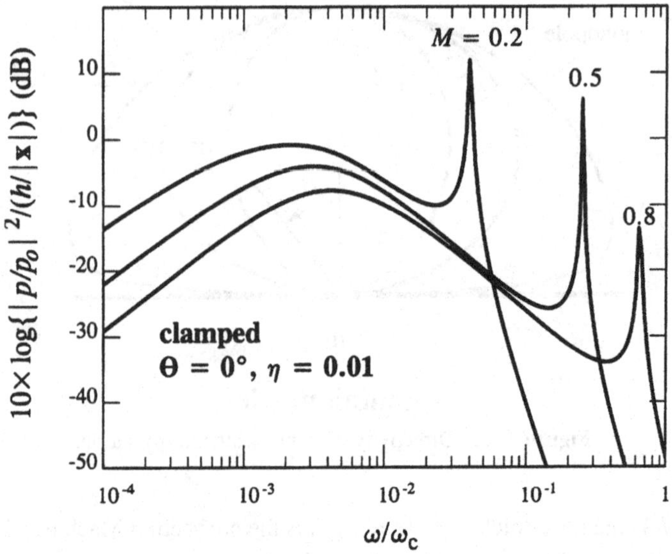

Figure 4.3.3. Efficiency of sound generation at a rib.

in calculating the motion of the vortex, at least during its interaction with the rib. Then the vortex exerts a steady, nonuniform blocked pressure on the plate that convects at speed v. Suppose the vortex passes over the rib (at $x_1 = 0$) at $t = 0$ and that the blocked pressure is the same as in incompressible flow (any differences due to compressibility will modify the overall level of the scattered sound but will not affect its general characteristics). Then [4]

$$p_s(x_1, t) = \rho_o v^2 \frac{4d^2[(x_1 - vt)^2 - d^2]}{[(x_1 - vt)^2 + d^2]^2},$$

$$p_s(k, \omega) = -2\rho_o d^2 |\omega| \mathrm{e}^{-|\omega| d/v} \delta(k - \omega/v). \tag{4.3.7}$$

Before calculating the sound from (4.3.4), let us examine the expected properties of the radiation by considering the flexural waves generated by the vortex, which produce the sound when they impinge on the rib. The flexural waves are given by the solution ζ_Γ of equation (1.4.9) when $F_2 = -p_s(x_1, t)$ and can be written

$$\zeta_\Gamma = 2\rho_o d^2 \int_{-\infty}^{\infty} \frac{|\omega| \mathrm{e}^{-|\omega| d/v - i\omega(t - x_1/v)}\, d\omega}{\left[\mathrm{B}\left(\frac{\omega}{v}\right)^4 - m\omega^2 - \frac{2i\rho_o\omega^2}{\gamma(\omega/v)}\right]}. \tag{4.3.8}$$

At large distances from the vortex, the value of this integral is dominated by residue contributions from poles near the ω-axis. These poles are located above the real axis at $\omega = \pm v^2\{m/[\mathrm{B}_o(1\mp i\eta)]\}^{1/2}$, provided ϵ is small. It follows that

$$\zeta_\Gamma \sim \zeta_\Gamma^o \sin[\Omega(t - x_1/v)]\mathrm{e}^{-\frac{1}{2}\eta\Omega(x_1/v-t)}, \quad x_1 - vt \to +\infty$$
$$\sim 0, \quad x_1 - vt \to -\infty,$$

where $\Omega = v^2\sqrt{m/\mathrm{B}_o} \equiv M^2\omega_c$ $(M = v/c_o)$, and ζ_Γ^o is a constant determined by the residues at the poles.

Ω is the frequency of the flexural wave resonantly excited by the vortex, whose phase velocity is v. However, the group velocity $\partial\omega/\partial k \approx 2k\sqrt{\mathrm{B}_o/m} \approx 2v$ when $k = \Omega/v$ so that energy imparted to the surface by the vortex runs *ahead* of the vortex. In the absence of structural damping this "forerunner" would extend infinitely far beyond the current location $x_1 = vt$ of the vortex, and its interaction with the rib would produce time-harmonic radiation of constant amplitude and frequency Ω at all times prior to the arrival of the vortex over the rib. Because the flexural wave energy in the wake of the vortex is negligible, the acoustic radiation should decay rapidly after the vortex passes the rib.

These qualitative statements are borne out by numerical evaluation of (4.3.4). The acoustic pressure signatures for clamped conditions at the rib and for different convection Mach numbers $M = v/c_o$ are plotted in Figure 4.3.4 for an aluminium plate in air when $h/d = 0.1$, $\eta = 0.05$, and $\Theta = 0°$. The radiation is dominated by the resonance frequency Ω; the acoustic amplitude increases exponentially with time until it is cutoff when the vortex passes over the rib. The growth rate is governed by the value of the structural loss factor: $\eta = 0.05$ is large for a homogeneous aluminium plate but is probably representative of damped fuselage structures of an aircraft. For a fixed value of η, the damping of the forerunner increases with frequency, and the low Mach number radiation therefore grows much more slowly than the higher Mach number waveforms.

Example 2: Green's function for a surface roughness element. A rigid, acoustically compact roughness element is attached to the surface $x_2 = +0$ of the elastic plate. Define the compact Green's function by the requirement that the diameter of the element is small compared to both the characteristic acoustic wavelength and the wavelengths of flexural waves. Show that in the absence of mean flow

$$G(\mathbf{x}, \mathbf{y}, t-\tau)$$
$$\approx \frac{1}{8\pi^2|\mathbf{x}|^2}\int_{-\infty}^{\infty}\left(\mathrm{e}^{i\kappa_o|\mathbf{x}-\mathbf{Y}|} + \mathcal{R}(\kappa_o\sin\Theta,\omega)\mathrm{e}^{i\kappa_o|\mathbf{x}-\bar{\mathbf{Y}}|}\right)\mathrm{e}^{-i\omega(t-\tau)}\,d\omega,$$

where $\mathcal{R}(\kappa_o\sin\Theta,\omega)$ is defined in (4.1.29), and $\mathbf{Y} = (Y_1, y_2, Y_3)$, $\bar{\mathbf{Y}} = (Y_1, -y_2, Y_3)$, Y_i being the velocity potential of flow at unit speed past the roughness

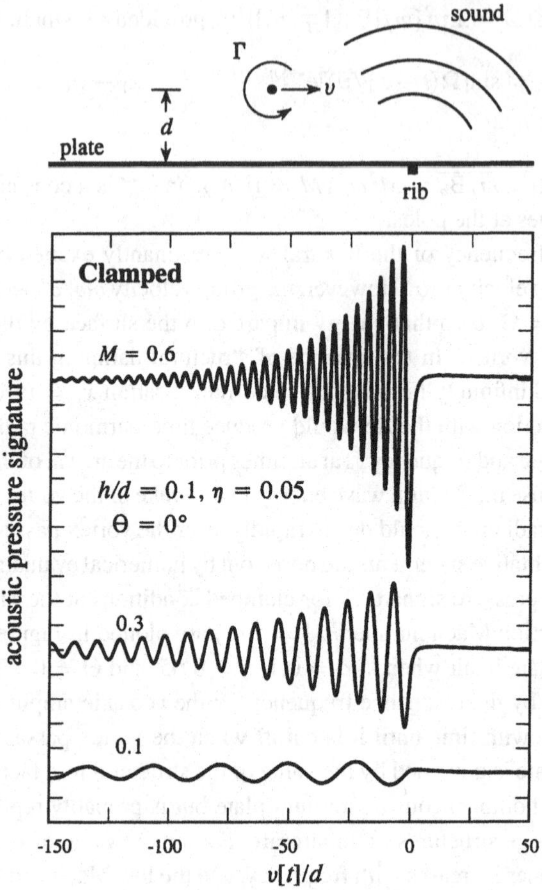

Figure 4.3.4. Pressure signatures (to the same nominal scale) of sound generated by vortex motion over a rib.

element in the i-direction ($i = 1$ and 3). The corresponding case of a roughness element on the face of an elastic half-space is discussed in [268].

Example 3: Green's function for a concentrated point mass. When the roughness element of Example 2 is replaced by a point mass m at the origin (the surface of the plate remaining aerodynamically smooth), show that (in the usual notation) the acoustic Green's function in the absence of mean flow is

$$G(\mathbf{x}, \mathbf{y}, t-\tau)$$

$$\approx \frac{-\mathrm{m}\cos\Theta}{2(2\pi)^4 c_o|\mathbf{x}|}\iint_{-\infty}^{\infty} \frac{\omega^3 \mathcal{T}(\kappa_o \sin\Theta, \omega) \mathrm{e}^{i(\mathbf{k}\cdot\mathbf{y}+\gamma(k)y_2-\omega[t])}}{\mathcal{Z}(\omega)\gamma(k)\mathcal{D}(k,\omega)}\, d^2\mathbf{k}\, d\omega,$$

$$|\mathbf{x}| \to \infty.$$

where $[t] = t - |\mathbf{x}|/c_o$, and

$$\mathcal{Z}(\omega) = 1 - \frac{\mathrm{m}K_o^2}{2\pi m}\int_0^\infty \frac{\lambda\, d\lambda}{[(1-i\eta)\lambda^4 - 1 - \sigma_2/\sqrt{\lambda^2-\mu^2}]}.$$

Example 4. Show that wall pressure fluctuations p_s applied to the infinite plate of Example 3 generate the acoustic far field

$$p(\mathbf{x},t) \approx \frac{-i\mathrm{m}\cos\Theta}{8\pi^2 c_o|\mathbf{x}|}\iint_{-\infty}^{\infty} \frac{\omega^3\mathcal{T}(\kappa_o\sin\Theta,\omega)p_s(\mathbf{k},\omega)\mathrm{e}^{-i\omega[t]}}{\mathcal{Z}(\omega)\mathcal{D}(k,\omega)}\, d^2\mathbf{k}\, d\omega.$$

4.4 Scattering of Flexural Waves at Edges

To understand the influence of surface compliance on trailing edge generated sound, it is necessary to examine two distinct types of interaction. The first involves the incidence of flexural waves on an edge, having been generated by the flow or other source of energy at locations far from the edge; the second considers the role of turbulence sources in the immediate vicinity of the edge.

The first problem is discussed in this section. Two idealized configurations are amenable to straightforward analysis, the results of which can be used to interpret more complicated situations. The simplest is an extension of our previous discussion and models an elastic edge by a semi-infinite, thin elastic plate; the case of a semi-infinite circular cylindrical elastic shell can be treated similarly [269], but it will not be discussed. These problems can both be solved by the Wiener–Hopf procedure, already outlined in Section 3.5 (Example 1) in connection with rigid trailing edge noise. The method is applicable generally to semi-infinite, wave-bearing structures and will be discussed in detail for the thin plate.

4.4.1 Edge Scattering of Bending Waves: General Solution [135, 247, 271–273]

We consider the basic problem of calculating the sound produced when a bending wave is incident on the edge of an elastic plate. The general solution will be shown to involve two arbitrary constants whose values are determined in any particular case by the mechanical restraints on plate motion at the edge.

Let the plate occupy $x_1 < 0$, $x_2 = 0$ and the bending wave (4.2.3) be incident on the edge (Figure 4.4.1). The phase velocity ω/κ is subsonic, and the flexural motion is accompanied by the evanescent pressures

$$p_\mathrm{I}(\mathbf{x},\omega) = -\frac{i\rho_o\omega^2\mathrm{sgn}(x_2)\zeta_o}{\gamma(\kappa)}\mathrm{e}^{i(\kappa x_1+\gamma(\kappa)|x_2|)}, \quad \mathbf{x} = (x_1, x_2). \qquad (4.4.1)$$

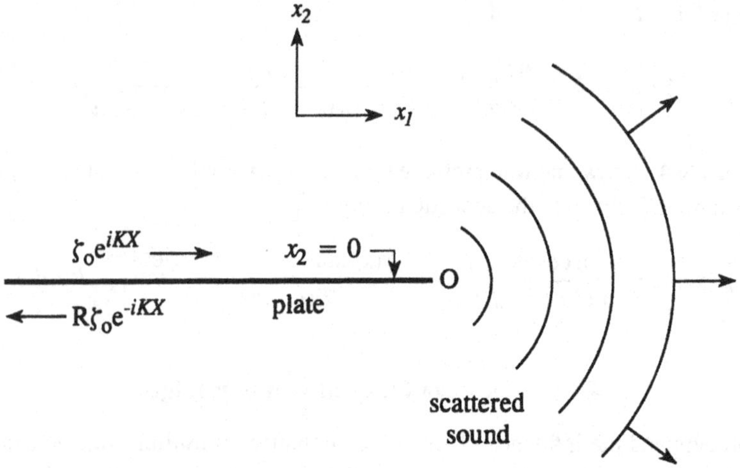

Figure 4.4.1. Scattering of a bending wave at an edge.

The edge-scattered displacements and pressure are written as Fourier integrals, and the total displacement and pressure are expressed in the form

$$\frac{\zeta(x_1,\omega)}{\zeta_o} = e^{i\kappa x_1} + \frac{i}{\kappa_o^2}\int_{-\infty}^{\infty} \gamma(k)\bar{\Im}(k)e^{ikx_1}\,dk, \tag{4.4.2}$$

$$p(\mathbf{x},\omega) = p_{\mathrm{I}}(\mathbf{x},\omega) + \mathrm{sgn}(x_2)\rho_o c_o^2\zeta_o \int_{-\infty}^{\infty} \bar{\Im}(k)e^{i(kx_1+\gamma(k)|x_2|)}\,dk, \tag{4.4.3}$$

where $\bar{\Im}(k)$ is an unknown dimensionless function of k, and $\gamma(k) = \sqrt{\kappa_o^2 - k^2}$ is defined in the usual way (see (1.13.9)). The integral in (4.4.3) is a solution of the Helmholtz equation with outgoing wave behavior and also satisfies (4.1.4) at $x_2 = \pm 0$. The most general form of $\bar{\Im}(k)$ is determined by requiring (4.4.2) and (4.4.3) to satisfy the bending wave equation (1.4.9) (with $F_2 = 0$) for $x_1 < 0$ and the condition that the pressure is finite and continuous everywhere.

Introduce the change of variable $k = \lambda K_o$ in the integrals (4.4.2) and (4.4.3) (where K_o is the vacuum wavenumber (4.1.2)), and substitute for ζ and $[p]$ in the bending wave equation (1.4.9) to obtain the first condition

$$\int_{-\infty}^{\infty} \mathcal{W}(\lambda)\Im(\lambda)e^{i\lambda X}\,d\lambda = 0, \qquad X \equiv K_o x_1 < 0, \tag{4.4.4}$$

where $\Im(\lambda) = \bar{\Im}(\lambda K_o)$ and

$$\mathcal{W}(\lambda) = (\lambda^4 - 1)\sqrt{\lambda^2 - \mu^2} - \sigma_2. \tag{4.4.5}$$

Branch cuts for $\sqrt{\lambda^2 - \mu^2}$ in the complex λ-plane are taken in the upper and

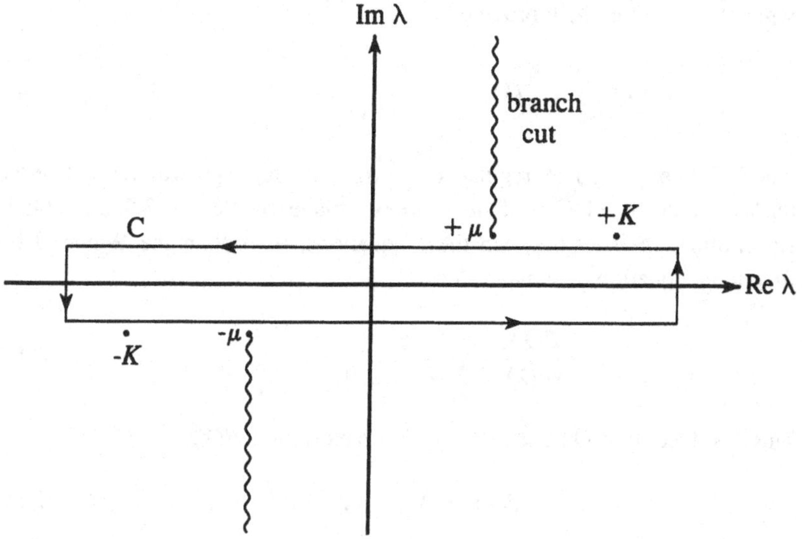

Figure 4.4.2. The λ-plane.

lower half-planes respectively from $\lambda = \pm\mu$ to $\pm i\infty$, such that $\sqrt{\lambda^2 - \mu^2}$ is positive on the real axis when $|\lambda| > \mu$ and *negative* imaginary for $|\lambda| < \mu$ (Figure 4.4.2). This choice is dictated by causality (Section 1.7.4) by assigning to ω (and therefore μ) a small positive imaginary part.

The pressure will be continuous throughout the fluid provided the limiting values as $x_2 \to \pm 0$ of the representation (4.4.3) are equal for $x_1 > 0$. Setting $K = \kappa/K_o$, and noting that

$$e^{iKX} = (1/2\pi i)\int_{-\infty}^{\infty} e^{i\lambda X} d\lambda/(\lambda - K - i0), \quad \text{for } X > 0,$$

this condition can be written

$$\int_{-\infty}^{\infty}\left(\Im(\lambda) - \frac{\beta}{(\lambda - K - i0)}\right)e^{i\lambda X} d\lambda = 0, \quad X > 0,$$

where

$$\beta = \frac{\mu^2}{2\pi i\sqrt{K^2 - \mu^2}}. \tag{4.4.6}$$

Equations (4.4.4) and (4.4.6) constitute dual integral equations for $\Im(\lambda)$, which can be solved by the Wiener–Hopf procedure [204–206], according to which they are satisfied, provided

$$\mathcal{W}(\lambda)\Im(\lambda) = \mathcal{L}(\lambda), \quad \Im(\lambda) - \frac{\beta}{(\lambda - K - i0)} = \mathcal{U}(\lambda), \tag{4.4.7}$$

where $\mathcal{L}(\lambda)$ and $\mathcal{U}(\lambda)$ are regular functions of algebraic growth, respectively in the lower and upper halves of the complex λ-plane (see Section 3.5, Example 1). By eliminating $\Im(\lambda)$ between these equations, we obtain the Wiener–Hopf functional equation in standard form:

$$\frac{\mathcal{L}(\lambda)}{\mathcal{W}(\lambda)} - \frac{\beta}{\lambda - K - i0} = \mathcal{U}(\lambda). \tag{4.4.8}$$

This is solved for $\mathcal{L}(\lambda)$ and $\mathcal{U}(\lambda)$ by first factorizing $\mathcal{W}(\lambda)$:

$$\mathcal{W}(\lambda) = \mathcal{W}_+(\lambda)\mathcal{W}_-(\lambda), \tag{4.4.9}$$

where $\mathcal{W}_\pm(\lambda)$ are respectively regular and nonzero in $\text{Im}\,\lambda \gtrless 0$. This is done in Example 1.

For *real* values of λ, we can then rewrite (4.4.8) as the following relation between $\mathcal{L}(\lambda)$ and $\mathcal{U}(\lambda)$

$$\begin{aligned} &\frac{\mathcal{L}(\lambda)}{\mathcal{W}_-(\lambda)} - \frac{\beta\mathcal{W}_+(K)}{(\lambda - K - i0)} \\ &\quad = \mathcal{U}(\lambda)\mathcal{W}_+(\lambda) + \frac{\beta}{(\lambda - K - i0)}[\mathcal{W}_+(\lambda) - \mathcal{W}_+(K)]. \end{aligned} \tag{4.4.10}$$

The left- and right-hand sides of this equation are respectively regular functions of λ in the lower and upper halves of the λ-plane. They accordingly define a function that is analytic everywhere and that, by Liouville's theorem [21], is a polynomial in λ. Either of equations (4.4.7) therefore supplies

$$\Im(\lambda) = \frac{\beta\mathcal{W}_+(K)}{\mathcal{W}_+(\lambda)}\left[\frac{1}{\lambda - K - i0} + \sum_{n\geq 0} A_n\lambda^n\right], \tag{4.4.11}$$

where the polynomial coefficients A_n remain to be determined.

The degree of the polynomial is found from the condition that the pressure loading $[p]$ of the plate must vanish as the edge is approached ($x_1 \to -0$). According to (4.4.1), $[p_{\mathrm{I}}]/\rho_o c_o^2 = -4\pi i\beta K_o\zeta_o e^{iKX}$. Thus, by recasting the integral in (4.4.3) in terms of nondimensional variables, and deforming the path of integration to pass *above* the pole at $\lambda = K$ of $\Im(\lambda)$, we find

$$\frac{[p]}{\rho_o c_o^2} = 2K_o \zeta_o \beta \mathcal{W}_+(K) \times \int_{-\infty}^{\infty} \frac{1}{\mathcal{W}_+(\lambda)} \left(\frac{1}{\lambda - K + i0} + \sum_{n \geq 0} A_n \lambda^n \right) e^{i\lambda X} d\lambda, \qquad (4.4.12)$$

where the notation $1/(\lambda - K + i0)$ is a reminder that the pole is now *below* the integration contour. When considered as a generalized function [19, 20] this expression for $[p]$ vanishes identically for $X > 0$. However, $|\mathcal{W}_+(\lambda)| \sim O(|\lambda|^{\frac{5}{2}})$ when $|\lambda| \to \infty$ (Example 1), which means that $[p]$ is singular at $X = 0$ unless the maximum value of the exponent $n \leq 1$. In this case, $[p]$ is continuous at $X = 0$ and vanishes as $X \to -0$.

Thus, when the integrand of (4.4.2) is expressed in terms of λ, and the integration contour is shifted to pass *above* the pole of $\Im(\lambda)$ at $\lambda = K$, the edge scattering problem for the displacement has the *general solution*

$$\frac{\zeta(X, \omega)}{\zeta_o} = \frac{-\beta \mathcal{W}_+(K)}{\mu^2} \int_{-\infty}^{\infty} \frac{\sqrt{\lambda^2 - \mu^2}}{\mathcal{W}_+(\lambda)} \times \left(\frac{1}{\lambda - K + i0} + A_0 + A_1 \lambda \right) e^{i\lambda X} d\lambda, \quad X < 0, \qquad (4.4.13)$$

where the arbitrary coefficients A_0 and A_1 remain to be determined by the mechanical conditions at the edge $X = 0$.

The Reflected Bending Wave

As $X \to -\infty$ on the plate, the displacement is dominated by the incident and reflected bending waves, and

$$\zeta / \zeta_o \approx e^{i\kappa x_1} + \mathrm{R} e^{-i\kappa x_1}, \qquad (4.4.14)$$

where the reflection coefficient R is determined by the residue contribution to (4.4.13) from the pole at $\lambda = -K$ (where $\mathcal{W}_+(\lambda) = 0$) just below the real axis. Substituting for β from (4.4.6) and calculating the residue, we find

$$\mathrm{R} = \frac{\mathcal{W}_+^2(K)}{\mathcal{W}'(K)} \left(\frac{1}{2K} - A_0 + A_1 K \right), \qquad (4.4.15)$$

where $\mathcal{W}'(\lambda) = \partial \mathcal{W}(\lambda)/\partial \lambda$.

The Edge-Generated Sound

The flexural wave power Π incident on unit length of the edge is given by (4.1.17). The reflected power is $|\mathrm{R}|^2\Pi$ so that the efficiency of sound generation at the edge of the plate is

$$\Pi_a/\Pi = 1 - |\mathrm{R}|^2, \tag{4.4.16}$$

where Π_a is the acoustic power radiated from unit length of edge.

The sound power can also be calculated by integrating the acoustic intensity $\langle p'^2\rangle/\rho_o c_o$ over the surface of a large circular cylinder of radius $|\mathbf{x}|$ centered on the edge of the plate, where p' is the scattered pressure, given by the integrated term in (4.4.3). The angle brackets denote a time average, which is evaluated after first restoring the time factor $\mathrm{e}^{-i\omega t}$ in (4.4.3) and taking the real part. As $|\mathbf{x}| \to \infty$, the integral may be replaced by its stationary phase approximation given by (1.13.13). Then

$$\Pi_a = \pi\rho_o c_o^2|\zeta_o|^2\omega \int_{-\pi}^{\pi} |\tilde{\Im}(\kappa_o\cos\theta)|^2 \sin^2\theta \, d\theta, \tag{4.4.17}$$

where θ is the radiation angle measured from the continuation of the plate (the positive x_1-axis); the integrand is proportional to the power radiated in the direction θ within the angular element $d\theta$. Replacing $\tilde{\Im}(\kappa_o\cos\theta)$ by its nondimensional representation (4.4.11), and using (4.1.17), we also have

$$\frac{\Pi_a}{\Pi} = \frac{\epsilon\mu|\mathcal{W}_+(K)|^2}{\pi K(5K^4 - 4K^2\mu^2 - 1)} \times \int_{-\pi}^{\pi} \frac{|A_0 + A_1\mu\cos\theta + 1/(\mu\cos\theta - K)|^2\sin^2\theta}{|\mathcal{W}_+(\mu\cos\theta)|^2} \, d\theta. \tag{4.4.18}$$

The acoustic efficiency can be calculated from either of equations (4.4.16) or (4.4.18). Equation (4.4.16) is more convenient for this purpose, but (4.4.18) also yields the directivity of the radiation because the integrand is proportional to the fraction of the incident power radiated in the direction θ.

Example 1: Factorization of $\mathcal{W}(\lambda)$

Let

$$\mathcal{W}(\lambda) = (\lambda^2 - K^2)(1+\lambda^2)^{\frac{3}{2}}\mathcal{Z}(\lambda),$$

where

$$\mathcal{Z}(\lambda) = \frac{(\lambda^4 - 1)\sqrt{\lambda^2 - \mu^2} - \sigma_2}{(\lambda^2 - K^2)(1+\lambda^2)^{\frac{3}{2}}}. \tag{4.4.19}$$

Then $\mathcal{Z}(\lambda)$ is regular and nonzero on the real λ-axis, and $\mathcal{Z} \to 1$ as $\lambda \to \pm\infty$.

Now, if $f(\lambda)$ is regular within and on the boundary of a closed contour C in the λ-plane, Cauchy's integral theorem [21] states that when λ lies inside C,

$$f(\lambda) = (1/2\pi i) \oint_C f(\xi)\, d\xi/(\xi - \lambda),$$

where the integration is around C in the anticlockwise sense. When $\mathrm{Im}\, \omega > 0$, the singularities of $\mathcal{Z}(\lambda)$ at the branch points $\lambda = \pm\mu$ lie *off* the real axis (Figure 4.4.2) (the points $\pm K$ are also displaced from the real axis, but $\mathcal{Z}(\lambda)$ is regular at these points). Hence, we can take $f(\lambda) = \ln \mathcal{Z}(\lambda)$, and let C be the rectangle shown in the figure with vertices $\pm L \pm i\delta$, where $0 < \delta < \mathrm{Im}\, \mu$. By considering the limits $L \to \infty$, $\delta \to +0$, it now follows that

$$\mathcal{Z}(\lambda) = \mathcal{Z}_+(\lambda)\mathcal{Z}_-(\lambda),$$

where

$$\mathcal{Z}_\pm(\lambda) = \exp\left(\pm\frac{1}{2\pi i}\int_{-\infty}^{\infty} \frac{\ln \mathcal{Z}(\xi)\, d\xi}{\xi - \lambda}\right) \quad \text{for } \mathrm{Im}\, \lambda \gtrless 0, \tag{4.4.20}$$

where the value of $\mathcal{Z}_\pm(\lambda)$ on the real axis is the limiting value of (4.4.20) as $\mathrm{Im}\, \lambda \to \pm 0$. The integral in (4.4.20) is readily evaluated numerically [205, 273], so that $\mathcal{Z}_\pm(\lambda)$ may now be regarded as known. Hence, also

$$\mathcal{W}_\pm(\lambda) = \mathrm{e}^{\frac{i\pi}{4}}(i \pm \lambda)^{\frac{3}{2}}(K \pm \lambda)\mathcal{Z}_\pm(\lambda), \quad \mathrm{Im}\, \lambda \gtrless, \quad \mathrm{Im}\, K = +0. \tag{4.4.21}$$

It is clear that $\mathcal{W}_\pm(\pm\lambda) = \mathcal{W}_\mp(\mp\lambda)$, and it may also be verified that $|\mathcal{W}_\mp(\lambda)| \sim \mathrm{O}(|\lambda|^{\frac{5}{2}})$ as $|\lambda| \to \infty$ [205, 206].

4.4.2 Scattering at a Free Edge

If the edge of the plate is free to vibrate without restraint, the coefficients A_0, A_1 in the general solution (4.4.13) are obtained from the conditions (Section 1.4)

$$\frac{\partial^2 \zeta}{\partial X^2} = \frac{\partial^3 \zeta}{\partial X^3} = 0, \quad \text{as } X \to -0. \tag{4.4.22}$$

Details of their straightforward but involved calculation are given in Example 2.

The acoustic efficiency (4.4.16) and the directivity of the scattered sound (the integrand of (4.4.18)) are now easily evaluated. To do this, it is necessary to perform the integration in (4.4.20) (required in the factorization of $\mathcal{W}(\lambda)$)

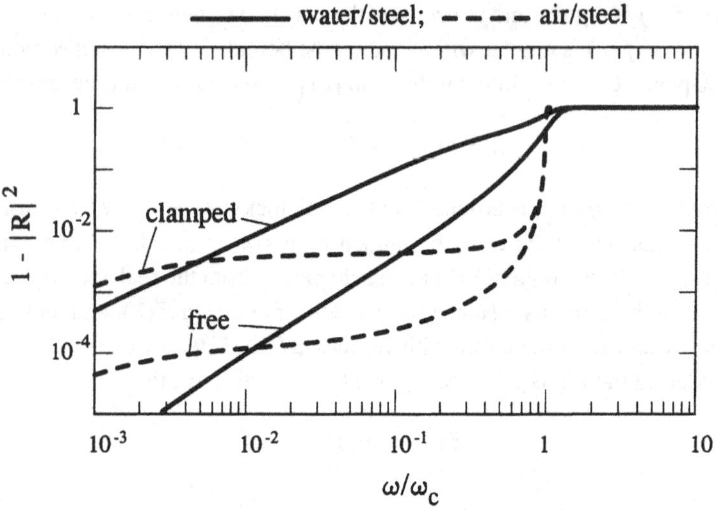

Figure 4.4.3. Acoustic efficiency of bending waves incident on the edge of a steel plate.

numerically for various values of λ [273]. When expressed as functions of ω/ω_c, the final results depend only on the value of the fluid loading parameter ϵ.

The acoustic efficiencies for steel plates in air and water are plotted against ω/ω_c in Figure 4.4.3. For light fluid loading ($\epsilon \ll 1$, air), the efficiency is very small when $\omega < \omega_c$ but rises precipitously as $\omega \rightarrow \omega_c$ as the reflection coefficient $\mathrm{R} \rightarrow 0$. The acoustic directivity in $x_2 > 0$ is shown in the left column of Figure 4.4.4 for a steel plate in water. The plots in this figure are of the integrand of (4.4.18) normalized such that the peak directivities are the same in all plots. At very low frequencies, the directivity $\sim \sin^2\theta$, which indicates that the edge is radiating like an acoustic dipole (oriented in the normal direction to the plate). This limiting behavior occurs when $\omega/\omega_c < \epsilon$, when A_0 and A_1 become independent of ω and $|\mathcal{W}_+(\mu\cos\theta)| \rightarrow K$ [273]. The peak directivity progressively rotates to higher angles with increasing frequency and flattens and develops two lobes. When ω exceeds the coincidence frequency, the phase velocity of the incident bending wave approaches the speed of sound in the fluid, and wave energy is conveyed principally in the evanescent, marginally subsonic motions on either side of the plate. At these frequencies $\mathrm{R} \rightarrow 0$, and ultimately the incident fluid-borne energy propagates directly across the edge into two "forward scattered" beams of sound, symmetrically on both side of the continuation of the plate; the null at $\theta = 0$ in Figure 4.4.4 is caused by destructive interference of the equal and opposite evanescent pressure waves incident on the edge on opposite sides of the plate. The residual peak for $\theta > 90°$ is produced by a leaky flexural wave excited at the edge, which, however, becomes negligible when $\omega \gg \omega_c$.

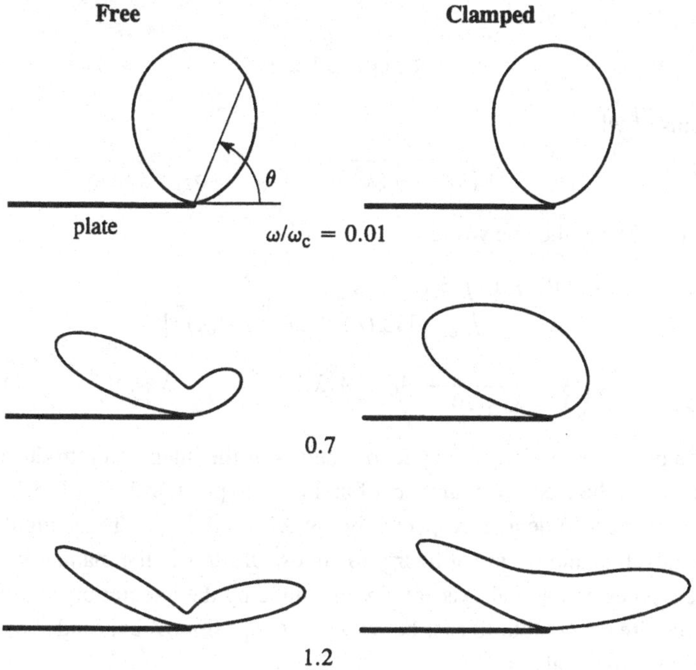

Figure 4.4.4. Directivity of sound produced when a bending wave impinges on the edge of a steel plate in water.

4.4.3 Scattering at a Clamped Edge

In this case,

$$\zeta = \frac{\partial \zeta}{\partial X} = 0, \quad \text{as } X \to -0. \tag{4.4.23}$$

A_0 and A_1 are determined in Example 3.

The plots in Figure 4.4.3 reveal that the radiation efficiency of a clamped edge is typically very much larger than for a free edge when $\omega < \omega_c$. The directivities (Figure 4.4.4) are similar; however, in particular, at low frequencies the field shape has the same characteristic dipole pattern.

Example 2: Scattering coefficients for a free edge. To apply the edge conditions (4.4.22) it is not permissible to differentiate under the integral sign in the formula (4.4.13) for ζ because the integrand $\sim O(1/\sqrt{\lambda})$ as $\lambda \to \pm\infty$. To overcome this difficulty the integrand is rearranged by adding and subtracting a term involving the nondimensional dispersion function of the plate in vacuo, namely the polynomial

$$\mathcal{P}(\lambda) \equiv \lambda^4 - 1,$$

in terms of which,

$$\mathcal{W}(\lambda) = \sqrt{(\lambda^2 - \mu^2\mathcal{P}(\lambda)} - \sigma_2,$$

and (4.4.13) can then be written

$$\frac{\zeta}{\zeta_o} = \frac{-\beta\mathcal{W}_+(K)}{\mu^2}\int_{-\infty}^{\infty}\left[\frac{\sigma_2}{\mathcal{W}_+(\lambda)\mathcal{P}(\lambda)} + \frac{\mathcal{W}_-(\lambda)}{\mathcal{P}(\lambda)}\right] \times\left(\frac{1}{\lambda - K + i0} + A_0 + A_1\lambda\right)e^{i\lambda X}\,d\lambda, \quad X < 0. \tag{4.4.24}$$

The presence of $\mathcal{P}(\lambda)$ in the denominators of the integrand introduces four new poles at the free wavenumbers of an in vacuo plate, at $\lambda = \pm 1, \pm i$, which it is convenient to denote, respectively, by $\pm\lambda_1$ and $\pm\lambda_2$. These singularities are removable, and we are at liberty to choose arbitrarily the manner in which the new poles on the real axis are circumvented by the integration contour; we shall require the contour to pass below λ_1 on the positive λ-axis and above $-\lambda_1$ on the negative real axis.

When $X < 0$, the part of the integral (4.4.24) involving the second term in the square brackets can be evaluated exactly as a sum of residues from the poles at $\lambda = -\lambda_1, -\lambda_2$ (there is no contribution from the pole at $\lambda = K - i0$ because $\mathcal{W}_-(K) \equiv 0$). The displacement then becomes

$$\frac{\zeta}{\zeta_o} = \frac{-\beta\mathcal{W}_+(K)}{\mu^2}\left\{\int_{-\infty}^{\infty}\frac{\sigma_2}{\mathcal{W}_+(\lambda)\mathcal{P}(\lambda)}\left(\frac{1}{\lambda - K + i0} + A_0 + A_1\lambda\right)e^{i\lambda X}d\lambda + \sum_j \frac{2\pi i\mathcal{W}_+(\lambda_j)}{\mathcal{P}'(\lambda_j)}\left(\frac{-1}{K + \lambda_j} + A_0 - A_1\lambda_j\right)e^{-i\lambda_j X}\right\}, \quad X < 0,$$

where $\mathcal{P}'(\lambda) = \partial\mathcal{P}(\lambda)/\partial\lambda$. The sum is over the zeros λ_1, λ_2 of $\mathcal{P}(\lambda)$, and in obtaining this formula it has been noted that $\mathcal{P}'(-\lambda) = -\mathcal{P}'(\lambda)$, $\mathcal{W}_-(-\lambda) = \mathcal{W}_+(\lambda)$. The remaining integrand is $O(1/\lambda^{11/2})$ as $\lambda \to \pm\infty$, and the integral is therefore absolutely convergent when differentiated under the integral sign up to four times, after which we can set $X = 0$ and evaluate the integral from its residues at λ_1, λ_2 *above* the integration contour.

This procedure yields the general formula,

$$\frac{\partial^n\zeta}{\partial X^n} = \frac{-2\pi i^{n+1}\zeta_o\beta\mathcal{W}_+(K)}{\mu^2}\left(A_0 S_n^{\pm} + A_1 S_{n+1}^{\mp} - T_n^{\mp}\right), \quad X \to -0, \quad 0 \le n \le 4, \tag{4.4.25}$$

where the upper and lower signs are taken accordingly as n is even or odd, and $S_n^\pm$ and $T_n^\pm$ are defined by

$$S_n^\pm = \sum_j \lambda_j^n \mathcal{C}_\pm(\lambda_j), \quad T_n^\pm = \sum_j \frac{\lambda_j^n[\lambda_j \mathcal{C}_\pm(\lambda_j) + K\mathcal{C}_\mp(\lambda_j)}{K^2 - \lambda_j^2},$$

where

$$\mathcal{C}_\pm(\lambda) = \frac{\sigma_2 \pm \mathcal{W}_+^2(\lambda)}{\mathcal{W}_+(\lambda)\mathcal{P}'(\lambda)},$$

and the summations are taken over the zeros λ_1 and λ_2 of $\mathcal{P}(\lambda)$.

Thus, the free-edge conditions yield the following linear equations for A_0 and A_1:

$$S_2^+ A_0 + S_3^- A_1 = T_2^-, \qquad S_3^- A_0 + S_4^+ A_1 = T_3^+.$$

Example 3: Scattering coefficients for a clamped edge. The edge conditions (4.4.23) and equation (4.4.25) give the following equations for A_0 and A_1

$$S_0^+ A_0 + S_1^- A_1 = T_0^-, \quad S_1^- A_0 + S_2^+ A_1 = T_1^+.$$

Example 4: Scattering at a spring-loaded edge. Consider the acoustic efficiency and directivity when the edge is loaded uniformly by a resonant mass-spring system. If m is the mass load per unit length of the edge, and Ω is the resonance frequency when the edge is clamped, take the edge conditions to be

$$\frac{\partial^3\zeta}{\partial X^3} - \frac{\mathrm{m}K_o/m\zeta}{1 - (\omega + i\chi)^2/\Omega^2} = 0, \quad \frac{\partial^2\zeta}{\partial X^2} = 0, \quad X \to -0,$$

where χ is a damping coefficient.

4.5 Aerodynamic Sound Generated at an Elastic Edge

Turbulent pressures acting on the edge of an elastic plate generate sound and structural vibrations. For large fluid loading (in water), the sound power produced at the edge at frequencies below coincidence tends to be smaller than for a nominally identical flow over a rigid edge [274, 275]. This decrease may be offset in practice by the sound subsequently generated at remote locations by the diffraction of structural waves produced by the turbulence at the edge and by sound generated when structural modes arriving from remote regions impinge on the edge.

We can expect fluid loading to modify the production of sound by turbulence near the edge when the frequency is small enough for the heavy fluid loading condition (4.1.12), $\omega/\omega_c < \epsilon^2$, to be satisfied. However, in order for the edge to be identified as a source of noise, as opposed to merely generating sound by scattering incident structural waves, it is also necessary that the edge should be sufficiently "rigid" to interact effectively with the turbulence sources. This means that, on the length scale U/ω of the local turbulence pressure fluctuations, the plate should appear to be rigid. The condition for this is $K_o \ll \omega/U$. Thus, the following frequency range will tend to be dominated by edge-generated sound,

$$K_o \ll \frac{\omega}{U} \Rightarrow M^2 = \left(\frac{U}{c_o}\right)^2 \ll \frac{\omega}{\omega_c},$$

and the effects of fluid loading will be significant in the subrange,

$$M^2 \ll \frac{\omega}{\omega_c} < \epsilon^2. \tag{4.5.1}$$

This frequency range can be large in underwater applications in which the Mach number is usually less than 0.01 and $\epsilon \sim 0.1$.

When this criterion is satisfied, the wavelengths of both the edge-generated sound and the structural waves are much larger than the turbulence scale, and the details of the motion near the edge may be calculated as if the plate were rigid and the fluid incompressible. The problem of calculating the flow noise source is then decoupled from that of calculating the radiation from the edge, and simple closed-form expressions for the generated sound can often be found in terms of a specified edge source. We shall do this by introducing a compact Green's function applicable to source flows whose distance from the edge is much smaller than both the characteristic acoustic and bending wavelengths.

4.5.1 Compact Green's Function

We consider flow at infinitesimal Mach number and define Green's function $G(\mathbf{x}, \mathbf{y}, t - \tau)$ (as in Section 4.1) to be the outgoing field generated by a point source $\mathcal{F}(\mathbf{x}, t) = \delta(\mathbf{x} - \mathbf{y})\delta(t - \tau)$ on the right of the pressure equation (1.6.8). The source at $\mathbf{y}$ generates both sound and bending waves, and in the compact approximation the source distance $r = \sqrt{y_1^2 + y_2^2}$ from the edge must be much smaller than both the acoustic and bending wavelengths (Figure 4.5.1). The observer at $\mathbf{x}$ may be assumed to be many acoustic wavelengths from the edge. In particular, we shall concentrate on cases in which $\omega/\omega_c < \epsilon^2$, where fluid loading effects are expected to be large.

The disturbed motion of the plate is governed by equations (4.1.25) for each component of frequency ω, and a Fourier integral representation of G can be

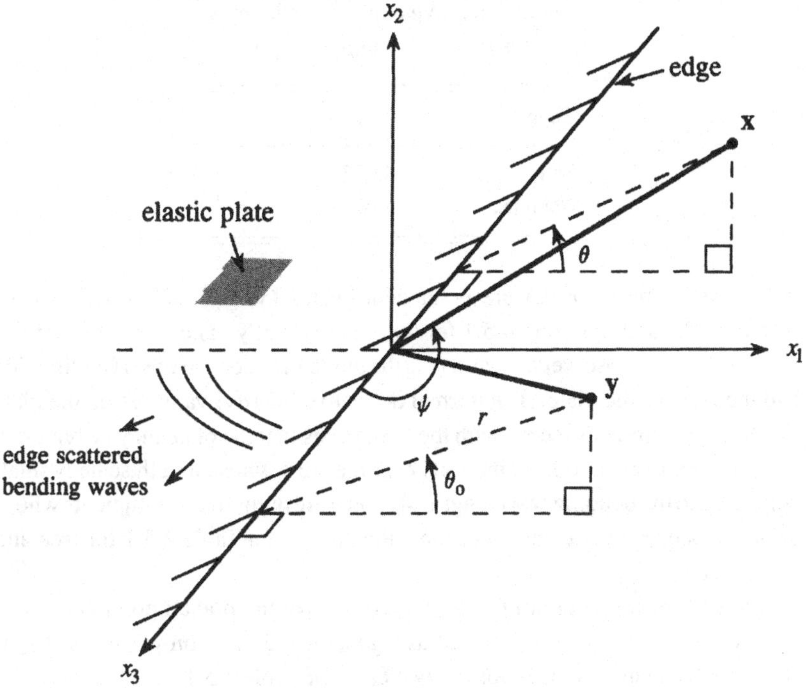

Figure 4.5.1. Coordinates defining the Green's function for an elastic plate.

derived by the Wiener–Hopf procedure, by extension of the method described in Section 4.4. As for the rigid plate (Section 3.2.4), the leading order term in the expansion of G in powers of r near the edge does not depend on r and determines the monopole sound from the source. This term is unimportant for turbulence quadrupoles at very low Mach numbers and may be discarded; the radiation produced by the edge interaction is governed by the second term in the expansion, which is of order $\sqrt{r}$. The details of the calculation of this term are too lengthy to be given here (see [275]), and it must suffice to quote the general expression (without the monopole term) in the limit of heavy fluid loading. The compact Green's function for $\omega/\omega_c < \epsilon^2$ is as follows:

$$G(\mathbf{x}, \mathbf{y}, t-\tau) \approx \frac{\mathcal{A}_n e^{-i(n+1)\frac{\pi}{2}}}{4\pi^2 \epsilon^{(2n+\frac{1}{2})}} \frac{\sin\theta \sin\psi \sin(\theta_o/2)}{|\mathbf{x} - \mathbf{i}_3 y_3|} \sqrt{\frac{\omega_c r}{2\pi c_o}}$$
$$\times \int_{-\infty}^{\infty} \left(\frac{\omega}{\omega_c}\right)^{n+1} e^{-i\omega(t-\tau-|\mathbf{x}-\mathbf{i}_3 y_3|/c_o)} d\omega,$$
$$K_o r \ll 1, \quad |\mathbf{x} - \mathbf{i}_3 y_3| \to \infty. \qquad (4.5.2)$$

Table 4.5.1. *Numerical coefficients for free and clamped edges*

Edge	$\mathcal{A}_n$	n
Free	1.097	$\frac{1}{5}$
Clamped	1.475	$-\frac{1}{5}$

In this formula, (r, θ_o) are polar coordinates $(y_1, y_2) = r(\cos\theta_o, \sin\theta_o)$ that are indicated in Figure 4.5.1 for the source point $\mathbf{y}$. The angle θ is defined similarly for the observer at $\mathbf{x}$, and ψ is the angle between the observer direction and the edge of the plate. The normal derivative $\partial G/\partial y_2$ vanishes on the plate $(\theta_o = \pm\pi)$. This is in accord with the hypothesis that the plate may be regarded as rigid with respect to fluid motions near the edge whose length scale is small compared to the bending wavelength. $\mathcal{A}_n$ and n are numerical coefficients whose values depend on the edge conditions; they are given Table 4.5.1 for free and clamped edges.

The path of integration in (4.5.2) passes above the branch point at $\omega = 0$, and the integral must be interpreted as a generalized function when $n + 1 \geq 0$ [19], in which case we have for the two cases of Table 4.5.1:

Free edge,

$$G(\mathbf{x}, \mathbf{y}, t - \tau) \approx \frac{\mathcal{A}_n}{2\pi^2\epsilon^{(2n+\frac{1}{2})}} \frac{\sin\theta \sin\psi \sin(\theta_o/2)}{|\mathbf{x} - \mathbf{i}_3 y_3|} \sqrt{\frac{\omega_c r}{2\pi c_o}} \times \frac{(n-1)! \sin(n\pi)}{\omega_c^{n+1}} \frac{\partial^2}{\partial t^2}\left(\frac{\mathrm{H}(t - \tau - |\mathbf{x} - \mathbf{i}_3 y_3|/c_o)}{(t - \tau - |\mathbf{x} - \mathbf{i}_3 y_3|/c_o)^n}\right); \quad (4.5.3)$$

Clamped edge,

$$G(\mathbf{x}, \mathbf{y}, t - \tau) \approx \frac{\mathcal{A}_n}{2\pi^2\epsilon^{(2n+\frac{1}{2})}} \frac{\sin\theta \sin\psi \sin(\theta_o/2)}{|\mathbf{x} - \mathbf{i}_3 y_3|} \sqrt{\frac{\omega_c r}{2\pi c_o}} \times \frac{n! \sin(|n|\pi)}{\omega_c^{n}} \frac{\partial}{\partial t}\left(\frac{\mathrm{H}(t - \tau - |\mathbf{x} - \mathbf{i}_3 y_3|/c_o)}{(t - \tau - |\mathbf{x} - \mathbf{i}_3 y_3|/c_o)^{n+1}}\right). \quad (4.5.4)$$

Note that $x_2/|\mathbf{x}| = \cos\Theta = \sin\theta \sin\psi$, where Θ is the angle between the observer direction $|\mathbf{x}|$ and the x_2-axis. Thus, the edge-generated sound has a dipole directivity with the dipole axis normal to the plate. This is characteristic of diffraction at the edge of an elastic plate at low frequencies [247], and also occurs when a bending wave is scattered at an edge (Figure 4.4.4).

Example 1: Two-dimensional Green's function. The pressure generated by an impulsive line source $\mathcal{F}(\mathbf{x}, t) = \delta(x_1 - y_1)\delta(x_2 - y_2)\delta(t - \tau)$ in (1.6.8) in the limit of heavy fluid loading is deduced from (4.5.2) by integrating over $-\infty < y_3 < \infty$. This can be done by the method of stationary phase (Section 1.13.2), when the observer is far from the edge, yielding

$$G(\mathbf{x}, \mathbf{y}, t - \tau) \approx \frac{\mathcal{A}_n \mathrm{e}^{-i(n+1)\frac{\pi}{2}}}{4\pi^2 \epsilon^{(2n+\frac{1}{2})}} \sin\theta \sin(\theta_o/2)\sqrt{\frac{r}{|\mathbf{x}|}}$$

$$\times \int_{-\infty}^{\infty} \left(\frac{\omega}{\omega_c}\right)^{n+\frac{1}{2}} \mathrm{e}^{-i\omega(t-\tau-|\mathbf{x}|/c_o)}\, d\omega,$$

$$= \frac{\mathcal{A}_n}{2\pi^2 \epsilon^{(2n+\frac{1}{2})}} \sin\theta \sin(\theta_o/2)\sqrt{\frac{r}{|\mathbf{x}|}}$$

$$\times \frac{\left(n - \frac{1}{2}\right)! \cos(n\pi)}{\omega_c^{n+\frac{1}{2}}} \frac{\partial}{\partial t}\left(\frac{\mathrm{H}(t - \tau - |\mathbf{x}|/c_o)}{(t - \tau - |\mathbf{x}|/c_o)^{n+1}}\right), \qquad (4.5.5)$$

where $|\mathbf{x}| = \sqrt{x_1^2 + x_2^2} \to \infty$.

4.5.2 Trailing Edge Noise

Let us apply these results to estimate the noise produced by low Mach number turbulent flow over an elastic trailing edge for frequencies in the range (4.5.1), where the influence of fluid loading is expected to be significant. The edge is modeled by the semiinfinite elastic plate $-\infty < x_1 < 0,\ x_2 = 0$, as in Figure 3.5.1, and a spanwise section of the edge of length L (centered on the origin) is wetted by the turbulence. The turbulent flow is assumed to be homentropic and confined to the "upper" surface ($x_2 > 0$) of the plate.

As in Section 3.5, we express the radiation in terms of the blocked wall pressure by considering the scattering of the hypothetical pressure p_{I} that would be produced by the same turbulent flow if the plate were absent. In the turbulence-free regions, where the mean flow is uniform at speed U in the x_1-direction and $M = U/c_o \ll 1$, convection of sound is ignored, and the pressure satisfies the homogeneous form of the wave equation (1.6.8) ($\mathcal{F}(\mathbf{x}, t) = 0$).

The scattered pressure $p'(\mathbf{x}, t)$ satisfies this equation everywhere so that Kirchhoff's formula (1.10.5) supplies the relation

$$p'(\mathbf{x}, t) = \int_{-\infty}^{\infty} d\tau \oint_S \left(p'(\mathbf{y}, \tau)\frac{\partial G}{\partial y_n}(\mathbf{x}, \mathbf{y}, t - \tau)\right.$$

$$\left. - G(\mathbf{x}, \mathbf{y}, t - \tau)\frac{\partial p'}{\partial y_n}(\mathbf{y}, \tau)\right) dy_1\, dy_3, \qquad (4.5.6)$$

where the surface integration is over both sides of the plate, with y_n measured outward from the plate.

The unknown pressure p' is eliminated from the surface integration by the following argument. A similar integral representation (1.10.5) can be written down for the total fluctuating pressure $p(\mathbf{x}, t) = p_I(\mathbf{x}, t) + p'(\mathbf{x}, t)$, in which the source term $\mathcal{F}(\mathbf{x}, t) \neq 0$ but represents the turbulence Reynolds stress sources in the boundary layer. However, G and $p(\mathbf{x}, t)$ both satisfy the elastic-plate boundary conditions, and the surface integral in this case must therefore vanish. Hence, observing also that $p_I(y_1, +0, y_3, \tau) = p_I(y_1, -0, y_3, \tau)$, and that $\partial p_I/\partial y_2$ is continuous across the plate, the representation (4.5.6) is actually equivalent to

$$p'(\mathbf{x}, t) = \iint_{-\infty}^{\infty} d\tau\, dy_3$$
$$\times \int_{-\infty}^{0} [G(\mathbf{x}, y_1, +0, y_3, t-\tau) - G(\mathbf{x}, y_1, -0, y_3, t-\tau)]$$
$$\times \left(\frac{\partial p_I}{\partial y_2}(\mathbf{y}, \tau)\right)_{y_2=0} dy_1. \tag{4.5.7}$$

The derivative $\partial p_I/\partial y_2$ at $y_2 = 0$ can be expressed in terms of the blocked pressure $2p_I(y_1, 0, y_3, \tau) = p_s(y_1, y_3, \tau)$ by differentiating the relation (3.5.2).

Thus, in the frequency range $M^2 \ll \omega/\omega_c < \epsilon^2$, $p'(\mathbf{x}, t)$ can be evaluated from (4.5.7) in terms of p_s by substituting for $G(\mathbf{x}, \mathbf{y}, t-\tau)$ from (4.5.2). When formally divergent integrals are interpreted as generalized functions, we find

$$p'(\mathbf{x}, t) \approx \frac{-A_n \sin\theta \sin\psi}{2\sqrt{2c_o}|\mathbf{x}|\epsilon^{2(n+\frac{1}{2})}(i\omega_c)^{(n+\frac{1}{2})}}$$
$$\times \int_{-\infty}^{\infty} \frac{\omega^{n+1}}{\sqrt{k_1 + i0}} p_s(k_1, \kappa_o \cos\psi, \omega) e^{-i\omega[t]}\, dk_1\, d\omega,$$
$$|\mathbf{x}| \to \infty, \tag{4.5.8}$$

where terms $\sim O(\kappa_o)$ have been neglected relative to the blocked pressure wavenumber k_1 (Section 3.4).

By assuming the blocked pressure is stationary random in both position and time (as in the discussion of trailing edge noise in Section 3.5), and by making use of the relation (3.5.11), the frequency spectral density of the sound $\Phi(\mathbf{x}, \omega)$ (defined in the second equation of (3.4.20)) is found to be

$$\Phi(\mathbf{x}, \omega) \approx \frac{A_n^2 \sin^2\theta \sin^2\psi}{8\pi |\mathbf{x}|^2 \epsilon^{(4n+1)}} \frac{\omega L}{c_o} \left(\frac{\omega}{\omega_c}\right)^{2n+1}$$
$$\times \int_{-\infty}^{\infty} \frac{P(k_1, \kappa_o \cos\psi, \omega)\, dk_1}{|k_1|}, \quad |\mathbf{x}| \to \infty, \tag{4.5.9}$$

where P(**k**, ω) is the blocked pressure wavenumber–frequency spectrum. This corresponds to the analogous formula (3.5.14) for a rigid edge.

The remaining integral is approximated in the usual way (Section 3.5.2) by taking the integrand to be small except in the vicinity of the convective peak of the wall pressure spectrum at $k_1 = \omega/U_c$, where U_c is the eddy convection velocity. Using the Corcos approximation (3.4.10), and setting $|k_1| = \omega/U_c$ in the denominator of the integrand, we find

$$\Phi(\mathbf{x}, \omega) \approx \frac{M_c L A_n^2 \sin^2\theta \sin^2\psi}{8\pi^2|\mathbf{x}|^2\epsilon^{(4n+1)}}\left(\frac{\omega}{\omega_c}\right)^{2n+1} \ell_3 \Phi_{pp}(\omega), \quad |\mathbf{x}| \to \infty, \tag{4.5.10}$$

where $M_c = U_c/c_o$. Because of the factor $(\omega/\omega_c)^{2n+1}$ in this expression, $\Phi(\mathbf{x}, \omega)$ decreases more rapidly with decreasing frequency than the corresponding spectrum (3.5.15) for a rigid edge. Furthermore, whereas the radiation intensity peaks in the upstream directions ($\theta = \pm\pi$) for the rigid edge, the low-frequency noise from the elastic edge has a dipole directivity, peaking in directions $\theta = \pm\pi/2$ normal to the surface.

A convenient comparison of the elastic- and rigid-plate radiated sound levels is obtained by calculating the spectral density $\Pi_a(\omega)$ of the total edge-generated sound (defined such that $\int_0^\infty \Pi_a(\omega)d\omega$ is the total edge-generated acoustic power). This is obtained by integrating $\Phi(\mathbf{x}, \omega)/\rho_o c_o$ over the surface of a large sphere of radius $|\mathbf{x}|$ enclosing the source region, which gives

$$\Pi_a(\omega) \approx \frac{M_c L A_n^2}{6\pi\rho_o c_o \epsilon^{(4n+1)}}\left(\frac{\omega}{\omega_c}\right)^{2n+1} \ell_3 \Phi_{pp}(\omega), \quad \omega \geq 0.$$

The sound power spectrum $\Pi_o(\omega)$ derived from (3.5.15) for a rigid edge is similarly found to be

$$\Pi_o(\omega) \approx \frac{M_c L}{4\rho_o c_o} \ell_3 \Phi_{pp}(\omega), \quad \omega \geq 0.$$

Thus, the order of magnitude of the influence of surface compliance on the edge noise is

$$\frac{\Pi_a}{\Pi_o} \sim \epsilon\left(\frac{\omega}{\epsilon^2\omega_c}\right)^{2n+1} = \left(\frac{\rho_o}{\rho_s}\right)\left(\frac{\omega}{\epsilon^2\omega_c}\right)^{2n+1} \Big/ \left(\frac{\omega_c h}{c_o}\right),$$

$$\text{for } M^2 \ll \frac{\omega}{\omega_c} < \epsilon^2, \tag{4.5.11}$$

where ρ_s is the density of the plate material. Table 4.5.1 shows that the

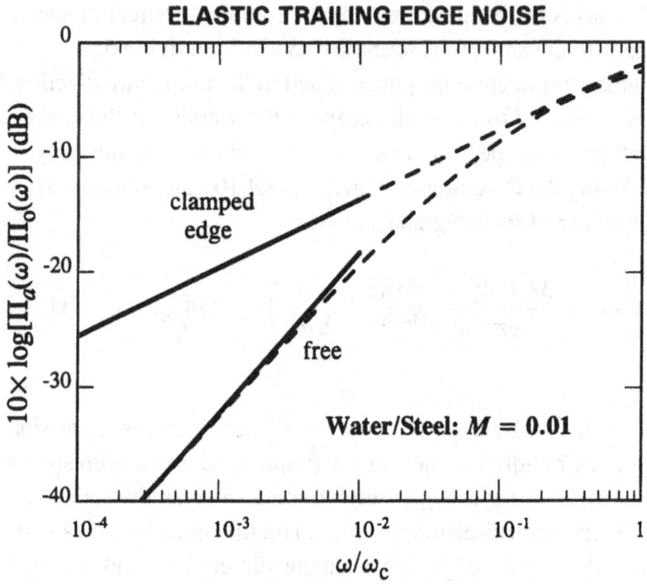

Figure 4.5.2. Edge noise sound relative to that from a rigid trailing edge.

clamped-edge noise exceeds that from the free edge by a factor of order $(\omega/\epsilon^2\omega_c)^{-\frac{4}{3}}$, which becomes large as ω/ω_c decreases. The solid portions of the curves in Figure 4.5.2 illustrate the low-frequency asymptotic behavior of $10 \log\, [\Pi_a(\omega)/\ \Pi_o(\omega)]$ (dB) in the interval $10^{-4} < \omega/\omega_c < 10^{-2}$ for a steel plate in water with free- or clamped-edge conditions. The fluid loading parameter $\epsilon \approx 0.135$, and the plots represent $\Pi_a(\omega)/\Pi_o(\omega) \equiv 2\mathcal{A}_n^2(\omega/\omega_c)^{(2n+1)}/3\pi\epsilon^{(4n+1)}$ with values of $\mathcal{A}_n$ and n taken from Table 4.5.1.

The identity

$$\frac{\omega}{\omega_c} = M\left(\frac{\omega\delta}{U}\right)\left(\frac{h}{\delta}\right)\Big/\left(\frac{\omega_c h}{c_o}\right),$$

where h and δ are the plate thickness and boundary layer thickness, implies that the peak in the rigid trailing edge noise spectrum (near $\omega\delta/U = 1$) occurs for $\omega/\omega_c < 10^{-2}$ when $M = 0.01$ and $h \leq \delta$, where according to Figure 4.5.2 the elastic-plate radiation is at least 15 dB smaller. More detailed calculations [274] indicate that $\Pi_a(\omega) < \Pi_o(\omega)$ for steel in water for all $\omega < \omega_c$, and the dashed curves in the figure are the estimated behaviors of $\Pi_a(\omega)/\Pi_o(\omega)$.

The contribution to the edge noise from resonant forcing of the plate by the boundary layer should occur at frequencies $\omega/\omega_c \leq O(M^2)$. In the absence of structural damping, the dominant flexural motions near the edge are then generated upstream of the edge by direct turbulence forcing of the surface.

4.5.3 Parallel Blade–Vortex Interaction Noise

A simple model for examining the influence of surface compliance on the sound generated by a leading-edge, parallel blade–vortex interaction is depicted in the upper part of Figure 4.5.3. A line vortex is aligned with the edge and convects at uniform velocity U in the x_1-direction parallel to the plate, where $M = U/c_o \ll 1$. The production of sound is governed in the low Mach number, linearized approximation by equation (3.3.1).

Let the vortex have circulation Γ about the x_3-axis (out of the plane of the paper in the figure), and take the coordinate origin at the edge. The vortex translates along a path at a fixed perpendicular distance ℓ from the plate, so that the linearized aeroacoustic source strength is

$$\boldsymbol{\omega} \wedge \mathbf{U} = \Gamma U \mathbf{i}_2 \delta(x_1 - Ut)\delta(x_2 - \ell),$$

where $\mathbf{i}_2$ is a unit vector in the x_2-direction, and time is measured from the instant at which the vortex crosses $x_1 = 0$, when its distance from the edge is the minimum value ℓ. If the plate were absent, this linear source would generate only the hydrodynamic near field of the vortex. When the vortex encounters the

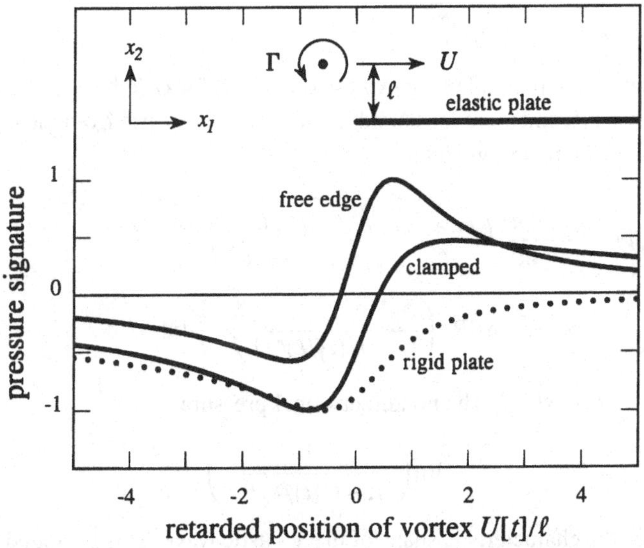

Figure 4.5.3. Parallel blade–vortex interaction acoustic pressure signatures.

plate, the interaction of its "upwash" with the edge generates both sound and bending waves. The characteristic frequency of the sound $\sim U/\ell$, and according to (4.1.12) the production of sound can be expected to be significantly affected by surface compliance when

$$M < \frac{\ell}{h}\frac{\omega_c h}{c_o}\epsilon^2,$$

where h is the plate thickness. If, in addition, the length scale ℓ is small compared to the characteristic bending wavelength ($K_o\ell \ll 1$), the motion of the vortex over the edge will be effectively the same as if the plate is rigid. Both of these conditions are satisfied when

$$M \ll \frac{h}{\ell}\bigg/\frac{\omega_c h}{c_o} < \frac{\epsilon^2}{M}, \tag{4.5.12}$$

which is also the condition of the validity of the two-dimensional version (4.5.5) of the compact Green's function. It is clear from Table 4.1.1 that (4.5.12) can be satisfied in water, provided h/ℓ is not too small, and we suppose this condition to be satisfied in the following discussion.

Then the acoustic pressure at large distances from the edge is given by (3.3.2) in the form

$$p(\mathbf{x}, t) \approx -\rho_o U\Gamma \int_{-\infty}^{\infty} \frac{\partial G}{\partial y_2}(\mathbf{x}, U\tau, \ell, t-\tau)\, d\tau,$$
$$|\mathbf{x}| = \sqrt{x_1^2 + x_2^2} \to \infty. \tag{4.5.13}$$

Evaluating the integral using expression (4.5.5) for G (where, in the present configuration θ_o and θ are measured in the clockwise sense from the upstream extension of the plate), we find

$$p(\mathbf{x}, t) = -\frac{A_n n!}{4\pi}\left(\frac{\rho_o U\Gamma}{\ell}\right)\sqrt{\frac{\ell}{\pi|\mathbf{x}|}}\left(\frac{U}{\ell\omega_c\epsilon^2}\right)^{n+\frac{1}{2}}$$
$$\times \sqrt{\epsilon}\sin\theta \operatorname{Re}\left(\frac{1}{(i - U[t]/\ell)^{n+1}}\right), \quad |\mathbf{x}| \to \infty, \tag{4.5.14}$$

where $[t] = t - |\mathbf{x}|/c_o$. The nondimensional pressure

$$-\operatorname{Re}\left(\frac{1}{(i - U[t]/\ell)^{n+1}}\right)$$

determines the characteristic shape of the acoustic wave. This is plotted in Figure 4.5.3 for free- and clamped-edge conditions (using the values of n from Table 4.5.1). In each case, the maximum absolute pressure is normalized to unity.

The integration with respect to τ in (4.5.13) and the functional form (4.5.5) of $G(\mathbf{x}, \mathbf{y}, t-\tau)$ show that the acoustic pressure received at $\mathbf{x}$ at time t depends on contributions to the integral from all times $\tau < t - |\mathbf{x}|/c_o$, which is typically the case for propagation in two dimensions (Section 1.7.3). An exception occurs, however, for a rigid plate where the compact Green's function assumes the special form (3.2.16) and which, in terms of the present coordinate system, becomes

$$G(\mathbf{x}, \mathbf{y}, t-\tau) = \sqrt{r}\,\sin(\theta_o/2)\frac{\sin(\theta/2)}{\pi\sqrt{|\mathbf{x}|}}\delta(t-\tau-|\mathbf{x}|/c_o), \quad |\mathbf{x}| \to \infty, \tag{4.5.15}$$

where θ_o and θ are measured in the clockwise sense from the upstream extension of the plate. In this limit, the radiation becomes

$$p(\mathbf{x}, t) = -\frac{1}{2\pi}\left(\frac{\rho_o U\Gamma}{\ell}\right)\sqrt{\frac{\ell}{|\mathbf{x}|}}\sin(\theta/2)\,\mathrm{Re}\left(\frac{1}{\sqrt{i - U[t]/\ell}}\right), \quad |\mathbf{x}| \to \infty.$$

The pressure signature $-\mathrm{Re}(1/\sqrt{i - U[t]/\ell})$ (again normalized to unit maximum amplitude) is plotted as the dotted curve in the Figure 4.5.3.

The first obvious effect of compliance, however, is the profound difference in the directivities of the sound radiated from the rigid and compliant edges. The dependence on $\sin\theta$ for the elastic edge under heavy fluid loading shows that the interaction radiation is equivalent to that produced by a dipole orientated normal to the plate. Second, in order of magnitude, the elastic-edge acoustic pressure is smaller than the rigid-edge pressure by a factor of order

$$\left(\frac{U}{\ell\omega_c\epsilon^2}\right)^{n+\frac{1}{2}}\sqrt{\epsilon},$$

which is the same as the estimate (4.5.11) for edge noise so that the relative mean square radiation pressures for the three cases will be much as indicated in Figure 4.5.2 for edge noise.

Example 2: Force on an oscillating airfoil. A rigid two-dimensional airfoil of chord $2a$ executes small-amplitude, vertical translational oscillations of amplitude $\zeta(\omega)$ of frequency ω in a uniform mean stream of speed U parallel to the plane of the airfoil (Figure 4.5.4). If the motion is incompressible, show that in the linearized approximation, the lift on unit span of the airfoil produced by the vorticity shed from the trailing edge to satisfy the Kutta condition is

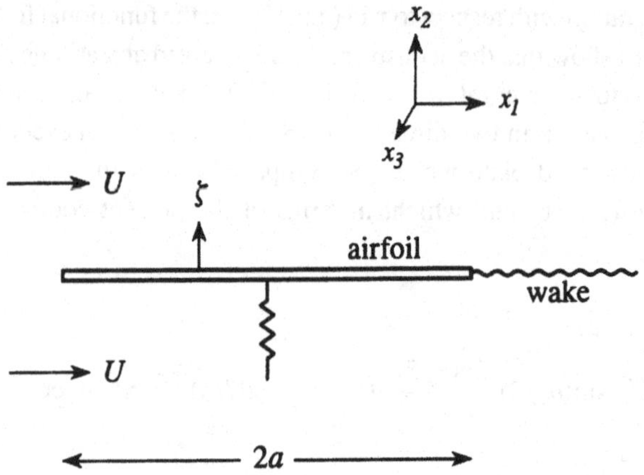

Figure 4.5.4. Airfoil vibrating in a mean stream.

given by

$$F_L = 2\pi i \rho_o U a \omega \zeta(\omega) \mathrm{C}\left(\frac{\omega a}{U}\right), \tag{4.5.16}$$

where

$$\mathrm{C}(x) = \frac{i\mathrm{H}_1^{(1)}(x)}{\mathrm{H}_0^{(1)}(x) + i\mathrm{H}_1^{(1)}(x)} \tag{4.5.17}$$

is called the *Theodorsen function* [276]. The real and imaginary parts of $\mathrm{C}(\omega a/U)$ are plotted in Figure 4.5.5 against real values of the reduced frequency $\omega a/U$.

Example 3: Elastic blade–vortex interaction noise in three dimensions. Use the airfoil of Example 2 to model an elastic airfoil of compact chord by assuming its motion is resisted by a linear restoring force according to the equation of motion

$$(\mathrm{m} + \mathrm{M}_{22})\left(\frac{d^2}{dt^2} + \omega_o^2\right)\zeta = F_L,$$

where m, $\mathrm{M}_{22}(\equiv \rho_o \pi a^2)$, F_L are, respectively, the airfoil mass, added mass, and lift per unit span, and ω_o is the natural frequency *in an ideal fluid* (no vortex shedding) [4, 17, 277].

In the notation of Section 3.3.2, show that when a gust of vorticity (3.3.10) is incident on the airfoil, equation (3.3.13) for the sound generated by a stationary

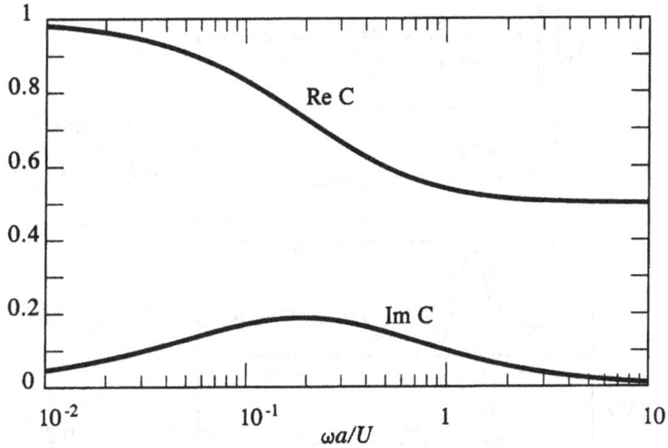

Figure 4.5.5. The Theodorsen function $C(\omega a/U)$.

airfoil takes the modified form [278]

$$p(\mathbf{x},t) \approx \frac{2i\rho_o U^2 \cos\Theta}{c_o|\mathbf{x}|}\frac{\partial}{\partial t}\iint_{-\infty}^{\infty}\left(\frac{\omega_o^2 - \omega^2 \mathrm{m}/(\mathrm{m}+\mathrm{M}_{22})}{\omega_o^2 - 2i\Delta\omega_o\omega - \omega^2}\right)$$
$$\times \frac{\Omega_3(\omega/U, k_2, 0)\mathrm{e}^{-i\omega[t]}\,dk_2 d\omega}{(\omega^2 + U^2k_2^2)\left(\mathrm{H}_0^{(1)}(\omega a/U) + i\mathrm{H}_1^{(1)}(\omega a/U)\right)}, \quad |\mathbf{x}| \to \infty, \tag{4.5.18}$$

where

$$\Delta = \frac{\mathrm{M}_{22}}{\mathrm{m}+\mathrm{M}_{22}}\frac{\mathrm{C}(\omega a/U)}{\omega_o a/U}$$

is a damping factor that accounts for losses caused by the production of kinetic energy in the wake by the vibrating airfoil.

When the gust is the axial vortex (3.3.15), deduce that the modified form of the acoustic pressure (3.3.17) can be written

$$p(\mathbf{x},t) \approx \frac{\rho_o U^2 \Gamma \cos\Theta}{8c_o|\mathbf{x}|}\frac{|\sin\phi|^{\frac{3}{2}}(a/\pi R)^{\frac{1}{2}}}{\tan\theta}\bar{\Im}(\alpha_+), \quad |\mathbf{x}| \to \infty,$$

where

$$\bar{\Im}(\alpha_+) = \frac{\mathrm{m}}{\mathrm{m}+\mathrm{M}_{22}}\left\{\Im(\alpha_+) + \frac{\mathrm{M}_{22}}{\pi\mathrm{m}}\sqrt{\frac{2i\omega_o R}{U|\sin\phi|}}\right.$$
$$\left.\times \int_{-\infty}^{\infty}\frac{\omega_o^{\frac{3}{2}}\exp(-(\omega R/2U\sin\phi)^2 - i\omega([t] + a/U))d\omega}{\sqrt{\omega}\left[\omega_o^2 - 2i\Delta\omega_o\omega - \omega^2\right]}\right\}.$$

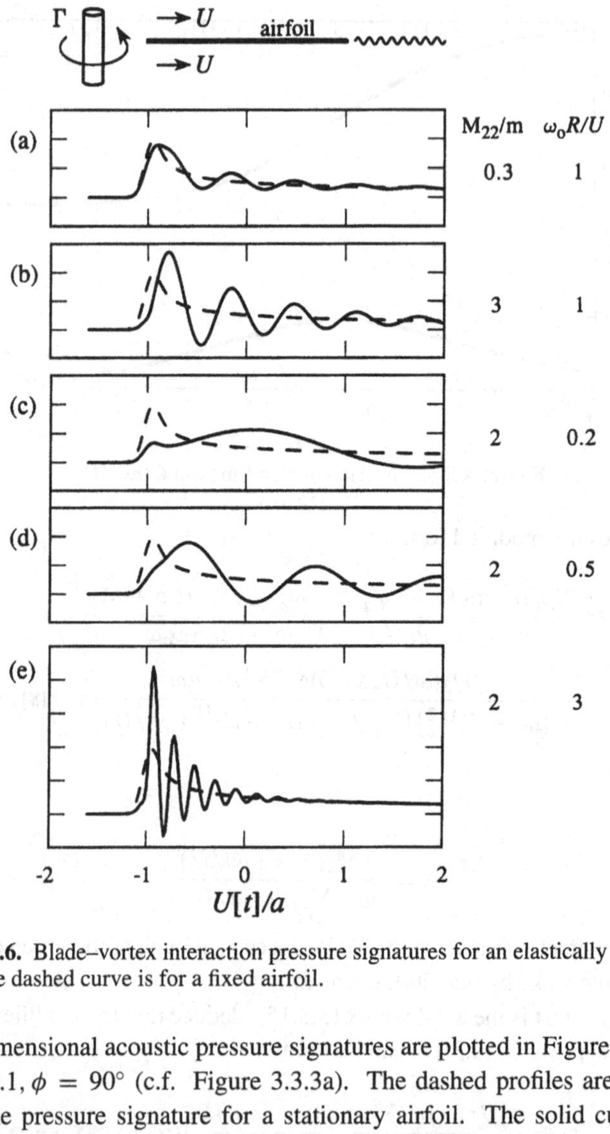

Figure 4.5.6. Blade–vortex interaction pressure signatures for an elastically supported airfoil. The dashed curve is for a fixed airfoil.

Nondimensional acoustic pressure signatures are plotted in Figure 4.5.6 for $R/a = 0.1, \phi = 90°$ (c.f. Figure 3.3.3a). The dashed profiles are plots of $\Im(\alpha_+)$, the pressure signature for a stationary airfoil. The solid curves are plots of $\tilde{\Im}(\alpha_+)$, the pressure signature for the vibrating airfoil for a nominal value $\Delta = 0.1$ of the damping coefficient. In case (a) the resonance frequency is comparable to the acoustic frequency $\sim U/R$ for the fixed airfoil, and the fluid loading $M_{22}/m = 0.3$ is small. When M_{22}/m is large, the radiation is dominated by the airfoil vibrations. In this case, substantial variations in the waveform occur with changes in $\omega_o R/U$.

5

Interaction of sound with solid structures

Fluid motion in the immediate vicinity of a solid surface is usually controlled by viscous stresses that cause an adjustment in the velocity to comply with the no-slip condition and by thermal gradients that similarly bring the temperatures of the solid and fluid to equality at the surface. At high Reynolds numbers, these adjustments occur across boundary layers whose thicknesses are much smaller than the other governing length scales of the motion. We have seen how the forced production of vorticity in boundary layers during convection of a "gust" past the edge of an airfoil can be modeled by application of the Kutta condition (Section 3.3). In this chapter, similar problems are discussed involving the generation of vorticity by sound impinging on both smooth surfaces and surfaces with sharp edges in the presence of flow. The aerodynamic sound generated by this vorticity augments the sound diffracted in the usual way by the surface. However, the near field kinetic energy of the vorticity is frequently derived wholly from the incident sound, so that unless the subsequent vortex motion is unstably coupled to the mean flow (acquiring additional kinetic energy from the mean stream as it evolves) there will usually be an overall decrease in the acoustic energy: The sound will be damped. General problems of this kind, including the influence of surface vibrations, are the subject of this chapter. We start with the simplest case of sound impinging on a plane wall.

5.1 Damping of Sound at a Smooth Wall

5.1.1 Dissipation in the Absence of Flow

The thermo-viscous attenuation of sound is greatly increased in the neighborhood of a solid boundary where temperature and velocity gradients are large. Let us examine this by considering the reflection of a time-harmonic wave from a plane wall at $x_2 = 0$. A sound wave of constant pressure amplitude $p_{\rm I}$

$$p(\mathbf{x}, \omega) = p_{\rm I} e^{i(n_1 x_1 - n_2 x_2)}, \quad \omega > 0$$

impinges on the wall from "above" ($x_2 > 0$), where

$$n_1 = \kappa_o \sin\Theta, \quad n_2 = \kappa_o \cos\Theta, \tag{5.1.1}$$

and Θ is the angle of incidence (see Figure 5.1.2). If $\mathcal{R}$ is an appropriate reflection coefficient, the total acoustic pressure above the wall is

$$p(\mathbf{x}, \omega) = p_{\rm I}(e^{-in_2 x_2} + \mathcal{R} e^{in_2 x_2}) e^{in_1 x_1}. \tag{5.1.2}$$

In an ideal fluid, the condition that the normal velocity vanishes on the surface implies that $v_2 = (-i/\rho_o\omega)\partial p/\partial x_2 = 0$ at $x_2 = 0$ and that $\mathcal{R} = 1$.

In a real fluid, momentum and thermal boundary layers are formed at the wall where the motion can no longer be regarded as irrotational and across which the velocity and temperature reduce to their respective values at the surface. The characteristic width of the layers is small compared to the acoustic wavelength for all frequencies of practical interest, but dissipation within them causes $\mathcal{R}$ to become complex with $|\mathcal{R}| < 1$. Motion within the boundary layers in the normal direction is constrained by the wall, which implies that the pressure is approximately uniform and equal to

$$p_s = p_{\rm I}(1 + \mathcal{R}) e^{in_1 x_1}. \tag{5.1.3}$$

Let δ be the larger of the thermal and momentum boundary layer thicknesses, and let v_δ denote the normal component of the acoustic particle velocity at the outer edge $x_2 = \delta$ of the boundary layers. The acoustic properties of the wall layers may then be expressed in terms of an *acoustic admittance* Y, defined by

$$Y \equiv Y_\nu + Y_\chi = \frac{-v_\delta}{p_s}, \tag{5.1.4}$$

where Y_ν and Y_χ are the *shear stress* and *thermal* admittances, respectively controlled by the viscous and thermal boundary layers. When they are known, the relation

$$v_\delta(x_1, \omega) = \frac{-n_2 p_{\rm I}}{\rho_o \omega}(1 - \mathcal{R}) e^{in_1 x_1}, \quad \kappa_o\delta \ll 1, \tag{5.1.5}$$

and (5.1.3) permit the reflection coefficient to be expressed in the form

$$\mathcal{R} = \frac{n_2/\rho_o\omega - Y}{n_2/\rho_o\omega + Y} = \frac{\cos\Theta - \rho_o c_o Y}{\cos\Theta + \rho_o c_o Y}. \tag{5.1.6}$$

The sound power Π dissipated per unit area of the wall is expressed in terms of these quantities by first restoring the time factor $e^{-i\omega t}$ to the complex representations of p_s and v_δ, taking their real parts, and forming the time average $-\langle p_s v_\delta \rangle$. This supplies

$$\Pi = \frac{1}{2}|p_s|^2 \operatorname{Re} Y = \frac{n_2|p_I|^2}{2\rho_o \omega}(1 - |\mathcal{R}|^2). \tag{5.1.7}$$

To calculate the admittance, we temporarily denote by p', ρ', and $\mathbf{v}'$ the perturbation pressure, density, and velocity in the *acoustic* field, which satisfy the ideal fluid equations in $x_2 > \delta$. However, these equations may be regarded as satisfied by the acoustic variables within the whole of $x_2 > 0$, including the boundary layers. Because both the exact and acoustic fields satisfy the linearized continuity equation and must vary like $e^{in_1x_1}$, we can then form the difference

$$in_1(v_1 - v_1') + \partial(v_2 - v_2')/\partial x_2 - (i\omega/\rho_o)(\rho - \rho') = 0,$$

and integrate over $0 < x_2 < \delta$. Because $v_2 \equiv v_2'$ at $x_2 = \delta$, $v_2 = 0$ at $x_2 = 0$, and $v_2' = v_\delta$ (since $\kappa_o\delta \ll 1$), we find that

$$v_\delta = -in_2 \int_0^\infty (v_1 - v_1')\, dx_2 + \frac{i\omega}{\rho_o} \int_0^\infty (\rho - \rho')\, dx_2, \tag{5.1.8}$$

where the upper limits of integration have been extended to ∞ because the integrands vanish for $x_2 > \delta$.

Very close to the wall the motion is tangential and retarded by viscous shear stresses, which causes fluid to be displaced outward relative to an ideal fluid at a rate equal to the first integral in (5.1.8). The second integral is the corresponding excess outward velocity produced by the volumetric expansion that occurs when a temperature gradient is established in the boundary layer to bring the wall and fluid into equilibrium following adiabatic heating or cooling by the incident sound. Hence, the definition (5.1.4) yields

$$Y_\nu = \frac{in_2}{p_s} \int_0^\infty (v_1 - v_1')\, dx_2, \quad Y_\chi = \frac{-i\omega}{\rho_o p_s} \int_0^\infty (\rho - \rho')\, dx_2. \tag{5.1.9}$$

Shear Stress Admittance

Y_ν is calculated by means of the *boundary layer approximation* [4, 17, 36, 178]. The *momentum* boundary layer of thickness δ_ν is driven by the acoustic pressure gradient $\partial p_s/\partial x_1$ and governed by the linearized, tangential component of the

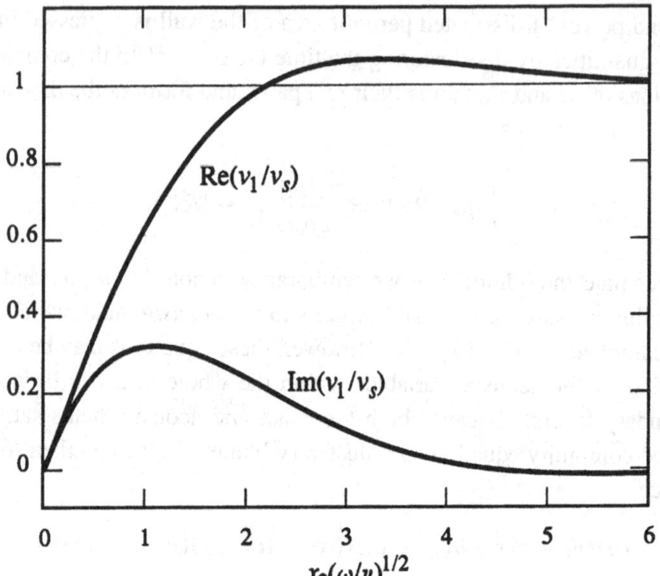

Figure 5.1.1. Kirchhoff–Stokes boundary layer.

momentum equation (1.2.9)

$$-i\omega(v_1 - v_s) = \nu\frac{\partial^2 v_1}{\partial x_2^2}, \quad x_2 < \delta_\nu, \tag{5.1.10}$$

where $v_s = (-i/\rho_o\omega)\partial p_s/\partial x_1 = n_1 p_s/\rho_o\omega$ is the acoustic particle "slip" velocity along the wall, which is uniform across the layer. The shear viscosity term on the right exceeds all other discarded viscous terms by a factor $\sim 1/(\kappa_o\delta_\nu)^2 \gg 1$, and the omitted compressive stresses (involving div $\mathbf{v}$) are smaller than the acoustic pressure gradient by a factor of order $\kappa_o\delta/(\delta c_o/\nu) \ll 1$. The solution of (5.1.10), which must remain finite at $x_2 = +\infty$ and satisfy the no-slip condition at the wall, is furnished by the *Kirchhoff–Stokes* formula

$$v_1 = v_s\left(1 - \mathrm{e}^{ix_2\sqrt{i\omega/\nu}}\right), \tag{5.1.11}$$

where $\mathrm{Im}\sqrt{i\omega} > 0$. The real and imaginary parts of v_1/v_s are plotted as functions of $x_2(\omega/\nu)^{1/2}$ in Figure 5.1.1; $v_1 \to v_s$ as x_2 becomes large, and $|v_1|$ is within about 94% of $|v_s|$ when $x_2 > \delta_\nu \approx 1.8\sqrt{\nu/\omega}$.

We may now set $v_1 - v_1' = v_1 - v_s \equiv v_1 - n_1 p_s/\rho_o\omega$ in the first integral of (5.1.9) to obtain from (5.1.11) and (5.1.10) the alternative representations

$$Y_\nu = \frac{n_1^2}{\rho_o\omega}\sqrt{\frac{\nu}{i\omega}} \equiv \frac{n_1\nu}{\omega p_s}\left(\frac{\partial v_1}{\partial x_2}\right)_{x_2=0}, \tag{5.1.12}$$

and from (5.1.7) the following alternative formulae for the acoustic power dissipated by viscous action per unit area of the wall:

$$\Pi_\nu = \frac{\sqrt{\nu/2\omega}}{2\rho_o\omega}\left|\frac{\partial p_s}{\partial x_1}\right|^2 \equiv \left\langle v_s\eta\left(\frac{\partial v_1}{\partial x_1}\right)_{x_2=0}\right\rangle. \tag{5.1.13}$$

The time average in the angle brackets (evaluated after restoration of the time factor $e^{-i\omega t}$ and taking real parts) associates the dissipation with the rate of working of the surface shear stress ("drag") on the *ideal* fluid.

Thermal Admittance

A similar calculation (Example 1), that assumes the heat capacity of the wall to be sufficiently large that the surface temperature can be regarded as constant, yields

$$Y_\chi = \frac{\omega\beta}{\rho_o c_p}\sqrt{\frac{\chi}{i\omega}} = \frac{\beta\chi}{p_s}\left(\frac{\partial T}{\partial x_2}\right)_{x_2=0}, \tag{5.1.14}$$

where β is the coefficient of expansion at constant pressure (Section 1.2.3), $\chi = \kappa/\rho_o c_p$ is the thermometric conductivity (κ being the thermal conductivity of Section 1.2.8), and T denotes temperature.

Combining (5.1.12) and (5.1.14), the overall acoustic admittance can be written

$$Y = \frac{e^{-\frac{i\pi}{4}}}{\rho_o\omega^{\frac{3}{2}}}\left(n_1^2\sqrt{\nu} + \frac{\beta\omega^2}{c_p}\sqrt{\chi}\right), \tag{5.1.15}$$

and the power dissipated per unit area of the wall is

$$\Pi = \frac{\sqrt{\omega}}{2\sqrt{2}\rho_o c_o^2}|p_s|^2\left(\sqrt{\nu}\sin^2\Theta + \frac{\beta c_o^2}{c_p}\sqrt{\chi}\right). \tag{5.1.16}$$

The Prandtl number $\mathrm{Pr} = \nu/\chi$ can be large in liquids (≈ 6.7 in water), but in gases $\mathrm{Pr} \sim O(1)$ and the shear stress and thermal contributions are of comparable importance. In an ideal gas of temperature T_o, $\beta = 1/T_o$, and $\beta c_o^2/c_p = \gamma - 1$, where γ is the ratio of the specific heats. For normally incident sound ($\Theta = 0$), the dissipation is due entirely to thermal losses.

Example 1. The thermal admittance is determined by the excess volumetric changes in the boundary layer produced by the conduction of heat to and from the wall over that produced by the adiabatic expansions and compressions caused by the sound. Let T_o denote the mean temperature of the fluid and wall. The thermodynamic formula

$$d\rho = dp/c_o^2 + (\partial\rho/\partial s)_p\, ds \equiv dp/c_o^2 - (\rho_o\beta T_o/c_p)\, ds$$

implies that $\rho - \rho' \approx -(\rho_o\beta T_o/c_p)s$ in (5.1.8), where the perturbation specific entropy $s = 0$ *except* within the thermal boundary layer. By setting $dp = p_s$ in the linearized relation $T_o\, ds = c_p\, dT - dp/\rho_o$, we find that

$$Y_\chi = \frac{i\omega\beta T_o}{c_p p_s}\int_0^\infty s\, dx_2 = \frac{i\omega\beta}{p_s}\int_0^\infty (T - T_s)\, dx_2, \qquad (5.1.17)$$

where T is the temperature perturbation from T_o, and $T_s = p_s/\rho_o c_p$ the adiabatic temperature change produced by the sound, both of which are equal at the outer edge $x_2 = \delta_\chi$ of the thermal boundary layer. By making the same substitution for s in the linearized form of the energy equation (1.2.34), and using the boundary layer approximation, show that

$$-i\omega(T - T_s) = \chi\frac{\partial^2 T}{\partial x_2^2}, \quad x_2 < \delta_\chi, \qquad (5.1.18)$$

with solution

$$T = T_s\left(1 - \mathrm{e}^{ix_2\sqrt{i\omega/\chi}}\right),$$

which vanishes on the wall. Hence, deduce (5.1.15) by substitution into (5.1.17).

Example 2. Show that the power dissipated in the viscous boundary layer can be written

$$\Pi_\nu = -\int_0^\infty \left\langle v_1\frac{\partial p}{\partial x_1}\right\rangle dx_2.$$

The integrand vanishes in $x_2 > \delta_\nu$, where v_1 is in quadrature with $\partial p/\partial x_1$; the dissipated power is therefore equal to the rate of working of the tangential pressure gradient force in the boundary layer. In complex notation, we also have

$$\Pi_\nu = -\frac{1}{2}|v_s|^2\int_0^\infty \mathrm{Im}\left(\frac{v_1(x_2)}{v_s}\right) dx_2, \qquad (5.1.19)$$

where $v_1(x_2)$ is given by (5.1.11). Figure 5.1.1 indicates that the main contribution to the integral is from that part of the boundary layer centered on $x_2(\omega/\nu)^{1/2} \approx 1$.

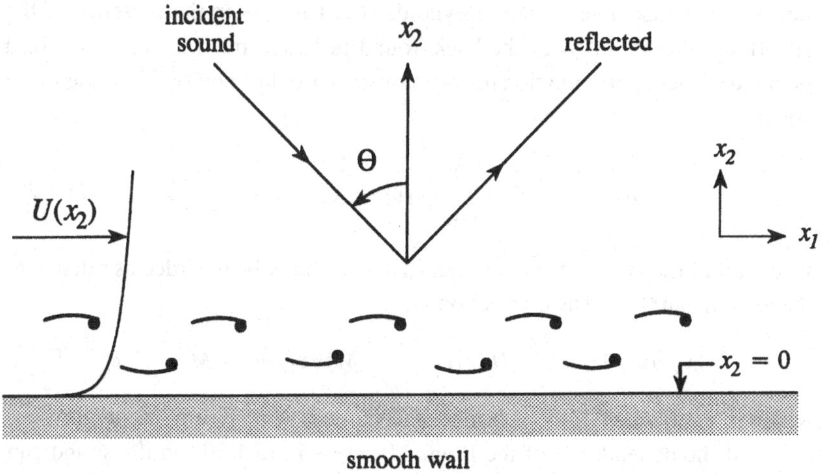

Figure 5.1.2. Sound incident on a turbulent boundary layer.

5.1.2 Dissipation by Wall Turbulence

The damping of sound impinging on a turbulent boundary layer is much larger than in free-field turbulence (Section 2.5) because the acoustic particle velocity gradients are increased by the presence of the wall [58, 279–286]. It is more important at lower frequencies, when the periodic straining of vortex lines by the sound becomes irreversible because there is then ample time in an acoustic cycle for energy to flow between the different degrees of freedom of the strained turbulence via their nonlinear couplings.

Let the turbulent boundary layer in $x_2 > 0$ be formed over a smooth rigid wall at $x_2 = 0$ (Figure 5.1.2), with mean velocity $U(x_2)$ in the x_1-direction. The generally slow dependence of U on x_1 is ignored [62], and the Mach number in the exterior mean flow is assumed to be small enough that convection of sound can be neglected. We proceed as in Section 5.1.1, by considering incident sound of wavelength $\gg \delta =$ the *turbulent boundary layer thickness*, and calculate the effective acoustic admittance of the boundary layer.

The generalization of the boundary layer approximation (5.1.10) for the tangential acoustic particle velocity is formed by averaging the momentum equation. When expressed in Reynolds' form (1.2.22), the term responsible for the transfer of x_1-momentum away from the wall is the Reynolds stress $\rho v_1 v_2 \approx \rho_o v_1 v_2$. For the sound wave (5.1.2), the mean rate of transfer of tangential acoustic momentum is equal to the difference

$$\rho_o \overline{v_1 v_2} - \rho_o \langle v_1 v_2 \rangle$$

between the ensemble average Reynolds stress $\rho_o\overline{v_1 v_2}$ in the presence of the sound and that $\rho_o\langle v_1 v_2\rangle$ of the background turbulent motion. The linearized boundary layer approximation for the acoustic particle velocity $v_1(x_2)$ therefore becomes

$$-i\omega(v_1 - v_s) = -\frac{\partial}{\partial x_2}(\overline{v_1 v_2} - \langle v_1 v_2\rangle) + \nu\frac{\partial^2 v_1}{\partial x_2^2}, \tag{5.1.20}$$

where all of the velocities on the right-hand side may be regarded as relative to the local mean flow. The convection term

$$U(x_2)\partial v_1/\partial x_1 = in_1 U(x_2)v_1 \sim \mathrm{O}(M)\partial v_1/\partial t, \quad M = U/c_o,$$

which is small when $M \ll 1$, has been discarded. A crude empirical representation of the dependence of the Reynolds stress in (5.1.20) on the sound can be obtained from the following extension of arguments originally proposed by Prandtl [62].

In equilibrium boundary layer flow (no sound), Prandtl equated the shear stress $-\rho_o\langle v_1 v_2\rangle$ (the rate of diffusion of x_1-momentum into the wall, which is *positive* if the mean drag is in the x_1-direction) to $\rho_o\ell_P^2(\partial U(x_2)/\partial x_2)^2$, where ℓ_P is a *mixing length* which he regarded as being analogous to the mean free path in the kinetic theory of gases [287], and over which distance a fluid particle is convected before being thoroughly mixed with its turbulent environment. This is equivalent to writing $-\rho_o\langle v_1 v_2\rangle = \rho_o\epsilon_m\partial U(x_2)/\partial x_2$, where $\epsilon_m \equiv \ell_P^2\partial U(x_2)/\partial x_2$ is an *eddy viscosity*. Experiments at low Mach numbers reveal that the mean shear stress is practically constant and equal to $\rho_o v_*^2$ over a substantial region of the boundary layer adjacent to the wall (v_* being the friction velocity, Section 3.4), and therefore $\epsilon_m\partial U/\partial x_2 \approx v_*^2$ within this region. By making the additional hypothesis that $\ell_P = \kappa_K x_2$, where κ_K is constant (the *von Kármán constant*), Prandtl deduced the following logarithmic formula for the mean velocity profile:

$$\frac{U(x_2)}{v_*} = \frac{1}{\kappa_K}\ln\left(\frac{x_2 v_*}{\nu}\right) + \text{constant}. \tag{5.1.21}$$

This agrees with measurements over a region of the boundary layer extending from about $x_2 v_*/\nu = 30$ to 2×10^3, provided $\kappa_K \approx 0.41$ and the constant ≈ 5 [62]. Prandtl's mixing length $\ell_P = \kappa_K x_2$ decreases linearly to zero at the wall, but it must actually vanish more rapidly than this within the "viscous sublayer" ($x_2 v_*/\nu < 7$), where the Reynolds stress is negligible. The sublayer motion is dominated by molecular diffusion, that is, $\nu\partial U/\partial x_2 = v_*^2$, where $U(x_2)/v_* \approx x_2 v_*/\nu$.

These ideas suggest that, in the presence of sound,

$$\overline{v_1 v_2} = -\ell_P^2 \left(\frac{\partial U}{\partial x_2} + \frac{\partial v_1}{\partial x_2} \right)^2,$$

in which case (5.1.20) becomes, when linearized,

$$-i\omega(v_1 - v_s) = \frac{\partial}{\partial x_2} \left([\nu + 2\epsilon_m(x_2)] \frac{\partial v_1}{\partial x_2} \right). \tag{5.1.22}$$

At low frequencies, the acoustic boundary layer defined by the solution of this equation will extend far out into the turbulent flow; the disturbed motion should be approximately quasi-static with $\epsilon_m \approx \kappa_K v_* x_2$ over most of the boundary layer. The turbulence relaxation time, during which the turbulence shear stress evolves to accommodate the change $v_1(x_2)$ in velocity profile, is then a small fraction of the acoustic period, and kinetic energy acquired by the turbulence by acoustic straining has ample time to flow between the different turbulence modes and is not, therefore, recovered by the acoustic field when the strain is released. In the opposite extreme of very high frequencies, the acoustic boundary layer is entirely within the viscous sublayer, and the turbulence has no influence on acoustic dissipation.

There must exist, however, a range of intermediate frequencies for which the principal acoustic–turbulence interactions occur in a near-wall region where both viscous and turbulence diffusion are significant, and where the simple linear model $\epsilon_m \approx \kappa_K v_* x_2$ is not satisfactory. This range is centered on a critical frequency ω_* whose value can be roughly estimated to correspond to the condition that the edge $x_2 \approx 7\nu/v_*$ of the sublayer should fall in the middle of the region of maximum slope of the acoustic particle velocity profile. This occurs at about $x_2(\omega_*/\nu)^{1/2} \approx \frac{3}{4}$ (Figure 5.1.1) when the profile is approximated by the Kirchhoff–Stokes formula (5.1.11), so that $\omega_* \nu / v_*^2 \approx 10^{-2}$. Experiments [285] also suggest that at this frequency the turbulence relaxation time near the wall is comparable to the acoustic period.

The following modification of Prandtl's linear approximation to ϵ_m can be used to model both the frequency dependence of the acoustic–boundary layer interaction and the rapid decrease of ϵ_m to zero within the sublayer:

$$\begin{aligned} \epsilon_m &= 0, && x_2 < \delta_v(\omega) \\ &= \kappa_K v_*(x_2 - \delta_v(\omega)), && x_2 > \delta_v(\omega). \end{aligned} \tag{5.1.23}$$

The frequency dependent length $\delta_v(\omega)$ must be approximately equal to the sublayer thickness $(7\nu/v_*)$ when $\omega/\omega_* \to 0$ and the interaction becomes quasi-static. At higher frequencies, δ_v must increase to account for the increased importance of viscous diffusion, which is dominant for $\omega > \omega_*$, and for the reduced efficiency of turbulent diffusion that occurs when the acoustic period decreases below the turbulence relaxation time. The following empirical formula fulfills these requirements and leads to predictions that are in good agreement with experiment [282]:

$$\frac{\delta_\nu v_*}{\nu} = 6.5\left(1 + \frac{1.7(\omega/\omega_*)^3}{1 + (\omega/\omega_*)^3}\right), \quad \omega_*\nu/v_*^2 \approx 0.01, \quad \omega > 0. \tag{5.1.24}$$

An additional modification is necessary at frequencies that are so small that the region of maximum slope of the acoustic boundary layer velocity profile extends beyond the domain of validity of the logarithmic mean velocity profile (5.1.21).

The Boundary Layer Admittance

When equation (5.1.22) is solved with ϵ_m defined as in (5.1.23) (Example 3), the shear stress admittance in the presence of the turbulent boundary layer can be calculated from (5.1.12) and expressed in the form

$$Y_\nu = \frac{n_1^2}{\rho_o\omega}\sqrt{\frac{\nu}{i\omega}}\mathcal{F}_o\left(\sqrt{\frac{i\omega\nu}{\kappa_K^2 v_*^2}}, \delta_\nu\sqrt{\frac{i\omega}{\nu}}\right), \tag{5.1.25}$$

where

$$\mathcal{F}_o(a, b) = \frac{i\left[H_1^{(1)}(a)\cos b - H_0^{(1)}(a)\sin b\right]}{H_0^{(1)}(a)\cos b + H_1^{(1)}(a)\sin b}. \tag{5.1.26}$$

The reader can easily verify that $\mathcal{F}_o \to 1$ as $v_* \to 0$ and that (5.1.25) then reduces to the no-flow formula (5.1.12).

The influence of turbulence mixing on the thermal admittance is obtained by applying the Reynolds analogy between the turbulence transport of momentum and heat [62]. This leads to a modified form of (5.1.18), in which χ is replaced by $\chi + 2\epsilon_T(x_2)$, where $\epsilon_T = \epsilon_m/\text{Pt}$ is an eddy thermal diffusivity and Pt is a turbulence Prandtl number. In air Pt ≈ 0.7 [62, 288], and by assuming it to be constant the thermal boundary layer equation can be solved by the method of Example 3, to give

$$Y_\chi = \frac{\omega\beta}{\rho_o c_p}\sqrt{\frac{\chi}{i\omega}}\mathcal{F}_o\left(\text{Pt}\sqrt{\frac{i\omega\chi}{\kappa_K^2 v_*^2}}, \delta_\nu\sqrt{\frac{i\omega}{\chi}}\right). \tag{5.1.27}$$

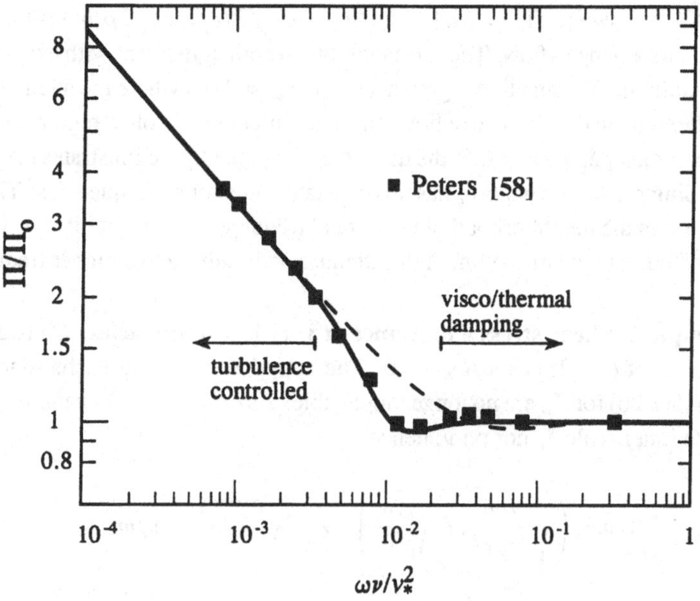

Figure 5.1.3. Turbulent boundary layer damping of sound in air at very low Mach numbers.

The total acoustic power dissipated per unit area of the wall is now found from (5.1.7):

$$\Pi = \frac{\sqrt{\omega}}{2\sqrt{2}\rho_o c_o^2}|p_s|^2 \operatorname{Re}\left(\sqrt{2}e^{-\frac{i\pi}{4}}\left[\sqrt{\nu}\sin^2\Theta\mathcal{F}_o\left(\sqrt{\frac{i\omega\nu}{\kappa_K^2 v_*^2}}, \delta_v\sqrt{\frac{i\omega}{\nu}}\right)\right.\right.$$
$$\left.\left.+\frac{\beta c_o^2\sqrt{\chi}}{c_p}\mathcal{F}_o\left(\mathrm{Pt}\sqrt{\frac{i\omega\chi}{\kappa_K^2 v_*^2}}, \delta_v\sqrt{\frac{i\omega}{\chi}}\right)\right]\right). \tag{5.1.28}$$

This formula is compared with measurements of the damping of axially propagating sound in a pipe with low Mach number turbulent air flow ($\nu = 1.5 \times 10^{-5}$ m²/s, $\chi = 2 \times 10^{-5}$ m²/s, Pt $= 0.7$) in Figure 5.1.3 [58, 286]. The predicted acoustic power dissipated per unit area of the pipe wall is given by (5.1.28) with $\Theta = 90°$, provided the acoustic boundary layer thickness is small compared to the pipe radius. It turns out that the mean flow has a significant effect on the acoustic wavenumber and also, therefore, on the attenuation, even when the Mach number is small (Example 7), and the experimental results in the figure are an extrapolation to zero Mach number obtained by averaging measurements for waves of the same frequency propagating in the upstream and downstream directions. The solid curve is the prediction (5.1.28) of Π/Π_o when δ_v is defined by (5.1.24), where $\Pi_o = \sqrt{\omega}|p_s|^2(\sqrt{\nu} + \beta c_o^2\sqrt{\chi}/c_p)/(2\sqrt{2}\rho_o c_o^2)$ is the dissipation for no mean flow. There is favorable overall agreement with experiment, especially in the transition region near $\omega_* \nu / v_*^2 \approx 0.01$, where the attenuation is *smaller* than in the absence of flow. In the turbulence controlled region, $\omega < \omega_*$, the parameter $\delta_\nu v_*/\nu \approx 6.5$; the dashed curve in the figure illustrates the corresponding prediction when δ_ν has this constant value for all frequencies. The dissipation in the neighborhood of the critical frequency is now significantly overestimated, and the minimum in the attenuation is shifted to a higher frequency.

Example 3: Shear stress admittance at very low frequencies. Consider the behavior of (5.1.25) as $\omega/\omega_* \to 0$. The hypotheses leading to the functional form (5.1.26) for $\mathcal{F}_o$ are no longer applicable. As $\omega/\omega_* \to 0$, Y_ν cannot depend on the inner scale δ_ν nor on ν; hence,

$$\sqrt{\nu}\mathcal{F}_o\left(\sqrt{\frac{i\omega\nu}{\kappa_K^2 v_*^2}}, \delta_\nu\sqrt{\frac{i\omega}{\nu}}\right) \to C'\sqrt{v_*^2/\omega}, \quad \omega/\omega_* \ll 1.$$

By examining the dissipation due to momentum transfer in the *quasi-static* limit, for which $\Pi_\nu \approx \langle v_s(\rho_o v_*^2)'\rangle$, where the prime denotes the perturbation in the shear stress when v_* is perturbed from its mean value by the tangential velocity field of the sound wave, deduce that $C' \sim v_*/U_\infty$ where $U_\infty = U(\infty)$ and therefore that $Y_\nu \sim v_*^2 n_1^2 e^{-\frac{i\pi}{4}}/\rho_o\omega^2 U_\infty$.

Example 4. The tangential velocity v_1 and the shear stress $(\nu + 2\epsilon_m(x_2))\,\partial v_1/\partial x_2$ are continuous across the momentum boundary layer. Show that the solution of (5.1.22) satisfying this and vanishing at the wall, when ϵ_m is defined by (5.1.23), can be written [289]

$$\begin{aligned}\frac{v_1(x_2)}{v_s} &= 1 - e^{ix_2\sqrt{i\omega/\nu}} \\ &\quad + \frac{H_1^{(1)}(a)\cos b + iH_0^{(1)}(a)\sin b}{H_0^{(1)}(a)\cos b + H_1^{(1)}(a)\sin b}\sin(x_2\sqrt{i\omega/\nu}), \quad 0 < x_2 < \delta_\nu, \\ &= 1 - \frac{H_0^{(1)}\left(a[1 + (2\kappa_K v_*/\nu)(x_2 - \delta_\nu)]^{\frac{1}{2}}\right)}{H_0^{(1)}(a)\cos b + H_1^{(1)}(a)\sin b}, \quad x_2 > \delta_\nu, \end{aligned} \tag{5.1.29}$$

where $a = \sqrt{i\omega\nu/\kappa_K^2 v_*^2}$, $b = \delta_\nu\sqrt{i\omega/\nu}$, and $\sqrt{i\omega/\nu}$ are taken to have positive real parts.

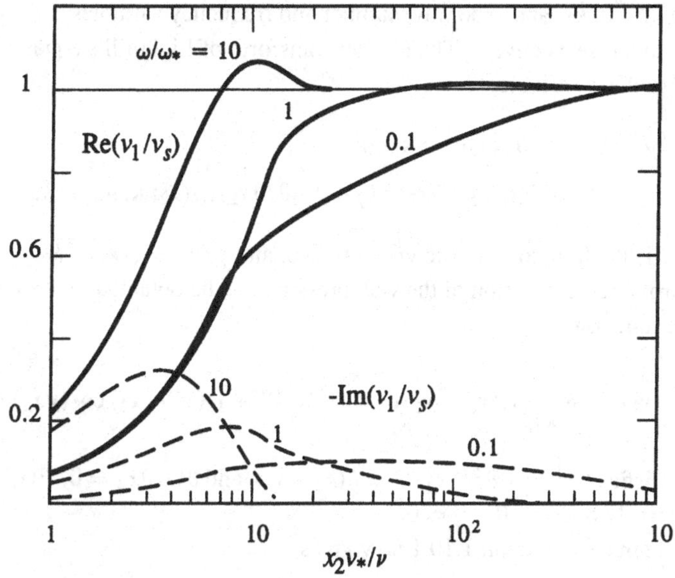

Figure 5.1.4. Acoustic momentum boundary layer (5.1.29) at low Mach numbers.

The solution is illustrated in Figure 5.1.4. When $\omega/\omega_* = 10$ the boundary layer is controlled by viscous diffusion in the sublayer at the wall (Figure 5.1.3), the dissipation integral (5.1.15) being dominated by the neighborhood of $x_2 v_*/\nu \approx 3$, where $-\mathrm{Im}(v_1/v_s)$ assumes its maximum value. At lower frequencies, the maximum of $-\mathrm{Im}(v_1/v_s)$ is much broader, and the region of high dissipation can extend far out into the turbulent flow. When $\omega/\omega_* < 1$, the profile of $\mathrm{Re}(v_1/v_s)$ resembles that of the mean velocity $U(x_2)$, increasing linearly out to about $x_2 v_*/\nu = 8$, and roughly logarithmically for larger values of $x_2 v_*/\nu$.

Example 5. Determine the behavior at low Mach numbers of the turbulent boundary layer wall pressure spectrum in the neighborhood of the acoustic wavenumber [188].

Let the boundary layer be formed on the rigid wall $x_2 = 0$ and occupy the region $x_2 > 0$. For low Mach number flow at uniform mean density, Lighthill's equation (2.1.5) can be taken in the form

$$(\partial^2/c_o^2\partial t^2 - \nabla^2)p = \partial^2\tau_{ij}/\partial x_i \partial x_j - \mathrm{div}\left(\eta\nabla^2\mathbf{v} + \frac{1}{3}\eta\nabla\mathrm{div}\,\mathbf{v}\right),$$

where $\tau_{ij} = \rho_o v_i v_j$. This equation is to be solved for the pressure p on the wall.

Let $\mathbf{k} = (k_1, 0, k_3)$ and ω be wavenumber and frequency variables conjugate to (x_1, x_3) and t, respectively. The Fourier transform of Lighthill's equation with respect to these variables is

$$\begin{aligned}&\left(\gamma^2(k) + \partial^2/\partial x_2^2\right) p(x_2; \mathbf{k}, \omega)\\ &\quad = -(ik_i + \delta_{i2}\partial/\partial x_2)(ik_j + \delta_{j2}\partial/\partial x_2)\tau_{ij}(x_2; \mathbf{k}, \omega) + Z,\end{aligned}$$

where Z is the transform of the viscous term, and $\gamma(k) = \sqrt{\kappa_o^2 - k^2}$.

A formal representation of the wall pressure can be obtained in terms of the Green's function

$$G(x_2, y_2) = \frac{-i}{\gamma(k)}\left(\mathrm{H}(x_2 - y_2)\cos(\gamma y_2)\mathrm{e}^{i\gamma x_2} + \mathrm{H}(y_2 - x_2)\cos(\gamma x_2)\mathrm{e}^{i\gamma y_2}\right),$$

which satisfies $(\gamma^2(k) + \partial^2/\partial x_2^2)G = \delta(x_2 - y_2)$, and $\partial G/\partial x_2 = 0,\ \partial G/\partial y_2 = 0$ respectively on $x_2 = 0,\ y_2 = 0$.

The method of Section 1.10.1 now gives

$$\begin{aligned}&p_s(\mathbf{k}, \omega) + \frac{\eta k_j}{\gamma(k)}\left(\frac{\partial v_j}{\partial x_2}(x_2; \mathbf{k}, \omega)\right)_{x_2=0}\\ &\quad = \frac{-i}{\gamma(k)}\int_0^\infty (k_i - \gamma\delta_{i2})(k_j - \gamma\delta_{j2})\tau_{ij}(y_2; \mathbf{k}, \omega)\mathrm{e}^{i\gamma y_2}\, dy_2, \qquad (5.1.30)\end{aligned}$$

where $p_s(\mathbf{k}, \omega) \equiv p(0; \mathbf{k}, \omega)$. In obtaining this result, the x_2-component of the momentum equation on $x_2 = 0$ has been used in the form

$$\partial p/\partial x_2 - (4\eta/3)\partial(\operatorname{div}\mathbf{v})/\partial x_2 = -\eta k_j \partial v_j/\partial x_2,$$

and viscous body stresses have been neglected in comparison with the Reynolds stress.

Sound is generated by nonlinear mechanisms in the boundary layer defined by the Reynolds stress integral, so that both the surface pressure $p_s(\mathbf{k}, \omega)$ and the surface shear stress $\eta(\partial v_j(x_2; \mathbf{k}, \omega)/\partial x_2)_{x_2=0}$ must be regarded as produced by the turbulence quadrupoles. In the acoustic domain $k < |\kappa_o|$, the length scale of the wall pressure fluctuations is the acoustic wavelength $\sim\delta/M \gg \delta$ (δ being the boundary layer thickness and $M \ll 1$ the mean flow Mach number). The sound received at a given point on the wall is dominated by quadrupole sources located over a very large expanse of the boundary layer; this impinging, long wavelength sound has propagated over the turbulent boundary layer through regions with which it is statistically independent, and the accompanying surface stress $\eta(\partial v_j(x_2; \mathbf{k}, \omega)/\partial x_2)_{x_2=0}$ is therefore related to the wall pressure p_s by

(5.1.12), which in the present notation becomes

$$k_j \eta (\partial v_j(x_2; \mathbf{k}, \omega)/\partial x_2)_{x_2=0} = \rho_o \omega Y_\nu(\mathbf{k}, \omega) p_s,$$

where Y_ν is the shear stress admittance of the boundary layer.

Let $S(\mathbf{k}, \omega)$ denote the integral in (5.1.30). When this can be regarded as the Fourier transform of a stationary random function of position and time [62, 63], we can write $\langle S(\mathbf{k}, \omega) S^*(\mathbf{k}', \omega') \rangle = \delta(\mathbf{k} - \mathbf{k}')\delta(\omega - \omega') Q(\mathbf{k}, \omega)$, where $Q > 0$ is the wavenumber–frequency spectrum of S. The wall pressure spectrum $\mathrm{P}(\mathbf{k}, \omega)$, defined as in Section 3.4, is then given in the acoustic domain by

$$\mathrm{P}(\mathbf{k}, \omega) = \frac{Q(\mathbf{k}, \omega)}{\left|\sqrt{\kappa_o^2 - k^2} + \rho_o \omega^2 Y_\nu(\mathbf{k}, \omega)\right|^2}, \quad k \leq |\kappa_o|.$$

Thus, when the interaction of sound with the turbulence through which it propagates is neglected (by setting $Y_\nu = 0$), the wavenumber–frequency spectrum $\mathrm{P}(\mathbf{k}, \omega)$ is singular at the edge $k = |\kappa_o|$ of the acoustic domain; the singularity is produced by the aggregate effect of sound waves incident from all sections of the boundary layer [172, 175, 179, 186, 187, 290]. The dominant boundary layer frequencies are in the range $\omega\delta/U = 0.01 - 1$, for which $\omega/\omega_* < 0.1$ when $U\delta/\nu \approx 10^6$. To a sufficient approximation, we can therefore use the quasi-static formula $Y_\nu \approx (v_*^2/\rho_o c_o^2 U_\infty)\mathrm{e}^{-\frac{i\pi}{4}}$ (Example 3) to estimate the peak level $\mathrm{P}_c \equiv \mathrm{P}(\mathbf{k}, \omega) \approx (U_\infty c_o/v_*^2)^2 \mathrm{P}(\mathbf{0}, \omega) \gg \mathrm{P}(\mathbf{0}, \omega)$ at $k = |\kappa_o|$, where $\mathrm{P}(\mathbf{0}, \omega) = S(\mathbf{0}, \omega)/\kappa_o^2$ is the wall pressure spectrum at $k = 0$ (which determines the sound radiated in the direction normal to the wall). This estimate of P_c is unconscionably large, and levels of this magnitude are not observed experimentally [173]. Bergeron [186, 290] argued that P_c would in practice be determined by the physical dimensions of the boundary layer; he obtained the much more modest estimate $\mathrm{P}_c \sim \kappa_o L \mathrm{P}(\mathbf{0}, \omega)$, where the length L is the characteristic size of the flow. Other mechanisms that might control the magnitude of this peak are discussed in [175].

Example 6. Use the method of Example 5 to obtain the following formula for the wavenumber–frequency spectrum of the surface shear stress $\sigma_{ij}(\mathbf{k}, \omega)$ beneath a low Mach number turbulent boundary layer as $k \to 0$:

$$\sigma_{ij}(\mathbf{k}, \omega) \approx \frac{\nu k_i k_j}{|\omega|} \left| \mathcal{F}_o\left(\sqrt{\frac{i\omega\nu}{\kappa_K^2 v_*^2}}, \delta_\nu \sqrt{\frac{i\omega}{\nu}}\right) \right|^2 \mathrm{P}(\mathbf{k}, \omega), \quad k\delta \ll 1.$$

Example 7: Attenuation in turbulent pipe flow. By averaging the momentum and continuity equations over the cross-section of uniform area $\mathcal{A}$ and perimeter

ℓ_p of a pipe conveying low Mach number fully developed turbulent flow, show that long wavelength sound propagating axially within the pipe satisfies

$$\frac{1}{c_o^2}\left(\frac{\partial}{\partial t}+U\frac{\partial}{\partial x}\right)^2 p-\frac{\partial^2 p}{\partial x^2}=\frac{-4\rho_o}{\mathrm{D}_p}\left(\frac{\partial}{\partial t}+U\frac{\partial}{\partial x}\right)v_\delta,$$

where U is the mean flow velocity, x is measured along the axis of the pipe, $\mathrm{D}_p = 4\mathcal{A}/\ell_p$ is the *hydraulic diameter*, and v_δ is the normal component of the acoustic particle velocity at the outer edge of the wall boundary layer (directed inward from the wall).

Deduce the dispersion equation

$$k^2-(\kappa_o-Mk)^2=-\frac{4i\rho_o c_o}{\mathrm{D}_p}(\omega-Uk)Y(k,\omega), \quad M=U/c_o$$

for plane waves proportional to $\mathrm{e}^{i(kx-\omega t)}$, $\omega>0$, $Y(k,\omega)$ being the admittance of the wall shear layer. For small damping, show that $k=\pm\kappa_o/(1\pm M)+k'_\pm$, respectively, for downstream and upstream propagating waves, where

$$k'_\pm\approx\frac{2i\rho_o c_o}{(1\pm M)\mathrm{D}_p}Y\left(\frac{\pm\kappa_o}{1\pm M},\omega\right).$$

The component $\mathrm{Im}\,k'_\pm$ governs the damping in the boundary layers. Deduce that $\mathrm{Im}\,k'_\pm=\pm\alpha_\pm$, where

$$\alpha_\pm=\frac{\sqrt{2\omega}}{c_o\mathrm{D}_p(1\pm M)}\mathrm{Re}\left(\sqrt{2}\mathrm{e}^{-\frac{i\pi}{4}}\left[\frac{\sqrt{\nu}}{(1\pm M)^2}\mathcal{F}_o\left(\sqrt{\frac{i\omega\nu}{\kappa_K^2 v_*^2}},\delta_v\sqrt{\frac{i\omega}{\nu}}\right)\right.\right.$$
$$\left.\left.+\frac{\beta c_o^2\sqrt{\chi}}{c_p}\mathcal{F}_o\left(\mathrm{Pt}\sqrt{\frac{i\omega\chi}{\kappa_K^2 v_*^2}},\delta_v\sqrt{\frac{i\omega}{\chi}}\right)\right]\right). \qquad (5.1.31)$$

This formula suggests that damping is strongly dependent on the mean flow Mach number. However, convection by the mean flow was neglected in the low Mach number acoustic boundary layer equation (5.1.22). The validity of (5.1.31) therefore requires that such effects, which are nominally $\mathrm{O}(M)$, should continue to be negligible at small but finite Mach numbers.

The mean flow causes the damping coefficients $\alpha_\pm$ to be different for waves of equal frequency propagating in the upstream and downstream directions. Figure 5.1.5 illustrates this for sound of frequency $\omega/2\pi=88$ Hz propagating in air in a hard-walled circular cylindrical pipe of radius $R=1.5$ cms: $a_\pm/a_o$ predicted by (5.1.31) (when δ_v is given by (5.1.24)) are plotted against

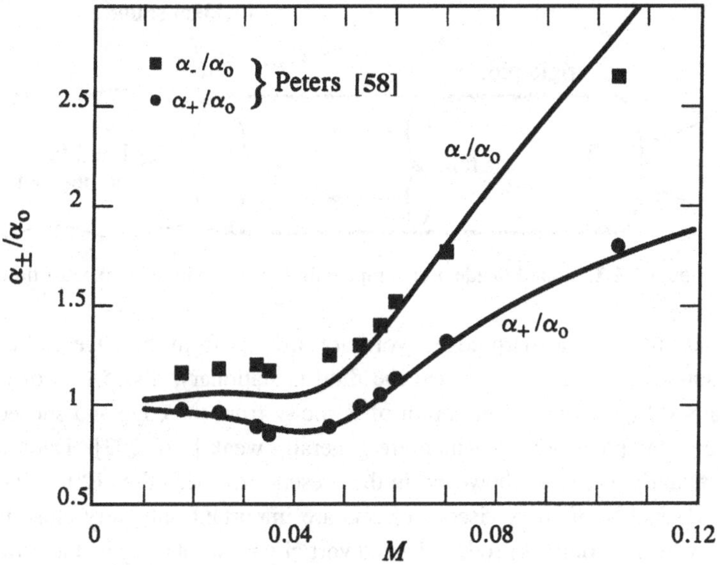

Figure 5.1.5. Attenuation of sound in turbulent pipe flow.

Mach number, where

$$\alpha_o = \frac{\sqrt{2\omega}}{c_o \mathrm{D}_p}\left(\sqrt{\nu} + \frac{\beta c_o^2}{c_p}\sqrt{\chi}\right), \tag{5.1.32}$$

is the damping in the absence of flow (see (5.1.16)). For each value of $M = U/c_o$, the friction velocity v_* is calculated by solving the empirical pipe flow formula [62]

$$U/v_* = 2.44\ln(v_* R/\nu) + 2.0.$$

The frequency ratio ω/ω_* becomes progressively smaller as the Mach number increases, with $\omega \approx \omega_*$ near $M = 0.05$, where the attenuations exhibit local minima, and $\omega/\omega_* \approx 0.3$ at $M = 0.1$. The acoustic boundary layer thickness also increases with Mach number, being equal to about $0.2R$ at $M = 0.1$. The experimental points shown in the figure are due to Peters [58, 286] at the same conditions and frequency.

5.2 Attenuation of Sound by Vorticity Production at Edges

Vorticity is produced by fluid motion over a solid surface, the rate of production being greatest in regions where the pressure and velocity in the primary flow change rapidly, such as at corners and sharp edges. The kinetic energy of the vortex motion is derived from the primary flow, and when this is a sound wave

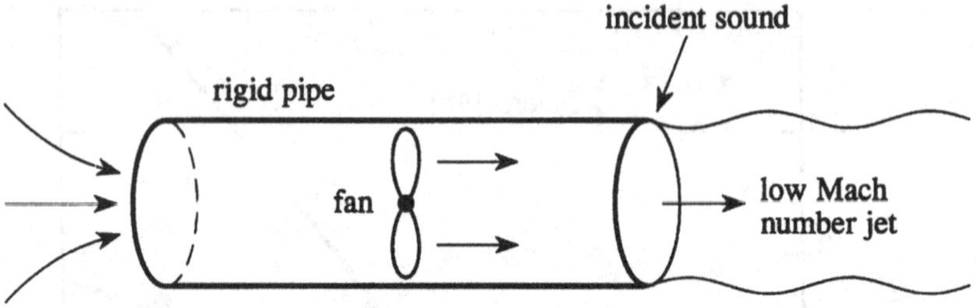

Figure 5.2.1. Sound incident on a pipe with low Mach number exhaust flow.

incident (e.g., on a sharp edge), vorticity diffuses from the edge and causes the sound to be damped. When the fluid is stationary, dissipation of sound is caused by nonlinear convection of vorticity from the edge and subsequent viscous damping, both of which are generally weak [291–293]. Damping is significantly increased, however, in the presence of mean flow [294–308]. At high Reynolds numbers viscous effects are important only very close to the edge, where vorticity is produced; shed vorticity is swept away by the flow, and its kinetic energy is permanently lost to the sound. This mechanism of energy transfer is the subject of this section.

5.2.1 Interaction of Sound with a Solid Surface

To fix ideas, consider the idealized situation of Figure 5.2.1, where sound is diffracted by a rigid pipe in the presence of a low subsonic, homentropic flow produced by a fan within the pipe. Let $\mathbf{v}_\omega(\mathbf{x}, t)$ denote the velocity induced by the vorticity $\boldsymbol{\omega}(\mathbf{x}, t)$ as if the flow were incompressible. In the absence of solid boundaries, $\mathbf{v}_\omega$ is given by the Biot–Savart formula (1.14.3). In the present case, the motion is augmented by the velocity field of an image system of the vorticity in the boundaries; it can always be assumed that this occurs in a manner that ensures that $\operatorname{div} \mathbf{v}_\omega = 0$ and that the normal component of $\mathbf{v}_\omega$ vanishes on the instantaneous boundary positions. The potential component of the motion produced by the fan is expressed in terms of a velocity potential ψ, which satisfies Laplace's equation $\nabla^2\psi = 0$, with normal velocity $\partial\psi/\partial x_n$ on a bounding surface equal to that of the surface. The total fluid velocity $\mathbf{v}$ may then be cast in the form

$$\mathbf{v} = \mathbf{v}_\omega + \nabla\psi + \nabla\varphi, \tag{5.2.1}$$

where the remaining part of the velocity, which accounts for the influence of compressibility, is represented by the velocity potential φ. This includes contributions from incident and diffracted sound waves and from compressible motions caused by the boundary motions and the production of aerodynamic sound.

When a typical flow Mach number $M \ll 1$, differences $\rho - \rho_o$ between the density and its mean value $\sim O(M^2)\rho_o$ (Section 1.6.5), and the total kinetic energy of the fluid is approximately

$$\begin{aligned} E_T &= \frac{1}{2}\rho_o \int (\mathbf{v}_\omega + \nabla(\psi + \varphi))^2 \, d^3\mathbf{x} \\ &\equiv \frac{1}{2}\rho_o \int \mathbf{v}_\omega^2 \, d^3\mathbf{x} + \frac{1}{2}\rho_o \int [\nabla(\psi + \varphi)]^2 \, d^3\mathbf{x}, \end{aligned} \quad (5.2.2)$$

where the integrations are over the fluid. The split into the final two terms follows from the divergence theorem and the condition that the normal component of $\mathbf{v}_\omega$ vanishes on the solid boundaries.

The kinetic energy of the fluid is therefore partitioned between the kinetic energy $E = 1/2\rho_o \int \mathbf{v}_\omega^2 \, d^3\mathbf{x}$ of the vortical motion, and the kinetic energy of the acoustic field and the incompressible, potential flow caused by boundary motion. The latter is localized to the neighborhood of the fan and represents energy temporarily stored in the fluid, in that ψ vanishes when the boundary motion ceases (Section 1.14). However, the continuous operation of the fan generates vortical kinetic energy in the jet exhausting from the pipe, which would ultimately become unbounded in the absence of dissipative processes, whereas the energy associated with ψ is always finite and localized.

The motion of the fan and the interaction of sound with the boundaries generates vorticity, which cause the vortical kinetic energy E to vary. To calculate this change, the momentum equation (1.2.18) for homentropic flow is first cast in the form

$$\rho_o \frac{\partial \mathbf{v}_\omega}{\partial t} + \nabla\left(\rho_o\left[B + \frac{\partial \psi}{\partial t} + \frac{\partial \varphi}{\partial t}\right]\right) = -\rho_o \boldsymbol{\omega} \wedge \mathbf{v} - \eta \,\mathrm{curl}\, \boldsymbol{\omega}, \quad (5.2.3)$$

where (for $M^2 \ll 1$) $B = p/\rho_o + 1/2v^2$ is the total enthalpy, and dissipation terms proportional to div $\mathbf{v}$ (including that associated with the bulk viscosity) are neglected, because viscous effects are important only near boundaries where the motion may be regarded as incompressible. An energy equation is now formed by taking the scalar product of this equation with $\mathbf{v}_\omega$. Noting that $\mathrm{div}\, \mathbf{v}_\omega = 0$, this yields

$$\begin{aligned} &\frac{\partial}{\partial t}\left(\frac{1}{2}\rho_o \mathbf{v}_\omega^2\right) + \mathrm{div}\left(\rho_o \mathbf{v}_\omega \left[B + \frac{\partial \psi}{\partial t} + \frac{\partial \varphi}{\partial t}\right]\right) \\ &\quad = \rho_o \boldsymbol{\omega} \wedge \mathbf{v} \cdot (\nabla \psi + \nabla \varphi) - \eta \mathbf{v}_\omega \cdot \mathrm{curl}\, \boldsymbol{\omega}. \end{aligned} \quad (5.2.4)$$

The argument of the divergence is the flux of vortical kinetic energy when convection by sound and unsteady surface motions is excluded. In irrotational regions, it reduces to the energy flux vector $-\rho_o \mathbf{v}_\omega \partial\Phi/\partial t$, where Φ is the velocity potential less the contributions from surface motions and sound (Section 1.2.8). The first term on the right is a kinetic energy *source* for the vortex field arising from the coupling of the vorticity with the surface motions and the sound.

By integrating over the fluid and using the relation

$$\int \partial f/\partial t \, d^3\mathbf{x} = \partial/\partial t \int f \, d^3\mathbf{x} + \oint_S f \nabla\psi \cdot d\mathbf{S},$$

where S is a solid boundary with surface element $d\mathbf{S}$ directed *into* the fluid, we find

$$\frac{\partial E}{\partial t} + 2\eta e_{ij} e_{ij} = \rho_o \int \boldsymbol{\omega} \wedge \mathbf{v} \cdot (\nabla\psi + \nabla\varphi) \, d^3\mathbf{x} - \frac{1}{2}\rho_o \oint_S \mathbf{v}_\omega^2 \nabla\psi \cdot d\mathbf{S}$$
$$+ \eta \oint_S (\boldsymbol{\omega} \wedge \mathbf{v}_\omega - 2(\mathbf{u}_S \cdot \nabla)\mathbf{v}) \cdot d\mathbf{S}, \qquad (5.2.5)$$

where $\mathbf{u}_S(\mathbf{x}, t)$ is the velocity of the moving boundary, and e_{ij} is the rate of strain tensor (1.2.7). The left-hand side is the sum of the power $\Pi_\omega = \partial E/\partial t$ entering the vortical flow and that dissipated by frictional forces. At high Reynolds numbers, the viscous dissipation occurs mainly in regions close to boundaries where the length scales of the motion are very small and in very small scales of the motion in the body of the fluid. In typical applications, the energy supplied by the source (first term on the right) therefore passes to the vortex motions and is not immediately dissipated by viscous action. The final surface integral is the energy flux associated with the viscous drag; this is usually small compared to the other source terms when the Reynolds number is large and is frequently neglected.

When the boundaries are stationary (fan not operating in Figure 5.2.1), $\psi = 0$ and vorticity production occurs solely as a result of incident sound waves. It follows that when surface losses (which are of the type studied in Section 5.1.1) are neglected, $\Pi_\omega \sim \mathrm{O}(\epsilon^3)$, where ϵ is a measure of the acoustic amplitude. When, however, boundary motions are sufficient to create a substantial mean flow (the jet in Figure 5.2.1), both $\mathbf{v}$ and the vorticity $\boldsymbol{\omega}$ can have large mean components that are independent of the incident sound. In this case, Π_ω is formally $\mathrm{O}(\epsilon^2)$, which suggests that the mean flow can increase considerably the rate of production of vortical energy at the expense of the sound. In many applications, the dominant interaction of the sound (or that component of the sound under investigation) with the vorticity occurs in regions where $\nabla\psi$ is negligible (e.g., near the nozzle exit in Figure 5.2.1 because, according to potential flow theory [4, 17], ψ decreases exponentially with axial distance from the fan). The energy supplied to the vortical flow must then be extracted directly from the sound in accordance with the acoustic dissipation formula:

$$\Pi_\omega \approx \rho_o \int \boldsymbol{\omega} \wedge \mathbf{v} \cdot \mathbf{u}\, d^3\mathbf{x}, \tag{5.2.6}$$

which determines the *rate of dissipation of acoustic energy*, where the acoustic particle velocity $\nabla\varphi$ is now denoted by $\mathbf{u}$. It will be seen below (and in Chapter 6) that Π_ω can be negative, for example, in systems where self-sustaining oscillations are maintained by "acoustic feedback," in which the phase of vorticity production enables a steady transfer of energy to the oscillations from a mean flow.

A special case occurs in the presence of a high-speed uniform mean flow parallel to a system of plane surfaces. For that part of the vorticity generated by the sound, and correct to second order in the acoustic amplitude, the velocity $\mathbf{v}$ in (5.2.6) can be replaced by the mean velocity and the vorticity assumed to convect at the mean stream velocity. The dissipation formula is then valid for *arbitrary* values of the mean flow Mach number, including supersonic! This is because vorticity swept out of the region of interaction with solid boundaries convects at precisely the velocity of the mean stream and produces neither pressure nor density fluctuations. The dissipated power must therefore be contained in the kinetic energy of a fluid of constant density ρ_o, the rate of increase of which must just equal the rate at which acoustic energy is dissipated.

5.2.2 Emission of Sound from a Jet Pipe [297, 301–304]

A simple illustration of the application of (5.2.6) is provided by the problem discussed in Section 3.2.3 of the radiation of internally generated sound from a circular cylindrical pipe in the presence of low Mach number nozzle flow (Figure 3.2.3). For time-harmonic waves of frequency ω, the free-space acoustic far field is given by (3.2.11) in terms of the amplitude p_{I} of the sound incident on the exit from within. In a first approximation the axial velocity perturbation in the nozzle exit plane $v_E \approx 2p_{\mathrm{I}}/\rho_o c_o$, and therefore, by integrating the average acoustic intensity $\langle p^2 \rangle/\rho_o c_o$ over the surface of a large sphere of radius $|\mathbf{x}|$, the acoustic power radiated from the nozzle is given approximately by

$$\Pi_a \approx \frac{\rho_o c_o \langle v_E^2 \rangle (\kappa_o A)^2 [1 + 10M^2/3]}{4\pi}, \tag{5.2.7}$$

where A is the pipe cross-section.

The acoustic power dissipated by the transfer of energy to the kinetic energy of the jet can be calculated from (5.2.6) by using the approximation of Section 3.2.3, where the free shear layer downstream of the nozzle is modeled by an axisymmetric cylindrical vortex sheet in the continuation of the pipe. Then $\mathbf{v}_\omega \approx (v_E, 0, 0)$ within the jet near the nozzle exit, vanishes outside the jet and varies significantly only over distance $\sim$O(U/ω) from the nozzle. Thus, when the nozzle diameter D satisfies $\omega D/U \ll 1$, $\int \langle \boldsymbol{\omega} \wedge \mathbf{v} \cdot \mathbf{u} \rangle \, d^3\mathbf{x} \approx \langle U v_E \oint \mathbf{u} \cdot d\mathbf{S} \rangle$, where the surface integral is over the cylindrical vortex sheet with surface element $d\mathbf{S}$ directed radially outward. However, $\oint \mathbf{u} \cdot d\mathbf{S} \approx A v_E$ for $\kappa_o D \ll 1$ (when the influence of compressibility is negligible) and therefore the acoustic power dissipated by kinetic energy production is

$$\Pi_\omega \approx \rho_o U A \langle v_E^2 \rangle. \tag{5.2.8}$$

The ratio $\Pi_a/(\Pi_a + \Pi_\omega)$ of the total radiated sound power to the total power supplied by the acoustic field within the pipe has been measured by Bechert [309]. His results are plotted against $\kappa_o D$ in Figure 5.2.2 for $M = U/c_o = 0.1$ and 0.3. The curves in this figure represent the prediction

$$\frac{\Pi_a}{\Pi_a + \Pi_\omega} \approx \frac{(\kappa_o D)^2}{(\kappa_o D)^2 + 16M/(1 + 10M^2/3)} \tag{5.2.9}$$

of equations (5.2.7) and (5.2.8) and are in close agreement with experiment when $\kappa_o D$ is small.

Example 1. The formula (1.13.3), $\mathbf{I}_a = (\rho\mathbf{v}' + \rho' U)B'$, for the acoustic intensity implies that the mean power emitted from the jet pipe of Section 5.2.2 is $\Pi = (A/\rho_o c_o)\langle p_I^2 \rangle \{(1+M)^2 - |\mathcal{R}|^2(1-M)^2\}$, where $\mathcal{R}$ is the pressure reflection coefficient, which determines the amplitude of the acoustic wave reflected back upstream within the pipe. Deduce that Π reduces to Π_ω given by (5.2.8) in the zeroth order approximation in which $\mathcal{R} = -1$.

Example 2. Similarly, if the perturbation pressure $p = 0$ in the nozzle exit plane (so that there are no acoustic losses),

$$\begin{aligned}\Pi &= A\langle(\rho\mathbf{v}' + \rho' U)B'\rangle = A\big\langle\big[\rho_o v_E + pU/c_o^2\big]\big[p/\rho_o + U v_E\big]\big\rangle \\ &= \rho_o U A \langle v_E^2 \rangle.\end{aligned}$$

Example 3. Consider a plane acoustic wave incident from upstream on the nozzle exit with a contraction (Figure 3.2.2). Let A, A_J denote the cross-section areas of the pipe upstream of the contraction and at the exit, respectively, and

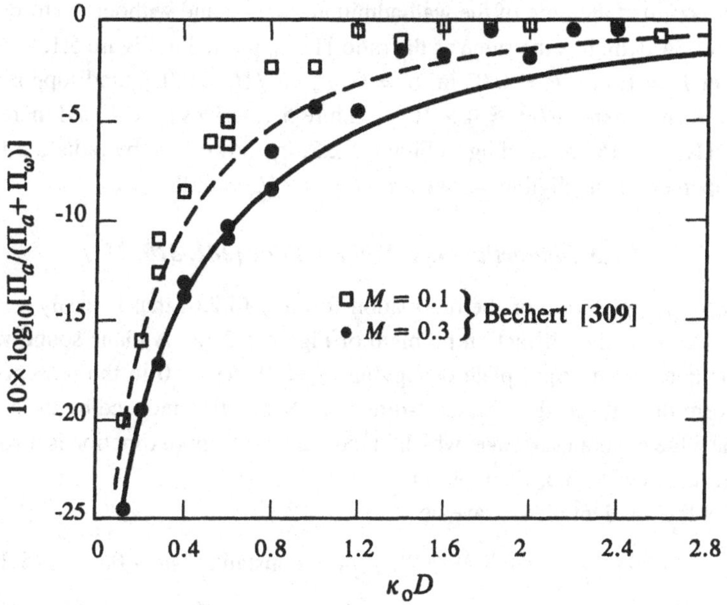

Figure 5.2.2. Absorption of sound in radiation from a jet pipe: The curves are predictions of (5.2.9).

let the contraction extend over a length L. Show that in low Mach number flow with exit Mach number M_J, the acoustic power absorbed by the jet is

$$\Pi_\omega \approx \frac{4\Pi_I M_J (A/A_J)}{1+(\kappa_o \ell)^2}, \quad M_J \to 0,$$

where $\ell \approx \frac{1}{2}L(A-A_J)/A_J$ is a nozzle "end correction" and Π_I is the acoustic power incident on the nozzle from within the pipe [10].

Example 4. A resonant acoustic wave $p = p_o \cos(\kappa_o x_1)\cos(\omega t)$ of amplitude p_o is maintained in a circular cylindrical pipe of length L and radius R. The pipe conveys a fully developed turbulent flow at a low Mach number mean velocity U, which exhausts into free space from the end at $x_1 = L$. The upstream end $x_1 = 0$ may be regarded as acoustically closed, so that the lowest order resonant mode has frequency $\omega \approx \pi c_o/2L$. The acoustic power Π_w dissipated within the pipe in the boundary layer $\approx \frac{\pi}{2} R L p_o^2 \operatorname{Re} Y$, where Y is the effective wall admittance (Section 5.1). The power absorbed by the flow at the pipe exit is $\Pi_E \approx \frac{\pi}{2} R^2 U p_o^2/\rho_o c_o^2$, so that

$$\Pi_w/\Pi_E \approx \alpha(\omega)(L/R\sqrt{2})(v_*/U)(\sqrt{\omega\nu/v_*^2} + (\gamma-1)\sqrt{\omega\chi/v_*^2}),$$

where $\alpha(\omega)$ is the ratio of the wall admittances with and without the turbulent mean flow, which is the same as the ratio Π/Π_o plotted in Figure 5.1.3.

Let $L = 1.5$ m, $R = 0.03$ m, $U = 0.3c_o$, $v_*/U \approx 0.035$, and suppose the fluid is air. Then $\omega\nu/v_*^2 \approx 4 \times 10^{-4}$, Figure 5.1.3 gives $\alpha \approx 4$, and therefore $\Pi_w/\Pi_E \approx 0.15$. According to Figure 5.2.2, the power lost by radiation from the open end is negligible, because $\kappa_o D = \pi R/L = 0.06$.

5.2.3 Interactions at a Trailing Edge [281, 310, 311]

A second application of the dissipation formula (5.2.6) that is easily treated analytically is the diffraction problem of Figure 5.2.3a. A plane sound wave is incident on the rigid plate occupying $x_1 < 0$, $x_2 = 0$ in the presence of uniform flow at speed U in the x_1-direction. Separation induced by the sound at the edge generates a wake, which in the linearized approximation is a vortex sheet (cf. Section 3.5, Example 1).

Let the incident plane wave be

$$p(\mathbf{x}, \omega) = p_{\mathrm{I}} e^{i(n_1 x_1 - n_2 x_2)}, \quad p_{\mathrm{I}} = \text{constant}, \quad \omega > 0. \tag{5.2.10}$$

This satisfies the homogeneous form of equation (1.7.13), with $M = U/c_o$, such that if θ is the angle between the plate and the normal to the incident wavefronts (surfaces of constant phase, see Figure 5.2.3a), then

$$(n_1, n_2) = \kappa_o(\cos\theta, \sin\theta)/(1 + M\cos\theta). \tag{5.2.11}$$

The solution of the diffraction problem is discussed in Example 5.

The wake vorticity $\boldsymbol{\omega} = (v_- - v_+)\delta(x_2)\mathbf{i}_3$ $(x_1 > 0)$, where $v_\pm$ are the x_1-components of velocity just above and below the sheet. Using the solution from Example 5, we find that

$$\boldsymbol{\omega}(x_1, \omega) = \frac{-2p_{\mathrm{I}}}{\rho_o U}\cos(\theta/2)\sqrt{2M(1 + M\cos\theta)}e^{i\kappa x_1}\delta(x_2)\mathbf{i}_3, \quad x_1 > 0, \quad M \le 1, \tag{5.2.12}$$

where $\kappa = \omega/U$ is the *hydrodynamic* wavenumber.

Correct to second order in the acoustic amplitude p_{I}, we can set $\mathbf{v} = U\mathbf{i}_1$ in (5.2.6) to express the dissipated acoustic power per unit length of trailing edge in the form

$$\Pi_\omega = \rho_o U \int_0^\infty \langle \Omega(x_1, t) u_2(x_1, 0, t)\rangle \, dx_1, \tag{5.2.13}$$

where Ω is the real part of the coefficient of $\delta(x_2)\mathbf{i}_3$ in (5.2.12) after restoration of the time factor $e^{-i\omega t}$, and the angle brackets denote a time average.

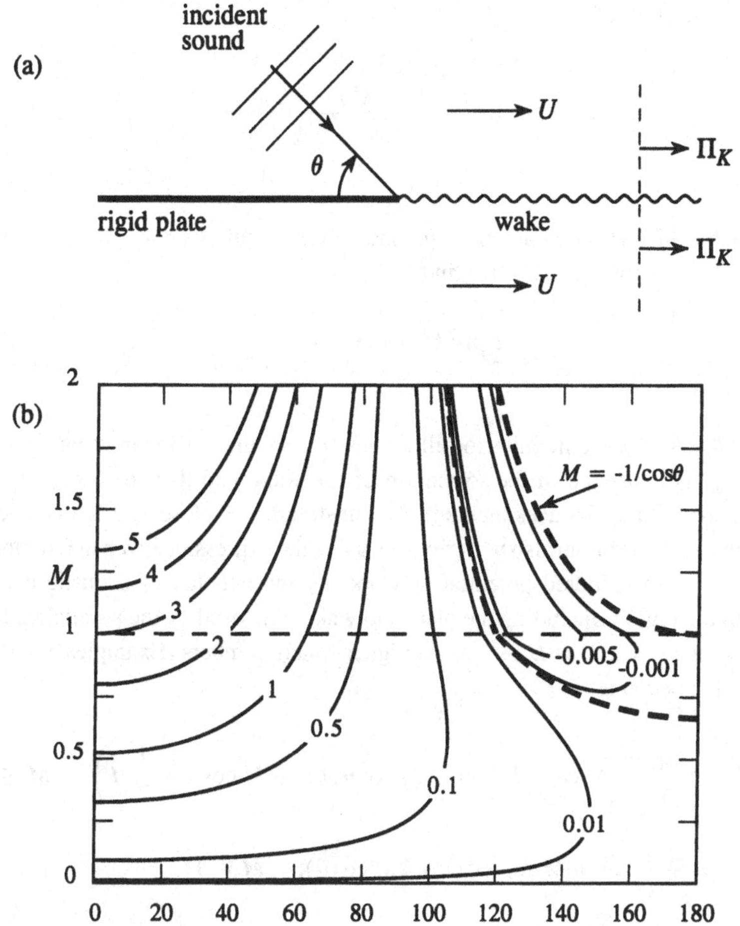

Figure 5.2.3. (a) Sound interacting with a trailing edge and (b) contours of constant values of $\rho_o\omega\Pi_\omega/2|p_I|^2$.

To evaluate this integral, one must determine the acoustic particle velocity $\mathbf{u} = \nabla\varphi$. The component $u_2(x_1, x_2, t)$ is made up of a diffracted velocity plus a component from the incident sound. It is easily confirmed, however, that the latter makes no contribution to (5.2.13), that is, that the dissipation is controlled by the coupling between the scattered acoustic field and the wake. Moreover, as $M \to 0$ it is only the *singular* part of the acoustic particle velocity u_2 very close to the edge that is important because the length scale of the vorticity is then infinitesimal on a scale of the acoustic wavelength and the integral is dominated by the contribution from near the edge. From Example 5, we find for $M \ll 1$

$$u_2(x_1, 0, t) \approx \mathrm{Re}\left(\frac{-p_\mathrm{I}\cos(\theta/2)}{\rho_o\omega}\sqrt{\frac{2\kappa_o}{\pi x_1}}\mathrm{e}^{i(\kappa x_1-\omega t)}\right),$$
$$x_1 > 0, \quad \kappa_o x_1 \ll 1, \tag{5.2.14}$$

which exhibits the characteristic potential flow singularity $\sim 1/\sqrt{x_1}$ at the edge. Substituting into (5.2.13), we find

$$\Pi_\omega \approx \frac{2|p_\mathrm{I}|^2 M\cos^2(\theta/2)}{\rho_o\omega}, \quad M \ll 1. \tag{5.2.15}$$

This is positive definite for all angles of incidence, showing that acoustic energy is expended in the formation of the wake and that the dissipation is greatest for sound waves incident from upstream at θ close to 0°. As a check, equation (5.2.15) can also be derived by using the expression given in Example 5 for the total scattered potential to work out the net flux of acoustic energy from the region (including the plate and wake) enclosed by the parallel control surfaces $x_2 = \pm L$ as $L \to \infty$. At higher Mach numbers (Examples 5 and 6), (5.2.15) becomes

$$\begin{aligned}\Pi_\omega &= \frac{2|p_\mathrm{I}|^2}{\rho_o\omega} M\cos^2(\theta/2)(1+M\cos\theta)\left(1+M\cos\theta-\frac{1}{2}M\right), \quad M \le 1,\\ &= \frac{|p_\mathrm{I}|^2}{2\rho_o\omega}(1+M\cos\theta)^2(1+2M\cos\theta), \quad M \ge 1.\end{aligned} \tag{5.2.16}$$

In the first of these expressions ($M \le 1$), the term $\frac{1}{2}M$ makes a *negative* contribution to the dissipation. The reader can verify that it corresponds to the contribution to the acoustic particle velocity $u_2(x_1, 0, t)$ in (5.2.13) produced by the interaction of the wake with the edge of the plate, that is, to the *generation of trailing edge noise by the wake*. Consequently, when M exceeds about $\frac{2}{3}$ the dissipation can be negative, resulting in a net transfer of energy from the mean flow to the acoustic field. A more detailed overall picture of the dependence of Π_ω on M and angle of incidence θ is shown in Figure 5.2.3b, which depicts contours of constant $\rho_o\omega\Pi_\omega/2|p_\mathrm{I}|^2$. Acoustic energy is dissipated except when (θ, M) lies within the region bounded by the broken curves, on which $\Pi_\omega = 0$. When $M > 1$, the incident wave cannot lie within the top right corner of the figure, bounded by the broken curve on which $\theta = \arccos(-1/M)$; θ cannot exceed this angle because of the convection of sound by the flow.

At large distances downstream of the plate, the motion on either side of the wake is dominated by the kinetic energy of the incompressible, evanescent wave that translates with the vorticity at its mean flow convection velocity U. At such points, the perturbation kinetic energy decays like $e^{-2\omega|x_2|/U}$ with distance from the wake. When $M < 1$, the flux of this kinetic energy is

$$\begin{aligned}\Pi_K &= \int_{-\infty}^{\infty} \langle \rho_o(v_1 + U)B' \rangle \, dx_2 \\ &\equiv -\rho_o \int_{-\infty}^{\infty} \langle \partial\Phi/\partial x_1 \partial\Phi/\partial t \rangle \, dx_2, \quad \text{as } x_1 \to \infty \\ &= (|p_I|^2/\rho_o\omega) M(1 + M\cos\theta)\cos^2(\theta/2), \end{aligned} \qquad (5.2.17)$$

where Φ is the velocity potential of the total diffracted field on either side of the wake. This is usually smaller than the total dissipated power (5.2.16), so that a portion $\Pi_\omega - \Pi_K$ of the dissipated power must be used to increase the kinetic energy of the mean stream, that is, $\Pi_\omega - \Pi_K = (\partial/\partial t) \int_{-\infty}^{\infty} \rho_o U v_{\omega 1} \, dx_1 \, dx_2$, where $v_{\omega 1}$ must be regarded as expanded to *second order* in p_I. There is accordingly a net gain of mean stream momentum in the flow direction which must be balanced by a second order thrust $(\Pi_\omega - \Pi_K)/U$ on the plate tending to accelerate it in the negative x_1-direction (Example 7) [312–314]. At very small Mach numbers, the dissipated energy is shared equally between the vortex wake and the accelerated mean stream.

Example 5. Solve the diffraction problem of Figure 5.2.3a by the Wiener–Hopf procedure described in Section 4.4.1. Show that for arbitrary subsonic Mach number M, the velocity potential Φ of the *total* scattered field satisfying the Kutta condition that the pressure load vanishes at the edge, is given by

$$\Phi(\mathbf{x}, \omega) = \frac{\operatorname{sgn}(x_2) p_I \sqrt{2\kappa_o(1 + M\cos\theta)}\cos(\theta/2)}{2\pi\rho_o\omega} \times \int_{-\infty}^{\infty} \left(\frac{1}{k - n_1 + i0} - \frac{1}{k - \kappa - i0} \right) \frac{e^{i(kx_1 + \Gamma(k)|x_2|)}}{\sqrt{\kappa_o + k(1 + M)}} \, dk,$$

where $\Gamma(k) = \sqrt{[\kappa_o - k(1 + M)][\kappa_o + k(1 - M)]}$. Similarly, show that the velocity potential Φ' of the *incompressible* motion generated by the vorticity (5.2.12) on either side of the wake and plate can be expressed in the form

$$\Phi'(\mathbf{x}, \omega) = \frac{-\operatorname{sgn}(x_2) p_I \sqrt{2M(1 + M\cos\theta)}\cos(\theta/2)}{2\pi\rho_o U\sqrt{\kappa}} \times \int_{-\infty}^{\infty} \frac{e^{ikx_1 - |kx_2|}}{(k - \kappa - i0)\sqrt{k + i0}} \, dk.$$

Observe that this solution has singular velocity $|\nabla\Phi'| \sim 1/\sqrt{|\mathbf{x}|}$ at the edge.

The acoustic velocity component u_2 of the scattered field is equal to $\partial(\Phi - \Phi')/\partial x_2$. The first line of (5.2.16) ($M \leq 1$) is obtained by substitution into (5.2.13). Care must be exercised in evaluating the integral because of the pole at $k = \kappa + i0$. The integration contours in the k-plane should first be displaced to run parallel to the real axis just above the pole (the residue contributions captured by this displacement make null contributions to (5.2.13)). The convergence of the integration with respect to x_1 is now assured for $\mathrm{Im}(k) > 0$, after performing which the k-integrals can be evaluated by residues.

Example 6. When the mean flow in Example 5 is supersonic, show that the scattered potential is

$$\Phi(\mathbf{x}, \omega) = \frac{p_{\mathrm{I}}(1 + M\cos\theta)\,\mathrm{sgn}(x_2)}{2\pi\rho_o\omega} \times \int_{-\infty}^{\infty}\left(\frac{1}{k - n_1 + i0} - \frac{1}{k - \kappa - i0}\right)\mathrm{e}^{i(kx_1 + \Gamma(k)|x_2|)}\,dk,$$

where $\Gamma(k) = \sqrt{\kappa_o - k(M+1)}\sqrt{\kappa_o - k(M-1)}$, and the path of integration runs below *both* of the branch cuts of $\Gamma(k)$, defined such that $\Gamma(k)$ is positive imaginary on the real axis for $\kappa_o/(M+1) < k < \kappa_o/(M-1)$, $\kappa_o > 0$. Show that the wake vorticity is

$$\omega = -2(p_{\mathrm{I}}/\rho_o U)(1 + M\cos\theta)\mathrm{e}^{i\kappa x_1}\delta(x_2)\mathbf{i}_3, \quad x_1 > 0, \quad M \geq 1,$$

which coincides with the subsonic result (5.2.12) when $M = 1$. Derive the dissipation formula (5.2.16) for $M \geq 1$.

Example 7. The sound wave (5.2.10) is incident on a rigid, two-dimensional, flat strip airfoil occupying $|x_1| < a$, $x_2 = 0$, of *compact* chord $2a$ at zero angle of attack to a mean flow of speed U and infinitesimal Mach number M in the x_1-direction.

When $\kappa_o a \ll 1$, the acoustic velocity potential close to the airfoil can be approximated by $\varphi(\mathbf{x}, \omega) = (-ip_{\mathrm{I}}/\rho_o\omega)(1 + i\kappa_o x_1\cos\theta - i\kappa_o X_2\sin\theta)$, $X_2 = \mathrm{Re}\{-i\sqrt{z^2 - a^2}\}$, $z = x_1 + ix_2$. If the wake vorticity $\omega = \gamma_o\mathrm{e}^{i\kappa x_1}\delta(x_2)\mathbf{i}_3$, ($x_1 > 0$, $\kappa = \omega/U$), where γ_o is the circulation per unit length, show by the method of conformal mapping [4, 17] that the velocity potential of incompressible flow generated by the wake is given by

$$\Phi = \frac{\gamma_o}{2\pi}\int_a^{\infty}\mathrm{Re}(i[\ln(\zeta - \xi) - \ln(\zeta - a^2/\xi)])\mathrm{e}^{i\kappa y_1}\,dy_1,$$

where $\zeta = z + \sqrt{z^2 - a^2}$, $\xi = y_1 + \sqrt{y_1^2 - a^2}$. Hence, apply the Kutta condition to obtain

$$\gamma_o = \frac{4ip_{\mathrm{I}} \sin\theta}{\rho_o c_o \left[\mathrm{H}_0^{(1)}(\kappa a) + i\mathrm{H}_1^{(1)}(\kappa a)\right]}.$$

Deduce that, to second order in p_{I}, the acoustic power dissipated per unit length of trailing edge is $\Pi_\omega = \pi \Pi_{\mathrm{I}} \operatorname{Re}\{\mathrm{C}(\kappa a)\} \sin\theta$, where $\mathrm{C}(\kappa a)$ is the Theodorsen function (4.5.17), and $\Pi_{\mathrm{I}} = a|p_{\mathrm{I}}|^2 \sin\theta/\rho_o c_o$ is the acoustic power incident on unit span of the airfoil. The kinetic energy flux in the wake (defined as in (5.2.17)) is

$$\Pi_K = p\Pi_{\mathrm{I}} M \sin\theta \big/ \left[\kappa a \left|\mathrm{H}_0^{(1)}(\kappa a) + i\mathrm{H}_1^{(1)}(\kappa a)\right|^2\right].$$

For large values of the reduced frequency κa, the absorbed sound power is shared equally between the wake and mean stream, but as $\kappa a \to 0$ the wake power becomes negligible. The sound induces a *leading edge suction force* and a tendency for the plate to be propelled against the stream.

Example 8. A two-dimensional, rigid-walled duct occupies the region $-\infty < x_1 < \infty$, $|x_2| < h$ and conveys a uniform flow at subsonic speed U in the x_1-direction. A rigid "splitter plate" extends upstream along the negative x_1-axis (Figure 5.2.4a). If the plane acoustic wave with velocity potential

$$\varphi(\mathbf{x}, \omega) = \mathrm{e}^{i\kappa_o x_1/(1+M)} (M = U/c_o,\ \omega > 0),$$

is incident on the edge of the plate from $x_1 = -\infty$ in the section $0 < x_2 < h$ above the splitter plate, show that for $M \leq 1$ the acoustic power dissipated by vorticity production at the edge is given by [315]

$$\begin{aligned}\frac{\Pi_\omega}{\Pi_{\mathrm{I}}} &= \frac{2M}{(1+M)^2} \\ &\times \prod_{n=1}^{\infty} \left[\left(\frac{1 + M\mu_{n-1/2}}{1 - M\mu_{n-1/2}}\right)\left(\frac{1 - M\mu_n}{1 + M\mu_n}\right)\left(\frac{1 - \mu_{n-1/2}}{1 + \mu_{n-1/2}}\right)\left(\frac{1 + \mu_n}{1 - \mu_n}\right)\right] \\ &\times \left(1 - M(1 - M^2) \sum_{n=1}^{\infty} \left[\frac{\mu_n}{1 - M^2\mu_n^2} - \frac{\mu_{n-1/2}}{1 - M^2\mu_{n-1/2}^2}\right]\right),\end{aligned}$$

where Π_{I} is the incident acoustic power, and $\mu_\nu = \sqrt{1 - (\nu\pi/\kappa_o h)^2(1 - M^2)}$ or 0 according as $(\nu\pi/\kappa_o h)\sqrt{1 - M^2} \lessgtr 1$. Because $\mu_\nu = 0$ for $\kappa_o < (\nu\pi/h)$ $\sqrt{1 - M^2}$, the infinite product and series actually involve only a finite number

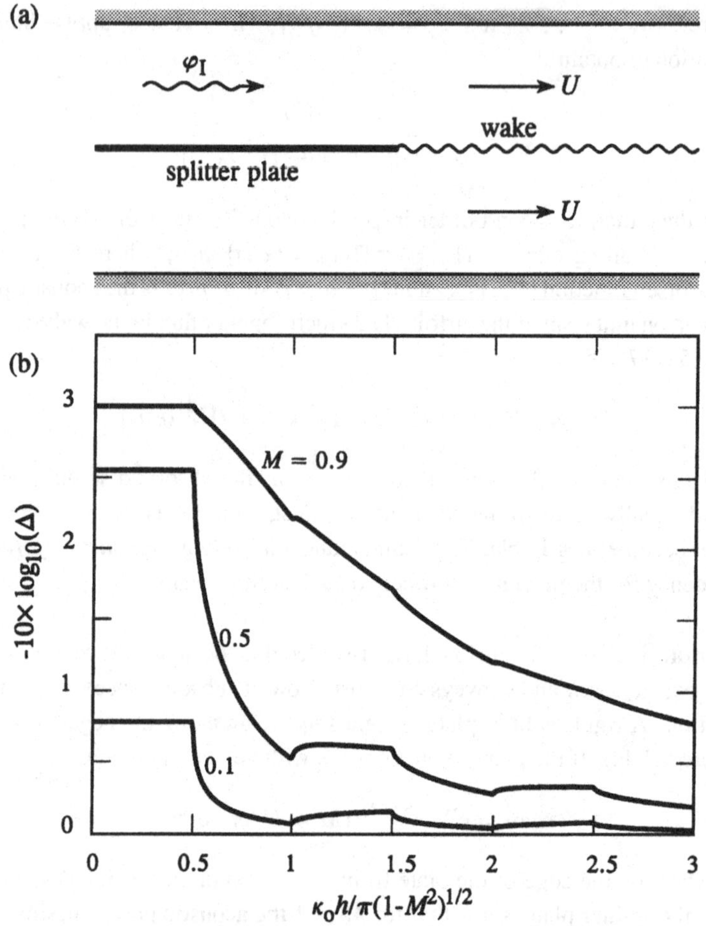

Figure 5.2.4. Dissipation at the edge of a splitter plate.

of terms determined by the acoustic frequency. The critical values at which $\kappa_o = ([n - \frac{1}{2}]\pi/h)\sqrt{1 - M^2}$ or $\kappa_o = (n\pi/h)\sqrt{1 - M^2}$ correspond to the "cut-on" frequencies of the wide and narrow sections of the duct respectively, above which transverse acoustic modes of order n are excited during the interaction at the edge of the splitter plate [316, 317]. Although Π_ω is a continuous function of frequency, this discontinuous excitation of higher order duct modes with increasing frequency produces discontinuities in the slope of the plot of Π_ω against frequency. Figure 5.2.4b exhibits this behavior in the plot of the absorption coefficient $\Delta = 1 - \Pi_\omega/\Pi_I$ against frequency.

At $M = 0.9$, about half the incident energy is dissipated at the edge for very low-frequency incident sound. However, the excitation of transverse modes

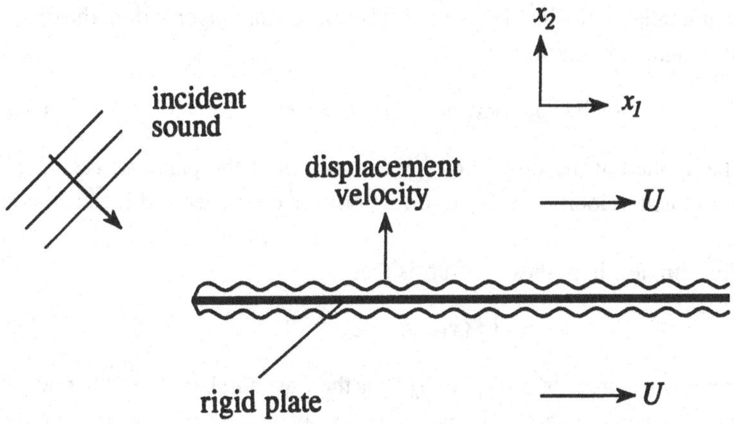

Figure 5.2.5. Interaction of sound with a leading edge.

with increasing frequency greatly reduces the efficiency of absorption by vorticity production, and there is typically a rapid decrease in Δ when the frequency exceeds the lowest cut-on frequency.

5.2.4 Leading-Edge Interactions [311]

According to linear theory there can be no exchange of energy between impinging sound and the mean flow at the leading edge of an airfoil in nominally steady potential mean flow (Figure 5.2.5). This is because vorticity generated at the leading edge remains on the airfoil, and its influence is canceled by that of equal and opposite image vorticity in the rigid surface. In practice, the leading edge motion excites *Tollmien–Schlichting* waves in the surface boundary layers [178] or causes flow separation and the appearance of discrete boundary layer structures. An estimate of the effect of such interactions on energy transfers between the flow and the sound can be made by assuming that the boundary layers remain thin and that the *displacement thickness velocity* fluctuation v_δ at the outer edges of the boundary layers is known in terms of the incident sound [318] (cf. Section 5.1, where the displacement velocity was determined by a surface admittance).

Let us consider the case in which the mean flow Mach number $M \ll 1$, and take the incident sound in the form (5.2.10) with $M = 0$ in (5.2.11). The plate occupies the semi-infinite domain $x_1 > 0,\ x_2 = 0$, and the unsteady motion outside the boundary layers is described by the velocity potential $\Phi + \varphi_{\mathrm{I}}$, where

$$\varphi_{\mathrm{I}}(\mathbf{x}, \omega) = (-ip_{\mathrm{I}}/\rho_o\omega)\mathrm{e}^{i(n_1x_1 - n_2x_2)}$$

is the potential of the incident sound. If the boundary layer is thin, the displacement boundary condition

$$\partial(\Phi + \varphi_I)/\partial x_2 = v_\delta^\pm, \quad x_2 = \pm 0, \quad x_1 > 0, \tag{5.2.18}$$

can be applied at the upper and lower surfaces of the plate, where $v_\delta^\pm$ is the displacement velocity (in the x_2-direction) for the upper and lower boundary layers.

The simplest hypothesis for $v_\delta^\pm$ is that

$$v_\delta^\pm(x_1, \omega) = v_\pm e^{i\kappa_c x_1}, \tag{5.2.19}$$

where $v_\pm$ are constants and $\kappa_c = \omega/U_c$ is the wavenumber of the "displacement thickness waves" generated in the boundary layer at the leading edge, and U_c is their phase speed. The internal dynamics of the boundary layer must determine the precise dependence of κ_c on ω.

The values of $v_\pm$ are found by imposing a *leading edge Kutta condition.* The displacement thickness waves cause fluid to be "pumped" around the edge of the plate. By rewriting (5.2.18) in the form

$$v_2(x_1, \pm 0, \omega) = \left[\frac{1}{2}(v_+ + v_-) \pm \frac{1}{2}(v_+ - v_-)\right] e^{i\kappa_c x_1}, \quad x_1 > 0,$$

it can be seen that the pumping action is controlled by the first terms in the square brackets, which produce an antisymmetric motion with respect to the plane $x_2 = 0$. The normal velocities determined by the second terms in the brackets cannot induce flow about the edge, but they generate an exterior potential flow that is finite everywhere. It follows that $v_+ - v_-$ cannot be determined from a leading-edge Kutta condition and must therefore vanish in the absence of other excitation mechanisms. This is equivalent to setting $v_+ = v_- = v_o$, so that the boundary layer motions on opposite sides of the plate are 180° out of phase, which means that the boundary layer cannot be a net source or sink of fluid. The determination of Φ from these conditions can be formulated as a standard Wiener–Hopf problem whose solution is discussed Example 9.

For a thin boundary layer, the dissipation integral (5.2.6) can be expressed in terms of the displacement thickness velocity v_o and the acoustic particle velocity component u_1 parallel to the plate. To see this, note first that $\boldsymbol{\omega}$ vanishes except within the boundary layers on the plate, where the motion may be regarded as incompressible when $M \ll 1$, and where the divergence of the momentum equation (1.2.18) for isentropic flow reduces to $\nabla^2 B = -\text{div}(\boldsymbol{\omega} \wedge \mathbf{v})$. Integrate this equation across the boundary layer and suppose the boundary layer thickness is small compared to the length scale of variation of B parallel to the plate. Then, because $\partial B/\partial x_2 = -\partial v_2/\partial t$ just outside the boundary layer, $v_\delta(x_1, \omega) = v_o e^{i\kappa_c x_1}$ where

$$\frac{\partial v_\delta}{\partial t} = \frac{\partial}{\partial x_1}\int_0^\infty (\boldsymbol{\omega}\wedge\mathbf{v})_1\, dx_2. \tag{5.2.20}$$

The normal component of the acoustic particle velocity vanishes on the plate; therefore, for a thin boundary layer only the tangential velocity u_1 makes a significant contribution to (5.2.6), the behavior of which near the leading edge is given by (Example 9)

$$u_1(x_1, \pm 0, \omega) \approx \pm(p_I/\rho_o\omega)\sqrt{2i\kappa_o/\pi x_1}\,\sin(\theta/2), \quad x_1 > 0, \tag{5.2.21}$$

when $M \ll 1$. Then (after restoring the time factor $e^{-i\omega t}$ and taking real parts, etc.), (5.2.6) yields

$$\Pi_\omega \approx 2\rho_o \int_0^\infty u_1(x_1, 0, t)\int_0^\infty (\boldsymbol{\omega}\wedge\mathbf{v})_1\, dx_2\, dx_1,$$

where the factor of two accounts for equal contributions from opposite sides of the plate. However, (5.2.20) implies $\int_0^\infty(\boldsymbol{\omega}\wedge\mathbf{v})_1(\mathbf{x},\omega)\,dx_2 \approx -(\omega v_o/\kappa_c)e^{i\kappa_c x_1}$, and Example 9 supplies

$$v_o = -(p_I/\rho_o U)\sqrt{2M}\,\sin(\theta/2), \tag{5.2.22}$$

so that

$$\Pi_\omega \approx \frac{-2|p_I|^2 M_c \sin^2(\theta/2)}{\rho_o\omega}, \quad M_c = U_c/c_o \ll 1.$$

This result can also be derived by calculating the net flux of acoustic energy through the control surfaces $x_2 = \pm L$ ($L \to \infty$) from the solution of Example 9.

Thus, in contrast to the trailing edge problem of Section 5.2.3, a leading-edge interaction at low Mach numbers results in the transfer of mean flow energy to the acoustic energy. It is a maximum when the impinging sound wave is incident from a downstream direction ($\theta = \pm\pi$); the wave produces a large perturbation of the flow near the edge, which is compensated by the generation of displacement thickness waves of proportionate amplitude. Experiments with laminar boundary layers [319] appear to confirm this direct proportionality between the amplitudes of the incident sound and boundary layer Tollmien–Schlichting waves.

It is not hard to see from the above discussion and Section 5.2.3 that the fundamental difference between leading and trailing edges is independent of the structure of the incident sound field. In fact, *any* mechanism that produces a reciprocating flow about an edge will bring about a net transfer of energy to or from the mean flow. This conclusion is important in the study of mean flow induced resonant oscillations (Chapter 6).

Example 9. In the diffraction problem of Figure 5.2.5, the velocity and velocity potential must be continuous across the upstream extension of the plate. In the notation of Section 5.2.3, the potential Φ of the scattered field can therefore be written

$$\Phi(\mathbf{x}, \omega) = \operatorname{sgn}(x_2) \int_{-\infty}^{\infty} \mathcal{L}(k) \mathrm{e}^{i(kx_1 + \gamma(k)|x_2|)}\, dk,$$

where $\mathcal{L}(k)$ is regular and of algebraic growth in $\operatorname{Im} k < 0$. Conditions (5.2.18), (5.2.19) (with $v_\pm = v_o$) supply a functional relation of the Wiener–Hopf type from which one finds

$$\mathcal{L}(k) = \frac{1}{2\pi\sqrt{\kappa_o - k}} \left(\frac{v_o}{(k - \kappa_c - i0)(\kappa_o + \kappa_c)} - \frac{\sqrt{2\kappa_o}(p_{\mathrm{I}}/\rho_o\omega)\sin(\theta/2)}{k - \kappa_o \cos\theta - i0} \right).$$

When the value (5.2.22) is chosen for displacement velocity v_o, then $\mathcal{L}(k) \sim \mathrm{O}(1/k^{5/2})$ as $|k| \to \infty$, and this is sufficient to ensure that the velocity and pressure are finite at the leading edge (Kutta condition). Thus,

$$\Phi(\mathbf{x}, \omega) = \frac{\operatorname{sgn}(x_2)\sqrt{\kappa_o/2}\, p_{\mathrm{I}} \sin(\theta/2)}{\pi \rho_o \omega} \int_{-\infty}^{\infty} \frac{1}{\sqrt{\kappa_o - k}} \times \left(\frac{1}{k - \kappa_c - i0} - \frac{1}{k - \kappa_o\cos\theta - i0} \right) \mathrm{e}^{i(kx_1 + \gamma(k)|x_2|}\, dk.$$

In the limit $M \to 0$, the first term in the brace brackets of the integrand yields the velocity potential of the incompressible velocity field $\mathbf{v}_\omega$ outside the boundary layers. The second term gives the scattered acoustic particle velocity, which is used to derive the approximation (5.2.21) for $u_1(x_1, \pm 0, \omega)$.

Example 10. The sound wave (5.2.10) impinges on a composite wall at $x_2 = 0$ in the presence of uniform flow at speed U in the x_1-direction (Figure 5.2.6a). The region $x_1 > 0$ of the wall is coated with a "pressure release" material (on which $p \equiv 0$), and the section $x_1 < 0$ is rigid. Acoustic energy is conserved at

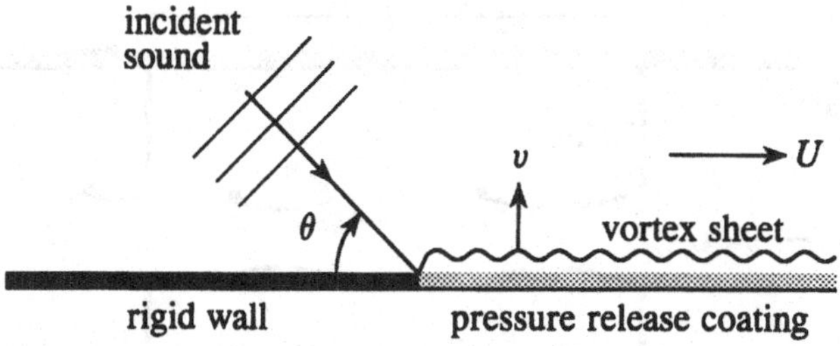

Figure 5.2.6. Interaction of sound with linear discontinuity.

the wall in the absence of flow. In the presence of flow, displacement thickness waves $v_\delta(x_1, \omega) = v_o e^{i\kappa x_1}$ ($\kappa_c = \omega/U_c$) are generated at the join $x_1 = 0$, where the magnitude of v_o ensures that the pressure and velocity remain finite at $x_1 = 0$ [320]. Linear theory requires that the displacement thickness phase velocity U_c should equal U if the pressure is to vanish over the pressure release surface. Show that

$$\Pi_\omega = \frac{4|p_{\mathrm{I}}|^2}{\rho_o \omega} M \cos^2(\theta/2)(1 + M\cos\theta)\left(1 + M\cos\theta - \frac{1}{2}M\right), \quad M \leq 1,$$

which is twice the trailing edge result (5.2.10).

5.2.5 Application to Resonant Oscillations

Vortex shedding and interactions with structures are sources of sound and can be responsible for the excitation of cavity resonances in ducts, rocket motor combustion chambers, heat exchangers, and so on. A resonance is usually strongly excited only at certain mean flow velocities because, as we have seen, sound can also be absorbed during vorticity production, and it is only when certain phase relations are satisfied that production mechanisms become dominant.

The dissipation integral (5.2.6) can often be used to predict the flow conditions for resonance. Figure 5.2.7a depicts a short baffled section of length L of a long duct of square cross-section conveying air at low Mach number. In an experimental arrangement of this kind [321] a longitudinal acoustic mode (of fixed frequency and wavelength equal approximately to the whole length of the duct) was excited at Strouhal numbers $fL/U \approx 4.7,\ 7.7$ and 11.0, where U is the mean velocity just upstream of the baffled section. For these conditions, respectively one, two, and three vortex pairs were observed within the baffled section, generated by shedding from the upstream baffles. By measuring the amplitude of the acoustic pressure fluctuations at the resonance frequency for

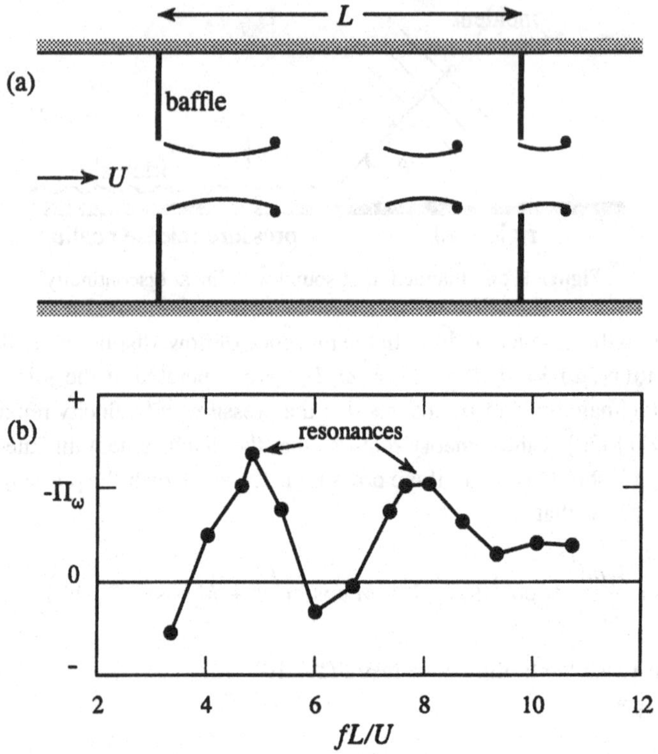

Figure 5.2.7. (a) Baffled section of a duct with flow and (b) mean power flux (in arbitrary units); adapted from [321].

a range of mean flow velocities, the following four-step numerical procedure can be used to determine whether the acoustic excitation can be attributed to the unsteady vortex flow in the baffled section of the duct:

1. The acoustic field is determined by solving the linearized acoustic equations at the observed resonance frequency f. The amplitude of the sound field is then set equal to measured value.
2. The calculated acoustic field is incorporated into an incompressible numerical solution of the Navier–Stokes equation (via the no-slip boundary condition and the contribution of the acoustic particle velocity to the convection of vorticity) to determine the velocity and vorticity distributions in the duct. This procedure locks the vortex shedding rate onto the acoustic frequency.
3. The average power $\langle -\Pi_\omega \rangle$ fed into the acoustic field is calculated using (5.2.6).

4. Steps 1–3 are repeated for a range of flow velocities U, and $\langle -\Pi_\omega \rangle$ is plotted as a function of fL/U. Those values of the Strouhal number where $\langle -\Pi_\omega \rangle$ is large and positive are identified with flow conditions at which resonant excitation is possible.

Figure 5.2.7b (adapted from [321]) shows calculated values of $\langle -\Pi_\omega \rangle$ plotted against fL/U. The two indicated resonances occur at Strouhal numbers close to the observed values of 4.7 and 7.7, but the higher resonance at $fL/U \approx 11$ is not well resolved. The principal contribution to the power flux integral (5.2.6) occurs when the large-scale vortex structures (idealized as vortex pairs in Figure 5.2.7a) encounter the downstream baffles. At resonance, the principal components of $\boldsymbol{\omega} \wedge \mathbf{v}$ tend to be locally parallel to the acoustic particle velocity $\mathbf{u}$ and to be in just the right phase to maximize the energy flux and ensure the continued maintenance of the acoustic field. Note that, unlike resonance phenomena of the edgetone type (Section 6.3), the resonance is not maintained by a feedback loop in which a pressure pulse generated at the downstream baffle triggers the shedding of a new vortex [322]. In the present case, the feedback is via the resonance: The motion in the acoustic field that induces vortex shedding from the upstream baffles is generated over many cycles and has been reflected back-and-forth many times between the ends of the duct [79].

5.3 Interactions with Perforated Screens

The absorption of acoustic energy by vorticity production can be exploited to enhance the effectiveness of acoustic liners and perforated screens: "Jetting" in surface apertures increases the rate of energy conversion, and the results of Section 5.2 suggest that the increase can be even larger in the presence of mean flow, because vorticity generated by sound at aperture edges is convected away and its energy is permanently lost to the sound [295, 296, 323]. Most practical screens used to "absorb" sound are perforated with small (usually circular) apertures. The apertures may still be large on the scale of the viscous motions, however, in which case the influence of viscosity is insignificant except near the edges. This is important if enhanced levels of dissipation are to be achieved by mean flow convection. In applications, screens may be of the "grazing" or "bias" flow type and are used in mean flow environments, such as in heat exchanger cavities [300, 305] to damp harmful acoustics resonances. The acoustic properties of the screen can usually be expressed in terms of the *Rayleigh conductivity* of the apertures, and this concept will be discussed before considering the influence of mean flow.

5.3.1 Rayleigh Conductivity

Consider the unsteady flow through an acoustically compact circular aperture of radius R in a rigid wall of thickness h produced by a uniform, time-dependent pressure differential $[p] = p_+(t) - p_-(t)$. The "upper" and "lower" faces of the wall are assumed to coincide respectively with $x_2 = \pm\frac{1}{2}h$, and p_+, p_- are the pressures above and below the wall. The motion near the aperture is regarded as incompressible, with volume flux $Q(t)$ through the aperture (in the x_2-direction), and fluctuating quantities are proportional to $\mathrm{e}^{-i\omega t}$. The Rayleigh conductivity K_R of the aperture is defined by [10, 324]

$$K_R = i\omega\rho_o Q/(p_+ - p_-). \tag{5.3.1}$$

When there is no mean flow on either side of the wall, this is equivalent to

$$K_R = Q/(\varphi_+ - \varphi_-), \tag{5.3.2}$$

where $\varphi_\pm$ are the *uniform* velocity potentials above and below the wall, which satisfy $p\pm = -\rho_o\partial\varphi_\pm/\partial t$. Because of the motion, the pressure and velocity potential cannot actually be constant close to the aperture; $p_\pm$ and $\varphi_\pm$ are the limiting values of these quantities at large distances ($\gg R$) from the aperture. Conductivity has the dimensions of length, and (5.3.2) is the analogue of *Ohm's law*, which determines the conductivity of an electrical circuit element as the ratio of the current to the potential difference.

It is sometimes useful to express the unsteady motion in terms of a length $\ell = A/K_R$, where A is the area of the aperture. Then, if V is the area averaged velocity in the aperture, equation (5.3.1) is equivalent to

$$\rho_o\ell\frac{dV}{dt} = -[p] \equiv p_- - p_+, \tag{5.3.3}$$

so that ℓ may be interpreted as the effective length of a slug of fluid of cross section A that participates in the motion through the aperture.

Let us determine K_R for an ideal (inviscid) fluid when $h \to 0$, in the absence of mean flow (Figure 5.3.1). The flow produced by the uniform potential difference $\varphi_+ - \varphi_-$ can be described by a velocity potential Φ; the total potential is then $\varphi = \varphi_\pm + \Phi$ above and below the wall, respectively. Φ is a solution of the Laplace equation for incompressible motion, and $\Phi \sim \mp Q/2\pi|x|$ as $x_2 \to \pm\infty$. Symmetry demands that the motion in the aperture must be in the normal direction, so that Φ must be constant there. The general solution (1.10.8) applied to incompressible flow (with $\rho = \rho_o$, $M = 0$, $\hat{\mathcal{F}} = 0$) can be

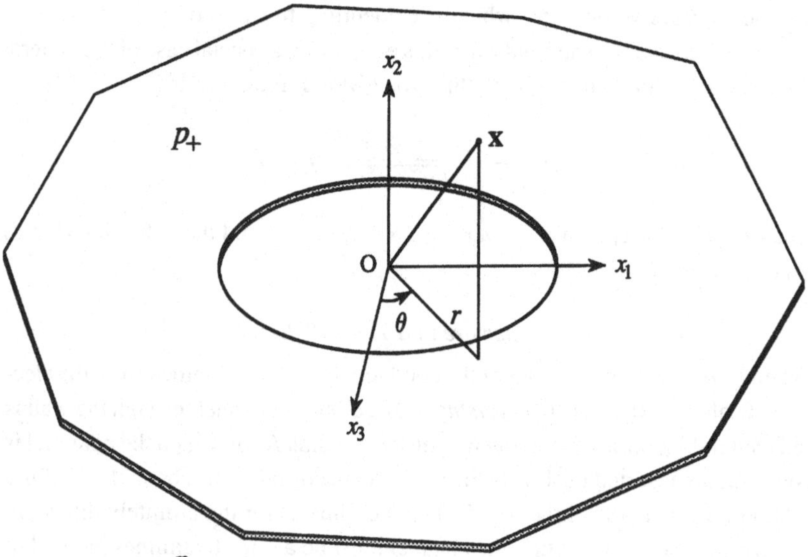

Figure 5.3.1. Circular aperture in a thin wall.

used to express Φ in terms of the velocity $v_2(\mathbf{x})$ at the wall and Green's function

$$G(\mathbf{x}, \mathbf{y}, \omega) = \frac{-1}{4\pi|\mathbf{x}-\mathbf{y}|} + \frac{-1}{4\pi|\mathbf{x}-\mathbf{y}'|}, \quad y' = (y_1, -y_2, y_3), \tag{5.3.4}$$

which has vanishing normal derivatives $\partial G/\partial x_2$, $\partial G/\partial y_2$ respectively on x_2 and $y_2 = 0$, that is

$$\varphi(\mathbf{x}) = \varphi_\pm - \frac{\operatorname{sgn}(x_2)}{2\pi}\int_{-\infty}^{\infty} \frac{v_2(y_1, \pm 0, y_3)}{|\mathbf{x}-\mathbf{y}|}\, dy_1\, dy_3, \quad (y_2 = 0), \tag{5.3.5}$$

where the integration is over the upper–lower surfaces $y_2 = \pm 0$ respectively for $x_2 \gtrless 0$. Because $v_2 \equiv 0$ on the rigid portions of the wall, the integration may be restricted to the region S occupied by the aperture provided v_2 has only integrable singularities at the aperture edge. The condition that φ (and the pressure) should be continuous across the aperture now supplies the following integral equation

$$\oint_S \frac{v_2(y_1, 0, y_3)}{|\mathbf{x}-\mathbf{y}|}\, dy_1\, dy_3 = \pi(\varphi_+ - \varphi_-),$$

$$\text{for } x_2 = y_2 = 0, \quad r = \sqrt{x_1^2 + x_3^2} < R \tag{5.3.6}$$

for the aperture velocity v_2, which is evidently a function of r alone, because $\varphi_+ - \varphi_-$ is constant. The solution of this equation is a special case of the general solution of integral equations of this type given in Example 1:

$$v_2(r) = \frac{(\varphi_+ - \varphi_-)}{\pi\sqrt{R^2 - r^2}}, \quad r < R.$$

Then $Q = \oint_S v_2(x_1, 0, x_3)\, dx_1\, dx_3 = 2R(\varphi_+ - \varphi_-)$, and the definition (5.3.2) yields $K_R = 2R$.

Aperture in a Thick Wall

When $h \neq 0$, an upper bound for the conductivity can be obtained from the electrical analogy [10, 324]. The *resistance* of a circular channel through the wall is reduced if thin, infinitely conducting disks of radius R (on which the potential is constant) are inserted flush with the wall over each end of the channel. The fluid motion on either side of the wall is then the same as for the infinitely thin wall. The hypothetical flow within the channel must be along streamlines parallel to the channel walls at uniform velocity $U \equiv Q/\pi R^2$; the potential difference between the ends of the channel is Uh and the volume flux is $\pi R^2 U$, so that the channel conductivity is $\pi R^2/h$. The total resistance to the flow through the wall is therefore equal to the sum of a part $1/2R$ due to the circular openings and a part $h/\pi R^2$ from the channel, and the actual conductivity of the channel therefore satisfies

$$K_R < \frac{\pi R^2}{h + \pi R/2}. \tag{5.3.7}$$

A *lower* bound for K_R can be obtained from the principle that the kinetic energy $\mathrm{T} = \frac{1}{2}\rho_o \int (\nabla\varphi)^2\, d^3\mathbf{x}$ of an ideal, incompressible flow through the aperture is smaller than any other possible motion satisfying the condition of vanishing normal velocity on the wall [17]. By performing the integration over the fluid on either side of the wall contained within a large sphere centered on the aperture (on the surface of which the velocity potential $\varphi = \varphi_\pm$ in $x_2 \gtrless 0$), we find by application of the divergence theorem that $\mathrm{T} = \frac{1}{2}\rho_o(\varphi_+ - \varphi_-)Q \equiv \frac{1}{2}\rho_o Q^2/K_R$. This is less than the kinetic energy calculated for any other motion that is consistent with continuity and the boundary conditions, and such a calculation would therefore provide a lower bound for K_R. For example, the introduction of a rigid obstacle or boundary into the flow will always produce an increase in T, because the new motion could have existed previously with the fluid displaced by the obstacle at rest. Rayleigh [10, 324] considered the case in which infinitely thin, rigid circular pistons are introduced at each end of the channel; the normal (x_{2-}) component of velocity at the openings is then constant and equal to $Q/\pi R^2$ (the motion within the channel being uniform, as previously). By calculating the corresponding velocity potential and kinetic energy T′ of the motion on either side of the wall, it was deduced that

$$K_R > \frac{\rho_o Q^2}{2\mathrm{T}'} = \frac{\pi R^2}{h + 16R/3\pi}. \tag{5.3.8}$$

It is instructive to express these conclusions in terms of the length ℓ of equation (5.3.3). The conductivity of the channel alone is $\pi R^2/h$, for which ℓ would equal h. Thus, (5.3.7) and (5.3.8) can be expressed in terms of an end correction ℓ', defined by

$$\ell \equiv A/K_R = h + \ell',$$

so that ℓ' is the amount by which h must be increased to account for the contributions from the openings. The end correction for a *single* opening therefore satisfies

$$\frac{\pi R}{4} < \ell' < \frac{8R}{3\pi}, \tag{5.3.9}$$

(or $0.785R < \ell' < 0.849R$). More detailed calculations [10] indicate that the correct value of $\ell' \approx 0.82R$. The end correction can be neglected when $h \gg R$.

Example 1: Copson's method [170, 325]. Let the right-hand side of the integral equation (5.3.6) be an arbitrary function $f(r,\theta)$ of the polar coordinates (r,θ) in the plane of the aperture. Suppose $f(r,\theta)$ has the expansion

$$f(r,\theta) = \sum_{n=0}^{\infty} f_n(r)\cos\{n(\theta-\theta_n)\}, \quad 0 \le r \le R, \quad 0 \le \theta \le 2\pi,$$

where $\theta_0, \theta_1, \ldots,$ are constants. Then

$$v_2(r,\theta) = \sum_{n=0}^{\infty} \mathcal{V}_n(r)\cos\{n(\theta-\theta_n)\}, \quad 0 \le r \le R,$$

where

$$\mathcal{V}_n(r) = -\frac{2r^{n-1}}{\pi R^n}\frac{\partial}{\partial r}\int_r^R \frac{t\chi_n(t)\,dt}{\sqrt{t^2-r^2}}, \quad \chi_n(r) = \frac{R^n}{2\pi r^{2n}}\frac{\partial}{\partial r}\int_0^r \frac{t^{n+1}f_n(t)\,dt}{\sqrt{r^2-t^2}}.$$

The volume flux is given by

$$Q = \oint_S v_2\, dx_1\, dx_3 = \frac{2}{\pi}\int_0^R \frac{t f_0(t)\, dt}{\sqrt{R^2 - t^2}}. \tag{5.3.10}$$

Example 2. The conductivity of an *elliptic* aperture of area A (with major and minor axes a and b) in a thin, rigid wall is given by [10, 324]

$$K_R = \frac{\sqrt{\pi A}}{\mathrm{K}(\epsilon)(1-\epsilon^2)^{1/4}} \approx 2\sqrt{\frac{A}{\pi}}\left(1 + \frac{\epsilon^4}{64} + \frac{\epsilon^6}{64} + \cdots\right),$$

where ϵ $(= \sqrt{1 - b^2/a^2})$ is the eccentricity of the ellipse, and $\mathrm{K}(x) = \int_0^{\pi/2} d\theta / \sqrt{1 - x^2 \sin^2\theta}$ is the complete elliptic integral of the first kind [289].

The circular aperture ($\epsilon = 0$) has the minimum conductivity for apertures of given area. The conductivity of the ellipse is less than 3% larger than the conductivity of the circular aperture of the same area provided $a/b < 2$ ($\epsilon < 0.866$). This suggests that the conductivity of any elongated aperture is approximately equal to that of a circular aperture of the same area. More generally, the conductivity is always increased if an aperture of given area is elongated or broken up into a set of smaller apertures that are sufficiently far apart to act independently. As $a/b \to \infty$ ($\epsilon \to 1$), $\mathrm{K}(\epsilon) \approx \ln\{4/\sqrt{1-\epsilon^2}\}$, so that $K_R \to \infty$.

Example 3: Conductivity of a nearly cylindrical tube of revolution [324]. Let $R(x)$ be the tube radius, where x is distance along the axis of symmetry. Introduce an indefinite number of infinitely thin, perfectly conducting planes perpendicular to the axis, on each of which the potential must be constant so that the resistance of the passage is reduced. Over an interval δx, the reduced value of the resistance is $\delta x / \pi R^2(x)$, and the conductivity of the tube therefore satisfies $1/K_R > \int_a^b dx/\pi R^2(x)$, where $x = a,\ b$ at the tube ends. This yields an upper bound for K_R.

The minimum kinetic energy principle is next applied to find a lower bound. By introducing rigid pistons perpendicular to the axis, the axial velocity $u = Q/\pi R^2(x)$ will be constant over a cross-section, and the radial velocity (obtained from the continuity equation for incompressible flow) becomes $w = -\frac{1}{2} r \partial/\partial x\{Q/\pi R^2(x)\}$. Calculating the kinetic energy, we obtain the upper bound

$$T \equiv \rho_o Q^2/2K_R < (\rho_o Q^2/2\pi)\int_a^b \left[\left\{1 + \frac{1}{2}(dR/dx)^2\right\}\Big/ \pi R^2(x)\right] dx.$$

The conductivity therefore satisfies

$$\int_a^b \frac{dx}{\pi R^2(x)} < \frac{1}{K_R} < \int_a^b \left[1 + \frac{1}{2}\left(\frac{dR}{dx}\right)^2\right] \frac{dx}{\pi R^2(x)}, \tag{5.3.11}$$

and either formula accordingly supplies an approximation for K_R with an error of order $\int_a^b [(dR/dx)^2/2\pi R^2(x)]\,dx$. This result can be combined with the upper and lower bounds (5.3.7), (5.3.8) to include the influence of flanged openings.

Example 4. Show that the conductivity of a conical tube of radius

$$R(x) = [R_1(b-x) + R_2(x-a)]/(b-a), \quad a < x < b,$$

with infinite flanged openings at $x = a$ and $x = b$ satisfies,

$$\frac{b-a}{\pi R_1 R_2} + \frac{1}{4}\left(\frac{1}{R_1} + \frac{1}{R_2}\right) < \frac{1}{K_R} < \frac{b-a}{\pi R_1 R_2} + \frac{(R_1 - R_2)^2}{\pi R_1 R_2 (b-a)} + \frac{8}{3\pi^2}\left(\frac{1}{R_1} + \frac{1}{R_2}\right).$$

Example 5: The Helmholtz resonator. Consider very low-frequency simple harmonic compressions and rarefactions of the air in a cavity of volume V that communicates with the ambient atmosphere through a narrow-necked channel of cross-sectional area A and length ℓ. The pressure and density within V may be regarded as uniform, and the kinetic energy neglected except within the neck, where the motion may be treated as incompressible. If the reciprocating volume flux from V is $Q = Q_o \cos(\omega t)$, the mean kinetic energy $T \approx \frac{1}{4}\rho_o Q_o^2/K_R$, where K_R is the conductivity of the channel. The pressure fluctuation within V is $p \approx -\rho_o c_o^2 A\zeta/\mathrm{V}$, where ζ is the outward displacement of fluid in the neck from the undisturbed state; the mean potential energy within V is therefore $\frac{1}{4}\rho_o c_o^2 Q_o^2/\omega^2 V$. The condition that the mean potential and kinetic energies are equal in simple harmonic motion then yields the formula for the resonant frequency:

$$\omega \approx c_o\sqrt{\frac{K_R}{\mathrm{V}}} = c_o\sqrt{\frac{A}{LV}}, \tag{5.3.12}$$

where $L = \ell + \ell'_+ + \ell'_-$, and $\ell'_\pm$ are end corrections for the outer and inner openings of the neck.

Example 6. According to the minimum kinetic energy principle, the end correction ℓ for the open end of a long, thin walled, circular cylindrical pipe of radius R is smaller than that ($\approx 0.82R$) for the same pipe with a flanged opening. Detailed calculation reveals that $\ell \approx 0.61R$ [205, 326].

5.3.2 Reflection and Transmission of Sound by a Perforated Screen

The reflection and transmission coefficients for long wavelength sound incident on a homogeneously perforated screen are easily expressed in terms of the conductivities of the component apertures, provided the local motion through an aperture may be assumed to be incompressible and is identical in structure to the flow in an isolated aperture in an infinite plane [327–329]. The latter condition is fulfilled when the distance between neighboring apertures is much larger than the aperture diameter.

Let the screen occupy the plane $x_2 = 0$, and suppose the apertures have equal conductivities K_R and that there are $\mathcal{N}$ per unit area of the screen. Consider a plane sound wave of the form (5.2.10) incident on the screen from above, where the acoustic wavelength is very much larger than the shortest distance d between neighboring apertures. Reflected and transmitted waves are generated at the screen, such that at distances $|x_2| \gg d$ from the screen the pressure can be written

$$\begin{aligned} p(\mathbf{x}, \omega) &= p_{\mathrm{I}}(\mathrm{e}^{-n_2 x_2} + \mathcal{R}\mathrm{e}^{in_2 x_2})\mathrm{e}^{in_1 x_1}, \quad && x_2 \gg d, \\ &= p_{\mathrm{I}}\mathcal{T}\mathrm{e}^{i(n_1 x_1 - n_2 x_2)}, && x_2 \ll -d. \end{aligned} \tag{5.3.13}$$

The more general situation involving mean flow on both sides of the screen is illustrated in Figure 5.3.2; for the moment any mean flow effects are ignored.

Because the acoustic wavelength is large compared to the aperture size and spacing, the velocity potentials $\varphi_\pm$ of the *acoustic field* above and below the wall may be regarded as constant over an aperture, in terms of which the volume flux Q through an aperture is then given by equation (5.3.2). Because there are $\mathcal{N}$ apertures per unit area, the total flux per unit area is $\mathcal{N}Q \equiv \mathcal{N}K_R(\varphi_+ - \varphi_-)$. Continuity requires $\partial\varphi_+/\partial x_2 = \partial\varphi_-/\partial x_2 = \mathcal{N}K_R(\varphi_+ - \varphi_-)$ at the screen, which is the same as

$$\partial p_+/\partial x_2 = \partial p_-/\partial x_2 = \mathcal{N}K_R(p_+ - p_-), \quad x_2 = 0, \tag{5.3.14}$$

where $p_\pm$ denotes the acoustic pressure above and below the screen, given by (5.3.13). The quantity $\mathcal{N}K_R$ is the effective *compliance* of the perforated

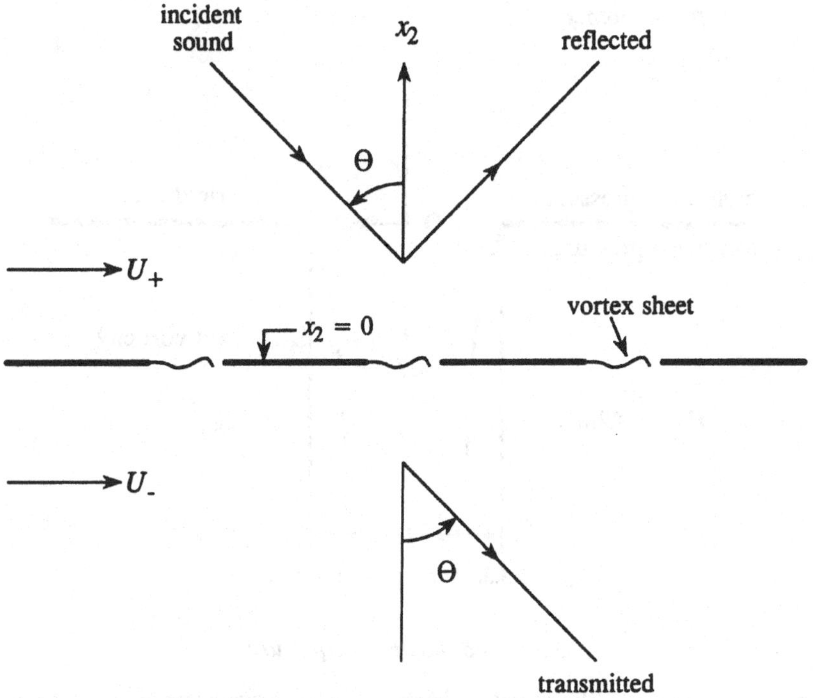

Figure 5.3.2. Sound incident on a perforated screen.

screen. By substituting from (5.3.13) into these equations, it now follows that

$$\mathcal{R} = \frac{\kappa_o \cos\Theta}{\kappa_o \cos\Theta + 2i\mathcal{N}K_R}, \quad \mathcal{T} = \frac{2i\mathcal{N}K_R}{\kappa_o \cos\Theta + 2i\mathcal{N}K_R}, \qquad (5.3.15)$$

where Θ is the angle of incidence shown in the figure (such that $n_2 = \kappa_o \cos\Theta$).

The absorption coefficient Δ is the fractional acoustic power absorbed by the screen and is given by

$$\Delta \equiv 1 - |\mathcal{R}|^2 - |\mathcal{T}|^2 = \frac{-4\mathcal{N}\kappa_o \cos\Theta \, \mathrm{Im}\, K_R}{|\kappa_o \cos\Theta + 2i\mathcal{N}K_R|^2}. \qquad (5.3.16)$$

When the aperture motion is *ideal*, the conductivity K_R determined by the procedures described in Section 5.3.1 is real, and there is no dissipation of acoustic energy at the screen ($\Delta = 0$).

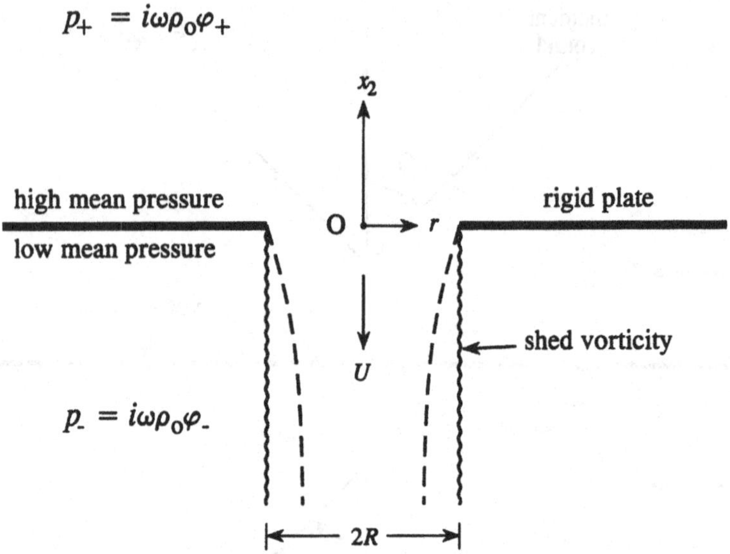

Figure 5.3.3. Aperture with bias flow jet.

5.3.3 The Bias Flow Aperture

A steady-pressure differential maintained across a wall containing a small aperture produces a nominally steady bias flow. When the Reynolds number $UD/\nu \gg 1$ (D being the aperture diameter), the motion is uninfluenced by viscosity except near the aperture edge, where the flow separates to form a jet. Vorticity generated at the edge by incident sound is swept away in the jet, and damping occurs by the transfer of acoustic energy to the kinetic energy of the jet, which implies that the Rayleigh conductivity K_R is now a complex function of frequency.

An approximate formula for K_R for a circular aperture with a bias flow can be derived using the model shown in Figure 5.3.3. The mean flow is from the upper to the lower surfaces, and the jet velocity U is assumed to have infinitesimal Mach number. According to *free streamline theory* [17], for an ideal fluid the jet contracts to an asymptotic cross section equal to about 0.62 times the aperture cross section, in the manner indicated by the dashed profile in the figure [17]; when the flow is turbulent, however, turbulence diffusion will tend to produce an expanding flow. The fluctuations in the flow produced by an applied time harmonic, uniform pressure differential $p_+ - p_-$ may be assumed to be axisymmetric; the additional vorticity generated at the edge of the aperture may be imagined to consist of a succession of vortex rings with infinitesimal cores, which convect at a velocity U_c and form an axisymmetric vortex sheet within the mean shear layer of the jet. To calculate the conductivity, the back-reaction of these vortex rings on the aperture flow must be determined. This is easily done if the variation in the radius of a ring is neglected, so that the vortex sheet is cylindrical, and equal to the aperture radius R [298]. This approximation yields predictions for the absorption of sound that are in remarkably good agreement with experiment [306].

Except very near the edges, the motion through the aperture may be regarded as inviscid, incompressible, and homentropic, for which case the divergence of the momentum equation (1.2.18) (with $F_i = 0$) reduces to $\nabla^2 B = -\text{div}(\boldsymbol{\omega} \wedge \mathbf{v})$. The vorticity $\boldsymbol{\omega}$ consists of a mean part associated with the undisturbed jet and a fluctuating part produced by the unsteady applied pressure. The dominant unsteady component $(\boldsymbol{\omega} \wedge \mathbf{v})'$ of $(\boldsymbol{\omega} \wedge \mathbf{v})$ is taken to correspond to the elementary vortex rings of radius R translating with velocity U_c parallel to the mean jet, that is,

$$(\boldsymbol{\omega} \wedge \mathbf{v})'(\mathbf{x}, \omega) = -\gamma_o U_c \delta(r - R)\mathbf{i}_r \mathrm{e}^{-i\kappa x_2}, \; x_2 < 0,$$
$$r = \sqrt{x_1^2 + x_3^2}, \tag{5.3.17}$$

where $\gamma_o \equiv \gamma_o(\omega)$ is the circulation of the perturbation vorticity per unit length of the jet, $\kappa = \omega/U_c$ is the hydrodynamic wavenumber appropriate for convection in the mean flow direction at velocity U_c, and $\mathbf{i}_r$ is a unit vector radially outward from the jet axis.

For points not on the vortex sheet, the unsteady motion can be described by an axisymmetric velocity potential $\varphi(x_2, r)$. On the side $x_2 < 0$ of the wall containing the jet, we set

$$\varphi = \varphi_- + \varphi_s + \varphi_w,$$

where $\varphi_- \equiv -ip_-/\rho_o\omega$, $\varphi_w(x_2, r)$ is the velocity potential attributable to the wake vorticity when the aperture is ignored (so that $\partial\varphi_w/\partial x_2 = 0$ at $x_2 = -0$, including the region occupied by the aperture), and $\varphi_s(x_2, r)$ accounts for the influence of the aperture. Similarly, in $x_2 > 0$ we write $\varphi = \varphi_+ + \varphi_s$, where $\varphi_+ \equiv -ip_+/\rho_o\omega$. The potential φ_s is given by the integrated term in (5.3.5), in terms of the perturbation normal velocity v_2 in the plane of the aperture. The condition that φ should be continuous across the plane of the aperture then yields the integral equation

$$\oint_S \frac{v_2(y_1, 0, y_3)}{|\mathbf{x} - \mathbf{y}|} dy_1\, dy_3 = \pi(\varphi_+ - \varphi_-) + \pi\gamma_o R \int_0^\infty \frac{\lambda \mathrm{H}_1^{(1)}(\lambda R) \mathrm{J}_0(\lambda r)\, d\lambda}{\kappa^2 + \lambda^2},$$
$$r < R, \quad x_2 = y_2 = 0, \tag{5.3.18}$$

for $v_2(x_1, 0, x_3) \equiv v_2(0, r)$, where the integrated term on the right-hand side is equal to $-\pi\varphi_w(0, r)$ (Example 7).

Solving this equation by Copson's method (Example 1), we find that

$$v_2(r) = \frac{\varphi_+ - \varphi_-}{\pi\sqrt{R^2 - r^2}} + \frac{\gamma_o R}{\pi}\int_0^\infty \frac{z\mathrm{H}_1^{(1)}(z)}{z^2 + \kappa^2 R^2} \times \left(\frac{\cos z}{\sqrt{R^2 - r^2}} + \int_r^R \frac{z\sin(zt/R)\,dt}{\sqrt{t^2 - r^2}}\right) dz, \quad r < R.$$

The strength γ_o of the vortex sheet is obtained by requiring $v_2(0, r)$ to remain finite as $r \to R$. The integrated term in the brace brackets vanishes in this limit, and the remaining terms are finite provided that

$$\gamma_o = -(\varphi_+ - \varphi_-)\Big/ R\int_0^\infty \frac{z\mathrm{H}_1^{(1)}(z)\cos z}{z^2 + \kappa^2 R^2}\,dz. \tag{5.3.19}$$

This formula can be used to evaluate the volume flux Q from (5.3.10), with $f_0(r)$ given by the right-hand side of (5.3.18). Forming the ratio $Q/(\varphi_+ - \varphi_-)$, the bias flow conductivity is then found to be

$$K_R = 2R(\Gamma_R - i\Delta_R), \tag{5.3.20}$$

where

$$\Gamma_R - i\Delta_R = 1 - \int_0^\infty \frac{\mathrm{H}_1^{(1)}(z)\sin z}{z^2 + \kappa^2 R^2}\,dz \Big/ \int_0^\infty \frac{z\mathrm{H}_1^{(1)}(z)\cos z}{z^2 + \kappa^2 R^2}\,dz$$
$$= 1 + \frac{\frac{\pi}{2}\mathrm{I}_1(\kappa R)\mathrm{e}^{-\kappa R} - i\mathrm{K}_1(\kappa R)\sinh(\kappa R)}{\kappa R\left[\frac{\pi}{2}\mathrm{I}_1(\kappa R)\mathrm{e}^{-\kappa R} + i\mathrm{K}_1(\kappa R)\cosh(\kappa R)\right]}. \tag{5.3.21}$$

Γ_R and Δ_R are both positive (for $\omega > 0$), and their dependence on the *Strouhal number* $\kappa R \equiv \omega R/U_c$ is illustrated in Figure 5.3.4. The following asymptotic approximations are easily derived:

$$K_R \approx 2R(1 - i/\kappa R), \qquad \kappa R \to \infty,$$
$$\approx 2R\left\{\frac{1}{3}(\kappa R)^2 - \frac{i\pi}{4}\kappa R\right\}, \quad \kappa R \to 0. \tag{5.3.22}$$

Thus, at very high frequencies vorticity production has a negligible influence on the fluctuating flow. This is because, when the vorticity length scale $\sim U_c/\omega$

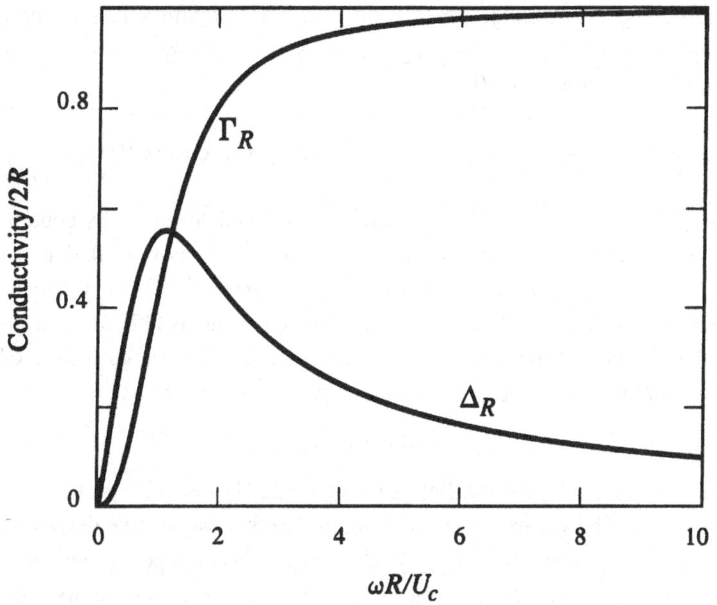

Figure 5.3.4. Conductivity of a bias flow circular aperture.

is small, the velocities induced by successive elementary vortex rings largely cancel except in the immediate vicinity of the edge of the aperture. At low Strouhal numbers, however, vorticity of one sign can stretch many aperture diameters downstream and produce a strong effect on the flow; in this case vortex shedding acts to block the aperture. The high frequency asymptotic formula is applicable for κR greater than about 5, but the low Strouhal number approximation is inapplicable unless $\kappa R \leq 0.04$, although the formula for Δ_R may be used for $\kappa R < 0.5$.

The Convection Velocity

If the bias flow jet emerging from the aperture is initially nonturbulent, the vorticity convection velocity U_c may be taken to be approximately equal to the mean velocity in the plane of the aperture. This value is consistent with the approximation in the derivation of K_R, as may be seen from the following argument.

Let p_o denote the steady excess pressure driving the flow through the aperture, V the asymptotic jet velocity far downstream, and A_∞ the asymptotic jet cross-sectional area. Then $V = \sqrt{2p_o/\rho_o}$, and the volume flux $Q = -A_\infty V$ (minus because our convention takes $Q > 0$ in the positive x_2-direction of Figure 5.3.3). The strength of the mean vortex sheet bounding the jet is $\gamma_o = -V$

because the pressure and jet velocity are constant just within the jet. The vortex strength may also be calculated by replacing $\varphi_+ - \varphi_-$ by $p_o/i\omega\rho_o$ in (5.3.19) and taking the limit $\omega \to 0$:

$$\gamma_o = -\left(\frac{p_o}{i\omega\rho_o R}\right) \bigg/ \left[\frac{\pi}{2}\mathrm{I}_1(\kappa R)\mathrm{e}^{-\kappa R} + i\mathrm{K}_1(\kappa R)\cosh(\kappa R)\right] \to \frac{-p_o}{\rho_o U_c}.$$

Hence, $U_c = p_o/\rho_o V = \frac{1}{2}V$. The value of A_∞ is determined by equating the volume flux $Q = -A_\infty V$ to the value given by (5.3.1) as $\omega \to 0$, that is, using (5.3.22), to $-\frac{\pi}{2}R^2 p_o/\rho_o U_c$, which gives $A_\infty/\pi R^2 = \frac{1}{2}$. This is identical with the minimum theoretical value obtained by other means [17] (experimentally observed values of the area contraction ratio are in the range 0.61–0.64). It follows that the mean jet velocity in the aperture is $\frac{1}{2}V \equiv U_c$.

Energy Dissipation

Work is performed by the oscillating pressure load $[p] = p_+ - p_-$ in generating the vorticity. The vortex power Π_ω supplied in this way can be determined by calculating the mean rate $-\oint_S \langle p\mathbf{v}\rangle \cdot d\mathbf{S}$ at which work is performed on a large spherical control surface S of radius $|\mathbf{x}| \gg R$ centered on the aperture (the surface element $d\mathbf{S}$ being radially outward). On S, we can take, after restoring the time factor $\mathrm{e}^{-i\omega t}$,

$$p = \mathrm{Re}\, p_\pm, \quad \text{and} \quad \mathbf{v} \approx \pm\mathrm{Re}\left(\frac{iK_R\mathbf{x}[p]}{2\pi|\mathbf{x}|^3}\rho_o\omega\right), \quad x_2 \gtrless 0,$$

and deduce that

$$\Pi_\omega = \frac{-\mathrm{Im}\{K_R\}|[p]|^2}{2\rho_o\omega} \equiv \frac{R\Delta_R|[p]|^2}{\rho_o\omega}. \tag{5.3.23}$$

Thus, $\Pi_\omega > 0$ because $\Delta_R > 0$.

Example 7. The velocity potential φ_w of the motion induced by the unsteady bias flow vorticity is determined by the solution of $\nabla^2 B_w = -\mathrm{div}(\boldsymbol{\omega} \wedge \mathbf{v})'$ (where $(\boldsymbol{\omega} \wedge \mathbf{v})'$ is given by (5.3.17)) subject to the condition that $\partial\varphi_w/\partial x_2 \equiv (1/i\omega)\partial B_w/\partial x_2 = 0$ for $x_2 = -0$. This condition can be satisfied by introducing image vortices in $x_2 > 0$. The equation for B_w is then extended to $-\infty < x_2 < \infty$, and the particular solution that vanishes at infinity can be found by first taking the Fourier transform with respect to x_2. The inverse transform yields

$$\varphi_w = \frac{-2\gamma_o R}{\pi}\int_0^\infty (\mathrm{H}(R-r)\mathrm{K}_1(kR)\mathrm{I}_0(kr)$$
$$- \mathrm{H}(r-R)\mathrm{I}_1(kR)\mathrm{K}_0(kr))\frac{k\cos(kx_2)\,dk}{(\kappa+i0)^2 - k^2},$$

where I_n and K_n are modified Bessel functions [24, 289], and H is the Heaviside step function. For $r < R$ in the plane $x_2 = 0$ of the aperture, the integration contour may be rotated onto the negative imaginary axis without encountering any singularities. This procedure leads to the final term on the right of (5.3.18).

Example 8. Show that *in the absence of vortex shedding* the velocity potential φ of the "acoustic particle velocity" produced by a time harmonic pressure load $p_+ - p_-$ applied across an acoustically compact, circular aperture in a thin rigid wall can be expressed in the form

$$\varphi = \frac{-\operatorname{sgn}(x_2)(p_+ - p_-)}{\pi i \rho_o \omega} \int_0^\infty \frac{\sin(kR)\mathrm{J}_0(kr)\mathrm{e}^{-k|x_2|}}{k}\, dk,$$

where the notation is defined as in Figure 5.3.3. Use this formula to confirm that the power Π_ω dissipated in the aperture in the presence of a bias flow (5.3.23) is given by the general formula (5.2.6), in which $\mathbf{u} = \nabla\varphi$ and $\boldsymbol{\omega} \wedge \mathbf{v}$ is replaced by $(\boldsymbol{\omega} \wedge \mathbf{v})'$ defined by (5.3.18) and (5.4.19).

5.3.4 The Bias Flow Perforated Screen

A steady pressure maintained across a homogeneously perforated screen can enhance the absorption of sound by vorticity production. The simplest configuration is illustrated schematically in Figure 5.3.5a. Vorticity generated in the apertures by impinging sound is convected away in the bias flow jets. The aperture conductivity determined by (5.3.21) is complex, and the absorption coefficient (5.3.16) is

$$\Delta = \frac{(8\alpha/\pi M_c)\Delta_R \kappa R \cos\Theta}{|\kappa R \cos\Theta + (4\alpha/\pi M_c)(\Delta_R + i\Gamma_R)|^2}, \tag{5.3.24}$$

where $\alpha = \mathcal{N}\pi R^2$ is the fractional open area of the screen and $M_c = U_c/c_o$. The dependence of Δ on Strouhal number $\omega R/U_c$ is shown in Figure 5.3.5b for different values of $\alpha/M_c \cos\Theta$. When $\omega R/U_c < 1$, the attenuation becomes maximal for $\alpha \approx M_c \cos\Theta$; as $\omega R/U_c \to 0$, this maximum approaches 0.5, one half of the incident sound being absorbed (Example 9).

Example 9. When $\omega R/U_c \to 0$, the absorption coefficient for a bias flow perforated screen becomes $\Delta = 2\alpha M_c \cos\Theta/(M_c\cos\Theta + \alpha)^2$, which has a maximum value of $\frac{1}{2}$ when $\alpha = M_c \cos\Theta$.

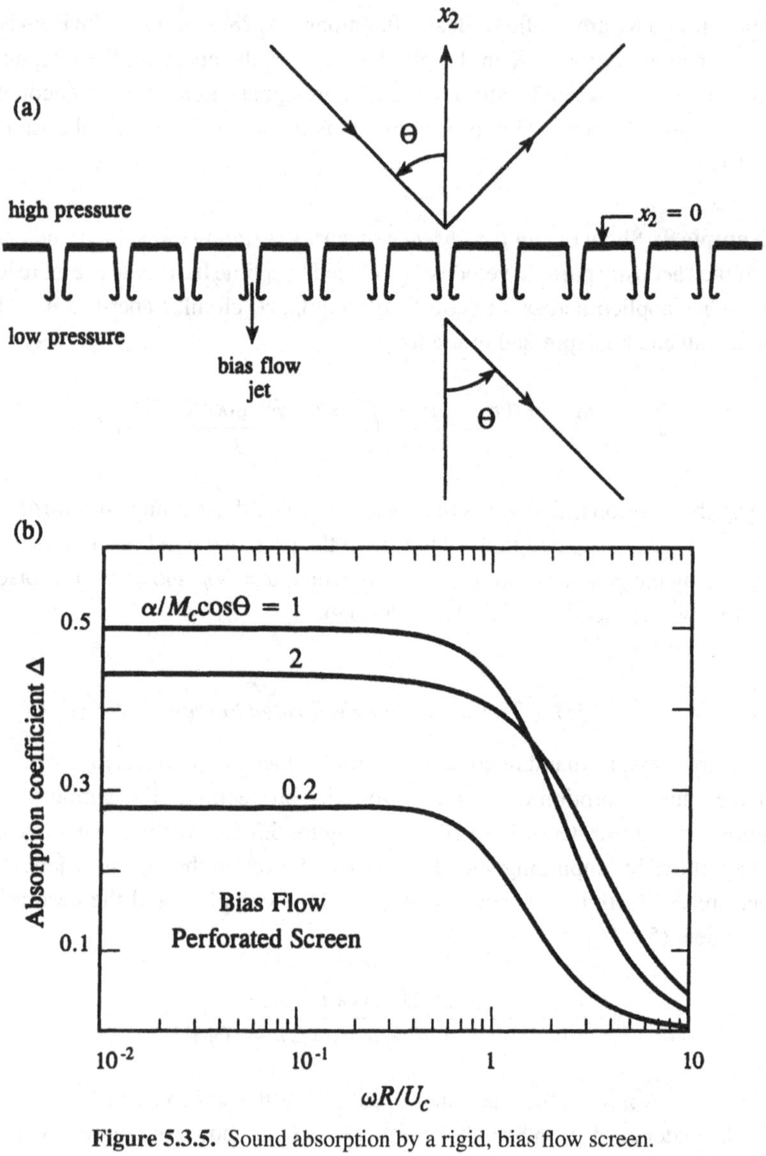

Figure 5.3.5. Sound absorption by a rigid, bias flow screen.

5.3.5 Cavity-Backed Screen

A rigid backing wall of the type illustrated in Figure 5.3.6a can greatly increase the efficiency with which sound is absorbed by a bias flow screen [306]. In practice the backing may be the wall of a wind tunnel, with fluid either injected or withdrawn from the main stream at the perforated surface, which behaves

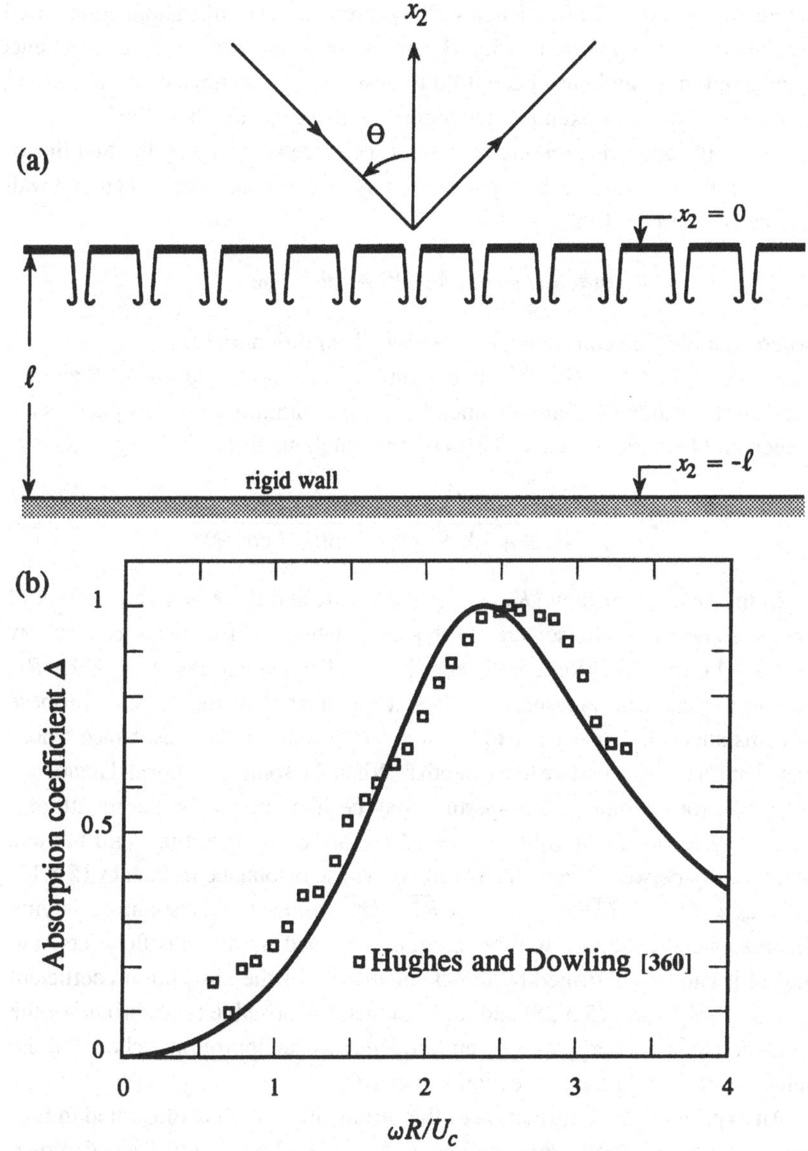

Figure 5.3.6. Sound absorption by a cavity backed, bias flow screen: $R = 0.15$ cm, $\ell = 2$ cm, $U_c = 4.76$ m/s, $\alpha = 0.024$. Experimental data adapted from Hughes and Dowling [306].

as an acoustic liner whose bias flow velocity can be adjusted to yield maximal attenuations at desired frequencies [294]. Predictions of this attenuation based on the conductivity formula (5.3.21) are in good agreement with experiments conducted in air and have been used to design practical devices for absorbing intense sound in, for example, jet engine combustion chambers [306].

When the acoustic pressure in front of the screen is given by the first line of (5.3.13), the pressure in the region $-\ell < x_2 < 0$ between the screen and wall can be taken in the form

$$p(\mathbf{x}, \omega)/p_{\mathrm{I}} = (\mathcal{A}\mathrm{e}^{-n_2 x_2} + \mathcal{B}\mathrm{e}^{i n_2 x_2})\mathrm{e}^{i n_1 x_1},$$

where $\mathcal{A}$ and $\mathcal{B}$ are constants. The rigid-wall condition $\partial p/\partial x_2 = 0,\ x_2 = -\ell$, implies that $\mathcal{A}\mathrm{e}^{i n_2 \ell} = \mathcal{B}\mathrm{e}^{-i n_2 \ell}$. Two additional equations relating $\mathcal{A}$, $\mathcal{B}$ and the reflection coefficient $\mathcal{R}$ are obtained from the continuity of $\partial p/\partial x_2$ across the screen and from the relation (5.3.14) from which we find

$$\mathcal{R} = \frac{\kappa_o \cos\Theta - i\mathcal{N}K_R\{1 - i\cot(\kappa_o \ell \cos\Theta)\}}{\kappa_o \cos\Theta + i\mathcal{N}K_R\{1 + i\cot(\kappa_o \ell \cos\Theta)\}}. \tag{5.3.25}$$

In the absence of flow ($K_R = 2R$) $|\mathcal{R}| = 1$, and there is no absorption of acoustic energy by the screen. In this case, when the frequency is very low ($\kappa_o \ell \ll 1$), the reflection coefficient $\mathcal{R} \to -1$ when $\kappa_o \ell \cos\Theta \to \sqrt{2\mathcal{N}R\ell}$, so that the acoustic pressure vanishes at the surface of the screen. In these circumstances, the cavity is excited at a very low-frequency resonance whose wavelength is large relative to the depth ℓ. When the sound is at normal incidence ($\Theta = 0$), for example, each aperture may be imagined to be the mouth of a *Helmholtz* resonator of volume $v = \ell/\mathcal{N}$, formed by inserting rigid barriers in the cavity between the wall and screen, whose resonance frequency (5.3.12) satisfies $\kappa_o \ell = \ell\sqrt{2R/v} \equiv \sqrt{2\mathcal{N}R\ell}$. This suggests that sound satisfying this resonance condition will be strongly absorbed by the bias flow jets, and indeed it can be confirmed by direct calculation of the absorption coefficient $\Delta = 1 - |\mathcal{R}|^2$ from (5.3.25) and (5.3.20) that it is possible to absorb *all* of the incident acoustic energy of a given frequency by an appropriate choice of the values of R, ℓ, and the convection velocity U_c.

An experimental confirmation of this strong absorption is illustrated in Figure 5.3.6b for normally incident sound on a bias flow, cavity backed screen in air for which $R = 0.15$ cm, $\ell = 2$ cm, $U_c = 4.76$ m/s, and the fractional open area $\alpha = 0.024$. The solid curve is the theoretical prediction of $\Delta = 1 - |\mathcal{R}|^2$ calculated from (5.3.25) and (5.3.20); the experimental data are adapted from measurements of Hughes and Dowling [306], where U_c is identified with the mean flow velocity in the apertures. Theory and

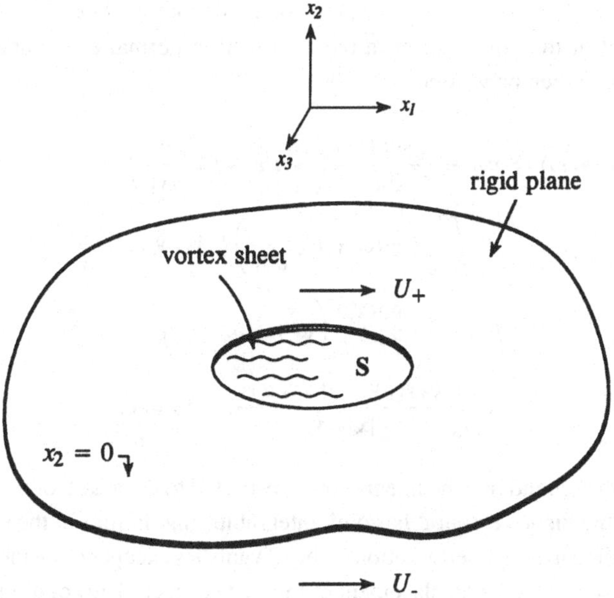

Figure 5.3.7. Tangential mean flow past an aperture in a thin wall.

experiment exhibit peak absorption in the neighborhood of the Helmholtz resonance frequency where $\kappa_o\ell = \sqrt{2\mathcal{N}R\ell} \equiv \sqrt{2\alpha\ell/\pi R} \approx 0.45$, that is, at $\omega R/U_c \approx 2.42$ ($\omega/2\pi \approx 1200$ Hz).

5.3.6 Circular Aperture in Grazing Mean Flow

In many practical situations, it is necessary to use a perforated screen in the presence of mean flow over one or both of its sides [300, 305] and in the absence of any significant bias flow. Let us determine the conductivity for a circular aperture in the presence of grazing flow in the practically important case of infinitesimal Mach number. Let the mean velocities be $U_\pm$ above and below the aperture (in the x_1-direction, Figure 5.3.7), and assume the Reynolds number is large enough that viscosity can be neglected except for its role in generating vorticity at sharp edges. Thus, in the steady state, when $U_+ \neq U_-$, the aperture is spanned by a vortex sheet across which the pressure, but not the velocity potential, is continuous.

When a uniform time-harmonic pressure load $[p] = p_+ - p_-$ is applied across the aperture, the velocity potential of the unsteady, incompressible motion on each side of the wall is given by equation (5.3.5), which can be transformed into an equivalent representation for the perturbation pressure

$p(\mathbf{x}, \omega) = -\rho_o(-i\omega + U_\pm \partial/\partial x_1)\varphi$. If we denote by $\zeta(x_1, x_3, \omega)$ the displacement of the vortex sheet in the x_2-direction normal to the aperture, the pressure can then be written

$$\begin{aligned} p(\mathbf{x}, \omega) &= p_\pm + \frac{\rho_o \operatorname{sgn}(x_2)}{2\pi}\left(-i\omega + U_\pm \frac{\partial}{\partial x_1}\right) \\ &\quad \times \int_{-\infty}^{\infty}\left(-i\omega + U_\pm \frac{\partial}{\partial y_1}\right)\zeta(y_1, y_3, \omega)\frac{dy_1\, dy_3}{|\mathbf{x}-\mathbf{y}|} \\ &= p_\pm - \frac{\rho_o \operatorname{sgn}(x_2)}{2\pi}\left(\omega + iU_\pm \frac{\partial}{\partial x_1}\right)^2 \\ &\quad \times \int_S \frac{\zeta(y_1, y_3, \omega)\, dy_1\, dy_3}{|\mathbf{x}-\mathbf{y}|}, \quad y_2 = 0, \end{aligned} \tag{5.3.26}$$

where in the second line the integration is restricted to the area S of the aperture, provided the displacement ζ has only integrable singularities at the edge.

According to linear perturbation theory ζ vanishes except within the aperture, where it represents the displacement of the vortex sheet. The condition that the pressure is continuous across the vortex sheet then yields the following equation for ζ:

$$\begin{aligned} &\left[\left(\omega + iU_+ \frac{\partial}{\partial x_1}\right)^2 + \left(\omega + iU_- \frac{\partial}{\partial x_1}\right)^2\right] \\ &\quad \times \int_S \frac{\zeta(y_1, y_3, \omega)}{2\pi\sqrt{(x_1-y_1)^2 + (x_3-y_3)^2}}\, dy_1\, dy_3 = \frac{p_+ - p_-}{\rho_o}, \end{aligned} \tag{5.3.27}$$

for $\sqrt{x_1^2 + x_3^2} < R$. By integrating this equation with respect to the second-order differential operator on the left, and noting that $(p_+ - p_-)/\rho_o$ is independent of $\mathbf{x}$, it is transformed into the integral equation

$$\begin{aligned} &\int_S \frac{\zeta'(y_1', y_3', \omega)\, dy_1'\, dy_3'}{\sqrt{(x_1' - y_1')^2 + (x_3' - y_3')^2}} + \lambda_1(x_3')e^{i\sigma_1 x_1'} + \lambda_2(x_3')e^{i\sigma_2 x_2'} = 1, \\ &\qquad \sqrt{x_1'^2 + x_3'^2} < 1, \end{aligned} \tag{5.3.28}$$

where

$$\zeta' = \frac{\rho_o \omega^2 R \zeta}{\pi(p_+ - p_-)}, \quad \mathbf{x}' = \frac{\mathbf{x}}{R}, \quad \text{and so on}$$

and σ_1, σ_2 are the dimensionless *Kelvin–Helmholtz* wavenumbers of the instability waves of the vortex sheet [4, 17],

$$\sigma_1 = \frac{\omega R(1+i)}{U_+ + iU_-}, \quad \sigma_2 = \frac{\omega R(1-i)}{U_+ - iU_-}. \tag{5.3.29}$$

The amplitudes $\lambda_1(x_3')$, $\lambda_2(x_3')$ of these waves are constant along any line drawn on the sheet in the streamwise direction but vary with transverse location x_3'. They are chosen to satisfy the Kutta condition, by requiring the vortex sheet to leave the upstream semicircular arc of the aperture edge tangentially. According to potential theory [4, 17, 170] the displacement ζ that satisfies these conditions has a mild singularity $\sim 1/\sqrt{R^2 - r^2}$ at the downstream edge $r \equiv \sqrt{x_1^2 + x_3^2} = R$, $x_1 > 0$ of the aperture. The singularity is integrable, however (so that the volume flux Q is finite), and is a physically acceptable, linear theory representation of the large amplitude and violent motions produced by vortex impingement on a leading edge [330]. In principle the singularity could be removed in a quasi-linear approximation, by introducing *displacement thickness waves* in the boundary layers downstream of the aperture as in Section 5.2.4. The amplitude of these waves can be adjusted to remove any singular behavior in the displacement at the trailing edge, but this does not significantly alter the conclusions of the present simpler approach [311, 331].

The volume flux is

$$Q = \int_{-\infty}^{\infty} v_2(x_1, 0, x_3)\, dx_1\, dx_3$$
$$\equiv \int_{-\infty}^{\infty} \left(-i\omega + U_\pm \frac{\partial}{\partial x_1}\right) \zeta(x_1, x_3, \omega)\, dx_1\, dx_3.$$

The derivative in the second integral makes a null contribution because ζ vanishes on the wall outside the aperture. The remaining integration may be confined to the region occupied by the aperture, and the definition (5.3.1) then supplies

$$K_R = \pi R \int_S \zeta'(x_1', x_3', \omega)\, dx_1'\, dx_3'. \tag{5.3.30}$$

The integral equation (5.3.28) must be solved numerically. The solution is used in this formula to calculate the real and imaginary parts of the conductivity K_R (defined as in (5.3.20)), which are functions of $\omega R/U_+$ and the velocity ratio U_-/U_+. The results of such calculations performed by Scott [332, 333]

Table 5.3.1. *Circular aperture in one-sided grazing flow:* $U_+ = U,\ U_- = 0$ [332]

$\omega R/U$	Γ_R	Δ_R	$\omega R/U$	Γ_R	Δ_R
0.00	0.000	0.000	1.50	1.949	0.623
0.10	−0.030	0.005	1.75	2.097	0.244
0.20	−0.119	0.041	2.00	2.233	−0.131
0.30	−0.253	0.145	2.25	2.409	−0.540
0.40	−0.395	0.360	2.50	2.708	−1.006
0.50	−0.464	0.710	2.75	3.277	−1.459
0.60	−0.350	1.140	3.00	4.184	−1.555
0.70	−0.016	1.503	3.25	4.887	−0.969
0.80	0.439	1.673	3.50	4.897	−0.300
0.90	0.872	1.652	3.75	4.628	−0.000
1.00	1.216	1.515	4.00	4.398	0.059
1.10	1.466	1.333	4.25	4.258	0.033
1.20	1.644	1.142	4.50	4.189	0.003
1.30	1.774	0.959	4.75	4.175	0.011
1.40	1.872	0.786	5.00	4.182	0.115

for the two important cases of (i) one-sided mean flow ($U_+ = U,\ U_- \equiv 0$) and (ii) two-sided uniform flow ($U_+ = U_- = U$) are shown in Figure 5.3.8. In solving (5.3.28) in case (ii), the λ−terms are replaced by

$$(\lambda_1(x_3') + \lambda_2(x_3')x_1')\mathrm{e}^{i\sigma x_1'}, \quad \text{where} \quad \sigma = \omega R/U.$$

Numerical values for one-sided flow are given in Table 5.3.1; for practical purposes the tabulated results can be approximated by cubic splines.

For one-sided grazing flow $\Gamma_R \approx 4$ for $\omega R/U > 5$, and the imaginary part Δ_R becomes very small. Δ_R is negative in the range $1.9 < \omega R/U < 3.9$ where, according to equation (5.3.23), negative work is performed by the applied pressure load $p_+ - p_-$ in generating vorticity in the aperture, that is, energy is actually extracted from the mean flow. The aperture then behaves as a net source of sound; in Section 5.3.8, it will be shown that the grazing flow is absolutely unstable and that sound generation by this means will occur even in the absence of a disturbing pressure load. When the mean velocity is the same on both sides of the wall, there is no vorticity in the absence of the applied pressures (at least for an ideal fluid). However, Figure 5.3.8 reveals the existence of periodically spaced frequency intervals (of period π when $\omega R/U > 3$) in which $\Delta_R < 0$, where energy is extracted from the mean flow, in spite of the fact (see Section 5.3.8) that the aperture motion is not unstable. This occurs because of the forced interaction of the vorticity with the downstream semicircular edge,

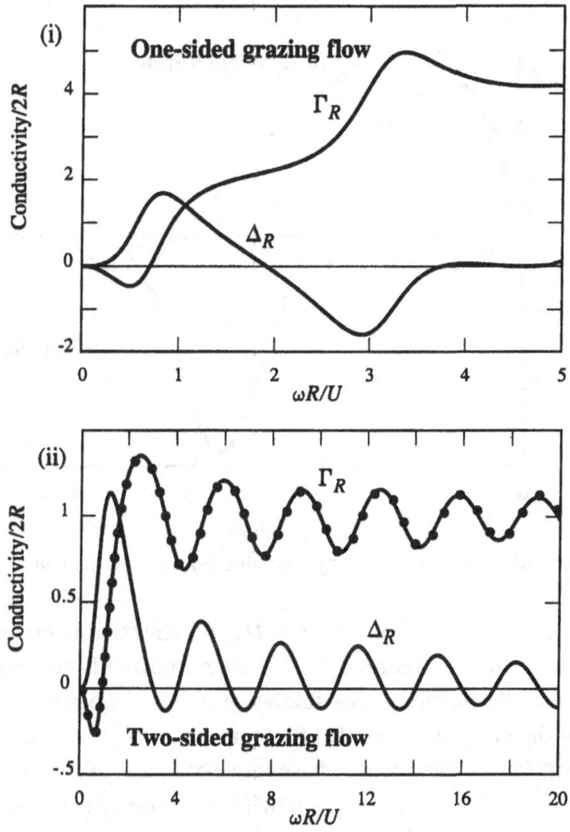

Figure 5.3.8. Conductivity of a circular aperture in grazing flow (i) $U_+ = U,\ U_- = 0$ and (ii) $U_+ = U_- = U$ [332].

on which it impinges after crossing the aperture (cf. Section 5.2.4). However, this idealized case of uniform two-sided flow is never realized in practice because of the small but finite thickness of a real wall. This causes vorticity to be present in the aperture even for nominally steady ideal flow; the motion is then unstable, and K_R turns out to be very similar in structure to the one-sided flow case (Section 5.3.9).

5.3.7 Grazing Flow Perforated Screen

The reflection and transmission coefficients for a homogeneously perforated screen in a grazing mean flow of infinitesimal Mach number are given by equations (5.3.15). Figure 5.3.9 illustrates the frequency dependence of the absorption coefficient Δ defined by (5.3.16) for one- and two-sided flows when the

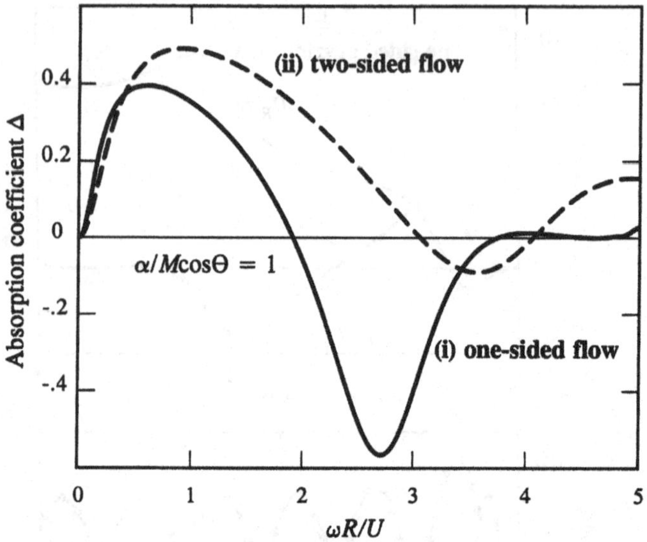

Figure 5.3.9. Absorption of sound by a grazing flow screen with circular apertures.

fractional open area $\alpha = M \cos\Theta$ ($M = U_+/c_o$) and the Rayleigh conductivity is given by the results in Figure 5.3.8. The greatest overall attenuations occur at low frequencies, the maximum possible being $\Delta = 1/2$ when $\alpha/M \cos\Theta = 0.5$ and 0.75 for the one- and two-sided flows, respectively. One-sided flow is absolutely unstable, and the screen behaves as a net source of sound in the interval $1.9 < \omega R/U < 3.9$ where $\Delta < 0$; much more modest gains in acoustic energy occur for two-sided flow within periodically recurring frequency ranges where $\Delta_R < 0$ in Figure 5.3.8. As $\omega R/U \to 0$, $\mathcal{T} \to 0$ and $\mathcal{R} \to 1$: The flow through the screen is then blocked by vortex shedding, and it becomes acoustically hard.

Cavity Backed Grazing Flow Screen

Two-sided grazing flow screens are used to attenuate low-frequency acoustic resonances in heat exchanger tube-bank cavities [305]; the relatively weak negative damping at higher frequencies appears to be unimportant. The strong instability associated with one-sided flow screens for $\omega R/U > 2$ is believed to be responsible for the "howling" of cavity-backed perforated duct liners in the presence of unsteady mean flow.

5.3.8 The Kramers–Kronig Formulae: Absolute and Conditional Instability

According to equation (5.3.1), the volume flux Q is determined by the applied pressure load $[p] \equiv p_+ - p_-$ and the conductivity K_R. All of these quantities are generally functions of the frequency ω, so that in real form we can write

$$\rho_o \frac{dQ}{dt} = -\int_{-\infty}^{\infty} K_R(\omega)[p(\omega)]\mathrm{e}^{-i\omega t}\, d\omega, \tag{5.3.31}$$

where $[p(\omega)] = \frac{1}{2\pi}\int_{-\infty}^{\infty}[p(t)]\mathrm{e}^{i\omega t}\, dt$. The integration is nominally taken over all real values of the frequency. However, if the integrand has singularities in the upper half of the ω-plane the integration contour must be deformed to pass *above* them to ensure that Q is causally related to the $[p]$. Because $[p(t)]$ is arbitrary and may be assumed to vanish prior to some initial instant, singularities in $\mathrm{Im}(\omega)>0$ must be associated with $K_R(\omega)$. When they are present, the system is *linearly* unstable, and an unsteady flux through the aperture can develop spontaneously. Also, because $Q(t)$ and $[p(t)]$ are real-valued quantities, it follows that $K_R(-\omega^*) = K_R^*(\omega)$, where the asterisk denotes complex conjugate.

Consider the stable case in which $K_R(\omega) = 2R\{\Gamma_R(\omega) - i\Delta_R(\omega)\}$ is regular in $\mathrm{Im}\,\omega > 0$. For real ω, we then have

$$\Gamma_R(-\omega) = \Gamma_R(\omega), \quad \Delta_R(-\omega) = -\Delta_R(\omega). \tag{5.3.32}$$

In general, $\Gamma_R(\omega) \to \Gamma_\infty$ = constant as $|\omega| \to \infty$ and $\Delta_R(\omega) \to 0$. Then, when regarded as a function of the complex variable ω, the function $f(\omega) = \Gamma_R(\omega) - \Gamma_\infty - i\Delta_R(\omega)$ is regular in $\mathrm{Im}\,\omega > 0$ and vanishes as $|\omega| \to \infty$. Cauchy's theorem [21, 334] applied to a closed contour consisting of the real axis and a semicircle in the upper half-plane whose radius grows without limit, then implies that $f(\omega) = (1/2\pi i)\int_{-\infty}^{\infty} f(\xi)\, d\xi/(\xi - \omega)$ for $\mathrm{Im}\,\omega > 0$, the integration being along the real axis. By taking the limit as ω approaches the real axis, the real and imaginary parts of this equation (and the relations (5.3.32)) yield the *Kramers–Kronig* formulae

$$\Gamma_R(\omega) - \Gamma_\infty = -\frac{2}{\pi}\int_0^\infty \frac{\xi\Delta_R(\xi)\, d\xi}{\xi^2 - \omega^2},$$

$$\Delta_R(\omega) = \frac{2\omega}{\pi}\int_0^\infty \frac{[\Gamma_R(\xi) - \Gamma_\infty]\, d\xi}{\xi^2 - \omega^2}, \quad (\omega \text{ real}), \tag{5.3.33}$$

where the integrals are principal values.

When a system is known to be stable, these equations can be used to determine, say, $\Gamma_R(\omega)$ from measured values of $\Delta_R(\omega)$. If experiment shows that $K_R(\omega)$ vanishes at zero frequency (as in the case of a bias flow aperture), the constant Γ_∞ is found by setting $\omega = 0$ in the first of (5.3.33).

Alternatively, when $\Gamma_R(\omega)$ and $\Delta_R(\omega)$ have both been found by experiment or numerical computation, equations (5.3.33) can be used to investigate the stability of the flow. If the equations are not satisfied $K_R(\omega)$ must have one or more singularities in $\mathrm{Im}\,\omega > 0$; the motion is then absolutely unstable, in the sense that the smallest perturbation of the flow can cause a spontaneous growth of large amplitude motions in the aperture. This is the case for an aperture with one-sided grazing flow: The Kramers–Kronig formulae (5.3.33) are not satisfied by the numerically derived values of Γ_R and Δ_R shown in Figure 5.3.8(i). However, the Kramers–Kronig formulae *are* satisfied by the numerical data for the two-sided flow of Figure 5.3.8(ii), even though Δ_R can be negative. The dotted curve in this figure represents Γ_R computed from the first of the Kramers–Kronig equations (5.3.33) using the numerical values of Δ_R shown in the figure (assuming that $\Delta_R = 0$ for $\omega R/U > 20$). In this case, the aperture motion is *conditionally unstable*: Energy extraction from the mean flow occurs only when an incident disturbance contains a component with frequency within an interval where $\Delta_R < 0$. Similarly, the real and imaginary parts of the bias flow conductivity (5.3.21) comply with the Kramers–Kronig formulae; because $\Delta_R > 0$ the motion is stable for all frequencies.

When the stability characteristics of $K_R(\omega)$ are known, it is sometimes useful to determine simple approximating formulae to the experimental or numerical representations of K_R that exhibit the appropriate regularity properties for complex ω, which can be used in analytical work. In several important cases this can be done by assuming that all singularities of the conductivity are simple poles occurring in pairs at

$$\kappa \equiv \omega R/U = \pm\kappa_n - i\beta_n, \quad n = 1, 2, \ldots s,$$

where κ_n and β_n are real, and $\beta_n > 0$ when the flow is stable. Then,

$$\Gamma_R(\omega) - i\Delta_R(\omega) \approx \sum_n \left(\frac{\alpha_n}{\kappa - (\kappa_n - i\beta_n)} - \frac{\alpha_n^*}{\kappa - (\kappa_n + i\beta_n)} \right) + \sum_n \frac{2(\alpha_{rn}\kappa_n - \alpha_{in}\beta_n)}{\kappa_n^2 + \beta_n^2}, \quad (5.3.34)$$

where α_{rn} and α_{in} are the real and imaginary parts of α_n.

For stable motion ($\beta_n > 0$), the coefficients $\alpha_n, \kappa_n, \beta_n$ are chosen to provide a good fit to $\Gamma_R - i\Delta_R$ for real values of ω. Thus, the bias flow conductivity given by (5.3.21) is well approximated by the *two-pole* model

$$\alpha_1 = -0.51i, \quad \kappa_1 = 0.40, \quad \beta_1 = 0.83; \quad \alpha_n = 0, \quad n > 1.$$

In this approximation, $K_R(\omega)$ is regular in $\mathrm{Im}\,\omega > 0$ and has simple poles at $\omega R/U = \pm 0.40 - 0.83i$ in the lower half-plane.

For unstable motion, the known behavior of $f(\omega) = \Gamma_R(\omega) - \Gamma_\infty - i\Delta_R(\omega)$ for real values of ω must first be used to express $f(\omega)$ as the sum of two functions $f^+(\omega) + f^-(\omega)$, which are regular respectively in $\mathrm{Im}\,\omega \gtrless 0$. This is done by means of the relations [205]

$$\Gamma_R^{\pm}(\omega) - \Gamma_{\infty}^{\pm} = \frac{\mp 1}{2\pi} \int_{-\infty}^{\infty} \frac{\Delta_R(\xi)\,d\xi}{\xi - \omega} + \frac{1}{2}(\Gamma_R(\omega) - \Gamma_\infty),$$
$$\Delta_R^{\pm}(\omega) = \frac{\pm 1}{2\pi} \int_{-\infty}^{\infty} \frac{[\Gamma_R(\xi) - \Gamma_\infty]\,d\xi}{\xi - \omega} + \frac{1}{2}\Delta_R(\omega), \tag{5.3.35}$$

where ω is real. The representation (5.3.34) is now obtained by approximating $f^{\pm}(\omega)$ by those terms in the summation respectively with poles in the lower and upper halves of the ω-plane.

Example 10. A four-pole approximation to the conductivity of a circular aperture in one-sided grazing is given by (5.3.34) when

$$\alpha_1 = 1.23, \quad \kappa_1 = 0.9, \quad \beta_1 = 1;$$
$$\alpha_2 = 0.82, \quad \kappa_2 = 2.9, \quad \beta_2 = -0.8; \quad \alpha_n = 0, \quad n > 2.$$

There are two poles in the upper half-plane at $\omega R/U = \pm 2.9 + 0.8i$ and two in the lower half-plane at $\omega R/U = \pm 0.9 - i$.

Example 11: Volume flux through an infinite slot. Determine the volume flux *per unit length* produced by a pressure load $[p] = p_0 e^{i\kappa_1 x_1}$ applied to an infinite slot of acoustically compact width $2s$ occupying the interval $-s < x_1 < s$ of the thin wall $x_2 = 0$ in the presence of grazing flow in the x_1-direction [331].

This two-dimensional problem is solved by the method of Section 5.3.6. The pressure above and below the wall is first expressed in the following form analogous to (5.3.26) using the two-dimensional Green's function (1.7.10)

$$p(\mathbf{x}, \omega) = p_{\pm} - \frac{i\rho_o \operatorname{sgn}(x_2)}{2} \left(\omega + iU_{\pm}\frac{\partial}{\partial x_1}\right)^2$$
$$\times \int_{-s}^{s} \mathrm{H}_0^{(1)}(\kappa_o|\mathbf{x} - \mathbf{y}|)\zeta(y_1, \omega)\,dy_1, \quad y_2 = 0. \tag{5.3.36}$$

This is used to formulate an equation of the form (5.3.27) for the displacement ζ of the vortex sheet. To do this, we first note that, when $x_1,\ y_1$ both lie within the slot, the compactness condition implies that $\kappa_o|x_1 - y_1| \ll 1$, and therefore (see (1.10.18))

$$\frac{-i}{2}\mathrm{H}_0^{(1)}(\kappa_o|x_1 - y_1|) \approx \frac{1}{\pi}\ln\left(\frac{|x_1 - y_1|}{s}\right) + a, \quad |x_1 - y_1| < s,$$

where

$$a = -\frac{i}{2} + \frac{1}{\pi}[\gamma_E + \ln(\kappa_o s/2)].$$

Make the change of variables

$$\sigma_\pm = \omega s/U_\pm, \quad \sigma_0 = \kappa_1 s, \quad \xi = x_1/s, \quad \eta = y_1/s, \quad \zeta'(\xi) = \zeta(x_1)/s,$$

and apply continuity of pressure to obtain the analogue of (5.3.27):

$$\left[\left(\sigma_+ + i\frac{\partial}{\partial \xi}\right)^2 + \mu^2\left(\sigma_- + i\frac{\partial}{\partial \xi}\right)^2\right]\int_{-1}^{1} \zeta'(\eta)\ln|\xi - \eta|\,d\eta$$
$$= \frac{-\pi p_0 e^{i\sigma_0\xi}}{\rho_o U_+^2} - \frac{2\pi i a\omega Q}{U_+^2}, \quad |\xi| < 1, \tag{5.3.37}$$

where $\mu = U_-/U_+$ and $Q(\omega) = -i\omega\int_{-s}^{s}\zeta(x_1, \omega)\,dx_1$ is the volume flux per unit length of the slot. Integrating with respect to the second-order differential operator on the left:

$$\int_{-1}^{1} \zeta'(\eta)\ln|\xi - \eta|\,d\eta = -\frac{\pi i a Q}{2\omega s^2} + \sum_{j=0}^{2}\lambda_j e^{i\sigma_j\xi}, \quad |\xi| < 1, \tag{5.3.38}$$

where λ_1, λ_2 are constants of integration,

$$\lambda_0 = -\pi p_o/\{\rho_o U_+^2[(\sigma_+ - \sigma_0)^2 + \mu^2(\sigma_- - \sigma_0)^2]\},$$

and σ_1, σ_2 are defined in (5.3.29) with R replaced by s.

The solution of equation (5.3.38) is [334]

$$\zeta'(\xi) = \frac{1}{\pi^2\sqrt{1-\xi^2}}\left(\fint_{-1}^{1}\frac{\sqrt{1-\eta^2}}{\eta - \xi}\chi'(\eta)\,d\eta - \frac{1}{\ln 2}\int_{-1}^{1}\frac{\chi(\eta)\,d\eta}{\sqrt{1-\eta^2}}\right),$$
$$|\xi| < 1, \tag{5.3.39}$$

where $\chi(\xi)$ denotes the right-hand side (5.3.38), $\chi'(\eta) = \partial\chi/\partial\eta$ and the first integral is a principal value. The integrals can be evaluated explicitly in terms of the Bessel function J_n to give

$$\zeta'(\xi) = \frac{1}{\pi\sqrt{1-\xi^2}}\left\{\frac{-\pi i a Q}{\omega s^2 \ln 2} + \sum_{j=0}^{3}\lambda_j\left[\frac{J_0(\sigma_j)}{\ln 2} + i\sigma_j\left(\xi J_0(\sigma_j) + iJ_1(\sigma_j)\right.\right.\right.$$
$$\left.\left.\left. - \sum_{n=1}^{\infty} 2i^n J_n(\sigma_j)\sin\vartheta\,\sin(n\vartheta)\right)\right]\right\} \tag{5.3.40}$$

where $\vartheta = \arccos\xi$.

This equation determines ζ' in terms of λ_0 (i.e., p_0) and the three unknown quantities λ_1, λ_2, and Q. The integration of (5.3.40) over the slot supplies the following relation between λ_1, λ_2, and Q:

$$Q = \frac{i\omega s^2}{\ln 2}\left(\frac{-\pi i a Q}{\omega s^2} + \sum_{j=0}^{2} \lambda_j \mathrm{J}_0(\sigma_j)\right). \tag{5.3.41}$$

The two further relations are obtained by application of the Kutta condition at the upstream edge $\xi = -1$ of the slot, where the vortex sheet is required to leave the edge tangentially: $\zeta' = \partial\zeta'/\partial\xi = 0$ as $\xi \to -1$. These are

$$\begin{aligned} &Q + \sum_{j=0}^{3} \omega s^2 \lambda_j \sigma_j [\mathrm{J}_0(\sigma_j) - i\mathrm{J}_1(\sigma_j)] = 0, \\ &\sum_{j=0}^{3} \lambda_j \sigma_j (\mathrm{J}_0(\sigma_j) - 2i\sigma_j[\mathrm{J}_0(\sigma_j) - i\mathrm{J}_1(\sigma_j)]) = 0. \end{aligned} \tag{5.3.42}$$

Solving these equations for Q:

$$Q = \frac{-ip_0}{\rho_o \omega} \Im(\sigma_0, \omega), \tag{5.3.43}$$

where

$$\Im(\sigma_0, \omega) = \frac{\pi\sigma_+^2}{[(\sigma_+ - \sigma_0)^2 + \mu^2(\sigma_- - \sigma_0)^2]} \times \frac{\sum_j \sigma_j[\mathrm{J}_0(\sigma_j) - 2\mathrm{W}(\sigma_j)][\mathrm{J}_0(\sigma_{j+1})\mathrm{W}(\sigma_{j+2}) - \mathrm{J}_0(\sigma_{j+2})\mathrm{W}(\sigma_{j+1})]}{\sigma_1[\mathrm{J}_0(\sigma_1) - 2\mathrm{W}(\sigma_1)][\mathrm{J}_0(\sigma_2) - \Psi\mathrm{W}(\sigma_2)] - \sigma_2[\mathrm{J}_0(\sigma_2) - 2\mathrm{W}(\sigma_2)][\mathrm{J}_0(\sigma_1) - \Psi\mathrm{W}(\sigma_1)]} \tag{5.3.44}$$

and $\Psi = \ln 2 - \pi a$, $\mathrm{W}(x) = ix[\mathrm{J}_0(x) - i\mathrm{J}_1(x)]$. The summation in the numerator is over $j = 0, 1, 2$, with suffixes evaluated modulo 3.

This solution is applicable for all values of κ_1. Also, unlike the case of a finite, compact aperture, the flux Q now exhibits a weak dependence on the ratio of the acoustic wavelength to the slot width $2s$ because Ψ involves $\ln(\kappa_o s)$.

Example 12: Slotted screen with tangential mean flow. Consider the reflection of sound from a rigid screen perforated with parallel slots of width $2s$, when the distance between the centerlines of adjacent slots is d, where $\kappa_o d \ll 1$.

Let the slots be orientated as in Example 11, and let the incident, reflected, and transmitted sound be defined as in (5.3.13). Show that the reflection and transmission coefficients are given by [335]

$$\mathcal{R} = \frac{-\frac{i}{\pi}\kappa_o d \cos\Theta[F(\omega s/U_+, \mu) + \ln(d/\pi s)]}{1 - \frac{i}{\pi}\kappa_o d \cos\Theta[F(\omega s/U_+, \mu) + \ln(d/\pi s)]},$$
$$\mathcal{T} = \frac{1}{1 - \frac{i}{\pi}\kappa_o d \cos\Theta[F(\omega s/U_+, \mu) + \ln(d/\pi s)]}, \tag{5.3.45}$$

where, in the notation of Example 11,

$$F(\omega s/U_+, \mu) = \frac{-[\sigma_1[\mathrm{J}_0(\sigma_1) - 2\mathrm{W}(\sigma_1)]\mathrm{J}_0(\sigma_2) - \sigma_2[\mathrm{J}_0(\sigma_2) - 2\mathrm{W}(\sigma_2)]\mathrm{J}_0(\sigma_1)]}{\sigma_1[\mathrm{J}_0(\sigma_1) - 2\mathrm{W}(\sigma_1)]\mathrm{W}(\sigma_2) - \sigma_2[\mathrm{J}_0(\sigma_2) - 2\mathrm{W}(\sigma_2)]\mathrm{W}(\sigma_1)}. \tag{5.3.46}$$

This function is related to $\Im(\sigma_0, \omega)$ by $\Im(0, \omega) = \frac{\pi}{2}/\{F(\omega s/U_+, \mu) + \Psi\}$.

The absorption coefficient $\Delta = 1 - |\mathcal{R}|^2 - |\mathcal{T}|^2$ determines the fractional acoustic power absorbed by the screen and is given by

$$\Delta = \frac{2(\kappa_o d/\pi)\cos\Theta \,\mathrm{Im}\, F(\omega s/U_+, \mu)}{|1 - i(\kappa_o d/\pi)\cos\Theta[F(\omega s/U_+, \mu) + \ln(d/\pi s)]|^2}.$$

Because $\kappa_o d \ll 1$, the denominator differs negligibly from unity except when the frequency is very small. The magnitude of the absorption is determined by the imaginary part of $F(\omega s/U_+, \mu)$. When this is negative, there is a net increase in the acoustic energy at the screen. This occurs for one-sided flow ($\mu = 0$) when $1.59 < \omega s/U < 3.49$. For two-sided, uniform flow ($\mu = 1$) $\mathrm{Im}\, F < 0$ for $2.4 + n\pi < \omega s/U < 3.7 + n\pi$ ($n = 0, 1, 2, \ldots$). At very small Strouhal numbers, $F(\omega s/U_+, \mu) \to \infty$, and the magnitude of the reflection coefficient increases toward unity and $|\mathcal{T}| \to 0$, and the screen becomes acoustically hard.

Example 13: The slotted bias flow screen. When a steady mean flow is maintained *through* the screen of Example 12, the reflection and transmission coefficients are given (5.3.45) with $F(\omega s/U_+, \mu)$ replaced by [308]

$$F(\sigma) = \frac{\pi \mathrm{e}^{-\sigma}\mathrm{I}_0(\sigma) + 2i\mathrm{K}_0(\sigma)\sinh\sigma}{\sigma[\pi \mathrm{e}^{-\sigma}\mathrm{I}_1(\sigma) + 2i\mathrm{K}_1(\sigma)\sinh\sigma]}, \qquad \sigma = \omega s/U_c,$$

where U_c is the mean convection velocity of the shed vorticity in the bias flow jets, which is approximately equal to half the asymptotic jet velocity. The reader may verify that $\mathrm{Im}\, F(\sigma) > 0$, so that the absorption coefficient $\Delta = 1 - |\mathcal{R}|^2 - |\mathcal{T}|^2$ is always positive.

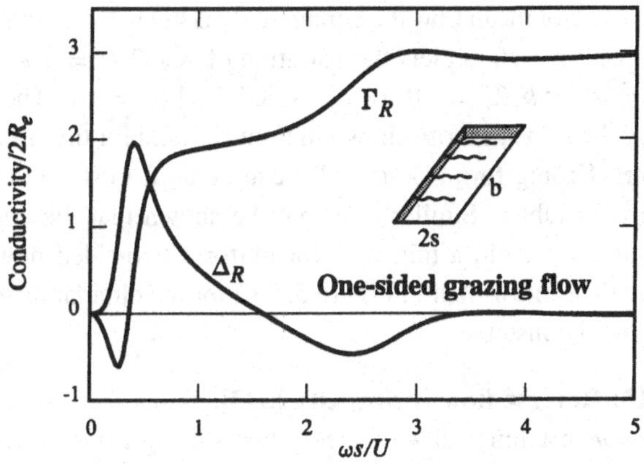

Figure 5.3.10. Conductivity of a rectangular aperture in a thin wall with one-sided grazing flow; $b/2s = 10$.

Example 14: Conductivity of a rectangular aperture in grazing flow [333]. Let the mean flow be in the x_1-direction with velocities $U_\pm$ above and below a thin rigid wall $x_2 = 0$, and suppose the aperture occupies the region $|x_1| < s$, $|x_3| < \frac{1}{2}b$ (Figure 5.3.10). The vortex sheet displacement ζ produced by a uniform, time-harmonic pressure load $p_+ - p_-$ is governed by the integro-differential equation (5.3.27). An approximate solution of this equation can be obtained when the *aspect ratio* $b/2s \gg 1$ by assuming the motion of the vortex sheet to be two-dimensional, that is, by neglecting the dependence of $\zeta(x_1, x_3, \omega)$ on the spanwise coordinate x_3. The integration with respect to y_3 over $(-\frac{1}{2}b, \frac{1}{2}b)$ may then be performed explicitly. If we also integrate over $-\frac{1}{2}b < x_3 < \frac{1}{2}b$, the resulting equation satisfies the condition of pressure continuity across the vortex sheet in a spanwise averaged sense. When $b \gg s$, the equation reduces to (5.3.37) with the right-hand side replaced by

$$\frac{-\pi(p_+ - p_-)}{\rho_o U_+^2} + \frac{2i\omega \ln(2b/s\mathrm{e})}{bU_+^2} Q,$$

where Q is the total volume flux through the aperture, and e is the exponential constant. The equation can be solved as in Example 11, and Q can be expressed in terms of the function $\Im(\sigma_0, \omega)$ with $\sigma_0 \equiv 0$ but with

$$\Psi = \ln(4b/s\mathrm{e}).$$

It then follows that

$$K_R \approx b\Im(0, \omega) = \frac{\pi b}{2[F(\omega s/U_+, \mu) + \ln(4b/s\mathrm{e})]}. \tag{5.3.47}$$

In the absence of mean flow the conductivity is given by setting $F = 0$ in this formula. Figure 5.3.10 depicts the variation of $K_R/2R_e = \Gamma_R - i\Delta_R$ where $R_e = \sqrt{2sb/\pi}$ for $b/2s = 10$ for one-sided flow ($\mu = 0$). These results are qualitatively the same as those shown in Figure 5.3.8i for the circular aperture. The Kramers–Kronig formulae (5.3.33) can be used to show that the motion is absolutely unstable. Similarly, it may be shown that the conductivity of a rectangular aperture in a thin wall for uniform two-sided flow ($\mu = 1$) is qualitatively the same as that in Figure 5.3.8ii for the circular aperture and that it is conditionally unstable.

Example 15: Reverse flow reciprocity [333]. Show that, for an aperture of *arbitrary shape* in a thin wall with one- or two-sided grazing flow, the Rayleigh conductivity $K_R(\omega)$ is unchanged when the directions of the mean flows are reversed, provided that in each case the Kutta condition is applied over that section of the aperture edge from which vortex shedding occurs.

5.3.9 Influence of Wall Thickness

It has already been mentioned that small but finite wall thickness is usually sufficient to make the conductivity of a wall aperture in uniform two-sided flow depart considerably from the case illustrated in Figure 5.3.8ii for a wall of infinitesimal thickness. To prove this, we make a minor extension of the theory of Section 5.3.6 (which will be discussed in more detail in Chapter 6) to deal with a wall of small but finite thickness h. The shear layers over the upper and lower faces of the aperture are modeled by vortex sheets; the fluid within the aperture (in $|x_2| < \frac{1}{2}h$) is assumed to be in a mean state of rest.

Denote by $\zeta_\pm(x_1, x_3, \omega)$ the respective displacements of the upper and lower vortex sheets from their undisturbed positions (Figure 5.3.11 illustrates the case $U_+ = U_- = U$). Then pressure fluctuations above and below the aperture are given by (5.3.26) with ζ replaced by $\zeta_\pm$ for $x_2 \gtrless \pm\frac{1}{2}h$, and the integrations for $x_2 \gtrless \pm\frac{1}{2}h$ being taken with $y_2 = \pm\frac{1}{2}h$.

For a *thin* wall ($h \ll R$, or $h \ll 2s$ for circular or rectangular apertures), and provided the wavelength of disturbances on the vortex sheets are large compared to h, the vertical displacement of fluid in the aperture is independent of x_2. We can then set $\zeta(x_1, x_3, \omega) \approx \zeta_\pm(x_1, x_3, \omega)$, and write the equation of motion of a fluid "column" within the aperture in the form

$$\rho_o h \frac{\partial^2 \zeta}{\partial t^2} = -[p], \qquad (5.3.48)$$

where $[p]$ is the difference in the pressures applied to the upper and lower ends of the column at $x_2 = \pm\frac{1}{2}h$. Substituting for $[p]$ from the modified form of (5.3.26), equation (5.3.48) becomes

$$\left[\left(\omega + iU_+\frac{\partial}{\partial x_1}\right)^2 + \left(\omega + iU_-\frac{\partial}{\partial x_1}\right)^2\right]\frac{1}{2\pi} \times \int \frac{\zeta(y_1, y_3, \omega)\, dy_1\, dy_3}{\sqrt{(x_1 - y_1)^2 + (x_3 - y_3)^2}} + h\omega^2\zeta(x_1, x_3, \omega) = \frac{p_+ - p_-}{\rho_o}, \tag{5.3.49}$$

where (x_1, x_3) lies within the aperture.

This equation must be solved numerically (see Section 6.1). The result of such a calculation is depicted in Figure 5.3.11, which shows the conductivity of a rectangular aperture of aspect ratio $b/2s = 2$ in a thin wall of thickness h when $h/2s = 0.1$ and when the flow is the same on both sides of the wall. The functional form of K_R is the same as for one-sided flow in Figures 5.3.8 and 5.3.10, and it may be verified from the Kramers–Kronig formulae that the motion is absolutely unstable.

5.4 Influence of Surface Compliance

The assumption in Sections 5.2 and 5.3 that a surface can be regarded as *rigid* during its interaction with sound is a useful approximation only for small fluid loading, such as for a steel plate in air. For elastic structures immersed in liquids, the loading is large enough that high acoustic intensities are accompanied by significant structural vibrations. The vorticity produced at the edge of a vibrating surface is modulated in strength by surface motions precisely as it is modulated by incident sound. A bias flow through a perforated elastic screen would therefore be expected to absorb structural vibrations of the screen as well as incident sound.

5.4.1 Interaction of Flexural Waves with a Small Aperture

Let us consider an infinite thin plate at $x_2 = 0$ containing a small circular aperture of radius R whose center is at the origin. Any mean flow, either parallel to the plate or through the aperture, will be assumed to have infinitesimal Mach number. Small-amplitude flexural motions are governed by the bending wave equation (1.4.9) (with $F_2 = 0$). For time-harmonic motions, we therefore have

$$(\mathrm{B}\nabla_2^4 - m\omega^2)\zeta_+ + [p] = 0, \quad \sqrt{x_1^2 + x_3^2} > R, \tag{5.4.1}$$

where ζ_+ denotes the plate displacement in the x_2-direction.

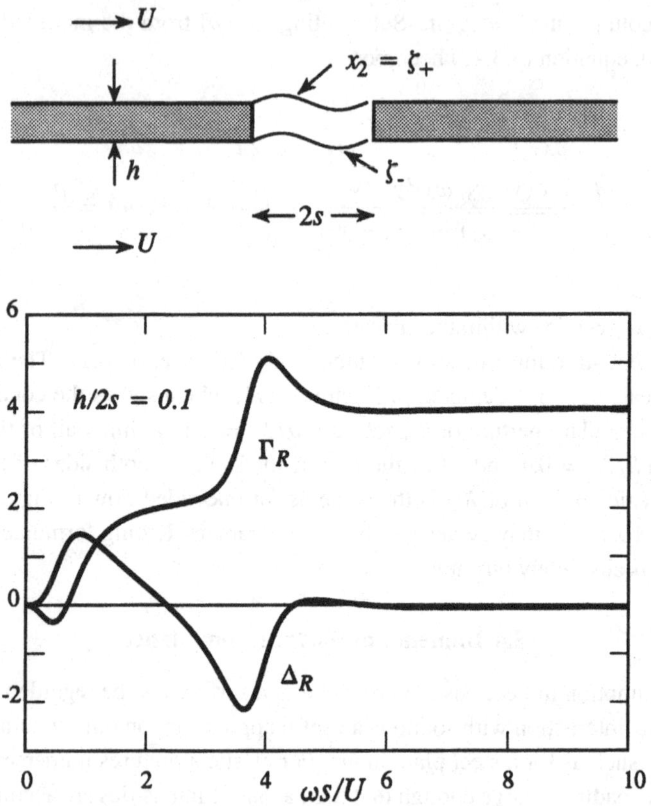

Figure 5.3.11. Conductivity of a rectangular aperture in a rigid wall of thickness $h = 0.2s$ for $b/2s = 2$.

A generalized bending wave equation containing source terms that represent the effect of the aperture can be derived by the method of Section 4.2.1, by setting

$$\zeta(x_1, x_3, \omega) = \zeta_+ \mathrm{H}(f - \varepsilon) + \zeta_- \mathrm{H}(-f - \varepsilon), \quad \varepsilon \to +0, \tag{5.4.2}$$

where H is the Heaviside step function,

$$f(x_1, x_3) \equiv r(x_1, x_3) - R, \quad r = \sqrt{x_1^2 + x_3^2},$$

and ζ_- is the perturbation displacement in the x_2-direction of fluid in the plane of the aperture. Multiplying (5.4.1) by $\mathrm{H}(f - \varepsilon)$ and rearranging, we find as

$\varepsilon \to +0$

$$\begin{aligned}(\mathrm{B}\nabla_2^4 - m\omega^2)\{\mathrm{H}(f)\zeta_+\} + \mathrm{H}(f)[p] \\ = \mathrm{B}\nabla_2^2(\nabla \cdot \{\zeta_+\nabla\mathrm{H}(f)\} + \nabla\mathrm{H}(f)\cdot\nabla\zeta_+) \\ + \mathrm{B}\big(\nabla\cdot\{(\nabla_2^2\zeta_+)\nabla\mathrm{H}(f)\} + \nabla\mathrm{H}(f)\cdot\nabla(\nabla_2^2\zeta_+)\big),\end{aligned} \quad (5.4.3)$$

where on the right-hand side ζ_+ and its derivatives are evaluated at the edge $r = R + 0$ of the aperture.

For a thin plate, continuity of pressure across the aperture implies that $[p] = 0$ for $r < R$, and we can therefore write

$$\begin{aligned}(\mathrm{B}\nabla_2^4 - m\omega^2)\{\mathrm{H}(-f)\zeta_-\} + \mathrm{H}(-f)[p] \\ = (\mathrm{B}\nabla_2^4 - m\omega^2)\{\mathrm{H}(-f)\zeta_-\}, \quad \varepsilon \to +0.\end{aligned}$$

Adding to (5.4.3), it follows that

$$\begin{aligned}(\mathrm{B}\nabla_2^4 - m\omega^2)\zeta + [p] \\ = (\mathrm{B}\nabla_2^4 - m\omega^2)\{\mathrm{H}(-f)\zeta_-\} + \mathrm{B}\nabla_2^2(\nabla\cdot\{\zeta_+\nabla\mathrm{H}(f)\} + \nabla\mathrm{H}(f)\cdot\nabla\zeta_+) \\ + \mathrm{B}\big(\nabla\cdot\{(\nabla_2^2\zeta_+)\nabla\mathrm{H}(f)\} + \nabla\mathrm{H}(f)\cdot\nabla(\nabla_2^2\zeta_+)\big),\end{aligned} \quad (5.4.4)$$

where the "sources" on the right depend on conditions at the edge of the aperture and the perturbation displacement of fluid through the aperture.

When $\kappa_o R,\ K_o R \ll 1$ (where K_o is the vacuum wavenumber (4.1.2), which is characteristic of bending wave motions), the waves scattered from the aperture must be axisymmetric. As $R \to 0$, the source terms vanish except at $r = 0$ and must ultimately reduce to combinations of the surface delta function $\delta_2(\mathbf{x}) \equiv \delta(x_1)\delta(x_3)$ and its axisymmetric derivatives [20, 336], where, however, the highest order derivatives of the terms in ζ_+ cannot exceed third order if the displacement of the plate is to remain finite as $r \to R + 0$. In the first term, $\{\mathrm{H}(-f)\zeta_-\} \to \hat{Q}\delta_2(\mathbf{x})$ as $R \to 0$, where $\hat{Q}$ $(= -Q/i\omega)$ is the perturbation volume *displacement* through the aperture. Thus, as $R \to 0$, equation (5.4.4) reduces to

$$(\mathrm{B}\nabla_2^4 - m\omega^2)\zeta + [p] = a_0\delta_2(\mathbf{x}) + a_2\nabla_2^2\delta_2(\mathbf{x}) + \hat{Q}(\mathrm{B}\nabla_2^4 - m\omega^2)\delta_2(\mathbf{x}), \quad (5.4.5)$$

where $a_0,\ a_2,\ \hat{Q}$ remain to be determined; they represent the strengths of the sources to which the aperture is equivalent as $R \to 0$.

Suppose now the bending wave $\zeta(x_1, \omega) \equiv \zeta_I(x_1, \omega) = \zeta_o e^{i\kappa x_1}$ is incident on the aperture, where $\omega > 0$, $\kappa R \ll 1$ and $\zeta_o =$ constant, and let $[p_I]$ denote the accompanying pressure jump across the plate. At large distances from the aperture, in an "outer region" where $|\mathbf{x}| \gg R$, the scattered surface displacement $\zeta'(x_1, x_3, \omega)$ and pressure $p'(\mathbf{x}, \omega)$ are given by the three-dimensional analogues of equations (4.2.4):

$$\zeta'(x_1, x_3, \omega) = \int_{-\infty}^{\infty} \zeta'(\mathbf{k}, \omega) e^{i\mathbf{k}\cdot\mathbf{x}}\, dk_1\, dk_3, \quad \mathbf{k} = (k_1, 0, k_3)$$

$$p'(\mathbf{x}, \omega) = -i\rho_o\omega^2 \operatorname{sgn}(x_2) \int_{-\infty}^{\infty} \frac{\zeta'(\mathbf{k}, \omega)}{\gamma(k)} e^{i(\mathbf{k}\cdot\mathbf{x} + \gamma(k)|x_2|)}\, dk_1\, dk_3, \tag{5.4.6}$$

where $\gamma(k) = \sqrt{\kappa_o^2 - k^2}$, and branch cuts for $\gamma(k)$ are defined in the usual way by (1.13.9).

By taking the Fourier transform of (5.4.5), these expressions can be used to determine ζ' in terms of the aperture sources. The net displacement at $x_2 = 0$ can then be cast in the form

$$\zeta(x_1, x_3, \omega) = \hat{Q}\delta_2(\mathbf{x}) + \frac{1}{(2\pi)^2} \times \int_{-\infty}^{\infty} \frac{\{a_0 - a_2 k^2 + 2i\hat{Q}\rho_o\omega^2/\gamma(k)\} e^{i\mathbf{k}\cdot\mathbf{x}}}{\mathcal{D}(k, \omega)} \times dk_1\, dk_3 + \zeta_o e^{i\kappa x_1}, \tag{5.4.7}$$

where $\mathcal{D}(k, \omega)$ is the bending wave dispersion function (4.1.6), and singularities of the integrand for real $\mathbf{k}$ are avoided by considering the limit $\omega = \omega + i0$ (Section 1.7.1). In this "outer" representation the δ-function corresponds to the component ζ_- of ζ that describes the volume displacement through the aperture. The remaining terms on the right are nonsingular at $\mathbf{x} = \mathbf{0}$ and determine the displacement ζ_+ of the plate (in $r > R$).

Two linear algebraic equations for a_0, a_2, $\hat{Q}$ are obtained by substituting for ζ_+ in the "free" edge condition of (1.4.16b), which, for axisymmetric motion, becomes

$$\nabla_2^2\zeta_+ - \frac{(1-\sigma)}{r}\frac{\partial\zeta_+}{\partial r} = 0, \quad \frac{\partial}{\partial r}\nabla_2^2\zeta_+ = 0, \quad r \to R + 0,$$

where σ is Poisson's ratio for the plate. To do this, the incident wave is first expanded in the form [24, 289]

$$\zeta_o e^{i\kappa x_1} \equiv \zeta_o e^{i\kappa r\cos\vartheta} = \zeta_o\left(\mathrm{J}_0(\kappa r) + \sum_{n=1}^{\infty} i^n \mathrm{J}_n(\kappa r)\cos(n\vartheta)\right), \quad x_1 = r\cos\vartheta,$$

where only the axisymmetric component $\zeta_o \mathrm{J}_0(\kappa r)$ need be retained. We then find that

$$\begin{aligned} &\mathcal{A}_{01}a_0 + \mathcal{A}_{21}a_2 + \mathcal{B}_1\hat{Q} = \frac{\zeta_o\kappa}{R}(\kappa R\mathrm{J}_0(\kappa R) - (1-\sigma)\mathrm{J}_1(\kappa R)) \\ &\mathcal{A}_{02}a_0 + \mathcal{A}_{22}a_2 + \mathcal{B}_2\hat{Q} = -\zeta_o\kappa^3\mathrm{J}_1(\kappa R), \end{aligned} \tag{5.4.8}$$

where the coefficients $\mathcal{A}_{0j}$, $\mathcal{A}_{2j}$, $\mathcal{B}_j$ are given in Example 1.

A third equation for a_0, a_2, $\hat{Q}$ is obtained by considering the fluid motion just above and below the aperture, where the pressure can be written

$$p(\mathbf{x}, \omega) = -\mathrm{sgn}(x_2)\frac{\rho_o\omega^2\zeta_o}{\sqrt{\kappa^2 - \kappa_o^2}} + \rho_o\omega^2\zeta_o x_2 + p'(\mathbf{x}, \omega).$$

The terms in ζ_o are the contributions from the incident bending wave when the influence of the aperture is ignored, for which the normal acceleration of the plate is just equal to $-\omega^2\zeta_o$. When the wavelength is much larger than R, the motion of the plate near the aperture will be identical to that of a rigid plane executing translational oscillations of amplitude ζ_o in the normal direction. For an acoustically compact aperture, the fluid motion in the immediate vicinity of the aperture may be regarded as incompressible, and the additional pressure $p'(\mathbf{x}, \omega)$ calculated as for an incompressible fluid subject to a *local* rigid surface condition $\partial p'/\partial x_2 = 0$ at $x_2 = \pm 0$.

The volume displacement can then be expressed in terms of the Rayleigh conductivity (5.3.1) by writing $\hat{Q} \equiv Q/(-i\omega) = K_R[p_\infty]/\rho_o\omega^2$, where p_∞ is the uniform component of the pressure at distances $|\mathbf{x}| \gg R$ from the aperture. This pressure is dominated by the incident bending wave, that is, $[p_\infty] \approx [p_\mathrm{I}] \equiv -2\rho_o\omega^2\zeta_o/\sqrt{\kappa^2 - \kappa_o^2}$ so that

$$\hat{Q} \approx \frac{K_R[p_\mathrm{I}]}{\rho_o\omega^2}.$$

The values of a_0 and a_2 are now obtained by substituting this expression for $\hat{Q}$ into (5.4.8). Expanding the solution in powers of $K_o R$ ($\ll 1$) to order $(K_o R)^2$

(the highest power that contains only axisymmetric modes), we find that

$$a_0 = -\pi R^2 \mathrm{B}\kappa^4\zeta_o - 2RK_R[p_\mathrm{I}], \quad a_2 = -\left(\frac{1+\sigma}{1-\sigma}\right)\pi R^2 \mathrm{B}\kappa^2\zeta_o.$$

Now $\kappa^4\zeta_o = \nabla_2^4\zeta_\mathrm{I}$, $\kappa^2\zeta_o = -\nabla_2^2\zeta_\mathrm{I}$, so that (5.4.5) can be written

$$\begin{aligned}(\mathrm{B}\nabla_2^4 - m\omega^2)\zeta + [p] \\ &= -\{\pi R^2 \mathrm{B}\nabla_2^4\zeta_\mathrm{I} + 2RK_R[p_\mathrm{I}]\}\delta_2(\mathbf{x}) + \left(\frac{1+\sigma}{1-\sigma}\right)\pi R^2 \mathrm{B}\nabla_2^2\zeta_\mathrm{I}\nabla_2^2\delta_2(\mathbf{x}) \\ &\quad + \frac{K_R[p_\mathrm{I}]}{\rho_o\omega^2}(\mathrm{B}\nabla_2^4 - m\omega^2)\delta_2(\mathbf{x}).\end{aligned} \tag{5.4.9}$$

In this form, the equation describes the scattering of an arbitrary incident displacement ζ_I and pressure field $[p_\mathrm{I}]$ whose characteristic length scale is much larger than the aperture radius. Note that the equation is applicable in the absence of fluid loading when the terms in $[p]$ and $[p_\mathrm{I}]$ are omitted.

Example 1. The coefficients $\mathcal{A}_{0j}$, $\mathcal{A}_{2j}$, $\mathcal{B}_j$ of equations (5.4.8) are given by

$$\begin{aligned}\mathcal{A}_{01} &= \frac{-i}{8\mathrm{B}}\Bigg(\mathrm{H}_0^{(1)}(K_oR) + \mathrm{H}_0^{(1)}(iK_oR) \\ &\quad - \frac{(1-\sigma)}{K_oR}\mathrm{H}_1^{(1)}(K_oR) + \frac{i(1-\sigma)}{K_oR}\mathrm{H}_1^{(1)}(iK_oR)\Bigg) \\ &\quad - \frac{i\rho_o\omega^2}{\pi R}\int_0^\infty \frac{k^2[kR\mathrm{J}_0(kR) - (1-\sigma)\mathrm{J}_1(kR)]\,dk}{\gamma(k)(\mathrm{B}k^4 - m\omega^2)\mathcal{D}(k,\omega)}, \\ \mathcal{A}_{21} &= \frac{-iK_o^2}{8\mathrm{B}}\Bigg(\mathrm{H}_0^{(1)}(K_oR) - \mathrm{H}_0^{(1)}(iK_oR) \\ &\quad - \frac{(1-\sigma)}{K_oR}\mathrm{H}_1^{(1)}(K_oR) + \frac{i(1-\sigma)}{K_oR}\mathrm{H}_1^{(1)}(iK_oR)\Bigg) \\ &\quad + \frac{i\rho_o\omega^2}{\pi R}\int_0^\infty \frac{k^4[kR\mathrm{J}_0(kR) - (1-\sigma)\mathrm{J}_1(kR)]\,dk}{\gamma(k)(\mathrm{B}k^4 - m\omega^2)\mathcal{D}(k,\omega)}, \\ \mathcal{B}_1 &= \frac{-i\rho_o\omega^2}{\pi R}\int_0^\infty \frac{k^2[kR\mathrm{J}_0(kR) - (1-\sigma)\mathrm{J}_1(kR)]\,dk}{\gamma(k)\mathcal{D}(k,\omega)}; \\ \mathcal{A}_{02} &= \frac{K_o}{8\mathrm{B}}\left(i\mathrm{H}_1^{(1)}(K_oR) - \mathrm{H}_1^{(1)}(iK_oR)\right) \\ &\quad + \frac{i\rho_o\omega^2}{\pi}\int_0^\infty \frac{k^4\mathrm{J}_0(kR)\,dk}{\gamma(k)(\mathrm{B}k^4 - m\omega^2)\mathcal{D}(k,\omega)},\end{aligned}$$

$$\mathcal{A}_{22} = \frac{-K_o^3}{8\mathrm{B}}\left(i\mathrm{H}_1^{(1)}(K_oR) - \mathrm{H}_1^{(1)}(iK_oR)\right)$$
$$- \frac{i\rho_o\omega^2}{\pi}\int_0^\infty \frac{k^6\mathrm{J}_1(kR)\,dk}{\gamma(k)(\mathrm{B}k^4 - m\omega^2)\mathcal{D}(k,\omega)},$$
$$\mathcal{B}_2 = \frac{i\rho_o\omega^2}{\pi\mathrm{B}}\left(\frac{1-\mathrm{e}^{i\kappa_oR}}{\kappa_oR} + \omega^2\int_0^\infty \frac{[m + 2i\rho_o/\gamma(k)]\mathrm{J}_1(kR)]\,dk}{\gamma(k)\mathcal{D}(k,\omega)}\right).$$

For $\omega > 0$, the integrations are taken along the positive real axis indented to pass below real singularities of the integrands. Because $\hat{Q} \approx K_R[p_\mathrm{I}]/\rho_o\omega^2$, the relative orders of magnitude in equations (5.4.8) of $\mathcal{B}_1\hat{Q}$ and that part of $\mathcal{B}_2\hat{Q}$ involving the integral in the definition of $\mathcal{B}_2$ are smaller than $\mathrm{O}\{(K_oR)^2\}$ and their contributions may be neglected.

Example 2. Show from (5.4.9) that the power dissipated (as opposed to *scattered*) when a bending wave $\zeta = \mathrm{Re}\{\zeta_o\mathrm{e}^{i(\kappa x_1 - \omega t)}\}$, $\omega > 0$, is incident on a small circular aperture of radius R is given by

$$\Pi \approx \frac{-2\rho_o\omega^3|\zeta_o|^2\mathrm{Im}\,K_R}{\kappa^2 - \kappa_o^2}, \quad \kappa R \ll 1.$$

This dissipation is due entirely to losses occurring at the aperture and does not include losses due to scattering into sound or into other bending waves. It arises from the *final* source term on the right of (5.4.9). The terms in ζ_I are responsible for scattering by the aperture edge (and are unchanged in the absence of fluid loading), and the contribution from first term on the right involving $[p_\mathrm{I}]$ is an order of magnitude smaller.

Example 3. When $K_oR \ll 1$, the far field acoustic mean square pressure generated by the interaction of Example 2 is given approximately by

$$\langle p^2(\mathbf{x},t)\rangle = \frac{\left(\rho_o\omega^2|\zeta_oK_R|\right)^2\mu^4\cos^2\Theta\left[\mu^4\sin^4\Theta - 1 - \frac{2R\rho_o}{m}\right]^2}{2(\pi K_o|\mathbf{x}|)^2(K^2-\mu^2)[\mu^4\cos^2\Theta(\mu^4\sin^4\Theta - 1)^2 + 4\epsilon^2]},$$
$$\mu = \sqrt{\frac{\omega}{\omega_c}}, \quad K = \frac{\kappa}{K_o},$$

where $\Theta = \arccos(x_2/|\mathbf{x}|)$ is the angle between the radiation direction and the normal to the plate, and the notation is the same as in Section 4.2.4, Example 8. We now see that the condition for the validity of the analogous formula (4.2.20) derived by a more heuristic argument is that there should be no dissipation in the aperture, and the fluid loading should be *light*:

$$K_R = 2R, \quad 2R\rho_o \ll m, \quad \text{that is,} \quad 2R\rho_o \ll h\rho_s.$$

Deduce that when $\omega/\omega_c \le \frac{1}{3}$ the power lost by acoustic radiation is given approximately by

$$\Pi_a \approx \frac{2\rho_o\omega^4|\zeta_o K_R|^2}{\pi c_o(\kappa^2-\kappa_o^2)}\left(1+\frac{2R\rho_o}{m}\right)\left(1-\frac{\arctan(\omega/2\epsilon\omega_c)}{(\omega/2\epsilon\omega_c)}\right). \tag{5.4.10}$$

Comparing this with the power Π dissipated within the aperture of Example 2, it may be concluded that, provided $\text{Im}\, K_R \ne 0$, $\Pi_a \sim \text{O}(\kappa_o R)\Pi \ll \Pi$.

5.4.2 The Perforated Elastic Screen

Let a thin elastic screen be uniformly perforated with $\mathcal{N}$ equal circular apertures of radius R per unit area, with fractional open area $\alpha = \mathcal{N}\pi R^2$. When $\alpha \ll 1$ the distance $(\sim 1/\sqrt{\mathcal{N}})$ between neighboring apertures is much greater than R, and provided $\kappa_o R$, $K_o R \ll 1$, the length scales of disturbances incident on an aperture (including those scattered by neighboring apertures) will be much larger than R. The dominant component of the scattered acoustic and flexural waves will then have radial symmetry with respect to that aperture, and equation (5.4.9) may be generalized by replacing the right-hand side by

$$\sum_n \Big\{ -\{\pi R^2 \text{B}\nabla_2^4\zeta_n + 2RK_R[p_n]\}\delta_2(\mathbf{x}-\mathbf{x}_n)$$
$$+\left(\frac{1+\sigma}{1-\sigma}\right)\pi R^2 \text{B}\nabla_2^2\zeta_n.\nabla_2^2\delta_2(\mathbf{x}-\mathbf{x}_n)$$
$$+\frac{K_R}{\rho_o\omega^2}[p_n]\{\text{B}\nabla_2^4 - m\omega^2\}\delta_2(\mathbf{x}-\mathbf{x}_n)\Big\}$$

where the summation is over all of the apertures. The center of the nth aperture is at $\mathbf{x}_n = (x_{n1}, 0, x_{n3})$, at which ζ_n and $[p_n]$ respectively denote the flexural displacement of the plate and pressure jump produced by incident sound or flexural waves, including collective contributions caused by scattering by the remaining apertures.

The sum can be simplified by averaging with respect to all possible aperture positions, which leads to a representation of the overall influence of the apertures, rather than one concerned with local details at each aperture. This procedure is frequently employed in studying wave propagation in dusty gases and bubbly media and in the classical *Lorentz–Lorenz* theory of a dielectric [337]. To determine the average contribution from the sources at $\mathbf{x} = \mathbf{x}_m$, say, we first obtain appropriate expressions for ζ_m and $[p_m]$ by the following argument. For a large plate of area A, the probability that $\mathbf{x}_n$ lies within the area element $d^2\mathbf{x}_n = dx_{n1}\,dx_{n3}$ is equal to $d^2\mathbf{x}_n/A$, and there are $\mathcal{N}A$ apertures in all. When $\alpha \ll 1$, the apertures may be regarded as independently distributed in a first approximation, and the average can be evaluated by first averaging over all $\mathbf{x}_n \neq \mathbf{x}_m$. The quantities ζ_m and $[p_m]$ must then be set equal to the corresponding averaged incident fields, following which we can average with respect to $\mathbf{x}_m$. When the total number of apertures is large, these incident field averages cannot differ significantly from the overall average plate displacement and pressure jump because the latter are hardly affected by the presence or absence of the aperture at $\mathbf{x}_m$ when $\mathcal{N}A \to \infty$. Thus, as $\mathcal{N}A \to \infty$, the contribution to the summation from $\mathbf{x}_m$ is obtained by writing

$$\zeta_m = \zeta_{\mathrm{P}}(x_{m1}, x_{m3}) \quad \text{and} \quad [p_m] = [p(x_{m1}, 0, x_{m3})],$$

where ζ_{P} and $[p]$ respectively denote the average displacement of the *plate* and the average pressure jump. If $\zeta_A(x_{m1}, x_{m3}) = \hat{Q}/\pi R^2 \equiv K_R[p]/(\rho_o\omega^2\pi R^2)$ is the corresponding average displacement in the apertures, then the average displacement of fluid on either side of the plate is

$$\zeta \equiv (1-\alpha)\zeta_{\mathrm{P}} + \alpha\zeta_{\mathrm{A}} = (1-\alpha)\zeta_{\mathrm{P}} + \mathcal{N}K_R[p]/\rho_o\omega^2.$$

This relation permits ζ_{P} to be expressed in terms of ζ and $[p]$.

Making these substitutions and performing the average (and discarding terms of order $\alpha^2 \ll 1$), we arrive at the generalized bending wave equation

$$\{\bar{\mathrm{B}}\nabla_2^4 - m\omega^2\}\zeta + \left(1 + 2\mathcal{N}RK_R\left[1 - \frac{\bar{\mathrm{B}}\nabla_2^4 - m\omega^2}{2R\rho_o\omega^2}\right]\right)[p] = 0,$$
$$\bar{\mathrm{B}} = \left(1 - \frac{2\alpha\sigma}{1-\sigma}\right)\mathrm{B}. \tag{5.4.11}$$

In this result, $\bar{\mathrm{B}}$ is the effective stiffness of the plate, which is reduced by the presence of the apertures; this is independent of the presence of the fluid. The influence of fluid motion in the apertures is given by the term involving the conductivity K_R.

Example 4. Show that the generalized bending wave equation (5.4.11) can be written

$$(1-\alpha)\left(\left[1 - \frac{2\alpha\sigma}{1-\sigma}\right]\mathrm{B} + m\frac{\partial^2}{\partial t^2}\right)\zeta_{\mathrm{P}} + [p] = 2\alpha R\rho_o\frac{\partial^2\zeta_{\mathrm{A}}}{\partial t^2}, \tag{5.4.12}$$

where ζ_P and ζ_A denote the displacement of the plate and the fluid displacement in the apertures, and all quantities are averaged over a region of the plate containing many apertures.

5.4.3 Absorption of Sound by a Perforated Elastic Screen

Let incident, reflected, and transmitted sound waves be defined as in (5.3.13). The reflection and transmission coefficients $\mathcal{R}$ and $\mathcal{T}$ are determined from the conditions that (1) the volume flux of fluid normal to the screen is the same on both sides, (2) the pressure and (generalized) displacement of the screen satisfy $\zeta = (1/\rho_o\omega^2)\partial p/\partial x_2$, $x_2 \to \pm 0$ (provided the mean flow Mach number is infinitesimal), and (3) ζ and $[p]$ satisfy the generalized bending wave equation (5.4.11). If Θ denotes the angle of incidence (defined as in Figure 5.3.5), then

$$\mathcal{R} = 1 - \mathcal{T}$$
$$= \frac{\left(\frac{[\bar{\mathrm{B}}\kappa_o^4 \sin^4\Theta - m\omega^2]}{\rho_o\omega^2}\right)\kappa_o\cos\Theta}{\left[\left(\frac{[\bar{\mathrm{B}}\kappa_o^4 \sin^4\Theta - m\omega^2]}{\rho_o\omega^2}\right)\kappa_o\cos\Theta - 2i\left(1 + 2\mathcal{N}RK_R\left[1 - \frac{(\bar{\mathrm{B}}\kappa_o^4\sin^4\Theta - m\omega^2)}{2R\rho_o\omega^2}\right]\right)\right]}. \quad (5.4.13)$$

This reduces to the rigid screen formulae (5.3.15) for an infinitely massive ($m/\rho_o R \to \infty$) or stiff ($\mathrm{B}/\rho_o c_o^2 R^3 \to \infty$) screen.

For a *bias flow* screen (Section 5.3.4), the absorption coefficient $\Delta = 1 - |\mathcal{R}|^2 - |\mathcal{T}|^2$ for sound incident normally ($\Theta = 0$) may be written

$$\Delta = \frac{\frac{8\alpha M_c S_t \Delta_R}{\pi}\left[1 + \frac{2\rho_o R}{\rho_s h}\right]}{S_t^2\left(M_c + \frac{4\alpha\Delta_R}{\pi S_t}\left[1 + \frac{2\rho_o R}{\rho_s h}\right]\right)^2 + 4\left(\frac{\rho_o R}{\rho_s h} + \frac{2\alpha\Gamma_R}{\pi}\left[1 + \frac{2\rho_o R}{\rho_s h}\right]\right)^2}, \quad (5.4.14)$$

where $S_t = \omega R/U_c$ is the Strouhal number based on the vorticity convection velocity U_c (the mean bias flow velocity in the plane of the aperture), $M_c = U_c/c_o$, and K_R has been replaced by $2R(\Gamma_R - i\Delta_R)$, Γ_R, and Δ_R being defined by (5.3.21).

Δ is plotted in Figure 5.4.1 (solid curves) for normal incidence on a perforated aluminium plate in air, when $M_c = 0.05$ and $R/h = 5$. The three cases shown of $\alpha/M_c = 1,\ 2$, and 0.2 correspond to the values in Figure 5.3.5 for a rigid screen. The maximum attenuation ($\Delta \approx \frac{1}{2}$) occurs when $\alpha = M_c$ for a range of Strouhal numbers centered on $S_t \approx 0.4$.

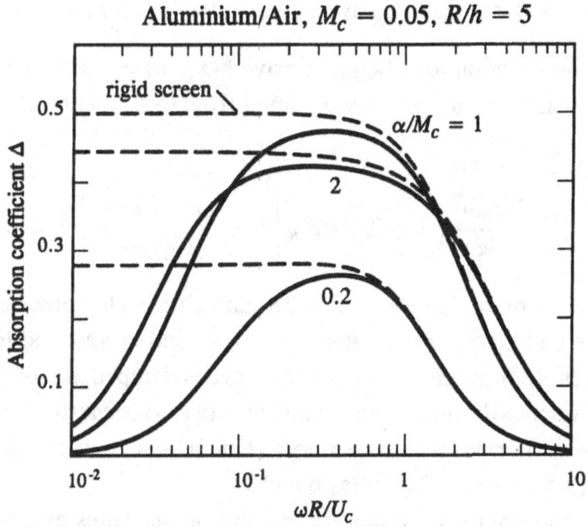

Figure 5.4.1. Absorption of normally incident sound by a bias flow perforated, elastic screen.

The corresponding attenuations for a rigid screen (obtained by setting $\rho_o R/\rho_s h = 0$ in (5.4.14)) are reproduced from Figure 5.3.5 as the broken curves in Figure 5.4.1. At low frequencies, the rigid screen absorption coefficient $\Delta \approx 2\alpha M_c/(M_c + \alpha)^2$ is independent of frequency. The elastic screen predictions depart from this asymptotic limit when the frequency is low enough for the screen to be regarded as acoustically transparent (Section 4.1); then the bias flow jets are not modulated by the sound, and there is no transfer of acoustic energy to the flow. The behavior illustrated in Figure 5.4.1 is typical of lightly loaded screens. When the fluid loading is large (which in practice means for $\rho_o/\rho_s > 10^{-2}$ and is exemplified by the case of a steel screen in water), the absorption of acoustic energy is negligible at all frequencies.

Example 5. By using the low Strouhal number approximation (5.3.22) for the bias flow conductivity, show that the normal incidence absorption coefficient becomes

$$\Delta = \frac{2\alpha M_c S_t^2\left[1 + \frac{2\rho_o R}{\rho_s h}\right]}{S_t^2\left(M_c + \alpha\left[1 + \frac{2\rho_o R}{\rho_s h}\right]\right)^2 + 4\left(\frac{\rho_o R}{\rho_s h}\right)^2}.$$

The damping is therefore negligible when $\omega R/U_c < R\rho_o/h\rho_s M_c$.

5.4.4 Bending Waves on a Bias Flow Plate

The dispersion equation for a bending wave $\zeta(x_1, \omega) = \zeta_o e^{i\kappa x_1}$ ($\kappa R \ll 1$) on a perforated elastic screen is readily obtained from (5.4.11) in the form

$$\bar{\mathcal{D}}(\kappa, \omega) \equiv \bar{\mathrm{B}}\kappa^4 - m\omega^2 - \frac{2i\rho_o\omega^2}{\gamma(\kappa)}\left(1 + 2\mathcal{N}RK_R\left[1 - \frac{\bar{\mathrm{B}}\kappa^4 - m\omega^2}{2R\rho_o\omega^2}\right]\right) = 0. \quad (5.4.15)$$

In the absence of dissipation, $K_R = 2R$, and $\bar{\mathcal{D}}(\kappa, \omega)$ has precisely two equal and opposite real zeros. They satisfy $|\kappa| > \kappa_o$ and correspond to undamped flexural waves. In practice, of course, the wave is damped, not only because of internal and viscous dissipative mechanisms, but also because of the generation of sound by scattering at the apertures (which is a higher order effect not included in (5.4.15)) (see Examples 6 and 7).

For a bias flow screen, dissipation within the apertures as a result of vorticity production causes the bending wave roots of (5.4.15) to acquire small imaginary parts. When $\omega > 0$, and for a wave traveling in the positive x_1-direction, the reduction in wave power over a propagation distance x_1 is equal to $20\,\mathrm{Im}(\kappa)x_1 \log(\mathrm{e})$ dB, so that the wave power absorbed per wavelength of propagation is

$$40\pi \log(\mathrm{e})\,\mathrm{Im}\{\kappa\}/\mathrm{Re}\{\kappa\} \approx 54.6\,\mathrm{Im}\{\kappa\}/\mathrm{Re}\{\kappa\}\,\mathrm{dB}.$$

This is plotted as a function of $\omega R/U_c$ in Figure 5.4.2 for a steel screen in water when $U_c = 7.5$ m/s, $R = 5h$ and for three different fractional open areas α. The peak attenuation of about 2 dB per wavelength of propagation for $\alpha = 10\%$ is comparable with that which can be achieved by coating the plate with elastomeric damping materials [26] and occurs near $\omega R/U_c = 0.8$, which corresponds to about 40 Hz for a screen 0.5 cm thick. The attenuation is usually smaller for a metallic screen in air and occurs at lower frequencies.

Example 6. When $K_o R \ll 1$, show that the bending wave roots of the dispersion equation (5.4.15) are close to those of

$$\mathcal{D}(\kappa, \omega) + \frac{4\rho_o\omega^2\mathcal{N}K_R}{\kappa^2 - \kappa_o^2} = 0,$$

where $\mathcal{D}(\kappa, \omega)$ is the dispersion function (4.1.6) for a thin homogeneous plate. Deduce that, for the flexural wave $\zeta(x_1, \omega) = \zeta_o e^{i\kappa x_1}$ ($\kappa R \ll 1$, $\omega > 0$), propagating in the direction of increasing x_1

$$\frac{\mathrm{Im}\,\kappa}{\mathrm{Re}\,\kappa} \approx \frac{8\alpha\rho_o\omega^2\Delta_R}{\pi\kappa R\left(\kappa^2 - \kappa_o^2\right)\partial\mathcal{D}(\kappa, \omega)/\partial\kappa}, \quad (5.4.16)$$

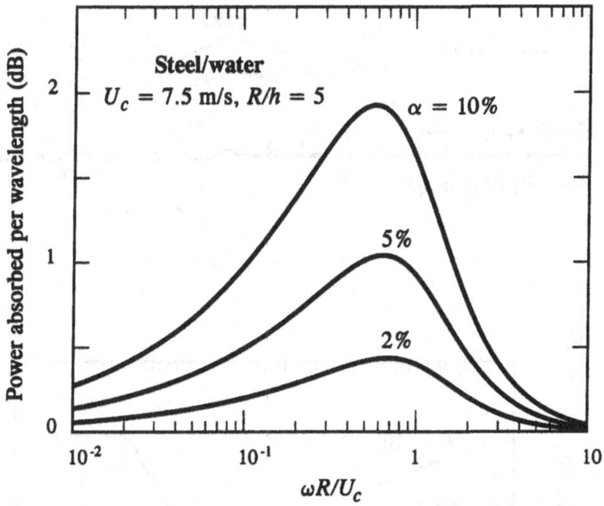

Figure 5.4.2. Damping of bending waves on a bias flow perforated screen.

where on the right-hand side κ denotes the real value of the wavenumber in the absence of dissipation.

Example 7. The contribution of acoustic scattering to the damping of a bending wave on a homogeneous perforated plate can be estimated from expression (5.4.10). That formula neglects the "short-circuiting" by neighboring apertures of the acoustic monopole to which the aperture is equivalent (except at very low frequencies) and therefore tends to overestimate the strength of the scattered sound, but it gives a sufficient estimate of the overall contribution of acoustic damping. The fraction of the incident bending wave power Π_{I}, say, scattered when the incident wave ($\zeta(x_1, \omega) = \zeta_o \mathrm{e}^{i\kappa x_1}$) travels a distance δx_1 is $\mathrm{Im}\{\kappa_a\}\delta x_1 \equiv \mathcal{N}\Pi_a \delta x_1/\Pi_{\mathrm{I}}$, where $\mathrm{Im}\{\kappa_a\}$ is the contribution to $\mathrm{Im}\{\kappa\}$ from acoustic dissipation. Using (5.4.10) and the approximation (4.1.21) $\Pi_{\mathrm{I}} = \frac{1}{4}\omega|\zeta_o|^2 \partial\mathcal{D}(\kappa, \omega)/\partial\kappa$, we find

$$\frac{\mathrm{Im}\,\kappa_a}{\mathrm{Re}\,\kappa} \approx \frac{32\alpha\rho_o\omega^2(\kappa_o/\kappa)\left(\Gamma_R^2 + \Delta_R^2\right)}{\pi\left(\kappa^2 - \kappa_o^2\right)\partial\mathcal{D}(\kappa, \omega)/\partial\kappa}\left(1 - \frac{\arctan(\omega/2\epsilon\omega_c)}{(\omega/2\epsilon\omega_c)}\right).$$

This is applicable for $\omega/\omega_c \leq \frac{1}{3}$, which covers most cases of practical interest (at higher frequencies the bending wavelength is comparable to the plate thickness, i.e., to the aperture diameter).

A comparison of this formula with (5.4.16) for the aperture dissipation shows that, at those frequencies for which damping in the apertures is significant, the contribution from acoustic losses is always at least $\mathrm{O}(\kappa_o R) \ll 1$ smaller.

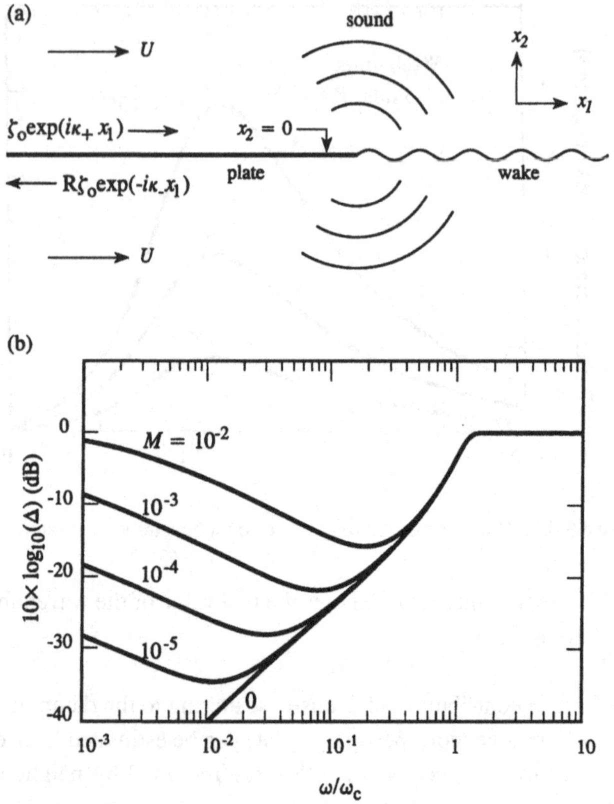

Figure 5.4.3. Bending wave power dissipated at a free edge.

5.4.5 Damping of Bending Waves at a Trailing Edge

Bending waves impinging on the trailing edge of an elastic plate generate sound and cause additional shedding of vorticity into the wake (Figure 5.4.3a). At low Mach number, linear perturbation theory applied to a thin plate suggests that this vorticity is generated at the expense of the structural energy. Energy losses caused by the radiation of sound from the edge in the absence of mean flow were examined in Section 4.4. Consider the situation of Figure 5.4.3a, in which the flexural wave $\zeta(x_1, \omega) = \zeta_o e^{i\kappa_+ x_1}$ is reflected at the edge of a semi-infinite plate in tangential mean flow at speed U in the x_1-direction. The reflected wave $\zeta = R\zeta_o e^{-i\kappa_- x_1}$ ($\kappa_- > 0$) emerges from the localized, complex fluid–structure interaction at the edge and propagates without attenuation to $x_1 = -\infty$. The wavenumbers κ_+ and $-\kappa_-$ are the two real roots of the dispersion equation

$$Bk^4 - m\omega^2 - \frac{2i\rho_o(\omega - Uk)^2}{\gamma(k)} = 0, \tag{5.4.17}$$

of bending waves in the presence of mean tangential flow at speed U ($M = U/c_o \ll 1$) on both sides of the plate.

The fractional loss of power at the edge, Δ, depends on the frequency, Mach number, and the mechanical constraints imposed on plate motion at the edge. The no-flow calculation of Section 4.4.1 is easily generalized to account for a low Mach number mean flow provided the shed vorticity is assumed to lie within a vortex sheet whose strength is determined in the usual way by application of the Kutta condition. The details are too lengthy to be given here [338], and it must suffice to give a brief summary of the general results for the case in which the edge of the plate is free to vibrate without constraint, when $\partial^2\zeta/\partial x_1^2 = \partial^3\zeta/\partial x_1^3 = 0$ at the edge (Section 1.4.2).

Δ is plotted in Figure 5.4.3b as a function of ω/ω_c for a steel plate in water for several values of M. When $M = 0$, the dissipation is caused only by acoustic radiation damping; this corresponds to the solid "free" edge curve of Figure 4.4.3. A significant increase in Δ occurs at low frequencies when $M \neq 0$ by the transfer of structural energy to the kinetic energy of the flow; the acoustic losses are hardly affected by the flow when $M \ll 1$. At high frequencies ($\omega \to \omega_c$), however, all of the curves collapse onto the no-flow curve, and dissipation is entirely a result of radiation losses; this is because, as $\omega \to \omega_c$, bending wave energy is contained principally in the evanescent pressure fields on either side of the plate and propagates with negligible plate motion at a velocity which is only slightly less than the speed of sound.

5.5 Tube Banks

Sound propagating within a tube bank cavity of an industrial heat exchanger [300, 305, 339, 344] is scattered by the tubes and dissipated by thermo-viscous action at the tube surfaces and by interaction with the turbulent flow. Flow–acoustic interactions occur in tube wakes or in the separated flows produced by "jetting" between adjacent tubes. When the frequency coincides with that of vortex shedding or tube vibration, or a "jet-switching" mechanism, resonance absorption or production of sound can occur, although unless the frequency also coincides with a cavity resonance the generated aerodynamic sound will be incoherent and essentially uncorrelated with the incident sound.

The scattering problem is too difficult to treat analytically in any great detail, except at low frequencies (when tube cross-sections are compact) when the tubes and their wakes are acoustically equivalent to a distribution of monopole and dipole sources whose strengths are related to the unsteady surface pressure. At high Reynolds numbers, the latter are dominated by turbulence Reynolds stress fluctuations, but there are also components correlated with the incident sound. These are associated with the unsteady potential flow around the tubes produced by the sound and by the modulation of shed vorticity by the sound.

5.5.1 Long Wavelength Sound

Let the tube bank consist of a uniform array of fixed, parallel, rigid cylinders, each of cross-sectional area A, whose axes are aligned with the x_3-direction. The mean flow is at right angles to the tubes, and the Mach number is sufficiently small that the convection of sound can be ignored. Conditions are homogeneous with respect to x_3 and we may therefore consider the propagation of time harmonic waves proportional to $e^{i(n_3x_3-\omega t)}$, where $n_3 = \kappa_o \cos\psi$, ψ being the angle between the propagation direction and the tube axes. When the "solidity" of the bank is small, that is, when the fractional cross-section of the tubes $\alpha = \mathcal{N}A \ll 1$, where $\mathcal{N}$ is the number of tubes per unit cross-section of the bank, the propagation of long sound waves (in the absence of vorticity production) is governed by the two-dimensional version of equation (1.11.11), which describes scattering by a compact rigid body. The tubes are noncompact in the x_3-direction, and the appropriate generalization of (1.11.11) is

$$\left(\frac{\partial^2}{\partial x_1^2} + \frac{\partial^2}{\partial x_2^2} + \kappa_o^2 \sin^2\psi\right) p(x_1, x_2, n_3, \omega)$$
$$= \sum_n \left\{ i\rho_o\omega q_n \delta(x_1 - x_{n1})\delta(x_2 - x_{n2}) \right.$$
$$\left. + F_{ni}\frac{\partial}{\partial x_i}(\delta(x_1 - x_{n1})\delta(x_2 - x_{n2}))\right\}, \tag{5.5.1}$$

where the repeated suffix i ranges over the x_1x_2-directions only,

$$q_n = \frac{A\sin^2\psi}{\rho_o c_o^2}\frac{\partial p_n}{\partial t},$$
$$F_{ni} = -(m_o\delta_{ij} + \mathrm{M}_{ij})\frac{\partial v_{nj}}{\partial t} \equiv (m_o\delta_{ij} + \mathrm{M}_{ij})\frac{1}{\rho_o}\frac{\partial p_n}{\partial x_j}, \quad i, j = 1, 2, \tag{5.5.2}$$

and p_n and $\mathbf{v}_n$ are the acoustic pressure and particle velocity of the sound incident on the nth tube, evaluated on its axis $\mathbf{x}_n = (x_{n1}, x_{n2})$. q_n is the equivalent monopole strength per unit length of the nth tube, $\mathbf{F}_n$ is the force exerted on the fluid by unit length of the tube, and m_o and M_{ij} respectively denote the mass of displaced fluid and the added mass tensor, both per unit length of cylinder. If the cylinders are elastic and can support flexural vibrations about their undisturbed axes (of wavelength $2\pi/n_3$), $\mathbf{v}_n$ must be replaced by $\mathbf{v}_n - \mathbf{U}_n$, where $\mathbf{U}_n$ is the flexural velocity of cylinder n. When vortex shedding occurs, $\mathbf{F}_n$ must include an additional component equal and opposite to vorticity-induced lift.

Example 1. Use the method of Section 1.11 (Examples 6 and 7) to derive equation (5.5.1). Take the compact Green's function for an isolated cylinder in the form

$$G(\mathbf{x}, \mathbf{y}; \omega) = -\frac{i}{4}\mathrm{H}_0^{(1)}\left(|\mathbf{X} - \mathbf{Y}|\sqrt{\kappa_o^2 - n_3^2}\right),$$

where $\mathbf{X} = (x_1 - \varphi_1^*, x_2 - \varphi_2^*)$, and so on.

5.5.2 Bank of Rigid Strips

An analytical treatment of the effect of flow can be derived very simply by taking the tubes to be flat, rigid strips set at zero angle of attack to the flow. We now have $A = 0$, $m_o = 0$, and only the component $\mathrm{M}_{22} = \rho_o a^2$ of the added mass tensor is nonzero. Thus, $q_n \equiv 0$, and in the absence of mean flow $\mathbf{F}_n = (0, a^2 \partial p_n / \partial x_2)$.

Equation (5.5.1) is simplified by averaging over strips, by invoking the argument of Section 5.4.2 to justify replacing the pressure p_n incident on the nth strip by the local mean acoustic pressure, which is henceforth denoted by p. The averaging can be performed formally by replacing $\sum_n(\bullet)$ by $\mathcal{N}\iint_{-\infty}^{\infty}(\bullet)\, d^2\mathbf{x}_n$, to obtain

$$\left(\nabla^2 - \alpha\frac{\partial^2}{\partial x_2^2}\right)p + \kappa_o^2 p = 0, \qquad (5.5.3)$$

where $\alpha = \mathcal{N}\pi a^2$ ($\ll 1$) is the effective solidity.

The dispersion equation for acoustic waves proportional to $\mathrm{e}^{i(\mathbf{n}.\mathbf{x}-\omega t)}$ is therefore

$$n^2 - \alpha n_2^2 = \kappa_o^2, \qquad (5.5.4)$$

which shows that, except for waves propagating parallel to the strips, the sound speed is reduced by the presence of the strips. This is because the effective

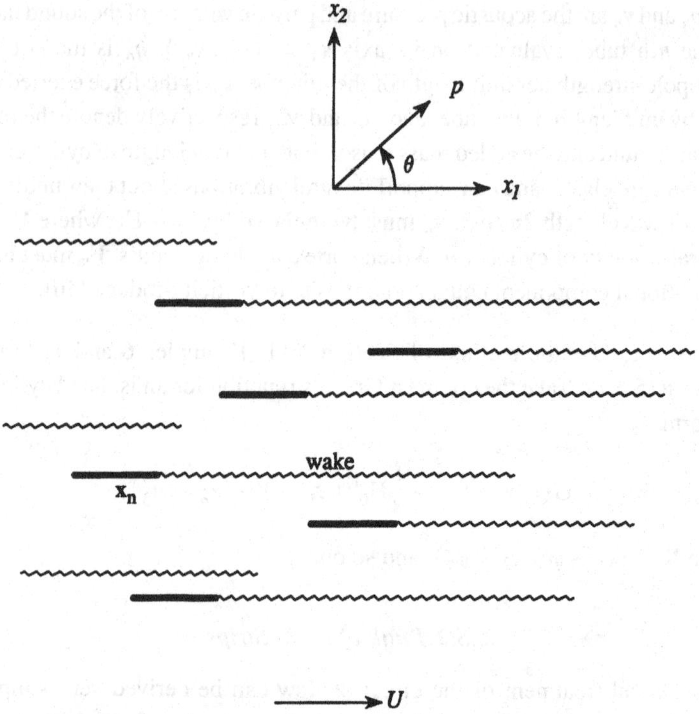

Figure 5.5.1. Sound propagation through a bank of rigid strips.

inertia of the fluid is increased by the normal reaction of the strips. When $n_3 = 0$, so that the sound propagates at right angles to the strip axes, the sound speed becomes

$$c = c_o\sqrt{1 - \alpha \sin^2 \theta}, \quad (5.5.5)$$

where θ is angle between the direction of propagation and the x_1-axis.

In this approximation (*Rayleigh* scattering), sound waves are not damped by their interactions with the strips. This is because the collective action of scattering by the strips is merely equivalent to a change in the inertia of the fluid [345–347]. To examine damping by vorticity production, we introduce a low Mach number mean flow at speed U in the x_1-direction, parallel to the strips (Figure 5.5.1).

Vortex shedding from the trailing edges is acoustically equivalent to a distribution of fluctuating dipoles sources whose strengths are equal and opposite to the lift fluctuations on the strips. Surface forces are also produced when a wake interacts with a neighboring, downstream strip. "Feedback" loops between different tubes caused by such interactions are frequently a source of unstable oscillations in real systems. When feedback cannot be eliminated by design of the tube bank cavity, its influence must be controlled by suitable damping procedures [300, 305].

These hydrodynamic interactions between strips cannot be modeled with any precision and will be ignored. The wakes will be replaced by vortex sheets, and the vortex sheet strength for the nth strip will be expressed in terms of the upwash velocity v_{n2} as in thin airfoil theory (Section 3.3). In a first approximation, the wake is identical with that produced by an oscillatory motion of the strip (in the x_2-direction) at velocity $-v_{n2}$, and the additional force exerted on the fluid per unit span of the nth strip is deduced from (4.5.16) to be

$$F_{n2} = \frac{2\pi i a^2 \mathrm{C}(S)}{S}\frac{\partial p_n}{\partial x_2}, \quad S = \frac{\omega a}{U}, \tag{5.5.6}$$

where $\mathrm{C}(S)$ is the Theodorsen function (4.5.17), provided interference from neighboring strips is negligible ($\alpha \ll 1$).

When this force is included on the right of (5.5.1) for each strip, and the result averaged as before, the acoustic equation becomes

$$\left(\nabla^2 - \alpha\left[1 + \frac{2i\mathrm{C}(S)}{S}\right]\frac{\partial^2}{\partial x_2^2}\right)p + \kappa_o^2 p = 0. \tag{5.5.7}$$

The dispersion equation is the same as (5.5.4) provided α is replaced by $\alpha\{1 + 2i\mathrm{C}(S)/S\}$; for real ω, it is satisfied only by complex values of the wavenumber $\mathbf{n}$. In particular, for sound propagating at right angles to the strip axes, at angle θ to the mean flow direction

$$n^2 = \frac{\kappa_o^2}{1 - \alpha\{1 + 2i\mathrm{C}(S)/S\}\sin^2\theta}, \quad S = \frac{\omega a}{U}, \mathbf{n} = n(\cos\theta, \sin\theta, 0). \tag{5.5.8}$$

The curves labeled *fixed strips* in Figure 5.5.2 show the variation with reduced frequency S of the sound speed ratio $c/c_o = \kappa_o/\mathrm{Re}\, n$ and the damping per wavelength of propagation ($\approx$54.6 Im(n)/Re(n) dB) for $\alpha = 0.1$ and $\theta = 90°$ (propagation in the x_2-direction). When S is large, there is no damping, but the sound speed is reduced. At lower frequencies, vortex shedding increases the effective "stiffness" as the sound speed increases above c_o. Damping progressively increases as the frequency decreases. However, the neglected hydrodynamic interactions between neighboring strips must be important when S becomes small, when the hydrodynamic length scale U/ω becomes comparable to the mean spacing $\sim 1/\sqrt{N}$ of the strips [348].

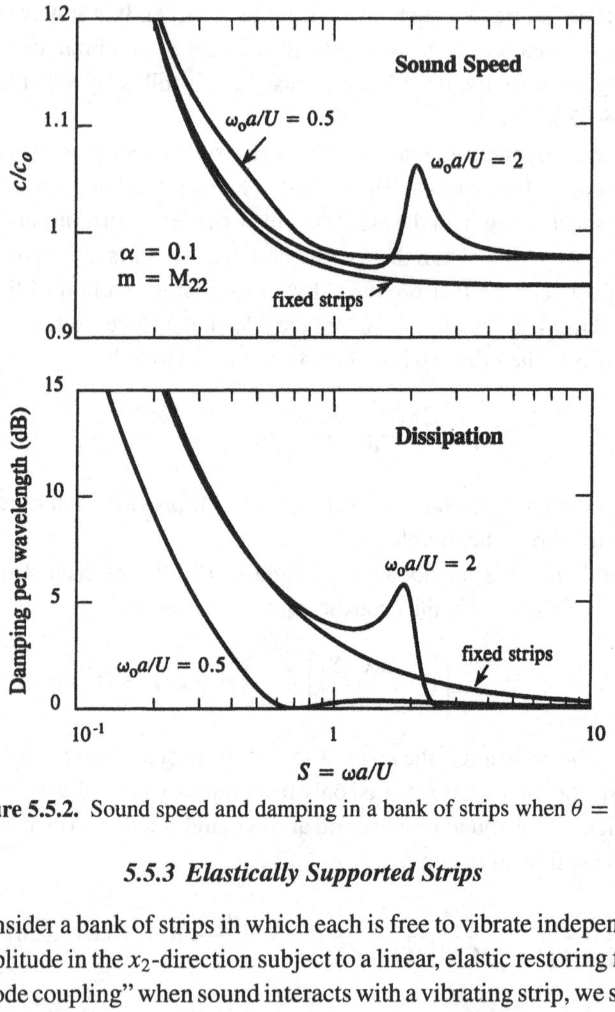

Figure 5.5.2. Sound speed and damping in a bank of strips when $\theta = 90°$.

5.5.3 Elastically Supported Strips

Let us consider a bank of strips in which each is free to vibrate independently at small amplitude in the x_2-direction subject to a linear, elastic restoring force. To avoid "mode coupling" when sound interacts with a vibrating strip, we shall consider only waves having $n_3 = 0$, that propagate at right angles to the strip axes.

The vibrations make two contributions to the dipole source F_{ni} of (5.5.2). First the velocity v_{n2} must be replaced by $v_{n2} - \partial\zeta_n/\partial t$, where ζ_n is the displacement of the nth strip, and secondly the wake force (5.5.6) must be modified to account for additional vorticity generated by the motion. The effect of both these corrections is to increase the net force on the fluid per unit span of the nth strip by

$$\mathrm{M}_{22}\frac{d^2\zeta_n}{dt^2}\left[1 + \frac{2i\mathrm{C}(S)}{S}\right].$$

We must now express ζ_n in terms of the local acoustic pressure p_n. Suppose that in the absence of the incident sound and of losses due to vortex shedding and radiation damping, each strip can vibrate with simple harmonic motion of frequency ω_o. The equation of motion of the nth strip then assumes the form (cf. Section 4.5.3, Example 3)

$$\frac{d^2\zeta_n}{dt^2} + 2\omega_o\Delta\frac{d\zeta_n}{dt} + \omega_o^2\zeta_n = \frac{-\pi a^2}{\mathrm{m}+\mathrm{M}_{22}}\left[1 + \frac{2i\mathrm{C}(S)}{S}\right]\frac{\partial p_n}{\partial x_2}, \tag{5.5.9}$$

where m is the mass of unit span of the strip, and

$$\Delta = \frac{\mathrm{M}_{22}}{\mathrm{m}+\mathrm{M}_{22}}\frac{\mathrm{C}(S)}{S_o}, \quad S_o = \frac{\omega_o a}{U}.$$

Because all quantities actually vary like $\mathrm{e}^{-i\omega t}$, equation (5.5.9) is readily solved, and the solution can be used to express the force exerted on the fluid by the nth strip in terms of p_n. The averaged equation for sound propagation at right angles to the bank of vibrating strips is then found in the form (5.5.7) with the term multiplied by α in square brackets replaced by

$$\Im(\omega) = \frac{\{1+2i\mathrm{C}(S)/S\}\{\mathrm{m}S^2/(\mathrm{m}+\mathrm{M}_{22}) - S_o^2\}}{S^2 + 2i\Delta S S_o - S_o^2}. \tag{5.5.10}$$

Thus, the dispersion equation for propagation in a direction making an angle θ to the flow direction is

$$n^2 = \frac{\kappa_o^2}{1 - \alpha\Im(\omega)\sin^2\theta}. \tag{5.5.11}$$

Figure 5.5.2 illustrates the effect of strip vibration when $\mathrm{m} = \mathrm{M}_{22}$ and the effective solidity $\alpha = 0.1$, for $S_o = 0.5$ and 2. When $S = S_o\sqrt{1+\mathrm{M}_{22}/\mathrm{m}} \approx 1.4S_o$, damping by vorticity production vanishes and the sound speed equals its free space value c_o because at this frequency the normal velocity of the strips and the x_2-component of the acoustic particle velocity are equal. The local peaks in the sound speed and damping at the resonance frequency $S = S_o = 2$ occur because Δ is small when $S_o > 1$. In such cases $\Im(\omega) \sim 1 + 2i\mathrm{C}(S)/S$, its value in the absence of strip vibrations, when $S < 1$, and the sound speed and damping curves are then close to those for the fixed strips. The fixed strip result is also recovered when $\mathrm{m} \to \infty$, or when the spring stiffness is so large that $S_o \gg S$ for all frequencies of interest.

Example 2. Show that the lift force $F_2(t)$ per unit span experienced by a vibrating strip is given by

$$F_2(t) = \mathrm{M}_{22}\frac{d}{dt}\int_{-\infty}^{\infty}\Im(\omega)v_2(\omega)\mathrm{e}^{-i\omega t}\,d\omega,$$

where $\Im(\omega)$ is defined as in (5.5.10) and v_2 is the x_2-component of the incident acoustic particle velocity. $\Im(\omega) \to \mathrm{m}/(\mathrm{m}+\mathrm{M}_{22})$ as $\omega \to \infty$, and has a simple pole at $\omega = 0$. Verify that

$$\Im(\omega) - \frac{\mathrm{m}}{\mathrm{m}+\mathrm{M}_{22}}\left(1+\frac{2i}{S}\right), \quad S = \frac{\omega a}{U},$$

is regular in $\operatorname{Im}\omega > 0$, and that its real and imaginary parts satisfy Kramers–Kronig formulas of the type (5.3.33). This shows that, when hydrodynamic interactions between the strips are neglected, the propagation of sound in the presence of mean flow is absolutely stable.

Example 3. Verify that in the regions of Figure 5.5.2 where the phase velocity $c > c_o$ and $S < 1$, the *group velocity* $\operatorname{Re}(\partial\omega/\partial k) < c_o$, so that the speed of propagation of acoustic energy does not exceed c_o. This latter statement remains true at the resonance peak $S = \omega_o a/U = 2$ where, however, the usual definition of group velocity is not a proper measure of the rate of energy transport [337].

5.5.4 Tubes of Circular Cross-Section

For tubes of circular cross-section the difficulty of modeling vortex shedding precludes the development of a general theory of sound propagation. In applications, only Strouhal numbers $S = \omega R/U$ in the range 0.1 to 10 are of interest, where R is the tube radius [340]. At the upper end of this range, it may be permissible to neglect hydrodynamic interactions between the tubes, provided the solidity $\alpha = \mathcal{N}\pi R^2$ is small ($\mathcal{N}$ being the number of tubes per unit cross-section). However, the blockage of the mean flow by the tubes and finite rates of heat transfer will typically produce substantial variations in mean pressure and temperature within a tube bank cavity. These effects will be neglected, as is the influence of tube vibration.

The added mass $\mathrm{M}_{ij} \equiv m_o\delta_{ij}$ when a tube has circular cross-section, where $m_o = \rho_o A$ is the fluid mass displaced by unit length of tube. The dipole of (5.5.2) is then

$$F_{ni} = 2\rho_o A \frac{\partial p_n}{\partial x_i},$$

and the averaged equation describing wave propagation *in the absence of flow* becomes

$$\left(\nabla^2 + \alpha\frac{\partial^2}{\partial x_3^2}\right)p + (1+\alpha)\kappa_o^2 p = 0, \quad \alpha \ll 1.$$

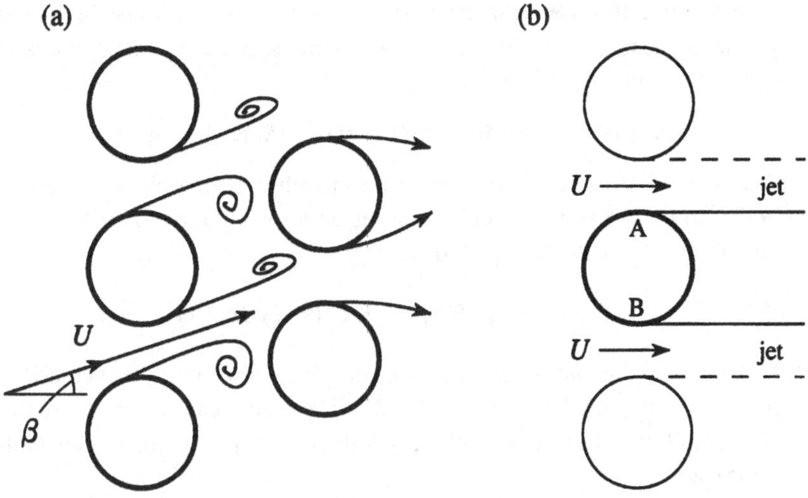

Figure 5.5.3. (a) "Jetting" in a tube bank and (b) idealized model.

Sound waves are not damped in this approximation. For propagation at right angles to the tube axes, the effective sound speed $c = \kappa_o/n \approx c_o(1 - \frac{1}{2}\alpha)$. This conclusion is strictly valid for tubes distributed "randomly" within the bank. The sound speed c generally exhibits a weak dependence on propagation direction when the tubes are arranged in certain specific, regular arrays [341, 349].

Consider next the influence of low Mach number flow through the bank in the x_1-direction. For densely packed tubes, there is usually no discernible structure to the turbulent wakes of the individual tubes [350], although there is often a distinct separation of the mean flow (Figure 5.5.3a) associated with "jetting" between tubes. Such configurations, usually involving tubes separated typically by one to four tube radii, often have small values of the solidity α. To model this case, we examine first the motion near a tube whose axis is temporarily taken as the x_3-axis, and assume the jets are parallel to the x_1-direction (this restriction is relaxed below). Suppose for simplicity that separation occurs for this tube at the 90° points A and B in Figure 5.5.3b, where the mean jet speed is U. When the tube radius is compact and the influence of neighboring tubes is neglected, the potential flow across the tube produced by a long wavelength incident sound wave is determined by the velocity potential

$$v_i X_i \equiv v_i x_i \left(1 + \frac{R^2}{x_1^2 + x_2^2}\right), \quad i = 1, 2,$$

where v_i is the incident acoustic particle velocity evaluated on the cylinder axis. Thus, the corresponding perturbation in the separation velocities at A and B are approximately equal to $2v_1$. These perturbations will be assumed to convect downstream in thin vortex sheets at the edges of the jets on either side of the cylinder at the mean velocity U. For time harmonic sound, the net perturbation in the wake vorticity is therefore

$$\boldsymbol{\omega}(\mathbf{x},\omega) \approx -2v_1\mathbf{i}_3(\delta(x_2-R)-\delta(x_2+R))\mathrm{e}^{i\kappa x_1}, \quad x_1>0,$$

where $\kappa = \omega/U$ is the hydrodynamic wavenumber of the wake vorticity. According to (1.14.13) this vorticity induces an unsteady force on the fluid per unit length of the cylinder equal to

$$F_i \approx -\rho_o \int (\nabla X_i \cdot \boldsymbol{\omega} \wedge \mathbf{U})(\mathbf{x},\omega)\, dx_1\, dx_2,$$

of which F_1 is the only nonzero component. This force must be included for each tube in the dipole source of (5.5.2). The reader can readily show that, when $\alpha \ll 1$, this leads to the following long wavelength approximation to the acoustic equation

$$\left(\nabla^2 - \alpha\left[\Psi(S)\frac{\partial^2}{\partial x_1^2} - \frac{\partial^2}{\partial x_2^2}\right]\right)p + (1+\alpha)\kappa_o^2 p = 0,$$

where

$$\Psi(S) = \frac{8i}{\pi S}\int_0^\infty \frac{\lambda \mathrm{e}^{i\lambda S}}{(1+\lambda^2)^2}\,d\lambda, \quad S = \frac{\omega R}{U}, \qquad (5.5.12)$$

(which can be evaluated in terms of exponential integrals [24]) determines the influence of vorticity production by the sound.

For a plane wave proportional $\mathrm{e}^{i(\mathbf{n}.\mathbf{x}-\omega t)}$ propagating *at right angles* to the tube axes, n satisfies

$$\frac{n}{\kappa_o} = 1 + \frac{\alpha}{2}(1+\Psi(S)\cos^2\theta), \quad \alpha \ll 1, \qquad (5.5.13)$$

where θ is the direction of propagation measured from the x_1-direction. This implies that transverse acoustic modes ($\theta = \pm 90°$) are not influenced by vortex shedding, which is a consequence of the assumption of Figure 5.5.3b that the jets are parallel to the direction of the mean flow through the bank. For a staggered array of tubes, the jets are usually inclined at finite angles $\pm\beta$, say, to the mean flow (Figure 5.5.3a) and switch spontaneously between these directions. This effect can be included in the present model by the expedient of replacing $\cos^2\theta$ in (5.5.13) by

$$\tfrac{1}{2}(\cos^2(\theta+\beta) + \cos^2(\theta-\beta)).$$

Predictions of this kind of theory are shown in Figure 5.5.4 for the mean propagation speed and damping of sound for two cases of waves propagating

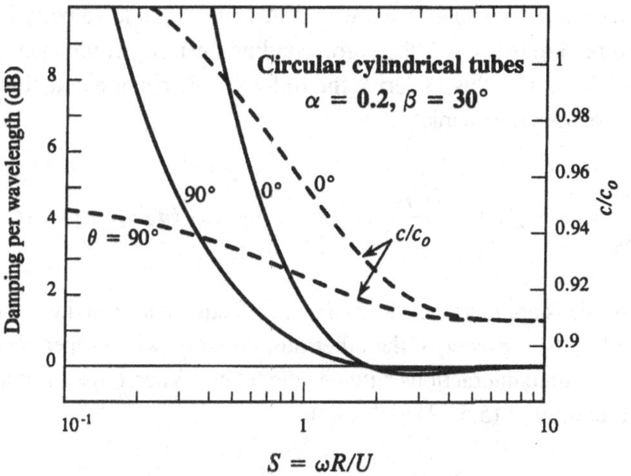

Figure 5.5.4. Predicted sound speed and damping in a tube bank.

parallel ($\theta = 0°$) and transverse ($\theta = 90°$) to the mean flow direction when $\beta = 30°$ and when the solidity $\alpha = 0.2$. Both the attenuation and sound speed increase with decreasing frequency, which is similar to the behavior in Figure 5.5.2 for transverse propagation in a bank of strips. At high frequencies the imaginary part of the wavenumber Im $n < 0$ (although very small in magnitude), which implies that the wave grows by extracting energy from the flow. The reduced sound speed and damping at higher frequencies can be responsible for the appearance of "bound" transverse resonances, that is, of large amplitude transverse waves ($\theta = \pm 90°$) trapped in a region of low sound speed [341].

Example 4. When the mean flow over a cylinder is in the x_1-direction, the unsteady force exerted on the fluid by unit length of cylinder is

$$F_1(t) = -\rho_o A U \frac{d}{dt} \int_{-\infty}^{\infty} v_1(\omega) \Psi(S) e^{-i\omega t} \, d\omega,$$

where v_1 is the component of the acoustic particle velocity in the mean flow direction. $\Psi(S)$, defined by (5.5.12), is regular in $\text{Im}\,\omega > 0$, so that, although Figure 5.5.4 implies that high-frequency sound waves can grow by extracting energy from the flow, the system is absolutely stable in the sense that sound cannot be generated spontaneously by this mechanism.

Example 5. The drag exerted on unit length of a cylinder of radius R immersed in a nominally steady, high Reynolds number cross flow of speed U is $\rho_o U^2 R C_D$, where C_D is a drag coefficient. When the mean flow is modulated by a low-frequency sound wave whose acoustic particle velocity in the mean stream direction is $v \ll U$, the corresponding unsteady component of the drag $\sim 2\rho_o v U R C_D$. Use this to derive the following dispersion equation for sound propagation in a tube bank

$$\frac{n}{\kappa_o} = 1 + \frac{\alpha}{2}\left(1 + \frac{iC_D}{\pi S}(\cos^2(\theta + \beta) + \cos^2(\theta - \beta))\right), \quad \alpha \ll 1.$$

This formula is applicable at arbitrarily low frequencies provided C_D has been measured *in the presence* of the other tubes, that is, with proper account taken of hydrodynamic interactions between the tubes. When $C_D = 2$, it agrees with asymptotic limit of (5.5.13) as $S \to 0$.

5.6 Nonlinear Interactions

When a sound wave impinges on a solid surface in the absence of mean flow, the dissipated energy is usually converted directly into heat through viscous action. At very high acoustic amplitudes, however, free vorticity may still be formed at edges, and dissipation may take place, as in the presence of mean flow, by the generation of vortical kinetic energy which escapes from the interaction zone by self-induction. This nonlinear mechanism can be important in small perforates or apertures; if the acoustic Reynolds number $\omega R^2/\nu$ is large ($R =$ aperture radius), viscosity has no control over the interaction except for its role in generating vorticity at the surface [293, 351–361]. Nonlinearity will cause sinusoidally varying sound incident on a perforated plate to produce periodic reflected and transmitted waves containing harmonics of the incident wave frequency. Problems of this kind are discussed in Sections 5.6.1–5.6.3.

Lighthill's theory of aerodynamic sound (Section 2.1) identifies fluctuations in the Reynolds stress as a source of sound. These stresses are associated with the presence of vorticity, and sound can itself generate vorticity when it interacts with a boundary. In Sections 5.6.4–5.6.5, we discuss the related nonlinear phenomenon where the mean Reynolds stress created by *sound* generates vorticity and a steady flow known as *acoustic streaming* [10, 36, 178, 362–369]. This occurs not only when sound impinges on solid boundaries, but also in intense, free-space beams of sound. Ultrasonic streaming over a surface exerts a small but steady surface shear stress that is often effective in removing loosely adhering surface contaminants. Streaming also causes fine dust sprinkled inside Kundt's tube to collect at the nodes of standing acoustic waves, where the acoustic particle velocity vanishes [10].

5.6.1 Nonlinear Aperture Flow

Consider a rigid wall of thickness h containing a circular cylindrical aperture of radius R. A uniform, time-periodic pressure load $[p(t)] = p_1 - p_2$ is applied across the wall as illustrated in Figure 5.6.1. The mean load $\langle [p] \rangle = 0$, so that there is no mean flow through the aperture, and the aperture radius and wall thickness are assumed to be much smaller than the wavelength of any sound. During the time intervals when $p_1 > p_2$, fluid is forced through the aperture from left to right. The Reynolds number is assumed to be sufficiently large that the flow separates and forms a jet, within which the motion can be regarded as irrotational, and on the free surface of which the pressure is equal to p_2. Let $V(t)$ denote the mean axial velocity of the jet where it enters the aperture (at O). An approximate equation determining $V(t)$ in terms of $[p(t)]$ can be derived by applying Bernoulli's equation (1.2.21) to the axial streamline within the jet, first between the point O and the point marked J just within the potential flow region at the head of the jet, where the velocity is V_J:

$$\frac{\partial}{\partial t}(\varphi_J - \varphi_O) + \frac{1}{2}V_J^2 - \frac{1}{2}V^2 = \frac{p_O - p_J}{\rho_o}, \tag{5.6.1}$$

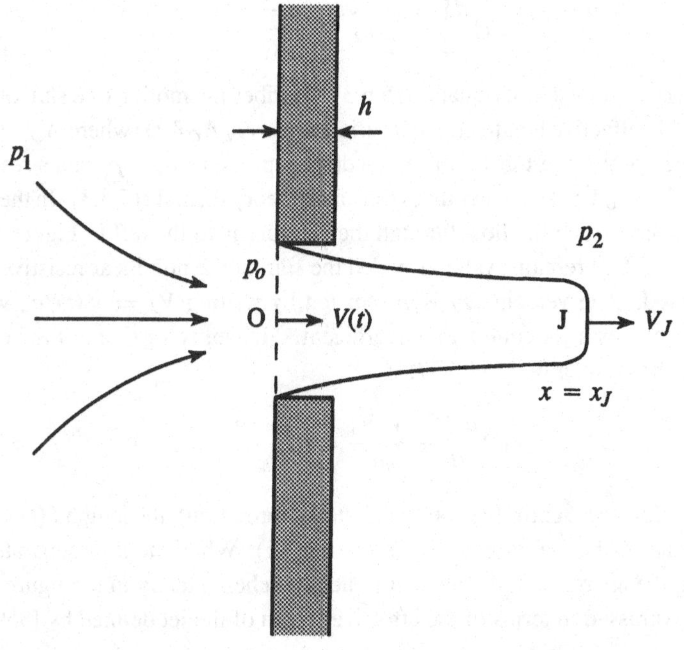

Figure 5.6.1. High Reynolds number separated flow through a wall aperture.

where φ_O and φ_J are the values of the velocity potential at O and J. Because the flow into the aperture is entirely irrotational, we can write similarly

$$\frac{\partial}{\partial t}(\varphi_O - \varphi_1) + \frac{1}{2}V^2 = \frac{p_1 - p_O}{\rho_o}, \tag{5.6.2}$$

where the amplitude of the motion at large distances from the aperture in the irrotational domain is assumed to be small enough to be governed by linearized equations, that is, $p_1 \approx -\rho_o \partial\varphi_1/\partial t$.

Rayleigh's formula (5.3.3) can be used to write $\varphi_O - \varphi_1 = \ell_1 V(t)$, where $\ell_1 \approx 0.82R$ is the end correction of the opening into the irrotational region. Similarly, we can write

$$\frac{\partial\varphi_J}{\partial t} - \frac{\partial\varphi_O}{\partial t} = \int_0^x \frac{\partial v(\xi, t)}{\partial t}\, d\xi,$$

where $v(x, t)$ is the axial jet velocity at distance x from the inlet plane. Assume that $p_J = p_2$, and denote the value of the integral in this formula by $\ell'(t)\partial V/\partial t$, where $\ell'(t)$ is a suitable, time-dependent length.

Inserting these expressions into equations (5.6.1) and (5.6.2) and adding, we find

$$\ell(t)\frac{dV}{dt} + \frac{1}{2}V_J^2 = \frac{p_1 - p_2}{\rho_o}, \tag{5.6.3}$$

where $\ell(t) = \ell_1 + \ell'(t)$. Equation (5.6.3) describes the motion of a slug of fluid of variable effective length $\ell(t)$ and variable mass $\rho_o A_O \ell(t)$ (where $A_O = \pi R^2$ is the area of the aperture) subject to a driving pressure $p_1 - p_2$ and a resistive force $-\frac{1}{2}\rho_o A_O V_J^2$, and generalizes the linear theory formula (5.3.3). In the half-cycle during which the flow through the aperture is to the left in Figure 5.6.1, equation (5.6.3) remains valid provided the sign of the nonlinear resistive term is reversed. The velocity V_J is eliminated by writing $V_J = V(t)/\sigma$, where σ here denotes a jet contraction coefficient. Hence, we arrive at the general aperture flow equation [358, 359]

$$\ell(t)\frac{dV}{dt} + \frac{V|V|}{2\sigma^2} = \frac{p_1 - p_2}{\rho_o}. \tag{5.6.4}$$

If the flow were entirely irrotational (no jet formation), the length $\ell(t)$ would be constant and equal to $h + 2\ell_1$ (Section 5.3.1). When the flow separates on entering the aperture, in the manner indicated schematically in the figure, $\ell(t)$ can be expressed in terms of the effective length of the jet defined by [359]

$$L(\tau) = \int_0^\tau |V(t)|\, dt, \tag{5.6.5}$$

where the time τ is measured from the beginning of the half cycle during which the sign of $V(t)$ is constant. From an examination of data from several experiments, in which the working fluid was either water or air [352–354], Cummings [359] proposed the following empirical relation between $\ell(t)$ and L,

$$\ell(t) = \ell_1 + \frac{h + \ell_1}{1 + \beta(L/2R)^{1.585}}, \quad \beta = \frac{1}{3}. \tag{5.6.6}$$

As L increases from the beginning of a half cycle, ℓ decreases because the inertia of the ideal potential flow within the aperture and on the jet side of the screen is progressively replaced by that of the jet.

There is no corresponding empirical formula for the dependence of the contraction ratio σ on τ. This would not be expected to vary significantly, however, except near the beginning of a half cycle. *Steady flow* experiment suggests that $\sigma \approx 0.61$–0.64. The assumption that σ is constant and equal to 0.75 is found to yield predictions that accord well with experiment [359], and this value is used below.

5.6.2 Transmission of High-Amplitude Sound Through an Orifice

The simplest application of equation (5.6.4) is to the problem of a plane sound wave $p = p_I \cos\{\omega(t - x_1/c_o)\}$ incident from $x_1 < 0$ on a thin rigid screen ($h = 0$) at $x_1 = 0$ containing a small aperture of radius R (Figure 5.6.2). When $\kappa_o R \ll 1$, the pressure at large distances on either side of the screen can be approximated by

$$\begin{aligned} p &= p_I(\cos\{\omega(t - x_1/c_o)\} + \cos\{\omega(t + x_1/c_o)\}) \\ &\quad - \frac{\rho_o A_O}{2\pi|\mathbf{x}|}\frac{\partial V}{\partial t}(t - |\mathbf{x}|/c_o), \quad x_1 < 0, \\ &= \frac{\rho_o A_O}{2\pi|\mathbf{x}|}\frac{\partial V}{\partial t}(t - |\mathbf{x}|/c_o), \quad x_1 > 0, \end{aligned} \tag{5.6.7}$$

where the coordinate origin is taken at the center of the aperture, and we can take $p_1 = 2p_I \cos(\omega t)$ and $p_2 = 0$.

By introducing a nondimensional aperture velocity $\bar{V} = V/(\omega R)$, and time $T = \omega t/2\pi$, and for a gaseous medium with ratio of specific heats γ, equations (5.6.4) and (5.6.5) may be cast in the form

$$\frac{\ell}{R}\frac{d\bar{V}}{dT} + \frac{\pi}{\sigma^2}\bar{V}|\bar{V}| = \frac{4\pi p_I \cos(2\pi T)}{\gamma p_o(\kappa_o R)^2}, \quad \frac{d}{dT}\left(\frac{L}{R}\right) = 2\pi|\bar{V}|, \tag{5.6.8}$$

where ℓ/R is determined by (5.6.6) with $h = 0$, and p_o is the undisturbed, mean pressure in the fluid. The numerical solution of these equations, starting

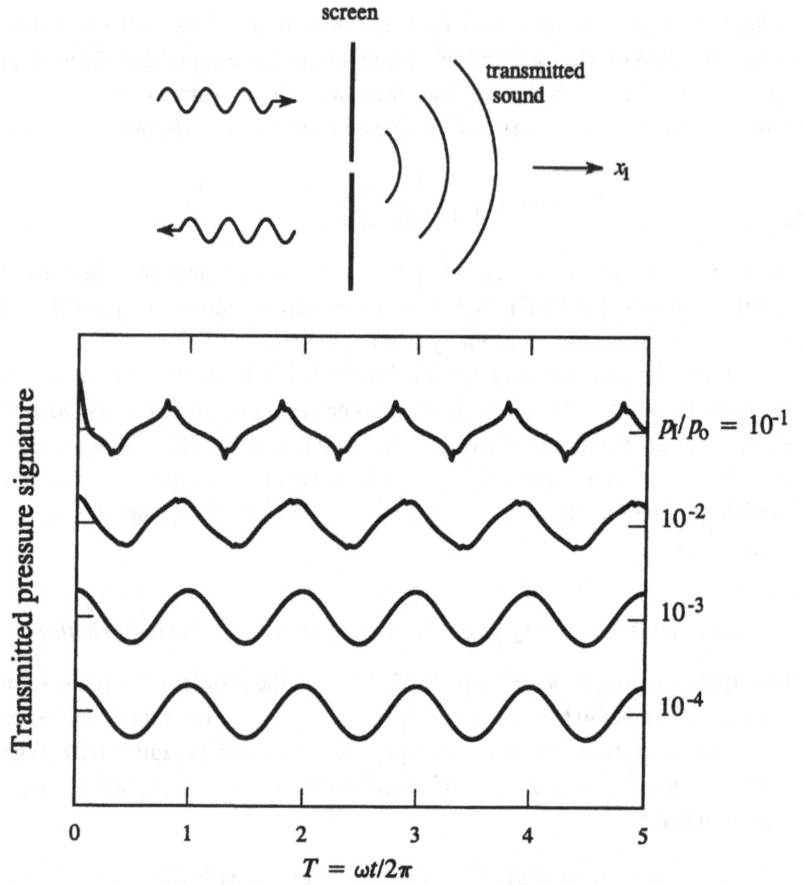

Figure 5.6.2. Periodic acoustic pressure transmitted through a small orifice for $\kappa_o R = 0.07$.

at $T = 0$ with the initial condition $\bar{V} = 0$ (the condition appropriate for linearized motion), can be used to determine the pressure scattered from the orifice (the terms in $\partial V/\partial t$ in (5.6.7)). After one or two cycles of integration, the scattered pressure becomes periodic with the period $2\pi/\omega$ of the incident sound. This is illustrated in Figure 5.6.2, which depicts the pressure signatures of the transmitted sound (not to equal scales) when $\kappa_o R = 0.07$, $\gamma = 1.4$ (air) and for several values of the incident wave amplitude p_I/p_o, which determines the degree of nonlinearity in the orifice. When $p_I/p_o = 10^{-4}$, the orifice motion is essentially linear; the transmitted pressure is in phase with p_I and has the same sinusoidal form. The wave profile becomes progressively distorted and retarded in phase as p_I/p_o increases, when orifice nonlinearity produces transmitted sound containing harmonics of the incident wave.

Example 1. An impulsive acoustic wave $p = P_I\delta(t - x_1/c_o)$ is incident on the screen of Figure 5.6.2. Show that the aperture velocity $V(t)$ (the solution of equation (5.6.4)) can be expressed in the form [359]

$$V(t) = V_o\mathrm{H}(t)\Big/\left(1 + \frac{V_o}{2\sigma^2}\int_0^t \frac{d\tau}{\ell(\tau)}\right), \quad \text{where } V_o = \frac{P_I}{\rho_o(h + 2\ell_1)},$$

where H is the Heaviside unit function. The transmitted sound accordingly consists of an impulsive spherical wave followed by a slowly decaying "tail":

$$p_2 \approx \frac{\rho_o A_O V_o}{2\pi|\mathbf{x}|}\left[\delta(t) - \frac{V_o\mathrm{H}(t)}{2\sigma^2\ell(t)}\Big/\left(1 + \frac{V_o}{2\sigma^2}\int_0^t \frac{d\tau}{\ell(\tau)}\right)^2\right]_{t-|\mathbf{x}|/c_o}.$$

5.6.3 The Perforated Screen

The influence of aperture nonlinearity on the transmission of sound through a perforated screen can be handled in a similar way to the single orifice problem, provided the average spacing d, say, of the apertures is small compared to the characteristic wavelength. This extends the linearized treatment of this problem given in Section 5.3.2. Let the fractional open area of the screen be α, and consider only the case of normal incidence (the results are easily generalized to arbitrary angles of incidence). In Figure 5.6.3, the plane acoustic wave $p = p_I\cos\{\omega(t - x_1/c_o)\}$ is incident from $x_1 < 0$ on a thin screen ($h = 0$) at $x_1 = 0$. The acoustic pressures at distances $\gg d$ from the screen are

$$\begin{aligned} p &= p_I\cos\{\omega(t - x_1/c_o)\} + p_R(t + x_1/c_o), \quad x_1 < 0, \\ &= p_T(t - x_1/c_o), \quad x_1 > 0. \end{aligned}$$

In the usual way (Section 5.3.2), conservation of mass at the screen supplies the following relations between $V(t)$ and the reflected and transmitted pressures:

$$\alpha V(t) = \{p_I\cos(\omega t) - p_R(t)\}/\rho_o c_o = p_T(t)/\rho_o c_o,$$

so that in the aperture equation (5.6.4) we can set

$$p_1 = 2p_I\cos(\omega t) - \alpha\rho_o c_o V(t) \quad \text{and} \quad p_2 = \alpha\rho_o c_o V(t).$$

The nondimensional equations determining the aperture motions are the same as for the single orifice problem of Section 5.6.2, except that the first of equations (5.6.8) is replaced by

$$\frac{\ell}{R}\frac{d\bar{V}}{dT} + \frac{\pi}{\sigma^2}\bar{V}|\bar{V}| + \frac{4\alpha\pi\bar{V}}{\kappa_o R} = \frac{4\pi p_I\cos(2\pi T)}{\gamma p_o(\kappa_o R)^2}. \tag{5.6.9}$$

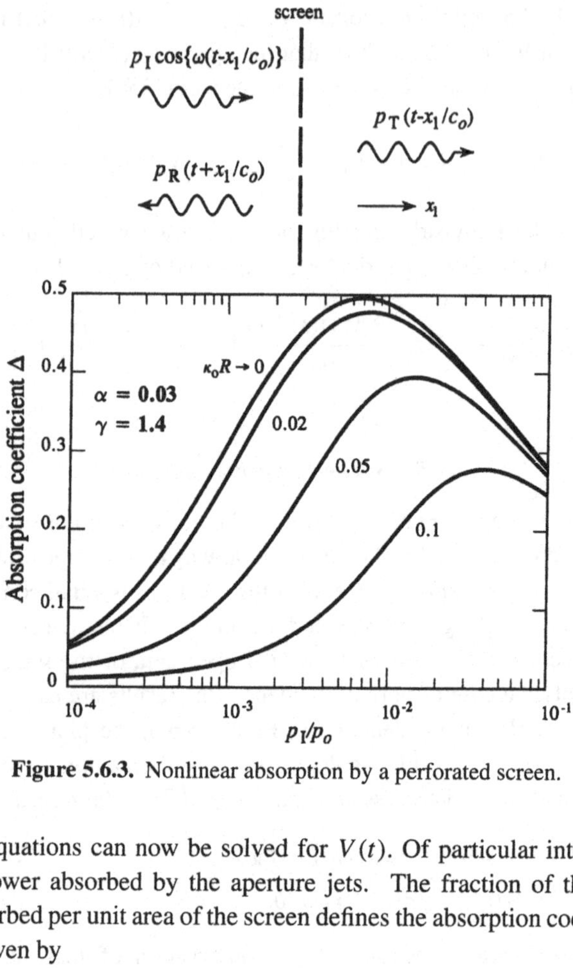

Figure 5.6.3. Nonlinear absorption by a perforated screen.

These equations can now be solved for $V(t)$. Of particular interest is the acoustic power absorbed by the aperture jets. The fraction of the incident power absorbed per unit area of the screen defines the absorption coefficient Δ, which is given by

$$\Delta = 1 - \frac{2}{(p_{\mathrm{I}}/p_o)^2}\int_0^1 \left[\left(\frac{p_{\mathrm{I}}}{p_o}\cos(2\pi T) - \alpha\gamma\kappa_o R\bar{V}(T)\right)^2 + (\alpha\gamma\kappa_o R\bar{V}(t))^2\right] dT, \qquad (5.6.10)$$

where the integral represents an average value calculated for one complete cycle $(2\pi/\omega)$ of the incident sound wave.

The curves in Figure 5.6.3 illustrate the dependence of Δ on the incident wave amplitude p_{I}/p_o relative to the mean pressure p_o for different values of $\kappa_o R$. The screen is in air ($\gamma = 1.4$) and has three percent fractional open area. The absorption can be considerable and comparable with the best achieved by a bias flow screen (Section 5.3) provided the frequency is small and $p_I/p_o \sim 0.01$, corresponding to an incident acoustic intensity of about 150 dB (re 2×10^{-5} N/m^2). For the bias flow screen, however, a proper choice of the bias flow velocity yields uniformly large absorption at low frequencies, irrespective of the amplitude of the sound.

Example 2. Show that as $\kappa_o R \to 0$, the solution of the perforated screen equation (5.6.9) is given by

$$\bar{V}(T)=\frac{2\alpha\sigma^2}{\kappa_o R}\operatorname{sgn}[\cos(2\pi T)]\left[\left(\frac{p_I|\cos(2\pi T)|}{p_o\gamma(\alpha\sigma)^2}+1\right)^{\frac{1}{2}}-1\right], \quad p_I/p_o>0.$$

Deduce that the absorption coefficient Δ defined in (5.6.10) is a function of $(p_I/p_o)/(\gamma\alpha^2\sigma^2)$ alone and attains a maximum value

$$\Delta_{\max} \approx 0.496 \text{ at } p_I/p_o \approx 9.75\gamma\alpha^2\sigma^2.$$

Example 3. Consider a plane sound wave $p = p_I(t - x_1/c_o)$ incident on a rigid screen at $x_1 = 0$ containing a small circular orifice forming the mouth of a Helmholtz resonator with rigid walls and volume V_R. Let $X(t)$ denote the displacement of fluid in the orifice, measured in the positive x_1-direction, such that the velocity $V(t)$ of Section 5.6.1 is equal to dX/dt. Show that when the characteristic acoustic wavelength is large relative to the diameter of the resonator [127]

$$\ell(t)\frac{d^2X}{dt^2}+\frac{1}{2\sigma^2}\frac{dX}{dt}\left|\frac{dX}{dt}\right|+\frac{c_o^2A_o}{V_R}X=\frac{2}{\rho_o}p_I(t).$$

Example 4: The perforated elastic screen. When the perforated screen of Figure 5.6.3 is a thin elastic plate of bending stiffness B and mass density m per unit area, use equation (5.4.12) to show that the sound transmission problem is governed by the nondimensional equations

$$\frac{\ell}{R}\frac{d\bar{V}}{dT}+\frac{\pi}{\sigma^2}\bar{V}|\bar{V}|+\frac{4\pi\bar{V}'}{\kappa_o R}=\frac{4\pi(p_I/p_o)}{\gamma(\kappa_o R)^2}\cos(2\pi T),$$

$$\frac{d\bar{V}'}{dT}+\frac{4\pi\bar{V}'}{\kappa_o R}\left(\frac{\rho_o R}{m}\right)-\alpha\left(1+\frac{2\rho_o R}{m}\right)\frac{d\bar{V}}{dT}$$

$$=\frac{4\pi(p_I/p_o)}{\gamma(\kappa_o R)^2}\left(\frac{\rho_o R}{m}\right)\cos(2\pi T),$$

$$\frac{d}{dT}\left(\frac{L}{R}\right)=2\pi|\bar{V}|,$$

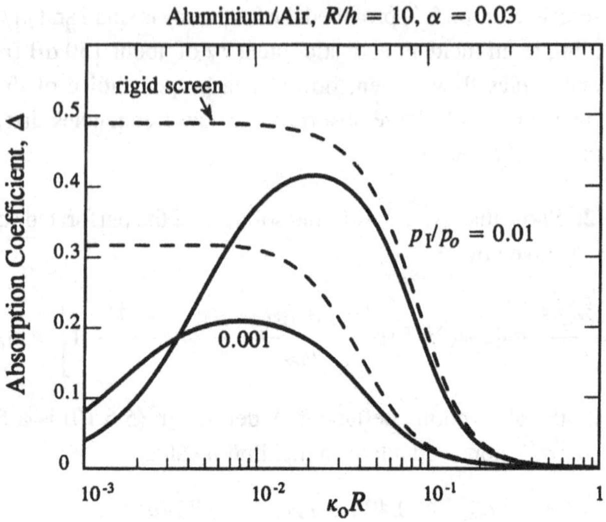

Figure 5.6.4. Nonlinear absorption by perforated rigid and elastic screens.

where, if V_P is the normal velocity of the plate, then $\bar{V}' = \{(1-\alpha)V_P + \alpha V\}/(\omega R)$. Deduce that the absorption coefficient can be expressed in the form

$$\Delta = \left(\frac{2\gamma\kappa_o R}{p_I/p_o}\right)^2 \int_0^1 \bar{V}'(T)\left(\frac{(p_I/p_o)}{\gamma\kappa_o R}\cos(2\pi T) - \bar{V}'(T)\right)dT.$$

At low frequencies, the elastic screen becomes progressively more transparent to the incident sound and, whereas the rigid-plate damping tends to a finite limit for each value of $\kappa_o R$, the elastic-plate damping ultimately decreases to zero. See Figure 5.6.4, where $p_I/p_o = 0.01,\ 0.001$.

5.6.4 Acoustic Streaming

A steady flow produced by an acoustic field is called *acoustic streaming* and is evidence of the generation of vorticity by the sound. To obtain a proper analytical understanding of streaming, explicit account must be taken of the consequent attenuation of the sound. Because acoustic quantities have vanishing mean values, streaming velocities are typically second order in the acoustic amplitude, and we shall see that the dominant cause of streaming is departures of the mean acoustic Reynolds stress $\rho_o\overline{v_i v_j}$ (where $\mathbf{v}$ is the acoustic particle velocity, and the overbar denotes a time average) from its value in the absence of attenuation. Henceforth, the first-order acoustic particle velocity is denoted by v and the streaming velocity by $\bar{v}$; time averages will be denoted by an overbar, but angle brackets will be used where necessary to avoid confusion.

The streaming velocity and mean pressure $\bar{p}$ satisfy the time-averaged equations of continuity and momentum. For *linear* acoustics, in the absence of an external mean flow $\langle \rho \mathbf{v} \rangle = \langle p' \mathbf{v} \rangle / c_o^2 \equiv \mathbf{I}_a / c_o^2$, where p' is the acoustic pressure perturbation and $\mathbf{I}_a$ the acoustic intensity (Section 13.1). Thus, the time-averaged continuity equation implies that

$$\operatorname{div} \bar{\mathbf{v}} = -\left(1/\rho_o c_o^2\right) \operatorname{div} \mathbf{I}_a. \tag{5.6.11}$$

Equation (1.13.1) (with $S_a = 0$) shows that the "source" on the right of this equation vanishes if the sound not attenuated.

The velocity $\bar{\mathbf{v}}$ is the *Eulerian* mean velocity, the average velocity at a fixed point in space. The right of (5.6.11) has a useful interpretation in terms of the difference $\bar{\mathbf{v}}_L - \bar{\mathbf{v}}$ where $\bar{\mathbf{v}}_L$ is the *Lagrangian* mean velocity, which is usually observed in experiments [36, 369]. By definition, the Lagrangian velocity is the weighted mean value $\langle \rho \mathbf{v} \rangle / \rho_o \equiv \bar{\mathbf{v}} + \mathbf{I}_a / \rho_o c_o^2$, that is,

$$\bar{\mathbf{v}}_L - \bar{\mathbf{v}} = \mathbf{I}_a / \rho_o c_o^2. \tag{5.6.12}$$

In practice, the mean Eulerian and Lagrangian velocities do not differ significantly, even when each is large, and this suggests that the source in the continuity equation is not normally an important cause of streaming.

To average the momentum equation (1.2.4) (with $F_i = 0$), observe that

$$\overline{\rho \frac{\partial v_i}{\partial t}} = \frac{\partial}{\partial t}\overline{(\rho v_i)} - \overline{v_i \frac{\partial \rho}{\partial t}} \equiv -\overline{v_i \frac{\partial \rho}{\partial t}} = \overline{v_i \frac{\partial (\rho v_j)}{\partial x_j}} \approx \rho_o \overline{v_i \frac{\partial v_j}{\partial x_j}},$$

where here and elsewhere in the averaged equation, density variations are neglected when multiplied by terms quadratic in the velocity. If the contribution from viscous terms involving $\operatorname{div} \bar{\mathbf{v}}$ is also discarded, we obtain the mean momentum equation in Lighthill's form [369]

$$\rho_o \bar{v}_j \frac{\partial \bar{v}_i}{\partial x_j} = -\frac{\partial \bar{p}}{\partial x_i} + F_i + \eta \nabla^2 \bar{v}_i, \tag{5.6.13}$$

$$F_i = -\frac{\partial (\rho_o \overline{v_i v_j})}{\partial x_j}. \tag{5.6.14}$$

Equation (5.6.13) balances the steady inertia of the streaming flow $\rho_o(\bar{v}_j \partial \bar{v}_i / \partial x_j)$ against driving forces consisting of the mean pressure gradient, the acoustic Reynolds stress force F_i, and the retardation produced by the mean viscous shear stress. If the sound is not damped the motion must be irrotational, and Bernoulli's equation expanded to second order implies that

$$\frac{\partial \bar{p}}{\partial x_i} + F_i = 0.$$

Accordingly, the acoustic driving terms in (5.6.13) vanish, and there can be no streaming motion from body forces. More generally, there is no streaming from acoustic sources in the momentum equation if F_i can be expressed as the gradient of a scalar because both $\bar{p}$ and F_i can then be eliminated by taking the curl of the momentum equation. The second-order body force F_i is then balanced by a mean pressure gradient $\partial \bar{p}/\partial x_i$ without the need for a streaming flow.

However, even when the nonlinear acoustic sources in the continuity and momentum equations are ignored, very weak streaming flows can still be produced by a vibrating surface [369]. A rapidly vibrating boundary radiates an acoustic beam. The Lagrangian mean velocity $\bar{v}_L$ must vanish near the vibrating surface, and equation (5.6.12) then shows that $\bar{\mathbf{v}} = -\mathbf{I}_a/\rho_o c_o^2$ at the surface, that is, the vibrating boundary behaves as a *sink*. Fluid is sucked into the sink from all directions (where mostly $\mathbf{I}_a = \mathbf{0}$) and is balanced by a net outflow at velocity $\bar{v}_L - \bar{v} \equiv \mathbf{I}_a/\rho_o c_o^2$ along the beam.

Equation (5.6.14) is the momentum equation for a steady flow driven by an external force F_i. Low velocity streaming can be assumed to be incompressible (div $\mathbf{I}_a = 0$ in (5.6.11)), and its inertia neglected. Equation (5.6.14) is then equivalent to the "creeping flow" equation of low Reynolds number hydrodynamics [4, 17]. The divergence and curl of this equation yield the following equations for the mean pressure and vorticity:

$$\nabla^2 \bar{p} = \operatorname{div} \mathbf{F}, \quad \eta \nabla^2 \bar{\omega} = -\operatorname{curl} \mathbf{F}. \tag{5.6.15}$$

The solutions of these equations are generally coupled through boundary conditions.

In two-dimensional problems where conditions are uniform in the x_3-direction and $\bar{\omega} = (0, 0, \bar{\omega}_3)$, the vorticity equation may be cast in terms of a stream function ψ in terms of which $\bar{\omega}_3 = -\nabla^2\psi$, and [4, 17]

$$\eta \nabla^4 \psi = \frac{\partial F_1}{\partial x_2} - \frac{\partial F_2}{\partial x_1} \quad \text{where} \quad (\bar{v}_1, \bar{v}_2) = \left(\frac{\partial \psi}{\partial x_2}, -\frac{\partial \psi}{\partial x_1}\right). \tag{5.6.16}$$

Similarly, when the streaming is axisymmetric with respect to the x-axis, the vorticity $\bar{\omega} = \bar{\omega}\mathbf{i}_\theta$, where $\mathbf{i}_\theta$ is a unit vector in the direction of increasing θ of the cylindrical polar coordinates (x, r, θ). Then

$$\bar{\omega} = \frac{-1}{r}\left(\frac{\partial^2}{\partial x^2} + \frac{\partial^2}{\partial r^2} - \frac{1}{r}\frac{\partial}{\partial r}\right)\psi,$$

and the vorticity equation becomes

$$\eta\left(\frac{\partial^2}{\partial x^2}+\frac{\partial^2}{\partial r^2}-\frac{1}{r}\frac{\partial}{\partial r}\right)^2\psi=-r\left(\frac{\partial F_x}{\partial r}-\frac{\partial F_r}{\partial x}\right),$$
$$(\bar{v}_x,\bar{v}_r)=\frac{1}{r}\left(\frac{\partial\psi}{\partial r},-\frac{\partial\psi}{\partial x}\right), \tag{5.6.17}$$

where $\bar{v}_x$, $\bar{v}_r$ are the velocity components in the x- and r-directions.

Example 5. Show that in the absence of attenuation the mean pressure in an acoustic field is given by $\bar{p}=\mathrm{V}-\mathrm{T}$, where $\mathrm{V}=\frac{1}{2}\overline{(p-p_o)^2}/\rho_o c_o^2$ and $\mathrm{T}=\frac{1}{2}\rho_o\overline{v^2}$ are the potential and kinetic energy densities of the acoustic field, and p_o is the undisturbed mean pressure. Verify that $\partial\bar{p}/\partial x_i=F_i$, that is, that no streaming is possible by momentum equation forcing.

5.6.5 Streaming Caused by Traveling and Standing Sound Waves

A narrow beam of intense very high frequency (1 MHz or more) sound can be generated by a piezoelectric vibrating disc. Early streaming experiments used quartz transducers to project sound into a liquid or gas, and the resulting streaming became known as the "quartz wind" [366]. If ℓ is a characteristic dimension of the streaming motion, it turns out that the streaming Reynolds number $\bar{v}\ell/\nu$ is usually large and that it is not permissible to neglect the inertia term on the left of (5.6.13). Calculation [369] indicates that streaming in free space is governed by the creeping flow equations only for transducer source powers less than about 10^{-6} Watts (Example 7). Free-jet streaming observed in practice must therefore be regarded as a high Reynolds number flow. At ultra-high frequencies, when the beam is rapidly damped, streaming can be modeled by taking the Reynolds stress force to be concentrated at a point (Examples 6 and 7). For high-powered beams at lower frequencies (<100 KHz), not only is inertia important, but the beam decays relatively slowly, the streaming flow forms a turbulent jet, and the attenuation of the sound tends to be dominated by the eddy viscosity of the jet. The reader should consult Lighthill's paper [369] for further details of this interesting case and for a comparison with experiment.

The Quartz Wind

The simplest configuration relating to early experimental realizations of the "quartz wind" is illustrated in Figure 5.6.5 [366]. A vibrating disc of radius a projects an axisymmetric beam of high-frequency sound of the same radius axially along an open, rigid tube of length L and radius R. Let us investigate low Reynolds number streaming in the tube (neglecting the inertia term in (5.6.14)),

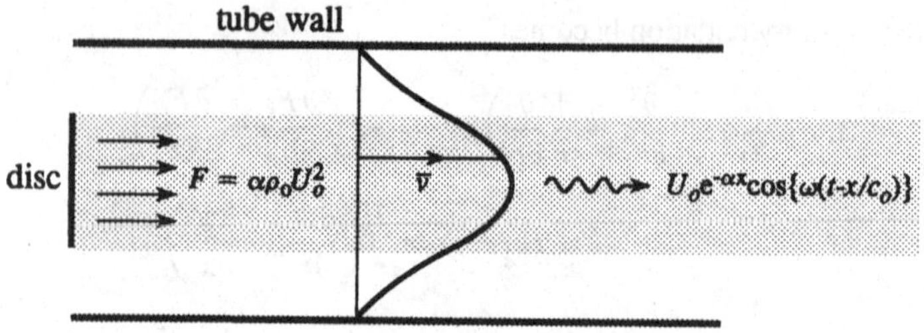

Figure 5.6.5. "Quartz wind" in an open tube.

when the flow is sufficiently slow to be regarded as incompressible (right-hand side of (5.6.11) is zero). Introduce cylindrical polar coordinates (x, r, θ), with x directed along the tube axis from the disc. The acoustic particle velocity v_o is parallel to the x-axis and is nonzero only within the beam, where it is assumed to be given by

$$v_o = U_o \mathrm{e}^{-\alpha x} \cos\{\omega(t - x/c_o)\}, \quad 0 < r < a, \tag{5.6.18}$$

where U_o is real, $\omega > 0$, and α (>0) is an attenuation coefficient.

Consider first a short, open tube in which $\bar{p} \equiv 0$ and $\alpha L \ll 1$. The Reynolds stress force F_i is nonzero only within the beam. It is parallel to the axis of the tube and equal to

$$F_o = \alpha\rho_o U_o^2 \mathrm{e}^{-2\alpha x} \approx \alpha\rho_o U_o^2 \quad \text{for } \alpha L \ll 1. \tag{5.6.19}$$

The streaming is parallel to the tube with velocity $\bar{v}$ determined by the axisymmetric form of (5.6.13)

$$\begin{aligned} \frac{1}{r}\frac{\partial}{\partial r}\left(r\frac{\partial \bar{v}}{\partial r}\right) &= \frac{-\alpha\rho_o U_o^2}{\eta}, \quad 0 < r < a, \\ &= 0, \quad a < r < R. \end{aligned}$$

The no-slip condition requires $\bar{v} = 0$ at $r = R$; $\bar{v}$ and the shear stress $\eta\partial\bar{v}/\partial r$ are continuous at the edge $r = a$ of the beam. The solution satisfying these conditions is

$$\begin{aligned} \bar{v} &= \frac{\alpha a^2 \rho_o U_o^2}{4\eta}(1 - (r/a)^2 + 2\ln(R/a)), \quad 0 < r < a, \\ &= \frac{\alpha a^2 \rho_o U_o^2}{2\eta}\ln(R/r), \quad a < r < R. \end{aligned} \tag{5.6.20}$$

The velocity profile is illustrated in the figure.

The acoustic attenuation coefficient can be expressed in the form

$$\alpha = \tfrac{1}{2}\kappa_o^2\delta/c_o,$$

where δ is called the *diffusivity* of sound [36], whose order of magnitude $\sim\eta/\rho_o$, although its value at very high frequencies is increased by molecular relaxation effects. The streaming velocity is proportional to $\rho_o\delta/\eta \approx \mathrm{O}(1)$, and $\bar{v}$ evidently remains significant even when $\eta \to 0$; additional dissipative mechanisms that raise the value of α produce an increase in the streaming velocity.

Equation (5.6.12) shows that the Lagrangian velocity $\bar{v}_L = \bar{v}$ outside the beam. Within the beam, $I_a/\rho_o c_o^2 = U_o^2/2c_o$, and

$$\frac{\bar{v}_L - \bar{v}}{\bar{v}} \approx \frac{2\eta}{\alpha\rho_o c_o a^2} \approx \mathrm{O}\left(\frac{1}{(\kappa_o a)^2}\right).$$

This is very small at ultrasonic frequencies, when the acoustic wavelength $2\pi/\kappa_o \ll a$, showing that $\bar{v}_L$ differs negligibly from $\bar{v}$.

When the sound beam completely fills the tube ($a = R$), it is necessary to take account of the acoustic boundary layer at the wall. At all relevant frequencies, the boundary layer thickness is a small fraction of the tube radius and very much smaller than the acoustic wavelength, and the variations across the boundary layer of both the acoustic and mean pressure $\bar{p}$ may be neglected. Outside the boundary layer, the acoustic particle velocity is given by (5.6.18). The mean pressure is not necessarily constant, however, and must be determined from the first of (5.6.15). Because $\operatorname{div}\mathbf{F} = -\rho_o\partial^2\bar{v}^2/\partial x^2$ depends only on axial position x, this yields

$$\bar{p} = A + Bx - \tfrac{1}{2}\rho_o U_o^2 \mathrm{e}^{-2\alpha x},$$

where A and B are constants, and when $\bar{p} = 0$ at both ends of the tube, we find

$$\bar{p} = \frac{1}{2}\rho_o U_o^2\left(1 - \mathrm{e}^{-2\alpha x} - \frac{x}{L}(1 - \mathrm{e}^{-2\alpha L})\right), \quad 0 < x < L. \qquad (5.6.21)$$

The pressure $\bar{p} = 0$ in the limit $\alpha L \to 0$, and we shall examine this case first. The Reynolds stress force is parallel to the tube and can be split into two components. The first

$$F_o = -\partial\left(\rho_o\overline{v_o^2}\right)/\partial x \approx \alpha\rho_o U_o^2 \quad (\alpha L \ll 1),$$

is the force considered previously, produced by the main acoustic beam. The remaining part $F - F_o$ is nonzero only within the acoustic boundary layer. The contribution to the streaming velocity $\bar{v}$ from each of these forces can be calculated separately. The component $\bar{v}_o$ generated by F_o is given by setting $a = R$ in (5.6.20). The axial velocity generated by the excess Reynolds stress force $F - F_o$ will be denoted by $\bar{v}_w$. Introduce local coordinates (x, z), with z measured into the fluid from the wall. When the terms determining $\bar{v}_o$ are canceled from the x-momentum equation of (5.6.13), the boundary layer approximation to the equation for $\bar{v}_w$ close to the wall is

$$\eta \frac{\partial^2 \bar{v}_w}{\partial z^2} = F_o - F,$$

where

$$F = -\frac{\partial}{\partial x}(\rho_o \overline{v^2}) - \frac{\partial}{\partial z}(\rho_o \overline{vw}), \qquad (5.6.22)$$

and (v, w) are the (x, z)-components of the acoustic particle velocity. These equations imply that $\bar{v}_w$ tends to a finite value $\bar{v}_s$ just outside the boundary layer (where $\partial \bar{v}_w/\partial z \to 0$). $\bar{v}_s$ is called the *slip velocity*; it is the component of the streaming velocity induced by the boundary layer stresses just outside the boundary layer and can be expressed in the form

$$\bar{v}_s = \frac{1}{\eta} \int_0^\infty z(F - F_o)\, dz. \qquad (5.6.23)$$

To evaluate the integrand, we use (5.6.18) and (5.1.11) to express the velocity $\mathbf{v}$ within the acoustic boundary layer in the form

$$v = U_o \mathrm{Re}\left(\left[1 - \mathrm{e}^{iz\sqrt{i\omega/\nu}}\right]\mathrm{e}^{-i\omega(t - x/c_o)}\right).$$

The normal velocity component w is calculated from the continuity equation which, because the pressure is uniform across the boundary layer, becomes

$$\partial v/\partial x + \partial w/\partial z = \partial v_o/\partial x,$$

provided thermal dissipation is ignored (Example 9). Then

$$w = -U_o \,\mathrm{Re}\left(\frac{\kappa_o}{\sqrt{i\omega/\nu}}\left[1 - \mathrm{e}^{iz\sqrt{i\omega/\nu}}\right]\mathrm{e}^{-i\omega(t - x/c_o)}\right). \qquad (5.6.24)$$

which satisfies $w = 0$ at the wall. These formulae for the velocity are now used in (5.6.22) to evaluate F. The first term on the right of (5.6.22) is found to make a negligible contribution to the slip velocity, which (5.6.23) determines to be

$$\bar{v}_s = \frac{U_o^2}{4c_o}.$$

By estimating the acoustic attenuation coefficient to be $\alpha \approx \kappa_o^2 \eta / \rho_o c_o$ (as before), and recalling that $\bar{v}_o$ is given by (5.6.20) with $a = R$, it follows that $\bar{v}_s/\bar{v}_o \approx O(1/(\kappa_o R)^2) \ll 1$. On the other hand, if α is given by (5.1.32), as it would be in a narrow tube when the damping is dominated by boundary layer losses, we find $\bar{v}_s/\bar{v}_o \approx O(\sqrt{\nu/\omega R^2})$, which is also small. In either case, the interior streaming is essentially unaffected by the contribution from the boundary layer.

This conclusion is changed completely when the tube is longer than the dissipation length $1/\alpha$. By letting $L \to \infty$ in (5.6.21), we find that the main beam force $F_o = \alpha \rho_o U_o^2 e^{-2\alpha x}$ is exactly balanced by the mean pressure gradient $\partial \bar{p}/\partial x$. The streaming is then very slow and entirely attributable to mean Reynolds stresses in the boundary layer. This also occurs in the case of a tube closed at one end, which is discussed next.

Tube Closed at One End

Consider a tube closed at the transducer end, when the beam radius and tube radius are equal. There is now no net mass flux through any cross section of the tube.

The pressure $\bar{p}$ is uniform over the cross section when $\alpha R \ll 1$, and its value is determined from the condition of mass conservation. In the interior of the tube, the axial streaming velocity $\bar{v}_o$ attributable to the beam force F_o of (5.6.19) satisfies

$$\frac{\eta}{r}\frac{\partial}{\partial r}\left(r\frac{\partial \bar{v}_o}{\partial r}\right) = \frac{\partial \bar{p}}{\partial x} - F_o$$

provided conditions vary slowly with x (i.e., $\alpha R \ll 1$), with solution

$$\bar{v}_o = \frac{1}{4\eta}\left(\frac{\partial \bar{p}}{\partial x} - F_o\right)(r^2 - R^2). \qquad (5.6.25)$$

The streaming generated by the boundary layer force $F - F_o$ is determined by the slip velocity $\bar{v}_s$ as before, which is equal to $(U_o^2/4c_o)e^{-2\alpha x}$. When αR is small, the corresponding motion in the interior of the pipe will vary slowly with x and can be calculated by applying (5.6.17) outside the boundary layer. When variations of w with x are neglected, the stream function satisfies $\{\partial^2/\partial r^2 - \partial/r\partial r\}^2\psi = 0$, and conditions of symmetry and finiteness at $r = 0$ imply that

$$\psi = f(x) r^2 (r^2 - R^2), \quad \bar{v}_w = 2f(x)(2r^2 - R^2), \qquad (5.6.26)$$

where $f(x)$ is a slowly varying function that must equal $(U_o^2/8R^2 c_o)e^{-2\alpha x}$ in order that $\bar{v}_w = \bar{v}_s$ at $r = R$.

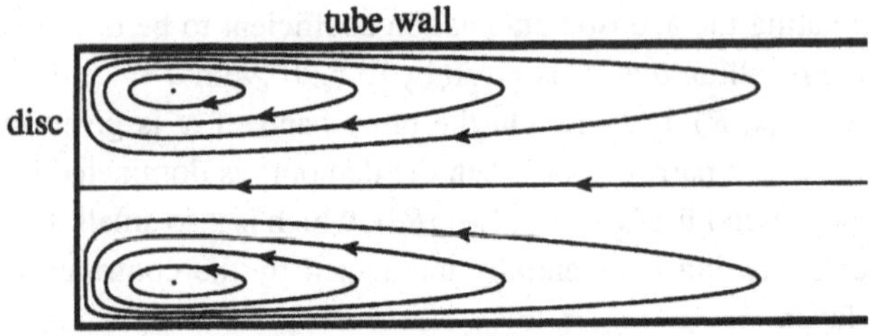

Figure 5.6.6. Recirculating streaming in a long tube.

The pressure gradient in (5.6.25) is found by equating to zero the net volume flux along the tube. This is determined by the Lagrangian axial velocity $\bar{v}_L = \bar{v}_o + \bar{v}_w + (U_o^2/2c_o)\mathrm{e}^{-2\alpha x}$ (see (5.6.12)). The stream function (5.6.26) defines a motion with no net volume flux through any cross section of the tube, and we accordingly find

$$\bar{p} = \frac{\rho_o U_o^2}{2}(\mathrm{e}^{-2\alpha L} - \mathrm{e}^{-2\alpha x})\left(1 + \frac{4\nu}{\alpha c_o R^2}\right).$$

The velocity $\bar{v}_o$ can now be found in terms of U_o by substituting this result into (5.6.25).

The streaming velocity is of order U_o^2/c_o everywhere, and an approximate Lagrangian description of the motion is easily seen to be represented by the stream function

$$\psi_L = \frac{3U_o^2 \mathrm{e}^{-2\alpha x}}{8c_o R^2} r^2(r^2 - R^2).$$

This is not applicable very close to the transducer, where $\bar{v}_L \to 0$ ($\psi_L \to 0$). Because also $\psi_L \to 0$ as $x \to \infty$, the stream tubes (on the surfaces of which ψ_L is constant) form a recirculating system of the type illustrated in Figure 5.6.6. The flow is in the beam direction close to the wall (where $r > R/\sqrt{2}$) and returns along the interior of the tube, that is, the flow in the central region is opposite to the quartz wind in a short, open tube. The stream tubes shown in the figure divide the motion into five equal regions of flow.

Kundt's Tube

Standing acoustic waves in a closed tube generate recirculating streaming patterns of the type shown in Figure 5.6.7. When the wavelength is much larger than the pipe diameter, the waves are one-dimensional, and damping is dominated by dissipation in the acoustic boundary layers at the wall. Then $F_o = \partial\bar{p}/\partial x$ outside the boundary layer. Let the acoustic particle velocity in the pipe be

$$v_o = U(x)\cos\omega t, \quad (\omega > 0),$$

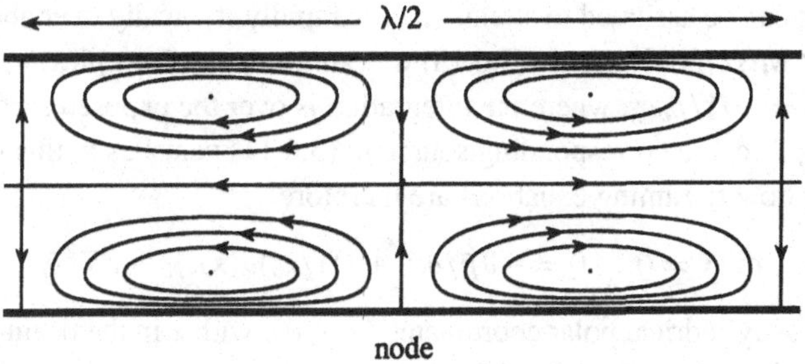

Figure 5.6.7. Kundt's tube: streaming produced by a standing acoustic wave.

where $U(x)$ varies slowly on a scale of the tube radius R. The slip velocity is again given by equations (5.6.22) and (5.6.23) (with $F_o \equiv 0$), where both terms on the right of (5.6.22) are important (because the axial variation of $\overline{v^2}$ is now determined by $U(x)$ and not the attenuation α), and

$$\begin{aligned} v &= U(x)\mathrm{Re}\left(\left[1 - \mathrm{e}^{iz\sqrt{i\omega/\nu}}\right]\mathrm{e}^{-i\omega t}\right), \\ w &= U'(x)\mathrm{Re}\left(\frac{1}{\sqrt{\omega/\nu}}\left[1 - \mathrm{e}^{iz\sqrt{i\omega/\nu}}\right]\mathrm{e}^{-i\omega(t-\pi/4)}\right), \end{aligned} \tag{5.6.27}$$

where $U'(x) = dU/dx$, and the expression for w neglects thermal dissipation in the boundary layer (Example 10).

The use of these results to evaluate the integral (5.6.23) leads to Rayleigh's formula for the slip velocity [10, 362]

$$\bar{v}_s = \frac{-3U(x)U'(x)}{4\omega}. \tag{5.6.28}$$

The direction of the streaming just outside the boundary layer is therefore toward the nodes of the standing acoustic wave, that is, toward those points in the tube where the acoustic particle velocity $v_o = U(x)\cos\omega t = 0$. The steady motion in the interior of the tube is defined by the stream function ψ of (5.6.26) with $f(x) = -3U(x)U'(x)/8R^2\omega$. Recirculating streamtubes are depicted in Figure 5.6.7 for $U(x) = U_o\cos(\kappa_o x)$, when the wavelength of the standing waves $\lambda = 2\pi/\kappa_o \gg R$, and $\bar{v}_s = 3U_o^2\sin(2\kappa_o x)/8c_o$. Fine dust deposited on the walls of the tube will tend to collect at the nodes, which are axial distances $\lambda/2$ apart.

Example 6: Low Reynolds number ultrasonic streaming in free space. A very high-frequency vibrating disc at the origin radiates a narrow beam of sound along the x_1-axis. If Π is the acoustic power of the source, the net force applied to the fluid is in the x_1-direction and equal to Π/c_o. At ultrasonic frequencies the beam can be assumed to attenuate very rapidly (typically over about 10 cm in air at 1 MHz [369]), and in (5.6.13) we can set $F_i = (\Pi/c_o)\delta(\mathbf{x})\delta_{i1}$. Because $\int_S \bar{\mathbf{v}} \cdot d\mathbf{S} = -\Pi/\rho_o c_o^2$, where the integration is over the exposed surface of the vibrating disc, the corresponding source in (5.6.11) vanishes in this limit. The creeping flow streaming equations are therefore

$$\operatorname{div} \bar{\mathbf{v}} = 0, \quad 0 = -\partial\bar{p}/\partial x_i + (\Pi/c_o)\delta(\mathbf{x})\delta_{i1} + \eta\nabla^2\bar{v}_i.$$

In terms of cylindrical polar coordinates (x, r, θ), with x in the beam direction, the streaming motion is defined by the stream function

$$\psi = \frac{\Pi}{8\pi\eta c_o}\frac{r^2}{\sqrt{x^2+r^2}}.$$

The stream tubes of this low Reynolds number approximation have the fore-and-aft symmetric structure shown in Figure 5.6.8a.

Example 7: High Reynolds number ultrasonic streaming in free space [369]. When inertia terms cannot be neglected in the problem of Example 6 (but the frequency is still high enough for the Reynolds stress force to be approximated by a point force), the stream function becomes [4, 370, 371]

$$\psi = \frac{2\eta r^2}{\rho_o[\vartheta\sqrt{x^2+r^2} - x]},$$

where ϑ (>1) is a constant, and x, r are the polar coordinates of Example 6. The constant ϑ is the solution of

$$F \equiv \frac{\Pi}{c_o} = \frac{16\pi\eta^2}{\rho_o}\vartheta\left[1 + \frac{4}{3(\vartheta^2-1)} - \frac{\vartheta}{2}\ln\left(\frac{\vartheta-1}{\vartheta+1}\right)\right].$$

When $\rho_o\Pi/\eta^2 c_o < 10$ (i.e., for $\Pi < 10^{-10}$ Watts in air), the solution of this equation is approximately $\vartheta \approx 16\pi\eta^2 c_o/\rho_o\Pi$, and the stream function reduces to the creeping flow result of Example 6. At higher source powers, the inertia of the streaming flow cannot be neglected (when $\rho_o\Pi/\eta^2 c_o$ is large, $\vartheta \to 1 + 32\pi\eta^2 c_o/3\rho_o\Pi$). The stream tubes illustrated in Figure 5.6.8b are for $\Pi = 10^{-4}$ Watts in air ($\rho_o\Pi/\eta^2 c_o = 1163$, $\vartheta = 1.0277$). The mean flow is now in the form of a narrow jet, with entrainment of slower fluid from the outside. The velocity on the jet axis at distance x from the force is equal to $4\eta/\rho_o x(\vartheta - 1)$, which is about 2 cm/sec when $x = 10$ cm.

Example 8. A time-harmonic sound wave propagating in the x_1-direction is incident on a compact rigid body, producing an acoustic boundary layer that

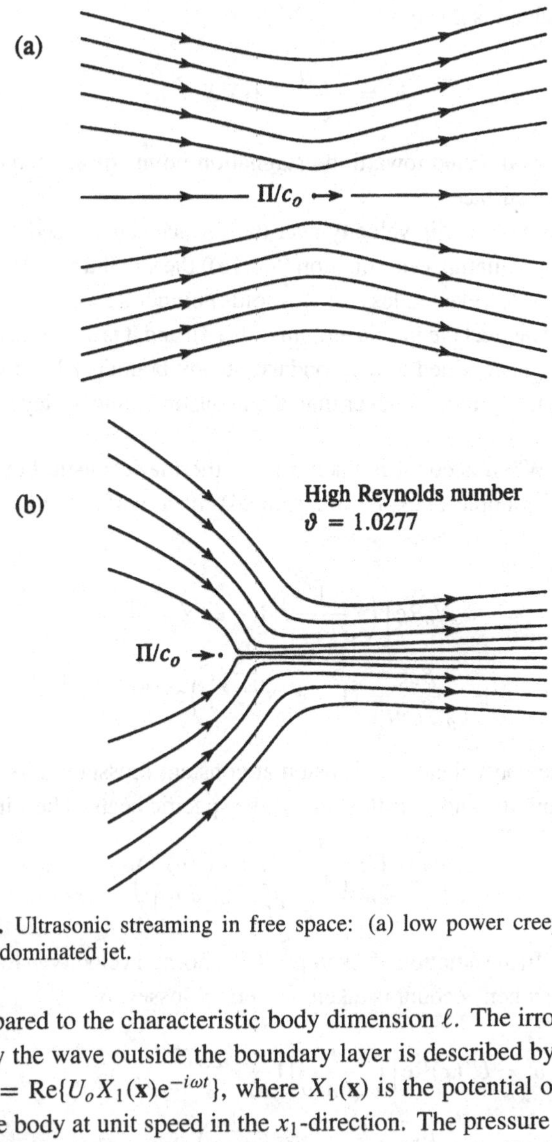

Figure 5.6.8. Ultrasonic streaming in free space: (a) low power creeping flow and (b) inertially dominated jet.

is thin compared to the characteristic body dimension ℓ. The irrotational flow produced by the wave outside the boundary layer is described by the velocity potential $\varphi = \mathrm{Re}\{U_o X_1(\mathbf{x})\mathrm{e}^{-i\omega t}\}$, where $X_1(\mathbf{x})$ is the potential of irrotational flow past the body at unit speed in the x_1-direction. The pressure gradient $\nabla \bar{p}$ within the boundary layer is equal to the Reynolds stress force $\mathbf{F}_o$ determined by φ. Let (x, z) and (v, w) be local coordinates and velocity components defining the acoustic boundary layer on the surface. Then $\mathbf{F}$ is given by (5.6.22), where v and w are given by (5.6.27) with $U(x) = U_o\partial X_1/\partial x$, and with the formula for w modified by the addition of $-zU'(x)\mathrm{e}^{-i\omega t}$ to the right-hand side (because the boundary layer motion is incompressible in the compact approximation). The additional boundary layer term is found to make no contribution to the slip velocity $\bar{v}_s$, which is given by

$$\bar{v}_s = \frac{-3U_o^2}{8\omega}\nabla((\nabla X_1)^2).$$

This velocity is directed toward the stagnation points of the acoustic particle velocity on the surface.

The distribution of slip velocity over the surface can be used as a boundary condition for calculating (via equation (5.6.13)) the streaming at large distances from the body. However, unless the Reynolds number Re$=U_o\ell/\nu \ll 1$, it is not permissible to neglect the inertia term in (5.6.13), and it is then found [367, 368] that streaming is confined to a secondary, steady boundary layer of thickness $\sim\ell/\sqrt{\text{Re}}$ (typically much thicker than the acoustic boundary layer).

Example 9. When account is taken of the thermal acoustic boundary layer (Section 5.1, Example 1) expression (5.6.24) for w within the boundary layer becomes

$$w = -\kappa_o U_o \text{Re}\left[\left(\frac{1}{\sqrt{i\omega/\nu}}\left[1 - \mathrm{e}^{iz\sqrt{i\omega/\nu}}\right] + \frac{\beta c_o^2}{c_p\sqrt{i\omega/\chi}}\left[1 - \mathrm{e}^{iz\sqrt{i\omega/\chi}}\right]\right)\mathrm{e}^{-i\omega(t-x/c_o)}\right],$$

where β is the coefficient of expansion at constant pressure, χ is the thermometric conductivity, and c_p is the ratio of the specific heats. The slip velocity is

$$\bar{v}_s = \frac{U_o^2}{4c_o}\left(1 + \frac{2\beta c_o^2(\chi/\nu)^2}{c_p(1+\chi/\nu)}\right).$$

Example 10. In the notation of Example 9, the normal velocity w for a *standing* acoustic wave when account is taken of thermal losses, is

$$w = U'(x)\text{Re}\left[\left(\frac{1}{\sqrt{\omega/\nu}}\left[1 - \mathrm{e}^{iz\sqrt{i\omega/\nu}}\right] + \frac{\beta c_o^2}{c_p\sqrt{\omega/\chi}}\left[1 - \mathrm{e}^{iz\sqrt{i\omega/\chi}}\right]\right)\mathrm{e}^{-i\omega(t-\pi/4)}\right],$$

and the slip velocity (5.6.28) becomes

$$\bar{v}_s = \frac{-3U(x)U'(x)}{4\omega}\left(1 + \frac{\beta c_o^2\sqrt{\chi/\nu}}{3c_p(1+\chi/\nu)}\right).$$

6

Resonant and unstable systems

Jets and shear layers are frequently responsible for the generation of intense acoustic tones. Instability of the mean flow over of a wall cavity excites "self-sustained" resonant cavity modes or periodic "hydrodynamic" oscillations, which are maintained by the steady extraction of energy from the flow. Whistles and musical instruments such as the flute and organ pipe are driven by unstable air jets, and shear layer instabilities are responsible for tonal resonances excited in wind tunnels, branched ducting systems, and in exposed openings on ships and aircraft and other high-speed vehicles. These mechanisms are examined in this chapter, starting with very high Reynolds number flows, where a shear layer can be approximated by a vortex sheet. We shall also discuss resonances where *thermal* processes play a fundamental role, such as in the Rijke tube and pulsed combustor. The problems to be investigated are generally too complicated to be treated analytically with full generality, but much insight can be gained from exact treatments of linearized models and by approximate nonlinear analyses based on simplified, yet plausible representations of the flow.

6.1 Linear Theory of Wall Aperture and Cavity Resonances

6.1.1 Stability of Flow Over a Circular Wall Aperture

The sound produced by nominally steady, high Reynolds number flow over an opening in a thin wall is the simplest possible system to treat analytically. Our approach is applicable to all linearly excited systems involving an unstable shear layer, and it is an extension of the method used in Section 5.3.6 to determine the conductivity of a circular aperture in a mean grazing flow. Consider one-sided flow (see Figure 5.3.7, with $U_+ = U,\ U_- = 0$) in the x_1-direction over an aperture in a thin, rigid wall at $x_2 = 0$, whose center is at the origin. The

Mach number $M = U/c_o$ is assumed to be small enough that secondary effects of convection of sound can be ignored. The aperture shear layer is modeled by a vortex sheet, as in Section 5.3.6. When the shear layer is disturbed by a uniform, time-dependent pressure $p_{\mathrm{I}}(t)$ applied in $x_2 > 0$, a volume flux $Q(t)$ is established through the aperture (in the x_2-direction) that results in the radiation of spherically spreading sound waves into the fluid on either side of the wall. When the frequency is small, the opening is acoustically compact, and the acoustic pressure at large distances is given by the volume monopole fields

$$p(\mathbf{x}, t) = \frac{\rho_o \mathrm{sgn}(x_2)}{2\pi |\mathbf{x}|} \frac{dQ}{dt}(t - |\mathbf{x}|/c_o), \quad |\mathbf{x}| \to \infty. \tag{6.1.1}$$

(cf. Section 1.8, Example 1; the divisor 4π is here replaced by 2π because the monopoles of strengths $\pm Q$ radiate into a solid angle 2π on either side of the wall).

According to Section 5.3.8, Q and p_{I} are related by

$$\rho_o \frac{dQ}{dt}(t) = -\int_{-\infty}^{\infty} K_R(\omega) p_{\mathrm{I}}(\omega) \mathrm{e}^{-i\omega t} \, d\omega, \tag{6.1.2}$$

where $p_{\mathrm{I}}(\omega) = (1/2\pi) \int_{-\infty}^{\infty} p_{\mathrm{I}}(t) \mathrm{e}^{i\omega t} dt$ is the Fourier transform of $p_{\mathrm{I}}(t)$, and $K_R(\omega)$ is the Rayleigh conductivity. It may be assumed that $p_{\mathrm{I}}(t)$ vanishes prior to some initial time t_{I}, which implies that $p_{\mathrm{I}}(\omega) \to 0$ as $\mathrm{Im}\,\omega \to +\infty$ and that it is regular in the upper complex frequency plane ($p_{\mathrm{I}}(\omega)$ is regular everywhere if $p_{\mathrm{I}}(t)$ is of finite duration [204]). To ensure that equation (6.1.2) determines the *causal* response of the volume flux to the applied pressure, that is, that $Q(t) = 0$ for $t < t_{\mathrm{I}}$, the path of integration must pass *above* any singularities of $K_R(\omega)$ (Section 1.7.4). For $t > t_{\mathrm{I}}$, the integral can be evaluated by displacing the integration contour downward toward the real axis, "wrapping" it around any singularities in $\mathrm{Im}\,\omega > 0$. The contributions from these singularities will grow exponentially with $t - t_{\mathrm{I}}$. Of course, unlimited growth cannot occur in reality: Although equation (6.1.2) is applicable quite generally, in a real flow the function $K_R(\omega)$ depends nonlinearly on Q, and the exponential increase, which is merely a mathematical artifact of linearization, would be suppressed.

The real and imaginary parts of $K_R(\omega)/2R = \Gamma_R - i\Delta_R$ for a circular aperture of radius R are displayed in Figure 5.3.8i for real ω. An approximate analytic continuation of K_R into the complex frequency plane is given by the *four-pole approximation* of Example 10, Section 5.3.8, according to which $K_R(\omega)$ has simple poles at $\omega R/U = \pm 2.9 + 0.8i$ in $\mathrm{Im}\,\omega > 0$. Thus, the volume flux at large times is determined by the residue contributions to the integral (6.1.2) from these poles, which yield the exponentially increasing monopole source strength

$$\rho_o \frac{dQ}{dt} \sim -8\pi U \alpha_2 |p_{\mathrm{I}}(\omega)| \cos\left(\frac{2.9Ut}{R} - \phi\right) \mathrm{e}^{0.8Ut/R},$$
$$\omega R/U = 2.9 + 0.8i, \quad \alpha_2 = 0.82,$$

where $\arg\{p_{\mathrm{I}}(\omega)\} = \phi$ (recall that $p_{\mathrm{I}}(-\omega^*) \equiv p_{\mathrm{I}}^*(\omega)$ when $p_{\mathrm{I}}(t)$ is real). This growth is ultimately controlled by nonlinearity, but this formula suggests that a self-sustaining acoustic mode of frequency Re $\omega \approx 2.9U/R$ could dominate the initial radiation, although the radiation may ultimately be "broadband" if the flow in the aperture becomes turbulent [372–397].

It has been argued [398] that nonlinear mechanisms that in practice cutoff unlimited growth do not necessarily alter the frequency of self-sustained oscillations because this is determined both in the linear and nonlinear regimes by the *convection* velocity of vorticity fluctuations across the mouth of the aperture, which, according to experiment [330, 399–401] is hardly influenced by vortex strength. The importance of the convection velocity is discussed in greater detail in Section 6.2; for the moment we shall proceed on the basis that the real part of the linear theory poles of $K_R(\omega)$ in Im $\omega > 0$ should be interpreted as possible frequencies of self-sustaining oscillations of the real system.

In practice, several additional self-sustained tones are observed, each of which "operates" when the flow velocity is within certain, possibly overlapping, ranges. The additional tones correspond to poles of $K_R(\omega)$ in Im $\omega > 0$ that are not included in the four-pole approximation. It is usual to express the frequencies as a Strouhal number fD/U, where $f =$ Re $(\omega)/2\pi$ and $D = 2R$, and the above-indicated calculation suggests the existence of a tone at $fD/U \approx 0.9$. The Strouhal number is frequently based on the *average* length of the opening in the streamwise direction: $L = \frac{\pi}{4}D$, in which case this tone occurs at $fL/U \approx 0.7$. A more accurate numerical simulation of the flow [402] has located poles in Im $\omega > 0$ at the four complex frequencies listed in Table 6.1.1.

Table 6.1.1. *Poles of $K_R(\omega)$ for a circular aperture [402]*

Pole: $\omega_r R/U + i\omega_i R/U$		Strouhal number
$\omega_r R/U$	$\omega_i R/U$	$fL/U = \frac{\pi}{4} fD/U$
±3.09	0.56	0.77
±4.99	1.98	1.25
±6.77	3.47	1.69
±8.50	4.98	2.13

These poles represent the linear theory prediction of the first four Strouhal numbers (of a nominally infinite sequence) of self-sustained oscillations of high Reynolds number flow over the circular aperture. Only the first three or four modes are usually observed in experiments.

6.1.2 The Rectangular Aperture

A simple and more complete picture of the linear stability of flow over an aperture, including the effects of wall thickness, can be given for rectangular apertures, of the type discussed in Section 5.3.9. Consider the case shown in profile in Figure 6.1.1 where the wall has thickness h. The mean flows "above" and "below" the wall are, respectively, at speeds $U_\pm$ in the x_1-direction. The midplane of the wall is taken to coincide with the plane $x_2 = 0$, and the origin is at the geometrical center of the aperture. The aperture is aligned with sides of length L parallel to the mean flow and of length b in the transverse (x_3-) direction so that the upper and lower openings occupy $|x_1| < s \equiv \frac{1}{2}L$, $x_2 = \pm\frac{1}{2}h$, $|x_3| < \frac{1}{2}b$. The shear layer over each face is modeled by a vortex sheet, and the fluid within the aperture (in $|x_2| < \frac{1}{2}h$) is in a mean state of rest.

Let us examine the motion in the *thin wall* approximation of Section 5.3.9, when the displacements $\zeta_\pm$ of the upper and lower vortex sheets can be assumed to equal ζ, say, which is governed by equation (5.3.48), that is, by (5.3.49) for an acoustically compact aperture. When the aperture is excited by the uniform pressure $p_I(\omega)$ of Section 6.1.1, the right-hand side of (5.3.49) is replaced by $p_I(\omega)/\rho_o$. Considerable simplification of the equation is possible by noting that vortex shedding from the straight leading edge (at $x_1 = -s$) tends to produce strongly correlated motions of the vortex sheets at different transverse locations

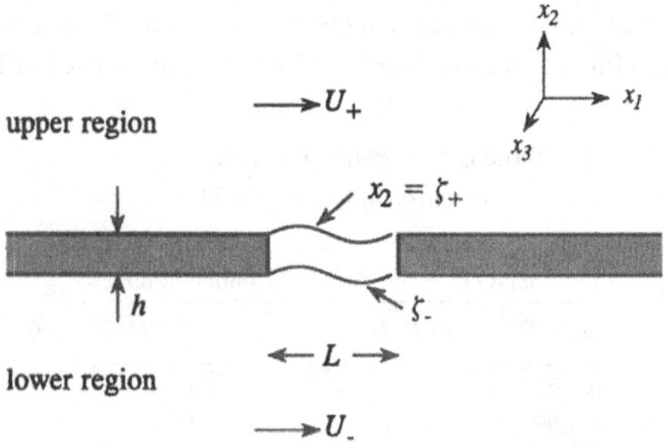

Figure 6.1.1. Rectangular wall aperture in two-sided flow.

x_3 so that in a first approximation ζ may be regarded as independent of x_3. The integration in (5.3.49) with respect to y_3 may then be performed explicitly; if the equation is also integrated over $-b/2 < x_3 < b/2$, it can be recast in the dimensionless form:

$$\left[\left(\sigma + i\frac{\partial}{\partial \xi}\right)^2 + \left(\sigma + i\mu\frac{\partial}{\partial \xi}\right)^2\right] \int_{-1}^{1} \zeta'(\eta)\{\ln|\xi - \eta| + \mathcal{L}(\xi, \eta)\}\, d\eta - \frac{2\pi\sigma^2 h}{L}\zeta'(\eta) = 2\sigma^2, \quad (6.1.3)$$

where

$$\zeta'(\xi) = \frac{-2\rho_o\omega^2 s\zeta(\xi)}{\pi p_I(\omega)} \quad (6.1.4)$$

is a dimensionless displacement, and

$$\sigma = \omega s/U_+, \quad \mu = U_-/U_+, \quad \xi = x_1/s, \quad \eta = y_1/s, \quad (6.1.5)$$

$$\mathcal{L}(\xi, \eta) = -\ln\{b/s + \surd[(b/s)^2 + (\xi - \eta)^2]\} + \surd\{1 + (s/b)^2(\xi - \eta)^2\} - (s/b)|\xi - \eta|. \quad (6.1.6)$$

Equation (6.1.3) is next integrated with respect to the second-order differential operator on the left by making use of the Green's function

$$G(\xi, \eta) = \frac{1}{2\sigma(1-\mu)}(\mathrm{H}(\xi - \eta)e^{i\sigma_1(\xi-\eta)} + \mathrm{H}(\eta - \xi)e^{i\sigma_2(\xi-\eta)}), \quad (6.1.7)$$

which is a particular solution of

$$\left[\left(\sigma + i\frac{\partial}{\partial \xi}\right)^2 + \left(\sigma + i\mu\frac{\partial}{\partial \xi}\right)^2\right] G(\xi, \eta) = \delta(\xi - \eta),$$

where $\sigma_{1,\,2}$ are the Kelvin–Helmholtz wavenumbers [17]

$$\sigma_1 = \sigma\left(\frac{1+i}{1+i\mu}\right), \quad \sigma_2 = \sigma\left(\frac{1-i}{1-i\mu}\right). \quad (6.1.8)$$

Then ζ' satisfies the integral equation

$$\int_{-1}^{1} \zeta'(\eta)\left[\ln|\xi - \eta| + \mathcal{L}(\xi, \eta) - 2\pi\sigma^2\frac{h}{L}G(\xi, \eta)\right] d\eta + \lambda_1 e^{i\sigma_1\xi} + \lambda_2 e^{i\sigma_2\xi} = 1, \quad |\xi| < 1, \quad (6.1.9)$$

where $\lambda_{1,2}$ are constants of integration (cf. equation (5.3.28) for the circular aperture).

This equation must be solved numerically. The values of $\lambda_{1,2}$ are fixed by the Kutta condition that

$$\zeta' = \partial\zeta'/\partial\xi = 0 \quad \text{as} \quad \xi \to -1,$$

and the solution then determines the conductivity from the formula $K_R = i\omega\rho_o Q/p_I(\omega)$, which is equivalent to

$$K_R = -\frac{\pi b}{2}\int_{-1}^{1}\zeta'(\eta)\,d\eta. \qquad (6.1.10)$$

In the special case of uniform, two-sided mean flow, where $U_- = U_+ \equiv U$, the wavenumbers $\sigma_{1,2}$ are both equal to $\sigma = \omega s/U$, and it is convenient to use the following degenerate Green's function instead of (6.1.7)

$$G(\xi,\eta) = -(\xi-\eta)\mathrm{H}(\xi-\eta)\mathrm{e}^{i\sigma(\xi-\eta)}.$$

The terms in $\lambda_{1,2}$ in equation (6.1.9) are then replaced by $(\lambda_1 + \lambda_2\xi)\mathrm{e}^{i\sigma\xi}$. The calculated motion of the vortex sheet is illustrated in Figure 6.1.2 at equal time intervals during a complete cycle for one-sided flow over a thin wall ($U_+ = U,\ U_- = 0,\ h = 0$) when the aperture aspect ratio is $b/L = 1.25$ and $fL/U = 1.1$.

It is now easy to prove that the singularities of $K_R(\omega)$ are all poles and, indeed, that they correspond to eigenvalues of the integral equation (6.1.9). They are a fundamental attribute of the shear layer and are independent of the assumption that p_I is uniform. To see this, consider the limiting process whereby the integral in (6.1.9) is discretized by using, for example, a Gauss integration scheme that expresses it as a sum of terms evaluated at N lattice points ξ_i $(1 \le i \le N)$, where $\xi_1 = -1+\delta,\ \xi_N = +1-\delta$, and $\delta \to +0$ as $N \to \infty$ [24]. The Kutta condition is imposed by setting $\zeta_1' = \zeta_2' = 0$, where $\zeta_i' \equiv \zeta'(\xi_i)$. The discrete form of (6.1.9) can then be written

$$\sum_{j=1}^{N} \mathcal{A}_{ij}\mathcal{Z}_j = 1,$$

where $\mathcal{Z}_1 = \lambda_1,\ \mathcal{Z}_2 = \lambda_2$, and $\mathcal{Z}_j = \zeta_j'$ for $j \ge 3$. For each fixed i,

$$\mathcal{A}_{i1} = e^{i\sigma_1\xi_i}, \quad \mathcal{A}_{i2} = e^{i\sigma_2\xi_i},$$

and the $\mathcal{A}_{ij}$ for $j \ge 3$ include similar terms, with coefficients dependent on the integration scheme used to approximate the integral. The eigenvalues of (6.1.9)

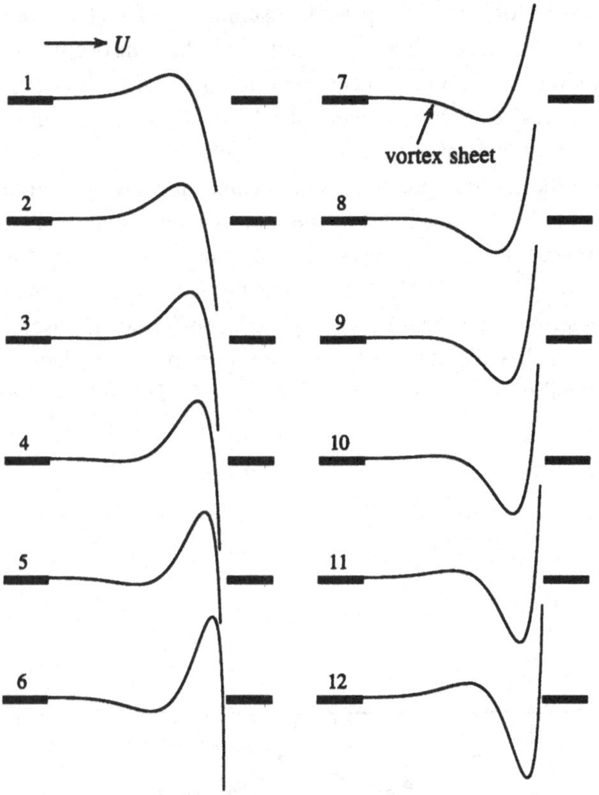

Figure 6.1.2. Vortex sheet profiles at twelve equally spaced time intervals during a complete cycle: $h/L = 0$, $b/L = 1.25$, and $fL/U = 1.1$.

are the (generally complex) roots $\omega s/U_+$ of

$$\det(\mathcal{A}_{ij}) = 0. \tag{6.1.11}$$

Because the integral (6.1.10) for $K_R(\omega)$ can be approximated by the same Gaussian integration formula, it follows from Cramer's rule [24], as N becomes large, that $K_R(\omega)$ is regular except for simple poles at the roots of (6.1.11).

Two-Sided Uniform Mean Flow

By solving equation (6.1.11) for the case $U_+ = U_-$, $h = 0$, we can verify the conclusion of Section 5.3.8 that uniform, two-sided flow past an aperture in a wall of zero thickness is stable. When $h = 0$ the roots all lie in $\text{Im}\,\omega < 0$. Of special interest are those with real parts close to $\omega_r s/U = \pm(n + \frac{1}{4})\pi, n = 1, 2, \ldots$. They lie just below the real axis and, because they are *poles* of K_R, are responsible for essentially periodic oscillations of $K_R(\omega)$ as ω varies along the real axis. (For the circular aperture, they produce the oscillatory behavior of K_R shown in Figure 5.3.8ii, successive values of $\omega_r R/U$ being approximately equal to the real Strouhal numbers at which Γ_R is a minimum.) When the wall thickness h increases from zero, unstable vortex sheets are formed over the faces of the aperture; the flow then becomes increasingly unstable because all of these poles eventually cross the real axis into $\mathrm{Im}\,\omega > 0$. Figure 6.1.3 illustrates how this happens for the first four modes $n = 1$ to 4 for an aperture of aspect ratio $b/L = 2$ [403]. Once a pole lies in the upper frequency plane, its real part determines the frequency of a possible self-sustained oscillation. High-frequency, small-scale instabilities (n large) are possible when h/L is small, but all instability modes exist when $h/L > 0.05$. The nth complex frequency

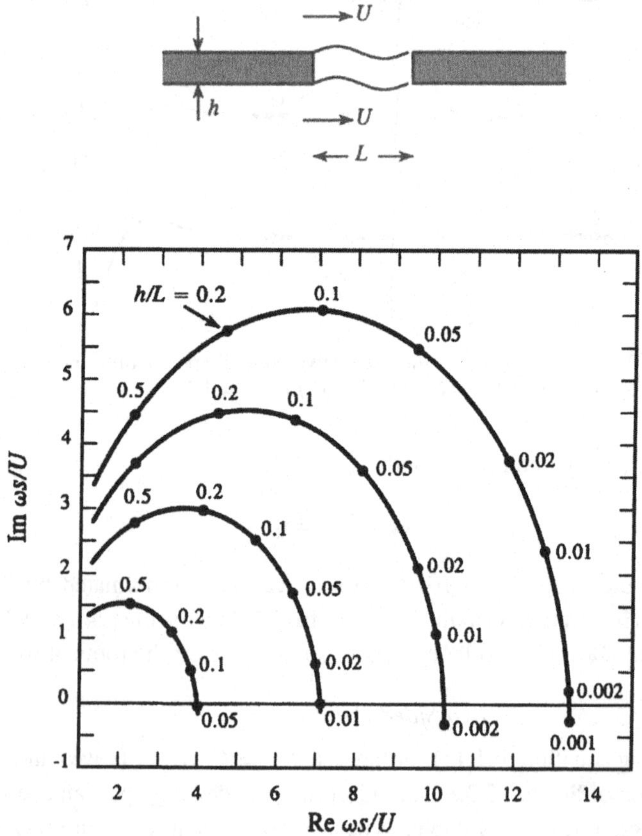

Figure 6.1.3. Poles of $K_R(\omega)$ for two-sided uniform flow for a rectangular aperture with $b/L = 2$ and variable h/L.

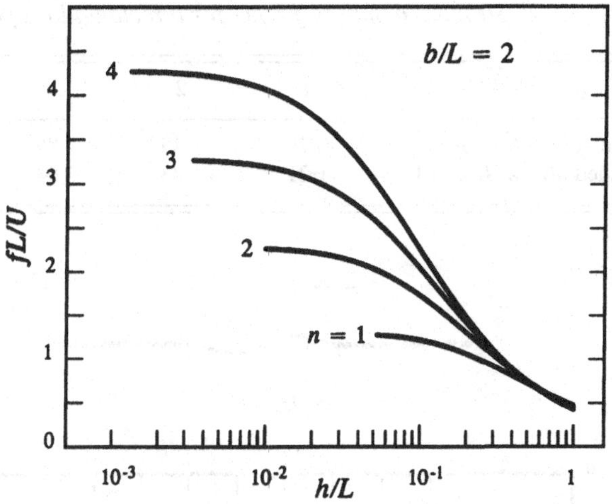

Figure 6.1.4. Strouhal number dependence on wall thickness: $U_+ = U_- = U, b/L = 2$.

is said to determine the nth "operating stage" of the system. In practice, only one or two modes radiate at any one time, and (for fixed h/L) higher frequency modes are successively excited as the flow velocity U increases. When h/L decreases from 0.05, the poles for $n = 1,\ 2,\ 3,$ and so forth successively cross into the lower half-plane; the first four stages are stable when h/L is less than about 10^{-3}. According to Figure 6.1.3 the poles converge onto the imaginary axis when h/L becomes large, although the thin-wall approximation is probably not applicable for h/L as large as 0.5. An alternative representation of these results is shown in Figure 6.1.4, where the dependence of the Strouhal number fL/U on h/L ($f = \text{Re}(\omega)/2\pi$) is shown for the first four operating stages. Each curve starts on the left at that finite, nonzero value of h/L at which the corresponding pole crosses into the upper frequency plane.

One-Sided Mean Flow

Instead of increasing wall thickness from $h = 0$, uniform two-sided flow can also be destabilized by gradually reducing the velocity ratio $\mu = U_-/U_+$ from one to zero, as illustrated in Figure 6.1.5 for the first four operating stages for a very large aspect ratio $b/L = 500$ (although the results are only very weakly dependent on b/L). The Strouhal numbers of the first four operating stages for one-sided flow ($\mu = 0$) are given in Table 6.1.2; they are close to those in Table 6.1.1 for the circular aperture. The Strouhal numbers of successive operating stages differ by about $\frac{1}{2}$. The table also lists the corresponding Strouhal numbers for two-sided flow when $h/L = 0.1$ (from Figure 6.1.4), which indicate that a relatively moderate wall thickness is sufficient to bring the stability

Table 6.1.2. *Strouhal numbers fL/U for a rectangular aperture*

Flow	1	2	3	4
One-sided flow $h = 0$	0.79	1.35	1.89	2.42
Two-sided flow $h/L = 0.1$	1.21	1.73	2.04	2.25

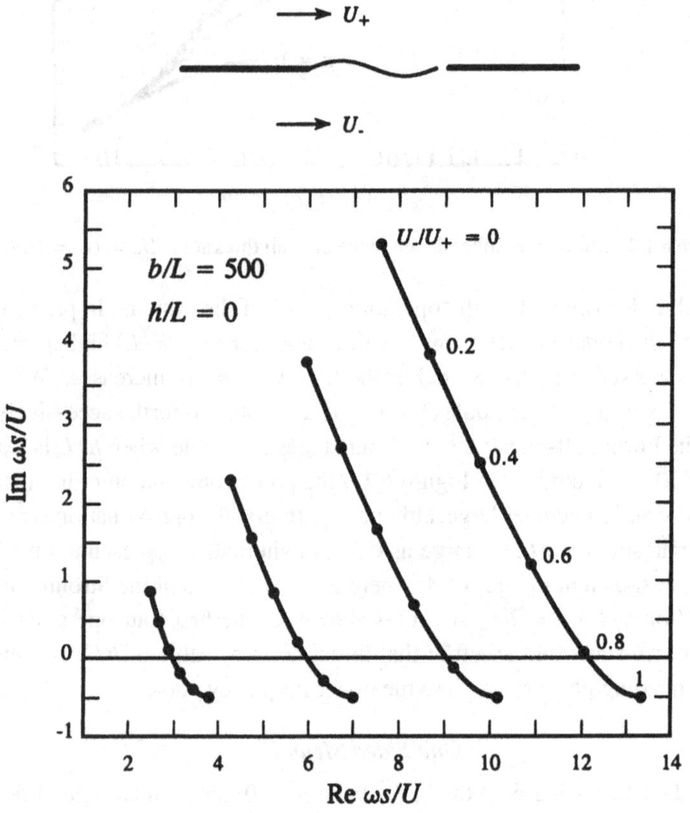

Figure 6.1.5. Poles of $K_R(\omega)$ for variable two-sided uniform flow over a rectangular aperture.

properties of the aperture close to those for one-sided flow, especially at smaller scales associated with larger values of n.

6.1.3 Cavity Depth Modes

When the wall aperture opens into a cavity (in $x_2 < 0$), as indicated in Figure 6.1.6, resonant acoustic modes of the cavity can be excited and maintained

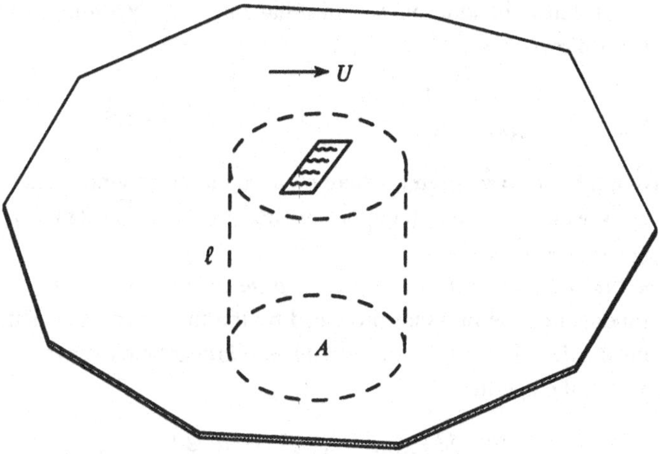

Figure 6.1.6. Cavity-backed wall aperture.

by the mean flow. As the mean velocity increases from zero, oscillations are observed over a range of increasing and discrete frequencies, different resonances progressively growing and decaying in amplitude. The resonances tend to be excited over a much larger range of velocities when the flow is turbulent, especially when the turbulence wall pressure spectrum has significant contributions at the cavity resonance frequencies. These frequencies generally exhibit a weak dependence on flow velocity, but certain other tones, analogous to wall aperture tones, show a marked dependence on velocity. When the resonances are regarded as triggered by an external pressure $p_I(t)$, as before, the volume flux Q through the mouth can again be expressed in the form (6.1.2) where now the conductivity is replaced by the *cavity conductivity* $K_c(\omega)$, which incorporates the influence of cavity modes on Q. Linear theory then associates the self-sustained cavity oscillations with poles of $K_c(\omega)$ in $\operatorname{Im}\omega > 0$.

Let us evaluate

$$K_c(\omega) \equiv \frac{i\omega\rho_o Q(\omega)}{p_I(\omega)} \tag{6.1.12}$$

when the cavity is rigid and *deep*, with depth ℓ and uniform cross-sectional area A, and when the shear layer is modeled as a vortex sheet. This can be done by first calculating the pressures $p_\pm$ and substituting into the definition (5.3.1) of the "bare" aperture conductivity K_R.

Low-frequency resonances are the first to be excited, corresponding to *depth modes*, where $p \approx p(x_2, \omega)$ varies only with x_2 $(-\ell < x_2 < 0)$. Mass conservation at the upper end of the cavity requires that $Q = A(-i/\rho_o\omega)\partial p/\partial x_2$ as

$x_2 \to -0$. Because, in addition, we must have $v_2 = (-i/\rho_o\omega)\partial p/\partial x_2 = 0$ at $x_2 = -\ell$, it follows that

$$p \approx \frac{-i\omega\rho_o Q}{Ak\sin(k\ell)}\cos\{k(x_2+\ell)\}, \quad -\ell < x_2 < 0.$$

The wavenumber $k = \kappa_o + i\alpha_o$, where α_o accounts for boundary layer losses at the cavity walls, and may be approximated by (5.1.32). Thus, $p_-(\omega) \approx (-i\omega\rho_o Q/Ak)\cot(k\ell)$.

Above the aperture, $p_+(\omega) = p_{\mathrm{I}}(\omega) + p'(\omega)$, where p' represents the uniform component of the pressure produced by the flux Q and is calculated by subtracting out the rapidly decaying *hydrodynamic* near field from the monopole radiation from the aperture:

$$p'(\omega) = \frac{-i\omega\rho_o Q}{2\pi|\mathbf{x}|}\left(\mathrm{e}^{i\kappa_o|\mathbf{x}|} - 1\right) \approx \frac{\omega\kappa_o\rho_o Q}{2\pi}, \quad \kappa_o|\mathbf{x}| \ll 1.$$

This locally uniform pressure generated by the aperture flux was ignored in Section 5.3, but it must be retained when account is to be taken of small effects of damping by radiation from the cavity.

Thus, substituting for $p_\pm$ into the definition (5.3.1) and rearranging, we find

$$\frac{p_{\mathrm{I}}(\omega)}{i\omega\rho_o Q} \equiv \frac{1}{K_c} = \frac{1}{K_R} + \frac{i\kappa_o}{2\pi} - \frac{\cot(k\ell)}{kA}. \tag{6.1.13}$$

This result is more conveniently expressed in terms of the reciprocal

$$Z_R(\omega) = \frac{1}{K_R(\omega)} \tag{6.1.14}$$

of the conductivity, which is an *impedance*.

Equation (6.1.13) then equates the impedance $Z_c = 1/K_c$ of the cavity-backed aperture to the sum of its *Rayleigh* impedance Z_R, the *radiation* impedance $i\kappa_o/2\pi$ of the flow, and the *cavity* impedance $-\cot(k\ell)/kA$. When the motion is incompressible ($c_o \to \infty$), the radiation impedance vanishes, but the cavity impedance becomes unbounded and $Q \to 0$.

Thus, we can write

$$Z_c(\omega) = Z_R(\omega) + Z_a,$$

where

$$Z_a = \frac{i\kappa_o}{2\pi} - \frac{\cot(k\ell)}{kA}$$

will be termed the *acoustic* impedance.

The frequencies of self-sustained depth modes are *zeros* of $Z_c(\omega)$ in $\mathrm{Im}\,\omega > 0$. These zeros fall into two sets. The first are close to the poles of $K_R(\omega) = 1/Z_R(\omega)$ and correspond to shear layer instabilities modified by the cavity; the amplitudes of these motions are usually small compared to modes coupled to cavity oscillations. The latter occur near the zeros of $Z_a(\omega)$ (close to the usual acoustic resonances of the cavity) and can attain considerable amplitudes by the accumulation of oscillatory energy within the resonator.

Low-order resonances of a very deep cavity can be calculated as follows. The resonance equation $Z_c(\omega) = 0$ is written

$$\cos(k\ell) - kA\left(Z_R(\omega) + \frac{i\kappa_o}{2\pi}\right)\sin(k\ell) = 0. \tag{6.1.15}$$

For a deep cavity, $k\ell \sim \mathrm{O}(1)$ for the low-order modes. The coefficient of $\sin(k\ell)$ is therefore small provided $AZ_R(\omega)/\ell \ll 1$, which is the case for low-order modes unless the aperture is very small because $Z_R \sim 1/L$, where L is representative of the aperture diameter. In this approximation, the cavity behaves as a pipe open at one end, with the approximate frequency equation

$$\cos\left(k\ell + kA\left(Z_R + \frac{i\kappa_o}{2\pi}\right)\right) = 0.$$

This result indicates that the effective length of the pipe is increased from ℓ by an end correction equal to $\mathrm{Re}(AZ_R)$ and that

$$\frac{\omega\ell}{c_o} \approx \pi\left(n - \frac{1}{2}\right)\left\{1 - \frac{i\alpha_o}{\kappa_o} - \frac{A}{\ell}\left(Z_R + \frac{i\kappa_o}{2\pi}\right)\right\}, \quad n = 1,\ 2, \ldots.$$

The terms in the brace brackets represent a correction to the (neutrally stable) open-pipe resonance frequency $\omega \approx (n - \frac{1}{2})\pi c_o/\ell$, and this value may be used to evaluate the right-hand side. Self-sustained oscillations are possible provided $\mathrm{Im}\,\omega > 0$, that is, provided

$$\mathrm{Im}\, Z_R(x) < -\frac{\kappa_o}{2\pi} - \frac{\ell\alpha_o}{\kappa_o A}, \quad \text{for} \quad \kappa_o\ell = \left(n - \frac{1}{2}\right)\pi. \tag{6.1.16}$$

This condition can be satisfied only if $\mathrm{Im}\, Z_R < 0$, and its implications are discussed in Section 6.1.4.

Example 1. When the cavity of Figure 6.1.6 is open at its lower end, show that the acoustic impedance for depth modes at the aperture exposed to the mean

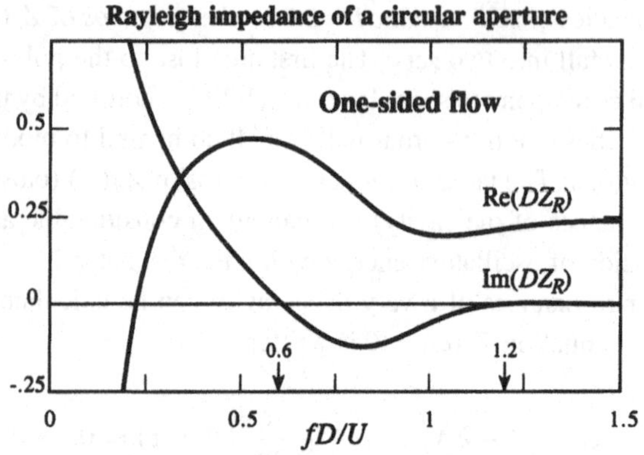

Figure 6.1.7. Rayleigh impedance of a circular aperture

flow is given approximately by

$$Z_a = \frac{i\kappa_o}{2\pi} + \frac{\tan\{k(\ell + \ell' + i\kappa_o A/4\pi)\}}{kA},$$

where ℓ' is the end correction for the lower open end.

Example 2: Rayleigh impedance of a circular aperture The values of $K_R(\omega)$ given in Table 5.3.1 for a circular aperture in a one-sided, high Reynolds number grazing flow permit the Rayleigh impedance $Z_R(\omega) = 1/K_R(\omega)$ to be plotted for real frequencies. This is done in Figure 6.1.7 as a function of the Strouhal number $fD/U \equiv \omega R/\pi U$. Im $Z_R < 0$ for $0.6 < fD/U < 1.2$, and according to (6.1.16) it is within this range of Strouhal numbers that depth modes can be excited in a cavity backing onto the aperture.

Example 3. Calculate the Rayleigh impedance for high Reynolds number flow over a deep rectangular cutout of depth ℓ and uniform rectangular cross section with side of length L ($\ll \ell$) parallel to the mean flow and transverse dimension b (Figure 6.1.8).

In the usual notation, with the shear layer modeled by a vortex sheet, the pressure on the upper surface $x_2 = +0$ of the vortex sheet is (Section 5.3.6)

$$p = p_+ - \frac{\rho_o}{2\pi}\left(\omega + iU\frac{\partial}{\partial x_1}\right)^2 \int_S \frac{\zeta(y_1, y_3, \omega)\, dy_1\, dy_3}{|\mathbf{x} - \mathbf{y}|},$$

$$|x_1| < s, \quad x_2 = +0, \quad |x_3| < b/2,$$

where the integration is over the mouth S, and $s = \frac{1}{2}L$. The motion in the cavity is uniform in the spanwise (x_3-) direction. If the vortex sheet is temporarily replaced by a rigid lid, the local incompressible flow produced by a spanwise

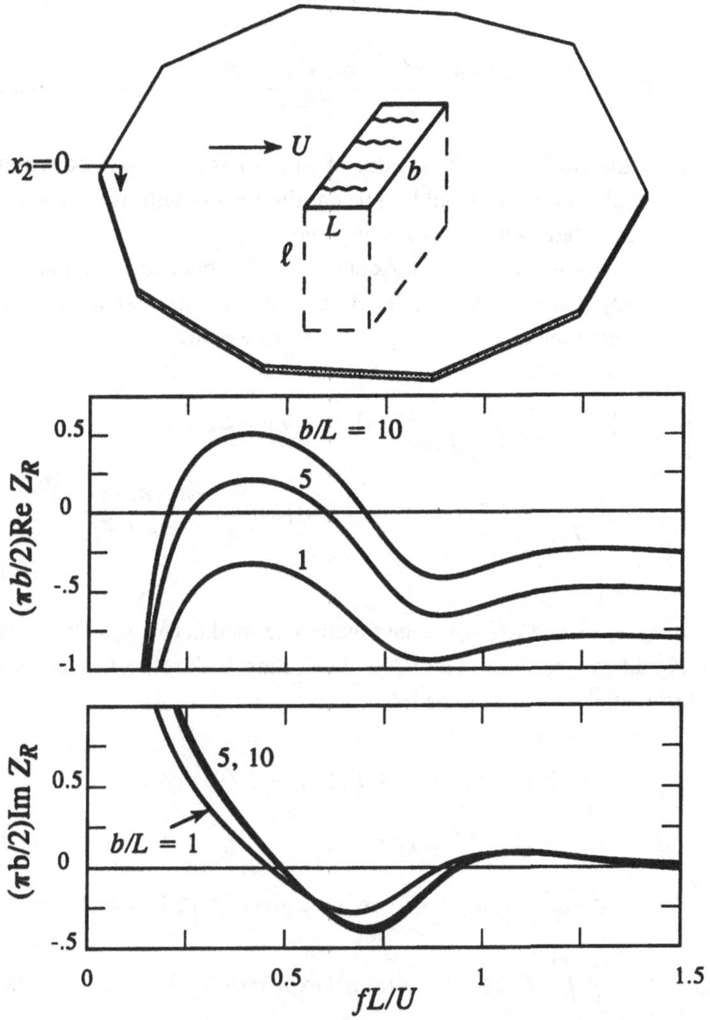

Figure 6.1.8. Rayleigh impedance of the mouth of a rectangular cavity.

line source of unit strength at $x_1 = y_1$, $x_2 = -0$, $|x_3| < b/2$ can be calculated from potential flow theory [4, 17, 404]. By representing the motion of the vortex by a distribution of these sources, the pressure on the lower surface $x_2 = -0$ of the vortex sheet can be found in the form

$$p = p_- - \frac{\rho_o \omega^2 s}{\pi} \int_{-1}^{1} \zeta(\eta)(\ln|\xi - \eta| + \mathcal{L}_-(\xi, \eta))\, d\eta,$$
$$\xi = x_1/s, \quad \eta = y_1/s,$$

where

$$\mathcal{L}_-(\xi,\eta) = \ln\left(\frac{4\sin\{\pi(\xi-\eta)/4\}\cos\{\pi(\xi+\eta)/4\}}{\xi-\eta}\right), \qquad (6.1.17)$$

where p_- is the uniform, limiting value of $p(\mathbf{x},\omega)$ as $x_2/L \to -\infty$ within the cavity. In applications p_- would represent the locally uniform pressure near the mouth generated by an acoustic depth mode.

The equation governing the vortex sheet is now obtained by equating pressures on its upper and lower surfaces. In the one-dimensional approximation, in which we integrate over $-\frac{1}{2}b < x_3,\ y_3 < \frac{1}{2}b$, we find

$$\left[\sigma^2 + \left(\sigma + i\frac{\partial}{\partial\xi}\right)^2\right]\int_{-1}^{1}\zeta(\eta)\{\ln|\xi-\eta| + \mathcal{L}(\xi,\eta)\}\,d\eta$$
$$+\sigma^2\int_{-1}^{1}\zeta(\eta)\mathcal{L}_-(\xi,\eta) - \mathcal{L}(\xi,\eta)\}\,d\eta = \frac{-\pi s(p_+ - p_-)}{\rho_o U^2},$$
$$|\xi| < 1, \qquad (6.1.18)$$

where $\sigma = \omega x/U$, and $\mathcal{L}(\xi,\eta)$ is the function defined in (6.1.6). This is cast in the standard form similar to (6.1.9) by integrating with respect to the second-order differential operator on the left:

$$\int_{-1}^{1}\zeta'(\eta)[\ln|\xi-\eta| + \mathcal{L}(\xi,\eta) + \mathcal{K}(\xi,\eta)]\,d\eta$$
$$+\lambda_1 e^{i\sigma_1\xi} + \lambda_2 e^{i\sigma_2\xi} = 1, \qquad |\xi| < 1, \qquad (6.1.19)$$

where $\zeta' = -2\zeta\rho_o\omega^2 s/\pi(p_+ - p_-)$, $\sigma_{1,2}$ are given by (6.1.8) with $\mu = 0$, and

$$\mathcal{K}(\xi,\eta) = \frac{\sigma}{2}\int_{-1}^{1}\{\mathcal{L}_-(\xi,\eta) - \mathcal{L}(\xi,\eta)\}\exp(i\sigma(\xi-\lambda) - \sigma|\xi-\lambda|)\,d\lambda.$$

Equation (6.1.19) is solved numerically subject to the Kutta condition $\zeta' = \partial\zeta'/\partial\xi = 0$ at $\xi = -1$. The Rayleigh impedance is then calculated from the formula

$$Z_R(\omega) \equiv \frac{1}{K_R(\omega)} = -\frac{\pi b}{2}\Big/\int_{-1}^{1}\zeta'(\eta)\,d\eta. \qquad (6.1.20)$$

$Z_R(\omega)$ is plotted in Figure 6.1.8 for three values of the aspect ratio b/L.

Example 4. Repeat the calculation of Example 3 for a side-branch opening in a rectangular duct, the opening being of rectangular cross-section of length

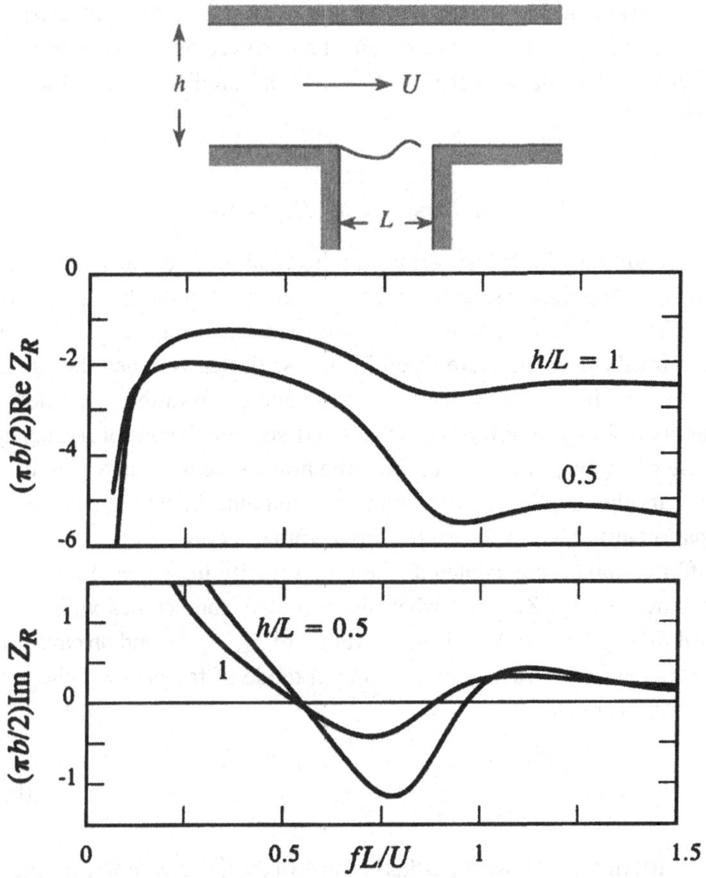

Figure 6.1.9. Rayleigh impedance of the mouth of a rectangular side branch.

L in the streamwise direction and span b. Assume the duct has height h and span b (Figure 6.1.9). Take the coordinate origin in the center of the side-branch mouth. When $p_+ - p_-$ is uniform across the span, the motion is entirely two dimensional, and the pressure on the upper surface of the vortex sheet can be written

$$p = p_+ - \frac{\rho_o \omega^2 s}{\pi}\left(\sigma + i\frac{\partial}{\partial \xi}\right)^2 \int_{-1}^{1} \zeta(\eta)(\ln|\xi - \eta| + \mathcal{L}_+(\xi, \eta))\, d\eta,$$

$$|\xi| = |x_1/s| < 1,$$

where

$$\mathcal{L}_+(\xi, \eta) = \ln\left(\frac{2\sinh\{(\pi L/4h)(\xi - \eta)\}}{\xi - \eta}\right).$$

The pressure on the lower surface of the vortex sheet is the same as in Example 3. The equation for ζ may now be obtained and solved as in Example 3. $Z_R(\omega)$ is independent of the aspect ratio b/L; typical predictions are illustrated in Figure 6.1.9.

6.1.4 Flow-Excited Depth Modes

In the condition (6.1.16) governing the excitation of depth modes by flow, the terms on the right respectively represent dissipation of the cavity mode by radiation losses into the flow and by damping in the interior cavity wall boundary layers. Im $Z_R(\omega)$ is negative when Im $K_R < 0$, that is, when the oscillatory motion of the boundary layer is extracting energy from the mean flow. The inequality (6.1.16) therefore states that a self-sustained mode of frequency ω is possible when the power derived from the flow exceeds the dissipation. When this occurs the amplitude of the motion and sound increases until nonlinear dissipation just balances the excess input from the flow.

Different modes are excited as the flow velocity increases. Figures 6.1.7–6.1.9 show that Im $Z_R < 0$ when the Strouhal number lies within a finite interval $S_1 < fL/U < S_2$, where the values of S_1, S_2 depend on cavity mouth geometry. Thus, in order to excite a depth mode of frequency f the velocity must lie in the range

$$\frac{fL}{S_2} < U < \frac{fL}{S_1}. \tag{6.1.21}$$

The actual range is narrower than this because of the linear damping mechanisms in (6.1.16). The formula shows that an increasing flow velocity progresses through distinct ranges where higher and higher order modes of the cavity can be excited, which is in qualitative accord with observation. Unfortunately, however, in most experiments [383] on depth modes the mean shear layer is too thick to be treated as a vortex sheet so that detailed quantitative comparisons are not possible.

6.1.5 The Shallow-Wall Cavity

At low Mach numbers, the mechanism of sound generation by flow over a cavity whose depth ℓ is very much smaller than the acoustic wavelength is similar to that governing aperture tones in a thin wall because the cavity is "too small" to store acoustic energy at the observed frequencies. Most experiments on wall cavities have been conducted at $M = U/c_o > 0.2$ and are relevant to vibration problems (at higher frequencies) experienced by exposed aircraft structures (Section 6.2).

Let the cavity have the dimensions shown in Figure 6.1.8, and take the coordinate origin in the center of the cavity mouth. At very small Mach numbers the acoustic pressure (6.1.1) at large distances must be taken in the more general form

$$p(\mathbf{x},t) \approx \frac{\rho_o}{2\pi|\mathbf{x}|}\frac{\partial}{\partial t}\int_S v_2(\mathbf{y}, t - |\mathbf{x}-\mathbf{y}|/c_o)\,dy_1\,dy_3, \text{ as } |\mathbf{x}| \to \infty, \tag{6.1.22}$$

where the integration is extended over any section S of the plane $y_2 = 0$ spanning the mouth. When the acoustic wavelength $\gg L$ or b, this expression would normally be representative of the monopole sound (6.1.1), where the volume flux $Q(t) = \int v_2(\mathbf{y},t)\,dy_1\,dy_3$. However, at very low frequencies the motion within the cavity becomes indistinguishable from that of an incompressible fluid, and there can be no net flux through the mouth ($Q \equiv 0$). The cavity must then radiate as a *dipole* whose pressure field is determined by expanding the integrand of (6.1.22) in powers of $\mathbf{y}/c_o$, which gives

$$p(\mathbf{x},t) \approx \frac{x_j}{2\pi c_o|\mathbf{x}|^2}\frac{\partial F_j}{\partial t}(t - |\mathbf{x}|/c_o) \quad |\mathbf{x}| \to \infty,$$

$$F_j = \rho_o\int_S y_j\frac{\partial v_2}{\partial t}(\mathbf{y},t)\,dy_1\,dy_3 \quad (j = 1 \text{ or } 3).$$

The axis of the dipole is parallel to the wall; the reader can show that its strength **F** is just equal to the unsteady *drag* exerted on the fluid by the cavity.

When $Q = 0$, a uniform pressure $p_I(t)$ applied in the vicinity of the cavity cannot exert a drag force nor produce any motion of an initially undisturbed shear layer over the mouth. The simplest disturbance that can induce drag is a *tangential pressure force* $-\partial p_I/\partial x_j$. Consider, in particular, a *uniform* pressure force applied in the direction of the mean flow; then $\mathbf{F} \equiv (F_1, 0, 0)$, and by analogy with equation (6.1.2) we can introduce a drag coefficient $D(\omega)$ defined such that

$$F_1(t) = -\int_{-\infty}^{\infty} D(\omega)p_I'(\omega)e^{-i\omega t}\,d\omega,$$

where $p_I'(\omega) \equiv \partial p_I(\omega)/\partial x_1$ is independent of $\mathbf{x}$. At high Reynolds numbers, when the shear layer is approximated by a linearly disturbed vortex sheet, the frequencies of self-sustained radiation from the cavity are determined by the poles of $D(\omega)$ in Im $\omega > 0$.

The motion of the vortex sheet is approximated, as in Section 6.1.2, by neglecting its dependence on the spanwise coordinate x_3, in which case

$$D(\omega) = \frac{\rho_o \omega^2 b}{p_o'(\omega)} \int_{-s}^{s} y_1 \zeta(y_1)\, dy_1, \tag{6.1.23}$$

where $\zeta(y_1)$ is the displacement produced by the uniform pressure gradient $p_{\mathrm{I}}'(\omega)$.

To simplify the calculation, the direct influence of the cavity base (at $x_2 = -\ell$) on the motion of the vortex sheet will be neglected by assuming that $\ell \gg L$. This is done by considering first *compressible* motion within the cavity (for which $Q \neq 0$) and subsequently obtaining the solution for incompressible cavity motion by taking the limit $\omega\ell/c_o \to 0$. When $\omega\ell/c_o$ is small, the motion of the sheet will excite depth modes, and the displacement ζ will satisfy equation (6.1.18) with

$$p_- = \frac{-i\rho_o c_o Q}{A} \cot(\kappa_o \ell), \qquad p_+ = x_1 p_o' + \omega\kappa_o\rho_o Q/2\pi,$$

where $A = bL$ is the cavity cross-section, thermo-viscous losses at the interior cavity walls may be ignored, and, without loss of generality, it is assumed that $p_{\mathrm{I}} = 0$ at $x_1 = 0$.

To solve equation (6.1.18), we set

$$\zeta = \frac{i\pi Q Z_a}{2\omega s}\zeta_1 - \frac{\pi p_{\mathrm{I}}'}{2\rho_o\omega^2}\zeta_2,$$

where $Z_a \equiv i\kappa_o/2\pi - \cot(\kappa_o\ell)/\kappa_o A$ is the acoustic impedance of the cavity. Then, ζ_1 is the solution of (6.1.19) satisfying the Kutta condition at $\xi = -1$, and ζ_2 is the corresponding solution when the right-hand side of (6.1.19) is replaced by $\xi - i/\sigma$. These solutions can be used to evaluate the drag coefficient $D(\omega)$ from the definition (6.1.23), and the volume flux from $Q = -i\omega b \int_{-s}^{s} \zeta(y_1)\, dy_1$. We find

$$D = \frac{\pi b s^2}{2} \frac{(\pi b/2) Z_a [\mathrm{I}_1 \mathrm{M}_2 - \mathrm{I}_2 \mathrm{M}_1] - \mathrm{M}_2}{1 - (\pi b/2) Z_a \mathrm{I}_1},$$

$$Q = \frac{(i\pi b s p_o'/2\rho_o\omega)\mathrm{I}_2}{1 - (\pi b/2) Z_a \mathrm{I}_1}, \tag{6.1.24}$$

where

$$\mathrm{I}_j = \int_{-1}^{1} \zeta_j(\eta)\, d\eta, \quad \mathrm{M}_j = \int_{-1}^{1} \eta\zeta_j(\eta)\, d\eta, \quad j = 1 \text{ or } 2.$$

The values of the moments I_j, M_j depend on the shape of the cavity and on the hydrodynamic flow in the mouth but are independent of fluid compressibility. Hence, because the acoustic impedance $Z_a \to \infty$ in the limit $\kappa_o \ell \to 0$ of incompressible cavity flow, it follows that $Q \to 0$, as expected, and that the drag coefficient D approaches the limiting value given by

$$\frac{D}{(\pi b s^2/2)} = \frac{M_1 I_2 - M_2 I_1}{I_1} \equiv \Gamma_D(\omega) - i\Delta_D(\omega), \qquad (6.1.25)$$

where Γ_D and $-\Delta_D$ are nondimensional real and imaginary parts of $D(\omega)$.

$\Gamma_D(\omega)$ and $\Delta_D(\omega)$ are plotted for real ω in Figure 6.1.10a for a cavity aspect ratio $b/L = 1$. It may easily be shown from the definition of D that the cavity absorbs energy from the applied pressure gradient p'_I when $\Delta_D(\omega) > 0$, for

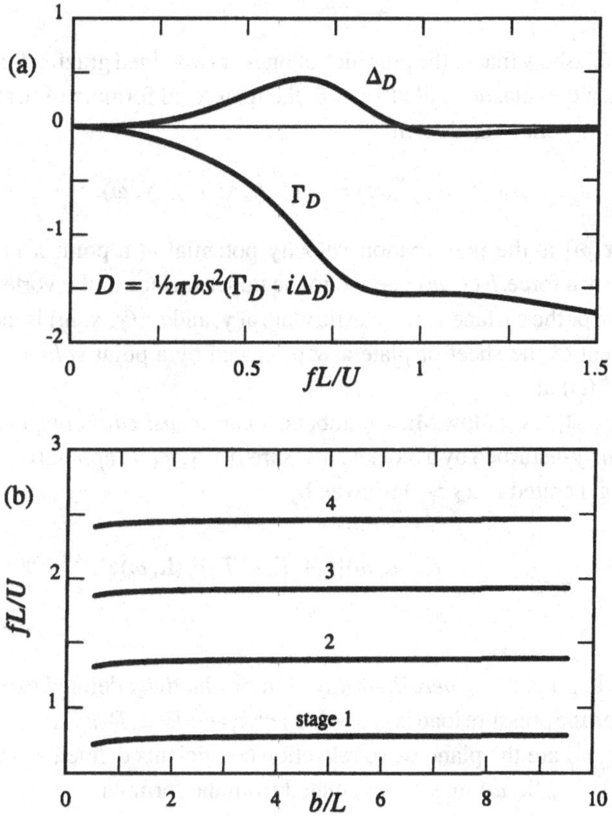

Figure 6.1.10. (a) Drag coefficient of a rectangular wall aperture and (b) shallow cavity Strouhal numbers for variable b/L.

$\omega s/U < 2.8$. The motion is unstable, however, because $D(\omega)$ has poles in $\mathrm{Im}(\omega) > 0$. The instability Strouhal numbers $fL/U \equiv \mathrm{Re}(\omega s/\pi)$ defined by these poles are illustrated in Figure 6.1.10b for the first four operating stages. fL/U increases very slowly with the aspect ratio, in broad agreement with experiment at higher Mach numbers [397], and jumps in value by about one-half between adjacent operating stages. These results are discussed further in Section 6.2.4.

Example 5. A steady incompressible flow over a shallow cavity is subject to a uniform, time-harmonic pressure gradient $\partial p_{\mathrm{I}}/\partial x_1 = \mathrm{Re}\{p_{\mathrm{I}}'(\omega)\mathrm{e}^{-i\omega t}$. Show that the power Π_ω absorbed from the incident field by the cavity flow is given by

$$\Pi_\omega = \frac{-|p_{\mathrm{I}}'|^2}{2\rho_o\omega}\mathrm{Im}\, D(\omega).$$

Example 6. Show that in the presence of one- or two-sided grazing flow over an aperture in a thin elastic wall at $x_2 = 0$, the reciprocal formula (4.3.1) assumes the following generalized form

$$\rho_o\varphi(\mathbf{x},\omega)q^R(\omega) = -\zeta^R(\mathbf{y},\mathbf{x};\omega)F(\mathbf{y},\omega),$$

where $\varphi(\mathbf{x},\omega)$ is the perturbation velocity potential at a point $\mathbf{x}$ in the fluid generated by a force $F(\mathbf{y},\omega)$ applied to the upper surface of the vortex sheet in the aperture or the surface of the elastic plate at $\mathbf{y}$, and $\zeta^R(\mathbf{y},\mathbf{x};\omega)$ is the flexural displacement of the sheet or plate at $\mathbf{y}$ produced by a point *volume* source of strength $q^R(\omega)$ at $\mathbf{x}$.

Deduce that, at very low Mach numbers, for an *acoustically compact* aperture that is *linearly* disturbed by a blocked pressure $p_s(x_1, x_3, t)$ applied on $x_2 = +0$, the sound generated in $x_2 > 0$ is given by

$$p(\mathbf{x},t) \approx \frac{-1}{4\pi|\mathbf{x}|}\iint_{-\infty}^{\infty} K_R(\mathbf{k},\omega)[1+\mathcal{R}-\mathcal{T}]p_s(\mathbf{k},\omega)\mathrm{e}^{-i\omega(t-|\mathbf{x}|/c_o)}\,d^2\mathbf{k}\,d\omega,$$
$$|\mathbf{x}| \to \infty,$$

where $K_R(\mathbf{k},\omega)$ is the *generalized Rayleigh conductivity* defined exactly as in (5.3.1) when the pressure load is $(p_+ - p_-)\mathrm{e}^{i\mathbf{k}\cdot\mathbf{x}}$, and $\mathcal{R} \equiv \mathcal{R}(\kappa_o \sin\Theta, \omega)$, $\mathcal{T} \equiv \mathcal{T}(\kappa_o \sin\Theta, \omega)$ are the plane wave reflection coefficients defined as in (4.1.29) and (4.1.30). $K_R(\mathbf{k},\omega)$ may be calculated from the formula

$$K_R(\mathbf{k},\omega) = \rho_o\omega^2\iint_{-\infty}^{\infty}\zeta^R(x_1,x_2,\omega)\mathrm{e}^{i\mathbf{k}\cdot\mathbf{x}}\,dx_1\,dx_2,$$

where $\zeta^R(x_1, x_2, \omega)$ is the displacement of the vortex sheet produced by a *spatially uniform* time-harmonic pressure load of *unit amplitude* when the *direction of the mean flow is reversed.* The displacement ζ^R is subject to a Kutta condition at the section of the aperture edge where shedding occurs in reverse flow.

Example 7. For a rectangular aperture in a thin wall, in the presence of one-sided flow ($U_+ = U,\ U_- = 0$), show that when aspect ratio $b/L \gg 1$

$$K_R(k_1, 0, \omega) \approx b\Im(\sigma_0, \omega),$$

where $\Im(\sigma_0, \omega)$ is given by (5.3.44) with $\sigma_+ = \omega s/U,\ \sigma_- = 0,\ \sigma_0 = k_1 s$, and $\Psi = \ln(8b/Le)$.

Example 8. Calculate the sound produced when the line vortex defined by the blocked pressure distribution (4.3.7) convects in a uniform mean flow at speed U in the x_1-direction over the aperture of Example 7 in a thin rigid plane (Figure 6.1.11). Assume the vortex translates at speed $U + v$ along a straight trajectory parallel to the x_1-axis at a distance d above the wall. Take $\mathcal{R} = 1,\ \mathcal{T} = 0$ in the formulae of Example 6, and ignore the contribution to the radiation from poles of the conductivity in $\text{Im}\,\omega > 0$.

The three typical acoustic pressure signatures in Figure 6.1.11 are calculated by making the following approximations for $K_R(\mathbf{k}, \omega)$: (i) $K_R(\mathbf{k}, \omega) = \pi b/2 \ln(8b/Le)$, its classical Rayleigh potential flow value in the absence of flow; (ii) $K_R(\mathbf{k}, \omega) = K_R(\mathbf{0}, \omega)$, its value in the presence of one-sided flow in the low wavenumber approximation $k_1 L \ll 1$; and (iii) its value given in Example 7. Because the aperture flow is unstable, the complete acoustic field will presumably consist of narrow band components at frequencies determined by the Strouhal numbers of the aperture operating stages, whose amplitudes are controlled by nonlinear mechanisms, together with the forced radiation field shown in the figure and determined by integration along the real ω-axis.

6.2 Nonlinear Theory of Cavity Resonances

To make predictions of flow-induced cavity resonances that more closely conform to observation it is necessary to account for both the thickening of the shear layer across the mouth and its ultimate breakup into discrete vortices. In practice, such a detailed knowledge of the hydrodynamic flow can be derived only from a full numerical simulation of the motion. However, an adequate understanding can be obtained from an analysis of simplified models, which we now proceed to describe.

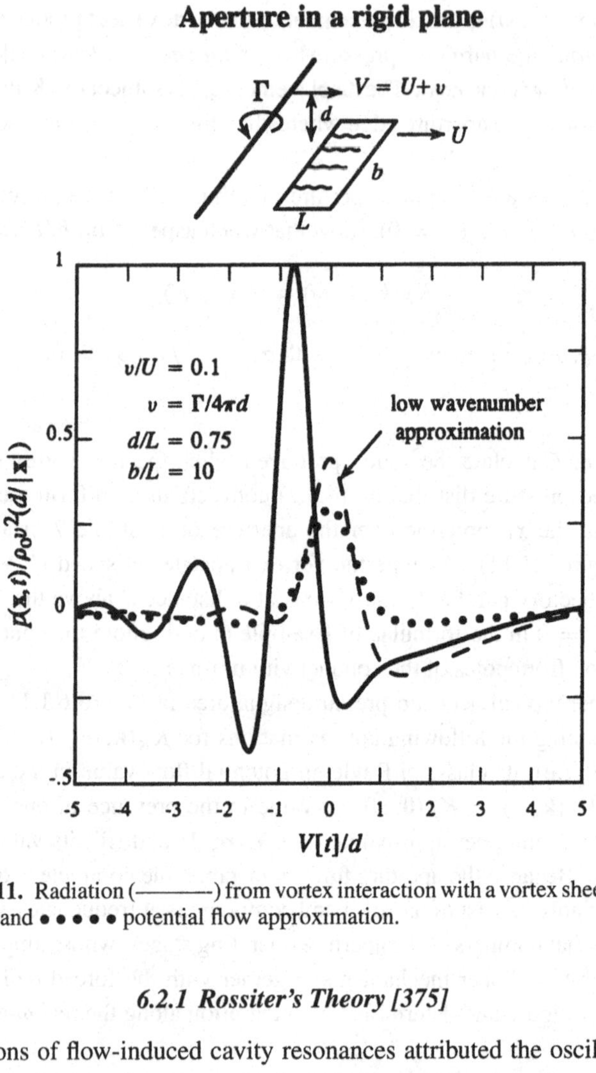

Figure 6.1.11. Radiation (———) from vortex interaction with a vortex sheet spanning an aperture and • • • • • potential flow approximation.

6.2.1 Rossiter's Theory [375]

Early notions of flow-induced cavity resonances attributed the oscillations to broadband excitation of cavity modes by turbulence in the shear layer. We have seen, however, that oscillations can be maintained when the flow is laminar and, indeed, it is often found that laminar flow resonances are more intense.

For shallow cavities ($\ell/L < 1$ in Figure 6.1.8), flow-excited tones do not generally correspond to acoustic modes of the cavity; they are not usually harmonically related and are more analogous to the shear tones and shallow cavity tones discussed in Sections 6.1.1, 6.1.2, and 6.1.5. Tonal Strouhal numbers fL/U are typically found to lie within certain well-defined bands when plotted against flow Mach number. The occurrence of these bands is consistent with an explanation of flow-maintained resonances in terms of a feedback loop [375], according to which discrete vortices are formed periodically just downstream of the leading edge of the cavity and proceed to convect toward the trailing edge. A sound pulse is generated when a vortex arrives at the trailing edge, which propagates upstream within the cavity and induces separation of the boundary layer just upstream of the edge. A vortex travels across the cavity in time L/U_c, where observation indicates that the convection velocity $U_c \approx 0.4U$ to $0.6U$, whereas the sound radiates back to the leading edge in time L/c_o. The returning sound will impinge on the leading edge at just the right instant to reinforce the periodic shedding provided the frequency f satisfies *Rossiter's equation*

$$\frac{L}{U_c} + \frac{L}{c_o} = \frac{n}{f}, \quad n = 1, 2, \ldots. \tag{6.2.1}$$

When predictions of this formula are compared with experiment [5], it is necessary to replace n by the factor $n - \beta$, where β is a constant "phase lag" that depends on cavity depth; β/f is the aggregate time delay between (i) the arrival of a vortex at the trailing edge and the emission of the main acoustic pulse and (ii) the arrival of the sound at the leading edge and the release of new vorticity.

Experiments at high subsonic Mach numbers [389, 397] actually relate measured Strouhal numbers to the following modified Rossiter's equation:

$$\frac{fL}{U} = \frac{n - \beta}{\left\{\frac{U}{U_c} + \frac{M}{\sqrt{1+(\gamma-1)M^2/2}}\right\}}, \quad n = 1, 2, \ldots, \tag{6.2.2}$$

where $M = U/c_o$, and γ is the ratio of specific heats of the fluid. The square root in the denominator accounts for an increased sound speed at the elevated fluid temperatures occurring at high Mach numbers. When $M > 0.2$, and for shallow, rectangular cavities with $\ell/L < 1$, predictions of Rossiter's equation agree well with observations when $\beta \approx \frac{1}{4}$ and $U_c/U \approx 0.6$, independently of the temperature of the free stream, the transverse cavity width b, and the Reynolds number UL/ν [397]. These tones are generated by the feedback cycle discussed earlier; *cavity acoustic modes* (whose frequencies are determined by the interior cavity dimensions) are unimportant unless $\ell/L > \frac{2}{5}$, when they can dominate the radiation, provided the Strouhal number satisfies Rossiter's equation (6.2.2). For shallow cavities at flow speeds up to and including the speed of sound, the radiation is usually dominated by the second and third Rossiter modes ($n = 2, 3$). Average-measured Strouhal numbers for cavities with $\frac{1}{6} < \ell/L < \frac{2}{3}$ (based on data from [397]) are compared in Figure 6.2.1 with the first four Rossiter modes when $\beta = 0.25$ and $U_c/U = 0.6$.

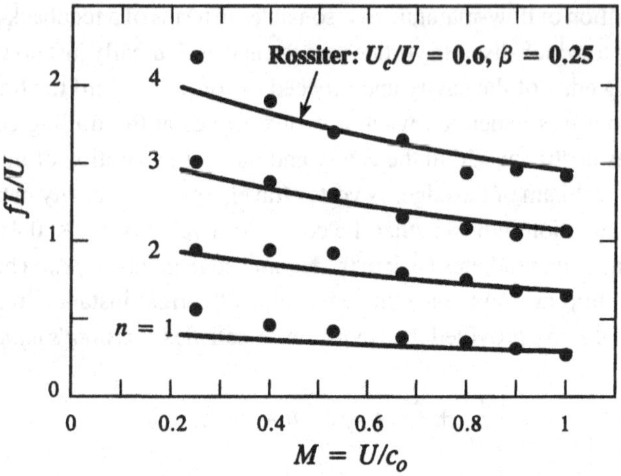

Figure 6.2.1. Shallow-cavity Strouhal numbers for $\frac{1}{6} < \ell/L < \frac{2}{3}$ [397].

6.2.2 Modeling Cavity Resonances at Low Mach Numbers

At low Mach numbers, vortex sound theory (Chapter 3) can be used to investigate the main properties of cavity resonances. To obtain realistic estimates, the unsteady flow over the cavity should be determined numerically. The procedure described in Section 5.2.5, involving the separate computation of hydrodynamic and acoustic modes, has been applied to evaluate the energy transfer integral (5.2.6) to identify Strouhal numbers at which the power flux $-\Pi_\omega$ from the hydrodynamic to the acoustic fields is a maximum [391, 392, 394, 395, 405]. This calculation can be done analytically for a cavity of rectangular cross-section by assuming that the shear layer has evolved into a periodic array of equal vortices of circulation Γ convecting across the mouth at a constant velocity U_c (Figure 6.2.2a). If the motion is periodic with frequency f, a discrete vortex is formed at the upstream edge of the mouth during time $1/f$. In a first approximation, when the flow at the edge is regarded as laminar, shed vorticity initially convects in a thin sheet of strength U at speed $1/2U$ so that the net circulation Γ around a contour enclosing all of the vorticity released in time $1/f$ is equal to $U \times \frac{U}{2} \times \frac{1}{f}$, that is,

$$\Gamma = \frac{U^2}{2f}. \tag{6.2.3}$$

Take coordinate axes with x_1 in the mean flow direction, x_2 normal to the wall, and let the origin be at the center of the cavity mouth. The vorticity

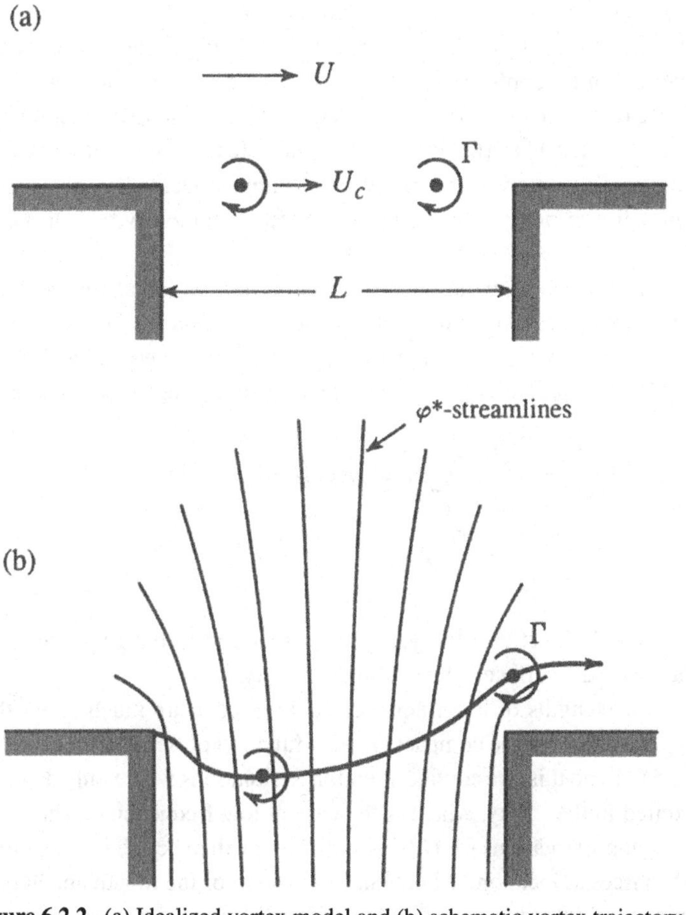

Figure 6.2.2. (a) Idealized vortex model and (b) schematic vortex trajectory.

distribution is then

$$\boldsymbol{\omega}(\mathbf{x}, t) \approx \sum_{n=-\infty}^{\infty} \frac{-U^2}{2f}\delta(x_2)\delta\big(x_1 + s - U_c(t - n/f)\big)\mathbf{i}_3, \quad |x_3| < \frac{1}{2}b, \tag{6.2.4}$$

where $\mathbf{i}_3$ is a unit vector in the x_3-direction (out of the paper in Figure 6.2.2) and $s = \frac{1}{2}L$. According to this highly simplified picture of the flow, the vortices lie in the plane of the wall; the velocities induced by those vortices lying upstream and downstream of the cavity are therefore cancelled by equal and opposite velocities induced by their images in the wall. Thus, the nth vortex effectively appears spontaneously at the upstream edge $x_1 = -s$ at time $t = n/f$ and disappears at time L/U_c later. In reality, vorticity is released smoothly, in a manner that must conform with the Kutta condition when the flow is regarded as inviscid (Section 3.2.5). A correction (to be discussed below) must be therefore applied when (6.2.4) is used to calculate the generated sound and local hydrodynamic motions in the cavity. The simple model (6.2.4) also ignores the tendency for vorticity to be violently ejected from the cavity near the trailing edge.

When M is small enough that the convection of sound can be neglected, the pressure fluctuations produced by the interaction of the vorticity (6.2.4) with the cavity is given at high Reynolds number by equation (3.3.2), with $\mathbf{U} = (U_c, 0, 0)$, where $B(\mathbf{x}, t) \approx p(\mathbf{x}, t)/\rho_o$ outside the source region; that is,

$$\frac{p(\mathbf{x}, t)}{\rho_o U^2} \approx \frac{U_c}{2f} \sum_n \int \delta(y_2)\delta(y_1 + s - U_c(\tau - n/f)) \times \frac{\partial G}{\partial y_2}(\mathbf{x}, \mathbf{y}, t - \tau)\, d^3\mathbf{y}\, d\tau\,, \qquad (6.2.5)$$

where Green's function $G(\mathbf{x}, \mathbf{y}, t - \tau)$ is required to have vanishing normal derivative on the surfaces of the wall and cavity.

The wavelengths of low-frequency cavity modes are much larger than the cavity mouth so that the compact Green's function of Section 1.11 can be used in (6.2.5). To do this, we confine attention to situations where only depth modes are excited in the cavity, which is the case at low frequencies. The following compact approximations for G (Example 1) may then be used for aerodynamic sound sources at locations $\mathbf{y}$ in the neighborhood of the mouth and at observer positions $\mathbf{x}$ far from the mouth, either within the cavity or in free space:

Cavity interior,

$$G(\mathbf{x}, \mathbf{y}, t - \tau) \approx \frac{-i\varphi^*(\mathbf{y})}{2\pi A}\frac{\partial}{\partial t}\int_{-\infty}^{\infty} \frac{\cos\{\kappa_o(x_2 + \ell)\}\mathrm{e}^{-i\omega(t-\tau)}\, d\omega}{\omega\cos\{\kappa_o(\ell + \ell' + i\kappa_o A/2\pi)\}}, \quad -\ell < x_2 < 0; \qquad (6.2.6)$$

Free space,

$$G(\mathbf{x}, \mathbf{y}, t - \tau) \approx \frac{\varphi^*(\mathbf{y})}{(2\pi)^2 c_o|\mathbf{x}|}\frac{\partial^2}{\partial t^2}\int_{-\infty}^{\infty} \frac{\sin(\kappa_o\ell)\mathrm{e}^{-i\omega(t-\tau-|\mathbf{x}|/c_o)}\, d\omega}{\omega\cos\{\kappa_o(\ell + \ell' + i\kappa_o A/2\pi)\}} + \frac{x_j Y_j(\mathbf{y})}{2\pi c_o|\mathbf{x}|^2}\delta'(t - \tau - |\mathbf{x}|/c_o), \quad |\mathbf{x}| \to \infty, \quad j = 1, 3, \qquad (6.2.7)$$

where the prime on the δ-function denotes differentiation with respect to the argument.

In these expressions, $A = bL$, ℓ' is the end correction of the mouth, calculated in the absence of flow and vortex shedding, and $\varphi^*(\mathbf{x})$ is the velocity potential of the ideal incompressible flow from the mouth produced by a uniform flow from the cavity interior that has unit speed in the x_2-direction inside the cavity. The function $Y_j(\mathbf{x})$ ($j = 1$ or 3) is the velocity potential of incompressible flow over the cavity having unit speed in the j-direction parallel to the wall and is exponentially small within the cavity [4, 17, 404].

The first term on the right of (6.2.7) determines the monopole response of the cavity to aerodynamic forcing, which dominates the radiation when characteristic source frequencies satisfy $\kappa_o\ell \geq 1$, for deep, resonant cavities. It is negligible for shallow cavities ($\kappa_o\ell \to 0$), when the radiation is governed by the second term, which is a dipole orientated in the j-direction. This dipole determines the radiation produced by the unsteady drag exerted on the flow by the cavity and is the only relevant source for shallow cavities at low Mach numbers. The interior Green's function (6.2.6) is valid only for deep cavities. In both formulae, damping due to boundary layer losses at cavity interior walls is neglected, but losses caused by radiation into the flow are included (and correspond to the imaginary term $i\kappa_o A/2\pi$ in the denominators of the integrands).

Example 1. Show that the velocity potential produced by a time-harmonic point source at $\mathbf{x}$ in the ambient fluid is given as a function of $\mathbf{y}$ in the neighborhood of the rectangular mouth of a *deep* cavity (of the type shown in Figure 6.2.2) by

$$G(\mathbf{y}, \mathbf{x}; \omega) \approx \frac{-e^{i\kappa_o|\mathbf{x}|}}{2\pi|\mathbf{x}|} + C\left(\varphi^*(\mathbf{y}) - \frac{i\kappa_o A}{2\pi}\right) + \frac{i\kappa_o x_i Y_j(\mathbf{y}) e^{i\kappa_o|\mathbf{x}|}}{2\pi|\mathbf{x}|^2},$$

$$\text{for} \quad \kappa_o\sqrt{A} \ll 1, \quad \kappa_o|\mathbf{x}| \to \infty,$$

where the constant C determines the amplitude of the monopole flow induced in the mouth of the cavity. For a deep cavity, the interior motion consists of the depth mode

$$G(\mathbf{y}, \mathbf{x}; \omega) \approx C' \cos\{\kappa_o(y_2 + \ell)\}.$$

The values of C, C' are found by matching the two representations of G and $\partial G/\partial y_2$ in the acoustically compact interval $\sqrt{A} \ll |y_2| \ll 1/|\kappa_o|$ just inside the cavity, where $Y_j(\mathbf{y})$ is exponentially small.

Deduce the compact free-space Green's function (6.2.7) from the reciprocal theorem (Section 1.9) and the Green's function Fourier inversion formula (1.7.19)

$$G(\mathbf{x}, \mathbf{y}, t-\tau) = (-1/2\pi)\int_{-\infty}^{\infty} G(\mathbf{x}, \mathbf{y}; \omega)\mathrm{e}^{-i\omega(t-\tau)}\, d\omega.$$

A similar procedure can be used to derive (6.2.6).

6.2.3 Deep Cavities: Base Pressure

We consider first the application of the formulae of Section 6.2.2 to determine the pressure fluctuations on the base of a deep cavity. This is important in practice because such pressures are representative of those developed in aircraft wheel-wells, and so forth and on surface structures of other moving vehicles with exposed openings. Because the vortices are aligned with the x_3-axis, it can be anticipated that the principal contributions to the integral (6.2.5) are from the vicinities of the leading and trailing edges of the cavity. The potential function $\varphi^*(\mathbf{y})$ may therefore be replaced by its value *averaged over the span* of the cavity. In particular, by using the interior Green's function (6.2.6) the pressure on $x_2 = -\ell$ becomes

$$\frac{p}{\rho_o U^2} \approx \frac{-ib}{4\pi f A}\frac{\partial}{\partial t}\int\int_{-\infty}^{\infty}\sum_n \frac{\cos\{\kappa_o(x_2+\ell)\}}{\omega\cos\{\kappa_o(\ell+\ell'+i\kappa_o A/2\pi)\}}$$
$$\times\left(\frac{\partial\varphi^*}{\partial y_2}\right)_0 \mathrm{e}^{-i\omega[t-(y_1+s)/U_c-n/f]}\, dy_1\, d\omega, \qquad (6.2.8)$$

where the notation $(\partial\varphi^*/\partial y_2)_0$ implies evaluation at $y_2 = 0$.

Causality (Section 1.7.4) requires the integration contour in the frequency plane to pass above the singularities of the integrand, which consist of a simple pole at $\omega = 0$ (determining a small-amplitude transient response of the cavity) and simple poles at the cavity resonance frequencies (all in $\mathrm{Im}\,\omega < 0$), which for the lower frequency modes are given by

$$\omega = \omega_m \approx \Omega_m(1 - i\varepsilon_m), \quad m = 0, \pm 1, \pm 2, \ldots,$$

where

$$\Omega_m = \frac{(2m-1)\pi c_o}{2(\ell+\ell')} \qquad \varepsilon_m = \frac{(2m-1)A}{4(\ell+\ell')^2} \ll 1.$$

On the cavity base $x_2 = -\ell$, the ω-integral in (6.2.8) vanishes identically except when n satisfies

$$n < f\{t - (y_1 + s)/U_c - (\ell + \ell')/c_o\},$$

that is, for those vortices in the infinite sequence (6.2.4) that are currently crossing the mouth or have already passed over the cavity. If $\mathcal{N}(y_1, t)$ is the largest value of n, which satisfies this condition, then for $n < \mathcal{N}$ the integration contour may be displaced to $\omega = -i\infty$ and the integral evaluated by the residues.

Consider a steady state in which vorticity is being shed at the frequency $f = \Omega_J/2\pi > 0$ of the Jth depth mode. The residue from the pole at $\omega = \omega_m$ is then found to involve the sum

$$\sum_{n=-\infty}^{\mathcal{N}} \exp\left(2\pi n\left[\frac{i\Omega_m}{\Omega_J} + \frac{\varepsilon_m\Omega_m}{\Omega_J}\right]\right),$$

where the nth term represents the contribution to the resonant mode Ω_m from the nth vortex. The sum is absolutely convergent because $\varepsilon_m\Omega_m > 0$. The separate contributions from the vortices are individually small and generally interfere destructively unless Ω_m happens to be an integral multiple of Ω_J, when their aggregate contribution is of order $1/\varepsilon_m \gg 1$, provided that the radiation losses are small. This is necessarily the case for the low-frequency cavity modes, and the principal contribution to the integral (6.2.8) is then supplied by the resonance mode of fundamental frequency f together with those harmonics Ω_m that are integral multiples of $2\pi f$. The particular contribution, $p(f)$, to the base pressure from the fundamental is

$$\frac{p(f)}{\rho_o U^2} \approx \frac{8(-1)^J(\ell + \ell')^2 b}{\pi(2J-1)^2 A^2} \int_{-s}^{s} \left(\frac{\partial\varphi^*}{\partial y_2}\right)_0 \sin\{\Omega_J[t - (y_1 + s)/U_c]\}\, dy_1. \tag{6.2.9}$$

When the Strouhal number $\Omega_J s/U_c > 1$, which is typically the case in practice except possibly for very deep cavities, the principal contributions to this integral are from the leading and trailing edges $y_1 = \pm s$ of the mouth, where $\partial\varphi^*/\partial y_2$ is singular. However, this integral gives an *incorrect* representation of the interaction of the vortices with the leading edge $y_1 = -s$ because, as noted in Section 6.2.2, when the unsteady flow is modeled by the vorticity distribution (6.2.4) no account is taken of the smoothing action of the Kutta condition. In our discussions of blade–vortex interaction noise and trailing edge noise in Chapter 3, we saw that when the characteristic scale of the vorticity is small, aerodynamic sound generation at an edge tends to be suppressed by the imposition of the Kutta condition. We shall assume this occurs also at the *leading edge* of the cavity and that application of the Kutta condition is equivalent to ignoring sound produced at that edge. When the Strouhal number exceeds unity, this can be done by discarding the asymptotic contribution to the integral (6.2.9) from the neighborhood of $y_1 = -s$. First the span-averaged velocity potential $\varphi^*(\mathbf{y})$ is approximated by the velocity potential of *two-dimensional* flow (independent of y_3) at unit speed from the interior of the cavity. By the method of conformal transformation (assuming the cavity to be of infinite depth [4, 17, 404]), we find

$$\varphi^*(\mathbf{y}) \approx \operatorname{Re}\left(\frac{2s}{\pi}\ln\{\zeta(z)\}\right), \quad z = s + \frac{2s}{\pi}\int_1^{\zeta}\frac{\sqrt{\xi^2-1}}{\xi}\,d\xi,$$
$$z = y_1 + iy_2.$$

Near the *trailing edge* ($\zeta \approx +1$)

$$\left(\frac{\partial\varphi^*}{\partial y_2}\right)_0 \approx \frac{(\sqrt{3}s/4\pi)^{\frac{1}{3}}}{(s-y_1)^{\frac{1}{3}}}, \quad y_1 \approx s. \tag{6.2.10}$$

Then (for $\Omega_J s/U_c > 1$) the contribution to (6.2.9) from the leading edge is suppressed by using this approximation for $(\partial\varphi^*/\partial y_2)_0$ in the integrand and by replacing the lower limit of integration by $-\infty$ [19, 405, 406]. This yields

$$\frac{p(f)}{\rho_o U^2} \approx \frac{0.9(-1)^J(\ell+\ell')^2}{(2J-1)^2 A}\,\frac{\sin\{\Omega_J[t-2s/U_c]+\frac{\pi}{3}\}}{(\Omega_J s/U_c)^{\frac{2}{3}}}.$$

The Strouhal number $\Omega_J s/U_c$ is typically of order unity, and this formula therefore implies that the base pressure $p(f)$ is substantially larger than $\rho_o U^2$ when the cavity depth $\ell \gg \sqrt{A}$, whereas measurements indicate that in practice $p(f) \sim \rho_o U^2$ [58, 391, 397, 408]. Two factors can be invoked to explain this discrepancy. First viscous and thermal losses within the cavity have been neglected. This might be important at low frequencies, but it is unlikely to be the main reason for the difference. The most important defect in the present model is the vorticity representation (6.2.4), according to which vorticity translates at constant speed across the plane of the mouth, passing directly over the trailing edge. A more realistic picture of the real motion is sketched in Figure 6.2.2b. Also shown are the "streamlines" of the cavity velocity potential $\varphi^*(\mathbf{x})$. As a vortex approaches the trailing edge and is ejected from the cavity, it will follow a trajectory closely parallel to these streamlines, especially for large-amplitude resonances, when the acoustic particle velocity in the mouth is comparable to U. The amplitude of the generated sound is determined by the rate at which these streamlines are cut by the vortices (Section 3.2), which in practice will be very much smaller than assumed in the calculation discussed above. Furthermore, additional vorticity with the *same* sign as Γ will tend to be shed into the exterior flow from the trailing edge. This vorticity will initially cut across the streamlines in a sense that is opposed to that of the primary vorticity and thereby reduces further the overall efficiency of sound production. These conclusions are supported by numerical simulations [58, 405].

Deep-Cavity Strouhal Numbers

Let us next use the method of Section 5.2.5 and the vortex model (6.2.4) to estimate the Strouhal numbers of flow-excited depth modes. Observations [391, 408] indicate that, except at very large acoustic amplitudes, a vortex may be regarded as formed at the instant at which the acoustic volume flux through the mouth begins to be directed *into* the cavity. For the mode of frequency f, the acoustic particle velocity $\mathbf{u}$ may therefore be taken in the form

$$\mathbf{u} = -u_o \nabla\varphi^*(\mathbf{x}) \sin(2\pi f t),$$

where $u_o > 0$ is a constant. At the formation time $t = n/f$ of the nth vortex, the vector $\mathbf{u}$ changes direction and begins the half-cycle during which the acoustic flux is into the cavity. The average power $\langle -\Pi_\omega\rangle = -\rho_o \int \langle \boldsymbol{\omega} \wedge \mathbf{v} \cdot \mathbf{u}\rangle \, d^3\mathbf{x}$ fed into the acoustic field by the flow (where $\langle\ \rangle$ denotes a time average, see Section 5.2.1) is then given by

$$\frac{\langle -\Pi_\omega\rangle}{\rho_o U^3 A} = \frac{-u_o U_c}{2fUL} \int_{-s}^{s} \Bigg\langle \sin(2\pi f t) \times \sum_n \delta(y_1 + s - U_c(t - n/f)) \Bigg\rangle \left(\frac{\partial \varphi^*}{\partial y_2}\right)_0 dy_1,$$

where, as before, $\varphi^*(\mathbf{y})$ is approximated by the potential of two-dimensional flow from the cavity. The time average in this formula can be evaluated by replacing the infinite sequence of δ-functions by its Fourier series expansion [19]

$$\sum_{n=-\infty}^{\infty} \delta(y_1 + s - U_c(t - n/f)) = \frac{f}{U_c} \sum_{n=-\infty}^{\infty} \exp(2\pi i n f[t - (y_1 + s)/U_c]), \tag{6.2.11}$$

in which case we find

$$\frac{\langle -\Pi_\omega\rangle}{\rho_o U^3 A} = \frac{-u_o}{2UL} \int_{-s}^{s} \left(\frac{\partial \varphi^*}{\partial y_2}\right)_0 \sin\{2\pi f (y_1 + s)/U_c\} \, dy_1.$$

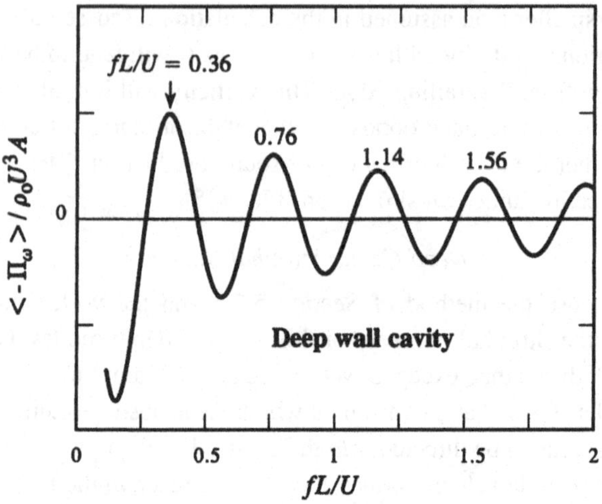

Figure 6.2.3. Deep-cavity Strouhal numbers.

The remaining integral is of precisely the form considered in (6.2.9) in calculating the base pressure. To evaluate it with proper account of the Kutta condition, we retain only the asymptotic contribution (for $2\pi f s/U_c > 1$) from the trailing edge $y_1 = s$. Using (6.2.10), we obtain

$$\frac{\langle -\Pi_\omega \rangle}{\rho_o U^3 A} \approx 0.17 \frac{u_o}{U} \frac{\cos(4\pi f s/U_c + \pi/6)}{(2\pi f s/U_c)^{\frac{2}{3}}}. \tag{6.2.12}$$

This function is plotted against the Strouhal number fL/U in Figure 6.2.3. Energy is transferred from the mean flow to maintain the acoustic oscillations when the Strouhal number lies within successive intervals in which $\langle -\Pi_\omega \rangle$ is positive. The acoustic gain is a maximum in the nth such interval when

$$\frac{fL}{U} \approx \frac{U_c}{U}\left(n - \frac{1}{12}\right), \quad n = 1, 2 \ldots. \tag{6.2.13}$$

For deep cavities, experiments and numerical simulations indicate that $U_c \approx 0.4U$. With this value, the principal Strouhal numbers for depth modes at very small Mach number are

$$fL/U \approx 0.36,\ 0.76,\ 1.16,\ 1.56, \ldots.$$

The first two members of this series are very close to values observed experimentally [394, 405], although for high-amplitude resonances (which occur typically in piping systems with side branches) where $u_o/U \sim O(1)$, the observed Strouhal numbers tend to be about 30% smaller.

Example 2. Use the free-space Green's function (6.2.7) to express the acoustic pressure radiated from a deep cavity at the fundamental frequency $f = \Omega_J/2\pi$ in the form

$$\frac{p(\mathbf{x},t)}{\rho_o U^2} \approx \frac{-2(\ell+\ell')b}{\pi(2J-1)A|\mathbf{x}|} \times \int_{-s}^{s} \left(\frac{\partial \varphi^*}{\partial y_2}\right)_0 \sin\{\Omega_J[t - |\mathbf{x}|/c_o - (y_1+s)/U_c]\}\, dy_1, \quad |\mathbf{x}| \to \infty.$$

When the Kutta condition is applied by means of the approximation (6.2.10),

$$\frac{p(\mathbf{x},t)}{\rho_o U^2} \approx \frac{-0.22(\ell+\ell')}{(2J-1)|\mathbf{x}|} \frac{\sin\{\Omega_J[t - |\mathbf{x}|/c_o - 2s/U_c] + \pi/3\}}{(\Omega_J s/U_c)^{\frac{2}{3}}}, \quad |\mathbf{x}| \to \infty.$$

Example 3. Show that the fundamental component $p(f)$ of the base pressure for a deep cavity of the type shown in Figure 6.1.6 can be calculated from (6.2.8) provided the area A in the argument of $\cos[\kappa_o(\ell + \ell' + i\kappa_o A/2\pi)]$ is replaced by $A' = Lb =$ the area of the rectangular aperture. By taking

$$\left(\frac{\partial \varphi^*}{\partial y_2}\right)_0 \approx \frac{A}{\pi b\sqrt{s^2 - y_1^2}}, \quad |y_1| < s$$

and applying the Kutta condition by expanding the integrand about $y_1 = s$, deduce that

$$\frac{p(f)}{\rho_o U^2} \approx \frac{1.02(-1)^J(\ell+\ell')^2}{(2J-1)^2 A'} \frac{\sin\{\Omega_J[t - 2s/U_c] + \pi/4\}}{(\Omega_J s/U_c)^{\frac{1}{2}}}.$$

Example 4. For a deep cavity of the type shown in Figure 6.1.6, show by using the approximation of Example 3 that the power supplied by the flow to the depth mode of frequency f is given by

$$\frac{\langle -\Pi_\omega \rangle}{\rho_o U^3 A} \approx \frac{u_o}{2\sqrt{2\pi}U} \frac{\sin(4\pi f s/U_c + 3\pi/4)}{(2\pi f s/U_c)^{\frac{1}{2}}},$$

with corresponding Strouhal number peaks $fL/U \approx (U_c/U)\{n - \frac{1}{8}\} \approx 0.35, 0.75, 1.15, \ldots$, for $n = 1, 2, 3, \ldots$, and $U_c = 0.4U$.

6.2.4 Shallow Cavity Resonances: The Sound Pressure

At low Mach numbers, the production of sound by flow over a shallow cavity is governed by the second, dipole, term in the Green's function (6.2.7). The far-field sound is given by (6.2.5) when the vortex sources are modeled by (6.2.4). There is evidently no contribution from the potential function $Y_3(\mathbf{y})$ so that the acoustic pressure is

$$\frac{p(\mathbf{x},t)}{\rho_o U^2} \approx \frac{bU_c \cos\theta}{4\pi f c_o|\mathbf{x}|}\frac{\partial}{\partial t}\int_{-s}^{s}\sum_n \left(\frac{\partial Y_1}{\partial y_2}\right)_0 \delta\{y_1 + s - U_c([t] - n/f)\}\, dy_1, \tag{6.2.14}$$

where θ is the angle between the radiation direction $\mathbf{x}$ and the positive x_1-axis, $[t] = t - |\mathbf{x}|/c_o$ is the retarded time, and the dependence of $Y_1(\mathbf{y})$ on the transverse coordinate y_3 is assumed to be averaged out.

Thus, $Y_1(\mathbf{y})$, the velocity potential of uniform incompressible flow at unit speed in the y_1-direction, can be approximated by the potential for two-dimensional flow over a rectangular cutout of the same section, which can be found by conformal transformation. Let the coordinate origin be at the center of the cavity mouth; the region above the cavity profile of Figure 6.2.4 is mapped onto the upper half of the ζ-plane by the transformation [409]

$$z = -i\ell + \frac{s}{\mathrm{E}(\mu)}\int_0^{\zeta}\left(\frac{1-\mu\xi^2}{1-\xi^2}\right)^{\frac{1}{2}} d\xi, \quad z = y_1 + iy_2, \tag{6.2.15}$$

where μ is the solution of the equation

$$\frac{\mathrm{K}(1-\mu) - \mathrm{E}(1-\mu)}{2\mathrm{E}(\mu)} = \frac{\ell}{L}, \tag{6.2.16}$$

$\mathrm{K}(m) \equiv \int_0^{\frac{\pi}{2}} d\xi/\sqrt{1 - m\sin^2\xi}$ and $\mathrm{E}(m) \equiv \int_0^{\frac{\pi}{2}} \sqrt{1 - m\sin^2\xi}\, d\xi\,(0 < m \le 1)$ being complete elliptic integrals [24]. The cavity corners $z = \pm s$ are mapped onto $\zeta = \pm 1/\sqrt{\mu}$, and the interior corners $z = \pm s - i\ell$ are mapped onto $\zeta = \pm 1$. It may then be verified that $Y_1(\mathbf{y}) = \mathrm{Re}\{s\sqrt{\mu}\zeta(z)/\mathrm{E}(\mu)\}$ and that near the trailing edge

$$\left(\frac{\partial Y_1}{\partial y_2}\right)_0 \approx \left(\frac{s(1-\mu)}{3\mathrm{E}(\mu)}\right)^{\frac{1}{3}} \frac{\sqrt{3}}{2(s-y_1)^{\frac{1}{3}}}, \quad y_1 \approx s.$$

By substituting this expression into (6.2.14), and extending the range of integration to $-\infty < y_1 < s$, we obtain the following approximation for the

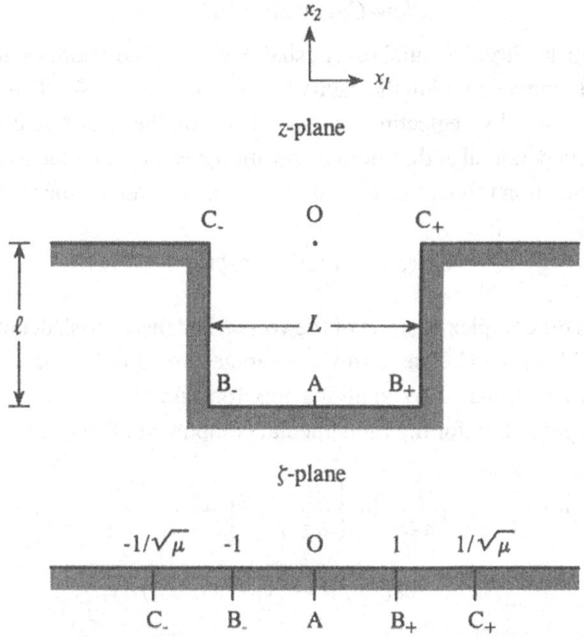

Figure 6.2.4. Mapping the shallow-cavity profile.

radiated sound with the satisfaction of the Kutta condition at $y_1 = -s$:

$$\frac{p(\mathbf{x},t)}{\rho_o U^2} \approx \frac{b(U_c/U)^2 M \cos\theta}{8\sqrt{3}\pi(fL/U)|\mathbf{x}|}\left(\frac{1-\mu}{6\mathrm{E}(\mu)}\right)^{\frac{1}{3}} \times \sum_{n=-\infty}^{\infty} \frac{\mathrm{H}\{L/U_c + n/f - [t]\}}{\{1-(U_c/L)([t]-n/f)\}^{\frac{4}{3}}},$$

where H is the Heaviside unit function. The radiation accordingly consists of a succession of identical pulses of period $1/f$, each rising to an infinitely large maximum as a vortex crosses the trailing edge. The singular behavior is a consequence of the assumption that the vortices pass directly by the trailing edge of the cavity; the pulses would be finite when allowance is made for the finite standoff distance occurring in practice. To be consistent with the assumption of compactness, the pulse width must be larger than the cavity length L, which is the case provided $U_c/c_o \ll 1$. The acoustic amplitude varies like $\rho_o U^2 M$, as for an aeroacoustic dipole produced when vorticity interacts with a compact body (Section 3.2). This dipole also contributes to the radiation from a deep cavity but was ignored in Section 6.2.3 because it is usually swamped by cavity resonances.

Shallow-Cavity Strouhal Numbers

To estimate the Strouhal numbers for shallow-cavity radiation, we first examine the velocity induced within the cavity by the vortices (6.2.4). This motion will be approximated by neglecting its dependence on the spanwise coordinate x_3. The velocity potential at the complex position $z = x_1 + ix_2$ due to a line vortex of unit circulation (about the x_3-direction) is then given by the real part of

$$\frac{-i}{2\pi}(\ln[\zeta(z) - \zeta(z_o)] - \ln[\zeta(z) - \zeta^*(z_o)]),$$

where z_o is the complex position of the vortex, and the asterisk denotes complex conjugate [4, 17, 404]. The spanwise averaged velocity fluctuation within the cavity can be estimated by applying this formula to each of the vortices of (6.2.4). In particular, for the fundamental component of frequency f, we find

$$v_2(z, f) = \frac{-U^2}{2\pi U_c} \operatorname{Im}\left\{\frac{d\zeta}{dz}\int_{-s}^{s} i\left[\frac{1}{\zeta - \zeta_o} - \frac{1}{\zeta - \zeta_o^*}\right] \times \cos(2\pi f[t - (y_1 + s)/U_c])\, dy_1\right\}, \tag{6.2.17}$$

where $\zeta = \zeta(z)$, $\zeta_o = \zeta(y_1)$.

Let us assume, as for a deep cavity, that a vortex is released from the upstream cavity edge at the beginning of the half-cycle during which the velocity v_2 in the neighborhood of that edge is directed *into* the cavity. Because vorticity is actually shed smoothly and continuously from the edge, the Kutta condition requires that the contribution from the discrete vortices near the leading edge must be excluded from (6.2.17), which then describes the velocity fluctuation produced by the interaction of the vortices with the trailing edge. The potential flow singularity of this velocity at the leading edge is cancelled by the shedding of a continuous stream of vorticity that subsequently develops into the discrete vortices that are regarded as released at times n/f $(n = 0, \pm 1, \pm 2, \ldots)$.

Provided $2\pi f L/U_c > 1$, the contributions from the leading edge vortices can be excluded by expanding the term in the large square brackets of (6.2.17) about $y_1 = s$ and by extending the range of integration to $-\infty < y_1 < s$. Then, near the upstream edge, the transformation (6.2.15) can be used to show that

$$v_2 \approx -\frac{3^{\frac{1}{3}}\Gamma\left(\frac{2}{3}\right)U^2}{16\pi U_c}(\mathrm{E}(\mu)\sqrt{1-\mu})^{\frac{4}{3}} \frac{s^{\frac{1}{3}}}{(2\pi f s/U_c)^{\frac{5}{3}}(s + x_1)^{\frac{1}{3}}} \times \sin\{2\pi f(t - 2s/U_c) + \pi/3\},$$

where here $\Gamma(x)$ denotes the Gamma function [24]. Hence, v_2 becomes negative

Table 6.2.1. *Shallow-wall cavity Strouhal numbers* fL/U

Stage	Equation (6.2.18) $U_c = 0.6U$	Linear theory of Section 6.1.5 $b/L = 5$	Data from [397] extrapolated to $M = 0$
1	0.7	0.78	0.7
2	1.3	1.37	1.1
3	1.9	1.92	1.7
4	2.5	2.45	2.5

(directed into the cavity) at the time $t = n/f$ of release of a vortex provided the argument of the sine is $2N\pi$, for some integer N. Admissible values of the Strouhal number must therefore satisfy

$$\frac{fL}{U} \approx \frac{U_c}{U}\left(n + \frac{1}{6}\right), \quad n = 1, 2, \ldots. \tag{6.2.18}$$

Taking $U_c = 0.6U$ for a shallow cavity, the Strouhal numbers predicted by this formula for the first four operating stages at infinitesimal mean flow Mach numbers are given in column one of Table 6.2.1. This table also includes the predictions of the linear theory of Section 6.1.5 (from Figure 6.1.10), which are seen to be in close agreement with (6.2.18). The final column represents the shallow-cavity experimental data in Figure 6.2.1 extrapolated to zero Mach number [397].

Example 5. Use equation (6.2.8) and the linear momentum equation $\partial v_2/\partial t = (-1/\rho_o)\partial p/\partial x_2$ to calculate the component $v_2(0)$ of the acoustic particle velocity in the mouth of a deep cavity (i.e., at $x_2 = -0$). By considering the contribution from the component of fundamental frequency $f = \Omega_J/2\pi$, and imposing the Kutta condition by using (6.2.10), deduce the Strouhal number formula (6.2.13) for a deep cavity from the requirement that $v_2(0)$ becomes negative at times $t = n/f$ (n an integer).

Example 6. The profile of the shallow-wall cavity is mapped onto the upper half of the complex ζ-plane by the transformation (6.2.15), with the correspondence of points shown in Figure 6.2.4. Table 6.2.2 lists parameter values that can be used to apply this transformation for a range of values of the depth-to-length ratio ℓ/L.

Example 7. The Helmholtz resonator illustrated in Figure 6.2.5 has resonance frequency Ω. Show that the compact Green's function for sources at points $\mathbf{y}$ in

Table 6.2.2. *Parameter values for mapping the shallow cavity onto* Im $\zeta > 0$

ℓ/L	μ	$\mathrm{E}(\mu)$	ℓ/L	μ	$\mathrm{E}(\mu)$
0.05	0.863	1.133	0.55	0.066	1.545
0.10	0.723	1.228	0.60	0.049	1.551
0.15	0.589	1.305	0.65	0.036	1.557
0.20	0.469	1.366	0.70	0.026	1.560
0.25	0.367	1.415	0.75	0.019	1.563
0.30	0.282	1.453	0.80	0.014	1.565
0.35	0.214	1.483	0.85	0.010	1.567
0.40	0.161	1.505	0.90	0.008	1.568
0.45	0.120	1.522	0.95	0.006	1.569
0.50	0.089	1.535	1.00	0.005	1.569

the neighborhood of the mouth, generating sound whose wavelength is much larger than the cavity diameter D, is given by [410]

$$\begin{aligned} G(\mathbf{x}, \mathbf{y}, t-\tau) \approx{}& \frac{1}{4\pi|\mathbf{x}|}\delta(t-s-|\mathbf{x}-\mathbf{Y}|/c_o) \\ &+ \frac{K_R\varphi^*(\mathbf{y})}{4\pi A|\mathbf{x}|}\delta(t-s-|\mathbf{x}|/c_o) - \frac{K_R\Omega\varphi^*(\mathbf{y})}{4\pi A|\mathbf{x}|} \\ &\times [\mathrm{H}(t)\sin(\Omega t)\mathrm{e}^{-\varepsilon t}]_{t-\tau-|\mathbf{x}|/c_o}, \quad |\mathbf{x}| \to \infty, \end{aligned} \tag{6.2.19}$$

where the origin is located in the vicinity of the mouth. In this formula, $\varphi^*(\mathbf{y})$ is the velocity potential of flow out of the neck of the resonator with the asymptotic properties

$$\begin{aligned} \varphi^*(\mathbf{y}) &\sim \frac{-A}{4\pi|\mathbf{y}|}, \quad |\mathbf{y}| \to \infty \quad \text{in free space} \\ &\sim \frac{-A}{K_R}, \quad \text{within the cavity} \end{aligned}$$

where A is the cross-sectional area of the neck, and K_R is its Rayleigh conductivity (Section 5.3.1). $Y_i(\mathbf{y})$ is the velocity potential of incompressible flow past the resonator that has unit speed in the i-direction at large distances from the resonator and may be assumed to be exponentially small inside the resonator. The factor $\varepsilon = K_R\Omega^2/8\pi c_o$ is the damping coefficient due to the radiation of sound by the resonator.

The first two terms on the right of (6.2.19) represent the transient response to a point source at $\mathbf{y}$ and are analogous to the compact Green's functions discussed in Section 3.2. The oscillatory term gives the resonant response and consists

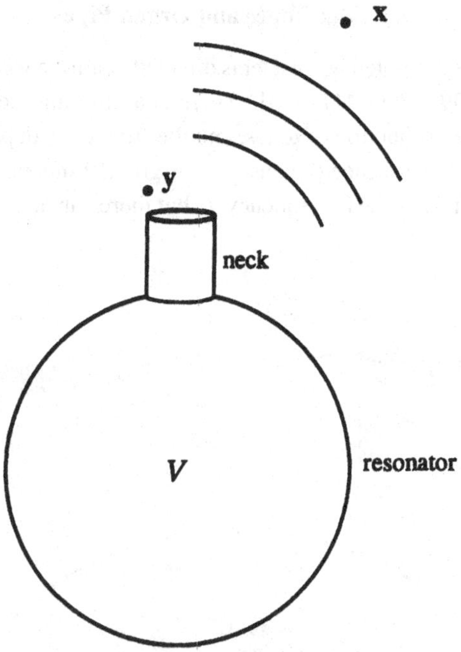

Figure 6.2.5. Helmholtz resonator.

of a sharp-fronted sinusoidal wave at the natural frequency of the resonator. The amplitude decays exponentially behind the wave front as the acoustic energy stored within the resonator is gradually dissipated by radiation. This term can be modified to include dissipation due to boundary layer losses in the resonator neck by suitably adjusting the value of ε. If the neck is a circular tube of radius R, show that when thermal and viscous boundary layer losses are included

$$\varepsilon = \frac{K_R \Omega^2}{8\pi c_o} + \frac{\sqrt{\Omega}}{R\sqrt{2}}\{\sqrt{\nu} + (\gamma - 1)\sqrt{\chi}\},$$

where ν and χ are respectively the kinematic viscosity and thermometric conductivity of the air in the neck, and γ is the ratio of the specific heats (Section 5.1). If the volume of the resonator is 10^3 cm^3, and the neck has an effective length $A/K_R = 6$ cm (including end corrections) and radius 0.75 cm, then for air ($c_o = 340$ m/s, $\nu \approx 1.5 \times 10^{-5}$ m^2/s, $\chi \approx 2.05 \times 10^{-5}$ m^2/s, $\gamma = 1.4$), we find $\Omega/2\pi \approx 93$ Hz (see (5.3.12)), and the ratio $\varepsilon_R/\varepsilon_\delta$ of the separate contributions from radiation and boundary layer damping is 0.009. That is, less than 1% of the damping is attributable to radiation losses.

6.3 Edge Tones and Organ Pipes

Edge tones are generated when a jet is directed against a wedge-shaped edge (Figure 6.3.1a) [399, 400, 411–413]. There is a minimum jet orifice-to-edge distance L for the production of tones, and the frequency depends on both the jet velocity and the jet orifice dimensions. A gradual increase in L is usually accompanied by a decrease in frequency f, but there can also be discontinuous

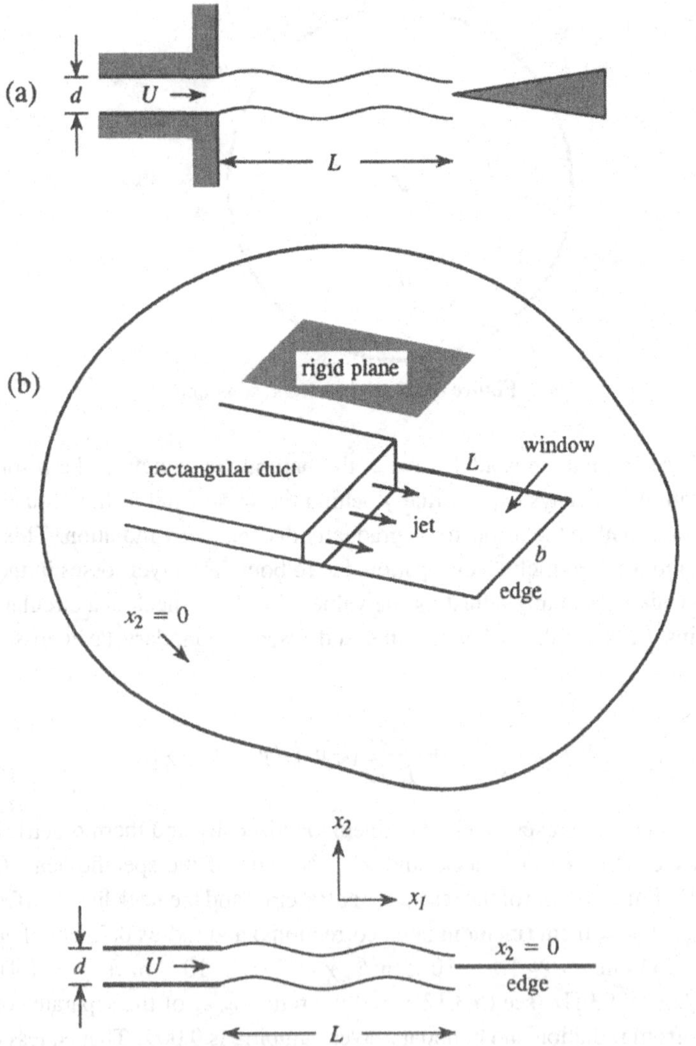

Figure 6.3.1. Jet–edge interaction.

increases in f at critical values of L; the edge tone is said to operate in different *stages* separated by these frequency jumps, and four or more stages are possible for a jet in the form of a thin *blade* of air directed at an edge. *Hysteresis* is observed when L decreases, when stages overlap, and the downward transitions from one stage to another occur at reduced critical values of L. The stages are analogous to the operating regimes of shallow-cavity tones (Section 6.2), and sound generation may be viewed in terms of the feedback to the orifice of disturbances produced by the interaction of vortices with the edge. When the jet–edge interaction occurs at the mouth of a resonant cavity, as in a flue organ pipe, the feedback and the operating stages are controlled by sound waves excited within the cavity.

6.3.1 Linear, Thin-Jet Theory of Edge Tones

Consider edge tones generated by a low Mach number stream of air issuing at velocity U from a thin-walled rectangular duct of height d and width b, where $b \gg d$ (see Figure 6.3.1). The duct is assumed to be symmetrically located within a semi-infinite rectangular slot of equal width in the rigid plane $x_2 = 0$, with its open end a distance $L \equiv 2s$ from the end of the slot, upon which the jet impinges. Take the origin on the centerline of the jet, midway between the orifice and the edge, with the x_1-axis in the flow direction so that the edge is at $x_1 = s, \ |x_3| < b/2$.

The edge-tone frequencies correspond to poles of the Rayleigh conductivity of the "window" $|x_1| < s, \ |x_3| < b/2$ connecting the upper and lower fluid regions. These poles are the eigenvalues of the equation of motion of the jet. We shall investigate them in the *thin-jet approximation*, when the wavelengths of disturbances on the jet are assumed to be much larger than the jet thickness d, and the equation determining the vertical displacement ζ of the jet (in the x_2-direction) may be approximated by

$$\rho_o d\left(\frac{\partial}{\partial t} + U\frac{\partial}{\partial x_1}\right)^2 \zeta = -[p], \quad |x_1| < s, \quad |x_3| < b/2, \tag{6.3.1}$$

where $[p]$ is the net pressure difference between the upper and lower surfaces $(x_2 \approx \pm\frac{1}{2}d)$ of the jet. It is assumed that the mean jet velocity U is uniform across the jet, and the gradual increase in jet thickness across the window is ignored.

The window is acoustically compact, and the motion near the window may be regarded as incompressible. In the absence of external forcing, the pressure perturbations in $x_2 \gtrless \pm\frac{1}{2}d$, just above and below the jet, can be expressed in the forms given in equation (5.3.26) with $U_+ = U_- = 0$ and $p_\pm = 0$, where now the integrations are over the planes $y_2 = \pm\frac{1}{2}d$, including the sections $|y_3| > \frac{1}{2}b$ of these planes to the sides of the main jet stream. When d is much smaller than either b or L, the pressures on the upper and lower surfaces of the jet can be approximated by setting $\zeta \equiv 0$ outside the window. If the spanwise variation of ζ is neglected (as in Section 6.2.1), transverse motions of the jet are then governed by the equation

$$\left(\sigma + i\frac{\partial}{\partial\xi}\right)^2 \zeta - \frac{2\sigma^2 s}{\pi d}\int_{-1}^{1} \zeta(\eta)(\ln|\xi - \eta| + \mathcal{L}(\xi, \eta))\,d\eta = 0, \qquad (6.3.2)$$

where $\sigma = \omega s/U$, $\xi = x_1/s$, $\eta = y_1/s$, and $\mathcal{L}(\xi, \eta)$ is defined as in (6.1.6).

A first integral of this equation is

$$\zeta(\xi) + \int_{-1}^{1} \mathcal{K}(\xi, \eta)\zeta(\eta)\,d\eta + (\lambda_1 + \xi\lambda_2)\mathrm{e}^{i\sigma\xi} = 0, \qquad (6.3.3)$$

where

$$\mathcal{K}(\xi, \eta) = \frac{2\sigma^2 s}{\pi d}\int_{-1}^{\xi} (\xi - \lambda)(\ln|\lambda - \eta| + \mathcal{L}(\lambda, \eta))\mathrm{e}^{i\sigma(\xi - \lambda)}\,d\lambda,$$

and λ_1, λ_2 are constants of integration that are determined by the thin-jet approximation to the Kutta condition: $\zeta = \partial\zeta/\partial\xi = 0$ as $\xi \to -1$.

The eigenvalues $\omega s/U$ of equation (6.3.3) are the roots of equation 6.1.11) as $N \to \infty$, where the matrix $\mathcal{A}_{ij}$ is defined by the procedure described in Section 6.1.2 for the wall aperture, with $\mathcal{A}_{i1} = e^{i\sigma\xi_i}$ and $\mathcal{A}_{i2} = \xi_i e^{i\sigma\xi_i}$. The roots have been found numerically [398] for $5 < L/d < 50$, for different fixed values of the "window" aspect ratio b/L. The real part of a root in $\mathrm{Im}(\omega) > 0$ is interpreted as the frequency $f = \mathrm{Re}(\omega)/2\pi$ of a possible edge tone. As L/d varies for each fixed value of b/L, the roots are found to lie along a succession of discrete bands in the complex plane, illustrated in Figure 6.3.2 for the quasi-two-dimensional case $b/L = 500$. Successive bands correspond to the edge-tone operating stages, in each of which the Strouhal number fL/U is a smoothly varying function of L/d. The jet–edge interaction is unable to operate in the first stage when L/d exceeds about fifteen, and there is ultimately a transition to the higher Strouhal number stage two and beyond. The solid circles in Figure 6.3.3 represent calculated values of the first stage Strouhal number plotted against L/d for the two aspect ratios $b/L = 0.5$ and 10. For fixed b/L, the points are collinear when L/d is large, and the solid lines in the figure are rectilinear

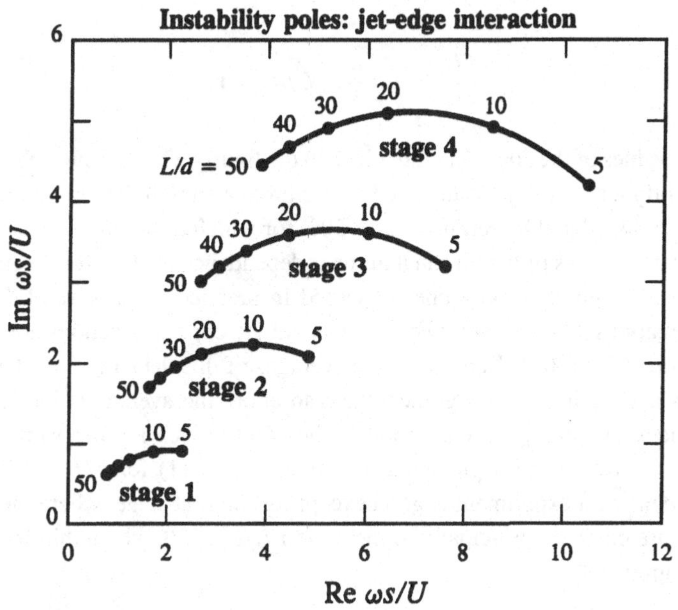

Figure 6.3.2. Thin jet–edge interaction poles for $b/L = 500$.

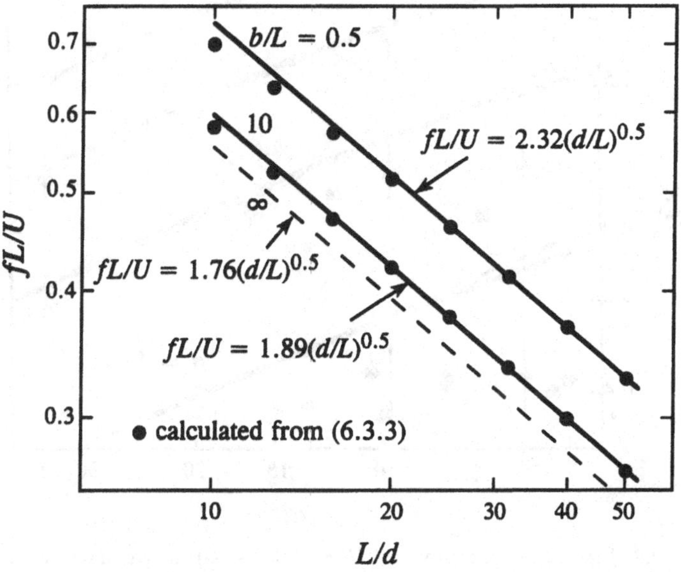

Figure 6.3.3. Strouhal number for first operating stage.

approximations defined by [322, 401]

$$\frac{fL}{U} = \alpha\sqrt{\frac{d}{L}}, \quad L/d \gg 1, \tag{6.3.4}$$

where values of the constant α are given in the figure. The limiting behavior for a "two-dimensional" jet is depicted by the broken straight line in Figure 6.3.3, which is calculated by setting $b/L = 500$, for which $\alpha \approx 1.76$.

Measurements of the Strouhal number dependence on L/d for the first four operating stages (the only ones observed in practice) from several different investigations of nominally two-dimensional jet–edge interactions are plotted in Figure 6.3.4 [401]. Representative averages of this data are plotted as solid circles and squares; any significant spread about the average is indicated by a vertical bar through a data point. The solid curves are the variations of $fL/U \equiv \mathrm{Re}(\omega s/\pi U)$ predicted by equation (6.1.11) for $b/L = 500$; the agreement with experiment is good except for the first stage, where most data points are confined principally to the region $L/d < 10$, where thin-jet theory is not applicable.

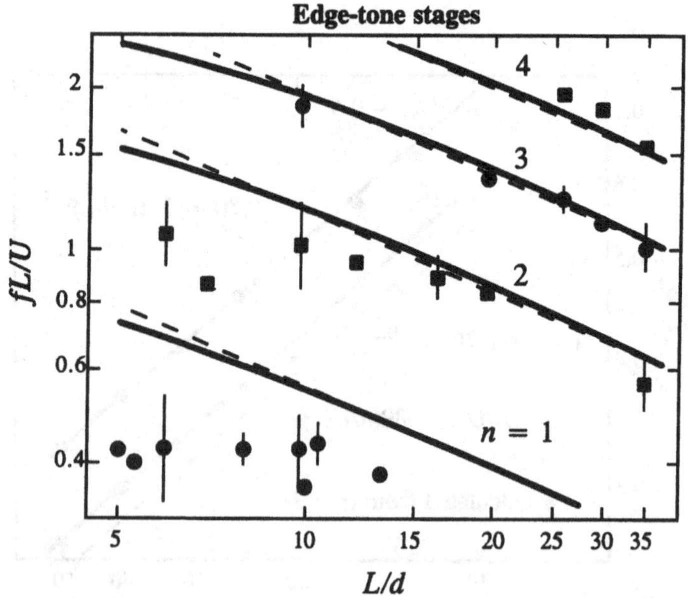

Figure 6.3.4. Linear thin-jet theory for $b/L = 500$ (——) compared with experiment [401]. The broken lines are predictions of (6.3.11).

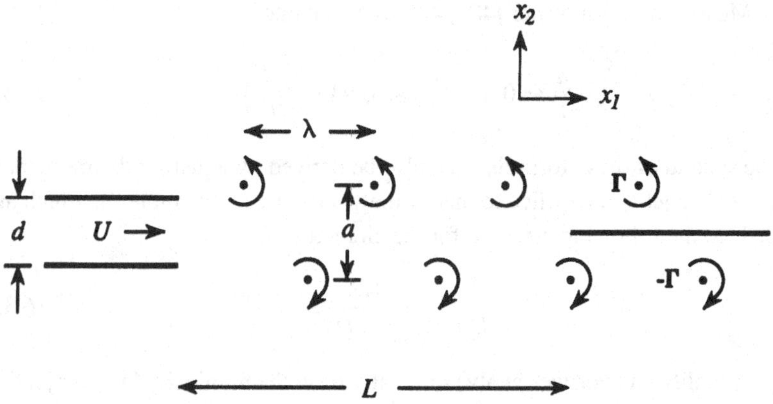

Figure 6.3.5. Vortex model of the jet–edge interaction.

6.3.2 Nonlinear Theory of Edge Tones

To estimate the amplitude of edge-tone radiation, it is necessary to consider a nonlinear theory of the kind discussed in Section 6.2.2. The jet is assumed to evolve nonlinearly into a Kármán vortex street [17, 404], consisting of a stable array of vortices of alternating sign convecting toward the edge at constant speed U_c (Figure 6.3.5). This model cannot be applicable when the Reynolds number Ud/ν is very large, when the flow is turbulent, but experiments suggest that discrete tones are generated only for Ud/ν smaller than about 2,000–3,000. In a locally two-dimensional approximation, the upper and lower rows of vortices in the figure are represented by

$$\omega(\mathbf{x}, t) \approx \sum_{n=-\infty}^{\infty} \Gamma \mathbf{i}_3(\delta(x_2 - a/2)\delta\{x_1 + s - U_c(t - n/f)\}$$
$$- \delta(x_2 + a/2)\delta\{x_1 + s - U_c[t - (n + 1/2)/f]\}),$$
$$x_1 > -s, \quad |x_3| < b/2, \qquad (6.3.5)$$

where a is the vertical distance between the rows, and $\lambda = U_c/f$ is the distance between neighboring vortices in the same row, which are released at frequency f. According to this model, vortices appear spontaneously in the neighborhood of the jet orifice and drift toward the edge at speed U_c. In fact, discrete vortices form at some distance downstream, typically between one and three hydrodynamic wavelengths from the orifice. Spurious contributions to the sound from vortices close to the orifice must therefore be suppressed by application of the Kutta condition in the manner discussed in Section 6.2.3.

Measurements in water [401, 411] indicate that

$$\frac{a}{\lambda} \approx 0.5, \quad \frac{U_c}{U} \approx 0.945\left(\frac{fd}{U}\right)^{\frac{1}{3}}. \tag{6.3.6}$$

The second of these formulae can also be derived by equating the momentum flux of the jet at the orifice to that of the vortex street (Example 2), which also leads to the following estimate for the circulation Γ

$$\frac{\Gamma}{Ud} \approx \frac{1.95}{(fd/U)^{\frac{1}{3}}}. \tag{6.3.7}$$

In reality, the vortices evolve gradually from the nominally laminar jet flow from the orifice and convey jet momentum that must be conserved as they move toward the edge. The momentum flux is equal to [401]

$$2\rho_o a\Gamma\left(1 - \frac{\lambda/a}{\tanh(\pi a/\lambda)}\right).$$

Viscous action causes the circulation Γ of each vortex to decay so that conservation of momentum flux results in the spreading of the jet, that is, in an increase in the distance a between the vortex rows with distance downstream [414].

We shall use the linear theory of Section 6.3.1 to relate the formation of the vortices to the phase of the transverse velocity v_2 induced near the orifice by the interaction of the vortices with the edge. Suppose there exists a phase lag $2\pi\theta$ between the formation of a vortex in the upper row and the beginning of the half-cycle during which v_2 is positive. For the fundamental mode of frequency f, we can then write

$$v_2 = v_o \sin[2\pi(ft - \theta)], \tag{6.3.8}$$

where $v_o > 0$ does not depend on time. This velocity can be calculated by the conformal mapping procedure described in Section 6.2.4 for a shallow cavity. To do this, we first observe (using Fourier series expansions of the type (6.2.11)) that the component $\omega(z, f)$ of the vorticity (6.3.5) of frequency f is

$$\begin{aligned}\omega(z, f) = {} & \frac{2\Gamma f \mathbf{i}_3}{U_c}(\delta(x_2 - a/2) + \delta(x_2 + a/2)) \\ & \times \cos\{2\pi f[t - (x_1 + s)/U_c]\}, \quad z = x_1 + ix_2.\end{aligned} \tag{6.3.9}$$

The corresponding component of v_2 is obtained from an integral of the type (6.2.17), where $\zeta \equiv \zeta(z)$ maps the z-plane with semi-infinite, rectilinear cuts along rays corresponding to the upper and lower walls of the jet duct and the edge onto the upper ζ-plane. For all modes of operation of the edge tone, the condition $2\pi f L/U_c > 1$ is usually satisfied, and the integral can be estimated asymptotically by expanding the nonoscillating part with respect to the integration variable about $y_1 = s$. This approximation suppresses contributions from vortex interactions with the orifice at $y_1 = -s$ and is equivalent to applying the Kutta condition. The behavior of $\zeta(z)$ near $z = s$ is well approximated by taking

$$\zeta = z/s + \sqrt{z^2/s^2 - 1}, \tag{6.3.10}$$

which maps the z-plane cut along the real axis from $\pm s$ to $\pm\infty$ into Im $\zeta > 0$ (neglecting the height d of the jet orifice relative to L).

The reader can verify that to calculate the *phase* of v_2 near the jet orifice, it is sufficient to take $a = 0$ in the δ-functions of expression (6.3.9) because a nonzero value of a affects only the amplitude. Then v_2 is given by (6.2.17) with the factor $-U^2/2\pi U_c$ replaced by $2\Gamma f/\pi U_c$. In the integrand, we expand $\zeta(z)$ about the jet orifice at $z = -s$ and $\zeta_o(y_1)$ about the edge $y_1 = +s$, to find, near the orifice, on the center line of the jet

$$\begin{aligned} v_2 &\approx \frac{-\sqrt{2}\Gamma f}{\pi U_c \sqrt{s}(s^2 - x_1^2)^{\frac{1}{2}}} \int_{-\infty}^{s} \sqrt{s - y_1} \cos\{2\pi f[t - (y_1 + s)/U_c]\}\, dy_1 \\ &\approx \frac{\Gamma}{4\pi^2 \sqrt{fs/U_c}(s^2 - x_1^2)^{\frac{1}{2}}} \sin[2\pi f(t - 2s/U_c) + \pi/4]. \end{aligned}$$

The integral in the first line is actually divergent, but only the finite part given in the second line from the vicinity of $y_1 = s$ is relevant asymptotically (formally derived by multiplying the integrand by $e^{\varepsilon y_1}$, where $\varepsilon > 0$ is subsequently allowed to vanish [19]). By comparing the phase of this result with that of v_2 in (6.3.8), it follows that

$$\frac{fL}{U_c} = n + \frac{1}{8} + \theta,$$

for integer values of n. Thus, using the second of (6.3.6), the jet–edge Strouhal numbers satisfy

$$\frac{fL}{U} \approx 0.92\sqrt{\frac{d}{L}}\left(n + \frac{1}{8} + \theta\right)^{\frac{3}{2}}, \quad n = 1, 2, \ldots.$$

We shall now argue that the Strouhal numbers predicted by this equation should coincide with those determined by the two-dimensional linear theory of

Section 6.3.1. This is because, although nonlinear mechanisms will ultimately control the amplitude of the periodic motions, they do not significantly change their frequencies, which are determined by the convection velocity, which varies slowly with vortex strength [399–401]. We can see by inspection of Figure 6.3.4 that linear theory is well approximated by the formula (6.3.4) for all operating stages when L/d is large. Linear and nonlinear theories will therefore be equivalent if $\alpha \equiv 0.92(n+\frac{1}{8}+\theta)^{\frac{3}{2}}$ for a unique value of the phase lag θ. When θ is chosen to make linear and nonlinear theories agree for $n = 1$ (by taking $\alpha = 1.76$; see Figure 6.3.3), we find $\theta \approx 0.42$, and therefore

$$\frac{fL}{U} \approx 0.92\sqrt{\frac{d}{L}}(n+0.54)^{\frac{3}{2}}. \tag{6.3.11}$$

The broken straight lines in Figure 6.3.4 are predictions of this formula, which agree with the thin-jet, linear theory for $L/d \gg 1$ for all of the observed stages. The choice $\theta = 0.42$ implies that a vortex with positive circulation (in the upper row of Figure 6.3.5) is formed nearly half a period before the beginning of the half-cycle during which the transverse velocity v_2 near the orifice is positive.

Example 1: The Kármán vortex street. The double array of line vortices in Figure 6.3.5 (equation (6.3.5)) is called a *vortex street.* The complex velocity potential $w(z)$ of the motion in an unbounded fluid, when the array extends to $x_1 = \pm\infty$ is given by [17, 404]

$$\begin{aligned} w = &-\frac{i\Gamma}{2\pi}\ln\left(\sin\left[\frac{\pi}{\lambda}(z-ia/2)\right]\right) \\ &+\frac{i\Gamma}{2\pi}\ln\left(\sin\left[\frac{\pi}{\lambda}(z-\lambda/2+ia/2)\right]\right), \quad z = x_1+ix_2, \end{aligned}$$

at the instant at which a vortex in the upper row ($x_2 = a/2$) passes above the coordinate origin. The street advances in the x_1-direction at constant speed

$$U_c = \frac{\Gamma}{2\lambda}\tanh\left(\frac{\pi a}{\lambda}\right). \tag{6.3.12}$$

The motion can be shown to be stable to small perturbations provided $a/\lambda \approx 0.281$ [17].

Example 2: Vortex strength deduced from conservation of momentum [401]. The strength Γ of the vortices in the vortex street (6.3.5) is estimated by equating the momentum flux in the free jet to that at the orifice. For most geometries, the velocity in the duct near the exit is controlled by viscous stresses and may be approximated by the Poiseuille profile [17] $v_1 = \frac{3}{2}U(1-4x_2^2/d^2)$, where U is the mean jet velocity. Therefore, the momentum flux per unit width of the jet is

$$\int_{-\frac{d}{2}}^{\frac{d}{2}} \rho_o v_1^2 \, dx_2 = \frac{6}{5}\rho_o U^2 d. \tag{6.3.13}$$

This is equated to the mean x_1-momentum flux in the vortex street $\langle \int_{-\infty}^{\infty}(p + \rho_o v_1^2)\, dx_2 \rangle$, where the angle brackets represent an average over the period $1/f = \lambda/U_c$ of the motion i.e., because conditions are periodic and steady in a frame convecting at speed U_c, to

$$\frac{1}{\lambda}\int_0^{\lambda} dx_1 \int_{-\infty}^{\infty} (p + \rho_o v_1^2)\, dx_2.$$

Suppose the vortices have circular cores of radius $R \ll a$ and uniform vorticity $\Omega \mathbf{i}_3$, where $\Omega = \pm\Gamma/\pi R^2$ in the upper and lower rows of Figure 6.3.5. From the inviscid, incompressible form of Crocco's equation (1.2.18) (with $\partial/\partial t = -U_c \partial/\partial x_1$), we can write

$$\left\langle \int_{-\infty}^{\infty} \left(p + \frac{1}{2}\rho_o(v_1^2 + v_2^2)\right) dx_2 \right\rangle$$
$$= \frac{\rho_o}{\lambda}\int_0^{\lambda} dx_1 \int_{-\infty}^{\infty} \left(U_c \frac{\partial \psi}{\partial x_2} - \sum_n \Omega(\psi - \psi_n)\mathrm{H}(R - r_n)\right) dx_2, \tag{6.3.14}$$

where the summation is over the two vortices in the region of integration, and r_n is the perpendicular distance from the axis of the nth vortex. ψ is a stream function, defined in the usual way such that $(v_1, v_2) = (\partial\psi/\partial x_2, -\partial\psi/\partial x_1)$; ψ_n is the stream function when the contribution from the nth vortex is excluded and may be regarded as constant across the core. Then [17]

$$\psi - \psi_n = \frac{1}{4}\Omega(R^2 - r_n^2) \quad \text{for} \quad r_n < R,$$

and observing that $\psi \to \pm a\Gamma/2\lambda$ as $x_2 \to \pm\infty$, the right-hand side of (6.3.14) evaluates to $\rho_o(aU_c\Gamma - \Gamma^2/4\pi)/\lambda$. Similarly,

$$\left\langle \int_{-\infty}^{\infty} \frac{1}{2}\rho_o(v_1^2 + v_2^2)\, dx_2 \right\rangle$$
$$\equiv \frac{\rho_o}{\lambda}\int_0^{\lambda} dx_1 \int_{-\infty}^{\infty} \sum_n \Omega v_1 x_2 \, dx_2 = \rho_o(aU_c\Gamma - \Gamma^2/4\pi)/\lambda.$$

Combining these results, the mean momentum flux in the vortex street per unit span is $2\rho_o(aU_c\Gamma - \Gamma^2/4\pi)/\lambda$, which with (6.3.13) implies that

$$\frac{3\lambda U^2 d}{5U_c} = a\Gamma - \frac{\Gamma^2}{4\pi U_c}. \tag{6.3.15}$$

The formula (6.3.12) now yields equations (6.3.6) and (6.3.7) in the form

$$U_c/U = F(\beta)(fd/U)^{\frac{1}{3}}, \quad \Gamma/Ud = 2\coth\beta\, F^2(\beta)/(fd/U)^{\frac{1}{3}},$$

where

$$F(\beta) = \left(\frac{3\pi\tanh^2\beta}{5(2\beta\tanh\beta - 1)}\right)^{\frac{1}{3}}, \quad \beta = \frac{\pi a}{\lambda}, \quad \lambda = \frac{U_c}{f}.$$

Experiments in water (with a configuration similar to that in Figure 6.3.5) over a large range of values of L/d confirm the breakup of the jet into a vortex street with $a/\lambda \approx 0.5$ [401], and this leads to the estimate (6.3.7).

Example 3. Practical edge-tone experiments use an arrangement of the type illustrated in Figure 6.3.6. Show that the fundamental component of the radiation at large distances from the edge is given by the dipole field

$$\frac{p(\mathbf{x},t)}{\rho_o} \approx \frac{-\Gamma f\cos\Theta}{\pi c_o|\mathbf{x}|}\frac{\partial}{\partial t}\int_{-\frac{b}{2}}^{\frac{b}{2}} dy_3 \times \int_{-s}^{s}\left(\frac{\partial Y_2}{\partial y_2}\right)_0 \cos(2\pi f\{[t] - (y_1+s)/U_c\})\, dy_1,$$

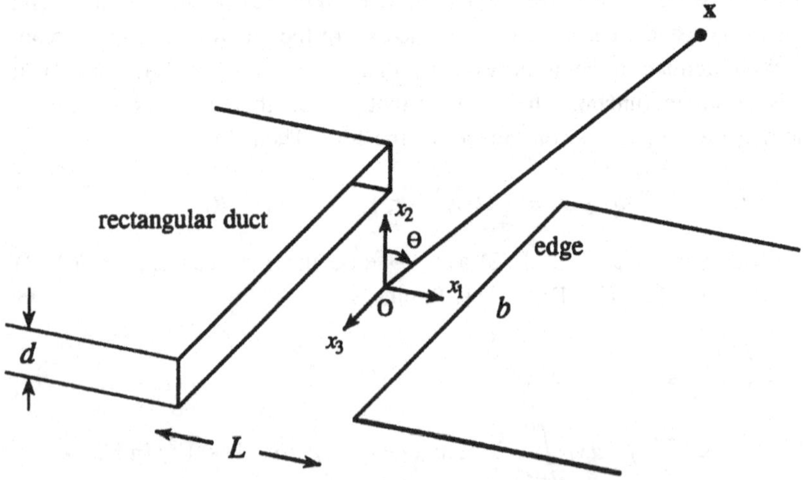

Figure 6.3.6. Edge-tone experimental arrangement.

where Θ is the angle between the normal to the jet and the radiation direction, $[t] = t - |\mathbf{x}|/c_o$, and the suffix 0 indicates evaluation at $y_2 = 0$, in the plane of the edge.

Because $fL/U_c > 1$, the Kutta condition is imposed at the jet exit by expanding $(\partial Y_2/\partial y_2)_0$ about the edge at $y_1 = s$ and extending the range of integration to $y_1 = -\infty$. The details of the behavior of Y_2 near the jet exit are then unimportant and, when $d/L \ll 1$, Y_2 may be approximated by the velocity potential of flow in the y_2-direction past the rigid strip with a transverse "slit" obtained by letting $d \to 0$ in Figure 6.3.6 [415, 416]:

$$Y_2 \approx \frac{\mathcal{M}(y_3, b, s)}{\pi} \operatorname{Re}(\zeta(z)), \quad z = y_1 + iy_2$$

where ζ is defined in (6.3.10), and

$$\begin{aligned} \mathcal{M}(y_3, b, s) &= \frac{\pi\sqrt{b^2/4 - y_3^2}}{\ln[(4/s)\sqrt{b^2/4 + \varepsilon^2 s^2 - y_3^2}]}, \quad |y_3| < b/2, \\ &= 0, \quad |y_3| > b/2, \end{aligned}$$

ε being an O(1) constant greater than $\frac{1}{4}$.

Deduce that the acoustic pressure in the nth stage is then given by

$$\begin{aligned} \frac{p(\mathbf{x}, t)}{\rho_o U^2} &\approx 0.6\mathcal{I}(b/s, \varepsilon) \cos\Theta \frac{b^2}{|\mathbf{x}|d}\left(\frac{d}{L}\right)^{\frac{5}{2}}\left(n + \frac{3}{8}\right)^2 \\ &\times \sin(2\pi f\{[t] - L/U_c\} + \pi/4), \quad |\mathbf{x}| \to \infty, \end{aligned}$$

where $M = U/c_o$, and $\mathcal{I}(b/s, \varepsilon) = (4/\pi b^2)\int_{-b/2}^{b/2} \mathcal{M}(y_3, b, s)\, dy_3$ is a dimensionless function of b/s and ε equal to about 0.2–0.4.

6.3.3 Whistles And Hole Tones

Hole tones are generated when a jet is directed at a sharp-edged orifice in a plate [380, 417]. In a whistling kettle, for example, a steam jet emerges from a hole at one end of a small cavity and impinges on a coaxial hole at the opposite end. Similarly, Rayleigh's *bird-call* [10] generates high-pitched tones when a stream of air issues from a circular hole in a thin plate and strikes symmetrically on a hole in a neighboring parallel plate. Rayleigh conjectured that the sound is maintained by a feedback mechanism in which pressure fluctuations generated at the second hole cause pulsations in the jet flow rate that excite axisymmetric instabilities of the jet. At large Reynolds numbers (typically for $UD/\nu > 10^4$, where D is the jet diameter), these instabilities grow into a succession of vortex rings that interact periodically with the second hole. The instabilities are symmetric, in contrast to the sinuous motions of a jet interacting with an edge. When the second plate is larger than the acoustic wavelength, the volume flux through the hole behaves as a monopole aerodynamic sound source, with an efficiency proportional to the jet Mach number M at low speeds (Section 3.2). For a small plate, or when the plate is replaced by a solid ring coaxial with the jet (producing a *ring tone* [417]), the sound is of dipole type, with an efficiency scaling as M^3; the dipole strength is equal to the unsteady force on the ring.

The monopole sound for the arrangement involving a large plate, shown in Figure 6.3.7a, can be calculated by using the compact Green's function (3.2.20). Assume the jet evolves into a succession of vortex rings of circulation Γ, radius R_o, with axes coincident with the x_1-axis, and having constant convection velocity U_c ($\sim 0.6U$). Suppose the vortices may be regarded as generated at the jet nozzle at $x_1 = -L$ at times $t = n/f$ $(n = 1, 2, \ldots)$ and that they impinge symmetrically on an orifice of radius $R > R_o$ in a thin, rigid plate at $x_1 = 0$.

Write

$$\omega(\mathbf{x}, t) = \sum_{n=-\infty}^{\infty} -\Gamma \mathbf{i}_\theta \delta(r - R_o)\delta(x_1 + L - U_c(t - n/f)),$$

where $\mathbf{i}_\theta$ is a unit vector in the direction of increasing θ in the cylindrical coordinate system (r, θ, x_1). At low Mach numbers, when the orifice is acoustically compact, the sound is given by (3.3.2) with $\mathbf{U} = U_c\mathbf{i}_1$. For the fundamental component of frequency f, the acoustic pressure is

$$\frac{p(\mathbf{x}, t)}{\rho_o} \approx \mp \frac{4f\Gamma R_o}{R|\mathbf{x}|} \times \int_{-\infty}^{\infty} \left(\frac{\partial \bar{Y}}{\partial r}\right)_{r=R_o} \cos(2\pi f\{[t] - (y_1 + L)/U_c\})\, dy_1, \quad |\mathbf{x}| \to \infty, \qquad (6.3.16)$$

where the upper–lower sign is taken according as $x_1 \gtrless 0$, the potential function $\bar{Y}$ is defined as in (3.2.20), and $[t] = t - |\mathbf{x}|/c_o$.

An integral representation of $\partial \bar{Y}(r, y_1)/\partial r$ can be derived from the formula given in Section 5.3.3, Example 8, which permits the acoustic pressure to be evaluated in the form

$$p(\mathbf{x}, t) \approx \mp \frac{2\rho_o f \Gamma R_o}{|\mathbf{x}|} \mathrm{I}_1(\kappa R_o) \mathrm{e}^{-\kappa R} \sin(2\pi f([t] - L/U_c)), \quad |\mathbf{x}| \to \infty, \qquad (6.3.17)$$

where $\mathrm{I}_1(\kappa R_o)$ is a modified Bessel function [24] ($\sim \mathrm{e}^{\kappa R_o}/\sqrt{\kappa R_o}$, $\kappa R_o \to \infty$), and $\kappa = 2\pi f/U_c$ is the hydrodynamic wavenumber. The Strouhal number

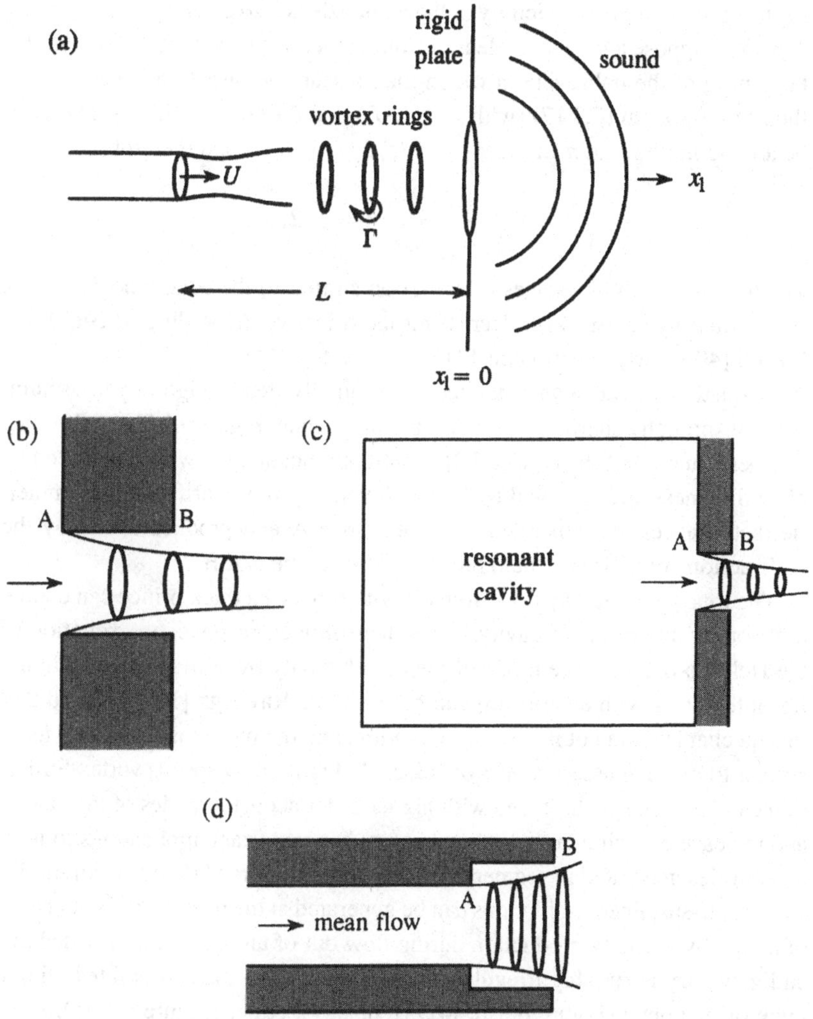

Figure 6.3.7. (a) Generation of hole tones, (b) flow through a thick orifice, (c) exit flow through a thick orifice, and (d) exit flow with a collar.

based on the jet diameter D ($\approx 2R_o$) and mean velocity U typically ranges from about 0.3 to 1 (which includes the Strouhal number $fD/U \approx$ 0.5–0.6 that characterizes low Mach number axisymmetric instabilities of a low–speed unimpeded jet [418]). The acoustic amplitude varies as $\rho_o U^2$, as expected for an aerodynamic monopole source.

The reader can readily verify that the compact Green's function (3.2.20) for a small aperture in a large rigid plane remains valid in the limit of incompressible flow ($c_o \to \infty$) provided $|\mathbf{x}| \gg R$. Thus, the fundamental component of the unsteady pressure in the vicinity of the jet nozzle is also given by (6.3.17) when $L \gg R$. Suppose a vortex is released from the jet orifice at time $\theta/2\pi$ *after* the beginning of the half-cycle in which the pressure perturbation is negative. It then follows from (6.3.17) (with $x_1 \approx -L$) that $2\pi f(t - L/U_c) - 2\pi\theta$ must be an odd multiple of π at times $t = 1/f, 2/f, 3/f, \ldots$ and therefore that

$$\frac{fL}{U_c} \approx n - \frac{1}{2} - \theta, \quad n = 1, 2, \ldots. \tag{6.3.18}$$

The hole tone exhibits stages of operation similar to the edge tone, including transitional hysteresis with decreasing jet velocity. According to Blake and Powell [400] early experiments [417] suggest that $\theta \approx -\frac{1}{4}$.

Acoustic tones are also generated by nominally steady, high Reynolds number flow through a sharp-edged orifice in a thick plate (Figure 6.3.7b). A Strouhal number equation of the type (6.3.18) is still applicable, but with L replaced by plate thickness and U_c equal to the mean velocity in the orifice. Axisymmetric disturbances are generated at the entrance A and produce sound by the "diffraction" of their near field pressure fields at the exit B.

The tone generated by flow from a cavity through a thick orifice can couple to resonant modes of the cavity. Thus, *whistling* corresponds to excitation of the Helmholtz resonance mode of the mouth cavity by vortices formed in the lips interacting with exterior lip geometry [418]. Rayleigh [10] observed that this mechanism did not involve lip vibration, and the mechanism appears to be similar to that discussed in Section 5.2.5. At higher flow speeds vortex formation can become synchronized with higher order acoustic modes of the cavity, and the case of a circular cylindrical cavity has important applications to flow instabilities generated in segmented solid rocket motors [419–421]. Similarly, strong self-sustained oscillations can be generated at the resonance frequencies of a pipe by vorticity production during flow out of an open end, provided the end geometry is suitably irregular, for example, when the end is fitted with a thick orifice plate (Figure 6.3.7c [422]), or has a collar (Figure 6.3.5d), or is flared like a horn. In all such cases the sound within the pipe causes vorticity production in the exit flow at points labeled A. Vorticity produced at such points tends to damp the oscillations (Section 5.2), but the subsequent interaction of the vortices, which can grow in strength by extracting energy from the mean flow, with the irregular geometrical features (labeled B) is an acoustic source, and the balance between the vortex damping at A and sound generation at B usually governs the amplitude of the oscillations.

Example 4. In some hole-tone experiments the vortices do not pass through the second hole but collide with the plate and expand radially. Assume that when the vortex radius R_o is greater than the hole radius R (Figure 6.3.7a) the vortices translate at uniform speed U_c until they reach the plate, where they are annihilated by images in the surface. The acoustic pressure at frequency f is then given by (6.3.16) with the integration restricted to the semi-infinite domain $-\infty < y_1 < 0$. In the limit $R_o \to R + 0$, show that the acoustic pressure is

$$p(\mathbf{x}, t) \approx \mp \frac{2\rho_o f \Gamma R_o}{|\mathbf{x}|} \mathcal{R}(\kappa R) \sin(2\pi\{f([t] - L/U_c) - \psi(\kappa R)\}),$$

$$|\mathbf{x}| \to \infty,$$

where $\lambda = 2\pi f/U_c$, the $\mp$ sign is taken according as $x_1 \gtrless 0$, and [289]

$$\mathcal{R}(x)e^{2\pi i\psi} = \frac{1}{2}\mathrm{I}_1(x)e^{-x} + \frac{i}{\pi}\mathrm{K}_1(x)\sinh(x).$$

The Strouhal number condition (6.3.18) becomes

$$\frac{fL}{U_c} \approx n - \frac{1}{2} - \theta - \psi, \quad n = 1, 2, \ldots.$$

The additional phase lag $\psi \equiv \psi(fD/U_c)$ is equal to $\frac{1}{4}$ when $fD/U_c = 0$, but it decreases rapidly to $\frac{1}{8}$ with increasing frequency and may be taken to be constant when $fD/U_c > 1.5$.

6.3.4 The Flue Organ Pipe

A flue organ pipe is depicted schematically in Figure 6.3.8. Our discussion will be confined to a simplified geometry, involving a uniform pipe of length ℓ and rectangular cross section $h \times b$ (the transverse dimension b being into the paper in the figure) open at both ends. Acoustic resonances of the pipe are generated by the impact of a nominally steady, thin blade of air issuing from the flue, which crosses the mouth and impinges on a sharp edge, often referred to as the *labium*. The distance L between the flue exit and the edge is much smaller than ℓ and is usually small compared to both h and b. The labium is actually wedge shaped, and the local geometry of the mouth is similar to that used in edge-tone experiments, but with the wedge offset from the centerline of the jet so that in steady, undisturbed flow the jet would not enter the pipe. Sinuous oscillations of the jet are forced by the acoustic volume flux through the mouth, but feedback from the labium is still important. It appears that two vortices are shed from the labium during each half-cycle of oscillation, once in response to the interruption of the jet flow as it deflects across the labium, and once more when the jet is clear of the edge in response to the acoustic flux [423–428].

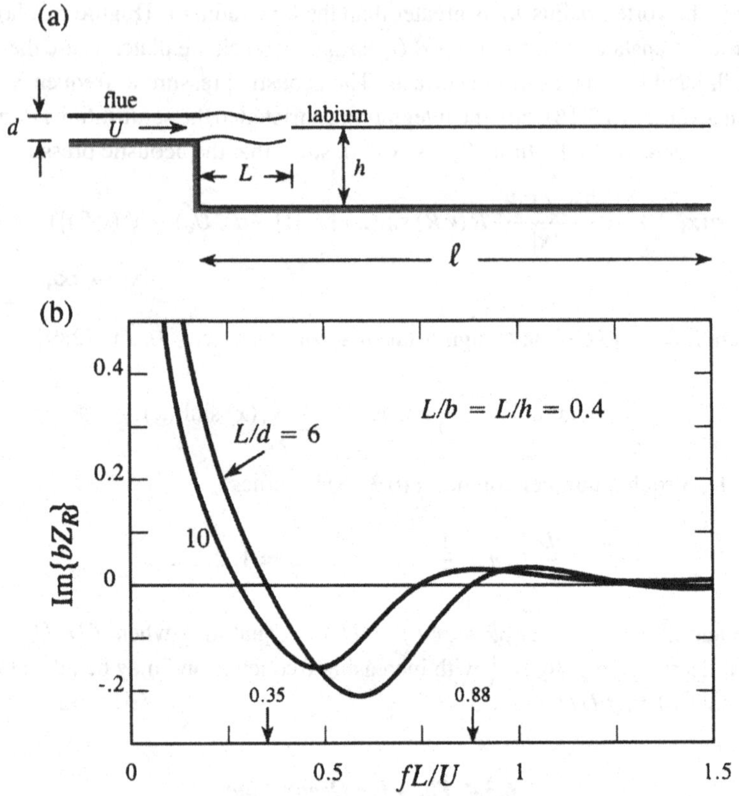

Figure 6.3.8. (a) Flue organ pipe, and (b) imaginary part of Rayleigh impedance.

The details of the interaction of the jet with the labium has traditionally been ignored [429–437], but the production of vorticity by the volume flux occurs at the expense of the acoustic energy stored in the pipe (Section 5.2), and this appears to determine the amplitude of the sound. The problem is analytically very complicated; numerical simulations are difficult and tend to give little insight [438]. It will suffice here to apply the thin-jet theory of Section 6.3.1 to examine the lowest order mode of oscillation of the jet, which is usually the most important operating stage in practice.

In the linearized approximation, the frequencies of self-sustained oscillations of the system formed by the jet and organ pipe correspond to the roots of the characteristic equation

$$Z_c(\omega) \equiv Z_R(\omega) + Z_a(\omega) = 0, \quad \operatorname{Im}\omega > 0, \qquad (6.3.19)$$

where $Z_R(\omega)$ and $Z_a(\omega)$ are, respectively, the Rayleigh impedance of the mouth and the acoustic impedance of the pipe at the mouth. Those roots that are close to the zeros of $Z_R(\omega)$ correspond to shear layer instabilities; they will be ignored, because they are not usually well coupled to acoustic modes of the pipe (which "stores" large amounts of acoustic energy at resonance); and the sound they produce tends to be random, of low amplitude, and relatively high frequency. For simplicity, the mouth will be assumed to have a rigid flange that is several times larger than the mouth (as in Figure 6.3.1b). This will permit the equation of motion of the jet to be formulated in the same way as for the edge tone. When the flange is acoustically compact, the acoustic impedance Z_a can be taken in the following modified form of that given in Example 1 of Section 6.1.1,

$$Z_a = \frac{i\kappa_o}{4\pi} + \frac{\tan(k(\ell + \ell' + i\kappa_o A/4\pi)}{kA}, \tag{6.3.20}$$

where $A = hb$ is the cross-sectional area of the pipe, and ℓ' is the end correction at the remote open end (at $x_1 = \ell$ in Figure 6.3.8). The wavenumber $k = \kappa_o + i\alpha_o$, where α_o is defined in (5.1.32), and accounts for boundary layer losses at the walls. The first term on the right of (6.3.20) is the radiation impedance at the mouth and is half of that in Example 1 of Section 6.1.1 because the flange is compact.

In a first approximation, the frequencies of the lower order pipe modes are close to the natural modes in the absence of the jet. Equation (6.3.19) may then be replaced by

$$\sin(k\{\ell + \ell' + A(Z_R + i\kappa_o/2\pi)\}) = 0.$$

It turns out that the magnitude of Z_R is comparable to $1/b$ so that this equation is valid provided $AZ_R/\ell \sim h/\ell \ll 1$, which is usually the case in organ pipes and recorderlike instruments. The equation is the "open end" analogue of the frequency equation for flow-excited depth modes (Section 6.1.3). The solutions are approximately

$$\frac{\omega(\ell + \ell')}{c_o} = n\pi\left(1 - \frac{i\alpha_o}{\kappa_o} - \frac{A}{\ell + \ell'}\left(Z_R + \frac{i\kappa_o}{2\pi}\right)\right), \quad n = 1, 2, \ldots,$$

and oscillations are self-sustaining when

$$\operatorname{Im} Z_R < \frac{-\kappa_o}{2\pi} - \frac{(\ell + \ell')\alpha_o}{A\kappa_o}, \quad \text{for} \quad \frac{\omega(\ell + \ell')}{c_o} = n\pi, \tag{6.3.21}$$

which is the condition for the energy supplied by the jet to exceed that dissipated by radiation from the ends of the pipe and by boundary layer losses inside the pipe.

Let the lateral width of the jet be equal to the width b of the pipe, and assume it impinges symmetrically on the labium, as in the edge-tone model of Section 6.3.1. Take the origin midway between the flue exit and the labium, on the axis of symmetry of the jet. The reader may then show, by the method of Section 6.3.1, that the integral equation governing the nondimensional displacement ζ' of the jet when forced by pressures $p_\pm$ above and below the mouth can be reduced to the following inhomogeneous form of the edge-tone equation (6.3.3)

$$\zeta'(\xi) + \int_{-1}^{1} \mathcal{K}(\xi, \eta)\zeta'(\eta)\, d\eta + (\lambda_1 + \xi\lambda_2)e^{i\sigma\xi} = 1,$$

where

$$\mathcal{K}(\xi, \eta) = \frac{2\sigma^2 s}{\pi d} \int_{-1}^{\xi} (\xi - \lambda) \times \left(\ln|\lambda - \eta| + \frac{1}{2}\{\mathcal{L}(\lambda, \eta) + \mathcal{L}_-(\lambda, \eta)\} \right) e^{i\sigma(\xi-\lambda)}\, d\lambda,$$

$\mathcal{L}(\lambda, \eta)$ is defined in (6.1.6), and

$$\mathcal{L}_-(\lambda, \eta) = \ln\left(\frac{4\sinh\{(\pi L/4h)(\xi - \eta)\}\sinh\{(\pi L/4h)(2 + \xi + \eta)\}}{\xi - \eta} \right).$$

The solution φ' satisfying the Kutta condition at $\xi = -1$ determines the Rayleigh impedance from the formula

$$Z_R = \frac{2d}{A\int_{-1}^{1} \zeta'(\eta)\, d\eta}.$$

The following parametric ratios are typical:

$$L/d = 6, \quad L/b = L/h = 0.4,$$

and Figure 6.3.8 illustrates the dependence of Im Z_R on Strouhal number $\omega L/2\pi U \equiv fL/U$, where U is the mean jet velocity. Acoustic modes of the pipe can be excited in this case provided

$$0.35 < \frac{fL}{U} < 0.88, \qquad (6.3.22)$$

where Im $Z_R < 0$.

If we set $f = nc_o/2(\ell + \ell')$, these inequalities determine the maximum range of jet speeds over which the nth mode can be excited. For a pipe whose effective length is 30 cm, with $L = 0.4$ cm, the fundamental frequency ($n = 1$) $f \approx 560$ Hz, which can be excited for jet speeds 2.55 m/s $< U <$ 6.4 m/s. In principle, condition (6.3.22) permits a mode and its first harmonic to speak simultaneously. The amplitude of the sound is governed by nonlinear mechanisms that are also generally responsible for the excitation of higher harmonics. In order to excite harmonics of the fundamental directly, the pipe must be "over blown" so that the Strouhal number for that mode is brought within the range (6.3.22). For example, the second harmonic can speak when 5.1 m/s $< U <$ 12.8 m/s (so that the fundamental and first harmonic can both speak when 5.1 m/s $< U <$ 6.4 m/s). When $L/d = 10$, the plot of Im Z_R, also shown in the figure, indicates that pipe tones are excited at lower Strouhal numbers (higher jet speeds) when the jet thickness decreases.

6.3.6 Vorticity Generation at The Labium

The acoustic particle velocity in the mouth of an organ pipe is typically of the same order of magnitude as the jet velocity U. Its equilibrium amplitude is controlled by *nonlinear* dissipation, which is additional to radiation and boundary layer losses. When the transversely oscillating jet has passed by the labium, the acoustic volume flux induces vortex shedding from the labium. The kinetic energy of this shed vorticity is derived from the acoustic energy stored in the pipe, and it seems likely that it is this nonlinear transfer of energy from the acoustic field that controls the amplitude of the pipe oscillations [425–428].

The equilibrium amplitude of the sound can be estimated by equating the rate of dissipation due to vorticity production to the rate of production of acoustic energy calculated according to thin-jet theory. Consider a situation in which the principal component of the volume flux from the mouth is $Q = Q_o \sin \omega t \equiv Q_o \mathrm{Re}\{i e^{-i\omega t}\}$. In a steady state, this volume flux is accompanied by a uniform pressure load $p = -\rho_o \omega Q_o \mathrm{Re}\{(Z_R + Z_a)e^{-i\omega t}\}$ on the mouth outside of the pipe. When self-sustaining oscillations occur, there is a net excess production of oscillatory energy by the jet at a rate

$$\Pi = \langle pQ \rangle \equiv -\frac{1}{2}\rho_o \omega Q_o^2 \,\mathrm{Im}\{Z_R + Z_a\}.$$

At resonance, $\omega(\ell + \ell')/c_o \approx n\pi$ ($n = 1,\ 2$ etc.) and $\Pi > 0$, provided the Strouhal number satisfies (6.3.21).

When $\Pi > 0$, the excess power is absorbed by vorticity generation at the labium. To calculate the absorption, consider the vorticity generated during the half-cycle in which $Q > 0$, which we can take as the interval $0 < t < \pi/\omega$.

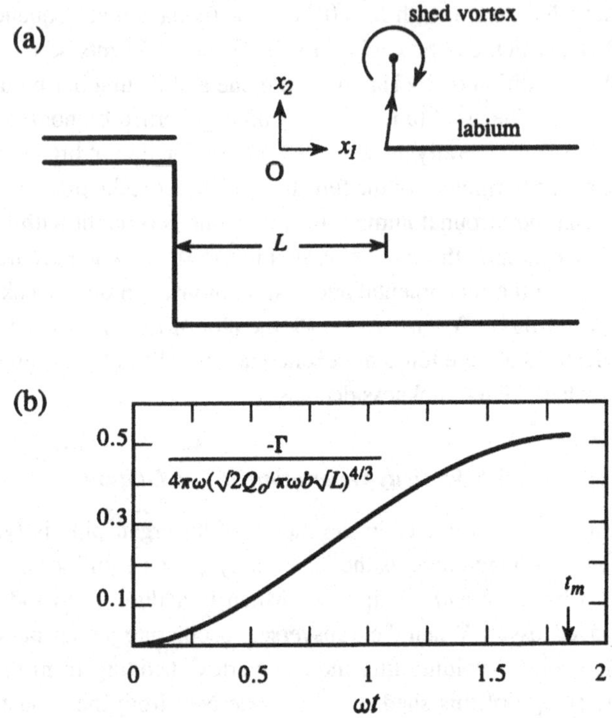

Figure 6.3.9. Dissipation by vortex shedding.

Measurements indicate that damping is significant over roughly the first quarter of this interval when the shed vorticity is still close to the labium. Suppose the motion can be regarded as locally two dimensional (Figure 6.3.9a) and that the continuous stream of shed vorticity rolls up into a tight core that can be modeled as a line vortex of variable circulation $\Gamma(t)$, whose value is determined by application of the Kutta condition at the edge of the labium. To calculate the motion near the labium produced by this vortex and the volume flux, the complex velocity potential w_Q of the flux through the mouth is approximated by that for an incompressible volume flux of strength $2Q$ through a slit of width $2L$:

$$w_Q = \frac{2Q}{\pi b}\ln(z'/L + \sqrt{z'^2/L^2 - 1}), \quad z' = x_1 + s + ix_2.$$

For this choice of w_Q, the normal velocity vanishes on the interior vertical wall $(x_1 = -s,\ x_2 < 0)$ of the pipe adjacent to the mouth.

If $z = z' - L$, the edge of the labium is at $z = 0$, where

$$w_Q \to \frac{2\sqrt{2}Qz^{\frac{1}{2}}}{\pi b\sqrt{L}}.$$

Let the complex position of the vortex $\Gamma(t)$ be $z_o(t)$ relative to the edge of the labium, then near the edge the potential due to the volume flux and the vortex can be approximated by

$$w(z) = \Psi(t)\zeta - \frac{i\Gamma(t)}{2\pi}(\ln(\zeta - \zeta_o) - \ln(\zeta - \zeta_o^*)), \quad \Psi(t) = \frac{2\sqrt{2}Q}{\pi b\sqrt{L}},$$

where $\zeta = \sqrt{z}$ maps the upper and lower sides of the labium (the positive real axis in the z-plane) onto the real axis in the ζ-plane, $\zeta_o(t) = \sqrt{z_o(t)}$, and the asterisk denotes the complex conjugate. The Kutta condition is satisfied at $z = 0$, provided $dw/d\zeta = 0$ at $\zeta = 0$, and this supplies the following equation for the circulation of the shed vortex

$$\Gamma(t) = \frac{4\sqrt{2}iQ|\zeta_o|^2}{b\sqrt{L}(\zeta_o^* - \zeta_o)}.$$

This model, in which vorticity is imagined to be shed continuously from the edge of the labium and is concentrated instantaneously in a tight core at $z = z_o(t)$, is actually inconsistent with the equations of motion and leads to the unphysical prediction of a force on the connecting sheet that joins the edge to the core and on the solid surface from which the vorticity is shed. As explained in Section 1.14.5, these forces can be canceled by requiring the translational velocity dz_o/dt of the core to be determined by the *Brown and Michael equation* (1.14.21). Proceeding as in Example 10 of Section 1.14, the equation of motion of the core is found to be

$$\frac{dZ^*}{dT} + \frac{(Z^* - |Z|)}{\Gamma}\frac{d\Gamma}{dT} = \frac{-|Z|\sin T}{(^*\sqrt{Z} - \sqrt{Z})}\left(\frac{1}{2Z} + \frac{1}{Z + |Z|}\right) + \frac{\sin T}{\sqrt{Z}}, \tag{6.3.23}$$

where $T = \omega t,\ Z = z_o\{\pi\omega b\sqrt{L}/\sqrt{2}Q_o\}^{\frac{2}{3}}$.

Equation (6.3.23) is integrated numerically in the interval $0 < T < \pi$, subject to the initial condition $Z = 0$ at $T = 0$. Vorticity of negative sign is initially generated, and the solution is physically acceptable provided $d\Gamma/dt$ is one signed. When this condition is violated, the absolute vortex strength would begin to decrease, and a more realistic model of the flow would require a second vortex to be released from the edge. The numerical solution is continued until

$d\Gamma/dt$ first changes sign at $t = t_m \approx 1.86/\omega$; the trajectory and the variation of $\Gamma(t)$ are depicted in Figure 6.3.9.

We shall ignore vorticity produced at later times $t_m < t < \pi/\omega$ of the half-cycle. Then the instantaneous power $\Pi_\Gamma(t)$ absorbed by vorticity production is given by (5.2.6), in which

$$\boldsymbol{\omega} = \Gamma(t)\mathbf{i}_3\mathrm{H}(b/2 - |x_3|)\delta(x_1 - x_{o1}(t))\delta(x_2 - x_{o2}(t)),$$

$$\mathbf{v} = (v_1, v_2), \quad \mathbf{u} = \Psi(t)\nabla\{\mathrm{Re}\sqrt{z}\},$$

where $v_1 - iv_2$ is equal to the right-hand side of (6.3.23) multiplied by $\omega(\sqrt{2}Q_o/\pi\omega b\sqrt{L})^{\frac{2}{3}}$. A similar calculation (leading to an identical prediction for Π_Γ) can be performed for the second half-cycle when the flux is into the mouth.

Thus, for equilibrium oscillations, we must have

$$\langle\Pi_\Gamma\rangle \equiv \frac{\omega}{\pi}\int_0^{t_m}\Pi_\Gamma(t)\,dt = -\frac{\rho_o\omega Q_o^2}{2}\,\mathrm{Im}\{Z_R + Z_a\}.$$

When the integral is evaluated numerically, we find that if $\bar{u} = Q_o/2bL$ is the root mean square acoustic particle velocity in the mouth for the nth mode ($\omega = 2\pi f \approx n\pi c_o/(\ell + \ell')$), then

$$\frac{\bar{u}}{U} \approx 12.8\,\frac{fL}{U}\left(\mathrm{Im}\{-bZ_R\} - \frac{\sqrt{2}(\ell + \ell')^{\frac{3}{2}}[\sqrt{\nu} + (\gamma - 1)\sqrt{\chi}]}{\sqrt{n\pi c_o}h^2}\right)^{\frac{3}{2}},$$

provided the Strouhal number is in the range where $\mathrm{Im}\{Z_R + Z_a\} < 0$, where self-excitation can occur.

In this formula, ν and χ are the kinematic viscosity and thermometric conductivity, and γ is the ratio of specific heats of the fluid (air); the relatively small contribution to Z_a from radiation damping has been discarded. For the case considered in Figure 6.3.8 (with $\ell + \ell' = 30$ cm, $L/d = 6$, $L/b = L/h = 0.4$), $\mathrm{Im}\{-bZ_R\}$ assumes a maximum positive value of 0.21 at $fL/U \approx 0.6$. For a pipe of square cross section with $h = 1.5$ cm, we then find $\bar{u}/U \sim 0.04$ for the fundamental mode ($n = 1$). The influence of boundary layer damping is smaller for a pipe of larger cross section, and the amplitude of the sound is larger: When $h = 2$ cm, we find $\bar{u}/U \sim 0.28$, which is typical of values observed experimentally.

6.4 Combustion Instabilities

There are two basic mechanisms by which sound is produced by entropy inhomogeneities (Sections 2.1 and 2.3). Monopole sound is produced by unsteady heating, which causes volumetric expansion; dipole sound is generated by the differential acceleration of "hot spots." In this section we examine monopole sources associated with *combustion*, an exothermic reaction involving a fuel and oxidant. The fuel is usually gas or liquid, and the oxidant is frequently air; solid fuel is used in some rocket motors. Sound is generated by unsteady heat release during burning and is often sufficiently intense to cause fatigue damage. In an enclosed space, such as the combustion chamber of a gas turbine engine, combustion mechanisms can couple to cavity resonances to produce oscillations capable of destroying the chamber [439–444]. Cavity resonances may also be *forced* by combustion, for example, when burning occurs as a sequence of explosions of roughly constant frequency (determined, for example, by a characteristic ignition time lag) that matches a resonance frequency of the chamber. Although these oscillations are generally undesirable, they are in some circumstances beneficial in causing increased rates of heat transfer.

Most combustion noise and vibration problems involve *deflagration*, that is, the propagation of a *flame* relative to an unburnt mixture of fuel and oxidant [442–445]. The flame is a thin, luminous region across which temperature and chemical species change very rapidly, the burnt products being of greatly elevated temperature and reduced density. The flame front forms an interface between combustion products and the unburnt mixture and usually propagates at less than about 1 m/s relative to the unburnt gases. The volumetric expansion produced by the propagating front generates sound; in a flow-through combustion chamber, the flame may be anchored to a flame holder or burner. In general the sound is either a broadband "roar" or a combustion driven oscillation at one or more discrete frequencies, involving acoustic feedback on combustion that enhances the conversion of chemical energy into sound. Pulsations associated with these feedback cycles can cause large-amplitude flexing of structures and early fatigue failure.

6.4.1 Flames

A flame is a luminous manifestation of highly exothermic oxidation. In a *diffusion* flame (such as a candle flame or the flame formed by a gas jet burning in air) burning occurs as the fuel makes contact with the air, mixing either by molecular diffusion or by turbulent convection. When the fuel and air are mixed prior to burning, the flame is called a *premixed flame*, which tends to give more intense combustion and higher effective heat transfer than the diffusion flame. In a gas turbine engine, a premixed flame is usually maintained at a fixed location within a combustion chamber, through which the combustible mixture of fuel and oxidant pass. When liquid fuel is injected upstream of the flame, droplets may also be present, producing a combined premixed-diffusive burning.

Simple arguments [63, 445, 446] indicate that the thickness of a flame front propagating through a stationary and uniform flammable mixture is proportional to $\sqrt{\chi\tau}$, where τ is the characteristic reaction time in the flame (which depends principally on temperature and is defined by the relation $de/dt = e/\tau$, where e is the thermal energy), and $\chi = \kappa/\rho_o c_p$ is the thermometric conductivity; the flame thickness is typically of order 1 mm. The normal velocity of the flame front U_L is called the *laminar burning velocity*. It is a fundamental property of the combustible mixture, of characteristic magnitude $\sqrt{\chi/\tau}$, and is determined by the reaction rate $1/\tau$ and the rate at which heat and mass are transferred from the flame to the unburnt gas.

A fuel–air mixture is said to be *stoichiometric* when there is just sufficient oxidant for complete combustion to occur. This typically occurs at values of the fuel–air mass ratio of less than 10%. To a good approximation, therefore, the gas constant R (Section 11.2.3) is effectively the same for the burnt and unburnt mixtures on either side of the flame, and this approximation is used in this section. The *equivalence ratio* ϕ is the value of the fuel–air mass ratio relative to stoichiometric (the mixture is "lean" or "rich" according as $\phi \lessgtr 1$). At atmospheric temperature and pressure, the laminar flame speed U_L attains a maximum when $\phi \approx 1.05$–1.1 (equal to about 45 cm/s for hydrocarbon–air mixtures); it can increase significantly if the proportion of oxygen in the air is increased. U_L increases with the temperature T_1 of the unburnt mixture; the variation of U_L is usually given for the *maximum* burning velocity by an empirical formula of the form

$$U_L \approx a + bT_1^n, \text{ m/s}, \tag{6.4.1}$$

where a, b, and n are constants, with $a \approx 0.1 - 0.3$, $b \approx 10^{-8} - 10^{-5}$, $1.75 < n < 2.9$ and, typically, $200 < T_1 < 600$°K. For methane, $a = 0.08$, $b = 1.6 \times 10^{-6}$, $n = 2.11$ [445, 447]. The flame speed is so very much smaller than the speed of sound that combustion is often assumed to occur at constant pressure. In a first approximation, changes of U_L with mean pressure may be neglected.

Turbulence can strongly influence flame speed [447]. For moderate turbulence levels (root mean square turbulence velocity $v \sim 2U_L$), the flame front is wrinkled by the larger eddies, and flame surface area is further increased by the entrainment of smaller eddies, both of which cause the *turbulent* flame speed U_T to be of order $2v$. The turbulence length scale is less important because it determines in this case only the relative contributions of the large- and small-scale eddies. For more intense turbulence, the scale of the dominant fluctuations

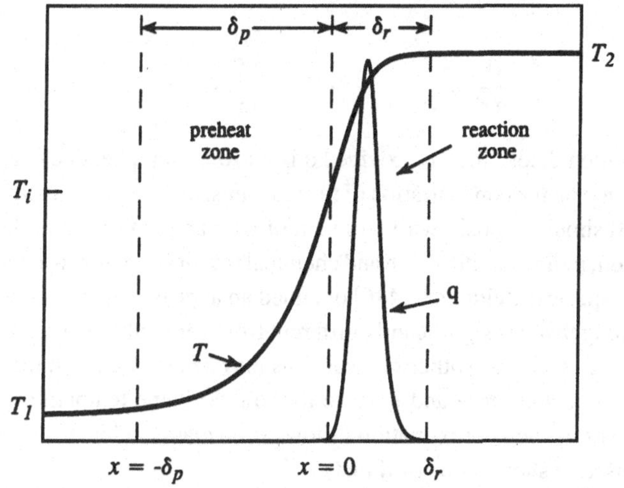

Figure 6.4.1. Temperature and heat release in a laminar flame.

decreases and the flame front ultimately becomes a diffuse layer of individually combusting eddies that advances at speed $U_T \approx \frac{1}{2}v\delta_L/\delta_K$, where δ_L is the *laminar flame thickness*, and $\delta_K = (\nu^3/\epsilon)^{\frac{1}{4}}$ is the *Kolmogorov dissipation length*, ϵ being the turbulence dissipation rate per unit mass [2, 62, 63].

Example 1: The premixed laminar flame. The temperature variation across a plane laminar flame is illustrated schematically in Figure 6.4.1. The coordinate system is at rest relative to the flame, and the unburnt mixture of fuel and oxidant at temperature T_1 and density ρ_1 flows at velocity U_L in the x-direction. Combustion and heat generation occur within the *reaction zone*, where the temperature increases to the final flame temperature T_2, and the density decreases to ρ_2; the magnitude of the flame velocity U_L depends on the rate at which this heat is transferred ahead of the flame. The temperature of the unburnt gases entering the flame increases rapidly in the *preheat region*, but very little chemical change occurs there.

At very low Mach numbers burning may be assumed to occur at constant pressure, and the left-hand side of the energy equation (1.2.34) can be written

$$\rho T \frac{Ds}{Dt} \approx \rho c_p \frac{DT}{Dt} \equiv \rho_1 c_p U_L \frac{dT}{dx}$$

because the mass flow rate is constant and the motion is steady relative to the coordinate frame. Let q denote the rate of heat generation per unit volume. Insert this on the right of (1.2.34) and neglect heat production by viscous stresses, to obtain

$$\frac{d}{dx}\left(\kappa\frac{dT}{dx}\right) - \rho_1 U_L c_p \frac{dT}{dx} + \mathrm{q} = 0. \tag{6.4.2}$$

This equation determines T provided q is known. But q depends on both temperature and on the concentration of the reactants, and the equation must strictly be solved simultaneously with a system of similar equations that describe the convection, molecular diffusion and chemical rate of formation (or depletion) of chemical species. Zeldovich [446] obtained an approximate solution of (6.4.2) by assuming that q is significantly different from zero only for $T > T_i$. T_i is the temperature at which exothermic reactions may be assumed to start; it is called the *ignition temperature* and is close to the final flame temperature T_2. That this is a reasonable approximation follows from classical kinetic theory, which may be used to show [446, 448] that

$$\mathrm{q} \sim \left(\frac{T_2 - T}{T_2 - T_1}\right)^n \mathrm{e}^{-E/\mathrm{R}T},$$

where R is the gas constant, and the *activation energy* E is large; n is a positive number determined by the type of reaction.

Thus, q is negligible in the preheat zone, where the temperature distribution is governed by a balance between molecular diffusion from the reaction zone and inward convection at speed U_L of the unburnt gases. Because $T \to T_1$ and $dT/dx \to 0$ as $x \to -\infty$, we find (neglecting the temperature dependence of κ and c_p)

$$\frac{dT}{dx} = \frac{\rho_1 U_L c_p}{\kappa}(T - T_1), \quad T - T_1 = (T_i - T_1)\exp\{\rho_1 U_L c_p x/\kappa\}, \tag{6.4.3}$$

where it is assumed that $x = 0$ at $T = T_i$. The temperature therefore rises exponentially in the preheat zone. In the reaction zone, convection by the flow may be neglected compared to molecular diffusion. Taking κ to be constant, writing $d^2T/dx^2 = d\{\frac{1}{2}(dT/dx)^2\}/dT$, and noting that $dT/dx = 0$ at $T = T_2$, we find

$$\frac{dT}{dx} = \left(\int_T^{T_2} \frac{2}{\kappa}\mathrm{q}(T)\,dT\right)^{\frac{1}{2}}, \quad T > T_i.$$

This expression for the temperature gradient must equal that given by (6.4.3) at the ignition temperature. In the integral, we can set the lower limit equal to

T_1 because q is negligible for $T < T_i$; in the first of (6.4.3), we can take $T_i \approx T_2$ because the temperature profile has an inflexion near T_i, and dT/dx varies very slowly. Hence, the flame speed is given by

$$U_L^2 \approx \frac{2\kappa \int_{T_1}^{T_2} \mathrm{q}(T)\, dT}{(\rho_1 c_p (T_2 - T_1))^2} = \frac{2\chi \bar{\mathrm{q}}}{\rho_1 c_p (T_2 - T_1)} \equiv \frac{\chi}{\tau},$$

where $\bar{\mathrm{q}}$ is the mean rate of heat release across the flame, and

$$\tau = \rho_1 c_p (T_2 - T_1)/2\bar{\mathrm{q}}$$

is the effective flame reaction rate. The thickness δ_p of the preheat zone is the distance $-x$ ahead of the ignition point at which $T - T_1 = 0.01(T_i - T_1)$; the second of (6.4.3) gives

$$\delta_p \approx 4.6\kappa/\rho_i U_L c_p,$$

which is typically of order 0.5 mm [445] in air at standard temperature and pressure. Except at very high flame speeds, the reaction zone thickness $\delta_r \sim \frac{1}{3}\delta_p$.

6.4.2 Combustion Noise

The direct sound produced by combustion is of monopole type. Most practical problems occur at relatively low Mach numbers, where the flow speed is small enough to maintain the integrity of flame fronts and combustion zones; density and temperature vary strongly with position and time in the source region, but the local motion resembles that of an incompressible fluid, where density changes are primarily a result of thermal expansion, and pressure variations are relatively small. In these circumstances, the acoustic analogy equation (2.3.13) can be taken in the simplified form (obtained by making use of (2.3.6) and (2.3.11))

$$\frac{\partial}{\partial t}\left(\frac{1}{\rho c^2}\frac{\partial p}{\partial t}\right) - \operatorname{div}\left(\frac{1}{\rho}\nabla p\right) = \frac{\partial^2}{\partial x_i x_j}(v_i v_j) + \frac{\partial}{\partial t}\left(\frac{\beta T}{c_p}\frac{Ds}{Dt}\right). \qquad (6.4.4)$$

For an ideal gas, the coefficient of expansion $\beta = 1/T$, and we shall make this substitution in the remainder of Section 6.4.

Equation (6.4.4) can be rearranged by replacing the density and sound speed on the left by their uniform ambient mean values ρ_o, c_o, respectively, whereupon it assumes the more familiar form

$$\frac{1}{c_o^2}\frac{\partial^2 p}{\partial t^2} - \nabla^2 p = \rho_o \frac{\partial}{\partial t}\left[\left(\frac{1}{\rho_o c_o^2} - \frac{1}{\rho c^2}\right)\frac{\partial p}{\partial t}\right] - \rho_o \mathrm{div}\left[\left(\frac{1}{\rho} - \frac{1}{\rho_o}\right)\nabla p\right]$$
$$+ \frac{\partial^2}{\partial x_i x_j}(\rho_o v_i v_j) + \frac{\partial}{\partial t}\left(\frac{\rho_o}{c_p}\frac{Ds}{Dt}\right). \tag{6.4.5}$$

In the absence of boundaries, the solution of this equation in the acoustic far field can be written in a form similar to (2.1.9) for an isentropic, inhomogeneous source flow, together with an additional contribution from the entropy source. The first two terms on the right describe the generation of sound by the scattering of source region pressure fluctuations by variations in fluid compressibility and density, as discussed in Section 2.1.3 – this is often called *indirect* combustion noise; the third source term is the Lighthill quadrupole. The entropic term is a *monopole* causing volumetric changes of the fluid due to heating by the flame.

If $\mathrm{q}(\mathbf{x}, t)$ is the rate of heat generation by combustion per unit volume, then

$$\frac{\partial}{\partial t}\left(\frac{\rho_o}{c_p}\frac{Ds}{Dt}\right) \approx \rho_o \frac{\partial}{\partial t}\left(\frac{\mathrm{q}(\mathbf{x}, t)}{c_p \rho T}\right), \tag{6.4.6}$$

provided thermal conduction and viscous dissipation are ignored. The monopole acoustic pressure is therefore

$$p(\mathbf{x}, t) \approx \frac{\rho_o}{4\pi|\mathbf{x}|}\frac{\partial}{\partial t}\int \left[\frac{\mathrm{q}(\mathbf{y}, t)}{c_p \rho T}\right] d^3\mathbf{y}, \quad |\mathbf{x}| \to \infty, \tag{6.4.7}$$

where the origin is within the source region, and the square brackets indicate evaluation at the retarded time $t - |\mathbf{x} - \mathbf{y}|/c_o$.

A turbulent flame may be viewed as a random distribution of burning elements of combustible mixture, each producing an increased volume of the heated gases and giving rise to a superposition of acoustic waves. Experiments indicate that when a given amount of combustible material burns in a given time the sound level can depend significantly on the geometry of the combustion process (Examples 2 and 3) [449]. Let us estimate the efficiency of sound generation by an anchored, premixed flame in a turbulent stream, such as that in a Bunsen burner (Figure 6.4.2). The undisturbed flame is approximately conical; the angle of the cone is determined by the condition that the normal component of velocity of the unburnt gases flowing into the flame is U_L. There is usually an outer luminous sheath, or "brush," that burns as a secondary diffusion

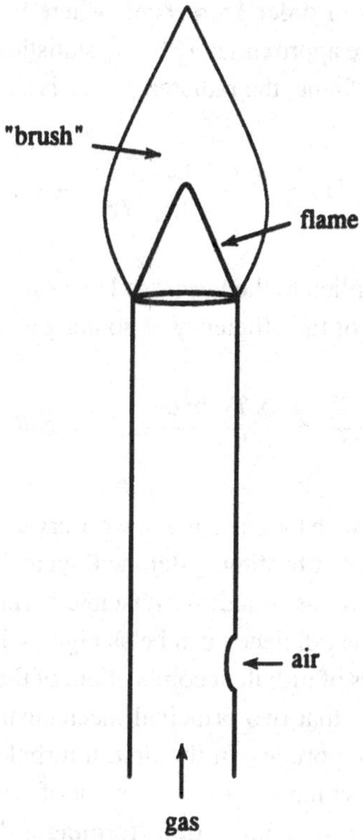

Figure 6.4.2. Bunsen burner.

flame, provided the initial gas–air mixture is rich enough. In the steady state (no turbulence) the heat production rate $\int q d^3\mathbf{x} = \rho_1 U_L c_p (T_2 - T_1) S_o$, where ρ_1 is the density of the unburnt gas, T_1 and T_2 are the temperatures before and after burning, and S_o is the mean surface area of the flame. The flame may be assumed to be acoustically compact, and according to (6.4.7) the acoustic pressure at large distances is

$$p(\mathbf{x}, t) \approx \frac{\rho_o U_L (T_2 - T_1)}{4\pi|\mathbf{x}|T_1} \left[\frac{dS}{dt}\right]_{t-|\mathbf{x}|/c_o},$$

where S is the flame area at time t. Thus, the mean square pressure depends on $\langle (dS/dt)^2 \rangle$, which is governed by the wrinkling of the flame front by the turbulent stream. Wrinkling occurs over a length scale determined by the correlation length ℓ of the turbulence (consistent with the practical observation that combustion noise is a low-frequency phenomenon), such that a local fluctuation in flame area in time δt is of order $\delta S \sim \ell v \delta t$, where v is a turbulence velocity. Hence, because there are approximately S_o/ℓ^2 statistically independent surface area fluctuations of the flame, the radiated power is

$$\Pi_a \sim \frac{\rho_o v^2 U_L^2 S_o (T_2 - T_1)^2}{4\pi c_o T_1^2}.$$

The thermal power supplied to the burner is $\Pi_T \sim \rho_1 U_L S_o c_p (T_2 - T_1)$ so that the order of magnitude of the efficiency of sound generation is

$$\frac{\Pi_a}{\Pi_T} \sim \frac{\rho_o T_2}{\rho_1 T_1} \frac{v^2 U_L}{c_o^3}, \quad T_2 \gg T_1.$$

This result is consistent with the experimental observation that, in the absence of feedback instabilities, a combusting system will typically radiate about 10^{-6} – 10^{-5} of the thermal power as sound; for unstable burning (with feedback) in a combustion chamber, the efficiency can be as high as 10^{-4}.

In more general cases of turbulent combustion, of the kind occurring in a gas turbine engine, it seems that two principal mechanisms of directly generated combustion noise are important. In the first, a turbulent eddy with entrained fuel vapor burns freely within a hot environment of air whose temperature has previously been elevated by combustion (forming a diffusion flame). In the second, an eddy consisting of premixed fuel and air burns in the manner of a laminar flame. The subsequent interaction of combustion products of nonuniform temperature with mean flow velocity gradients and structural boundaries generates indirect combustion noise (Section 2.1.3), whose relative importance depends on the particular flow in question.

Example 2: Spherically expanding flame [449]. A stationary, spherical soap bubble of radius r_o in air contains a combustible mixture of ethylene and air at uniform temperature T_1 ($= T_o$) and density ρ_1 (Figure 6.4.3a). The mixture is ignited centrally by a spark at time $t = 0$, producing a spherically expanding flame front. The flame speed is low enough that the steady pressures on either side of the flame are equal to $p_o = \rho_2 R T_2 \equiv \rho_1 R T_1$, where ρ_2 and T_2 are the density and temperatures respectively, of the burnt gases, and R is the gas constant.

The heat release is confined to the thin flame front of radius $r(t)$ so that the volumetric rate of production of burnt gases is $4\pi r^2\, dr/dt$. Because $T_2/T_1 =$

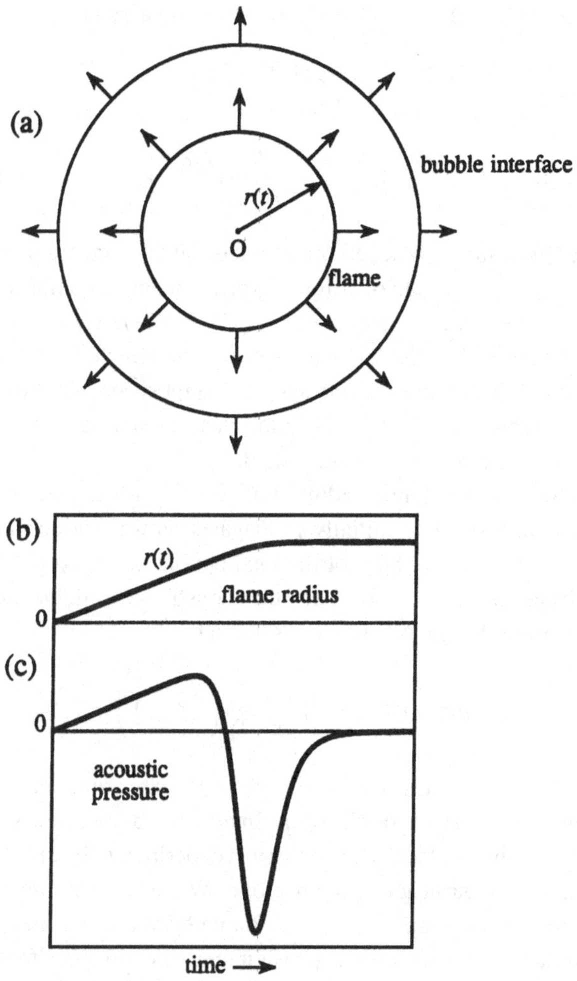

Figure 6.4.3. (a) Spherical flame in a bubble, (b) flame radius, and (c) the acoustic pressure.

ρ_1/ρ_2, the volumetric rate of consumption of the fuel–air mixture is

$$Q_o = \frac{T_1}{T_2} 4\pi r^2 \frac{dr}{dt},$$

and when the coordinate origin is taken at the center of the bubble,

$$\mathrm{q}(\mathbf{x}, t) \approx \rho_1 c_p (T_2 - T_1) Q_o(t) \delta\{|\mathbf{x}| - r(t)\}.$$

The radiation produced by the flame is therefore obtained from (6.4.7) to be

$$p(\mathbf{x},t) \approx \frac{\rho_o}{4\pi|\mathbf{x}|}\left(\frac{T_2}{T_1}-1\right)\left[\frac{dQ_o}{dt}\right]$$
$$\equiv \frac{\rho_o}{|\mathbf{x}|}\left(1-\frac{T_1}{T_2}\right)\left[2r\left(\frac{dr}{dt}\right)^2+r^2\frac{d^2r}{dt^2}\right], \qquad |\mathbf{x}|\to\infty,$$

where the terms in square brackets are evaluated at the retarded time $t-|\mathbf{x}|/c_o$. $(T_2/T_1-1)Q_o$ is the net rate of volume expansion due to combustion and corresponds to a radial fluid velocity $v_r = (1-T_1/T_2)\,dr/dt$ just outside the flame front. Except when the flame approaches the bubble interface, the burning is steady and the flame propagates at a constant speed $U_L = dr/dt - v_r \equiv (T_1/T_2)dr/dt$ relative to the unburnt fluid. Near the interface the flame decelerates, and the flame is finally extinguished.

The variation of the flame radius with time is illustrated schematically in Figure 6.4.3b. The flame initially propagates outward at constant velocity $U_L(T_2/T_1)$, until r is about 80% of the final bubble radius $r_o(T_2/T_1)^{1/3}$. During this initial period, $d^2r/dt^2=0$, and the acoustic pressure increases linearly with retarded time (Figure 6.4.3c) and is given by

$$p(\mathbf{x},t) \approx 2\rho_o U_L^2\frac{[r]}{|\mathbf{x}|}\frac{T_2}{T_1}\left(\frac{T_2}{T_1}-1\right).$$

When the flame decelerates, $r^2\,d^2r/dt^2$ becomes large and negative, and the radiation ends with a steep rarefaction. Small oscillations (not shown in the figure) following the rarefaction are observed experimentally and are attributed to pulsations of the burnt gaseous envelope. When the burning velocity U_L exceeds about 1 m/s, the initially smooth flame front can become wrinkled so that the flame area and volumetric expansion rate are no longer simple functions of the radius; this produces an increased rate of consumption of combustible gases and causes the peak positive pressure to exceed that predicted by the simple theory of Figure 6.4.3.

Example 3: Inward burning flame [449]. When the fuel–air mixture of Example 2 is ignited simultaneously at all points of the bubble interface

$$p(\mathbf{x},t) \approx \frac{\rho_o}{|\mathbf{x}|}\left(\frac{T_2}{T_1}-1\right)\left[2r\left(\frac{dr}{dt}\right)^2+r^2\frac{d^2r}{dt^2}\right], \qquad |\mathbf{x}|\to\infty,$$

the flame converges on the bubble center at constant speed $dr/dt = -U_L$, which is smaller by a factor $T_1/T_2 \ll 1$ than for the expanding flame of Example 2 because combustion products are now able to expand freely outside the bubble without disturbing the unburnt gas. The acoustic amplitude is therefore smaller by a factor $T_1/T_2 \approx 0.1$.

6.4.3 Rayleigh's Criterion

Large-amplitude, combustion-induced oscillations are observed in most systems involving continuous flow, such as aeroengine afterburners, gas boilers, and rocket motors. They occur when the back reaction of the sound on the thermoacoustic sources is correctly phased to promote the growth of the oscillations and were first explained in general terms by Rayleigh [10, 450, 451]. The simplest form of *Rayleigh's criterion* for the onset of instability is applicable at low Mach numbers where the mean pressure is uniform, and it may be derived from the acoustic analogy equation (6.4.4) (or alternatively from the continuity equation in the form (2.3.8)). Discard the quadrupole source, replace $(1/\rho)\nabla p$ in the second term on the left by $-\partial \mathbf{v}/\partial t$ from the linearized momentum equation, and integrate once with respect to time. The equation is then multiplied by p, and the momentum equation is used again to form the energy equation

$$\frac{\partial}{\partial t}\left(\frac{p^2}{2\rho c^2} + \frac{\rho v^2}{2}\right) + \operatorname{div}(p\mathbf{v}) = \frac{p\mathrm{q}}{c_p \rho T}, \tag{6.4.8}$$

where p and $\mathbf{v}$ are the perturbation pressure and velocity, respectively, q is the heat release rate per unit volume, and all other quantities assume their local mean values. The left-hand side of (6.4.8) coincides with the general energy equation (1.13.2) in the absence of mean flow.

This equation is applicable at low Mach numbers to any combusting system, confined or not; the energy source on the right determines the radiated acoustic energy and represents the work done on the exterior fluid by the pressure during volumetric expansion induced by the thermal source. For a free-burning flame, the pressure on the right hand side would be determined by the near-field part of the solution of (6.4.4). For a confined flame or heat source, however, the pressure at the flame involves contributions from a standing acoustic field within the combustion chamber. Large-amplitude, resonant oscillations are therefore possible only when this pressure satisfies

$$\int \langle p(\mathbf{x}, t)\mathrm{q}(\mathbf{x}, t)\rangle \, d^3\mathbf{x} > 0, \tag{6.4.9}$$

where the mean value $\langle p\mathrm{q}\rangle$ is evaluated over a period of the oscillations, and the integration is over the entire thermal source region. For a confined region of heat release, equation (6.4.8) shows that this result can be expressed in the form

$$\frac{1}{c_p\rho T}\int \langle p(\mathbf{x},t)\mathrm{q}(\mathbf{x},t)\rangle\, d^3\mathbf{x} = \oint \langle p(\mathbf{x},t)\mathbf{v}(\mathbf{x},t)\cdot d\mathrm{S}\rangle > 0,$$

where the second integral is over a surface that just encloses the combustion zone. The surface element $d\mathbf{S}$ is directed into the ambient fluid, and the second integral is just the rate of working of the source region on the ambient acoustic medium. Rayleigh's criterion is necessary, but not sufficient for the maintenance of the sound, because it takes no account of dissipation: The power supplied by the heat source must also be sufficient to overcome losses in boundary layers, due to vorticity production at edges, acoustic radiation from openings, and so forth.

The inequality (6.4.9) can be satisfied in two opposite extremes: first when heat is supplied at a time of compression, or alternatively, when heat is *withdrawn* during a time of expansion. If the heat source is at a pressure node of a particular acoustic mode of the chamber, that mode cannot be maintained by acoustic feedback to the source. There are many ways in which feedback can result in the favorable phasing of heat input and the pressure so that (6.4.9) is satisfied; for example, acoustic pressure fluctuations can induce pulsations in fuel supply rate, flame area, the production of fuel droplets, burning rates associated with periodic mixing of fuel and air, and so forth. A simple illustration is the "singing" diffusion flame, formed at the exit of a gas supply line inserted along the axis of an open-ended, vertical tube; the fundamental mode of the tube can be excited when the flame is near the center of the tube. The operation depends on the details of the supply line configuration, and so forth, but typically the fuel supply rate is modulated by the excitation of waves in the supply line by the standing wave in the tube; when singing occurs, the integral (6.4.9) turns out to be positive because of the existence of a phase lag between a change in the fuel supply rate and the heat release by burning [451, 452].

The Sondhauss tube [10, 450, 453] is a glass tube about 15 cm long open at one end and terminated at the other end by a bulb about 2 cm in diameter (Figure 6.4.4a). The air within the tube can be caused to vibrate when a steady, localized source of heat is applied to the bulb. The frequency of the sound corresponds closely to that of the lowest order resonance of the hot air in the tube (the acoustic wavelength is about four times the tube length), provided the dimensions of the bulb are not too great in relation to the tube length. Rayleigh [450] pointed out that just before the maximum compression of the air in the bulb, the flow is into the bulb but from a *colder* part of the tube so that, provided

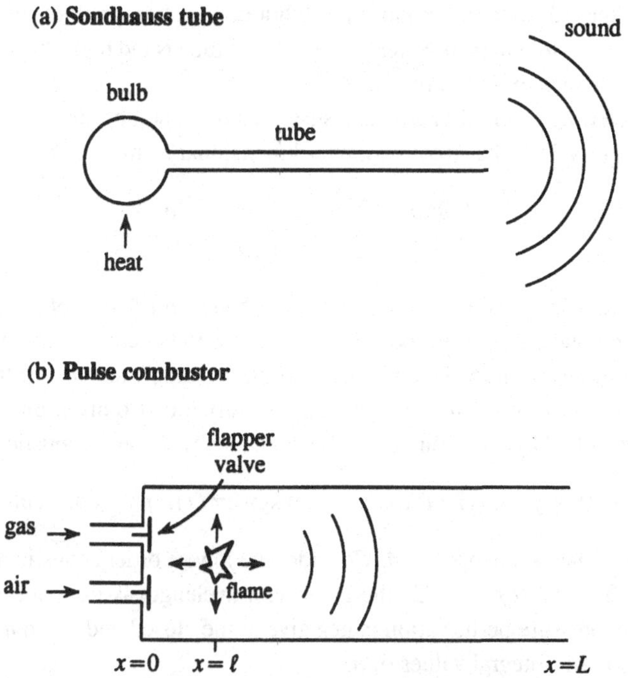

Figure 6.4.4. Sandhauss tube, and (b) pulse combustor.

the temperature variation along the glass tube is large, the air entering the bulb, although raised in temperature by virtue of compression, is initially cooler than the glass bulb, and there is therefore a net transfer of heat to the fluid at a time of compression. Similarly, when the air in the bulb expands, hot air leaving the bulb loses heat to the cooler tube wall. In these circumstances condition (6.4.9) is satisfied and a steady oscillation will ensue. Oscillations attributable to large thermal gradients are also observed in narrow tubes open at one end when the open end is immersed in liquid helium and the closed end is at room temperature.

The pulse combustor consists of a long tube of length L open at the end $x = L$, illustrated schematically in Figure 6.4.4b. Natural gas and air enter at the closed end ($x = 0$) through inlets controlled by "flapper" valves, which are the only moving parts in the combustor. Under favorable conditions, combustion occurs periodically under the control of a low-frequency acoustic resonance of the tube. The valves open when the acoustic pressure at $x = 0$ falls below a certain level; the fuel and air mix and subsequently burn explosively near the closed end (ignition usually being initiated by the remnants of the preceding flame that is swept back into the closed-end region during the rise in pressure), causing the valves to close and forcing hot combustion products to flow from the open end. In practical systems, the exhaust tube is fed into a heat exchanger where it divides to form a tube bundle.

To determine conditions for self-sustained operation of the combustor, take equation (6.4.5) in the simplified, one-dimensional form

$$\frac{1}{c_2^2}\frac{\partial^2 p}{\partial t^2} - \frac{\partial^2 p}{\partial x^2} = \frac{\rho_2}{A c_p \rho_1 T_1}\frac{\partial \mathrm{q}}{\partial t}, \tag{6.4.10}$$

where $\mathrm{q} \equiv \mathrm{q}(x, t)$ is the rate of heat release *per unit length of tube*; A is the cross-sectional area of the tube, which is filled with hot combustion products of density ρ_2, temperature T_2, and sound speed c_2; and ρ_1 and T_1 are the density and temperature of the unburnt fuel and air mixture. Combustion is assumed to be controlled by the acoustic mode of frequency f, with acoustic pressure

$$p(x, t) \approx p' \sin\{(2\pi f/c_2)(x - L)\} \sin(2\pi f t), \quad p' = \text{constant} > 0,$$

which vanishes at the open end. Consider the lowest order acoustic resonance, for which $2\pi f L/c_2 \approx \pi/2$ (the acoustic wavelength is then approximately $4L$). The pressure perturbation is negative at the closed end when $n/f < t < (n + \frac{1}{2})/f$, for integral values of n.

Write the periodic heat source as a succession of localized explosions

$$\mathrm{q} = \rho_1 c_p V_o (T_2 - T_1) \sum_n \delta(t - \tau - n/f)\delta(x - \ell),$$

where V_o is the volume of unburnt gases consumed per cycle; the argument of the δ function $\delta(x - \ell)$ vanishes at $x = \ell$ very near the closed end of the tube, where combustion is assumed to occur at time τ after the acoustic pressure goes negative at the closed end. The value of τ is determined by fuel characteristics and details of the mixing process. The component of q of frequency f is

$$\mathrm{q}(x, f) = 2 f \rho_1 c_p V_o (T_2 - T_1)\delta(x - \ell)\cos\{2\pi f(t - \tau)\}.$$

To simplify the calculation, assume that fuel and air enter the combustor only when $p < 0$ at the flapper valves. If losses due to vorticity production, and so forth, at the valves are ignored, the volume $V_o \equiv V(\tau)$ is determined by the equation

$$\rho_1 \frac{d^2 V}{dt^2} = -K_R\, p, 0 \quad \text{according as } p \lessgtr 0,$$

where $K_R > 0$ is a real-valued, "lumped" conductivity of the flapper valves. Taking $p = -p' \sin(2\pi f L/c_2) \sin(2\pi f t)$, which is negative when $0 < t <$

$1/2f$, we find

$$V_o = \frac{p' K_R \sin(2\pi f L/c_2)}{4\pi^2 \rho_1 f^2}(2\pi f t_m - \sin(2\pi f t_m)),$$

where

$$t_m = \min\{\tau, 1/2f\}.$$

We may now evaluate the source term in (6.4.10) for oscillations of frequency f. However, in this linearized approximation, instability will cause the amplitude of the pressure fluctuations to grow exponentially with time, and this has been neglected in evaluating $\mathrm{q}(x, f)$. The approximation will be satisfactory, however, provided the growth rate is small, so that the change in source strength per cycle is also small. The effect of this slow increase in amplitude is incorporated into the left of (6.4.10) by writing p in the form

$$p = \frac{p'}{2}\{i \sin\{\kappa_2(x - L)\}\mathrm{e}^{-i\omega t} + c.c.\}, \quad f = \mathrm{Re}\,\omega/2\pi,$$

where ω is complex, and $\kappa_2 = \omega/c_2$. This expression is strictly applicable only to the right ($x > \ell$) of the combustion zone, which acts as a *piston* forcing the acoustic oscillations.

Instability occurs when $\mathrm{Im}\,\omega > 0$, and the condition for this is obtained by substituting $\mathrm{q}(x, f)$ for q in (6.4.10) and integrating over a small interval $0 < x < \ell + \delta\,(\delta > 0)$ just enclosing the combustion zone. This can be done for each component of the solution proportional to $\exp(\pm 2\pi i f t)$. Because $\omega\ell/c_2 \sim \ell/L \ll 1$, only $\partial^2 p/\partial x^2$ (which has a δ-function singularity at $x = \ell$) contributes to this integral on the left-hand side: $\int_0^{\ell+\delta}(\partial^2 p/\partial x^2)\,dx = (\partial p/\partial x)_{\ell+0}$ as $\delta \to 0$. Hence, for $\ell \ll L$ we can write

$$\cos\{\kappa_2 L + \varepsilon \mathrm{e}^{2\pi i f \tau}\} \approx 0,$$

provided

$$\varepsilon \equiv \frac{2}{\pi^2}\left(\frac{K_R L}{A}\right)\left(1 - \frac{T_1}{T_2}\right)\{2\pi f t_m - \sin(2\pi f t_m)\} \ll 1.$$

We are at liberty to assume this condition to be satisfied so that the lowest order complex resonance frequency is

$$\frac{\omega L}{c_2} \approx \frac{\pi}{2} - \varepsilon \exp(2\pi i f \tau).$$

Because $\varepsilon > 0$, this will have positive imaginary part, causing the oscillations to grow and become self-sustaining, if the lag τ satisfies $1/2f < \tau < 1/f$, that is if combustion occurs during the time of high acoustic pressure at the closed end, in agreement with the Rayleigh criterion (6.4.9).

Example 4. In the enunciation of the criterion (6.4.9), Rayleigh [10, 450] also pointed out that "if heat is given at the moment of greatest rarefaction, or abstracted at the moment of greatest condensation [compression], the pitch [frequency] is not affected."

"... the pitch is raised if heat be communicated to the air a quarter period before the phase of greatest condensation; and the pitch is lowered if heat is communicated a quarter period after the phase of greatest condensation."

6.4.4 The Rijke Tube

A time lag in heat transfer also plays a crucial part in the excitation of oscillations in the Rijke tube [28, 454–458], which consists of a vertical cylindrical tube of length ℓ, open at both ends with a wire gauze stretched across it at a distance L from the lower end (Figure 6.4.5a). The gauze is maintained at a constant high temperature T_w by a steady electric current. In a typical demonstration, the tube is circular with diameter between 5–15 cm, and ℓ is between one and two meters. The hot gauze induces a steady upward convection flow through the tube at speed U, say, usually less than one meter per second. Large-amplitude acoustic oscillations are excited in the tube when the gauze is in the *lower half*

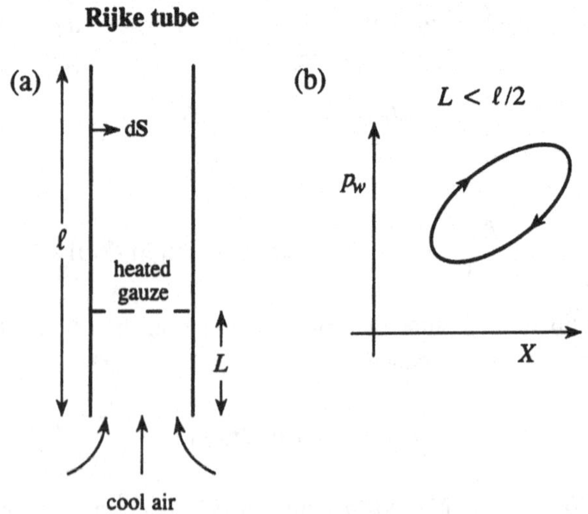

Figure 6.4.5. Rijke tube.

of the tube ($L < \frac{1}{2}\ell$); although all resonances are excited, the contribution from the fundamental (of wavelength 2ℓ) is usually dominant.

The convection of sound by the mean flow can be neglected. The temperature rise across the gauze is usually small enough that the jumps in mean sound speed and fluid density have a negligible impact on tube resonances, which may be regarded as subject to small-amplitude forcing at the gauze. Thus, the perturbation pressure $p(x,t)$ and velocity $v(x,t)$ for a longitudinal mode of frequency $\omega > 0$ may be taken in the real forms

$$p = p' \sin(\kappa_o x)\cos(\omega t), \quad v = \frac{-p'}{\rho_o c_o}\cos(\kappa_o x)\sin(\omega t), \quad p' = \text{constant},$$

where x is measured along the tube from the lower end, and the pressure vanishes at the open ends, provided $\kappa_o \ell \equiv \omega\ell/c_o = n\pi$ ($n = 1, 2,$ etc.).

If E is the acoustic energy within the tube, integration of the energy equation (6.4.8) yields

$$\frac{dE}{dt} - \oint p\mathbf{v}\cdot d\mathbf{S} = \frac{p_w \mathrm{Q}}{c_p \rho_o T_o},$$

where $\mathrm{Q} = \int \mathrm{q}\, d^3\mathbf{x}$ is the total rate of unsteady heat input from the gauze, and p_w is the pressure evaluated at the gauze. The surface integral is over the interior of the tube and the end planes, representing, respectively, energy losses in the wall boundary layers and radiation from the open ends. Unsteady heat input is controlled by the fluctuating velocity $v_w(t) = v(L,t)$ at the gauze [459, 460], and is a function of time of the form

$$\mathrm{Q} = \beta_w v_w(t - \tau),$$

where the constant coefficient $\beta_w = L_w(T_w - T_o)\sqrt{\pi\kappa c_v \rho_o D/2U}$, L_w being the total length of wire in the gauze, D the wire diameter, and κ the thermal conductivity of the air. The transfer of heat between the fluid and gauze depends on both molecular diffusion and convection by the flow, and τ represents a time lag between the heat release and the acoustic velocity v_w at the gauze. When the wire Reynolds number UD/ν is large and the Strouhal number $\omega D/U < 20$ (the usual case in practice), the time lag $\tau \approx 0.2D/U$ [460], and the mean rate of supply of thermal energy to the acoustic field is

$$\frac{\langle p_w \mathrm{Q}\rangle}{c_p \rho_o T_o} = \frac{\beta_w p'^2}{4\rho_o^2 c_o c_p T_o}\sin(2\kappa_o L)\sin(\omega\tau).$$

For the fundamental mode, $\omega\ell/c_o = \pi$, $0 < \omega\tau < \pi$ so that $\sin(\omega\tau) \equiv$

$\sin(\pi\tau c_o/\ell) > 0$. Thus, self-excitation is possible provided $L < \frac{1}{2}\ell$, the energy input to the sound being a maximum when $L = \frac{1}{4}\ell$.

More generally, introducing the explicit form of β_w, the condition for excitation is

$$(T_w - T_o)\sin\left(\frac{2\pi L}{\ell}\right) > 0,$$

which reveals that a *cold* gauze ($T_w < T_o$) can also excite the fundamental resonance, provided it is placed in the *upper* half of the tube. This happens in the *Bosscha* tube, which can be realized by feeding warm air into the tube, when resonant oscillations are excited until the gauze is fully heated to the air temperature.

Heating in the Rijke tube produces an effective volume outflow from the gauze equivalent to an axial velocity jump $[v] = (\beta_w/c_p\rho_o T_o A)v_w(t-\tau)$, where A is the cross-sectional area of the tube. The system may therefore be regarded as a *heat engine* in which a piston does an amount of work $Ap_w dX$ on the fluid in time dt, where $dX = [v]\,dt$, that is,

$$X = (\beta_w p'/\omega\rho_o^2 c_o c_p T_o A)\cos(\pi L/\ell)\cos\{\omega(t-\tau)\}.$$

In one complete cycle the piston does work $A\oint p_w dX$, where the integration is over a cycle, the work performed is positive, and the acoustic oscillations are thereby maintained, provided heat is added when the gas is compressed (Rayleigh's criterion). During the cycle, the point (X, p_w) in the Xp-plane must then traverse a closed (elliptic) loop in the *clockwise* direction, whose enclosed area is equal to the energy ceded to the sound, as indicated in Figure 6.4.5b. The ellipse has maximum area when $L = \frac{1}{4}\ell$; it would collapse to a straight line if the lag τ between the velocity and heat input at the gauze vanishes, in which case the phase point moves back and forth along the line and no work is performed. When the gauze is placed in the upper half of the tube, the ellipse is traversed in the counterclockwise sense, and energy is absorbed by the gauze.

Example 5. Show that when account is taken of damping in *wall* boundary layers and free-field radiation from the open ends, the condition for instability of the Rijke tube can be written

$$\frac{A_w}{A}\left(\frac{T_w}{T_o}-1\right)\sqrt{\frac{\chi}{2\pi\gamma U D}}\sin(2\pi L/\ell)\sin(\omega\tau)$$
$$> \frac{A}{\ell^2} + \frac{\ell}{R}\sqrt{\frac{2}{\pi}}\frac{[\sqrt{\nu}+(\gamma-1)\sqrt{\chi}]}{\sqrt{c_o\ell}},$$

where $A_w = L_w D$ is the frontal area of the wire gauze, R is the tube radius, and $\chi = \kappa/\rho_o c_p$ is the thermal diffusivity.

6.4.5 Flame Propagation in a Tube [461–466]

Let us consider the sound generated when a flame propagates in a long straight tube, closed at one end. The tube is initially filled with a combustible gas mixture that is ignited at the open end. The level and the intensity of the sound are in practice critically dependent on the fuel–air mixture, a 1 or 2% change in composition is often sufficient to inhibit completely the production of sound. The flame speed is very small relative to the speed of sound, and the steady mean pressure is effectively constant throughout the tube, although the changes in temperature and density across the flame region are large. The flame will be regarded as a hot flexible sheet whose thermal output is modulated periodically by acoustic waves excited in the tube. There is considerable debate concerning the precise nature of these modulations, and several theoretical models have been proposed.

Suppose the flame occupies a narrow region translating at constant speed U in the positive x-direction. Let the tube have total length $\ell = \ell_1 + \ell_2$, where ℓ_2 is the current distance of the flame from the open end (Figure 6.4.6a), and denote the respective mean densities, sound speeds and temperatures of the unburnt and burnt gases by ρ_1, c_1, T_1 and ρ_2, c_2, T_2. Thermally excited, one-dimensional sound waves in the tube are described by the following simplified form of equation (6.4.4)

$$\frac{1}{\rho c^2}\frac{\partial^2 p}{\partial t^2} - \frac{\partial}{\partial x}\left(\frac{1}{\rho}\frac{\partial p}{\partial x}\right) = \frac{1}{Ac_p\rho T}\frac{\partial \mathrm{q}}{\partial t},$$

where q is the unsteady rate of heat release per unit length of the tube, and A is the cross-sectional area. The acoustic pressure does not change across the flame, but by integrating this equation across the flame zone and by using the momentum equation on either side, it is seen that there is jump $v_1 - v_2$ in the acoustic particle velocity v, given by

$$v_1 - v_2 = \frac{\mathrm{Q}}{Ac_p\rho T}, \tag{6.4.11}$$

where $\mathrm{Q} = \int \mathrm{q}\,dx$ is the net heat release rate; $\rho T \equiv p_o/\mathrm{R}$ may be regarded as constant throughout the tube.

Translation of the flame causes the resonance frequencies to change slowly in time because of the changes in the lengths ℓ_1 and ℓ_2 of the acoustic domains on

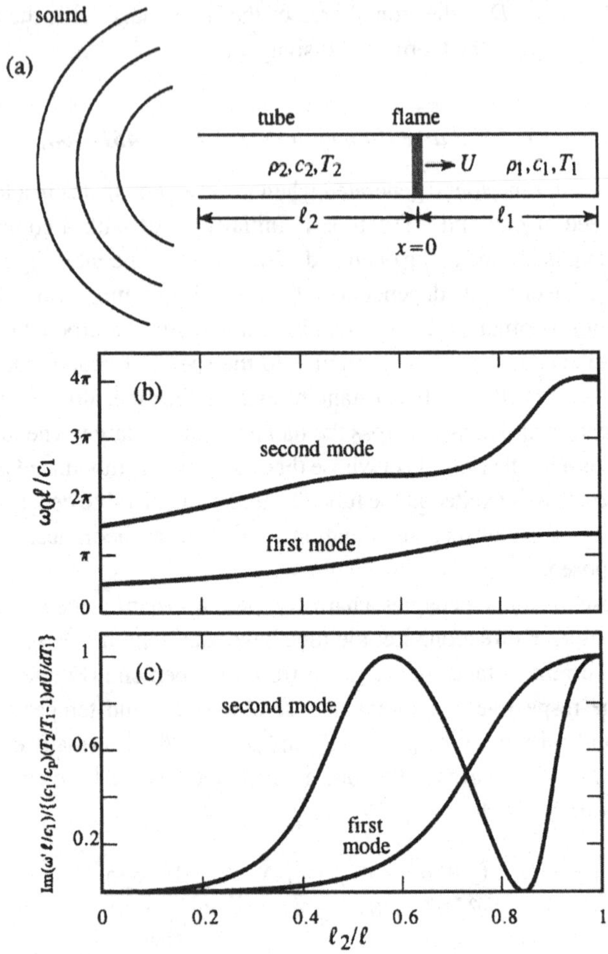

Figure 6.4.6. (a) Flame propagation in a tube, (b) and (c) dependence of frequency and growth rate on flame position.

either side of the flame. However, the fractional change during a characteristic period of oscillation $\delta\ell_1/\ell_1 \approx M \ll 1$ because $M = U/c_1 \sim 10^{-3}$, where $U \approx U_L S/A$ is the flame propagation velocity, and S is the area of the flame front. If the acoustic pressure in the tube is $pe^{-i\omega t}$, we can set

$$p = p_2 \sin\{\kappa_2(x+\ell_2)\}, \qquad -\ell_2 < x < 0, \quad \kappa_2 = \omega/c_2,$$
$$= p_1 \sin\{\kappa_1(x-\ell_1)\}, \qquad 0 < x < \ell_1, \quad \kappa_1 = \omega/c_1,$$

where p_1 and p_2 are constants, and the coordinate origin is taken at the current position of the flame. These expressions satisfy the open- and closed-end conditions, respectively, at $x = -\ell_2$ and ℓ_1. Continuity of pressure and the jump condition (6.4.11) at the flame ($x = 0$) require further that

$$p_1 \cos(\kappa_1 \ell_1) - p_2 \sin(\kappa_2 \ell_2) = 0,$$
$$\frac{p_1 \sin(\kappa_1 \ell_1)}{\rho_1 c_1} - \frac{p_2 \cos(\kappa_2 \ell_2)}{\rho_2 c_2} = \frac{i\mathrm{Q}'}{Ac_p \rho T},$$

where Q′ is the complex amplitude of the heat release $\mathrm{Q} = \mathrm{Q}' \mathrm{e}^{-i\omega t}$.

By eliminating p_2 between these two equations and by using the ideal gas relation $\rho_2 c_2 / \rho_1 c_1 = c_1 / c_2$, we find

$$\cot(\kappa_1 \ell_1) \cot(\kappa_2 \ell_2) - c_1/c_2 = \frac{-ic_2(\mathrm{Q}'/p_1)}{Ac_p T_2 \sin(\kappa_1 \ell_1)}. \tag{6.4.12}$$

If $\mathrm{Q}' = 0$, this would be the frequency equation for acoustic modes in the pipe in the presence of jumps in the mean density and sound speed at a distance ℓ_2 from the open end. The unsteadiness in the actual heat release rate Q is caused by the interaction of the sound with the flame; we can verify a posteriori that the magnitude of this effect is small and that under favorable circumstances it causes only a small shift in the value of the resonance frequencies from their real values obtained from (6.4.12) when $\mathrm{Q}' = 0$. Thus, we can set $\omega = \omega_0 + \omega'$, where ω_0 is a solution of (6.4.12) for $\mathrm{Q}' = 0$, and $\omega' \ll \omega_0$. In a first approximation, we then find

$$\frac{\omega' \ell}{c_1} \approx \frac{ic_1(\mathrm{Q}'/p_1) \sec \vartheta}{Ac_p T_1[1 + (\ell_1/\ell + \ell_2 T_1/\ell T_2) \tan^2 \vartheta]}, \quad \vartheta = \frac{\omega_0 \ell_1}{c_1}, \tag{6.4.13}$$

so that oscillations are excited at frequency ω_0 when $\mathrm{Im}\, \omega' > 0$, provided the growth rate is large enough to overcome radiation and wall boundary layer damping, neither of which has been included in the present calculation (cf. Example 5).

The first- and second-resonance frequencies $\omega_0 \ell / c_1$ of the tube are plotted in Figure 6.4.6b as a function of the fractional propagation distance ℓ_2/ℓ of the flame from the open end for the case $T_2/T_1 = 8$; the frequencies progressively increase as more of the tube is filled with the hot, burnt gas of elevated sound speed c_2. To calculate the growth rate, it is necessary to express the heat release Q′ in terms of the sound. There is no consensus on how this should be done, and to illustrate the procedure we outline a simple model due to Jones [462], who assumes that fluctuations in the specific heat release $U_L c_p(T_2 - T_1)$ from the flame depend only on changes in the laminar flame speed U_L, which in turn varies with the temperature of the unburnt gases. This hypothesis involves the assumption that the flame front surface area S is constant (see below). For an ideal gas, satisfying $p = \rho RT$, adiabatic fluctuations in temperature due to the sound satisfy $\partial T/\partial t = (1/c_p\rho)\partial p/\partial t$, and we can now write

$$Q' \equiv A(T_2 - T_1)\frac{dU}{dT_1}\, p_1 \cos(\kappa_1\ell_1), \quad U = U_L S/A,$$

whereupon (6.4.13) becomes

$$\frac{\omega'\ell}{c_1} \approx \frac{ic_1(T_2 - T_1)\, dU/dT_1}{c_p T_1[1 + (\ell_1/\ell + \ell_2 T_1/\ell T_2)\tan^2\vartheta]}. \tag{6.4.14}$$

An acoustic mode of nondimensional frequency $\omega_0\ell/c_1$ is excited and maintained by the flame, provided Im ω' is sufficiently large and positive. Both the frequency and growth rate vary with the position of the flame, as indicated in Figure 6.4.6 for the first two modes. According to Figure 6.4.6c, the second harmonic is excited first, initially when about half of the available combustible gases have been consumed, but it subsequently gives way to the lowest order mode. As the flame approaches the closed end, both modes (and higher order modes) are excited, and experiment indicates that the acoustic field consists of a multitude of different harmonics [463]. In the quasi-static approximation $U = U_L S/A$, where U_L is given empirically by a formula of the type (6.4.1), in which case ω' is positive imaginary, and $\omega'/\omega_0 \approx (\gamma - 1)(T_2/T_1 - 1)M \approx 10^{-3}$, where $M = U/c_1$. This estimate justifies the approximations leading to (6.4.14).

However, the approximation neglects possible phase lags between changes in U_L and the acoustic temperature fluctuation [464], which must become important at very high frequencies. There are, in addition, alternative hypotheses regarding the fluctuating heat output [461, 465–468]. It has been argued, for example, that the oscillations cause an elongation of the flame close to the wall and a fluctuation in heat output resulting from the change in flame surface area. Furthermore, observations reveal that the flame front usually becomes rippled with increasing flame velocity, developing a "cellular" structure with a curved flame front. Because the density changes discontinuously, the front becomes unstable in the acceleration field of the sound wave, causing the flame area, and therefore the heat output, to fluctuate.

6.4.6 Anchored Flames

In engineering applications, a flame in a duct with a mean flow is usually stabilized on a bluff body or "flame holder," from which the flame profile may be

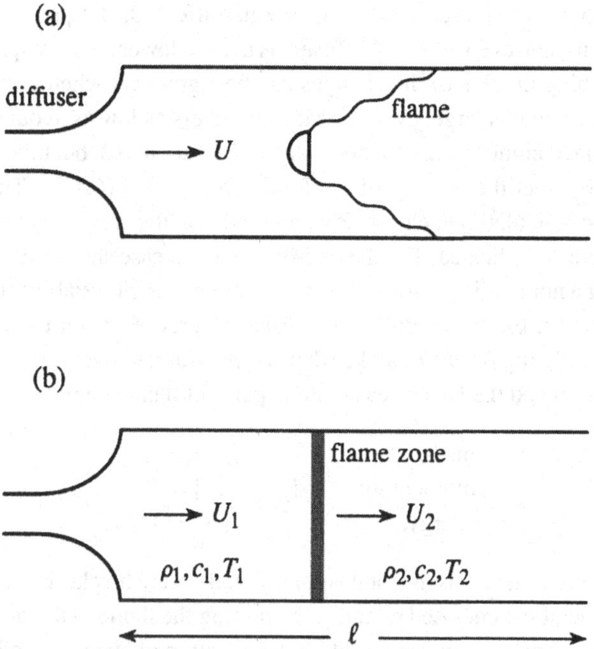

Figure 6.4.7. (a) Anchored flames and (b) anchored plane flame.

pictured as trailing downstream to maintain a normal flame front velocity equal to U_L (Figure 6.4.7a). When the flow is turbulent, the flame becomes wrinkled. The shape and effective burning surface area of the flame then fluctuate in time, and such motions can couple to a resonant acoustic mode in the duct. The simplest case is furnished by a wire gauze across the duct, on which an anchored flame can generate sound as in the Rijke tube [451, 469–471]. More generally, the impingement of sound on the flame holder causes fluctuations in the flame separation point, which in turn produces a wavelike rippling of the flame that propagates along the front at a phase velocity roughly equal to the component of the mean velocity tangential to the flame; this may cause fluctuations in the heat release rate necessary to maintain resonant oscillations [472]. This can occur, for example, in the after burner of a jet engine, giving rise to a phenomenon known as *reheat buzz* [473–475]. The flame often extends a considerable distance downstream of the holder, and this distributed heat source can affect the values of the resonance frequencies as well as the stability characteristics. In flow-through combustion chambers, fuel is usually injected just upstream of the flame. The presence of the flame holder introduces an unsteady drag (vortex shedding) that modifies the natural frequencies and combustion stability; vortex shedding at a downstream exit may be an additional source of damping

(Section 5.2). The presence of mean flow also affects the frequencies of coupled oscillations, and even when a "diffuser" is used to lower the flow speed (to avoid extinguishing the flame) the changes can be significant when the temperature gain at the flame is large, even for Mach numbers as low as about 0.15.

The Mach number does not normally exceed about 0.3, but this may be large enough to affect the validity of the Rayleigh criterion (6.4.9). The analytical problem of calculating the unsteady motion due to the presence of sound is now much more complicated. For the one-dimensional case shown in Figure 6.4.7b involving a nominally stationary, thin planar flame region stabilized in a steady mean flow (in the x-direction), it is usual to proceed from the conservation equations relating ρ, p, T, and v (density, pressure, temperature, and velocity, respectively) and the heat release rate Q per unit flame area:

$$\begin{aligned} &\text{mass,} && [\rho v] = 0 \\ &\text{momentum,} && [p + \rho v^2] = 0 \\ &\text{energy,} && [B] = \mathrm{Q}/\rho v, \end{aligned}$$

where $B = c_p T + \frac{1}{2}v^2$ is the total enthalpy, and square brackets denote the jump in the value of the enclosed quantity in crossing the flame in the flow direction. These equations are supplemented by the equation of state $p = \rho \mathrm{R} T$.

The motion in the uniform regions on each side of the flame is irrotational. If we let primed and unprimed variables, respectively, denote perturbation and mean quantities (the latter being constant on each side of the flame), then $B' - Ts'$ satisfies the convected wave equation

$$\left(\frac{1}{c_j^2}\left(\frac{\partial}{\partial t} + U_j\frac{\partial}{\partial x}\right)^2 - \frac{\partial^2}{\partial x^2}\right)(B' - Ts') = 0, \quad j = 1, 2,$$

where the suffixes $j = 1, 2$ refer, respectively, to conditions ahead and to the rear of the flame, and U_j is the mean flow velocity (see (2.3.22)). According to the momentum equation, the perturbation velocity $v' = \partial\varphi'/\partial x$, where the velocity potential φ' satisfies

$$\partial\varphi'/\partial t = -(B' - Ts'),$$

which is continuous across the flame because s' must vanish ahead of the flame, and energy conservation requires that $B_2' - B_1' = T_2 s_2'$. An acoustic energy equation can be formed, as in Section 1.13.1, on each side of the flame, from which it is readily deduced that the acoustic power generated by unit area of flame is

$$\Pi_a = -\frac{\partial\varphi'}{\partial t}\left[\rho\frac{\partial\varphi'}{\partial x} + \frac{Up'}{c^2}\right],$$

where the square brackets denote the jump in the enclosed quantity across the flame. Conservation of mass implies that $[\rho \partial \varphi'/\partial x] + [\rho' U] = 0$, where

$$\rho_1' = p_1'/c_1^2, \quad \rho_2' = p_2'/c_2^2 - \rho_2 s_2'/c_p \equiv p_2'/c_2^2 - \mathrm{Q}'/c_p U_2 T_2,$$

and Q' is the unsteady component of the heat release rate per unit flame area. Thus, the condition for the net production of sound becomes $\Pi_a \equiv (-\partial \varphi'/\partial t)(\mathrm{Q}'/c_p T_2) > 0$, which may also be written

$$\left(p_2' + \rho U \frac{\partial \varphi_2'}{\partial x} \right) \frac{\mathrm{Q}'}{c_p \rho_2 T_2} > 0.$$

This reduces to Rayleigh's criterion (6.4.9) when $U = 0$ (i.e., when the Mach number $M = U/c \ll 1$), in which case $\rho_2 T_2 \equiv \rho T =$ constant.

Example 6. In the absence of the flame, show that the natural frequencies of longitudinal modes of the combustion chamber of Figure 6.4.7b (of uniform sound speed c_o, mean density ρ_o, and flow velocity U) are given by

$$\frac{\omega \ell}{c_o(1 - M^2)} = \left(n + \frac{1}{2} \right) \pi - i \ln \left(\frac{1 + M}{1 - M} \right), \quad M = \frac{U}{c_o},$$

when the inlet is "choked" (mass flux = constant [104]). Assume the acoustic pressure vanishes at the open-end $x = \ell$ and apply the choking condition in the form $\rho'/\rho_o + v'/U = 0$ at the inlet. The frequencies have negative imaginary part showing that oscillations are damped by vorticity production at the outlet (see Section 5.2.2).

6.4.7 Vortex Shedding

The fuel in solid propellant rocket motors usually burns outward toward the rocket motor casing. The hot combustion products exhaust at high speed from the rocket nozzle and often experience unsteady oscillations when high-amplitude sound is generated within the combustion chamber; the unsteady pressures can generate strong unsteady loads on the rocket wall casing and other components. The heat release occurs over a substantial length of the combustion chamber, and simple notions such as the Rayleigh criterion often do not readily supply conditions for stability. When the solid fuel is in the form of axially segmented rings, as indicated in Figure 6.4.8, there can be strong coupling of vortex shedding from the "restrictors" (separating the fuel segments) and the acoustic field within the chamber. This instability is analogous to that

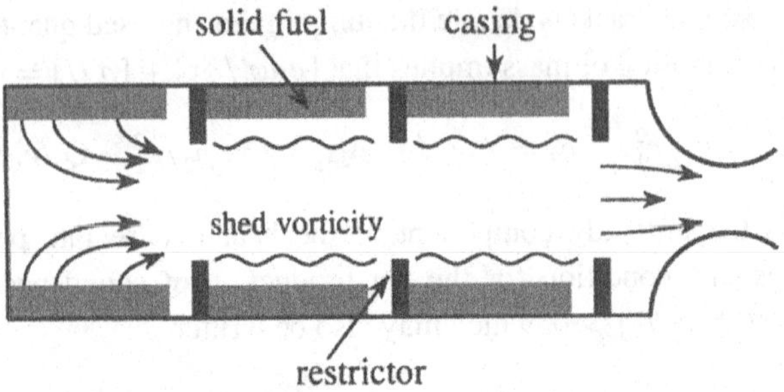

Figure 6.4.8. Segmented solid rocket motor.

discussed in Section 5.2.5 [419–421, 476–478], the feedback being essentially hydrodynamic rather than thermal.

Example 7: Generalized Rayleigh criterion. Derive the following generalization of Rayleigh's criterion (6.4.9) for low Mach number rotational flows:

$$\frac{1}{c_p \rho T} \int \langle p\mathrm{q} \rangle \, d^3\mathbf{x} - \int \rho \langle \boldsymbol{\omega} \wedge \mathbf{v} \cdot \mathbf{u} \rangle \, d^3\mathbf{x} > 0,$$

where **u** is the acoustic particle velocity.

6.5 Thermoacoustic Engines and Heat Pumps

The Sondhauss tube (Section 6.4.3) is the simplest example of a *thermoacoustic engine* [453, 479–485]. In 1962, it was discovered that the intensity of the sound could be greatly increased by inserting a coaxial glass tube, closed at one end, in the resonator near the bulb (Figure 6.5.1a). A similar arrangement using a resonator without a bulb was also found to be effective in generating sound; in this case, the closed end was maintained at a high temperature by an electrical heater (Figure 6.5.1b). The efficiency of sound production was further increased when the glass insert was replaced by a bundle of glass tubes (Figure 6.5.1c). The acoustic power supplied to the gas can be converted into useful work by closing the open end with a piston or other suitable transducer (Figure 6.5.1d), which transforms acoustic energy to mechanical or electrical energy. Of course, the presence of the transducer changes the acoustic boundary condition at that end; typically the wavelength of the lowest order resonance is now reduced to about twice the tube length.

Practical thermoacoustic engines are similar in principle to that shown in Figure 6.5.1d, except that the tube bundle is usually replaced by a "stack" of parallel plates aligned with the resonator axis, and heat is supplied via a heat

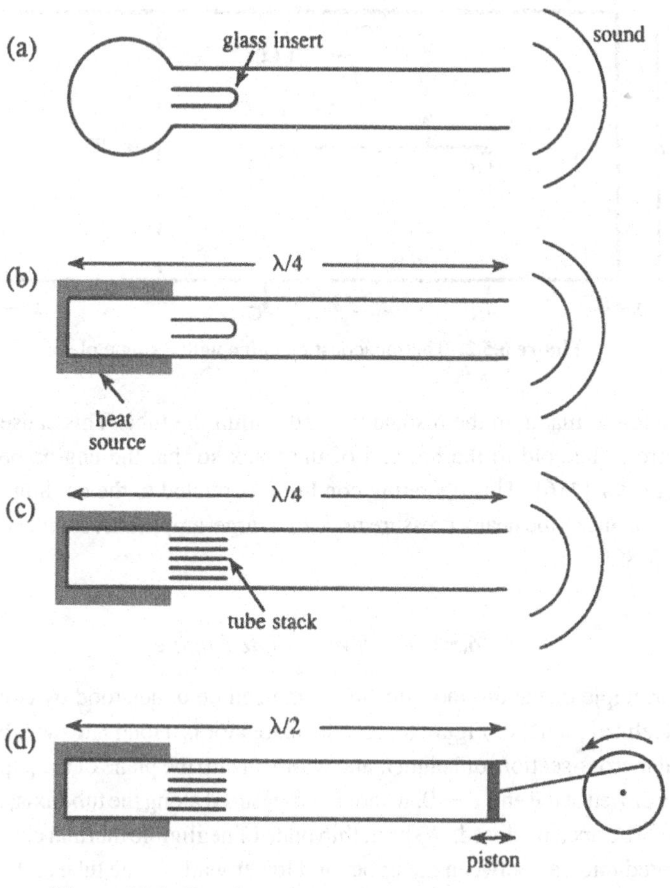

Figure 6.5.1. Development of the thermoacoustic engine.

exchanger at the leftmost end of the stack. In simple configurations, the heat exchanger consists of a set of evenly spaced short plates parallel to the resonator, that are made of highly conducting metal such as copper and are connected to a suitable source of heat. According to Rayleigh's criterion (Section 6.4.3), it is desirable to have a large temperature gradient along the stack, which is therefore constructed of a poorly conducting material (typically, stainless steel). The temperature of the "cold" end of the stack is controlled by a second set of copper strips in thermal contact with a heat sink, such as circulating cold water. The engine can function only when the temperature gradient exceeds a certain critical value determined by the geometry and material construction of the resonator. Below this value, sound cannot be excited and no useful work is performed, although the engine can then be run in reverse, by actuating

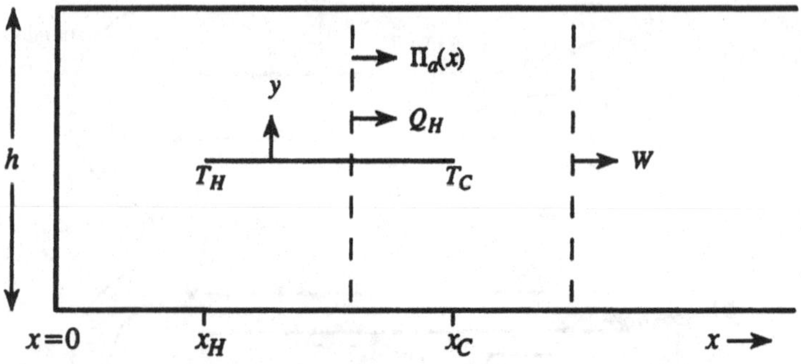

Figure 6.5.2. Thermoacoustic engine with a single plate.

the piston to maintain the resonant sound within the tube. This causes heat to flow from the cold to the hot end of the stack so that the engine becomes a *refrigerator* [486]. This operating condition is related to the cooling observed in a resonance tube near a pressure node of a large-amplitude, standing acoustic wave [480].

6.5.1 The Single-Plate Engine

The principle of the thermoacoustic engine can be understood by considering the idealized model of Figure 6.5.2. The resonator is a long narrow tube of rectangular cross-section, of height h and width d (into the plane of the paper). The tube is closed at the end $x = 0$, where x is measured along the tube axis; the other end may be open or closed. A short, thin plate of negligible thermal conductivity is located midway between the upper and lower walls of the tube and spans the resonator in the interval $x_H < x < x_C$ near the closed end. The ends of the plate are maintained at temperatures T_H, T_C ($T_H > T_C$) by heat exchangers. In the steady state, there is a standing acoustic wave of pressure $\text{Re}\{p(x, \omega)\text{e}^{-i\omega t}\}$ and acoustic particle velocity $\text{Re}\{u(x, \omega)\text{e}^{-i\omega t}\}$ parallel to the x-axis. There are small normal components of velocity near the solid surfaces of the plate and tube associated with viscous and thermal boundary layers (Section 5.1). The boundary layers on the resonator walls are responsible solely for the dissipation of sound and will be temporarily ignored. Then, except within the boundary layers of the plate, we can write

$$p(x, \omega) = p_0 \cos(\kappa_o x). \tag{6.5.1}$$

The plate length $\ell = x_C - x_H$ is assumed to be very small compared to the acoustic wavelength and is located sufficiently near the closed end so that $\kappa_o x_C \ll 1$.

The time-harmonic form of equation (6.4.4), with the quadrupole source omitted, can be used to relate the standing wave (6.5.1) to conditions on the plate. Periodic expansion and contraction of fluid near the plate occur because of the heat exchange between the fluid, plate, and heat exchangers. These exchanges can be expressed in terms of the temperature gradient within the fluid by means of the linearized energy equation (1.2.34), which permits the entropy source to be replaced by $-i\omega(\beta/\rho c_p)\mathrm{div}(\kappa\nabla T)$, where κ is the thermal conductivity of the fluid. We consider only the case of an ideal gas, for which $\beta = 1/T$, and integrate (6.4.4) over the region of fluid bounded by the closed end at $x = 0$ and a plane of constant $x > x_H$ (illustrated by either of the dashed lines in the figure). Then the divergence theorem supplies

$$\int_0^{h/2} \{u(x, y) - u_0(x)\}\, dy = -\int_{x_H}^{x} \left(\frac{\kappa}{c_p \rho_m T_m}\frac{\partial T}{\partial y} + \frac{i}{\omega\rho_m}\frac{\partial p}{\partial y}\right) dx, \tag{6.5.2}$$

where $u_0(x)$ is the (axial) particle displacement velocity for an ideal fluid *in the absence of the plate*, and y is measured in the normal direction from the upper surface of the plate. The left-hand side of (6.5.2) is the excess volume flux through a plane of constant x due to the presence of the plate; the integral on the right is taken along the upper surface of the plate, where the mean density and temperature are $\rho_m(x)$ and $T_m(x)$, respectively. The reader may verify from the boundary layer approximation to the momentum equation (1.2.9) that $\partial p/\partial y = -\rho_m \nu \partial^2 u/\partial x \partial y$ at $y = 0$. Thus, the two terms in the integrand on the right of (6.5.2), respectively, account for the influence on the sound of heat transfer and viscous drag at the surface of the plate. They can be expressed in terms of the acoustic variables by making use of boundary layer theory (Section 5.1), as described in Example 1, whereupon we find

$$\begin{aligned}\int_0^{h/2} &\{u(x, y) - u_0(x)\}\, dy \\ &= i\sqrt{i\omega}\int_{x_H}^{x}\left\{\frac{[\sqrt{\nu} + (\gamma - 1)\sqrt{\chi}]p(x,\omega)}{\rho_m c_m^2}\right. \\ &\qquad \left. - \frac{\chi}{\omega(\sqrt{\nu} + \sqrt{\chi})}\frac{i\bar{u}(x,\omega)}{T_m}\frac{dT_m}{dx}\right\} dx,\end{aligned} \tag{6.5.3}$$

where $\bar{u}(x, \omega)$ is the velocity just outside the thermal boundary layer on the plate, and γ is the ratio of the specific heats of the fluid. The first term in the brace brackets on the right would be present in the absence of the applied temperature gradient dT_m/dx, and represents the dissipation of sound in the thermal and viscous boundary layers (Section 5.1).

Equation (6.5.3) can be used to evaluate the acoustic power flux in the x-direction, namely,

$$\Pi_a(x) = 2d \int_0^{h/2} \langle \mathrm{Re}\{(u - u_0)\mathrm{e}^{-i\omega t}\}\mathrm{Re}(p\mathrm{e}^{-i\omega t})\rangle \, dy,$$

where the factor of two accounts for equal contributions above and below the plate. $\Pi_a(x)$ increases steadily with x over the interval $x_H < x < x_C$ occupied by the plate and becomes constant for $x > x_C$, where its value will be denoted by $W \equiv \Pi_a(x_C)$. W is the rate at which work is performed by the thermoacoustic engine, and it can be calculated from (6.5.3) by noting that the pressure is effectively uniform when $\kappa_o x \ll 1$ so that

$$W \approx \frac{d\sqrt{\omega}}{2\sqrt{2}\rho_o c_o^2} \int_{x_H}^{x_C} \left\{ \frac{\chi}{(\sqrt{\nu} + \sqrt{\chi})\kappa_o^2 T_m} \frac{\partial |p|^2}{\partial x} \frac{dT_m}{dx} \right.$$
$$\left. - 2[\sqrt{\nu} + (\gamma - 1)\sqrt{\chi}]|p|^2 \right\} dx.$$

Now $\partial|p|^2/\partial x < 0$ near the closed end of the resonator, and the integrand is therefore positive only when the mean temperature gradient dT_m/dx is sufficiently large and negative, that is, for $T_H > T_C$. A more explicit approximation can be made by using the representation (6.5.1) of the pressure. Because $\kappa_o x_C \ll 1$, we find

$$W \approx \frac{d\sqrt{\omega}|p_o|^2}{\sqrt{2}\rho_o c_o^2} \int_{x_H}^{x_C} \left\{ \frac{\chi}{(\sqrt{\nu} + \sqrt{\chi})} \frac{x}{T_m} \left| \frac{dT_m}{dx} \right| \right.$$
$$\left. - [\sqrt{\nu} + (\gamma - 1)\sqrt{\chi}] \right\} dx, \qquad (6.5.4)$$

which is positive, provided $|dT_m/dx|$ exceeds the critical value

$$\left| \frac{dT_m}{dx} \right|_W = (1 + \sqrt{\mathrm{Pr}})(\gamma - 1 + \sqrt{\mathrm{Pr}}) \left\langle \frac{T_m}{x} \right\rangle, \qquad (6.5.5)$$

where the angle brackets $\langle \; \rangle$ denote a suitable mean value over the interval occupied by the plate, and $\mathrm{Pr} = \chi/\nu$ is the Prandtl number.

When this critical temperature gradient is surpassed, the power supplied by the heat sources exceeds that dissipated in the boundary layers on the plate, which thus becomes a net source of acoustic energy, and the system can in principle function as a thermoacoustic engine. This is only a necessary condition for the generation of sound because the resonant oscillations are maintained only if the source is strong enough to overcome other losses, which include boundary layer damping at the resonator walls and radiation losses from any openings. When $|dT_m/dx|$ is subcritical, the plate absorbs acoustic energy.

Heat Transfer Along the Plate

To obtain a clearer understanding of the significance of the two cases $|dT_m/dx| \gtrless |dT_m/dx|_W$, let us examine the transfer of heat between the heat exchangers at the two ends of the plate. By hypothesis, the conduction of heat through the material of the plate may be neglected so that thermal contact is via the fluid alone. The principal mechanism of heat transfer is a second-order effect, within the thermal boundary layers on the plate, called *thermoacoustic streaming* [480, 481], whose magnitude is determined by evaluating the energy flux vector **I** of equation (1.2.31) correct to second order. At resonance the axial perturbation velocity is large (often exceeding several meters per second) and the component I_x of **I** parallel to the resonator is well approximated by $\rho v_x B$, where $v_x = \mathrm{Re}\{u(x,\omega)e^{-i\omega t}\}$. To evaluate I_x to second order, it is sufficient to take $B' \approx T_m s' + p'/\rho_m$, where the prime denotes a perturbation quantity, and s is the specific entropy.

The energy flux in the x-direction is then

$$\Pi \approx 2d \int_0^{h/2} \rho_m T_m \langle v_x s' \rangle \, dy + 2d \int_0^{h/2} \langle v_x p' \rangle \, dy,$$

where the factor of two accounts for equal contributions from both sides of the plate. The first integral is the heat flux due to convection by the fluid, and the second is the acoustic power $\Pi_a(x)$. In the steady state,

$$\begin{aligned} \Pi &= Q_H \quad \text{for } x_H < x < x_C, \\ &= W \quad \text{for } x > x_C, \end{aligned} \tag{6.5.6}$$

where Q_H is the rate of heat input from the heat exchanger at x_H. Q_H is found by making use of formulae derived in Example 1, which yield

$$Q_H \approx \frac{d\sqrt{\chi}(1-\mathrm{Pr}^{\frac{3}{2}})}{\sqrt{2}(1+\mathrm{Pr})\rho_o\omega^{\frac{3}{2}}} \left(\frac{|\partial p/\partial x|^2}{\kappa_o^2(\gamma-1)(1-\mathrm{Pr})T_m} \left| \frac{dT_m}{dx} \right| + \frac{1}{2}\frac{\partial |p|^2}{\partial x} \right).$$

In steady-state operation, the mean temperature gradient dT_m/dx adjusts itself to make Q_H independent of x in the region $x_H < x < x_C$ occupied by the plate. For a short plate near the closed end ($\kappa_o x_C \ll 1$), we have approximately

$$Q_H \approx \frac{xd|p_0|^2\sqrt{\omega\chi}\,(1-\mathrm{Pr}^{\frac{3}{2}})}{\sqrt{2}(1+\mathrm{Pr})\rho_o c_o^2}\left(\frac{1}{(\gamma-1)(1-\mathrm{Pr})}\frac{x}{T_m}\left|\frac{dT_m}{dx}\right|-1\right), \tag{6.5.7}$$

and a comparison of this with (6.5.4) indicates that $Q_H > 0$, whenever $W > 0$. Also,

$$Q_H - Q_C = W,$$

where Q_C is the rate of heat extraction at the cold end of the plate.

The efficiency of the thermoacoustic engine is W/Q_H. Because dT_m/dx must in practice be large enough to overcome all losses, the efficiency can be estimated by retaining only terms involving dT_m/dx in (6.5.4) and (6.5.7), in which case

$$\frac{W}{Q_H} \approx \frac{(1+\mathrm{Pr})}{(1+\sqrt{\mathrm{Pr}}+\mathrm{Pr})}(\gamma-1)\frac{\ell}{x_m}, \quad x_m \approx \frac{1}{2}(x_H + x_C),$$

where ℓ is the length of the plate. Apart from the dependence on γ and Pr, it is remarkable that the efficiency depends only on the size and position of the plate in the resonator and not, for example, on the temperature gradient dT_m/dT. The factor involving Pr varies very slowly for gases, a good average value being 0.67. Thus, $W/Q_H \approx 0.3\ell/x_m$ for air, and it is slightly larger for a monatomic gas such a helium.

Experiments suggest that unlike, for example, sound generated by shear layer instabilities, the equilibrium amplitude of the sound within the resonator is not determined by nonlinear properties of the motion (which are ignored in the simple theory outlined earlier) but rather by the rate at which heat can be supplied and extracted by the heat exchangers. It appears that the limiting value of the amplitude of the acoustic pressure $|p_o| \sim 0.1\rho_o c_o^2 \approx 0.1\gamma p_m$, where ρ_o and p_m are, respectively, the mean density and pressure in the resonator. Equations (6.5.4) and (6.5.7) accordingly indicate that W and Q_H are both increased when the engine is operated at high mean pressure and at high frequency.

Example 1: Viscous and thermal boundary layers. For time-harmonic oscillations, the analogue of equation (5.1.10) for the viscous boundary layer on the upper surface of the plate is

$$\partial^2 u/\partial y^2 + (i\omega/\nu)(u - \bar{u}) = 0,$$

where $\bar{u}$ is the acoustic particle velocity parallel to the plate just outside the boundary layer. Then

$$u = \bar{u}\left\{1 - e^{iy\sqrt{i\omega/\nu}}\right\}, \quad y > 0 \tag{6.5.8}$$

so that

$$\partial p/\partial y = -\rho_m \nu \partial^2 u/\partial x \partial y = -\rho_m \nu \{\omega\sqrt{i\omega/\nu}/\rho_m c_m^2\} p(x,\omega), \quad y = 0,$$

where $p(x,\omega)$ is defined in (6.5.1), and the adiabatic continuity equation $(1/\rho_m c_m^2)\partial p/\partial t + \partial \bar{u}/\partial x = 0$, applicable just outside the boundary layer, has been used (c_m being the local sound speed).

Because of the mean temperature gradient at the plate, the thermal boundary layer equation is coupled to the unsteady velocity. In the linearized approximation, near the plate, we can set

$$\rho T \frac{Ds}{Dt} = \rho c_p \frac{DT}{Dt} - \frac{Dp}{Dt} \approx \rho_m c_p \left(u(x) \frac{dT_m}{dx} - i\omega (T' - \bar{T}) \right) e^{-i\omega t},$$

where $T = T' e^{-i\omega t} + T_m$, T' being the temperature perturbation due to the sound, and $\bar{T} = p(x,\omega)/\rho_m c_p$ is the local adiabatic temperature fluctuation. Then, the boundary layer approximation to the heat equation (1.2.34) assumes the form

$$\partial^2 T'/\partial y^2 + (i\omega/\chi)(T' - \bar{T}) = (u(x)/\chi)\, dT_m/dx, \quad \chi = \kappa/\rho_m c_p,$$

where $T' \to \bar{T}$ at the outer edge of the boundary layer. It will be assumed that the thermal capacity of the plate is large enough that the temperature fluctuation $T' = 0$ at $y = 0$. Hence, using (6.5.8), we find

$$T' = \bar{T}\left\{1 - e^{iy\sqrt{i\omega/\chi}}\right\} - \frac{i\bar{u}}{\omega}\frac{dT_m}{dx}\left(1 - \frac{\nu e^{iy\sqrt{i\omega/\nu}} - \chi e^{iy\sqrt{i\omega/\chi}}}{\nu - \chi}\right),$$
$$y > 0, \tag{6.5.9}$$

where the second term on the right-hand side gives the contribution from the interaction of the sound with the mean temperature gradient. Thus,

$$\frac{\partial T}{\partial y} = -i\sqrt{i\omega/\chi}\left(\bar{T} - \frac{i\bar{u}}{\omega}\frac{dT_m}{dx}\frac{\sqrt{\chi}}{\sqrt{\nu} + \sqrt{\chi}}\right), \quad y \to 0.$$

Example 2. Establish the following formulae for the single-plate engine of Figure 6.5.2, which are applicable for $(x_C/\ell)\ln(T_H/T_C) \gg (\gamma-1)(1-\text{Pr})$:

$$\frac{1}{T_m}\frac{dT_m}{dx} = \frac{-x_H x_C \ln(T_H/T_C)}{x^2\ell}$$

$$W = \frac{d\sqrt{\omega\chi}|p_0|^2}{\sqrt{2}\rho_o c_o^2(1+\sqrt{\text{Pr}})}\left(\frac{x_H x_C}{\ell}\right)\ln(T_H/T_C)\ln(x_C/x_H)$$

$$Q_H = \frac{d\sqrt{\omega\chi}|p_0|^2(1-\text{Pr}^{\frac{3}{2}})}{\sqrt{2}\rho_o c_o^2(1+\sqrt{\text{Pr}})(\gamma-1)}\left(\frac{x_H x_C}{\ell}\right)\ln(T_H/T_C)$$

$$\frac{W}{Q_H} = \frac{(\gamma-1)(1+\text{Pr})}{(1+\sqrt{\text{Pr}}+\text{Pr})}\ln(x_C/x_H) \approx \frac{(\gamma-1)(1+\text{Pr})}{(1+\sqrt{\text{Pr}}+\text{Pr})}\frac{\ell}{x_m},$$

$$\text{for } \ell \ll x_m \equiv \tfrac{1}{2}(x_H+x_C).$$

6.5.2 Thermoacoustic Refrigerator

When the heat flux Q_H of (6.5.6) is negative, heat energy flows out of the system at the hot end x_H of the single-plate resonator. In this case, equations (6.5.4) and (6.5.7) show that acoustic energy is *absorbed* at the plate. The resonator now behaves as a *heat pump* or refrigerator, the heat flow being from the cold end to the hot end of the plate; steady-state operation is possible only by the extraction of work from the standing wave field (which, for example, must be maintained by an external force applied to the piston of Figure 6.5.1d).

From equation (6.5.7) for the short plate, we see that a second critical temperature gradient can be defined by

$$\left|\frac{dT_m}{dx}\right|_Q = (1-\text{Pr})(\gamma-1)\left\langle\frac{T_m}{x}\right\rangle.$$

Comparison with the definition (6.5.5) of $|dT_m/dx|_W$ reveals that $|dT_m/dx|_Q < |dT_m/dx|_W$ and that the resonator functions as (i) a heat engine for $|dT_m/dx| > |dT_m/dx|_W (W > 0, Q_H > 0)$ and (ii) a refrigerator for $|dT_m/dx| < |dT_m/dx|_Q (W < 0, Q_H < 0)$.

The engine is nonfunctional in the intermediate range $|dT_m/dx|_Q < |dT_m/dx| < |dT_m/dx|_W$, where heat is absorbed at the hot end and rejected at the cold end of the plate, and acoustic energy is absorbed.

Example 3: Thermoacoustic streaming [485, 486]. The mechanism of heat transfer along the plate can be understood by considering the limiting case in which the Prandtl number Pr $\rightarrow$ 0. This limit is not applicable for a gas (where Pr $\approx \frac{2}{3}$), but it can be realized in practice if the working fluid is a liquid metal such as sodium, for which Pr $\approx$ 0.004. In this case the reader can show that the critical temperature gradients

$$|dT_m/dx|_Q = |dT_m/dx|_W = \delta T/\delta x,$$

where $\delta T > 0$ is the amplitude of the adiabatic temperature change experienced by a fluid particle in the thermal boundary layer whose displacement amplitude caused by the sound is $\delta x > 0$.

Let us consider the cyclic thermal process experienced by a small particle of fluid in reciprocating motion within the thermal boundary layer on the plate, that is, at a distance from the plate not exceeding $\delta_\chi \sim \sqrt{\chi/\omega}$. In one complete cycle the particle may be imagined to move between the extreme locations $a \equiv x - \delta x \leq x \leq x + \delta x \equiv a'$ and back (Figure 6.5.3) on either side of the equilibrium position of the particle. To simplify the discussion, we suppose the motion between a and a' is rapid enough that all heat exchanges between the particle and plate may be assumed to occur only at a and a'. There are three cases, determined by the value of the mean temperature gradient dT_m/dx; in the steady state, the mean temperatures of both the plate and adjacent fluid are the same and equal to $T_m(x)$ so that the undisturbed temperatures at a and a' may be written $T_m + \delta T_m$ and $T_m - \delta T_m$, where $\delta T_m = |dT_m/dx|\delta x$.

1. $|dT_m/dx| > \delta T/\delta x$: When the particle moves to a, it is compressed, and its temperature rises by the adiabatic amount δT to a value that is less than

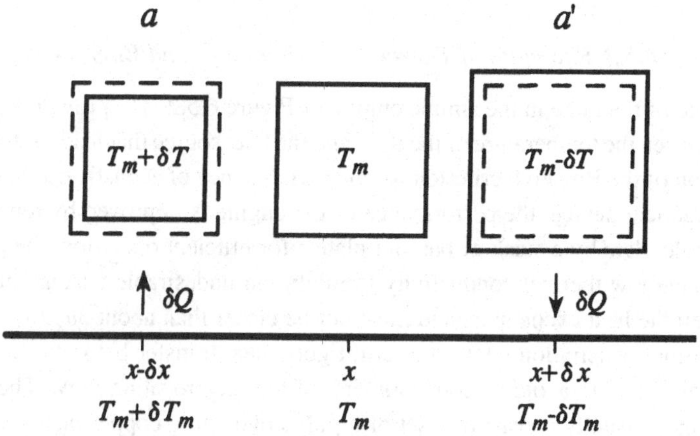

Figure 6.5.3. Thermoacoustic streaming; the indicated heat flow is for Case (i) of Example 3.

that of the plate at a by $\delta T_m - \delta T$ so that an amount of heat δQ, say, flows from the plate to the particle. This process satisfies the Rayleigh criterion and favors the excitation of sound. When the particle arrives at a', it expands in the reduced acoustic pressure, its temperature exceeds that of the plate at a' by the same amount $\delta T_m - \delta T$, and the heat δQ flows into the plate, again in a phase satisfying the Rayleigh criterion. Heat is transferred incrementally along the plate by this mechanism by all fluid particles in the thermal boundary layers, resulting in a net transfer of heat from the hot to the cold heat exchanger, the plate being a temporary store of heat that facilitates the passage of heat between neighboring fluid particles. Each elementary system consisting of a fluid particle and the adjacent section of the plate constitutes an infinitesimal heat engine, each of which generates sound in accordance with the Rayleigh criterion. Because of this, the magnitude of δQ decreases along the plate, as a certain fraction of the thermal energy is ceded to the sound at each step. It follows that the power supplied to the sound is proportional to the length of the plate.

2. $|dT_m/dx| = \delta T/\delta x$: No heat is transferred along the plate and no energy is exchanged with the sound wave because $\delta T_m = \delta T$.
3. $|dT_m/dx| < \delta T/\delta x$: At a, the particle temperature exceeds that of the plate by $\delta T - \delta T_m$; at a', the plate temperature exceeds that of the fluid particle by $\delta T - \delta T_m$. Heat transfer is therefore in the direction from the cold to the hot ends of the plate, and energy must be supplied by the sound wave to maintain a steady state. Evidently the ratio $\delta T/\delta x$ defines the maximum possible temperature gradient along the plate when the system is operating as a refrigerator.

6.5.3 Examples of Thermoacoustic Pumps and Engines

The role of the plate in the simple engine of Figure 6.5.2 is to provide a phase lag between the temperature of the fluid and the heat source that leads to the satisfaction of the Rayleigh criterion for the maintenance of acoustic oscillations. In a practical device, the performance of the engine is improved by replacing the single plate by a stack of parallel plates; for efficient operation, the plates must have low thermal conductivity to inhibit an undesirable leakage of heat between the heat exchangers and must not be closer than about $3\delta_\chi$ (δ_χ being the thermal penetration depth) to permit good heat transfer between the fluid and plates, and to avoid viscous blocking of the reciprocating flow. The heat exchangers usually consist of a set of highly conducting copper plates, parallel to the stack. According to Example 3, the heat exchanger plates should be no longer than about twice the fluid particle displacement amplitude because longer ones merely increase dissipation, and for shorter plates fluid particles in reciprocating motion over the plate will not be in thermal contact sufficiently long to effect efficient heat transfer. Acoustic losses at the walls of the resonator can be reduced by suitably shaping the resonator cavity; thus, large boundary layer losses associated with a tubelike resonator are considerably reduced by using a shorter tube with a large bulb; losses at the bulb walls are negligible (the tube now effectively behaves as a quarter-wavelength resonator, just as if the entrance to the bulb were an open end) [485].

A schematic thermoacoustic refrigerator is illustrated in Figure 6.5.4a. It is driven by a vibrating "high impedance" piston at the "hot" end of the resonator, which generates a standing acoustic wave with a pressure antinode at the hot end. The copper heat exchanger at the hot end of the stack is in thermal contact with water at room temperature circulating in pipe wrapped around the resonator. In a typical experimental setup [484], the resonance tube is about 30–40 cm long (excluding the bulb) and operates at a frequency of about 500 Hz; the working fluid is helium at a mean pressure of ten atmospheres, with peak acoustic amplitude of about one-third of an atmosphere. In steady-state operation, this

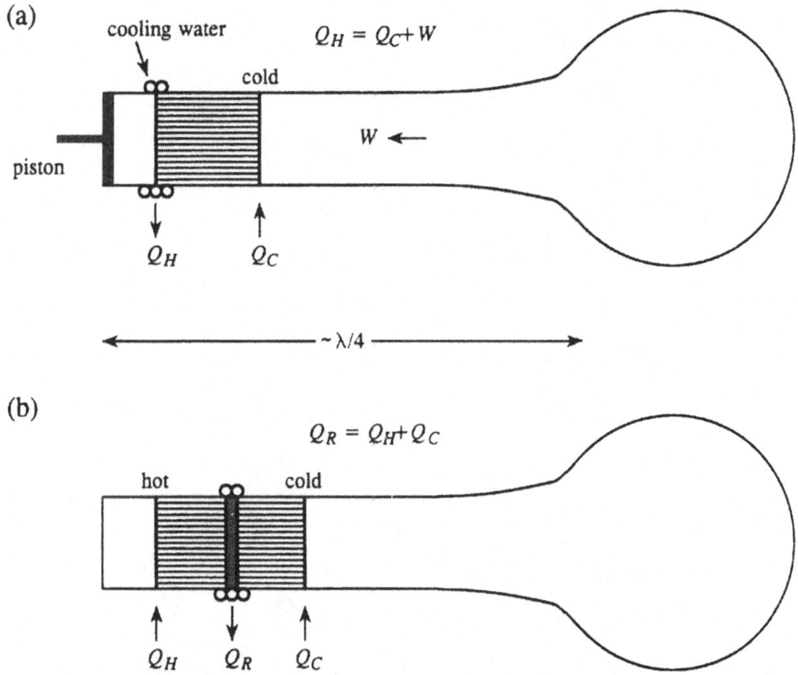

Figure 6.5.4. (a) Schematic thermoacoustic refrigerator (after [484]) and (b) heat driven refrigerator (after [485]).

simple device can reduce the temperature at the cold heat exchanger to as much as 20–40°C below room temperature.

Figure 6.5.4b shows a modification in which the refrigerator is driven by a heat source. The resonator now contains two stacks and three heat exchangers. The hot heat exchanger supplies heat at rate Q_H. Both stacks are connected to a central heat exchanger that is maintained at room temperature by water circulating in a collar outside the resonator, where heat is extracted at rate Q_R. When the temperature at Q_H is sufficiently high, the engine stack between Q_H and Q_R excites a standing wave in the resonator. The second stack functions as a refrigerator driven by the sound, pumping heat at a rate Q_C from the cold heat exchanger, which cools below freezing. The net rate of extraction of heat by the water is $Q_R \approx Q_H + Q_C$ [485].

References

1. Bhatia, A. B. (1967). *Ultrasonic Absorption*. Oxford: Clarendon Press.
2. Landau, L. D., & Lifshitz, E. M. (1987). *Fluid Mechanics* (2nd ed.). Oxford, England: Pergamon Press.
3. Lighthill, M. J. (1963). Introduction: Boundary layer theory. In *Laminar Boundary Layers* (L. Rosenhead, ed.). Oxford University Press.
4. Batchelor, G. K. (1967). *An Introduction to Fluid Dynamics*. Cambridge University Press.
5. Morton, B. R. (1984). The generation and decay of vorticity. *Geophys. Astrophys Fluid Dynamics* **20**, 277–308.
6. Lighthill, M. J. (1956). Viscosity effects in sound waves of finite amplitude. In *Surveys in Mechanics* (G. K. Batchelor & R. M. Davies, eds.). Cambridge University Press.
7. Graff, K. F. (1975). *Wave Motion in Elastic Solids*. Ohio State University Press.
8. Landau, L. D., & Lifshitz, E. M. (1986). *Theory of Elasticity* (3rd ed.). Oxford, England: Pergamon Press.
9. Morse, P. M., & Feshbach, H. (1953). *Methods of Theoretical Physics* (Vol. 1). New York: McGraw-Hill.
10. Rayleigh, Lord. (1894). *The Theory of Sound* (two volumes). London: Macmillan. (Reprinted by Dover, New York, 1945).
11. Gelfand, I. M., & Fomin, S. V. (1963). *Calculus of Variations*. Englewood Cliffs, NJ: Prentice Hall.
12. Dowell, E. H. (1973). Theoretical vibration and flutter studies of point supported panels. *J. Spacecraft and Rockets* **10**, 389–395.
13. Cremer, L., Heckl, M., & Ungar, E. E. (1988). *Structure-Borne Sound* (2nd ed.). New York: Springer-Verlag.
14. Mindlin, R. D. (1951). Influence of rotary inertia and shear on flexural motions of isotropic, elastic plates. *J. Appl. Mech.* **18**, 31–38.
15. Kraus, H. (1967). *Thin Elastic Shells*. New York: Wiley.
16. Love, A. E. H. (1944). *The Mathematical Theory of Elasticity*. New York: Dover.
17. Lamb, H. (1932). *Hydrodynamics* (6th ed.). Cambridge University Press.
18. Garrick, I. E. (1957). *Nonsteady Wing Characteristics, Section F of Aerodynamic Components of Aircraft at High Speeds* (A. F. Donovan & H. R. Lawrence, eds.). Princeton University Press.

19. Lighthill, M. J. (1958). *An Introduction to Fourier Analysis and Generalised Functions*. Cambridge University Press.
20. Jones, D. S. (1982). *The Theory of Generalised Functions* (2nd ed.). Cambridge University Press.
21. Titchmarsh, E. C. (1952). *Theory of Functions* (2nd corrected ed.). Oxford University Press.
22. Landau, L. (1946). On the vibrations of the electronic plasma. *J. Phys. USSR* **10**, 25–40.
23. Lighthill, M. J. (1960). Studies on magneto-hydrodynamic waves and other anisotropic wave motions. *Phil. Trans. Roy. Soc.* **A252**, 397–430.
24. Abramowitz, M., & Stegun, I. A. (1970). (eds.). *Handbook of Mathematical Functions* (9th corrected printing), U.S Dept. of Commerce, National Bureau of Standards Applied Mathematical Services No. 55.
25. Kempton, A. J. (1976). The ambiguity of acoustic sources–A possibility for active control? *J. Sound Vib.* **48**, 475–483.
26. Beranek, L. E., & Vér, I. L. (eds.) (1992). *Noise and Vibration Control Engineering*. New York: Wiley.
27. Möhring, W. (1979). Modelling low Mach number noise. *Proceedings Symposium Mechanics of Sound Generation in Flows*, Göttingen, August 28–31 (ed., E.-A. Müller, pp. 85–96). Berlin: Springer.
28. Howe, M. S. (1975). The generation of sound by aerodynamic sources in an inhomogeneous steady flow. *J. Fluid Mech.* **67**, 579–610.
29. van Dyke, M. (1964). *Perturbation Methods in Fluid Mechanics.* New York: Academic Press.
30. Crow, S. C. (1970). Aerodynamic sound emission as a singular perturbation problem. *Studies in Appl. Math.* **49**, 21–44.
31. Lesser, M. B., & Crighton, D. G. (1976). Physical acoustics and the method of matched asymptotic expansions. *Physical Acoustics* **11**, 69–149.
32. Taylor, K. (1978). A transformation of the acoustic equation with implications for wind tunnel and low speed flight tests. *Proc. R. Soc. Lond.* **A363**, 271–281.
33. Dowling, A. (1976). Convective amplification of real simple sources. *J. Fluid Mech.* **74**, 529–546.
34. Cantrell, R. H., & Hart, R. W. (1964). Interaction between sound and flow in acoustic cavities: mass, momentum and energy considerations. *J. Acoust. Soc. Am.* **36**, 697–706.
35. Myers, M. K. (1991). Transport of energy disturbances in arbitrary steady flows. *J. Fluid Mech.* **226**, 383–400.
36. Lighthill, J. (1978). *Waves in Fluids*. Cambridge University Press.
37. Felsen, L., & Marcuvitz, N. (1973). *Radiation and Scattering of Waves.* Englewood Cliffs, NJ: Prentice-Hall.
38. Havelock, T. H. (1914). *The propagation of disturbances in dispersive media* (Cambridge Tracts in Mathematics and Physics, No. 17). Cambridge University Press.
39. Wu, J. C. (1981). Theory for aerodynamic force and moment in viscous flows. *AIAA J.* **19**, 432–441.
40. Howe, M. S. (1989). On unsteady surface forces, and sound produced by the normal chopping of a rectilinear vortex. *J. Fluid Mech.* **206**, 131–153.
41. Howe, M. S. (1991). On the estimation of the sound produced by complex fluid–structure interactions, with application to a vortex interacting with a shrouded rotor in a duct. *Proc. Roy. Soc. Lond.* **A433**, 573–598.
42. Howe, M. S. (1995). On the force and moment exerted on a body in an

incompressible fluid, with application to rigid bodies and bubbles at high and low Reynolds numbers. *Q. J. Mech. Appl. Math.* **48**, 401–426.
43. Levich, V. (1962). *Physic-Chemical Hydrodynamics.* Englewood Cliffs, NJ: Prentice Hall.
44. Howe, M. S. (1992). A note on the Kraichnan–Phillips theorem. *J. Fluid Mech.* **234**, 443–448.
45. Brown, C. E., & Michael, W. H. (1954). Effect of leading edge separation on the lift of a delta wing. *J. Aero. Sci.* **21**, 690–706.
46. Brown, C. E., & Michael, W. H. (1955). On slender delta wings with leading-edge separation. *NACA Tech. Note* **3430**.
47. Rott, N. (1956). Diffraction of a weak shock with vortex generation. *J. Fluid Mech.* **1**, 111–128.
48. Mangler, K. W., & Smith, J. H. B. (1959). A theory for the flow past a slender delta wing with leading edge separation. *Proc. Roy. Soc. Lond.* **A251**, 200–217.
49. Smith, J. H. B. (1959). A theory of the separated flow from the curved leading edge of a slender wing. *Reports and Memoranda of the Aeronautical Research Council*, **No. 3116**.
50. Smith, J. H. B. (1968). Improved calculations of leading edge separation from slender, thin, delta wings. *Proc. Roy. Soc. Lond.* **A306**, 67–90.
51. Dore, B. D. (1966). Nonlinear theory for slender wings in sudden plunging motion. *The Aeronautical Quarterly*, **17** 187–200.
52. Lowson, M. V. (1963). The separated flow on slender wings in unsteady motion. *Reports and Memoranda of the Aeronautical Research Council*, **No. 3448**.
53. Randall, D. G. (1966). Oscillating slender wings with leading edge separation. *The Aeronautical Quarterly*, **17**, 311–331.
54. Clements, R. R. (1973). An inviscid model of two-dimensional vortex shedding. *J. Fluid Mech.* **57**, 321–336.
55. Graham, J. M. R. (1980). The forces on the sharp-edged cylinders in oscillatory flow at low Keulegan–Carpenter numbers. *J. Fluid Mech.* **97**, 331–346.
56. Cortelezzi, L., & Leonard, A. (1993). Point vortex model for the unsteady separated flow past a semi-infinite plate with transverse motion. *Fluid. Dyn. Res.* **11**, 263–295.
57. Cortelezzi, L., Leonard, A., & Doyle, J. C. (1994). An example of active circulation control of the unsteady separated flow past a semi-infinite plate. *J. Fluid Mech.* **260**, 127–154.
58. Peters, M. C. A. M. (1993). *Aeroacoustic sources in internal flows.* Ph.D. thesis, Eindhoven University of Technology.
59. Tavares, T. S., & McCune, J. E. (1993). Aerodynamics of maneuvering slender wings with leading edge separation. *AIAA J.* **31**, 977–986.
60. Lighthill, M. J. (1952). On sound generated aerodynamically. Part I: General theory. *Proc. Roy. Soc. Lond.* **A211**, 564–587.
61. Lighthill, M. J. (1954). On sound generated aerodynamically. Part II: Turbulence as a source of sound. *Proc. Roy. Soc. Lond.* **A222**, 1–32.
62. Hinze, J. O. (1975). *Turbulence* (2nd ed.). New York: McGraw-Hill.
63. Batchelor, G. K. (1953). *The theory of homogeneous turbulence.* Cambridge University Press.
64. Morfey, C. L. (1973). Amplification of aerodynamic noise by convected flow inhomogeneities. *J. Sound Vib.* **31**, 391–397.
65. Crighton, D. G., & Ffowcs Williams, J. E. (1969). Sound generation by turbulent two-phase flow. *J. Fluid Mech.* **36**, 585–603.

66. Ffowcs Williams, J. E. (1969). Hydrodynamic noise. *Ann. Rev. Fluid Mech.* **1**, 197–222.
67. Davies, H. G., & Ffowcs Williams, J. E. (1968). Aerodynamic sound generation in a pipe. *J. Fluid Mech.* **32**, 765–778.
68. Basset, A. B. (1961). *A Treatise on Hydrodynamics* (Vol. 2). New York: Dover Publications.
69. Ffowcs Williams, J. E., & Hawkings, D. L. (1969). Sound generation by turbulence and surfaces in arbitrary motion. *Phil. Trans. Roy. Soc.* **A264**, 321–342.
70. Curle, N. (1955). The influence of solid boundaries upon aerodynamic sound. *Proc. Roy. Soc. Lond.* **A231**, 505–514.
71. Ffowcs Williams, J. E. (1974). Sound production at the edge of a steady flow. *J. Fluid Mech.* **66**, 791–816.
72. Ffowcs Williams, J. E. (1963). The noise from turbulence convected at high speed. *Phil. Trans. Roy. Soc. Lond.* **A255**, 469–503.
73. Lush, P. A. (1971). Measurement of subsonic jet noise and comparison with theory. *J. Fluid Mech.* **46**, 477–500.
74. Ffowcs Williams, J. E. (1977). Aeroacoustics. *Ann. Rev. Fluid Mech.* **9**, 447–468.
75. Dowling, A. P., Ffowcs Williams, J. E., & Goldstein, M. E. (1978). Sound production in a moving stream. *Phil. Trans. Roy. Soc. Lond.* **A288**, 321–349.
76. Goldstein, M. E. (1984). Aeroacoustics of turbulent shear flows. *Ann. Rev. Fluid Mech.* **16**, 263–285.
77. Powell, A. (1961). *Vortex Sound* (Rept. No. 61–70), Department of Engineering, University of California, Los Angeles.
78. Powell, A. (1963). *Mechanisms of Aerodynamic Sound Production.* AGARD Rept. 466.
79. Howe, M. S. (1975). Contributions to the theory of aerodynamic sound, with application to excess jet noise and the theory of the flute. *J. Fluid Mech.* **71**, 625–673.
80. Obermeier, F. (1985). Aerodynamic sound generation caused by viscous processes. *J. Sound Vib.* **99**, 111–120.
81. Westervelt, P. J., & Larson, R. S. (1973). Laser-excited broadside array. *J. Acoust. Soc. Am.* **54**, 121–122.
82. Minnaert, M. (1933). On musical air bubbles and the sounds of running water. *Phil Mag. S. 7* **16**, 235–249.
83. Devin, C. (1959). Survey of thermal, radiation, and viscous damping of pulsating air bubbles in water. *J. Acoust. Soc. Am.* **31**, 1654–1667.
84. Eller, A. I. (1970). Damping constants of pulsating bubbles. *J. Acoust. Soc. Am.* **47**, 1469–1470.
85. Carstensen, E. L., & Foldy, L. L. (1947). Propagation of sound through a liquid containing bubbles. *J. Acoust. Soc. Am.* **19**, 481–501.
86. Batchelor, G. K. (1967). Compression waves in a suspension of gas bubbles in liquid. *Fluid Dynamics Trans.* **4**, 425–445.
87. van Wijngaarden, L. (1979). Sound and shock waves in bubbly liquids. pp 127–140, In *Cavitation and Inhomogeneities in Underwater Acoustics* (ed. W. Lauterborn); Berlin: Springer.
88. Kraichnan, R. H. (1953). The scattering of sound in a turbulent medium. *J. Acoust. Soc. Am.* **25**, 1096–1104.
89. Lighthill, M. J. (1953). On the energy scattered from the interaction of turbulence with sound or shock waves. *Proc. Camb. Phil. Soc.* **49**, 531–551.
90. Batchelor, G. K. (1957). Wave scattering due to turbulence. Chapter 16, *Naval Hydrodynamics*. Washington, DC: National Academy Science (Publication No. 515 of the National Research Council).

91. Noir, D. T., & George, A. R. (1978). Absorption of sound by homogeneous turbulence. *J. Fluid Mech.* **86**, 593–608.
92. Howe, M. S. (1971). On wave scattering by random inhomogeneities, with application to the theory of weak bores. *J. Fluid Mech.* **45**, 785–804.
93. Stratonovich, R. L. (1963). *Topics in the Theory of Random Noise* (Vol. 1). New York: Gordon and Breach.
94. Crow, S. C. (1968). Viscoelastic properties of fine-grained incompressible turbulence. *J. Fluid Mech.* **33**, 1–20.
95. Howe, M. S. (1983). On the scattering of sound by a vortex ring. *J. Sound Vib.* **87**, 567–571.
96. Broadbent, E. G. (1977). Acoustic ray theory applied to vortex refraction. *J. Inst. Maths. Applics.* **19**, 1–27.
97. Williams, F. A. (1985). *Combustion Theory* (2nd ed.). Menlo Park, CA: Benjamin/Cummings.
98. Temkin, S., & Dobbins, R. A. (1966). Measurements of attenuation and dispersion of sound by an aerosol. *J. Acoust. Soc. Am.* **40**, 1016–1024.
99. Marble, F. E., & Wooton, D. C. (1970). Sound attenuation in condensing vapor. *Phys. Fluids* **13**, 2657–2664.
100. Marble, F. E. (1975). Acoustic attenuation of fans and ducts by vaporization of liquid droplets–Application to noise reduction in aircraft power plants. *Cal. Tech. Guggenheim Jet Propulsion Center Rept. AFOSR-TR 75-0511.*
101. Vincenti, W. G., & Kruger, C. H. (1977). *Introduction to Physical Gas Dynamics.* Huntington, NY: Krieger.
102. Clarke, J. F. (1978). Gas dynamics with relaxation effects. *Rep. Prog. Phys.* **41**, 807–864.
103. Society of Automotive Engineers. (1985). *Gas Turbine Jet Exhaust Noise Prediction.* (SAE ARP 876C. Society of Automotive Engineers, Inc, 400 Commonwealth Drive, Warrendale, PA 15096).
104. Liepmann, H. W., & Roshko, A. (1957). *Elements of Gas Dynamics.* New York: Wiley.
105. Lighthill, M. J. (1962). Sound generated aerodynamically. *Proc. Roy. Soc. Lond.* **A267**, 147–182.
106. Stone, J. R. (1977). Prediction of in-flight exhaust noise for turbojet and turbofan engines. *Noise Control Engineering* **10**(1), 40–46.
107. Powell, A. (1977). Flow noise: A perspective on some aspects of flow noise, and of jet noise in particular: Part 2. *Noise Con. Eng.* **8**(3), 108–119.
108. Powell, A. (1953). On the mechanism of choked jet noise. *Proc. Phys. Soc.* **66**, 1039–1056.
109. Harper-Bourne, M., & Fisher, M. J. (1973). Noise mechanisms. The noise from shock waves in supersonic jets. *Paper No. 11, AGARD-CP-131.*
110. Tam, C. K. W. (1991). Chapter 6 of *Aeroacoustics of Flight Vehicles: Theory and Practice: Vol. 1: Noise Sources. Jet Noise Generated by Large-Scale Coherent Motion.* NASA Ref. Publ. 1258 (vol. 1).
111. Ross, D. (1976). *Mechanics of Underwater Noise.* Oxford, England: Pergamon.
112. Powell, A. (1960). Aerodynamic noise and the plane boundary. *J. Acoust. Soc. Am.* **32**, 962–990.
113. Crighton, D. G. (1975). Scattering and diffraction of sound by moving bodies. *J. Fluid Mech.* **72**, 209–227.
114. Ffowcs Williams, J. E., & Lovely, J. (1975). Sound radiated into moving flow by compact surface vibrations. *J. Fluid Mech.* **71**, 689–700.
115. Blokhintsev, D. I. (1946). Acoustics of nonhomogeneous moving media. *NACA Technical Memorandum 1399.*

116. Phillips, O. M. (1956). The intensity of Aeolian tones. *J. Fluid Mech.* **1**, 607–624.
117. Goldstein, S. (ed.) (1965). *Modern Developments in Fluid Dynamics* (Vol. 2). New York: Dover.
118. Roshko, A. (1961). Experiments on the flow past a cylinder at very high Reynolds number. *J. Fluid Mech.* **10**, 345–356.
119. Ffowcs Williams, J. E., & Howe, M. S. (1975). The generation of sound by density inhomogeneities in low Mach number nozzle flows. *J. Fluid Mech.* **70**, 605–622.
120. Ribner, H. S., & Tucker, M. (1953). Spectrum of turbulence in a contracting stream. *NACA Rept. No. 1113.*
121. Batchelor, G. K., & Proudman, I. (1954). The effect of rapid distortion of a fluid in turbulent motion. *Q. J. Mech. Appl. Math.* **7**, 83–103.
122. Candel, S. M. (1972). *Analytical studies of some acoustic problems of jet engines.* Ph.D. thesis, California Institute of Technology.
123. Marble, F. E., & Candel, S. M. (1977). Acoustic disturbance from gas non-uniformities convected through a nozzle. *J. Sound Vib.* **55**, 225–243.
124. Leppington, F. G. (1971). Aeronautical Research Council Working Party on novel aerodynamic noise source mechanisms at low jet speeds: *ARC 32 925 N. 742* (Chapter 5): Scattering of quadrupole sources near the end of a rigid semi-infinite circular pipe.
125. Crow, S. C., & Champagne, F. H. (1971). Orderly structure in jet turbulence. *J. Fluid Mech.* **48**, 547–591.
126. Crighton, D. G. (1972). The excess noise field of subsonic jets. *Aeronautical Research Council, Noise Research Committee. Paper ARC 33 714.*
127. Howe, M. S. (1979). Attenuation of sound in a low Mach number nozzle flow. *J. Fluid Mech.* **91**, 209–230.
128. Pinker, R. A., & Bryce, W. D. (1976). The radiation of plane-wave duct noise from a jet exhaust statically and in flight. *National Gas Turbine Establishment, Technical Note No. NT 1024.*
129. Munt, R. M. (1977). The interaction of sound with a subsonic jet issuing from a semi-infinite cylindrical pipe. *J. Fluid Mech.* **83**, 609–640.
130. Ffowcs Williams, J. E. (1971). Aeronautical Research Council Working Party on novel aerodynamic noise source mechanisms at low jet speeds: *ARC 32 925 N. 742* (Chapter 2): Transmission of low frequency jet pipe sound through a nozzle flow.
131. Ffowcs Williams, J. E., & Hall, L. H. (1970). Aerodynamic sound generation by turbulent flow in the vicinity of a scattering half-plane. *J. Fluid Mech.* **40**, 657–670.
132. Crighton, D. G., & Leppington, F. G. (1971). On the scattering of aerodynamic noise. *J. Fluid Mech.* **46**, 577–597.
133. Jones, D. S. (1964). *The Theory of Electromagnetism.* Oxford, England: Pergamon Press.
134. Bowman, J. J., Senior, T. B. A., & Uslenghi, P. L. E. (1987). *Electromagnetic and Acoustic Scattering by Simple Shapes* (revised printing). New York: Hemisphere.
135. Crighton, D. G., & Leppington, F. G. (1970). Scattering of aerodynamic noise by a semi-infinite compliant plate. *J. Fluid Mech.* **43**, 721–736.
136. Kambe, T., Minota, T., & Ikushima, Y. (1985). Acoustic wave emitted by a vortex ring passing near the edge of a half-plane. *J. Fluid Mech.* **155**, 77–103.
137. Crighton, D. G. (1972). Radiation from vortex filament motion near a half plane. *J. Fluid Mech.* **51**, 357–362.
138. Leverton, J. W. (1989). Twenty-five years of rotorcraft aeroacoustics: Historical prospective and important issues. *J. Sound Vib.* **133**, 261–287.
139. Schmitz, F. H. (1991). Rotor Noise. Chapter 2 of *Aeroacoustics of Flight Vehicles:*

Theory and Practice. Vol. 1: Noise Sources. NASA Special Publication 1258, Volume 1 (ed. H. H. Hubbard).

140. Widnall, S. (1971). Helicopter noise due to blade–vortex interactions. *J. Acoust. Soc. Am.* **50**, 354–365.
141. Filotas, L. T. (1973). Vortex induced helicopter blade loads and noise. *J. Sound Vib.* **27**, 387–398.
142. Lee, D. J., & Roberts, L. (1985). Interaction of a turbulent vortex with a lifting surface. *AIAA Paper* **85-0004**.
143. Hardin, J. C., & Lamkin, S. L. (1984). Aeroacoustic interaction of a distributed vortex with a lifting Joukowski airfoil. *AIAA Paper* **84-2287**.
144. Poling D. R., & Dadone, L. (1987). *AIAA J.* **27**, 694–699. Blade–vortex interaction.
145. Goldstein, M. E. (1978). Unsteady vortical and entropic distortions of potential flows round arbitrary obstacles. *J. Fluid Mech.* **89**, 433–468.
146. Baeder, J. D. (1987). *Computation of nonlinear acoustics in two-dimensional blade–vortex interactions.* Paper presented at the Thirteenth European Rotorcraft Forum, Arles, France, Sept. 8–11.
147. Tadghighi, H. (1989). An analytical model for prediction of main rotor/tail rotor interaction noise. *AIAA Paper* **89-1130**.
148. Atassi, H., & Dusey, M. (1990). Acoustic radiation from a thin airfoil in nonuniform subsonic flows. *AIAA Paper* **90-3910**.
149. Atassi, H., & Subramaniam, S. (1990). Acoustic radiation from a lifting airfoil in nonuniform subsonic flows. *AIAA Paper* **90-3911**.
150. Goldstein, M. E., & Atassi, J. (1976). A complete second order theory for the unsteady flow about an airfoil due to a periodic gust. *J. Fluid Mech.* **74**, 741–765.
151. Atassi, H. M. (1984). The Sears problem for a lifting airfoil revisited—New results. *J. Fluid Mech.* **141**, 109–122.
152. Sears, W. R. (1966). Aerodynamics, noise and the sonic boom. *AIAA J.* **7**, 577–586.
153. Ashley, H. and Landahl, M. (1965). *Aerodynamics of Wings and Bodies.* Reading, MA: Addison-Wesley.
154. Sears, W. R. (1941). Some aspects of non-stationary airfoil theory and its practical applications. *J. Aero. Sci.* **8**, 104–108.
155. Howe, M. S. (1976). The influence of vortex shedding on the generation of sound by convected turbulence. *J. Fluid Mech.* **76**, 711–740.
156. Amiet, R. K. (1975). Acoustic radiation from an airfoil in a turbulent stream. *J. Sound Vib.* **41**, 407–420.
157. Amiet, R. K. (1986). Airfoil gust response and the sound produced by airfoil–vortex interaction. *J. Sound Vib.* **107**, 487–506.
158. Myers, M. R., & Kerschen, E. J. (1986). Effect of airfoil camber on convected gust interaction noise. *AIAA Paper* **86-1873**.
159. Tsai, C., & Kerschen, E. (1990). Influence of airfoil nose radius on sound generated by gust interactions. *AIAA Paper* **90-3912**.
160. Ffowcs Williams, J. E., & Guo, Y. P. (1988). Sound generated from the interruption of a steady flow by a supersonically moving airfoil. *J. Fluid Mech.* **195**, 113–135.
161. Howe, M. S. (1988). Contributions to the theory of sound production by vortex–airfoil interaction, with application to vortices with finite axial velocity defect. *Proc. Roy. Soc. Lond.* **A420**, 157–182.
162. Schlinker, R. H., & Amiet, R. K. (1983). Rotor–vortex interaction noise. *AIAA Paper* **83-0720**.
163. Ahmadi, A. R. (1986). An experimental investigation of blade–vortex interaction at normal incidence. *J. Aircraft* **23**, 47–55.

164. Amiet, R. K. (1986). Intersection of a jet by an infinite span airfoil. *J. Sound Vib.* **111**, 499–503.
165. Basu, B. C., & Hancock, G. J. (1978). The unsteady motion of a two-dimensional aerofoil in incompressible inviscid flow. *J. Fluid Mech.* **87**, 159–178.
166. Hsu, A. T., & Wu, J. C. (1988). Vortex flow model for the blade–vortex interaction problem. *AIAA J.* **26**, 621–623.
167. Renzoni, P., & Mayle, R. E. (1991). Incremental force and moment coefficients for a parallel blade–vortex interaction. *AIAA J.* **29**, 6–13.
168. Amiet, R. K. (1990). Gust response for flat plate airfoils and the Kutta condition. *AIAA J.* **28**, 1718–1727.
169. Howe, M. S. (1989). On sound generated when a vortex is chopped by a circular airfoil. *J. Sound Vib.* **128**, 487–503.
170. Sneddon, I. N. (1966). *Mixed Boundary Value Problems in Potential Theory.* Amsterdam: North-Holland.
171. Blake, W. K. (1986). *Mechanics of Flow-Induced Sound and Vibration (Vol. 2): Complex Flow–Structure Interactions.* New York: Academic Press.
172. Chase, D. M. (1987). The character of the turbulent wall pressure spectrum at subconvective wavenumbers and a suggested comprehensive model. *J. Sound Vib.* **112**, 125–147.
173. Leehey, P. (1988). Structural excitation by a turbulent boundary layer: An overview. *J. Vib. Stress and Reliability in Design* **110**, 220–225.
174. Zorumski, W. E. (1987). Fluctuating pressure loads under high speed boundary layers. *NASA Tech. Memo.* 100517.
175. Howe, M. S. (1991). Surface pressures and sound produced by turbulent flow over smooth and rough walls. *J. Acoust. Soc. Am.* **90**, 1041–1047.
176. Schewe, G. (1983). On the structure and resolution of wall pressure fluctuations associated with turbulent boundary layer flow. *J. Fluid Mech.* **134**, 311–328.
177. Lauchle, G. C., & Daniels, M. (1987). Wall pressure fluctuations in turbulent pipe flow. *Phys. Fluids* **30**, 3019–3024.
178. Schlichting, H. (1979). *Boundary Layer Theory.* (7th ed.). New York: McGraw-Hill.
179. Chase, D. M. (1980). Modeling the wavevector–frequency spectrum of turbulent boundary layer wall pressure. *J. Sound Vib.* **70**, 29–67.
180. Choi, H., & Moin, P. (1990). On the space–time characteristics of wall pressure fluctuations. *Phys. Fluids A* **2**, 1450–1460.
181. Morrison, W. R. B., & Kronauer, R. R. (1969). Structural similarity for fully developed turbulence in smooth tubes. *J. Fluid Mech.* **39**, 117–141.
182. Corcos, G. M. (1964). The structure of the turbulent pressure field in boundary layer flows. *J. Fluid Mech.* **18**, 353–378.
183. Kraichnan, R. H. (1956). Pressure fluctuations in turbulent flow over a flat plate. *J. Acoust. Soc. Am.* **28**, 378–390.
184. Phillips, O. M. (1956). On the aerodynamic surface sound from a plane turbulent boundary layer. *Proc. Roy. Soc. Lond.* **A234**, 327–335.
185. Sevik, M. M. (1986). Topics in hydroacoustics. In *Proceedings of IUTAM Symposium Aero- and Hydroacoustics*, Lyon, July 3–6, 1985. Berlin: Springer-Verlag.
186. Howe, M. S. (1987). On the structure of the turbulent boundary layer wall pressure spectrum in the vicinity of the acoustic wavenumber. *Proc. Roy. Soc. Lond.* **A412**, 389–401.
187. Ffowcs Williams, J. E. (1982). Boundary-layer pressures and the Corcos model: A development to incorporate low wavenumber constraints. *J. Fluid Mech.* **125**, 9–25.
188. Howe, M. S. (1979). The role of surface shear stress fluctuations in the generation of boundary layer noise. *J. Sound Vib.* **65**, 159–164.

189. Laganelli, A. L., Martellucci, A., & Shaw, L. L. (1983). Wall pressure fluctuations in attached boundary layer flow. *AIAA J.* **21**, 495–502.
190. Laganelli, A. L., & Wolfe, H. (1989). Prediction of fluctuating pressure in attached and separated turbulent boundary layer flow. *AIAA Paper* **89-1064**.
191. Efimtsov, B. M. (1982). Characteristics of the field of turbulent wall pressure fluctuations at large Reynolds numbers. *Sov. Phys. Acoust.* **28**, 289–292.
192. Hersh, A. S. (1983). Surface roughness generated flow noise. *AIAA Paper* **No. 87-2663.**
193. Blake, W. K., & Gershfeld, J. L. (1989). *The aeroacoustics of trailing edges.* In Lecture Notes in Engineering 46 (ed., M. Gad-el-Hak), Frontiers in Experimental Fluid Mechanics.
194. Brooks, T. F., Pope, D. S., & Marcolini, M. A. (1989). Airfoil self-noise and prediction. *NASA Reference Publication No. 1218.*
195. Crighton, D. G. (1991). Airframe Noise. Chapter 7 of *Aeroacoustics of Flight Vehicles: Theory and Practice. Vol. 1: Noise Sources.* NASA Reference Publication 1258 (edited by H. H. Hubbard).
196. Chase, D. M. (1972). Sound radiated by turbulent flow off a rigid half-plane as obtained from a wavevector spectrum of hydrodynamic pressure. *J. Acoust. Soc. Am.* **52**, 1011–1023.
197. Chandiramani, K. L. (1974). Diffraction of evanescent waves, with applications to aerodynamically scattered sound and radiation from unbaffled plates. *J. Acoust. Soc. Am.* **55**, 19–29.
198. Chase, D. M. (1975). Noise radiated from an edge in turbulent flow. *AIAA J.* **13**, 1041–1047.
199. Amiet, R. K. (1976). Noise due to turbulent flow past a trailing edge. *J. Sound Vib.* **47**, 387–393.
200. Amiet, R. K. (1978). Effect of the incidence surface pressure field on noise due to turbulent flow past a trailing edge. *J. Sound Vib.* **57**, 305–306.
201. Olsen, W., & Boldman, D. (1979). Trailing edge noise data with comparison to theory. *AIAA Paper* **79-1524**.
202. Goldstein, M. E. (1979). Scattering and distortion of the unsteady motion on transversely sheared mean flows. *J. Fluid Mech.* **91**, 601–632.
203. Howe, M. S. (1981). The displacement-thickness theory of trailing edge noise. *J. Sound Vib.* **75**, 239–250.
204. Titchmarsh, E. C. (1948). *Introduction to the theory of Fourier integrals* (2nd edition). Oxford University Press.
205. Noble, B. (1958). *Methods Based on the Wiener–Hopf Technique*. London: Pergamon Press.
206. Clemmow, P. C. (1966). *The plane wave spectrum representation of electromagnetic fields*. Oxford, England: Pergamon Press.
207. Howe, M. S. (1988). The influence of surface rounding on trailing edge noise. *J. Sound Vib.* **126**, 503–523.
208. Crighton, D. G. (1972). Radiation properties of the semi-infinite vortex sheet. *Proc. Roy. Soc. Lond.* **A330**, 185–198.
209. Dassen, T., Parchen, R., Bruggeman, J., & Hagg, F. (1996). *Results of a wind tunnel study on the reduction of airfoil self-noise by the application of serrated blade trailing edges.* Paper presented at the European Union Wind Energy Conference and Exhibition, Gothenburg.
210. Howe, M. S. (1991). Aerodynamic noise of a serrated trailing edge. *J. Fluids and Structures* **5**, 33–45.
211. Howe, M. S. (1991). Noise produced by a sawtooth trailing edge. *J. Acoust. Soc. Am.* **90**. 482–487.

212. Hayden, R. E. (1976). Reduction of noise from airfoils and propulsive lift systems using variable impedance systems. *AIAA Paper* **76-500**.
213. Howe, M. S. (1979). On the added mass of a perforated shell, with application to the generation of aerodynamic sound by a perforated trailing edge. *Proc. Roy. Soc. Lond.* **A365**, 209–233.
214. Gutin, L. (1948). On the sound field of a rotating blade. *NACA TM* **1195**.
215. Ozawa, S., Morito, Y., Maeda, T., & Kinosita, M. (1976). Investigation of the pressure wave radiated from a tunnel exit. *Railway Technical Research Institute Rept.* (in Japanese) 1023.
216. Ozawa, S. (1979). Studies of the micro-pressure wave radiated from a tunnel exit. *Railway Technical Research Institute Rept.* (in Japanese) 1121.
217. Hawkings, D. L. (1979). Noise generation by transonic open rotors. *Proceeding of IUTAM Symposium on Mechanics of Sound Generation in Flows* (ed., E.-A. Müller, pp. 294–300). Berlin: Springer-Verlag.
218. Hanson, D. B., & Fink, M. R. (1979). The importance of quadrupole sources in prediction of transonic tip speed propeller noise. *J. Sound Vib.* **62**, 19–38.
219. Schmitz, F. H., & Yu, Y. H. (1986). Helicopter impulsive noise: Theoretical and experimental status. *J. Sound Vib.* **109**, 361–422.
220. Magliozzi, B., Hanson, D. B., & Amiet, R. K. (1991). Propeller and Propfan Noise. Chapter 1 of *Aeroacoustics of Flight Vehicles: Theory and Practice. Vol. 1: Noise Sources*. NASA Reference Publication 1258 (edited by H. H. Hubbard).
221. Schmitz, F. H., & Yu, Y. H. (1981). Transonic rotor noise–Theoretical and experimental comparisons. *Vertica* **5**, 55–74.
222. Ardavan, H. (1991). The breakdown of linearised theory and the role of quadrupole sources in transonic rotor acoustics. *J. Fluid Mech.* **226**, 591–624.
223. Brentner, K. S., & Farassat, F. (1994). Helicopter noise prediction: The current status and future direction. *J. Sound Vib.* **170**, 79–96.
224. Parry, A. B., & Crighton, D. G. (1989). Asymptotic theory of propeller noise—Part 1: Subsonic single rotation propeller. *AIAA J.* **27**, 1184–1190.
225. Myers, M. K., & Wydeven, R. (1991). Asymptotic/numerical analysis of supersonic propeller noise. *AIAA J.* **29**, 1374–1382.
226. Hawkings, D. L., & Lowson, M. C. (1974). Theory of open supersonic rotor noise. *J. Sound Vib.* **36**, 1–20.
227. Farassat, F. (1981). Linear acoustic formulas for calculation of rotating blade noise. *AIAA J.* **19**, 1122–1130.
228. Watson, G. N. (1944). *A Treatise on the Theory of Bessel Functions* (2nd ed.). Cambridge University Press.
229. Morfey, C. L. (1973). Rotating blades and aerodynamic sound. *J. Sound Vib.* **28**, 587–617.
230. Simonich, J. C., Amiet, R. K., Schlinker, R. H., & Greitzer, E. M. (1990). Rotor noise due to atmospheric turbulence ingestion–Part 1: Fluid mechanics. *J. Aircraft* **27**, 7–14.
231. Amiet, R. K., Simonich, J. C., & Schlinker, R. H. (1990). Rotor noise due to atmospheric turbulence ingestion–Part 2: Aeroacoustic results. *J. Aircraft* **27**, 14–22.
232. Goldstein, M. (1974). Unified approach to aerodynamic sound generation in the presence of solid boundaries. *J. Acoust. Soc. Am.* **56**, 497–509.
233. Amiet, R. K. (1988). Thickness noise of a propeller and its relation to blade sweep. *J. Fluid Mech.* **192**, 535–560.
234. Peake, N., & Crighton, D. G. (1991). Lighthill quadrupole radiation in supersonic propeller acoustics. *J. Fluid Mech.* **223**, 363–382.
235. Crighton, D. G., & Parry, A. B. (1991). Asymptotic theory of propeller noise Part 2: Supersonic single-rotation propeller. *AIAA J.* **29**, 2031–2037.

236. Crighton, D. G., & Parry, A. B. (1992). Higher approximations in the asymptotic theory of propeller noise. *AIAA J.* **30**, 23–28.

237. Ozawa, S., Maeda, T., Matsumura, T., Uchida, K., Kajiyama, H., & Tanemoto, K. (1991). Countermeasures to reduce micro-pressure waves radiating from exits of Shinkansen tunnels. *In Aerodynamics and Ventilation of Vehicle Tunnels* **(253–266)** Amsterdam: Elsevier Science.

238. Liang, M., Kitamura, T., Matsubayashi, K., Kosaka, T., Maeda, T., Kudo, N., & Yamada, S. (1994). Active attenuation of low frequency noise radiated from tunnel exit of high speed train. *J. Low Frequency Noise and Vibration* **13**(2), 39–47.

239. Maeda, T., Matsumura, T., Iida, M., Nakatani, K., & Uchida, K. (1993). Effect of shape of train nose on compression wave generated by train entering tunnel. *Proceedings of the International Conference on Speedup Technology for Railway and Maglev Vehicles* pp. 315–319 Yokohama, Japan 22–26 November.

240. Howe, M. S. (1998). The compression wave produced by a high-speed train entering a tunnel. *Proc. Roy. Soc. A* **454**.

241. Whitham, G. B. (1961). Group velocity and energy propagation for three-dimensional waves. *Comm. Pure Appl. Math.* **14**, 675–691.

242. Chandiramani, K. L. (1986). The role of spectral derivatives of structure reactance in vibration problems. *Bolt Beranek and Newman Technical Information Report No. 105.*

243. Lighthill, M. J. (1965). Group velocity. *J. Ins. Maths. Applics.* **1**, 1–28.

244. Whitham, G. B. (1974). *Linear and Nonlinear Waves.* New York: Wiley.

245. Cairns, R. A. (1979). The role of negative energy waves in some instabilities of parallel flows. *J. Fluid Mech.* **92**, 1–14.

246. Crighton, D. G. (1979). The free and forced waves on a fluid loaded elastic plate. *J. Sound Vib.* **63**, 225–235.

247. Crighton, D. G. (1989). The 1988 Rayleigh Medal Lecture: Fluid loading—The interaction between sound and vibration. *J. Sound Vib.* **133**, 1–27.

248. Feit, D. (1966). Pressure radiated by a point excited elastic plate. *J. Acoust. Soc. Am.* **40**, 1489–1494.

249. Nayak, P. R. (1970). Line admittance of infinite isotropic fluid loaded plates. *J. Acoust. Soc. Am.* **47**, 191–201.

250. Innes, D., & Crighton, D. G. (1988). Power radiated by an infinite plate subject to fluid loading and line drive. *J. Sound Vib.* **123**, 437–450.

251. Ungar, E. E. (1961). Transmission of plate flexural waves through reinforcing beams: Dynamic stress concentrations. *J. Acoust. Soc. Am.* **33**, 633–639.

252. Smith, P. W. (1962). Response and radiation of structural modes excited by sound. *J. Acoust. Soc. Am.* **34**, 640–647.

253. Lyon, R. H. (1962). Sound radiation from a beam attached to a plate. *J. Acoust. Soc. Am.* **34**, 1265–1268.

254. Kouzov, D. P. (1963). Diffraction of a plane hydroacoustic wave at a crack in an elastic plate. *Prikladnaia Matematiki I Mekhanika* (English Translation) **27**, 1593–1601.

255. Lyapunov, V. T. (1969). Flexural wave propagation in a liquid-loaded plate with an obstruction. *Sov. Phys. Acoust.* **14**, 352–355.

256. Maidanik, G., Tucker, A. J., & Vogel, W. H. (1976). Transmission of free waves across a rib on a panel. *J. Sound Vib.* **49**, 445–452.

257. Chandiramani, K. L. (1977). Vibration response of fluid loaded structures to low speed flow noise. *J. Acoust. Soc. Am.* **61**, 1460–1470.

258. Belinski, B. P. (1978). Radiation of sound by a plate reinforced with a projecting beam. *Sov. Phys. Acoust.* **24**, 183–187.

259. Leppington, F. G. (1978). Acoustic scattering by membranes and plates with line constraints. *J. Sound Vib.* **58**, 319–332.
260. Stepanishen, P. R. (1978). The acoustic transmission and scattering characteristics of a plate with line impedance discontinuities. *J. Sound Vib.* **58**, 257–272.
261. Howe, M. S. (1986). Attenuation and diffraction of bending waves at gaps in fluid loaded plates. *IMA J. Appl. Math.* **36**, 247–262.
262. Guo, Y. P. (1993). Effects of structural joints on sound scattering. *J. Acoust. Soc. Am.* **93**, 857–863.
263. Howe, M. S. (1994). Scattering of bending waves by open and closed cracks and joints in a fluid loaded elastic plate. *Proc. Roy. Soc. Lond.* **A444**, 555–571.
264. Hayden, R. E., Murrey, B. M., & Theobald, M. A. (1983). A study of interior noise levels, noise sources and transmission paths in light aircraft. *NASA CR-172152*.
265. Mixson, J. S., & Wilby, J. F. (1991). Interior Noise. Chapter 16 of *Aeroacoustics of Flight Vehicles: Theory and Practice* (Vol. 2). NASA Ref. Pub. No. 1258.
266. Graham, W. R. (1993). Boundary layer noise and vibration, Doctoral Thesis, Cambridge University Engineering Department.
267. Shah, P. L., & Howe, M. S. (1996). Sound generated by a vortex interacting with a rib-stiffened elastic plate. *J. Sound Vib.* **197**, 103–115.
268. Howe, M. S. (1990). Scattering by a surface inhomogeneity on an elastic half-space, with application to fluid–structure interaction noise. *Proc. Roy. Soc. Lond.* **A429**, 203–226.
269. Howe, M. S. (1994). On the sound produced when flexural waves are reflected at the open end of a fluid loaded elastic cylinder. *J. Acoust. Soc. Am.* **96**, 265–276.
270. Cannell, P. A. (1975). Edge scattering of aerodynamic sound by a lightly loaded elastic half-plane. *Proc. Roy. Soc. Lond.* **A347**, 213–238.
271. Cannell, P. A. (1976). Acoustic edge scattering by a heavily loaded elastic half-plane. *Proc. Roy. Soc. Lond.* **A350**, 71–89.
272. Crighton, D. G., & Innes, D. (1984). The modes, resonances and forced response of elastic structures under heavy fluid loading. *Phil. Trans. Roy. Soc. Lond.* **A312**, 291–341.
273. Howe, M. S. (1992). Sound produced by an aerodynamic source adjacent to a partly coated, finite elastic plate. *Proc. Roy. Soc. Lond.* **A436**, 351–372.
274. Howe, M. S. (1993a). Structural and acoustic noise produced by turbulent flow over an elastic trailing edge. *Proc. Roy. Soc. Lond.* **A442**, 533–554.
275. Howe, M. S. (1993b). The compact Green's function for the semi-infinite elastic plate, with application to trailing edge noise and blade–vortex interaction noise. *J. Acoust. Soc. Am.* **94**, 2353–2364.
276. Theodorsen, T. (1935). General theory of aerodynamic instability and the mechanism of flutter. *NACA Report* **496**.
277. Newman, J. N. (1977). *Marine Hydrodynamics.* MIT Press.
278. Howe, M. S. (1994). Elastic blade–vortex interaction noise. *J. Sound Vib.* **177**, 325–337.
279. Ronneberger, D., & Ahrens, C. D. (1977). Wall shear stress caused by small amplitude perturbations of turbulent boundary layer flow: An experimental investigation. *J. Fluid Mech.* **83**, 433–464.
280. Howe, M. S. (1979). The interaction of sound with low Mach number wall turbulence, with application to sound propagation in turbulent pipe flow. *J. Fluid Mech.* **94**, 729–744.
281. Howe, M. S. (1984). On the absorption of sound by turbulence and other hydrodynamic flows. *IMA J. Appl. Math.* **32**, 187–209.
282. Howe, M. S. (1995). The Damping of Sound by Wall Turbulent Shear Layers. *J. Acoust. Soc. Am.* **98**. 1723–1730.

283. Ahrens, C., & Ronneberger, D. (1971). Acoustic attenuation in rigid and rough tubes with turbulent air flow. *Acustica* **25**, 150–157.
284. Ingard, U., & Singhal, V. K. (1974). Sound attenuation in turbulent pipe flow. *J. Acoust. Soc. Am.* **55**, 535–538.
285. Mankbadi, R. R., & Liu, J. T. C. (1992). Near-wall response in turbulent shear flows subjected to imposed unsteadiness. *J. Fluid Mech.* **238**, 55–72.
286. Peters, M. C. A. M., Hirschberg, A., Reijnen, A. J., & Wijnands, A. P. J. (1993). Damping and reflection coefficient measurements for an open pipe at low Mach and low Helmholtz numbers. *J. Fluid Mech.* **256**, 499–534.
287. Chapman, S., & Cowling, T. G. (1970). *The Mathematical Theory of Non-Uniform Gases* (3rd ed.). Cambridge University Press.
288. Blom, J. (1970). *An experimental determination of the turbulent Prandtl number in a temperature boundary layer.* Doctoral thesis, Eindhoven University of Technology.
289. Gradshteyn, I. S., & I. M. Ryzhik (1994). *Table of Integrals, Series and Products* (5th ed.). New York: Academic Press.
290. Bergeron, R. F., Jr. (1973). Aerodynamic sound and the low-wavenumber wall-pressure spectrum of nearly incompressible boundary layer turbulence. *J. Acoust. Soc. Am.* **54**, 123–133.
291. Zinn, B. T. (1970). A theoretical study of nonlinear damping by Helmholtz resonators. *J. Sound Vib.* **13**, 347–356.
292. Melling, T. H. (1973). The acoustic impedance of perforates at medium and high sound pressure levels. *J. Sound Vib.* **29**, 1–65.
293. Cummings, A. (1984). Acoustic nonlinearities and power losses at orifices. *AIAA J.* **22**, 786–792.
294. Dean, P. D., & Tester, B. J. (1975). Duct wall impedance control as an advanced concept for acoustic suppression. *NASA CR-134998*.
295. Bechert, D., Michel, U., & Pfizenmaier, E. (1977). Experiments on the transmission of sound through jets. *AIAA Paper* **77-1278.**
296. Bechert, D. W. (1979). Sound absorption caused by vorticity shedding, demonstrated with a jet flow. *AIAA Paper* **79-0575**.
297. Howe, M. S. (1980). The dissipation of sound at an edge. *J. Sound Vib.* **70**, 407–411.
298. Howe, M. S. (1979). On the theory of unsteady high Reynolds number flow through a circular aperture. *Proc. Roy. Soc. Lond.* **A366**, 205–233.
299. Howe, M. S. (1980). On the diffraction of sound by a screen with circular apertures in the presence of a low Mach number grazing flow. *Proc. Roy. Soc. Lond.* **A370**, 523–544.
300. Vér, I. L. (1982). *Perforated baffles prevent flow-induced acoustic resonances in heat exchangers.* Paper presented at 1982 meeting of the Federation of the Acoustical Societies of Europe, Göttingen, September 1982.
301. Rienstra, S. W. (1980). On the acoustical implications of vortex shedding from an exhaust pipe. *ASME Paper 80-WA/NC-16*.
302. Cargill, A. M. (1982). Low frequency sound radiation and generation due to the interaction of unsteady flow with a jet pipe. *J. Fluid Mech.* **121**, 59–105.
303. Cargill, A. M. (1982). Low frequency acoustic radiation from a jet pipe–A second order theory. *J. Sound Vib.* **83**, 339–354.
304. Rienstra, S. W. (1983). A small Strouhal number analysis for acoustic wave–jet flow–pipe interaction. *J. Sound Vib.* **86**, 539–556.
305. Vér, I. L. (1990). Practical examples of noise and vibration control: Case history of consulting projects. *Noise Control Eng. J.* **35** (Nov/Dec issue), pp. 115–125.
306. Hughes, I. J., & Dowling, A. P. (1990). The absorption of sound by perforated linings. *J. Fluid Mech.* **218**, 299–336.

307. Fukumoto, Y., & Takayama, M. (1991). Vorticity production at the edge of a slit by sound waves in the presence of a low Mach number bias flow. *Phys. Fluids* **A3**, 3080–3082.
308. Dowling A. P., & Hughes, I. J. (1992). Sound absorption by a screen with a regular array of slits. *J. Sound Vib.* **156**, 387–405.
309. Bechert, D. (1979). Sound sinks in flows, a real possibility? *Proceedings IUTAM/ICA/AIAA Symposium: Mechanics of Sound Generation in Flows* (edited by E.-A. Müller), pp. 26–34. Berlin: Springer-Verlag.
310. Rienstra, S. W. (1981). Sound diffraction at a trailing edge. *J. Fluid Mech.* **108**, 443–460.
311. Howe, M. S. (1981). On the role of displacement thickness in hydroacoustics, and the jet-drive mechanism in flue organ pipes. *Proc. Roy. Soc. Lond.* **A374**, 543–568.
312. Lighthill, Sir J. (1975). *Mathematical Biofluiddynamics*. Philadelphia: Society of Industrial and Applied Mathematics.
313. Childress, S. (1981). *Mechanics of Swimming and Flying*. Cambridge University Press.
314. Fung, Y. C. (1993). *An Introduction to the Theory of Aeroelasticity.* New York: Dover.
315. Howe, M. S. (1986). Attenuation of sound due to vortex shedding from a splitter plate in a mean flow duct. *J. Sound Vib.* **105**, 385–396.
316. Morse, P. M., & Ingard, K. U. (1968). *Theoretical Acoustics.* New York: McGraw-Hill.
317. Pierce, A. D. (1989). *Acoustics: An Introduction to its Principles and Applications.* American Institute of Physics.
318. Liepmann, H. W. (1954). On the acoustic radiation from boundary layers and jets. *Guggenheim Aero. Lab. Rept.* Pasadena, CA: California Institute of Technology.
319. Shapiro, P. J. (1977). The influence of sound upon laminar boundary layer instability. *MIT Acoustics and Vibration Lab. Rept. 83458-83560-1.*
320. Quinn, M. C., & Howe, M. S. (1984). On the production and absorption of sound by lossless liners in the presence of mean flow. *J. Sound Vib.* **97**, 1–9.
321. Hourigan, K., Welsh, M. C., Thompson, M. C., & Stokes, A. N. (1990). Aerodynamic sources of acoustic resonance in a duct with baffles. *J. Fluids and Structures* **4**, 345–370.
322. Crighton, D. G. (1992). The jet edge-tone feedback cycle: Linear theory for the operating stages. *J. Fluid Mech.* **234**, 361–392.
323. Barthel, F. (1958). Investigations on nonlinear Helmholtz resonators. *Frequenz* **12**, 1–11.
324. Rayleigh, Lord. (1870). On the theory of resonance. *Phil. Trans. Roy. Soc. Lond.* **161**, 77–118.
325. Copson, E. T. (1947). On the problem of the electrified disc. *Proc. Edinburgh Math. Soc.* **8**, 14–19.
326. Levine, H., & Schwinger, J. (1948). On the radiation of sound from an unflanged circular pipe. *Phys. Rev.* **73**, 383–406.
327. Ffowcs Williams, J. E. (1972). The acoustics of turbulence near sound-absorbent liners. *J. Fluid Mech.* **51**, 737–749.
328. Leppington, F. G., & Levine, H. (1973). Reflexion and transmission at a plane screen with periodically arranged circular or elliptical apertures. *J. Fluid Mech.* **61**, 109–127.
329. Leppington, F. G. (1977). The effective compliance of perforated screens. *Mathematika* **24**, 199–215.
330. Rockwell, D. (1983). Oscillations of impinging shear layers. *AIAA J.* **21**, 645–664.
331. Howe, M. S. (1981). On the theory of unsteady shearing flow over a slot. *Phil. Trans. Roy. Soc. Lond.* **A303**, 151–180.

332. Scott, M. I. (1995). *The Rayleigh conductivity of a circular aperture in the presence of a grazing flow.* Master's thesis, Boston University.
333. Howe, M. S., Scott, M. I., & Sipcic, S. R. (1996). The influence of tangential mean flow on the Rayleigh conductivity of an aperture. *Proc. Roy. Soc. Lond.* **A452**, 2303–2317.
334. Carrier, G. F., Krook, M., & Pearson, C. E. (1966). *Functions of a Complex Variable*. New York: McGraw-Hill.
335. Howe, M. S. (1981). The influence of mean shear on unsteady aperture flow, with application to acoustical diffraction and self-sustained cavity oscillations. *J. Fluid Mech.* **109**, 125–146.
336. Gel'fand, I. M., & Shilov, G. E. (1964). *Generalized Functions. Vol. 2: Spaces of Fundamental and Generalized Functions*. New York: Academic Press.
337. Brillouin, L. (1960). *Wave propagation and Group Velocity*. New York: Academic Press.
338. Howe, M. S. (1992). On the damping of structural vibrations by vortex shedding. *J. d'Acoustique* **5**, 603–620.
339. Walker, E. M., & Reising, G. F. S. (1968). Flow-induced vibrations in cross-flow heat exchangers. *Chem. Proc. Engng.* **49**, 95–103.
340. Blevins, R. D. (1977). *Flow Induced Vibration*. New York: Van Nostrand Reinhold.
341. Parker, R. (1978). Acoustic resonances in passages containing banks of heat exchanger tubes. *J. Sound Vib.* **57**, 245–260.
342. Bolleter, U., & Blevins, R. D. (1982). Natural frequencies of finned heat exchanger tubes. *J. Sound Vib.* **80**, 367–371.
343. Blevins, R. D. (1986). Acoustic modes of heat exchanger tube bundles. *J. Sound Vib.* **109**, 19–31.
344. Byrne, K. P. (1983). The use of porous baffles to control acoustic vibrations in crossflow tubular heat exchangers. *Trans. AMSE, J. Heat Trans.* **105**, 751–758.
345. Lyon, R. H., & Maidanik, G. (1962). Power flow between linearly coupled oscillators. *J. Acoust. Soc. Am.* **34**, 623–639.
346. Ungar, E. E. (1967). Statistical energy analysis of vibrating systems. *J. Eng. Ind., Transactions ASME Ser.* **B87**, 629–632.
347. Lyon, R. H. (1975). *Statistical energy analysis of dynamical systems: Theory and application.* MIT Press.
348. Quinn, Margaret C., & Howe, M. S. (1984). The influence of mean flow on the acoustic properties of a tube-bank. *Proc. Roy. Soc. Lond.* **A396**, 383–403.
349. Burton, T. E. (1980). Sound speed in a heat exchanger tube bank. *J. Sound Vib.* **71**, 157–160.
350. Blevins, R. D. (1984). Review of sound induced by vortex shedding from cylinders. *J. Sound Vib.* **92**, 455–470.
351. Sivian, J. L. (1935). Acoustic impedance of small orifices. *J. Acoust. Soc. Am.* **7**, 94–101.
352. Ingard, U. (1953). On the theory and design of acoustic resonators. *J. Acoust. Soc. Am.* **25**, 1037–1061.
353. Thurston, G. B., Hargrove, L. E., & Cook, W. D. (1957). Nonlinear properties of circular orifices. *J. Acoust. Soc. Am.* **29**, 992–1001.
354. Ingard, U., & Ising, H. (1967). Acoustic nonlinearity of an orifice. *J. Acoust. Soc. Am.* **42**, 6–17.
355. Panton, R. L., & Goldman A. L. (1976). Correlation of nonlinear orifice impedance. *J. Acoust. Soc. Am.* **60**, 1390–1396.
356. Hersh, A. S., & Walker, B. (1977). Fluid mechanical model of the Helmholtz resonator. *NASA Contractor Report CR-2904*.
357. Salikuddin, M., & Plumlee, H. E. (1980). Low frequency sound absorption of orifice plates, perforated plates and nozzles. *AIAA Paper* **80-0991**.

358. Cummings, A., & Eversman, W. (1983). High amplitude acoustic transmission through duct terminations: Theory. *J. Sound Vib.* **91**, 503–518.
359. Cummings, A. (1986). Transient and multiple frequency sound transmission through perforated plates at high amplitude. *J. Acoust. Soc. Am.* **79**, 942–951.
360. Salikuddin, M. (1990). Acoustic behaviour of orifice plates and perforated plates with reference to low-frequency sound absorption. *J. Sound Vib.* **139**, 361–382.
361. Salikuddin, M., & Brown, W. H. (1990). Nonlinear effects in finite amplitude wave propagation through orifice plate and perforated terminations. *J. Sound Vib.* **139**, 383–406.
362. Rayleigh, Lord. (1883). On the circulation of air observed in Kundt's tubes, and on some allied acoustical problems. *Phil. Trans. Roy. Soc. Lond.* **175**, 1–21.
363. Nyborg, W. L. (1953). Acoustic streaming due to attenuated plane waves. *J. Acoust. Soc. Am.* **25**, 68–75.
364. Westervelt, P. J. (1953). The theory of steady rotational flow generated by a sound field. *J. Acoust. Soc. Am.* **25**, 60–67.
365. Stuart, J. T. (1963). Unsteady boundary layers. Chapter 7 of *Laminar Boundary Layers* (ed., L. Rosenhead). Oxford University Press.
366. Nyborg, W. L. (1965). Acoustic streaming. *Physical Acoustics* **IIB**, 266–331.
367. Stuart, J. T. (1966). Double boundary layers in oscillatory viscous flows. *J. Fluid Mech.* **24**, 673–687.
368. Riley, N. (1967). Oscillatory viscous flows: Review and extension. *J. Inst. Math. Applics.* **3**, 419–434.
369. Lighthill, Sir J. (1978). Acoustic streaming. *J. Sound Vib.* **61**, 391–418.
370. Landau, L. D. (1944). A new exact solution of the Navier–Stokes equations. *C. R. Acad. Sci. URSS*, **43**, 286–287.
371. Squire, H. B. (1951). The round laminar jet. *Q. J. Mech. Appl. Math.* **4**, 321–329.
372. Krishnamurty, K. (1955). Acoustic radiation from two-dimensional rectangular cut-outs in aerodynamic surfaces. *NACA TN* **3487**.
373. Roshko, A. (1955). Some measurements of flow in a rectangular cut-out. *NACA TN* **3488**.
374. Dunham, W. H. (1962). *Fourth Symposium on Naval Hydrodynamics–Propulsion Hydroelasticity* **ACR-92**, 1057–1081.
375. Rossiter, J. E. (1962). The effect of cavities on the buffeting of aircraft. *Royal Aircraft Establishment Technical Memorandum* **754**.
376. Spee, B. M. (1966). Wind tunnel experiments on unsteady cavity flow at high subsonic speeds. *AGARD CP* **4**, 941–974.
377. East, L. F. (1966). Aerodynamically induced resonance in rectangular cavities. *J. Sound Vib.* **3**, 277–287.
378. Heller, H. H., Holmes, D. G., & Covert, E. E. (1971). Flow induced pressure oscillations in shallow cavities. *J. Sound Vib.* **18**, 545–553.
379. Ronneberger, D. (1972). The acoustical impedance of holes in the wall of flow ducts. *J. Sound Vib.* **24**, 133–150.
380. Chanaud, R. C. (1970). Aerodynamic whistles. *Scientific American* **221**, 40–46.
381. Bilanin, A. J., & Covert, E. E. (1973). Estimation of possible excitation frequencies for shallow rectangular cavities. *AIAA J.* **11**, 347–351.
382. Block, P. J. W. (1976). Noise response of cavities of varying dimensions at subsonic speeds. *NASA TN* **D-8351**.
383. DeMetz, F. C., & Farabee, T. M. (1977). Laminar and turbulent shear flow induced cavity resonances. *AIAA Paper* **77-1293**.
384. Tam, C. K. W., & Block, P. J. W. (1978). On the tones and pressure oscillations induced by flow over rectangular cavities. *J. Fluid Mech.* **89**, 373–399.
385. Robins, A. J. (1979). A theoretical and experimental study of flow-induced cavity resonances. *Admiralty Marine Technology Establishment Rept. AMTE(N) R79102*.

386. Ronneberger, D. (1980). The dynamics of shearing flow over a cavity–A visual study related to the acoustic impedance of small orifices. *J. Sound Vib.* **71**, 565–581.
387. Kompenhans, J., & Ronneberger, D. (1980). The acoustic impedance of orifices in the wall of a flow duct with a laminar or turbulent boundary layer. *AIAA Paper* **80-0990**.
388. Elder, S. A., Farabee, T. M., & DeMetz, F. C. (1982). Mechanisms of flow-excited cavity oscillations at low Mach number. *J. Acoust. Soc. Am.* **72**, 532–549.
389. Komerath, N. M., Ahuja, K. K., & Chambers, F. W. (1987). Prediction and measurement of flows over cavities–A survey. *AIAA Paper* **87-022**.
390. Jungowski, W. M., Botros, K. K., Studzinski, W., & Berg, D. H. (1987). Tone generation by flow past confined, deep cylindrical cavities. *AIAA Paper* **87-2666**.
391. Bruggeman, J. C. (1987). *Flow induced pulsations in pipe systems*, Ph.D. thesis, Eindhoven University of Technology.
392. Bruggeman, J. C., Hirschberg, A., van Dongen, M. E. H., Wijnands, A. P. J., & Gorter, J. (1989). Flow induced pulsations in gas transport systems: analysis of the influence of closed side branches. *J. Fluids Eng.* **111**, 484–491.
393. Panton, R. L. (1990). Effect of orifice geometry on Helmholtz resonator excitation by grazing flow. *AIAA J.* **28**, 60–65.
394. Bruggeman, J. C., Hirschberg, A., van Dongen, M. E. H., & Wijnands, A. (1991). Self-sustained aero-acoustic pulsations in gas transport systems: Experimental study of the influence of closed side branches. *J. Sound Vib.* **150**, 371–394.
395. Hirschberg, A., & Rienstra, S. W. (1994). Elements of aero-acoustics. *von Karman Institute Lecture Notes*.
396. Burroughs, C. B., & Stinebring, D. R. (1994). Cavity flow tones in water. *J. Acoust. Soc. Am.* **95**, 1256–1263.
397. Ahuja, K. K., & Mendoza, J. (1995). Effects of cavity dimensions, boundary layer, and temperature on cavity noise with emphasis on benchmark data to validate computational aeroacoustic codes. *NASA Contractor Report: Final Report Contract NAS1-19061, Task 13*.
398. Howe, M. S. (1997). Edge, cavity and aperture tones at very low Mach numbers. *J. Fluid Mech.* **330**, 61–84.
399. Powell, A. (1961). On the edgetone. *J. Acoust. Soc. Am.* **33**, 395–409.
400. Blake, W. K., & Powell, A. (1986). The development of contemporary views of flow-tone generation. In *Recent Advances in Aeroacoustics* (eds. A. Krothapali & C. A. Smith), pp. 247–345. New York: Springer.
401. Holger, D. K., Wilson, T. A., & Beavers, G. S. (1977). Fluid mechanics of the edgetone. *J. Acoust. Soc. Am.* **62**, 1116–1128.
402. Grace, S. M., Wood, T. H., & Howe, M. S. (1998). The stability of high Reynolds number flow over a circular aperture. *Proc. Roy. Soc. Lond. A* (in press).
403. Howe, M. S. (1997). Influence of wall thickness on Rayleigh conductivity and flow-induced aperture tones. *J. Fluids and Structures* **11**, 351–366.
404. Milne-Thomson, L. M. (1968). *Theoretical Hydrodynamics* (5th ed.). London: Macmillan.
405. Kriesels, P. C., Peters, M. C. A. M., Hirschberg, A., Wijnands, A. P. J., Iafrati, A., Riccardi, G., Piva, R., & Bruggeman, J. C. (1995). High amplitude vortex induced pulsations in gas transport systems. *J. Sound Vib.* **184**, 343–368.
406. Copson, E. T. (1965). *Asymptotic Expansions*. Cambridge University Press.
407. Jeffreys, H. (1968). *Asymptotic Approximations* (2nd ed.). Oxford University Press.
408. Nelson, P. A., Halliwell, N. A., & Doak, P. E. (1983). Fluid dynamics of a flow excited resonance, Part 2: Flow acoustic interaction. *J. Sound. Vib.* **91**, 375–402.
409. Ivanov, V. I., & Trubetskov, M. K. (1995). *Handbook of Conformal Mapping with Computer-Aided Visualization*. Boca Raton, FL: CRC Press.

410. Howe, M. S. (1976). On the Helmholtz resonator. *J. Sound Vib.* **45**, 427–440.
411. Brown, G. B. (1937). The vortex motion causing edge tones. *Proc. Phys. Soc. Lond.* **49**, 493–520.
412. Jones, A. T. (1942). Edge tones. *J. Acoust. Soc. Am.* **14**, 131–139.
413. Nyborg, W. L., Burkhard, M. D., & Schilling, H. K. (1952). Acoustical characteristics of jet-edge and jet-edge-resonator systems. *J. Acoust. Soc. Am.* **24**, 293–304.
414. Birkhoff, G. (1953). Formation of vortex streets. *J. Applied Physics* **24**, 98–103.
415. Dunne, R. C. (1995). Sound produced by a vortex interacting with a blade tip of a ducted rotor. MS Thesis, College of Engineering, Boston University.
416. Dunne, R. C., & Howe, M. S. (1997). Wall-bounded blade–tip vortex interaction noise. *J. Sound Vib.* **202**, 605–618.
417. Chanaud, R. C., & Powell, A. (1965). Some experiments concerning the hole and ring tone. *J. Acoust. Soc. Am.* **37**, 902–911.
418. Wilson, T. A., Beavers, G. S., DeCoster, M. A., Holger, D. K., & Regenfuss, M. D. (1971). Experiments on the fluid mechanics of whistling. *J. Acoust. Soc. Am.* **50**, 366–372.
419. Brown, R. S., Dunlap, R., Young, S. W., & Waugh, B. G. (1981). Vortex shedding as a source of acoustic energy in segmented solid rockets. *J. Spacecraft and Rockets* **18**, 312–319.
420. Nomoto, H., & Culick, F. E. C. (1982). An experimental investigation of pure tone generation by vortex shedding in a duct. *J. Sound Vib.* **84**, 247–252.
421. Culick, F. E. C., & Yang, V. (1992). Prediction of the stability of unsteady motions in solid-propellant rocket motors, Chapter 18 of Nonsteady burning and combustion stability of solid propellants. (*AIAA Progress in Astronautics and Aeronautic Series*, ed. L. de Luca, E. W. Price, & M. Summerfield.)
422. Anderson, A. B. C. (1955). Structure and velocity of the periodic vortex-ring flow pattern of a Primary Pfeifenton (Pipe tone) jet. *J. Acoust. Soc. Am.* **27**, 1048–1053.
423. Kakayoglu, R., & Rockwell, D. (1986a). Unstable jet–edge interaction. Part 1: Instantaneous pressure field at a single frequency. *J. Fluid Mech.* **169**, 125–149.
424. Kakayoglu, R., & Rockwell, D. (1986b). Unstable jet–edge interaction. Part 2: Multiple frequency pressure fields. *J. Fluid Mech.* **169**, 151–172.
425. Hirschberg, A., Wijnands, A. P. J., van Steenbergen, A., de Vries, J. J., Fabre, B., & Gilbert, J. (1990). *Jet drive in a flue organ pipe: Flow visualization of stationary and transient behaviors.* Paper presented at Modele Physiques Creation Musicale et Ordinateur Colloquium, Grenoble.
426. Verge, M. P., Fabre, B. Mahu, W. E. A., Hirschberg, A., van Hassel, R. R.,Wijnands, A. P. J., de Vries, J. J. and Hogendoorn, C. J. (1994a). Jet formation and jet velocity in a flue organ pipe. *J. Acoust. Soc. Am.* **95**, 1119–1132.
427. Verge, M. P., Caussé, R., Fabre, B., Hirschberg, A., Wijnands, A. P. J., & van Steenbergen, A. (1994b). Jet oscillations and jet drive in recorder-like instruments. *Acta Acustica* **2**, 403–419.
428. Fabre, B., Hirschberg, A., & Wijnands, A. P. J. (1996). Vortex shedding in steady oscillation of a flue organ pipe. *Acustica* **82**, 863–877.
429. Coltman, J. W. (1968). Acoustics of the flute. *Physics Today* **21**, 25–32.
430. Coltman, J. W. (1976). Jet drive mechanisms in edge tones and organ pipes. *J. Acoust. Soc. Am.* **60**, 725–733.
431. Coltman, J. W. (1992). Time-domain simulation of the flute. *J. Acoust. Soc. Am.* **92**, 69–73.
432. Coltman, J. W. (1992). Jet behavior in the flute. *J. Acoust. Soc. Am.* **92**, 74–83.
433. Elder, S. A. (1992). The mechanism of sound production in organ pipes and cavity resonators. *J. Acoust. Soc. Japan (E)* **13**,11–24.

434. Fletcher, N. H. (1976). Sound production by organ flue pipes. *J. Acoust. Soc. Am.* **60**, 926–936.
435. Fletcher, N. H. and Rossing, T. D. (1991). *The Physics of Musical Instruments.* New York: Springer-Verlag.
436. Fletcher, N. H. (1993). Autonomous vibration of simple pressure-controlled valves in gas flows. *J. Acoust. Soc. Am.* **93**, 2172–2180.
437. Yoshikawa, S., & Saneyoshi, J. (1980). Feedback excitation mechanism in organ pipes. *J. Acoust. Soc. Japan* **1**, 175–191.
438. Skordos, P. A. (1995). *Modeling flue pipes: Subsonic flow, lattice Boltzmann, and parallel distributed computers.* Doctoral thesis, Department of Electrical Engineering and Computer Science, MIT.
439. Crocco, L., & Cheng, S.-I. (1956). *Theory of Combustion Instability in Liquid Propellant Rocket Motors.* London: Butterworths Scientific Publications.
440. Bragg, S. L. (1963). Combustion noise. *J. Inst. Fuel* (January issue), 12–16.
441. Kuo, K. K. and Summerfield, M. (ed.) (1984). *Fundamentals of Solid-Propellant Combustion, Vol. 90, Progress in Astronautics and Aeronautics.* New York: American Institute of Aeronautics and Astronautics.
442. Mahan, J. R., & Karchmer, A. (1991). Combustion and core noise. *Aeroacoustics of Flight Vehicles: Theory and Practice* (Vol. 1). NASA Ref. Pub. No. 1258.
443. Strahle, W.C. (1977). Combustion noise. *Stanford University Report: JIAA TR-7.*
444. Strahle, W. C. (1985). A more modern theory of combustion noise. In *Recent Advances in Aerospace Sciences* (ed. C. Casci) New York: Plenum.
445. Gaydon, A. G., & Wolfhard, H. G. (1970). *Flames: Their Structure, Radiation and Properties* (3rd ed.). London: Chapman and Hall.
446. Frank-Kamenetskii, D. A. (1955). *Diffusion and Heat Exchange in Chemical Kinetics.* Princeton University Press.
447. Lefebvre, A. H. (1983). *Gas Turbine Combustion.* New York: Hemisphere.
448. Pauling, L. (1988). *General Chemistry.* New York: Dover.
449. Thomas, A., & Williams, G. T. (1966). Flame noise: Sound emission from spark-ignited bubbles of combustible gas. *Proc. Roy. Soc. Lond.* **A294**, 449–466.
450. Rayleigh, Lord (1878). The explanation of certain acoustical phenomena. *Nature* **18**, 319–321.
451. Putnam, A. A. (1971). *Combustion Driven Oscillations in Industry.* New York: Elsevier.
452. Jones, A. A. (1945). Singing flames. *J. Acoust. Soc. Am.* **16**, 254–266.
453. Feldman, K. T. Jr. (1968a). Review of the literature on Sondhauss thermoacoustic phenomena. *J. Sound Vib.* **7**, 71–82.
454. Feldman, K. T. Jr. (1968b). Review of the literature on Rijke thermoacoustic phenomena. *J. Sound Vib.* **7**, 83–89.
455. Lighthill, M. J., & Ffowcs Williams, J. E. (1971). *Bulletin of the Institute of Mathematics and its Applications* 7 (part 4) 3–10. Demonstration of a shared lecture.
456. Heckl, M. A. (1988). Active control of the noise from a Rijke tube. *J. Sound Vib.* **124**, 117–133.
457. Nicoli, C., & Pelce, P. (1989). One dimensional model for the Rijke tube. *J. Fluid Mech.* **202**, 83–96.
458. Heckl, M. A. (1990). Nonlinear acoustic effects in the Rijke tube. *Acustica* **72**, 63–71.
459. King, L. V. (1914). On the convection of heat from small cylinders in a stream of fluid. *Phil. Trans. Roy. Soc. Lond.* **A214**, 373–432.
460. Lighthill, M. J. (1954). The response of laminar skin friction and heat transfer to fluctuations in the stream velocity. *Proc. Roy. Soc. Lond.* **A224**,1–23.

461. Markstein, G. H. (ed.) (1964). *Non-Steady Flame Propagation.* Oxford, England: Pergamon.
462. Jones, H. (1977). The mechanics of vibrating flames in tubes. *Proc. Roy. Soc. Lond.* **A353**, 459–473.
463. Jones, H. (1979). The generation of sound by flames. *Proc. Roy. Soc. Lond.* **A367**, 291–309.
464. Clavin, P., Pelce, P. & He, L. (1990). One-dimensional vibratory instability of planar flames propagating in tubes. *J. Fluid Mech.* **216**, 299–322.
465. Searby, G., & Rochwerger, D. (1991). A parametric acoustic instability in premixed flames. *J. Fluid Mech.* **231**, 529–543.
466. Pelce, P., & Rochwerger, D. (1992). Vibratory instability of cellular flames propagating in tubes. *J. Fluid Mech.* **239**, 293–307.
467. Markstein, G. H., & Squire, W. (1955). On the stability of a plane flame front in oscillating flow. *J. Acoust. Soc. Am.* **27**, 416–424.
468. Clavin, P. (1994). Premixed combustion and gasdynamics. *Ann. Rev. Fluid Mech.* **26**, 321–352.
469. Joos, F., & Vortmeyer, D. (1986). Self-excited oscillations in combustion chambers with premixed flames and several frequencies. *Combust. Flame* **65**, 253–262.
470. Putnam, A. A., & Dennis, W. R. (1954). Burner oscillations of the gauze-tone type. *J. Acoust. Soc. Am.* **26**, 716–725.
471. Putnam, A. A., & Dennis, W. R. (1956). Survey of organ-pipe oscillations in combustion systems. *J. Acoust. Soc. Am.* **28**, 246–269.
472. Boyer, A. L. & Quinard, J. (1990). On the dynamics of anchored flames. *Combust. Flame* **82**, 51–65.
473. Bloxsidge, G. J., Dowling, A. P., & Langhorne, P. J. (1988). Reheat buzz: An acoustically coupled combustion instability. Part 2. Theory. *J. Fluid Mech.* **193**, 445–473.
474. Macquisten, M. A. & Dowling, A. P. (1995). Combustion oscillations in a twin-stream afterburner. *J. Sound Vib.* **188**, 545–560.
475. Dowling, A. P. (1995). The calculation of thermoacoustic oscillations. *J. Sound Vib.* **180**, 557–581.
476. Flandro, G. A. (1995). Effects of vorticity on rocket combustion stability. *J. Propulsion and Power* **11**, 607–625.
477. Vuillot, F. (1995). Vortex shedding phenomena in solid rocket motors. *J. Propulsion and Power* **11**, 626–639.
478. Roh, T.-S., Tseng, I.-S., & Yang, V. (1995). Effects of acoustic oscillations on flame dynamics of homogeneous propellants in rocket motors. *J. Propulsion and Power* **11**, 640–650.
479. Carter, R. L., White, M., & Steele, A. M. (1962). Research performed at the Atomics International Division of North American Aviation Inc.
480. Merkli, P., & Thomann, H. (1975). Thermoacoustic effects in a resonant tube. *J. Fluid Mech.* **70**, 161–177.
481. Rott, N. (1980). Thermoacoustics. *Adv. Appl. Mech.* **20**, 135–175.
482. Rott, N. (1984). Thermoacoustic heating at the closed end of an oscillating gas column. *J. Fluid Mech.* **145**, 1–10.
483. Rott, N. (1986). A simple theory of the Sondhauss tube, In *Recent Advances in Aeroacoustics* (eds. A. Krothapali and C. A. Smith, pp. 327–338), New York: Springer.
484. Wheatley, J., Hofler, T., Swift, G. W., & Migliori, A. (1985). Understanding some simple phenomena in thermoacoustics with applications to acoustical heat engines. *Am. J. Phys.* **53**, 147–162.
485. Swift, G. W. (1988). Thermoacoustic engines. *J. Acoust. Soc. Am.* **84**, 1145–1180.
486. Gifford, W. E., & Longsworth, R. C. (1966). Surface heat pumping. *Adv. Cryog. Eng.* **11**, 171–179.

Index